Models of the Visual System

TOPICS IN BIOMEDICAL ENGINEERING INTERNATIONAL BOOK SERIES

Series Editor: Evangelia Micheli-Tzanakou
Rutgers University
Piscataway, New Jersey

Signals and Systems in Biomedical Engineering:
Signal Processing and Physiological Systems Modeling
Suresh R. Devasahayam

Models of the Visual System
Edited by George K. Hung and Kenneth J. Ciuffreda

A Continuation Order Plan is available for this series. A continuation order will bring delivery of each new volume immediately upon publication. Volumes are billed only upon actual shipment. For further information please contact the publisher.

Models of the Visual System

Edited by

George K. Hung
Rutgers University
Piscataway, New Jersey

and

Kenneth J. Ciuffreda
State University of New York
New York, New York

Kluwer Academic / Plenum Publishers
New York, Boston, Dordrecht, London, Moscow

Library of Congress Cataloging-in-Publication Data

Models of the visual system/edited by George K. Hung and Kenenth J. Ciuffreda.
p. cm. — (Topics in biomedical engineering international book series)
ISBN 0-306-46715-1
1. Vision—Research—Methodology. 2. Vision—Computer simulation. 3. Vision—Mathematical models. 4. Visual pathways—Research—Methodology. 5. Visual cortex—Research—Methodology. I. Hung, George K. II. Ciuffreda, Kenneth J., 1947- III. Series.

QP475 .M634 2001
612.8′4—dc21

2001053924

ISBN 0-306-46715-1

233 Spring Street, New York, New York 10013

http://www.wkap.nl/

10 9 8 7 6 5 4 3 2 1

A C.I.P. record for this book is available from the Library of Congress

Printed in the United States of America

To the memory of my father, Ernest K. Hung, who brought his five children to the United States with $40 in his pockets and a desire to improve our lives; and to my wife, Xiaosi Yang, for her love, devotion and support.

—G.K.H.

To my daughters, Gabrielle and Marcelline, for their love, affection and support.

—K.J.C.

and

To our mentor, Larry Stark, for introducing us to bioengineering models of the visual system.

—G.K.H. and K.J.C

FOREWORD

The co-editors and former students of mine, George K. Hung and Kenneth J. Ciuffreda, have made an explicit commitment to describing phenomena in terms of modeling and detailed quantitative analyses of the visual system in this important new book, "Models of the Visual System".

Philosophers, like Plato and Kant, have long speculated that we can have no knowledge of the external "real" world, the chaos outside of our minds. We can only introspect and consider various levels of models and by surviving convince ourselves that these schemas describe the so-called "real world". Bioengineering requires explicit models stated in mathematical form and often addressing cybernetic aspects of, for example, visual mechanisms, which is the topic of this volume.

More than fifty years ago, cybernetics embodied the spirit of the times. Norbert Wiener's book "Cybernetics: Control and Communication in Animals and Machines" was published in the same year, 1948, that Claude Shannon's great paper "A Mathematical Theory of Communication" came out in the Bell Systems Journal. Earlier, in the 1940's, McCulloch and Pitts constructed the formal neuron, 'poverty stricken' yet capable of logical computations; indeed von Neumann used their schema to design the logical units of the Edvac computer, the first stored program digital computer. McCulloch, together with Pitts and Lettvin, went on to invent the neural net and wrote such great papers on sensory organization and visual modeling as, "How We Know Universals" and "What the Frog's Eye Tells the Frog's Brain".

To give one a flavor of how we continue to use and think in terms of the cybernetic revolution, let us consider the amount of information capacity in a digital image. It is limited by its resolution or number of pixels, each considered as a discrete informational symbol. Each pixel in turn is specified by a certain number of gray levels (perhaps to 8 bits precision) or of trichromatic color levels (perhaps to 24 bits) entered into Shannon's formula as a logarithmic term. The product of the logarithm to the base two of the number of gray levels times the number of pixels gives the total informational content of a digital image in bits. If the sampling is sufficient, any analog image is completely transformed into a digital image – clearly in the noisy universe in which we are immersed, there cannot be an infinite amount of information in any analog value. The informational capacity is clearly not equivalent to the lesser amount of information with which an image is usually coded.

As well, the information received by the retinal and cortical mechanisms is also much less than in the original image. The "jnd's" or just noticeable differences in the shade of a pixel and the resolution available in the retinal-cortical receivers strongly limits the amount of information about an image that can be sensed. This is now an active area in engineering image compression and indeed in the study of visual mechanisms. Yet another aspect of perception appeared in 1971 when Noton and I formulated an explicit hypothesis, the "scanpath theory" concerning internal control of perception. This postulated that top-down internal spatial cognitive models controlled not only perception, but also active scanning eye movements that sequentially shift the focus of attention to confirm the expected loci and detailed features of important elements of a viewed picture. The scanpath sequence of eye movements could be objectively measured and modeled, and provided for the first time scientific evidence that supported earlier philosophical speculations.

This Hung-Ciuffreda volume explicitly contributes to these important areas of sensory organization in Section 2 on the Retina and the Visual Cortex and implicitly in the sensory component descriptions of the various motor control systems discussed. Importantly, perceptual issues are covered in Section 4 with chapters on Motion and Texture appreciation, on Visual Attention and on Cognitive Processing.

Next we turn to the motor mechanisms that can aid vision – accommodation minimizing blur, the pupil dynamically optimizing light flux and aperture size. Here Norbert Wiener's formulation of control, filter and identification theory, using his mathematical expertise in harmonic analysis and stochastic theory, changed the shape of these aspects of engineering in the forties. Wiener, with his widely-circulated 'classified' memorandum on fire control, with his linear filter theory, and with his Wiener functionals contributed a major component of the cybernetic evolution. Wiener once said to me, "Stark, the Wiener functionals are ideal for complex systems; biological systems are complex systems. Stark, you'll go far using Wiener functionals!" This later statement, of course, violated Mark Twain's admonition; "Never make predictions, especially about the future!"

Wiener functionals, a general method for non-linear sytem analysis, were the subject of Professor George Hung's thesis at Berkeley some years ago. His insights into their representation and the unique understanding that they provide have stood him in good stead. His many contributions on the expansion of non-linear theory into neurological and especially visual-motor control systems have formed an essential nucleus of the present volume.

Professor Ken Ciuffreda has also worked on eye movement control systems in the service of vision. He brought his clinical acumen to bear upon the problems of strabismus and amblyopia. Both the disparity vergence mechanism and the versional eye movement systems impaired by their dependence upon the reduced vision of an amblyopic eye demonstrate syndromes of deficits that highlighted the elegant control mechanisms underlying their normal function. These have been modeled explicitly and have been defined and clarified. Examples in this book are provided in Section 3 with chapters that cover Models of Accommodation, Vergence and Accommodation, and Eye Tracking as well as in Section 5 with clinical systems models covering the topics of Reading Disability and of Nystagmus.

Models come in various shapes and levels; perhaps the earliest were models as parables, perhaps somewhat more 'workable' than purely verbal descriptions. Mathematics as model is well illustrated in the famous Hodgkin-Huxley equations for the non-linear feedback explosion of the nerve impulse. Linear and non-linear systems models use the mathematics of control theory and Voltera-Wiener series, and these have been applied to visual mechanisms like the pupil. The machinery of model simulation has played an important supportive role for cybernetics and bioengineering over the last fifty years: from analogue computers, their formulation as digital programs from Selfridge (1955), to Simulink. Using this latter tool, the heuristic differential-integral model has come to dominance, and this excellent volume is replete with examples. My own scientific thrust has been self-described as, " I see models everywhere". Although it has been stated that, "statistics is a way of being wrong with confidence", we modelers must humbly add to that epigram the addendum, "modeling is a way of being wrong with certainty".

I can enthusiastically recommend this volume to the many vision neuroscientists, clinicians, bioengineers, psychologists, and students interested or deeply involved in the quantitative functioning of vision.

Dr. Lawrence W. Stark
Emeritus Professor of Physiological Optics and Engineering Sciences
School of Optometry
University of California at Berkeley
Berkeley, CA

PREFACE

The visual system has evolved to perform exquisitely complex sensory, motor, and perceptual tasks. It is able to focus on a range of objects in depth, track moving targets, attend to embedded figures, and perceive form in detailed contours. Yet, it does so effortlessly and continuously in daily life. This is due to the visual system's organizational structure, which permits automatic performance of these complex tasks. Understanding this structure requires more than qualitative or simple analytical approaches. Indeed, models are essential for assessing quantitatively its various components and their interactions. Over the past thirty years, with improvements in electronics and computer technology, great strides have been made in model-based quantitative analysis of the visual system. There is a need to synthesize and summarize these important modeling efforts. It is with such considerations in mind that we have compiled this updated, comprehensive, and quantitative model-based edited volume encompassing various components of the visual system. Some of the best vision scientists in the world in their respective fields have contributed to chapters in this book. They have expertise in a wide variety of fields, including bioengineering, basic and clinical vision science, medicine, neurophysiology, optometry, and psychology.

The book is organized into five sections: (1) Optical systems - cornea and the lens; (2) Neurosensory systems - retina and visual cortex; (3) Oculomotor systems - accommodation, vergence, and eye tracking; (4) Perceptual systems - texture, motion, visual attention, and cognitive processing; and (5) Clinical systems - refractive error development, clinical vergence testing, reading disability, and nystagmus.

The book is aimed first towards college seniors and beginning graduate students in biomedical engineering, neurophysiology, optometry, and psychology, who will gain a broad understanding of quantitative analysis of the visual system. In addition, it has considerable depth and important new information in each area to be useful as an updated reference and tutorial for graduate and post-doctoral students, as well as established vision scientists.

George K. Hung, Ph. D.
Piscataway, New Jersey

Kenneth J. Ciuffreda, O.D., Ph.D.
New York City

CONTENTS

Cognitive Processing

Perceptual Space

Vergence Model Parameters and Clinical Vergence Tests

Refractive Error Development

Reading Disability

Dysfunction

I

OPTICAL SYSTEM MODELS

Chapter 1

Physiological System Models of the Cornea

Jeffrey W. Ruberti[1], Ph.D.; Stephen D. Klyce[2], Ph.D.

[1]*Cambridge Polymer Group,11 Ward St., Suite 100, Somerville, MA 02143, PH: (617) 864-5566, FX: (617) 864-5577, EM: jeff@campoly.com*

[2]*Dept. of Ophthalmology, Louisiana State University Eye Center, 2020 Gravier St., Suite B,New Orleans, LA 70112-2234, PH: (504) 412-1329, FX: (504) 412-1322, EM: sklyce@klyce.com*

1.1 INTRODUCTION

The cornea is the primary refractive component of the visual system. It provides the crystal-clear window through which our world is observed. Given the importance of vision to everyday life and the accessibility of the cornea to investigation, it is easy to understand the magnitude and intensity of the research effort that has been focused on this apparently simple tissue. The mystery of its clarity and the dependence of this property on transport physiology ignited the initial frenzy of investigation in the late1940's while the development of contact lenses, proliferation of refractive surgeries and corneal transplantation spur it on today. Mathematical modeling has and will continue to complement the experimental investigations that have guided our understanding of the interaction between corneal structure and function.

1.1.1 Corneal Structure

To understand physiology through modeling or experimentation, it is essential to characterise the structures that sustain the molecular gradients necessary for life to exist. With regard to modeling, it is still more important to identify those aspects of the tissue that critically influence the spatial or temporal distribution of dependent variables of interest such as pressure, hydration or concentration. For practical purposes, the modeller is often forced to simplify the structure and functions of the tissue that is to be modelled; the cornea is no exception. Fortunately, lacking vasculature and muscle, the cornea already comprises a fairly simple structure that makes the development and testing of models less difficult than it is in most other physiological systems.

1.1.1.1 Gross Anatomy

The cornea makes up one-sixth of the tough outer tunic of the ocular globe. It is a well-ordered, transparent tissue that joins the more disorganized, opaque sclera at the limbus. Though not perfectly circular, the diameter of the cornea is approximately 12 mm in the human. The average radius of curvature of the central anterior surface is 7.8 mm and it is roughly 520 microns thick at the center and 650 microns thick in the periphery. Note that we are already beginning to develop model parameters just by laying out the anatomy of the tissue.

1.1.1.2 Microscopic Structure

Transversely, there are three layers important to the transport physiology of the cornea; they are the epithelium, the stroma and the endothelium. Fig. 1.1 shows the orderly structure of the tissue and the relative dimensions of the three major layers.

Epithelium. The corneal epithelium is a multicellular "tight" stratified squamous epithelium comprising three distinct functional layers. The deepest cellular layer is the stratum germinatum, this is the only layer capable of undergoing mitosis. The middle layer comprises the daughter cells (wing cells) of the basal layer which are pushed anteriorly. The surface layer comprises the squamous cells that form the complete tight junctions which generate the primary barrier to transport in the cornea and are critical to understanding its vegetative physiology.

Stroma. The stroma is a relatively acellular extracellular matrix comprising hydrated type I collagen fibrils (15% wet weight; Maurice (1969)) of uniform diameter (approximately 30 nm) in humans (Meek and Leonard, 1993; Daxer et al., 1998); glycosaminoglycans (GAGs) keratan sulfate and dermatan sulfate (1% wet weight; Meyer et al, 1953; Anseth, 1961); various proteoglycan (PG) core proteins (Axelsson and Heinegard, 1975) and miscellaneous other protein constituents including fibronectin, laminin and type VI collagen. The collagen fibrils are packed in parallel arrays that make up the approximately 300 to 500 lamellae (Hamada et al., 1972) of the stromal tissue.

Stromal uniformity - transverse section. Traditionally thought of as uniform in cross section, there are known spatial gradients in the distribution of stromal keratocytes (Muller et al. 1995), GAGs (Bettelheim and Goetz, 1976) and their associated PG core proteins, water (Bettelheim, 1975; Turss et al.1971; Bauer et al., 1998) and even swelling pressure (Kikkawa and Hirayama, 1970; Lee and Wilson, 1981). For the distribution of GAGs at least, it has been suggested that gradients in the local oxygen concentration may influence the ultimate tissue architecture (Scott, 1994). It is also likely that the remaining spatial differences in ultrastructure and keratocyte population are linked to similar gradients in the local chemical make up of the stroma. In spite of this, most corneal models treat the tissue as a uniform structure that contributes about 90% of the total thickness to the cornea.

Figure 1.1. The cornea (approximately 500 microns thick in humans) is a layered structure comprising the epithelium (E), Bowman's layer (Bw), stroma (St), Descemet's membrane (De), and the endothelium. Reprinted from Klyce and Beuerman (1997), Fig. 1-1, with permission of Butterworth-Heinemann.

Stromal uniformity - lateral section. Though there is some evidence to the contrary (Ling, 1987), most models of the cornea treat stromal tissue as structurally uniform from the center to the periphery. In general, however, there is a gradual thickening as the limbus is approached, a change in GAG distribution and a gradient in collagen fibril thickness. With the exception of Wiig's (1989) pressure data, the importance of the structural changes toward the periphery on stromal transport physiology has not been demonstrated. Thus, for transport, the cornea is typically modelled as a *one-dimensional* tissue in the anterior-to-posterior direction.

Endothelium. The "endothelium" is a "leaky" monolayer of approximately 400,000 hexagonal cells of 20 microns in diameter and 4-6 microns in height. The length of the paracellular pathway of this primary active transport layer is approximately 10 times the height of the cells themselves.

1.1.2 Importance of Mathematical Models to the Basic Corneal Scientist and Clinician

The transparency of the corneal stroma depends critically on the regular, short-order spacing of its constituent collagen fibrils (Maurice, 1957; Benedek, 1971). Isolated stromal tissue will swell freely in isotonic saline and become opaque, but *in vivo* corneas remain thin and clear. How the steady-state corneal thickness is maintained, given biological variability and naturally varying ambient conditions, remains one of the most perplexing questions in corneal physiology.

Modeling of the *whole* tissue system and its interaction with the environment is critical to our understanding its function and how to alter that function in a therapeutic manner. In his 1971 paper in Nature entitled "Application of computer experimentation to the cornea" Friedman succinctly summarizes the utility of mathematical and computer models in physiology:

- *It is often impractical or impossible to precisely measure all of the relevant dependent variables in the physiological system of interest during an experiment*
- *Computers and theoretical models offer perfect reproducibility for a given set of parameters and input values, unlike biological samples*
- *Modeling reduces the questions associated with artefact and trauma that may arise experimentally*
- *Models aid in the identification of critical experiments whose results can further illuminate the problem at hand.*

Thus there is a great wealth of information that can be gleaned from well-posed theoretical problems as long as the resulting data is treated with the caution it deserves. Basic scientists have used models of the cornea to extract the critical transport properties from data that would otherwise be difficult to interpret, while clinicians gain insight to the properties of the tissue for the administration of therapy.

1.2 EVOLUTION OF CORNEAL MODELS

In 1873, Leber demonstrated that the cornea swells and becomes opaque when placed in aqueous solution. Since that time, physiologists and engineers have examined the transport properties that control the thickness (and thus transparency) of this hydrophilic tissue and its limiting membranes. True quantitative experimental investigations of corneal physiology have been ongoing for 5 decades, while more complete mathematical and computational models of the transport properties of the tissue began in the mid 60s. All of this effort has led to a better understanding of the mechanisms that maintain the stroma in the state of relative dehydration necessary to allow the transmission of light (Maurice, 1957; Benedek, 1971).

1.2.1 Membrane Transport Models

When considering transport of solute and solvent across living membranes, modellers typically choose either a microscopic or a thermodynamic approach (Skadhauge, 1977). Microscopic models attempt to incorporate the microstructure of the membrane into the definition of the transport system. They require detailed geometric descriptions as well as knowledge of the transport coefficients of structures internal to the membrane. In their famous work on standing gradients, Diamond and Bossert (1967) formulated the first of these microscopic models. However, because of the complexity of the membrane cellular structure and our inability to observe the most important parameters (e.g. concentration gradients) in the relevant locations (e.g. lateral intercellular spaces (LIS)), microscopic models of membrane transport have evolved slowly and uncertainly (Diamond, 1979). The continuing avalanche of membrane solute channel data and the discovery of highly specific water channels in the cornea (Fischbarg, 1995) have also contributed to the slow development of a comprehensive microscopic membrane model. In addition to this, mathematical descriptions of the mechanics of isotonic fluid transport across epithelia have met with only limited success (Sackin and Boulpaep, 1975; Hill, 1975; Hill, 1980; Fischbarg, 1997). It is certainly not beyond the state of the art to include an extensive set of membrane transport parameters into microscopic models of the corneal epithelia (Leibovitch and Weinbaum, 1981). However, the predictions of such models ultimately depend highly on the choice of coefficients whose values have not been fully validated. Consequently, the most successful (Klyce and Russell, 1979) and the latest (Bryant and McDonnell, 1998) models of the full corneal transport system include very simple thermodynamic mathematical representations of the limiting layers, but complicated representations of the stroma.

Thermodynamic models of epithelia do not require detailed knowledge of the cellular microstructure and typically do not solve for spatially dependent concentrations or pressures. Such models are based on the work of Kedem and Katchalsky (1958) who employed the formalism of irreversible thermodynamics and Onsager's (1931) reciprocal relations to describe flows of solute and water through biological membranes. Thermodynamic models eventually became quite complex and have included basolateral membrane transport, multiple ions, and paracellular pathways (Sackin and Boulpaep, 1975). Models that utilize this formalism rely on phenomenological transport coefficients.

1.2.1.1 Corneal Epithelial Physiology

A stratified epithelium lines the outer aspect of the cornea and consists of four to six layers of cells some 50 microns in total thickness in the human. With respect to the maintenance of corneal hydration, attention has been focused on its ion transport and permeability properties. A net excretion of solute from the central stromal connective tissue compartment, for example, could prevent edema and maintain corneal thickness.

The fundamental electrophysiological characteristics of the corneal epithelium were established with microelectrode experiments with frogs, rabbits and cats and permit the modeling of the tissue with electrical analogs (Kikkawa, 1964; Ehlers 1970; Fee and Edelhauser, 1970; Akaike and Hori, 1970; Akaike, 1971; Klyce, 1971; Klyce 1972; Klyce and Wong 1977; Festen and Slegers, 1979; Klyce and Marshall 1982; Marshall and Klyce, 1983). The superficial squamous cells and the wing cells which comprise the outer half of the epithelium are well coupled together electrically by means of desmosomal interconnections, which permit electrochemical cross-talk and allow the group to be treated as an electrical syncytium. The innermost basal cells are the germinative cells that give rise to the outer cells and are less well coupled to them. Overall, the cell layer generates a trans-epithelial potential of 20-30 mV, which is negative on the tear side, and arises from the activities of ion transport systems and asymmetries in outer and inner epithelial membrane ion permeabilities. The outer membrane of the epithelium forms a high resistance barrier due to the presence of continuous apical tight junctions around the cells and a low conductance cell membrane. These structural features give the cornea its barrier properties and help protect the eye from the penetration of tear-borne pathogens. While the primary function of the corneal epithelium is that of a barrier, it does exhibit trans-epithelial ion transport systems and their neuropharmacology is fairly well understood.

Active Na^+ Transport. Initially, a tears-to-stroma transport of Na^+ had been demonstrated in the rabbit and frog (Donn et al., 1959a, 1959b; Green, 1965; Candia and Askew, 1968; Langham et al., 1969; Klyce et al., 1973; Van der Heyden et al., 1975; Klyce, 1975). This absorptive transport of Na^+ presented somewhat of an anomaly inasmuch as an inward solute transport to the stroma would increase its osmotic pressure and presumably favor stromal edema.

The presence of the specific energy-supplying enzyme for Na^+ transport, namely Na-K-activated ATPase, has been demonstrated histochemically (Kaye and Tice, 1968). Furthermore, inward Na+ transport can be enhanced several fold by increasing the cation permeability of the superficial epithelial cell membrane with amphotericin B or Ag^+ (Shapiro and Candia, 1973; Klyce and Marshall, 1982) and can be inhibited by ouabain, a specific inhibitor of Na-K-ATPase.

However, further studies (see below) indicate that the epithelial transport of Na^+ is most likely secondary to the K^+/Na^+ exchange activity of the deeper cell's membranes, which maintains the high K^+ and low Na^+ characteristic of the cell's interiors.

Active Cl^- Transport. Subsequent to the findings on Na+ transport, the corneal epithelium was found to transport Cl^- in the opposite secretory direction, that is, from stroma to tears (Zadunaisky, 1966; Klyce et al., 1973; Wiederholt, 1980). It has been shown that the Cl^- transport dominates, such that in the living eye there will be a net movement of NaCl from the stroma to the tears (Klyce, 1975). Corneal epithelial transport of Cl^- is regulated by the β-adrenergic receptor/adenylate cyclase complex. Stimulation of Cl^- transport is accompanied by an elevation of intracellular cyclic AMP and can be accomplished by the addition of membrane-permeable forms of cyclic AMP (Klyce et al., 1973). Catecholamines, such as epinephrine, stimulate Cl^- secretion, unless blocked by β-adrenoreceptor antagonists, such as propranolol or timolol. This stimulation is enhanced by aminophylline and theophylline, which are compounds that inhibit phosphodiesterase and slow the breakdown of cyclic AMP.

The active step in Cl^- transport appears to be Cl^- uptake by the epithelial cells from the stroma in conjunction with a passive diffusional barrier at the corneal surface, which limits the exit of Cl^- from the superficial epithelial cells to the tear film. The action of epinephrine in the regulation of the overall net secretion of Cl^- appears to be the formation of functional Cl^- conductance channels in the apical membrane of the superficial epithelial cells (Klyce and Marshall, 1982).

Neural Control. The likely physiologic source of catecholamines in the cornea appears to be the sympathetic fibres that innervate the tissue. The population of the β-adrenoreceptor receptors is dynamic; their density decreases after topical administration of epinephrine (Candia and Neufeld, 1978) and increases after superior cervical ganglionectomy (Klyce et al., 1985).

Other epithelial regulatory receptors have been identified. Both serotonin (Klyce et al., 1985) and dopamine (Klyce et al., 1982) have been found to stimulate Cl^- secretion in the rabbit cornea and their specific receptor antagonists block this action. Serotonin increases cyclic AMP levels in the epithelium (Neufeld et al., 1982) and evokes membrane conductance changes similar to those of epinephrine, including increased Cl^- conductance in the outer membrane of the superficial cells (Marshall and Klyce, 1984). Curiously, the nonselective, β_1- and β_2-adrenoreceptor antagonist, timolol, was found to block the activity of serotonin and dopamine. However, it was shown that the corneas of animals that had undergone superior cervical ganglionectomy (Klyce et al., 1985) and corneal epithelium in tissue culture (Jumblatt and Neufeld, 1983) lacked responsiveness to serotonin. Hence, it was proposed that the serotonin receptors and, potentially, the dopamine receptors, are located on the sympathetic fibres themselves. In this scheme, activation of serotonin or dopamine receptors located on the sympathetic fibers causes the release of catecholamines (presumably norepinephrine), which, in turn, activates epithelial β-adrenoreceptors (Fig. 1.2).

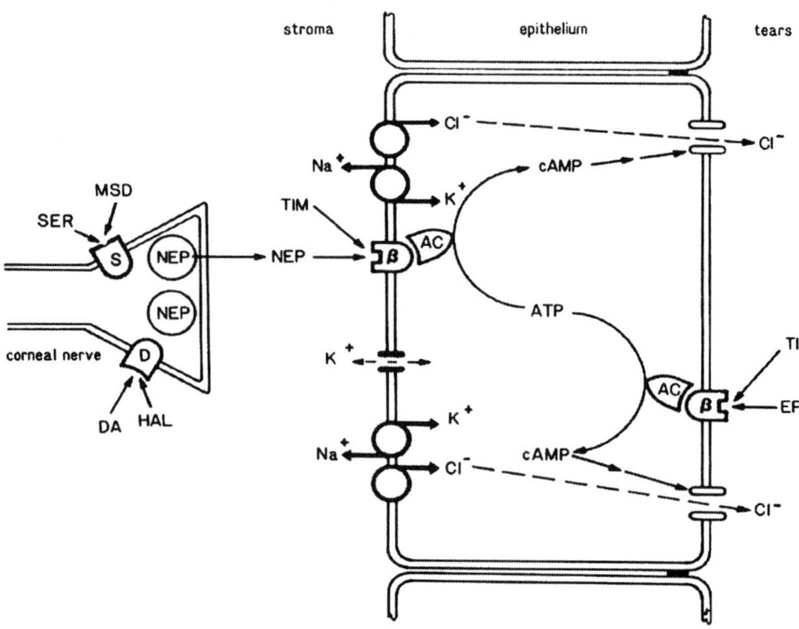

Figure 1.2. (See previous page). Model for the neuroregulation of Cl^- transport in the corneal epithelium. Serotonin and/or dopamine evoke the release of norepinephrine from sympathetic nerve fibers in the cornea. In turn, norepinephrine activates adenylate cyclase via β-adrenoreceptors increasing cell levels of cyclic AMP, and increasing the chloride conductance of the apical epithelial membrane. *EPN*, epinephrine; *NEP*, norepinephrine; *TIM*, timolol; *SER*, serotonin; *MSD*, methysergide; *DA*, dopamine; *HAL*, haloperidol; β, β-adrenoreceptor; *S*, serotonin receptor; *D*, dopamine receptor; *AC*, adenylate cyclase. Reprinted from Klyce and Crosson (1985), pg. 326, Fig. 2, with permission of Swets and Zeitlinger.

1.2.1.2 Corneal Epithelial Transport Models

It is instructive to briefly consider a classic work on the modeling of the rabbit corneal epithelium. Friedman (1978) constructed a multi-ionic, three-layer membrane transport model based on thermodynamics. The epithelium was represented as a single cell layer with a paracellular shunt. The trans-cellular route for transport comprised two membranes in series with an intervening cytoplasmic phase. The lumped resistance of the model membranes represented the general characteristics of the anterior portion or the posterior portion of the multi-cellular epithelium respectively. This was considered the simplest representation of the epithelium that captured all of the salient characteristics that determine the transport properties of the tissue. Specifically, Friedman modelled the transport of monovalent ions across a charged membrane using the Nernst-Planck equations while neglecting solute-solute interactions and solvent flow. Donnan equilibria at the membrane interfaces and electroneutrality were employed. Even with this "simplified" model, the number of dependent variables numbered 34 just for

σ_{nacl}	Lp^a	$\omega_{NaCl}{}^b$	$J_a{}^c$	Source
-	-	0.025 (Na)	-	Maurice (1951)
-	-	0.014 (Cl)	0.03	Klyce (1975)
0.79	6.1	0.019 (NaCl)	0.08	Klyce & Russell (1979)
-	-	0.025 (Cl)	0.055	Friedman (1978)
1.0*	6.9	-	-	Mishima & Hedbys (1967)
1.0*	6.5	-	-	Stanley et al. (1966)
0.8	0.4	-	-	Green & Green (1969)
-	0.9	-	-	Green & Downs (1976)
-	9.59	-	-	Fischbarg&Montoreano

Table 1.1. Corneal Epithelial Membrane Transport Parameters (σ_{NaCl}-reflection coefficient to NaCl; *Lp*-hydraulic conductivity; ω_{NaCl}-solute permeability; *Ja*-active pump rate). Units: ^{a}x10^{-12} cm^3/(dyne sec); ^{b}x10^{-5} cm/sec; ^{c}x10^{-10} mole/sec.

sodium and chloride transport. This large number of variables (already considered a "reduced" set) did not take into account the ability of the membrane parameters to change with time. Consequently, it is easy to see why simpler methods with experimentally verifiable phenomenological parameters have been used in models of corneal hydration control. Table 1.1 represents a list of the thermodynamic membrane transport coefficients found for the epithelium by various investigators.

1.2.1.3 Corneal Endothelial Physiology

The physiology of the corneal endothelium has been investigated vigorously for the last 5 decades. Driving this active period, investigations centred on the search for the properties of the tissue that maintain stromal hydration and transparency. It turns out that both the passive and active properties of the endothelium are implicated in stromal hydration control. Davson (1949) and Maurice (1951) were the first to suggest that active processes located in the corneal membranes were instrumental in keeping the cornea deturgesced. Davson (1955) and Harris & Nordquist (1955) provided supporting evidence by demonstrating that the corneas of enucleated rabbit eyes swell on refrigeration and return to their original thicknesses when rewarmed. Trenberth and Mishima (1968) implicated the endothelium in this process by showing that oubain inhibits transport across that layer in the *in vitro* cornea. Conversely, subsequent electrical studies of the endothelium demonstrated a low trans-membrane potential (~ 0.5 mV; tear side negative; Fischbarg, 1972), which suggests low levels of active transport. Further, the discovery of active movement of sodium from the tears to the stroma by the epithelium (Donn et al., 1959; Green, 1965) led to a hypothesis implicating the epithelium and not the endothelium as the primary site of hydration control (Green, 1968). However, in 1971, Hodson proposed the existence of a bicarbonate-dependent sodium pump in the endothelium that moves solute into the aqueous humour. In 1972, in a classic experiment Maurice demonstrated stromal deturgescence without the epithelium, indicating that the true site of active corneal hydration control is located primarily in the endothelium. Finally, Hodson & Miller (1976) measured the rate of solute transport in the endothelium and identified bicarbonate as the major solute transported. Later studies clarified the details of the sodium-potassium ATPase (Riley, 1977) and its relation to the active translocation of bicarbonate (Riley and Peters, 1981; Lim and Ussing, 1982).

Knowing the passive membrane transport properties is critical to the development of any model of stromal deturgescence. Table 1.2 lists endothelial transport parameters found by various investigators. However, the exact mechanics of isotonic "fluid" transport across the endothelium remain obscure. As we shall see, modellers are forced to make simplifying assumptions that have limited the validity of their constructs of fluid transporting epithelia.

σ_{nacl}	Lp^a	ωRT^b	J_a^c	Source
-	-	2.0 (Na)	-	Maurice (1951)
0.6	15.8	-	-	Mishima & Hedbys (1967)
0.6*	20.0	-	-	Stanley *et al.* (1966)
-	-	2.1 (urea)	-	Trenberth & Mishima (1968)
0.4	1.4	2.0 (NaCl)	-	Green & Green (1969)
-	-	2.2 (Cl)	-	Kim *et al.* (1971)
-	-	-	2.9	Hodson & Miller (1976)
-	3.0	-	-	Green & Downs (1976)
0.6*	8.0	-	-	Fischbarg *et al.*(1977)
-	-	-	6.97	Hull *et al.* (1977)
0.45	42	8.0 (NaCl)	4.7	Klyce & Russell (1979)
0.42	14.2	1.6 (NaCl)	4.2	Liebovitch & Weinbaum (1981)
1.0	-	-	-	Hodson & Lawton (1987)
-	-	5.0	-	Hodson & Wigham (1983)
-	2.45	-	-	Liebovitch *et al.* (1981)
-	49.77	-	-	Fischbarg & Montoreano (1982)

Table 1.2 Corneal Endothelial Membrane Parameters. Units: ^{a}x10^{-12} cm^3/(dyne sec); ^{b}x10^{-5} cm/sec; ^{c}x10^{-10} mole/sec.

1.2.1.4 Corneal Endothelial Transport Models

Because of its central role in the active dehydration of the corneal stroma, the endothelium has been modelled extensively (Shapiro and Candia, 1973; Lim and Fischbarg, 1976; Rehm and Spangler, 1977; Liebovitch and Weinbaum, 1981; Fischbarg et al., 1985). In 1976, Lim and Fischbarg examined the validity of the standing-gradient model of osmotic flow across the endothelium. In this microscopic model they present an extension of the perturbation analysis of Segel (1970) and demonstrate that the numerical results of Diamond and Bossert (1967) could be matched analytically. However, they determine that the standing gradient model of Diamond and Bossert (1967) cannot generate isotonic emergent fluid for even outlying values of paracellular pathway dimensions and osmotic permeability. Later measurements of high endothelial membrane hydraulic conductance (Klyce and Russell, 1979; Fischbarg and Montoreanno, 1982) prompted Fischbarg et al. (1985) to revise their model and altered their assessment of the standing gradients to a more favourable one, but isotonicity was still not achieved. In addition to such microscopic models, phenomenological models based on irreversible thermodynamics have also been applied to the corneal endothelium (Shapiro and Candia, 1973; Rehm and Spangler, 1977, Candia and Reinach, 1982).

Liebovitch and Weinbaum (1981) generated a comprehensive hybrid model of corneal endothelial transport. This highly detailed model was then applied to the corneal endothelium and is similar in complexity to the model of Friedman (1978) for the corneal epithelium. However, unlike Friedman's model (1978), Liebovitch and Weinbaum (1981) synthesize the thermodynamic with the microstructural approach. They used the thermodynamic approach for fluxes across cell membranes and a microstructural approach to determine the concentration and hydrostatic pressure *gradients* in the LIS. A schematic of their model is shown in Fig. 1.3.

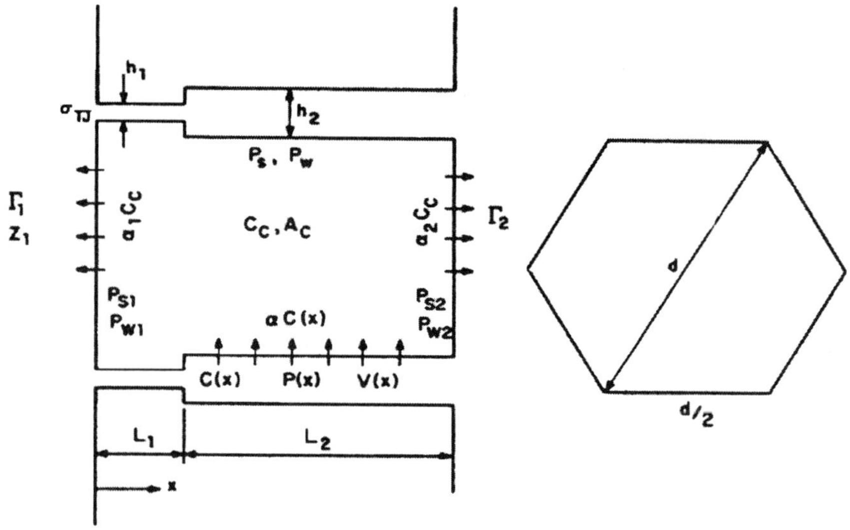

Figure 1.3. Transverse and face on views of the corneal endothelium comprising a single cell layer and showing the relevant parameters in the model. h_1 and h_2 are the dimensions of the LIS at the junction and between the cells, P_S and P_W are the solute and water membrane permeability (subscript definitions: no subscript - lateral membrane, *1* - apical membrane, *2* is basal membrane), Γ is the concentration of permeant solute (*1* - aqueous humour, *2* - stroma), Z_1 is the concentration of the impermeant solute in the aqueous humour, α is the pump rate proportionality constant (subscript definitions: no subscript - lateral membrane, *1* - apical membrane, *2* is basal membrane), σ_{TJ} is the reflection coefficient of the tight junction to the solute, *P(x)* is the hydrostatic pressure in the LIS, *C(x)* is the concentration of solute in the LIS, *V(x)* is the fluid velocity in the LIS, and L is the length of either the tight junction (subscript 1) or the remainder of the LIS (subscript 2). Reprinted from Liebovitch and Weinbaum (1981), Fig. 3, with permission of the Biophysics Society.

Liebovitch and Weinbaum (1981) prescribe the cell dimensions, the pump proportionality constants, the cell membrane permeabilities to both water and solute, the pressure on the apical and basal side of the endothelium, and the concentrations of the solutes on the apical and basal sides of the membrane. They solved for the intracellular concentration of solutes and proteins and the spatial distributions of concentration, hydrostatic pressure and solvent velocity in the LIS. Appropriately, they wrote down five equations to solve for these unknowns.

Their specific advancement over contemporary modes was that the cells were not allowed to continuously imbibe fluid without accountability for their volume changes. Their equations were solved using a perturbation series, which extended the work of Segel (1970). A detailed account of the solution is provided in the appendix of Liebovitch and Weinbaum (1981).

To apply their general formulation to the corneal endothelium (a "backwards" pumping membrane), they parameterised it in the following way:

- LIS width: 30 nm; length: 12 μm
- Tight junction width: 4 nm; length: 1 μm
- Bounding solute concentrations: 300 mOsm
- Diffusion coefficient in channel: 1×10^{-5} cm^2/sec
- Reflection coefficient across tight junction: 0.0
- Membrane osmotic/hydraulic conductance: 0.275 cm^4/mole/sec
- Membrane solute permeability: 3.0×10^{-7} cm/sec
- Lateral membrane pump strength (inward) ~ 2×10^{-6} cm/sec
- Apical solute pump strength ~ 4.4×10^{-6} cm/sec
- Pressure difference *in vivo*: -70 torr; *in vitro* (swollen): 0.0 torr

The active pump strength was adjusted to fit the net fluid transport to the experimentally determined *in vitro* values of 2-6 μl/h/cm^2 (Maurice, 1972). The model was used to examine the gradients in the LIS, the concentrations inside the cell and the net fluxes for water and solute across the whole membrane.

For the *in vitro* case corresponding to the deswelling experiment of Maurice (1972), the model is set to generate a net solvent flux across the membrane of 2.5 μl/hr/cm^2. For the *in vivo* case, a trans-membrane pressure difference of 70 torr (stromal side negative) is applied to just halt the trans-membrane fluid flux (see Fig. 1.4 for details).

In vivo, the model predicts that the stromal swelling pressure forces water through the tight junctions towards the stroma so that there is no net flow across the membrane. The model also predicts a lower concentration of solutes in the cell, which results in the transfer of equal amounts of water out of the cell through both the stromal and aqueous humor faces. Water lost through the membranes must be resupplied to the cell through the lateral membrane lining the LIS (which is maintained at a low concentration by the active pumps). Because it is more difficult for fluid to enter the LIS through the apical tight junction, a sharp pressure drop occurs across it, which ultimately balances the stromal swelling pressure. Using the result of their model, Liebovitch and Weinbaum (1981) go on to calculate trans-endothelial values for the permeability to their electroneutral "solute", the hydraulic conductance and reflection coefficient to their solute. The values obtained are presented in Table 1.2 and are in relatively good agreement with the literature.

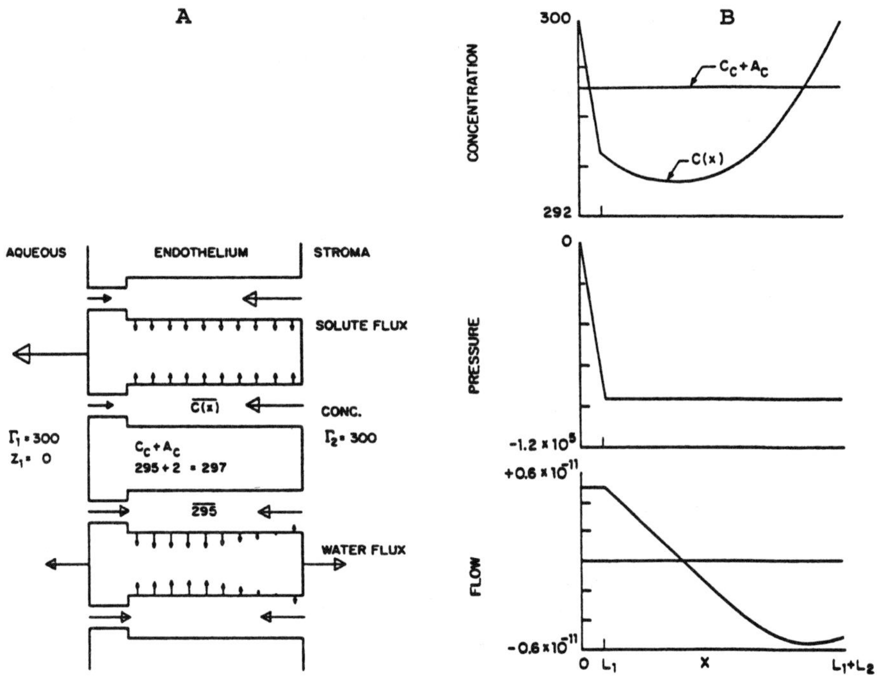

Figure 1.4. Flows of solute and water in the *in vivo* case (trans-membrane pressure difference of 70.0 mmHg). (A) There is no net fluid or solute flux from across the membrane. The overbar indicates average concentration in the LIS. (B) Concentration (mOsm), pressure (dyne/cm2) and volume flux (cm^3/sec/cell) profiles along the LIS. Reprinted from Liebovitch and Weinbaum (1981), Figs. 6 and 7, with permission of the Biophysics Society.

The model of Leibovitch and Weinbaum (1981) demonstrates how complex mathematical models can be designed to represent physiological systems fairly accurately. The one adjustable parameter used, the pump rate, turned out to be close to the values found by other methods (Table 1.2). However, the predicted value of the emergent osmolarity of the fluid pumped by this system was at least twice that required for isotonicity. Thus, the model does not describe isotonic transport of fluid adequately.

It appears that standing gradients and thermodynamic models cannot alone explain isotonic fluid transport. Perhaps a more complicated model, such as the sodium recycling mechanism proposed by Ussing and Eskesen (1989) is necessary.

1.2.2 Models of the Corneal Stroma

Since the early observations of stromal swelling by Leber (1873) there have been many investigations that attempt to determine the physicochemical properties of the corneal stroma. Accurate assessments of the forces that drive the imbibition of fluid are critical to the generation of models of the whole corneal transport system. Such swelling forces must be countered by the pump and barrier functions of the limiting epithelia to maintain corneal transparency (Maurice, 1957; Benedek, 1971). The success of the mathematical modeling of the stroma is probably due to the small number and low activity level of the stromal keratocytes. Without the confounding behavior of a large cell population, the stroma can be reduced to a stable, well-organized biological polyelectrolyte gel. This conception of the tissue has worked extremely well for most investigations of the stroma performed on a short time scale. Only recently has the short-term stability of stromal properties been called into question (Ruberti et al. 2000). Consequently, the corneal stroma has been extensively investigated and described with regard to its swelling pressure and transport properties with a high degree of success.

1.2.2.1 Empirically-Based Models

Empirical models require experimental data. For the corneal stroma, equations that relate the swelling pressure and the permeability to hydration, H, (mass of water/mass of dry tissue) proved essential for the development of empirically based dynamic models of the stromal tissue. The elegant work of Hedbys and Mishima (1962) relates the stromal flow conductivity, k/n, in both the in-plane and transverse direction to hydration (Fig. 1.5). Hedbys and Dohlman (1963) performed an additional series of experiments that related the swelling pressure of the corneal stroma also to hydration (Fig. 1.6). These data remain the gold standard.

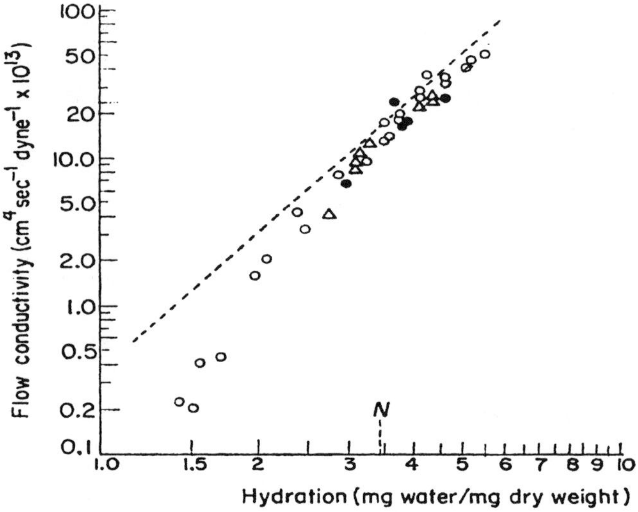

Figure 1.5. Flow conductivity versus hydration across the cornea. The symbols refer to measurements taken with (closed circles) and without (open circles) Descemet's membrane in 0.9% NaCl, or without Descemet's membrane in destilled water (triangles). The broken line represents the flow conductivity versus hydration relationship for flow along the cornea. Reprinted from Hedbys and Mishima (1962), Fig. 12, with permission of Harcourt, Inc.

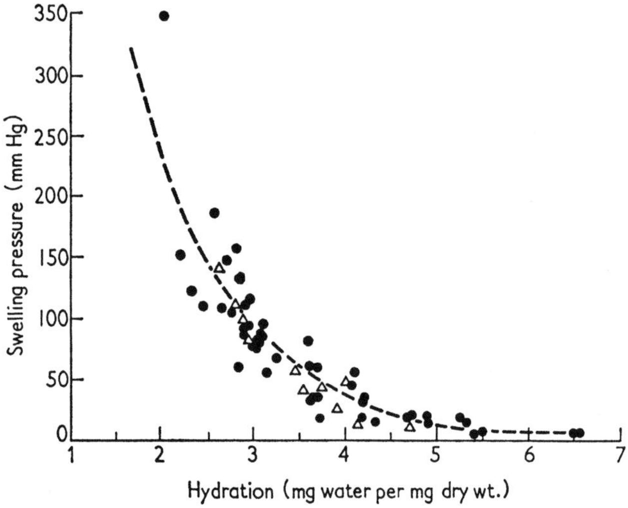

Figure 1.6. Swelling pressure rabbit (filled circles) and human (open triangles) stroma plotted against hydration. The broken curve represents steer swelling pressure. Reprinted from Hedbys and Dohlman (1963), Fig. 6, with permission of Harcourt, Inc.

In their seminal work, Fatt and Goldstick (1965) produced one of the first general theoretical descriptions of the physicokinetics of swelling tissue. Their analysis focused on the general mathematical description of the free swelling of membranes and applied this theory to corneal stroma.

Fatt and Goldstick considered stromal swelling to occur one-dimensionally in the anterior-to-posterior direction. The absence of cross-linking in this direction and the high modulus of collagen along the lamellae made this assumption a reasonable one. In support, the work of Hedbys and Mishima (1966) had demonstrated that during swelling, the thickness and the hydration of the tissue were linearly related. The assumption allowed Fatt and Goldstick to treat the dry tissue mass in a unit differential element to be a constant while the dimensional changes during swelling were ascribed to the unidirectional arrival and retention of solvent. To cast the problem in terms of familiar a concept, they chose to make the hydration, H, a dependent variable in the equation set. Their one-dimensional conservation equation was:

$$\frac{\partial H}{\partial t} = -A\frac{\partial f}{\partial m} \tag{1.1}$$

where t is time, f is the mass flow rate of solvent through unit area, A, and m is the unchanging mass of the *dry* tissue. Thus, the spatial coordinate, x, which is variable in time and attached to the tissue boundary can be recovered from the following:

$$dx = \frac{dm}{\delta A} + \frac{dm_w}{\gamma A} \tag{1.2}$$

where the δ is the density of the dry stromal tissue, γ is the density of water and m_w is the mass of water in the tissue. Using the definition of H and introducing ε (ratio of the density of water to the density of tissue, δ/γ) and ψ (the time invariant volume of the dry tissue, $m/\delta A$) the time variant spatial coordinate can be written:

$$dx = \left[1 + \frac{\delta}{\gamma}H\right]d\psi \tag{1.3}$$

Fatt and Goldstick then define the volume flow rate though a segment of tissue based on Darcy's law, which is converted to mass flow rate using the density of water to:

$$f = -\frac{k\gamma}{\eta}\frac{\partial P}{\partial x} \tag{1.4}$$

where k is the permeability constant, η, is the viscosity of water and P is the hydrostatic pressure at any point. Using the definition provided by Eq 1.3:

$$f = -\frac{k\lambda\varepsilon}{\eta\,(\varepsilon + H\,)}\frac{\partial P}{\partial \psi} \tag{1.5}$$

which is Darcy's law cast in terms of the hydration, the local pressure and the volume of the dry of the tissue. Note that the moving spatial coordinate has been removed as an independent variable at the cost of the introduction of a new dependent variable, H and the new independent time invariant coordinate, ψ. Finally, their initial conservation equation can be recast in light of the above equation as:

$$\frac{\partial H}{\partial t} = \frac{1}{\delta}\frac{\partial\left[\frac{k\gamma\varepsilon}{\eta\,(\varepsilon+H)}\frac{\partial P}{\partial \psi}\right]}{\partial \psi} \tag{1.6}$$

To make use of Eq. 1.6, one needs to know the hydrostatic pressure as a function of position inside a piece of swelling tissue. Based on the work of Hedbys et al. (1963) and Hedbys and Dohlman (1963), Fatt and Goldstick assumed that the only other relevant pressure, the swelling pressure, p, in the tissue was a unique function of position. Under equilibrium conditions, Hedbys and Dohlman (1963) measured the swelling pressure and found it to be the negative of the mechanical pressure necessary to keep tissue from swelling (i.e. no discernible solid stress). Given that osmotic pressure in the tissue was "negligible" (Green and Green, 1969) Fatt and Goldstick wrote the hydrostatic pressure in the following simple way:

$$P = -\lambda\, p \tag{1.7}$$

where λ is an empirical parameter. Now that the hydrostatic pressure is written in terms of the swelling pressure, p, the fact that Hedbys and Dohlman (1963) found the swelling pressure to be a monotonic function of hydration can be put to use. Equation 1.7 is then differentiated with respect to space to give:

$$\frac{\partial P}{\partial \psi} = -\lambda\frac{dp}{dH}\frac{\partial H}{\partial \psi} \tag{1.8}$$

Combining Eqs. 1.8 and 1.6 gives:

$$\frac{\partial H}{\partial t}=-\frac{\lambda k\varepsilon^{2}}{\eta(\varepsilon+H)}\frac{dp}{dH}\frac{\partial^{2}H}{\partial\psi^{2}} \tag{1.9}$$

where the products of the higher power derivatives are omitted. This is a general expression for the free swelling of gels. It has the form of an unsteady-state diffusion equation with a non-linear transport coefficient, *D(H)*.

$$D(H)=-\frac{\lambda k\varepsilon^{2}}{\eta(\varepsilon+H)}\frac{dp}{dH} \tag{1.10}$$

This term makes the conservation equation non-linear and reflects the balance of the tissue resistance and pressure gradient in the generation of fluid movement. Fatt and Goldstick linearized Eq. 1.9 by assuming that the coefficient *D(H)* could be replaced by an average value, $\underline{D(H)}$. The solutions to the resulting equation, in terms of $\underline{D(H)}$, were provided using the methods of Crank (1956). The results are shown in Fig. 1.7.

Fatt and Goldstick applied their theory to the corneal stroma. However, because of the strong dependence of both the swelling pressure and the permeability on hydration, Fatt and Goldstick's linearized equation could not be tested against corneal data, which was available only over large ranges of hydration. In any case, they correlated the swelling pressure vs hydration data of Hedbys and Dohlman (1963) and the permeability vs hydration data of Hedbys and Mishima (1962) to calculate an *average* transport coefficient. However, it is clear from the data of Hedby's and his collaborators that neither the swelling pressure nor the permeability of the corneal stroma can be represented over even a small range of hydrations by an average value.

In general, the value of the transport coefficient, *D(H),* can be obtained for all values of *H* only if *p*, k/η, and λ are known. Fatt and Goldstick characterize the first two by fitting available data but are only marginally successful with the empirical constant λ that relates the swelling pressure to the hydrostatic pressure during fluid imbibition (non-equilibrium). It will subsequently be shown that $\lambda \sim 1$ (Friedman, 1971b). Figure 1.8 shows the value of the transport coefficient as a function of hydration for the stromas of three different animals (variable ε).

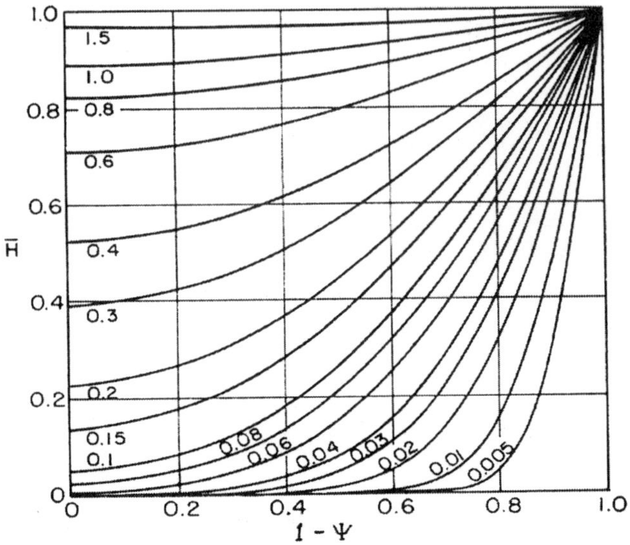

Figure 1.7. Relative hydration, $\underline{H}=(H-H_{initial})/(H_{final}-H_{initial})$, as a function of relative depth. The tissue slab is imbibing fluid from both sides (only half is slab is shown). The numbers on each curve represent dimensionless time, τ, scaled off of the transport coefficient. Reprinted from Fatt and Goldstick (1965), Fig. 2, with permission of J. Colloid and Interface Science.

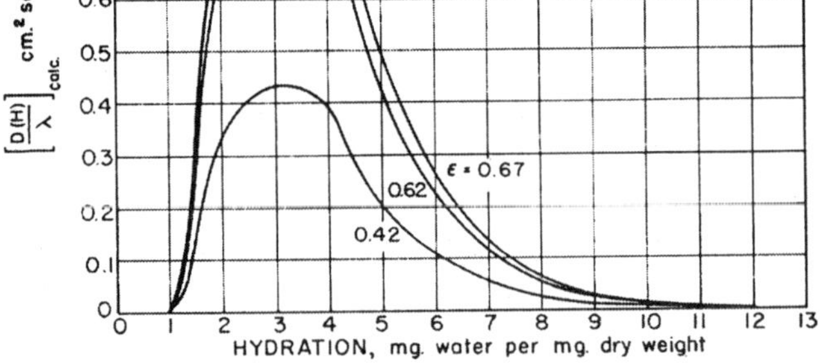

Figure 1.8. Fatt and Goldstick's transport coefficient calculated from their correlates of the equilibrium data of Hedbys and Mishima (1962) and Hedbys and Dohlman (1963). Steer ε=0.67, large rabbit ε=0.62 and small rabbit ε=0.42. Reprinted from Fatt and Goldstick (1965), Fig. 8, with permission of J. Colloid and Interface Science.

For all three types of cornea examined, the transport coefficient is maximal at a space-averaged hydration near the normal physiological hydration for corneal stroma. Clearly, using an average value to represent the transport coefficient over the entire range of hydrations is a serious limitation of the analysis.

In general, the work of Fatt and Goldstick is seminal with regard to the theoretical treatment of swelling tissue, even though their governing equation (Eq. 1.9) is not tractable analytically due to its inherent non-linear transport coefficient. However, this will not be a limitation for subsequent numerical investigations

Fatt (1968a) attempted to test his linearized theory of stromal swelling over a very small range of hydrations. In this way, the error introduced by using an average value for the transport coefficient was reduced. He used Eq. 1.9, his own measurements of the dynamic response of the stroma to loading and static swelling pressure measurements over a small range of H to estimate the permeability of the corneal stroma. His estimations of permeability matched the data of Hedbys and Mishima (1962) fairly well, indicating that his linearized theory accurately modelled the swelling response of the stromal tissue for small ranges of hydration.

Friedman (1971b) extended the work of Fatt and Goldstick (1965) with a more rigorous treatment that attempted to remove the parameter that relates the swelling pressure to the hydrostatic pressure, λ. He starts with the Fatt and Goldstick's conservation equation, Eq. 1.1, and then relates the flux of fluid to the hydration and the hydrostatic pressure using conservation of momentum. Friedman writes 5 equations in five dependent variables: hydration, mass flux of water, velocity of the local stromal element, swelling pressure and hydrostatic pressure. In formulating the relationship between swelling pressure and hydration, he assumes that the swelling pressure is only a function of H (as did Fatt and Goldstick) and ends up with the same form of the equation Fatt and Goldstick used when correlating p to H. For permeability, he uses the data of Hedby's and Mishima (1962) in conjunction with a theoretically based *form* of the dependence of Darcy's law on hydration. The resulting correlation equation is:

$$\frac{k}{\eta} = C_1 \frac{H^3}{H + \varepsilon} \tag{1.11}$$

where C_1 is an empirical constant extracted from the data of Hedbys and Mishima (1962). Friedman then numerically solves his reduced non-linear equation set using the method of characteristics for the case of free stromal swelling (which includes a plane of symmetry). Here conditions at the free swelling boundary require that the first element of stroma be swollen maximally throughout the simulation while the rest of the tissue swells to follow. From his simulation, inertia is found to be negligible and thus the swelling equations simply reduce to a single equation similar to that found by Fatt and Goldstick (1965):

$$\frac{\partial H}{\partial t} = -\frac{k\varepsilon^2}{\eta(\varepsilon + H)}\frac{\partial}{\partial \psi}\left[\frac{dp}{dH}\frac{\partial H}{\partial \psi}\right] \tag{1.12}$$

which is Fatt and Goldstick's Eq. 1.9 without the parameter, λ, and with the products of derivatives that Fatt and Goldstick excluded "arbitrarily". Friedman then solved Eq. 1.12 numerically for the condition of free swelling. A comparison of swelling curves for the average hydration of steer stroma indicates that the Friedman's model swells more slowly than actual stroma. Friedman suggests that Fatt and Goldstick's results which predict a faster swelling rate are due to their different correlations of the permeability data of Hedbys and Mishima (1962). Other than the differences in the correlations of the swelling pressure and permeability data, the work of Friedman (1971b) and Fatt and Goldstick (1965) are ostensibly the same.

The solutions to Friedman's (1971b) swelling equations were obtained numerically but were not fully testable given the limited available data. Friedman and Green (1971) produced an experimental and theoretical paper that examined free swelling curves of rabbit and steer stroma. The goal of the investigation was to verify Friedman's (1971b) original stromal swelling model and to re-evaluate the empirical transport coefficients that relate stromal swelling pressure and conductivity to hydration. Their swelling pressure equation was:

$$p = \gamma \exp(-\beta H) \tag{1.13}$$

where Friedman and Green use $\gamma = 886$ and $\beta = 0.809$ and Fatt and Goldstick used $\gamma = 1810$ and $\beta = 1$. The reason for the different correlation is not clearly outlined. For the permeability Friedman and Green use the same form as Friedman (1971b). These relationships were placed into an integral form of the original Friedman differential swelling equation:

$$\frac{d\overline{H}}{dt} = \frac{2\varepsilon^2 \overline{p}}{2\Psi}\frac{\overline{k}/\eta}{H + \varepsilon} \tag{1.14}$$

where, Ψ, is the Lagrangian distance from the free stromal surface and the overbar implies the value of the parameter is taken at the spatially averaged hydration. This serves to soften the gradients and keeps the empirical relationships within the limits of the available data. This equation is integrated and used to fit the swelling behavior of both steer and rabbit stroma. The results show that the swelling rate of the tissue versus time is well characterized by the integral equation if the coefficient for the permeability relationship, C_1, is set to 22.0 x 10^{-14} cm^4/dyne sec for steer stroma and 33.0 x 10^{-14} cm^4/dyne sec for rabbit stroma. With these values, the empirical relationship for stromal hydraulic conductivity slightly overestimates the empirical data of Hedbys and Mishima (1962) and Fatt (1968a).

The work of Fatt and Goldstick (1965), Friedman (1971b) and Friedman and Green (1971) characterized the free swelling properties of the corneal stroma fairly accurately. Their approaches were general and resulted in an unsteady-state diffusion equation with a non-linear transport coefficient. Only the characterization of the transport coefficients required empirical data.

1.2.2.2 Moving Towards Models Based On First Principles

Ideally, from the details of the structure and composition of a tissue and the nature of the transported solutes and solvents, one should be able to write down a set of equations that accurately describes the relevant stromal transport dynamics. Using this approach, attempts have been made to estimate both the permeability and the swelling pressure of the corneal stroma based on both biochemical and ultrastructural data.

Estimates of stromal water permeability

Attempts to estimate the stromal hydraulic conductivity from biochemical data have resulted in overestimates of more than an order of magnitude (Hedbys and Mishima, 1962). Basing his work on biochemical and structural information, Levick (1987) outlines how the main constituents of extracellular matrix generate low hydraulic conductivities even though estimates of their individual contributions give very high values. Though collagen is the major stromal component, the principle contribution of hydrated collagen fibrils to the resistance of the tissue is due to their exclusion effects. They not only reduce the available area for flow, but also compress the GAGs and the core proteins into the remaining flow space. This is why the contribution of both GAGs and proteoglycans to the tissue resistance appears greater than would be expected given their respective concentrations of 10.3 mg/ml and 27.5 mg/ml (Hedbys, 1963).

In a direct demonstration of this effect, Overby et al. (2000) have applied the transmission electron microscopic preparation technique of quick freeze/deep etch to corneal stroma. This method relies on quick freezing of the tissue to preserve the nanostructure while deep etching of the exposed well-frozen matrix removes water from the relevant hydrodynamic spaces. The etched surface is coated with a thin layer of platinum and carbon. The resulting replicas were examined on a TEM equipped with a goniometer. Using the stereo images to estimate the replica thickness, a sophisticated stereological technique to extract the void volume and the wetted surface area, and the Carmen-Kozeny equation, Overby et al. (2000) were able to estimate to reasonable accuracy the specific hydraulic conductivity of the stroma directly from a micrograph.

Estimates of stromal swelling pressure

Because of the tendency of the stroma to imbibe fluid and lose its transparency when either limiting cellular membrane is damaged, the forces driving the swelling have been extensively investigated. The result has been a long history of debate over the details of mechanisms responsible for the total swelling pressure of the stroma. Until they are fully resolved, models of corneal swelling based only on physical properties will remain elusive.

In 1961, Hedbys implicated electrostatic repulsion between the fixed charges on the polysaccharides of the stroma as the primary driving mechanism for swelling. In the same year these polysaccharides were isolated from the stromal tissue (Anseth, 1961). Indeed, in the stroma of the ox, the "fixed" negative charge of the tissue of 47 meq/liter (Hodson, 1971 - at physiological pH and NaCl concentration) is adequate to explain the observed corneal swelling pressure (Hodson et al., 1992). In the early 1970s, Green et al. (1971) postulated that the stromal GAGs have the ability to bind sodium ions. They coupled this with the fact that the corneal epithelium was found to actively transport sodium into the stroma and developed a theory for stromal hydration control. The theory held that the corneal epithelium would alter the sodium transport rate to control the stromal swelling pressure. Increasing sodium transport would collapse the GAGs via ion binding, and reducing the sodium transport allows dissociation of the sodium and a concomitant increased swelling pressure. This proposition led to a lively exchange in the literature, but ultimately succumbed to Maurice's (1972) unequivocal demonstration that the endothelium controls stromal hydration and to the demonstration (Klyce et al., 1973) that the net flow of ions across the epithelium was to the tears and not into the stromal tissue.

Though there does appear to be a significantly lower sodium activity in the stroma (Stiemke et al. 1992) than the total sodium concentration there would suggest (Otori, 1967), stromal Na^+ does not modulate stromal hydration.

There has also been a significant amount of effort exerted to examine the binding of chloride ions in the corneal stroma (Hodson et al., 1992; Elliott et al., 1980). In his attempt to explain changes in the cationic exchange capacity of the stromal tissue with concentration, Elliott (1980) proposed the existence of stromal chloride ion binding ligand. Elliott and Hodson (1998) solidified their idea that transient "fixed" charges, attributable to a chloride binding ligand, are responsible for the majority of the stromal swelling pressure at physiological concentrations. If true, any model based on first physical principles must include the reaction kinetics associated with this putative ligand. However, there is considerable evidence against Cl^- binding, Maroudas (1980) has shown ion binding of GAGs to be small and Eisenberg and Grodzinsky (1985, 1987) have adequately described stromal swelling stress without invoking any putative ion binding ligand.

In their continuum model, Eisenberg and Grodzinsky (1987) look at isometric stress relaxation in the corneal stroma and in articular cartilage following changes in external ionic concentration. Their model is electro-mechano-chemical in nature and extends the KLM biphasic theory of Mow et al. (1980) and Myers et al. (1984) to include the effects of the presence of ionic species (in this case NaCl). Specifically, the aggregate modulus of the tissue, H_A and the chemical stress, β were considered functions of ionic concentration, c. The uniaxial compressive stress equation for this poroelastic model of stroma in confined compression was:

$$\sigma = H_A(c)\varepsilon + \beta(c) + P_f \tag{1.15}$$

where σ is the compressive stress, ε is the strain in the z-direction and P_f is the fluid pressure in the tissue. The mechanical stress is that necessary to keep the tissue from swelling and is equal in magnitude to the swelling pressure of the tissue:

$$p(c,\varepsilon) = H_A(c)\varepsilon + \beta(c) \tag{1.16}$$

where $p(c,\varepsilon)$ is the swelling pressure term. Conservation of mass is written in terms of the fluid velocity, U, and the tissue strain:

$$\frac{\partial U}{\partial z} = \frac{\partial \varepsilon}{\partial t} \tag{1.17}$$

The fluid velocity is determined by the permeability, k, and the local pressure gradient (Darcy's Law). Momentum is conserved by assuming there are no inertial effects and setting the compressive stress spatial gradient to zero. The mechanical stresses in the problem are defined in terms of the local concentration. The flux of the ions through the tissue is defined by:

$$\Gamma_i = -D_i \frac{\partial \bar{c}_i}{\partial z} + \frac{z_i}{|z_i|} \bar{c}_i \bar{\mu}_i E + \bar{c}_i \upsilon \tag{1.18}$$

where Γ_i is the flux of the ith ion, D_i is its diffusion coefficient, z_i is the valence, μ_i is the intratissue ion mobility, E is the electric field and v is the local fluid velocity in the z-direction. The overall current is zero under the open circuit conditions thus the sum of the fluxes of all ions must be zero. Neglecting chemical reactions or ion binding (Maroudas, 1980) and using the ambipolar diffusion, $\underline{D}_A$, coefficient modified for tortuous matrices (Mackie and Meares, 1955), the unsteady diffusion equation governing ionic flux becomes:

$$\frac{\partial \bar{c}_\pm}{\partial t} = \frac{\partial}{\partial z}\left(\underline{D}_A \frac{\partial \bar{c}_\pm}{\partial z}\right) \tag{1.19}$$

The dimensionless swelling stress equation that results from the combination of Eqs. 1.15 through 1.17 and conservation of momentum is:

$$\frac{\tau_m}{\tau_c}\frac{\partial p}{\partial \underline{t}} = \frac{\underline{H}_A}{\pi^2}\frac{\partial^2 p}{\partial \underline{z}^2} + \frac{\tau_m}{\tau_c}\left[\frac{(p-\beta)}{\underline{H}_A}\frac{\partial \underline{H}_A}{\partial \bar{c}_+} + \frac{\partial \beta}{\partial \bar{c}_+}\right]\frac{\partial \bar{c}_+}{\partial \underline{t}} \tag{1.20}$$

where t is normalized by the characteristic chemical time constant, $\tau_c=4\delta^2/\pi^2\underline{D}^i_A$, τ_m is the characteristic mechanical time constant, $\delta^2/\pi^2 H^i_a k$, the superscript indicates that the value is taken prior to the addition of the salt and the underbar indicates a normalized parameter. The aggregate modulus, H_A, is normalized to the modulus prior to the addition of NaCl. Because the tissue is in confined compression, its permeability is considered constant, unlike that in the free swelling descriptions of Fatt and Goldstick (1965) and Friedman (1971b).

Equations 1.19 and 1.20 represent a coupled non-linear system of differential equations. To be solved, they require constitutive relationships for the aggregate modulus and the swelling pressure as functions of concentration. These are provided *empirically* via data taken in previous work (Eisenberg and Grodzinsky, 1985). The aggregate modulus is a linear function of concentration. The chemical stress in the tissue was determined to be a decaying exponential dependent on concentration. The total swelling stress in the corneal stroma as a function of NaCl concentration is given by $H_A(c) + \beta_A(c)$ and is shown in Fig. 1.9.

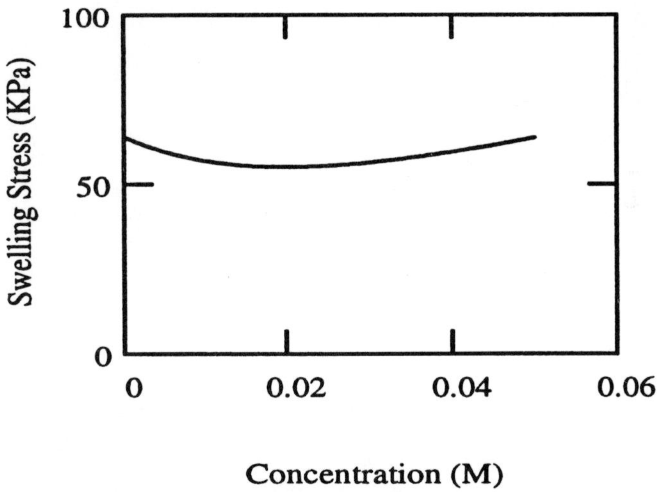

Figure 1.9. The composite empirical swelling stress ($H_A(c) + \beta(c)$) vs NaCl concentration curve used for corneal stroma.

Equation 1.15 was analytically solved to give the transient concentration profile in the tissue as function of time for boundary conditions consistent with application of a step function change in concentration at one surface of a stroma in confined compression. The result was then used in conjunction with the empirical relationships for $H_A(c)$ and $\beta(c)$ to numerically determine the evolution of the stress profile in the tissue and to calculate the swelling stress at the surface of the tissue.

The results show that when the τ_m/τ_c is near 1.0, the calculated evolution of the swelling pressure transient matches available data (Fig. 1.10). This work demonstrates that the drop in the stromal swelling pressure following a decrease in the external bath NaCl concentration is initially rapid, but then plateaus.

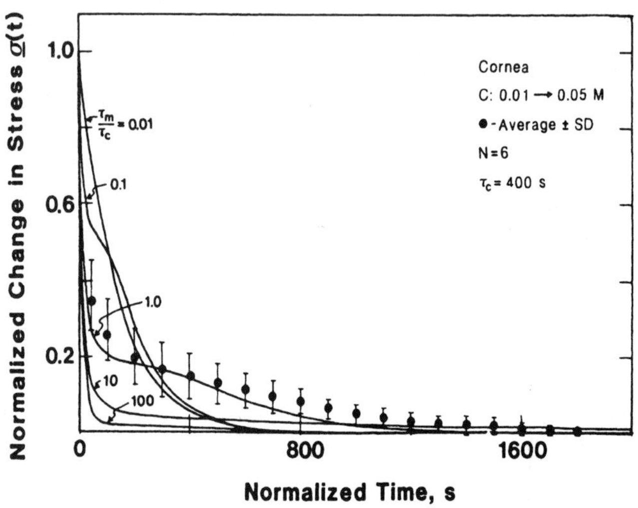

Figure 1.10. Measured stress transient (black dots) for six samples of corneal stroma undergoing an NaCl concentration step change from 0.01 to 0.05 M. Solid lines are the prediction of the numerical model parameterized by the ratio τ_m/τ_c with τ_c = 400 secs. Reprineted from Eisenberg and Grodzinsky (1987), Fig. 5, with permission of Am. Soc. Mech. Engin.

That the work of Eisenberg and Grodzinsky describes the swelling stress during the non-equlibrium diffusion of NaCl into the stroma refutes to a degree the assertion of Elliot and Hodson (1998) that an ion binding ligand is necessary to explain stromal swelling. For a more detailed array of arguments against ion binding to specific sites in the stroma, see the review of Grodzinsky (1983). More recently, Buschmann and Grodzinsky (1995) develop a microcontinuum Poisson-Boltzman model of the electrostatic forces associated with collagen. This type of model has not been applied to the cornea but will likely further reduce the dependence on empirical data.

1.2.3 Whole Corneal Transport Models

Once all of the necessary parts of the cornea have been described mathematically with some success, the next logical step is to couple the components together to see if the whole system can be accurately portrayed. There are both steady-state and dynamic models of the cornea that have been pieced together in this manner (Klyce and Russell, 1979; Bryant and McDonnell, 1998). In this section we describe in detail models that have revealed important information about the cornea and its limiting membranes.

1.2.3.1 Steady-State Whole Corneal Models

Steady-state transport models of the cornea, typically have been developed along two lines:

1. Fixed boundary models
2. Evolved boundary models

Fixed boundary models typically assume a set of tissue dimensions and then calculate the internal concentration profiles of various solutes or the flux through the tissue. Though more numerous and important for their determination of critical corneal parameters and physiological operation they will not be considered in detail. As our goal is to establish the most relevant transport parameters that, when placed into a model, generate normal corneal hydration, the evolved boundary models will be given a more thorough review.

Fixed boundary models

Fixed boundary models of the corneal transport system have provided a wealth of useful data. Stromal oxygen concentration of the open eye in air was estimated by Fatt and Bieber (1968) and of the open eye in nitrogen by Fatt (1968b). The important dynamics of oxygen diffusion through contact lenses was originally considered theoretically by Fatt et al. (1969) and revised recently by Harvitt and Bonanno, (1999). McCarey and Schmidt, (1990) generated two-dimensional models of glucose distribution in the cornea to explain anterior stromal melting following the implantation of an intrastromal lens. In addition, there are numerous examples of fixed boundary models of the cornea that have been used to investigate corneal drug permeation with good success (Grass et al., 1988; Friedrich et al., 1993; Edwards and Prausnitz, 1998).

Evolved boundary models

Evolved boundary models are used to determine the resting hydration of the stroma, given a fixed set of boundary conditions and stromal properties. Friedman (1972) constructed a one-dimensional model of the *in vivo* cornea for conditions of the time· averaged steady-state. His hypothesis was that corneal hydration was controlled purely by the effect of evaporation from the anterior surface of the eye. By his theory, the induced fluid flow through the cornea generates a drag force on the endothelium that is exactly equal to the intrinsic stromal swelling force. Thus after setting up his mathematical model and applying the appropriate physical transport properties to the various layers, he constrained the tissue such that there was no divergence in either the solute or the solvent flux.

Averaging the unsteady-state conditions was necessary to account for the diurnal variations in the concentration of the tear film (open vs. closed eyes). To simplify his model, Friedman makes the reasonable assumption that the anterior-most lamellae of the corneal stroma carry the load transmitted by the IOP and that the endothelium floats freely in mechanical equilibrium at interface of the stroma and the aqueous humour.

Friedman's rigorous treatment begins at the bounding membranes with the Kedem and Katchalsky (1961) general description for the electrochemical potential gradient in a general "membrane":

$$-\frac{d\mu_j}{dx} = \sum_{j \neq k} f_{jk}\left(v_j - v_k\right) \tag{1.21}$$

where μ_j is the electrochemical potential of the *jth* species, f_{jk} is the friction coefficient relating the drag force on a mole of *j* caused by the flow the *kth* species. After applying Onsager (1931) reciprocity, electroneutrality and making some assumptions about the similarity of the sodium and chloride permeability to the ratio of their mobility in free solution Friedman generates two equations specific for the limiting membranes of the stroma. The *solute* flux equation:

$$-\frac{2RT\Delta c_s(k)}{\bar{c}_{sk}\Delta x_k} = (K_{sk}J_s - J^a_{+k})\frac{f_{Tk}}{\bar{c}_{sk}} - K_{sk}f_{+k}\bar{V}_0(J_0 - J^a_{0k}) \tag{1.22}$$

where *R*, is the gas constant, *T* is the temperature, $\Delta c_s(k)$ is concentration difference across the *kth* membrane, c_{sk} is the average membrane concentration, Δx_k, is the membrane thickness, K_{sk} is a measure of the resistance of the membrane to chloride flux, J_s is the chloride flux, $J^a{}_{+k}$ is the active sodium flux, f_{tk} is the drag coefficient of the salt on the membrane, f_{+k} is the drag coefficient of sodium on the membrane, J_0 is the steady-state solvent flux, and $J^a{}_{0k}$ is the is the active water flux. The description of *solvent* flow through the either membrane was:

$$-\frac{\Delta P_k}{\Delta x_k}=\frac{RT\Delta c_I(k)}{\Delta x_k}+(f_{Tk}-f_{+k})(K_{sk}J_s-J_{+k}^a)+f_{0k}(J_0-J_{0k}^a) \quad (1.23)$$

where ΔP_k is the pressure drop across the kth membrane, $\Delta C_I(k)$ is the concentration of impermeant molecules and f_{0k} is the drag coefficient for water on the membrane. Friedman assumes that the interaction of the water and solutes occurs in the intercellular spaces between the cells of epithelium and endothelium. To obtain the hydration of the corneal stroma, Friedman writes the following equation:

$$\frac{dH}{dt}=(dp/dH)^{-1}(k/\eta)^{-1}(J_s\overline{V}_s+J_0\overline{V}_0) \quad (1.24)$$

where k/η is the permeability, J_s is the solute flux, J_0 is the water flux, V_s is the specific volume of the NaCl, and V_0 is the specific volume of water. As discussed previously, both the swelling pressure and the permeability of the stroma are purely functions of hydration. Friedman repeats his earlier (Friedman, 1971b) analysis and ties the variable spatial coordinate to the dry tissue thickness. He applies his own empirical correlations of the data of Hedbys and Mishima (1962) and Hedbys and Dohlman (1963) and integrates Eq. 1.24 to generate an explicit expression for hydration (water concentration).

For the solute concentration, Friedman corrects the diffusion coefficient for the available solvent in the stroma, assumes that the concentration gradient is negligible in the tissue and integrates from the epithelium to the endothelium. Eight equations with eight unknowns are the result. There are four membrane flux equations, two stromal flux equations and two mechanical equilibrium conditions (endothelial pressure balance and stromal pressure balance). The difference between the averaged, steady-state tear film concentration (177.0 mM) and the aqueous humor concentration (149.0 mM) provided the majority of the driving force for fluid transport in this model.

Friedman finds that his model can account for the stromal steady-state thickness without invoking an endothelial pump (which was his goal). He obtained a trans-corneal flow rate of 11 microns/sec. This flow rate generated a drag force on the endothelium adequate to compress the stromal tissue to a thickness of about 370 microns, which is near the normal rabbit stromal thickness. The major finding, though incorrect, was that no endothelium pump is necessary to maintain the stroma at its normal thickness. It is interesting to consider why Friedman's analysis predicts this, in light of the established fact that an endothelium fluid pump is necessary (Maurice, 1972). The error may reside in the low value used for endothelial hydraulic conductivity, which was based on the data of Green and Green (1969). This may have resulted in an excessively large solvent drag force across the endothelium, which compressed the stroma.

Bryant and McDonnell, (1998) produced the most recent and advanced model of corneal transport and swelling for both the intact eye and the stroma in confined compression. Their goal was to unify the "pump-leak" hypothesis (Klyce and Russell, 1979; Maurice, 1984) with the Donnan view of stromal swelling (Hodson, 1971). In the "pump-leak" hypothesis, the hydrostatic pressure in the stroma is lower than that in the aqueous, but is balanced by the osmotic pressure across the endothelium. Unfortunately, the total concentration of ions in the stroma appears to be higher than it is in the aqueous humor (Otori, 1967; Stiemke et al, 1992) making a "balance" impossible. To reconcile this apparent inconsistency, Bryant and McDonnell include the interaction of the solute ions with the fixed negative charges on the stromal proteoglycans. In this triphasic analysis of corneal swelling and hydration control, they generate a steady-state model designed to determine the swelling pressure and thickness of the corneal stroma under a variety of conditions (including confined compression). They start with the theory of Lai et al (1991) and add appropriate boundary conditions that represent the active and passive properties of the endothelium. Their model is "two-dimensional" but is reduced using symmetry. In the stroma, they assume that the number of fixed negative charges remains constant and that there is no net velocity of the matrix, the fluid or the ions. The solid phase of the tissue is modelled as linearly elastic and transversely isotropic and the mixture stress is written for small strains as:

$$\underline{\sigma} = \underline{\underline{C}}\underline{\varepsilon} + \underline{\sigma}^0 - p\underline{I} \tag{1.25}$$

where $\underline{\sigma}$ is the mixture stress, $\underline{C}$ is the material property tensor, $\underline{\varepsilon}$ is the strain tensor, $\underline{\sigma}^0$ is the initial stress tensor, p is the hydrostatic pressure and $\underline{I}$ is the identity matrix. The equilibrium expressions for fluid and ions in the stroma are written as:

$$\nabla p - RT(2\nabla c_- + \nabla c_f) = 0 \tag{1.26}$$

$$RT\frac{\nabla c_+}{c_+} + F\nabla\psi = 0 \tag{1.27}$$

$$RT\frac{\nabla c_-}{c_-} + F\nabla\psi = 0 \tag{1.28}$$

where c is the concentration of the ion or the fixed charge (indicated as +, - or f subscript), F is Faraday's constant, and ψ is the electrical potential. For the intact eye, which is modelled as a shell with spherical symmetry, a stress balance equation is written that relates the displacement in the radial direction to the gradients in the radial stress and hydrostatic pressure and momentum equations are written that relate the concentration gradient to the hydrostatic pressure gradient.

At the boundaries, the stress (IOP) is applied directly to the mixture with the assumption that the membranes perfectly transmit applied pressure. It is assumed that the stress distribution in the cornea can be adequately modelled using the standard elasticity solution for a spherical shell. The endothelium boundary condition is modelled in a manner consistent with Katchalsky and Curran (1965) and includes the passive permeability to ions and solvent and an active ion transporter. For the solutes, the active and passive ion fluxes at the endothelium were set to generate zero net flux. The endothelial ion flux balance equation was cast in terms of the cation concentrations (using Donnan theory) of the stroma and the aqueous humour, the fixed negative charge concentration of the stroma and the endothelial pump rate:

$$c_{+*}^2 - c_{+*}c_{f*} - c_a^2 + (c_{+*} + c_a - c_{f*})\frac{J_a}{2P_m} \tag{1.29}$$

where J_a is the active pump rate, P_m is the endothelial permeability to sodium, the subscript, a, refers to the aqueous humour, and the * indicates the value in the stroma adjacent to the membrane. For the solvent, the net flux was also assumed to be zero. An expression that balanced the stromal hydrostatic pressure, the IOP, and the Donnan osmotic pressure difference across the endothelium was written:

$$p - IOP - RT[2(c_{+*} - c_a) - c_{+*}] = 0 \tag{1.30}$$

The stromal stress and transport equations were solved numerically. The value of the endothelial pump rate was modulated to generate the proper pressure in the stroma. The final value of the pump rate was 7.0 x 10^{-11} moles/cm^2 sec. The results of this analysis demonstrate that the *trends* in stromal hydration and thickness can be duplicated with a model that reconciles the Donnan swelling pressure theory with the "pump-leak" hypothesis (Fig. 1.11) giving a stromal hydrostatic pressure that is higher than the IOP and a stromal sodium concentration that is higher than that in the aqueous humour.

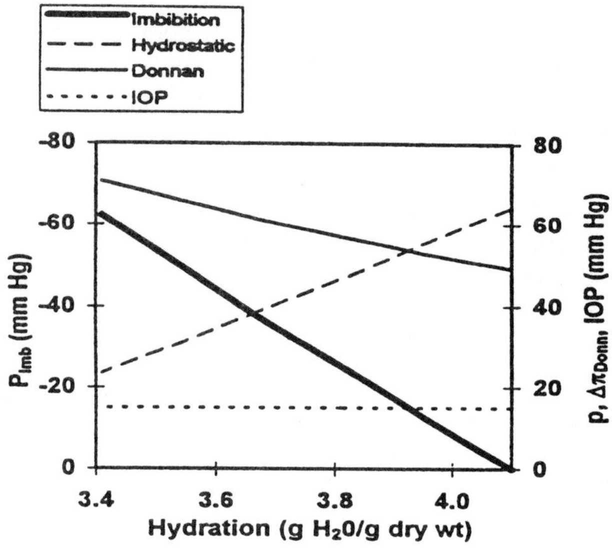

Figure 1.11. Components of the imbibition pressure, P_{imb}, in the intact cornea as functions of hydration. The imbibition pressure comprises the Donnan osmotic pressure, Δp_{Donn}, the hydrostatic fluid pressure, p, and the intraocular pressure, *IOP*. Specifically, $P_{imb} = -(\Delta p_{Donn} - p) - IOP$. Reprinted from Bryant and McDonnell (1998), Fig. 5, with permission of Am. Soc. Mech. Engin.

Note that the results of Bryant and McDonnell's work qualitatively agree with data in the literature (Fig. 1.12). In addition, the model predicts that the corneal steady-state thickness will decrease if the endothelial bathing solution is made hypertonic and increase if the bathing solution is made hypotonic. However, Ruberti (1998) has demonstrated, in rabbit corneas with epithelial transport blocked, that small perturbations in the endothelial perfusate (±15 mOsm), hypertonicity will slightly swell the cornea and hypotonicity will slightly thin the cornea. This implies that more data is necessary to ensure models can capture the true behavior of the whole corneal transport system. The primary limitations of Bryant and McDonnell's model are its inability to handle transients and steady-state flows (zero-flux condition), the simple endothelial description, and the neglect of the epithelium.

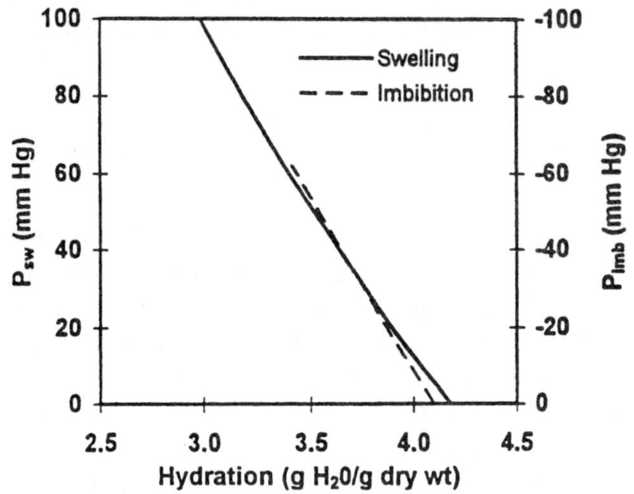

Figure 1.12. Calculated swelling pressure for confined compression and imbibition pressure in the intact cornea (generated by modulating the pump rate). The two pressures agree closely as functions of the corneal hydration. Reprinted from Bryant and McDonnell (1998), Fig. 9, with permission of Am. Soc. Mech. Engin.

Non steady-state whole corneal models

Non steady-state models of corneal transport attempt generate mathematical descriptions of the corneal transport system that are capable of responding realistically to environmental stimuli. Typically the purpose of these non steady-state corneal models is to reproduce corneal dynamics given a set of experimentally determined transport coefficients (Friedman, 1973) or to determine the values of the transport coefficients themselves by fitting the model to data from actual corneal system perturbations (Stanley et al., 1966; Klyce and Russell, 1979).

Non steady-state model development

In an investigation of the latter type, Stanley et al. (1966) attempted to determine the values of the epithelial and endothelial osmotic permeability ($Lp \times \sigma$) in the *in vivo* rabbit. A chamber was placed over the eye of an anaesthetized rabbit and was filled with isotonic perfusate. To induce a change in stromal thickness, the tonicity of the solution was increased by either 131 or 197 mOsm. The time course of the central stromal thickness change was monitored with a specially designed pachometer. To interpret the resulting thickness time course, a simple three-compartment, one-dimensional model of the corneal transport system was constructed. The membranes were modelled as homogeneous thin films impermeable to NaCl, but not to water. The volume change of stromal compartment, $\underline{C}_2$ was written as:

$$\frac{dV}{dt} = -\sigma_1 Lp_1 RT(C_1 - \underline{C}_2) + \sigma_2 Lp_2 RT(\underline{C}_2 - C_3) - \sigma_1 Lp_1 RT(\Delta C) \tag{1.31}$$

where the subscript 1 refers to the epithelium, the subscript 2 refers to the endothelium, *Lp* is hydraulic conductivity, σ is the reflection coefficient and *C* is the concentration of solute. Equation 1.31 was integrated to give an expression for the stromal thickness as a function of time. To fit the *in vivo* stromal response to a concentration change applied to the epithelium, the epithelial parameter, $Lp_1\sigma_1$, was manipulated to give the best fit to the data. Then, using the best fitting $Lp_1\sigma_1$, the endothelial parameter, $Lp_2\sigma_2$, was manipulated to give the final fit. The results of this approach are given in Tables 1.1 and 1.2. In this model, the compartments were all considered to be well mixed, the stromal hydraulic conductivity was neglected as was its swelling pressure during the hydration changes, the values of the reflection coefficient of the membranes were assumed not calculated and the transport of solutes though the model were not included.

Friedman also numerically investigated non-steady transport and hydration dynamics in the cornea (Friedman, 1973). This model was an extension of his 1972 effort and incorporated solute and solvent accumulation in the stroma and a non-steady state tear film concentration boundary condition to simulate the sleep wake cycle of rabbits. With his model he demonstrates that stromal thickness slowly oscillates around a nominal value between sleep and wake cycles. The oscillations and the general maintenance of hydration are due primarily to tear film evaporation in this model. The argument that no endothelial fluid pump is necessary to explain normal corneal hydration is again reiterated. This assertion was made in spite of recent experimental evidence that the endothelium did indeed contribute substantially to stromal hydration control (Maurice, 1972).

In 1979, Klyce and Russell generated a numerical model of corneal hydration dynamics. Their model provided a one-dimensional description of the transport of solutes and water through the corneal stroma with the attached limiting membranes. Consistent with the Kedem and Katchalsky (1958) descriptions of irreversible thermodynamics in biological systems, the epithelial and endothelial membranes were modelled as thin uniform sheets characterised by four transport parameters: hydraulic conductance, L_p, solute permeability, ω, reflection coefficient, σ, and active transport rate, J_a. A schematic of their model is shown in Fig. 1.13.

The tissue was divided artificially into compartments and membranes. Klyce and Russell (1979) recognized the generality of the Kedem and Katchalsky (1958) equations and used them as the governing flux relationships for the "membranes" both internal and external to the stroma:

$$J_v = L_p(\Delta p - \sigma\, RT\Delta C) \tag{1.32}$$

$$J_s = (1-\sigma)J_v\underline{C} + \omega\, RT\Delta C + J_a \tag{1.33}$$

where J_v is the volume flux through a unit area of tissue/membrane, J_s is the solute flux through the same unit area, ω is the membrane permeability to NaCl, ΔP is the local pressure difference, ΔC is the local NaCl concentration difference and $\underline{C}$ is "membrane" concentration.

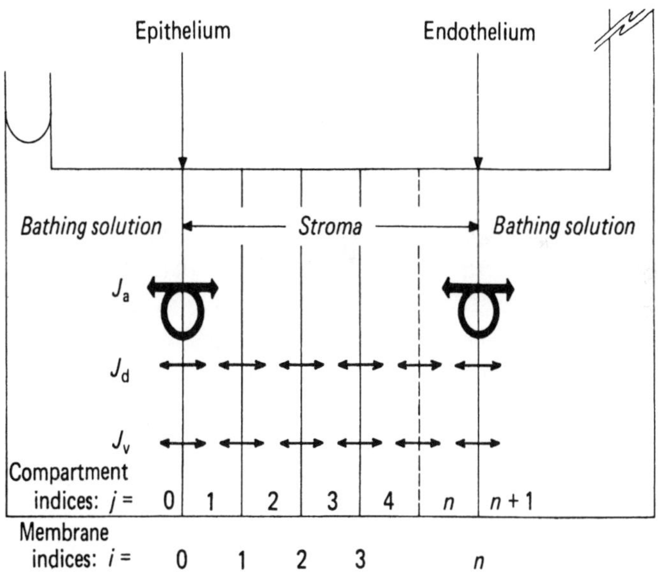

Figure 1.13. Schematic representation of Klyce and Russell (1979) corneal model. J_V is the volume flow, J_d is the passive neutral salt flux, J_a is the active NaCl flux. Reprinted from Klyce and Russell (1979), Fig. 1, with permission of J. Physiol.

To differentiate between the limiting membranes, the values of hydraulic conductance, reflection coefficient, pump rate and solute permeability were changed to match those values in the epithelium or the endothelium in their integral based numerical procedure. In the stromal "membranes", the reflection coefficient is assumed to be zero for small solutes (Green and Green, 1969), and both the diffusion coefficient, ω and the hydraulic conductivity, L_p were modified to act over the thickness of the adjacent compartments. In that case, Eq. 1.32 reduces to an integral based form of Darcy's law and Eq. 1.33 reduces to an integral based form of Fick's law. In the "compartments" between the "membranes", the local hydrations and concentrations were integrated based on the fluxes into and out of each compartment (where no gradients were assumed). The Fatt and Goldstick (1965) correlations of the data of Hedbys and Dohlman (1963) and Hedbys and Mishima (1962) were used to calculate the local pressure and flow conductivity in each compartment given the integrated local hydration. The total thickness of the stroma was estimated as the sum of the thickness of the local compartments. The equations were solved numerically on a "minicomputer".

The goal of the model was to estimate the values of the active and passive membrane transport parameters by adjusting them to fit the model's predicted response to the actual swelling response of *in vitro* rabbit corneas to osmotic NaCl perturbations applied to the epithelium. In their first series of experiments, endothelial transport was blocked with silicone oil. The epithelial perfusate osmolarity was increased by 30 mOsm or decreased by 23 mOsm. The resulting curves were fit (by eye) to the data by adjusting the four epithelial transport parameters. These values were retained for the next experiment, which required osmotic perturbations of the epithelium with the endothelium intact and functioning. This time the model parameters for the endothelium were adjusted to fit the swelling curves (Fig. 1.14). The resulting model parameters that provide the best fits of the stromal thickness time course induced by NaCl osmotic perturbations are presented in Tables 1.1 and 1.2. It is interesting to note that the large value for endothelial hydraulic conductance ($L_p = 42.0 \times 10^{-12}$ cm^3/dyne-sec) found by Klyce and Russell is eventually confirmed as the time resolution of other methods improves (Liebovitch et al., 1981; Fischbarg and Montoreano, 1982). To lend credibility to their results, Klyce and Russell then used the model with its fitted parameters successfully to test its performance against other known experimental results such as the "temperature reversal" experiment and ouabain inhibition of the endothelium pump. In addition to single solute transport, their model was capable of simulating the transport of multiple non-interacting solutes. They repeated numerically the endothelial perturbation experiment of Fischbarg (1973). The values of the endothelial sucrose permeability and reflection coefficient were adjusted to reproduce the data (Fig. 1.15).

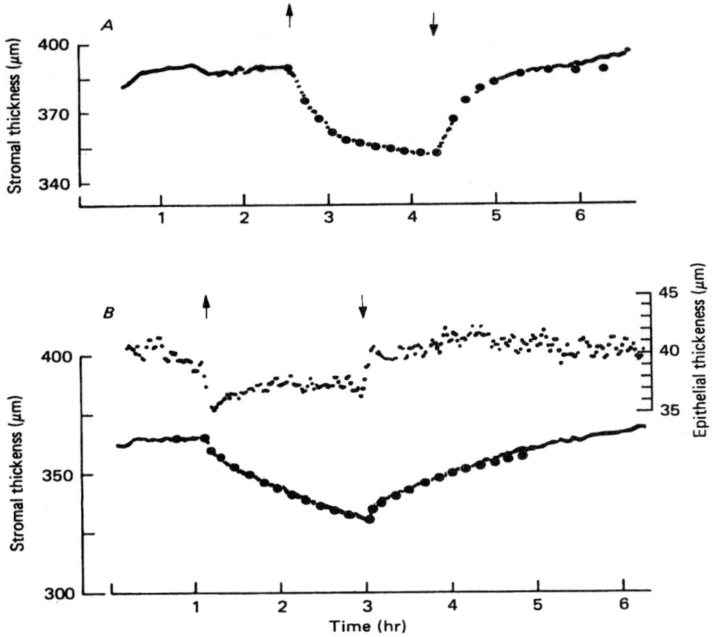

Figure 1.14. Stromal thickness response (solid lines) to a 30.0 mOsm perturbation and reversal applied to the corneal epithelium with the endothelium blocked (A) and with the endothelium intact and functioning (B). Fit of model (black dots). The inset in graph B is the epithelial thickness. (from Klyce and Russell, 1979). Reprinted from Klyce and Russell (1979), Fig. 5, with permission of J. Physiol.

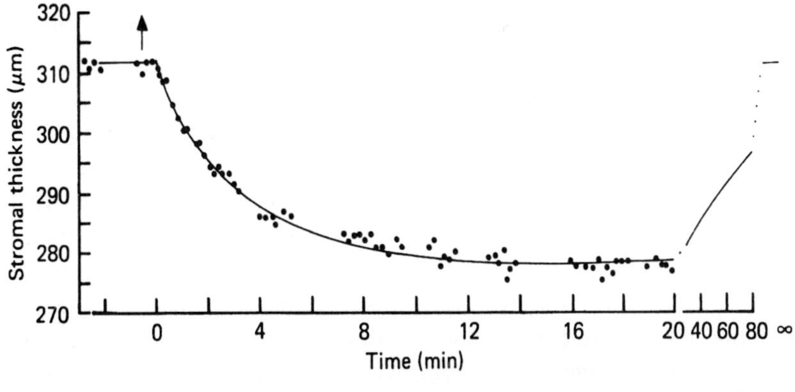

Figure 1.15. Model fit (solid line) of the stromal thickness response to a sucrose osmotic perturbation applied to the endothelial surface (at arrow). (Data points from Fischbarg, 1973; figure from Klyce and Russell, 1979). Reprinted from Klyce and Russell (1979), Fig. 9, with permission of J. Physiol.

Though detailed, thorough, and advanced for its time, the results of Klyce and Russell's model did not match the endothelial NaCl permeability data provided by tracer experiments (Maurice, 1951). The model overestimated tracer permeability by four times. In an attempt to resolve the latter question, very small direct NaCl perturbations of the endothelium have been carried out. These experiments did not resolve the issue, but did demonstrate dependence of the endothelial transport properties on the concentration of the applied perturbation (Ruberti, 1998).

Unsteady model applications

To resolve one of the more important questions in corneal transport physiology, Klyce (1981) extended his model (Klyce and Russell, 1979) to explain the swelling of the cornea induced by anterior surface hypoxia (which could be produced by wearing contact lenses). The specific model extensions included the addition of mathematical representations of unstirred diffusion boundary layers to the epithelial and endothelial surfaces. Klyce performed a series of physiological experiments demonstrating that compromised endothelial transport could not account for the swelling observed during epithelial hypoxia. Instead, he showed that increased lactate production (and its coincident osmotic load) by the stressed epithelial cells was the likely cause.

The most recent use of an unsteady model of corneal hydrodynamics was by Ruberti et al. (2000), who attempted to explain both the total and intrastromal fluid increases observed following mechanical debridement in the rabbit corneas. They demonstrated (on *in vitro* rabbit corneas, perfused on the endothelial side) that when the epithelium was mechanically debrided under silicone oil, a significant increase (30 microns) in stromal thickness resulted. Using a sensitive automatic specular microscope (Klyce and Russell, 1978), they discovered that there was also an unusual, concomitant internal shift of fluid from the posterior stroma to the anterior stroma.

In an attempt to explain both the local and the total stromal fluid dynamics, they modified the Klyce and Russell (1979) model to include rapid release of osmotic particles in the anterior half of the stroma. The resulting total and local corneal hydrodynamics *qualitatively* matched the experimental data quite nicely (Fig. 1.16). The results clearly indicate that the epithelial injury induced a large swelling pressure increase in the anterior stroma. The model proved adept not only in simulating the stromal response to injury, it was also capable of ruling out a potential swelling artefact due to inadvertent endothelial damage.

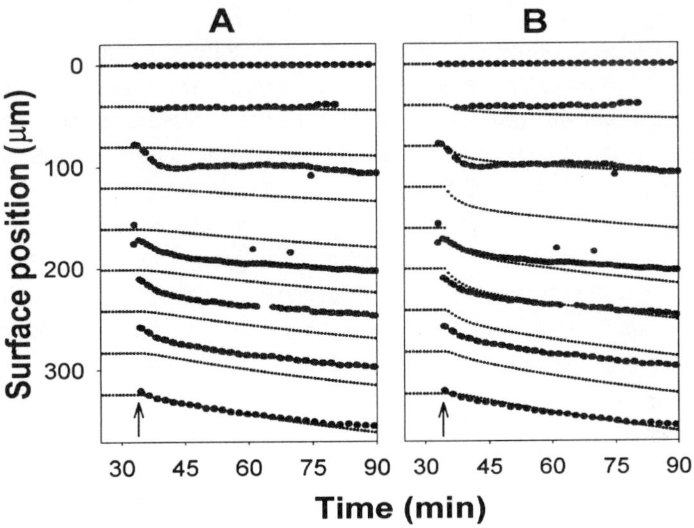

Figure 1.16. Numerical simulation of rapid stromal edema. The results of two different simulations (small dots) are compared with expiermental data (large dots). In *A*, possible artefactual damage to the endothelium was modelled as a three-fold increase in permeability. Though the total stromal swelling matched the data fairly well, the internal fluid shifts were not duplicated. In *B*, to test the hypothesis that osmotic particles due to lysis of cells or matrix are generated following debridement, an impermeant solute (3.75 mOsm) was added as a step function to the anterior half of the stromal tissue. Both the total and the local swelling a fairly well matched by this maneuver. Reprinted from Ruberti et al. (2000), with permission of Invest. Ophthal. Vis. Sci.

1.2.4 Corneal Hydration Control

Understanding corneal hydration control ultimately comes down to examination of the force balance between the hydrophillic stroma and its limiting membranes. The magnitude and dependence of the stromal swelling pressure on hydration has been correlated (Fatt and Goldstick, 1965) as a negative exponential with a normal *in vivo* swelling pressure of ~57 torr. Because of the small degree of cross-linking, this is the pressure that must be opposed by combined active and passive transport properties of the corneal membranes. We now briefly consider how these membranes may comprise a controller for stromal hydration.

The role of the highly innervated epithelium has been identified primarily as that of a tight barrier, although a small, net secretion of NaCl has been reported under certain conditions (Klyce, 1975). It is therefore unlikely that the epithelium can have a major active role in the control of stromal hydration except through homeostatic modulation of its barrier function which has not been observed. Because the endothelium does not exhibit a high degree of innervation, it appears that stromal hydration is not likely to be under neural control. How then, is the stromal thickness maintained?

The steep relationship between swelling pressure and hydration suggests that large changes in the ability of the endothelial "fluid pump" to generate an opposing pressure result in only small changes in stromal thickness. Indeed, a loss of 20 mmHg of endothelial fluid pump pressure (~30% of the total capacity) will result in an increase in the thickness of a rabbit cornea by only 40 microns (~10%). This demonstrates that a forgiving relationship exists between pump rate and stromal thickness. The implication of this is that the mechanism in control of hydration could be purely the result of intrinsic properties of the stroma, the epithelium and the endothelium. Interestingly, Ruberti (1998) found that numerous combinations of the solute permeability, reflection coefficient, hydraulic conductance and pump rate would give nearly identical thickness response curves to the same magnitude osmotic perturbation using the Klyce and Russell (1979) corneal model. In particular, it appeared that the active pump rate and solute permeability produced offsetting effects. Though this makes it difficult for the modeller to define a unique set of parameters, it is actually beneficial to this physiological control system and implies that a reasonable corneal hydration can be maintained by manipulating either the endothelial pump rate or the solute permeability.

It is known that the endothelial pump site density is maintained at a constant level throughout life in healthy eyes (Geroski et al, 1985), even as endothelial cell density decreases. If pump site density is assumed to be constant, then any active control of hydration might reside in other membrane transport parameters. Indeed, Rilcy ct al. (1998) has demonstrated that solute permeability can be modulated with adenosine. Additionally, Ruberti (1998) has provided data that imply a hydration-controlling permeability change follows very small NaCl concentration changes in the endothelial perfusate. Thus it is possible that hydration control is effected on a minute by minute basis through the action of adenosine (or other regulatory agent) or in response to changes in osmolarity. However, neither of these investigations identifies a sensing mechanism capable of monitoring a ambient parameters (i.e pressure, transparency etc). Thus, in spite of these investigations, we are left with little solid evidence for the existence of a homeostatic mechanism that controls stromal hydration. In the end, we suspect that corneal transparency is at the mercy of the forgiving relationship in the intrinsic physico-chemistry of its three layers and the fastidious maintenance of a constant pump site density in the endothelium throughout life.

1.3 BASIC CLINICAL IMPLICATIONS OF THE MODELS AND FUTURE DIRECTIONS

Basic clinical implications. The evolution of models of corneal hydration control and transport has been a long process that is far from complete. The complex physiology of the corneal limiting membranes, which comprise layers of extremely active cells in constant regulation, have uncertain dynamic transport properties and confound our ability to represent them accurately. In addition, the microscopic details of the mechanism of isotonic solute transport from the stroma to the aqueous humor remains obscure. Further, there is little data available regarding the transport properties of corneas with intact tear films. This has made models of *in vivo* transport difficult to produce. For all of these reasons, models of hydration control have not had a large impact in the clinic. However, the model of Klyce and Russell (1979) is a notable exception. It has been used to implicate the osmotic effect of lactic acid build-up as the main driver for stromal swelling during epithelial hypoxia (Klyce, 1981). A two-dimensional version of it has been used to suggest that reduced transport of metabolites is responsible for anterior stromal melting induced by impermeable intracorneal lenses (Klyce and Beuerman, 1997). More recently, it has been used to estimate the magnitude of an acute anomalous swelling response in the rabbit anterior stroma following epithelial debridement (Ruberti et al. 2000). All of these results would have been difficult to achieve experimentally.

Future directions. The cornea is a fairly simple physiological system that can be maintained *ex vivo* for extended periods without noticeable degradation. Consequently, modellers have access to an impressive amount of information for model validation. All of this has lead to highly developed representations of the tissue. However, this process is far from finished. Ultimately, it is important that a complete, valid model of the vegetative state of the cornea be produced. There are two reasons for this statement.

1. *If a complete model of a tissue system cannot be constructed, then the system is not understood.*

2. *The vegetative state of a tissue represents the simplest physiological configuration. It is the quintessential starting point for any physiological model.*

The above statements are universally true for modeling physiology and are somewhat tautological. However, reality has permitted only extremely small steps in this direction. For full models of the cornea to improve, the reliance on empirical data and phenomenological formulations should be reduced. A movement to first principles is necessary.

Membranes. In particular, the membranes need attention. Both the Klyce and Russell (1979) and the Bryant and McDonnell (1998) whole corneal models use simple, infinitely thin, homogeneous representations based on irreversible thermodynamics (Kedem and Katchalsky, 1958). In spite of the success of such phenomenological descriptions there are concerns about how they are applied and the number of adjustable parameters needed. Additionally, the assumption that both solvent and solute cross the membrane via the *same* route is likely to be a limitation. The presence of highly specific channels for both solutes and solvent call that assumption into question and may ultimately result in the decoupling of the solvent from the solute flux (Ussing and Eskesen, 1989). This will eliminate some of the frictional cross-coefficients and will introduce an entirely new set of channel-based transport parameters. Once the process of defining channel-based transport has begun, phenomenological models will not be adequate.

Ultimately membrane models should become microscopic in nature and will be based on first principles. Such models will necessarily dwarf that of Liebovitch and Weinbaum (1981) in complexity. They will need to include multi-ionic transport, the effects of membrane fixed charge density, ion and water channel hydrodynamic and electro-chemical properties, intracellular mechano-electro-chemical properties, and more. Although micro-scale fluid and solute transport has advanced rapidly in the last few decades, for the reasons outlined by Diamond (1979) long ago, their application to physiological membranes may not be a rapid process.

The final improvement to membrane models will be the addition of transient behavior in response to various perturbations and injury. This will require the coupling of detailed physiological and structural information in the microscopic model of the membrane. Such a complex addition is barely at the empirical level for the cornea.

Stroma. For the stroma, mathematical modeling has met with relative success due to its low cellular content and its "stability" *ex vivo* (see Ruberti et al., 2000). There are promising micro-scale models developed for cartilage that produce realistic swelling stresses (Buschmann and Grodzinsky, 1995). This type of model could be modified and applied to the corneal stroma in the future to mathematically describe stromal swelling pressure. With regard to hydraulic conductivity the stroma will likely require an approach based on its nano-structure similar to that of Overby et al. (2000). This too will require some time to incorporate. An excellent *next* step in modeling of corneal transport and hydration control will likely entail the coupling of simplified microscale membrane models with an advanced triphasic representation of the corneal stroma similar to the triphasic model of Bryant and McDonnell (1998).

Advanced modeling. In the future, corneal models will be expected to handle more than transport mechanics and hydration control. They will need to predict the behavior of the cornea in the short and long-term following perturbations and insults of all types. For instance, to predict the outcome of refractive surgical procedures, incorporation of the biomechanical properties of the tissue in three-dimensions is necessary. Many previous investigations have attempted to do just that for radial keratotomy (Bryant et al., 1987; Hanna et al., 1989a,b; Vito et al., 1989; Pinsky and Datye, 1991). However, because of the difficulty in performing precisely controlled experiments of refractive surgical procedures, these models have not been adequately validated (Bryant and McDonnell, 1996). More recent attempts to predict LASIK surgical outcomes have also met with only limited success (Bryant et al. 2000). In addition to mechanical properties, incorporating dynamic remodeling of the cornea following injury (surgical or traumatic) will be important to the clinician. This will require that modellers understand the behavior of the activated stromal keratocytes during the wound healing process.

Final note. We have seen the results of the effort that has been invested in modeling a fairly simple physiological system. The cornea's accessibility, its importance to vision, and its curious hydration control system have resulted in intense investigation by physiologists, clinicians and mathematical modellers alike. What should be impressed on the reader at this point is the continued reliance of both the clinician and the modeller on the experimentalist. So many aspects of corneal physiology, mechanics and cellular biology remain unknown that a complete model of this simple tissue continues to remain at large. In the light of the current atmosphere of proliferating refractive surgeries it is instructive to remember the message of first tautological statement above: *if we cannot model it, we don't understand it.*

1.4 ACKNOWLEDGMENTS

The authors would like to thank Dr. Alan Grodzinsky for his help clarifying the role of ion binding in the corneal stroma. Dr. Thomas Goldstick for his kind discussions of the original work that helped create the field of modeling swelling in soft tissue. Finally, the authors would like to pay tribute to the late Dr. Irving Fatt. Without his extraordinary insight and energy our understanding of ocular transport mechanics would certainly not have risen to the level that we enjoy today.

1.5 REFERENCES

Akaike, N., 1971, The origin of the basal cell potential in frog corneal epithelium, *J. Physiol. (Lond).* **219**: 57-75.

Akaike, N. and Hori, M., 1970, Effect of anions and cations on membrane potential of rabbit corneal epithelium, *Am. J. Physiol.* **219**: 1811-1818.

Anseth, A., 1961, Studies on corneal polysaccharides. III. Topographic and comparative biochemistry, *Exp. Eye Res.* **1**: 106-115.

Axelsson, I. and Heinegard, D., 1975, Fractionation of proteoglycans from bovine corneal stroma, *Biochem. J.* **145**: 491-500.

Bauer, N. J., Wicksted, J. P., Jongsma, F. H., et al., 1998, Noninvasive assesment of the hydration gradient across cornea using confocal Raman spectroscopy, *Invest. Ophthal. Vis. Sci.* **39**: 831-835.

Benedek, G. B., 1971, Theory of transparency of the eye, *Applied Optics.* **10**: 459-473.

Bettelheim, F. A., 1975, The hydration of proteoglycans of bovine cornea, *Biochim. Biophys. Acta.* **381**: 203.

Bettelheim, F. A. and Goetz, D., 1976, Distribution of hexosamines in bovine cornea, *Invest. Ophthal. Vis. Sci.* **15**: 301-304.

Bryant, M. R., Marchi, V. and Juhasz, T., 2000, Mathematical models of picosecond laser keratomileusis for high myopia, *J. Refract. Surg.* **16**: 155-162.

Bryant, M. R. and McDonnell, P. J., 1996, Constitutive laws for biomechanical modeling of refractive surgery, *J. Biomech. Eng.* **118**: 473-481.

Bryant, M. R. and McDonnell, P. J., 1998, A triphasic analysis of corneal swelling and hydration control, *J. Biomech. Eng.* **120**: 370-381.

Bryant, M. R., Velinsky, S. A., Plesha, M. E. and Clarke, G. P., 1987, Computer-aided surgical design in refractive keratotomy, *CIAO J.* **13**: 238-342.

Buschmann, M. D. and Grodzinsky, A. J., 1995, A molecular model of proteoglycan-associated electrostatic forces in cartilage mechanics, *J. Biomech. Eng.* **117**: 179-192.

Candia, O. A. and Askew, W. A., 1968, Active sodium transport in the isolated bullfrog cornea, *Biochim. Biophys. Acta.* **163**: 262-265.

Candia, O. A. and Neufeld, A. H., 1978, Topical epinephrine causes a decrease in density of beta-adrenergic receptors and cathecholamine-stimulated chloride transport in the rabbit cornea, *Biochim. Biophys. Acta.* **543**: 403-408.

Candia, O. A. and Reinach, P. S., 1982, Thermodynamic analysis of active sodium and potassium transport in the frog corneal epithelium, *Am. J. Physiol.* **242**: F690-698.

Crank, J., 1956, *The Mathematics of Diffusion*, Oxford University Press, London.

Davson, H., 1949, Some considerations on the salt content of fresh and old ox corneae, *Br. J. Ophthalmol.* **33**: 175-182.

Davson, H., 1955, The hydration of the cornea, *Biochem. J.* **59**: 25-28.

Daxer, A., Misof, K., Grabner, B., Ettl, A. and Fratzl, P., 1998, Collagen fibrils in the human corneal stroma: structure and aging, *Invest. Ophthal. Vis. Sci.* **39**: 644-648.

Diamond, J. M., 1979, Topical review. Osmotic water flow in leaky epithelia, *J. Membr. Biol.* **51**: 195-216.

Diamond, J. M. and Bossert, W. H., 1967, Standing-gradient osmotic flow : A mechanism for coupling of water and solute transport in epithelia, *J. Gen. Physiol.* **50**: 2061-2083.

Donn, A., Maurice, D. and Mills, N., 1959, Studies on the living cornea in vitro. I. Method and physiologic measurements, *Arch. Ophthalmol.* **62**: 741-747.

Donn, A., Maurice, D. and Mills, N., 1959, Studies on the living cornea in vitro. II. The active transport of sodium across the epithelium, *Arch. Ophthalmol.* **62**: 748-757.

Edwards, A. and Prausnitz, M. R., 1998, Fiber matrix model of sclera and corneal stroma for drug delivery to the eye, *AIchE J.* **44**: 214-225.

Ehlers, N., 1970, Some comparative studies on the mammalian corneal epithelium, *Acta. Ophthalmol.* **48**: 821-828.

Eisenberg, S. R. and Grodzinsky, A. J., 1985, Swelling of articular cartilage and other connective tissues: electromechanochemical forces, *J. Orthop. Res.* **3**: 148-159.

Eisenberg, S. R. and Grodzinsky, A. J., 1987, The kinetics of chemically induced nonequilibrium swelling of articular cartilage and corneal stroma, *J. Biomech. Eng.* **109**: 79-89.

Elliott, G. F., 1980, Measurement of the electric charge and ion binding of the protein filaments in intact muscle and cornea, with implications for filament assembly., *Biophys. J.* **32**: 95-97.

Elliott, G. F., Goodfellow, J. M. and Woolgar, A. E., 1980, Swelling studies of bovine corneal stroma without bounding membranes, *J. Physiol. (Lond).* **298**: 453-470.

Elliott, G. F. and Hodson, S. A., 1998, Cornea, and the swelling of polyelectrolyte gels of biological interest, *Rep Prog Phys.* **61**: 1325-1365.

Fatt, I., 1968, Dynamics of water transport in the corneal stroma, *Exp. Eye. Res.* **7**: 402-412.

Fatt, I., 1968, Steady-state distribution of oxygen and carbon dioxide in the in vivo cornea. II. The open eye in nitrogen and the covered eye, *Exp. Eye. Res.* **7**: 413-430.

Fatt, I. and Bieber, M. T., 1968, The steady-state distribution of oxygen and carbon dioxide in the in vivo cornea. I. The open eye in air and the closed eye, *Exp. Eye Res.* **7**: 103-112.

Fatt, I., Bieber, M. T. and Pye, S. D., 1969, Steady state distribution of oxygen and carbon dioxide in the in vivo cornea of an eye covered by a gas-permeable contact lens, *Am. J. Optom. Arch. Am. Acad. Optom.* **46**: 3-14.

Fatt, I. and Goldstick, T., 1965, Dynamics of water transport in swelling membranes, *J. Colloid Sci.* **20**: 962-989.

Fee, J. P. and Edelhauser, H. F., 1970, Intracellular electrical potentials in the rabbit corneal epithelium, *Exp. Eye Res.* **9**: 233-40.

Festen, C. M. and Slegers, J. F., 1979, The influence of ions, ouabain, propranolol and amiloride on the transepithelial potential and resistance of rabbit cornea, *Exp. Eye Res.* **28**: 413-426.

Fischbarg, J., 1972, Potential difference and fluid transport across rabbit corneal endothelium, *Biochim. Biophys. Acta.* **288**: 362-366.

Fischbarg, J., 1973, Active and passive properties of the rabbit corneal endothelium, *Exp. Eye Res.* **15**: 615-638.

Fischbarg, J., 1995, A rapidly emerging field: water channel proteins in the eye, *Invest. Ophthal. Vis. Sci.* **36**: 758-763.

Fischbarg, J., 1997, Mechanism of fluid transport across corneal endothelium and other epithelial layers: a possible explanation based on cyclic cell volume regulatory changes, *Br. J. Ophthalmol.* **81**: 85-89.

Fischbarg, J., Hernandez, J., Liebovitch, L. S. and Koniarek, J. P., 1985, The mechanism of fluid and electrolyte transport across corneal endothelium: critical revision and update of a model, *Curr Eye Res.* **4**: 351-360.

Fischbarg, J. and Montoreano, R., 1982, Osmotic permeabilities across corneal endothelium and antidiuretic hormone-stimulated toad urinary bladder structures, *Biochim. Biophys. Acta.* **690**: 207-214.

Fischbarg, J., Warshavsky, C. R. and Lim, J. J., 1977, Pathways for hydraulically and osmotically-induced water flows across epithelia, *Nature.* **266**: 71-74.

Friedman, M., 1972, A quantitative description of equilibrium and homeostatic thickness regulation in the in vivo cornea. I. Normal cornea, *Biophys. J.* **12**: 648-665.

Friedman, M., 1973, Unsteady transport and hydration dynamics in the in vivo cornea, *Biophys. J.* **13**: 890-910.

Friedman, M., 1978, Mathematical modeling of transport in structured tissues: corneal epithelium, *Am. J. Physiol.* **234**: F215-F224.

Friedman, M. H., 1971, Application of computer experimentation to the cornea, *Nature.* **233**: 553-555.

Friedman, M. H., 1971, General theory of tissue swelling with application to the corneal stroma, *J Theor Biol.* **30**: 93-109.

Friedman, M. H. and Green, K., 1971, Swelling rate of corneal stroma, *Exp. Eye Res.* **12**: 239-250.

Friedrich, S. W., Cheng, Y. L. and Saville, B. A., 1993, Theoretical corneal permeation model for ionizable drugs, *J. Ocul. Pharmacol.* **9**: 229-49.

Geroski, D. H., Matsuda, M., Yee, R. W. and Edelhauser, H. F., 1985, Pump function of the human corneal endothelium, *Ophthalmol.* **92**: 759-763.

Grass, G. M., Cooper, E. R. and Robinson, J. R., 1988, Mechanisms of corneal drug penetration. III: Modeling of molecular transport, *J. Pharm. Sci.* **77**: 24-26.

Green, K., 1965, Ion transport in isolated cornea of the rabbit, *Am. J. Physiol.* **209**: 1311-1316.

Green, K., 1968, Relation of epithelial ion transport to corneal thickness and hydration, *Nature.* **217**: 1074-1075.

Green, K. and Downs, S. J., 1976, Corneal membrane water permeability as a function of temperature, *Invest. Ophthalmol.* **15**: 304-307.

Green, K. and Green, M., 1969, Permeability to water of rabbit corneal membranes, *Am. J. Physiol.* **217**: 635-641.

Green, K., Hastings, B. and Friedman, M. H., 1971, Sodium ion binding in isolated corneal stroma, *Am. J. Physiol.* **220**: 520-525.

Grodzinsky, A. J., 1983, Electromechanical and physicochemical properties of connective tissue, *Crit. Rev. Biomed. Eng.* **9**: 133-199.

Hamada, R., Giraud, J. P., Graf, B. and Pouliquen, Y., 1972, Etude analytique et statistique des lamelles, des keratocytes, des fibrillesdel collagene de la region centrale de la cornee humaine normale, *Arch. Ophthalmol. (Paris).* **32**: 563-570.

Hanna, K. D., Jouve, F. E. and Waring, G. O. d., 1989, Preliminary computer simulation of the effects of radial keratotomy, *Arch. Ophthalmol.* **107**: 911-918.

Hanna, K. D., Jouve, F. E., Waring, G. O. d. and Ciarlet, P. G., 1989, Computer simulation of arcuate and radial incisions involving the corneoscleral limbus, *Eye.* **3**: 227-239.

Harris, J. and Nordquist, L., 1955, The hydration of the cornea. I. Transport of water from the cornea, *Am. J. Ophthalmol.* **40**: 100-110.

Harvitt, D. M. and Bonanno, J. A., 1999, Re-evaluation of the oxygen diffusion model for predicting minimum contact lens Dk/t values needed to avoid corneal anoxia, *Optom. Vis. Sci.* **76**: 712-719.

Hedbys, B. and Dohlman, C., 1963, A new method for the determination of the swelling pressure of the corneal stroma in vitro, *Exp. Eye Res.* **2**: 122-129.

Hedbys, B. and Mishima, S., 1962, Flow of water in the corneal stroma, *Exp. Eye Res.* **1**: 262-275.

Hedbys, B. and Mishima, S., 1966, The thickness-hydration relationship of the cornea, *Exp. Eye Res.* **5**: 221-228.

Hedbys, B., Mishima, S. and Maurice, D., 1963, The imbibition pressure of the corneal stroma, *Exp. Eye Res.* **2**: 99-111.

Hedbys, B. O., 1961, The role of polysaccharides in corneal swelling, *Exp. Eye Res.* **1**: 81.

Hedbys, B. O., 1963, Corneal resistance to the flow of water after enzymatic digestion, *Exp. Eye Res.* **2**: 112-121.

Hill, A., 1980, Salt-water coupling in leaky epithelia, *J. Membr. Biol.* **56**: 177-182.

Hill, A. E., 1975, Solute-solvent coupling in epithelia: a critical examination of the standing-gradient osmotic flow theory, *Proc. R. Soc. Lond. B. Biol. Sci.* **190**: 99-114.

Hodson, S., 1971, Why the cornea swells, *J. Theor. Biol.* **33**: 419-427.

Hodson, S., Kaila, D., Hammond, S., Rebello, G. and al-Omari, Y., 1992, Transient chloride binding as a contributory factor to corneal stromal swelling in the ox, *J Physiol (Lond).* **450**: 89-103.

Hodson, S. and Lawton, D., 1987, The apparent reflection coefficient of the leaky corneal endothelium to sodium chloride is about one in the rabbit, *J. Physiol. (Lond).* **385**: 97-106.

Hodson, S. and Miller, F., 1976, The bicarbonate ion pump in the endothelium which regulates the hydration of rabbit cornea, *J. Physiol. (Lond).* **263**: 563-577.

Hodson, S. and Wigham, C., 1983, The permeability of rabbit and human corneal endothelium, *J. Physiol. (Lond).* **342**: 409-419.

Hull, D., Green, K., Boyd, M. and Wynn, H., 1977, Corneal endothelium bicarbonate transport and the effect of carbonic anhydrase inhibitiors on endothelial permeability and fluxes and corneal thickness, *Invest. Ophthal. Vis. Sci.* **16**: 883-892.

Jumblatt, M. M. and Neufeld, A. H., 1983, Beta-adrenergic and serotonergic responsiveness of rabbit corneal epithelial cells in culture, *Invest. Ophthal. Vis. Sci.* **24**: 1139-1143.

Katchalsky, A. and Curran, P. F., 1965, *Non-equilibrium thermodynamics in biophysics*, Harvard University Press, Cambridge.

Kaye, G. I. and Tice, L. W., 1968, Studies on the cornea. V. Electron microscopic localization of adenosine triphosphate activity in the rabbit cornea in relation to transport., *Invest. Ophthalmol.* **5**: 22.

Kedem, O. and Katchalsky, A., 1958, Thermodynamic analysis of the permeability of biological membranes to non-electrolytes, *Biochim. Biophys. Acta.* **27**: 229-246.

Kedem, O. and Katchalsky, A., 1961, *J. Gen. Physiol.* **45**: 143.

Kikkawa, Y., 1964, The intracellular potential of the corneal epithelium, *Exp. Eye Res.* **3**: 132.

Kikkawa, Y. and Hirayama, K., 1970, Uneven swelling of the corneal stroma, *Invest. Ophthal. Vis. Sci.* **9**: 735-744.

Kim, J., Green, K., Martinez, M. and Paton, D., 1971, Solute permeability of the corneal endothelium and Descemet's membrane, *Exp. Eye Res.* **12**: 231-238.

Klyce, S., 1975, Transport of Na, Cl and water by the rabbit corneal epithelium at resting potential, *Am. J. Physiol.* **228**: 1446-1452.

Klyce, S. and Russell, S., 1978, System for monitoring the thickness of transparent layered structures, *Rev. Sci. Instrum.* **49**: 1318-1321.

Klyce, S. and Russell, S., 1979, Numerical solution of coupled transport equations applied to corneal hydration dynamics, *J. Physiol. (Lond).* **292**: 107-134.

Klyce, S. D. (1971). Electrophysiology of the corneal epithelium. Department of Physiology. New Haven, Yale University.

Klyce, S. D., 1972, Electrical profiles in the corneal epithelium, *J. Physiol. (Lond).* **226**: 407-429.

Klyce, S. D., 1981, Stromal lactate accumulation can account for corneal oedema osmotically following epithelial hypoxia in the rabbit, *J. Physiol. (Lond).* **321**: 49-64.

Klyce, S. D. and Beuerman, R. W., 1997, Structure and function of the cornea, H. E. Kaufman, M. B. McDonald and B. A. Barron. *The Cornea.* Butterworth-Heineman Medical, London, pp. 3-50.

Klyce, S. D., Beuerman, R. W. and Crosson, C. E., 1985, Alteration of corneal epithelial ion transport by sympathectomy, *Invest. Ophthal. Vis. Sci.* **26**: 434-42.

Klyce, S. D. and Crosson, C. E., 1985, Transport processes across the rabbit corneal epithelium: a review, *Curr. Eye Res.* **4**: 323-331.

Klyce, S. D. and Marshall, W. S., 1982, Effects of Ag+ on ion transport by the corneal epithelium of the rabbit, *J. Membr. Biol.* **66**: 133-144.

Klyce, S. D., Neufeld, A. H. and Zadunaisky, J. A., 1973, The activation of chloride transport by epinephrine and Db cyclic-AMP in the cornea of the rabbit, *Invest. Ophthalmol.* **12**: 127-139.

Klyce, S. D., Palkama, K. A., Harkonen, M., et al., 1982, Neural serotonin stimulates chloride transport in the rabbit corneal epithelium, *Invest. Ophthal. Vis. Sci.* **23**: 181-192.

Klyce, S. D. and Wong, R. K. S., 1977, Site and mode of adrenaline action on chloride transport across the rabbit corneal epithelium, *J. Physiol. (Lond).* **266**: 777-799.

Lai, W. M., Hou, J. S. and Mow, V. C., 1991, A triphasic theory for the swelling and deformation behaviors of articular cartilage, *J. Biomech. Eng.* **113**: 245-258.

Langham, M. E., Hart, R. W. and Cox, J., 1969, The interaction of collagen and mucopolysaccharides, M. Langham. *The Cornea. Macromolecular Organization of a Connective Tissue.* Johns Hopkins Press, Baltimore: 157.

Leber, T., 1873, Studien uber den Flussigkeitswechsel im Auge, *v. Graefes Arch. Ophthalmol.* **19**: 87.

Lee, D. and Wilson, G., 1981, Non-uniform swelling properties of the corneal stroma, *Curr. Eye. Res.* **1**: 457-461.

Levick, J. R., 1987, Flow through interstitium and other fibrous matrices, *Q. J. Exp. Physiol.* **72**: 409-437.

Liebovitch, L. and Weinbaum, S., 1981, A model of epithelial transport. The corneal endothelium, *Biophys. J.* **35**: 315-338.

Liebovitch, L. S., Fischbarg, J. and Koatz, R., 1981, Osmotic water permeability of rabbit corneal endothelium and its dependence on ambient concentration, *Biochim. Biophys. Acta.* **646**: 71-76.

Lim, J. J. and Fischbarg, J., 1976, Standing-gradient osmotic flow. Examination of its validity using an analytical method, *Biochim. Biophys. Acta.* **443**: 339-347.

Lim, J. J. and Ussing, H. H., 1982, Analysis of presteady-state Na+ fluxes across the rabbit corneal endothelium, *J. Membr. Biol.* **65**: 197-204.

Ling, T., 1987, Osmotically induced central and peripheral corneal swelliing in the cat. *Am. J. Opt. Phys. Optics.* **64**: 674-677.

Mackie, J. S. and Mears, P., 1955, The diffusion of electrolytes in a cation-exchange resin membrane, *Proc. R. Soc.*. **A232**: 498-509.

Maroudas, A., 1980, Physical chemistry of articular cartilage and the intervertebral disk, L. Sokoloff. *The Joints and Synovial Fluid.* Academic Press, New York, **II**.

Marshall, W. S. and Klyce, S. D., 1983, Cellular and paracellular pathway resistances in the "tight" Cl- secreting epithelium of rabbit cornea, *J. Membr. Biol.* **73**: 275-282.

Marshall, W. S. and Klyce, S. D., 1984, Cellular mode of serotonin action on Cl- transport in the rabbit corneal epithelium, *Biochim. Biophys. Acta.* **778**: 139-43.

Maurice, D., 1951, The permeability to sodium ions of the living rabbit's cornea, *J. Physiol. (Lond).* **112**: 367-391.

Maurice, D., 1957, The structure and transparency of the cornea, *J. Physiol. (Lond).* **136**: 263-286.

Maurice, D., 1972, The location of the fluid pump in the cornea, *J. Physiol. (Lond).* **221**: 43-54.

Maurice, D. M., 1969, The cornea and sclera, in: H. Davson. *The Eye,* vol. 1, Academic Press, New York - London, pg. 489.

Maurice, D. M., 1984, The cornea and sclera: 1b Vegetative physiology and biochemistry, in: H. Davson. *The Eye,* vol. 1, Academic Press, New York.

McCarey, B. E. and Schmidt, F. H., 1990, Modeling glucose distribution in the cornea, *Curr. Eye Res.* **9**: 1025-1039.

Meek, K. M. and Leonard, D. W., 1993, Ultrastructure of the corneal stroma: a comparative study, *Biophys. J.* **64**: 273-280.

Meyer, K., Linker, A., Davidson, E. A. and Weissmann, B., 1953, The mucopolysaccharides of the bovine cornea, *J. Biol. Chem.* **205**: 611-616.

Mishima, S. and Hedbys, B., 1967, The permeability of the corneal epithelium and endothelium to water, *Exp. Eye Res.* **6**: 10-32.

Mow, V. C., Kuei, S. C., Lai, W. M. and Armstrong, C. G., 1980, Biphasic creep and stress relaxation of articular cartilage in compression? Theory and experiments, *J. Biomech. Eng.* **102**: 73-84.

Muller, L. J., Pels, L. and Vrensen, G., 1995, Novel aspects of the ultrastructural organization of human corneal keratocytes, *Invest. Ophthal. Vis. Sci.* **36**: 2557-2567.

Myers, E. R., Lai, W. M. and Mow, V. C., 1984, A continuum theory and an experiment for the ion-induced swelling behavior of articular cartilage, *J. Biomech. Eng.* **106**: 151-158.

Neufeld, A. H., Ledgard, S. E., Jumblatt, M. M. and Klyce, S. D., 1982, Serotonin-stimulated cyclic AMP synthesis in the rabbit corneal epithelium, *Invest. Ophthal. Vis. Sci.* **23**: 193-198.

Onsager, L., 1931, Reciprocal Relations in Irreversible Processes, I., *Physiol. Rev.* **37**: 405.

Otori, T., 1967, Electrolyte content of the rabbit corneal stroma., *Exp. Eye Res.* **6**: 356-367.

Overby, D., Ruberti, J. W., Gong, H., Freddo, T. F. and Johnson, M., Specific hydraulic conductivity of corneal stroma as seen by quick-freeze/deep-etch, *J. Biomech. Eng.*, in press.

Pinsky, P. M. and Datye, D. V., 1991, A microstructurally-based finite element model of the incised human cornea, *J. Biomech.* **24**: 907-922.

Rehm, W. S. and Spangler, S. G., 1977, A theory of endothelial control of corneal hydration, *Am. J. Optom. Physiol. Opt.* **54**: 439-444.

Riley, M. V., 1977, Anion-sensitive ATPase in rabbit corneal endothelium and its relation to corneal hydration, *Exp. Eye Res.* **25**: 483-494.

Riley, M. V. and Peters, M. I., 1981, The localization of the anion-sensitive ATPase activity in corneal endothelium, *Biochim. Biophys. Acta.* **644**: 251-256.

Riley, M. V., Winkler, B. S., Starnes, C. A., Peters, M. I. and Dang, L., 1998, Regulation of corneal endothelial barrier function by adenosine, cyclic AMP, and protein kinases, *Invest. Ophthal. Vis. Sci.* **39**: 2076-2084.

Ruberti, J. W. (1998), Experimental and computational investigation of corneal transport properties, Dept. Biomedical Engineering, New Orleans, Tulane University.

Ruberti, J. W., Klyce, S. D., Smolek, M. K. and Karon, M. D., 2000, Anomalous acute inflammatory response in rabbit corneal stroma, *Invest. Ophthal. Vis. Sci.* **41**: 2523-2530.

Sackin, H. and Boulpaep, E. L., 1975, Models for coupling of salt and water transport. Proximal tubular reabsorption in Necturus kidney, *J. Gen. Physiol.* **66**: 671.

Scott, J. E., 1994, Keratan sulphate--a 'reserve' polysaccharide?, *Eur. J. Clin. Chem. Clin. Biochem.* **32**: 217-223.

Segel, L. A., 1970, Standing-gradient flows driven by active solute transport, *J. Theor. Biol.* **29**: 233-250.

Shapiro, M. P. and Candia, O. A., 1973, Corneal hydration and metabolically dependent transcellular passive transfer of water, *Exp. Eye Res.* **15**: 659-666.

Skadhauge, E., 1977, Analysis of computer models in transport of ions and water in animals, B. L. Gupta, J. L. Oschman, R. M. Moreton and B. J. Wall. Academic Press, New York.

Stanley, J., Mishima, S. and Klyce, S. D., 1966, In vivo determination of endothelial permeability to water, *Invest. Ophthal. Vis. Sci.* **7**: 371-377.

Stiemke, M. M., Roman, R. J., Palmer, M. L. and Edelhauser, H. F., 1992, Sodium activity in the aqueous humor and corneal stroma of the rabbit, *Exp. Eye Res.* **55**: 425-433.

Trenberth, S. and Mishima, S., 1968, The effect of ouabain on the rabbit corneal endothelium, *Invest. Ophthal. Vis. Sci.* **7**: 44-52.

Turss, R., Friend, J., Riem, M. and Dohlman, C. H., 1971, Glucose concentration and hydration of the corneal stroma, *Opthalmol. Res.* **2**: 253-260.

Ussing, H. H. and Eskesen, K., 1989, Mechanism of isotonic water transport in glands, *Acta. Physiol. Scand.* **136**: 443-454.

Van der Heyden, C., Weekers, J. F. and Schoffeniels, E., 1975, Sodium and chloride transport across the isolated rabbit cornea, *Exp. Eye Res.* **20**: 89-96.

Vito, R. P., Shin, T. J. and McCarey, B. E., 1989, A mechanical model of the cornea: the effects of physiological and surgical factors on radial keratotomy surgery, *Refract. Corneal Surg.* **5**: 82-88.

Wiederholt, M., 1980, Physiology of epithelial transport in the human eye, *Klin. Wochenschr.* **58**: 975-984.

Wiig, H., 1989, Cornea Fluid Dynamics I : Measurement of Hydrostatic and Colloid Osmotic Pressure in Rabbits, *Exp. Eye Res.* **49**: 1015-1030.

Zadunaisky, J. A., 1966, Active transport of chloride in frog cornea, *Am. J. Physiol.* **211**: 506-512.

Chapter 2

Models of the Lens and Aging Effects

Jane Koretz, Ph.D.

Center for Biophysics and Dept. of Biology, Rensselaer Polytechnic Institute Science Center, Troy, NY 12180-3590, PH: (518)276-6492, FX: (518) 276-2344, EM:koretj@rpi.edu

2.1 INTRODUCTION

The crystalline lens of the human visual system provides the variable refractive power needed for focus by the eye at all distances. Of the major refractive tissues, only the normal lens exhibits a uniquely specific and reproducible development path with increasing age. Optical modeling of the process of image formation on the retina thus relies on accurate and complete descriptions of the age dependence of lens shape, curvature, placement within the globe relative to the cornea and retina, and refractive index gradient. Other contexts in which an optical model of the lens may be important include studies of the posterior of the globe, and especially the retina, where changes in lens transparency, color, and refractive power could affect images and spectra in this region. Finally, the recent interest by both vision scientists and refractive surgeons in improving visual resolution to its theoretical limit (also known as "super-vision"), using the Shack-Hartmann method for collection of the wavefront and Zernike polynomials for its analysis, necessitates a more detailed examination of the changing contribution of the crystalline lens to overall image formation and image quality.

Up to this time, optical modeling of the crystalline lens, either as a component in overall image formation or as an independent refractive element, has been beset by questions and/or inappropriate assumptions about its properties. There are four fundamental problems in developing an optical model of the lens. The first, and arguably the most serious, problem is to determine what aspects of lens properties are age-dependent and which are not; for the former, the quantitative relationship of the property with age also needs to be defined. A second problem comes from the difficulties in accurate measurement of lens location relative to other important optical elements of the eye, although the increased accuracy of ultrasonographic and slit-lamp photographic data has partially solved this. The third problem is characterizing the shape of the lens as a function of age and accommodation state; these data are now becoming more available, but general trends and appropriate approximations need to be defined. Finally, the problem of the refractive index properties of the lens, including potential changes with age, remains a contentious one; a variety of optical refractive gradients have been proposed, but it is difficult to discriminate between different representations, especially since any one of a number of them appears to satisfy system constraints.

This chapter is devoted to a description of the normal human crystalline lens, its relationship to other elements of the globe, the changes that occur in its properties with increasing age, and how accommodation – the internal focusing process of the eye – is itself dependent on these changes. The changing ultrastructure of the lens during development and aging will be briefly reviewed and correlated with changes in overall lens shape and transparency. Lens orientation within the anterior segment will also be considered for both the emmetropic and ametropic eye. Finally, models of the lens as a refracting component of the visual system will be described, with emphasis placed on different representations of the gradient refractive indices. The information summarized here can ultimately be used as the basis for construction of a quantitative optical model of the lens, for critical evaluation of models in the biomedical literature, and for assessment of significant directions in future lens-related studies.

2.2 LENS DEVELOPMENT

The lens arises embryonically from the ectoderm as an invagination of this tissue, consisting of a single epithelial layer turned inside-out (Davson 1990). The epithelial cells from this layer then begin to differentiate, generating the lens capsule – a collagenous basement membrane covering that serves as the mechanism for attachment between the lens and the surrounding ciliary body – and the initial lens fiber cells that form the embryonic lens nucleus. Because of this inversion of the embryonic epithelial layer, the differentiating cells of the lens are on its surface and continue the process of adding lens fiber cells throughout life.

The post-natal lens can be considered to consist of two general regions, the central nuclear region and the surrounding cortical region. The nucleus consists of the original embryonic lens embedded in subsequent layers of lens fiber cells that define the fetal lens. The cortex appears to be laid down after birth, and becomes a larger percentage of total lens volume with development and aging. It is very difficult to discern the boundary between these regions visually in the normal lens, although they are easily distinguished upon development of a nuclear cataract, but there are clear morphological differences apparent at the microscopic level (Kuszak 1995; Taylor et al 1996). The size of the lens fiber cells, the appearance of the membranes separating adjacent cells, and the organization of membrane proteins within these layers all differ as a function of lens region. There are also differences between regions in the relative percentages of the crystallins – proteins of the lens which, due to their very high concentration in the lens fiber cell cytoplasm, provide the increased refractive index necessary for the lens to contribute to image formation (Bessems et al 1986; Bessems et al 1983; Fu et al 1984; Garland et al 1996). Although these compositional differences do not contribute to visible boundaries between the nucleus and cortex, it is likely that their variations contribute to the lens refractive index gradient, and age-related changes in this gradient.

Lens growth stems from the continuing differentiation of the lens epithelial cells, which are located on the anterior surface of the lens. The parent cells look like any other epithelial cell, being roughly cuboidal, but as they differentiate, they become elongated and flattened. Their altered size and shape enable them to form layers that evenly cover the lens surface, and in fact they participate in the development of the superficial layers of the lens cortex. Their location during differentiation is peripheral to and outside the central optical region, so that their cell nuclei and organelles do not scatter light under normal conditions. With continued cell differentiation and formation of new outer layers, the previous cells become more buried, further away from the lens surface. The cell nucleus and organelles, including mitochondria, ribosomes, etc., are gradually lost, preventing them from interfering with lens transparency. The fully differentiated lens fiber cells are therefore little more than membranous bags containing high (>300mg/ml) concentrations of the lens crystallins, as well as some cytoskeletal structures and ions.

The fiber cells are highly organized (Taylor et al 1996; Willekens & Vrensen 1981) within the overall lens structure (Fig. 2.1a). A cross-section of the lens in a plane that contains the lens pole (anterior to posterior axis of symmetry) will resemble the trunk of a tree, with concentric layers of cells circling the embryonic nucleus (Fig. 2.1b). A cross-section through the lens equator, separating the anterior and posterior halves of the lens, shows the close packing of the fibers (Fig. 2.1c). Each cell has a cross-section resembling a flattened hexagon, and the cell layers pack such that each fiber cell connects with four others circumferentially, while stacking on the upper and lower surfaces radially. In addition, every one of these six surfaces makes specific interlocking connections with the adjacent lens fiber cell surfaces. These connections may be tongue-in-groove, ball-in-socket, or other morphologies; their number and characteristics appear to be related to the region within which the cells are packed. Other connections between the cells are involved in intercellular transport through aligned gap junctions and other membrane inclusions. Taken together, these interconnections between each fiber cell and its six nearest neighbors strongly suggest that the lens fiber cells cannot slide past each other during accommodation (the focusing process). Changes in lens shape and thickness must occur through the movement of lens cytoplasm within each cell's volume, thereby limiting the nature and extent of the deformations that might be possible (Koretz et al 1984).

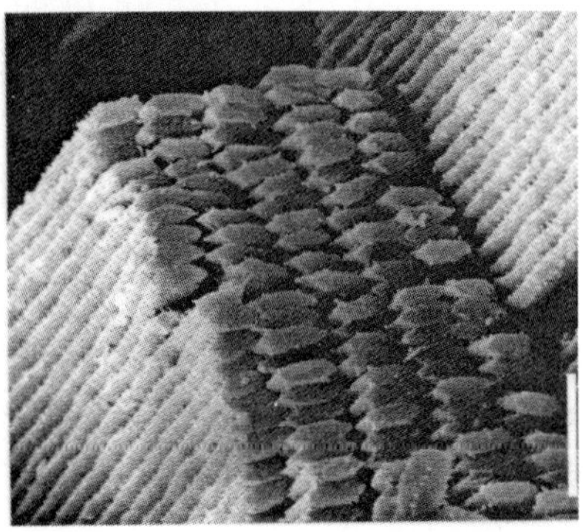

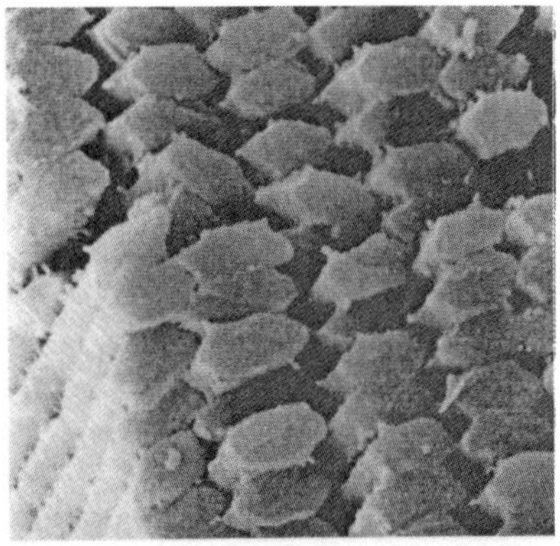

Figure 2.1. Scanning electron micrograph of a small internal region of a human lens. Clockwise from top-left: (a) .Transverse and radial packing of the fiber cells. (b) Magnified section showing side-by-side packing of the fiber cells, with small projections from their surfaces providing the ball-and-socket or tongue-in-groove connections between adjacent cells. (c) Magnified cross-section showing the cells' flattened hexagonal shape and close packing with nearest neighbors. Some of the stabilizing projections are visible. (Micrograph courtesy of Dr. Jer R. Kuszak, Rush-Presbyterian St. Lukes Medical Center.)

There is an additional constraint on lens development that has an impact on its structure, transparency, and deformability – the lens sutures. These are linear zones extending out from the poles on the anterior and posterior lens surfaces, where the lens fiber cell terminations interleave to form a closed seam. In the older literature on the human lens, it was believed that the lens fiber cells could grow long enough to span the circumference from anterior to posterior, but in fact their length is limited. This means that, with new fiber cells continually being laid down and the surface area expanding, there must be an alternative mechanism for generating closed surfaces using seam-like connections (Kuszak et al 1991; Kuszak et al 1994; Koretz et al 1994). The human eye solves this topological problem by increasing the number of suture branches with increasing lens size/age (Fig. 2.2a-e). In the fetal and young post-natal lens, the sutures are in the form of the three arms of a "Y" around each lens pole, with the posterior "Y" (Fig. 2.2a) offset from the anterior (Fig. 2.2b) by 60° to minimize reduction in image quality. When a surface defined by these sutures is filled, a new surface or shell is initiated, this one consisting of six arms radiating from the poles, and offset from both each other by 30° and from the underlying "Y" (Fig. 2.2c). The increased number of sutures enables the surface to be covered by a larger number of lens fiber cells without necessitating further elongation of each. A third shell (Fig. 2.2d) contains nine arms for each surface, offset appropriately, while the fourth (and generally final) shell is generated by twelve arms on each surface (Fig. 2.2e). Furthermore, each shell is offset relative to the other shells in order to prevent the increased image distortion that suture alignment would generate. It has been suggested that the symmetry of the last set of sutures, which begins to be laid down around age 40, can easily be distorted, and that this misalignment may be correlated

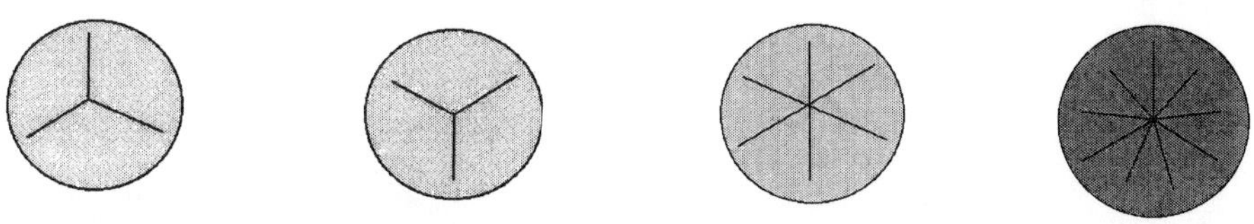

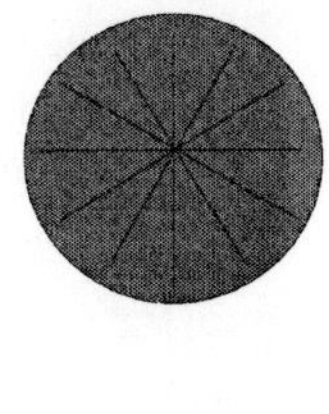

Figure 2.2. Schematic representations of the lens sutures, viewed down the sagittal axis, with each lens shell laid down in a precisely programmed sequence. Clockwise from top-left: Y-suture, with posterior surface (a) rotated relative to the anterior surface (b). (c) Six-star suture, anterior surface. (d) Nine-star suture, anterior surface. (e) Twelve-star suture, anterior surface. Each shell encompasses the previous one, with the sutures rotationally displaced to prevent reinforcement of optical distortions due to suture organization. The posterior lens half is organized similarly, and rotated relative to the anterior as well.

with the development of cortical cataracts; certainly their wedge shape is directly related to the inter-suture surface organization.

2.3 OPTICAL EFFECTS ASSOCIATED WITH LENS DEVELOPMENT AND AGING

The normal development and aging of the crystalline lens result in a number of changes in its optical properties that can be directly correlated with its morphology. Many of these changes were first observed and/or described using optical visualization methods such as slit-lamp photography, although some have been seen by other methods and then characterized biochemically. One such change – the shape of the gradient refractive index – will be considered separately because of its paramount importance in optical model-building; other developmental changes exhibit either no or little effect on image formation, although they may strongly influence aspects of image quality.

2.3.1 Nuclear Compaction and the Youthful Lens

The crystalline lens at birth exhibits dimensions of about 6.5 mm equatorially and about 3 mm sagitally. By about age 20 yr, the equatorial diameter has increased to about 9-10 mm, where it remains essentially unchanged throughout the adult lifespan (Strenk et al 1999), while sagittal lens thickness differs little from its initial value, although it will increase considerably after age 20. Recent Scheimpflug slit-lamp photographic studies of the anterior segment in youthful subjects in the age range of 5-20 yr provide information on the internal changes in lens ultrastructure during this time; they indicate that dynamic alterations in internal organization take place during this developmental period. The lens nuclear region essentially fills almost the entire volume of the developing lens, with the cortex only marginally visible in the very young lenses. As the individual grows to age 20 yr, the size of the nucleus shrinks, with the growth of the lens cortex almost exactly matching the rate of nuclear shrinkage; the two processes lead to the maintenance – or even slight shrinkage – of the lens in the sagittal direction (Al-Ghoul et al 2001; Bron et al 2000; Brown et al 1988).

It is believed that nuclear compaction occurs through the developmentally programmed loss of water (and perhaps lens crystallins as well), at an exponentially declining rate that essentially goes to zero by 20 yr of age. The growth of the cortical region is also exponential, first at the same rate as nuclear diminution, but then leveling off by age 20 yr to a constant rate that leads to continued sagittal lens growth through the human lifespan. It remains to be determined whether: (a) the nuclear compaction process is indeed compaction, due primarily to a loss of water and leading to an increased nuclear refractive index; (b) the reduction in nuclear size results from equivalent losses of both water and proteins for maintenance of a relatively constant nuclear refractive index; or (c) the process is intermediate between these two extremes. Since other aspects of the lens, and indeed of the globe itself, are essentially unchanged through this developmental period, it is likely that any nuclear refractive index changes are slight, but will compensate for the decreased contribution of the lower refractive index of the cortex.

2.3.2 Zones of Discontinuity

When adult lenses are visualized using slit-lamp photography, a set of internal curves in the anterior and posterior cortical regions is visible (Fig. 2.3). Each curve appears to be a boundary that scatters light more than surrounding regions. The anatomical basis for such a boundary was initially unknown, and these so-called "zones of discontinuity" were considered to be artifacts of the photography process. However, these zones are photographically reproducible in location in the same lens at different times; furthermore, when looking at a photographic sequence of an accommodating eye, these zones change curvature in concert with the lens nucleus and lens itself. These observations indicate that these zones are not artifactual, but arise from some physical cause (Brown 1973a; Koretz et al 1984).

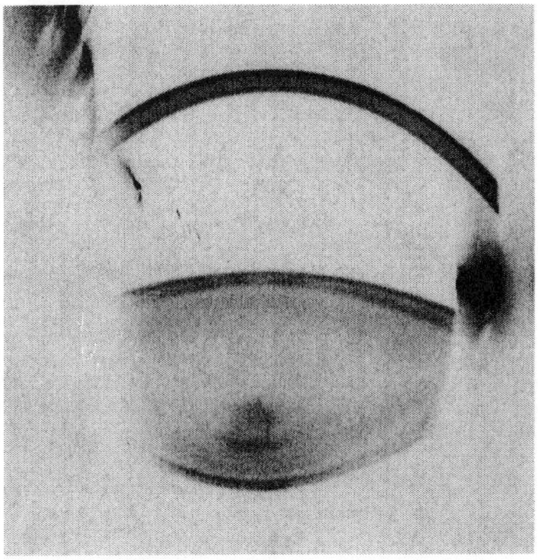

Figure 2.3. Scheimpflug photograph of a young subject's right eye. Note the presence of several distinct zones of discontinuity, and a central clear region that appears to bisect the lens nucleus.

Measurements of the location of these zones relative to the center of the lens nucleus along the polar axis indicate that their locations fall into specific groupings, with the number of groupings related to lens age. Since human lenses continue to grow throughout life, and appear to grow at about the same annual rate under normal conditions, it is possible to correlate the location of each zone with the age in years when the zone began to appear. In the adult lens, up to four zones of discontinuity can be clearly discerned in the anterior cortical region. When their location is correlated to the rate of cortical growth, taking into account the change in rate from youthful development to adult steady state, it is apparent that each zone of discontinuity signifies the initiation of a new suture shell (Koretz et al 1994; Kuszak et al 1994). The lens structural development can be followed non-invasively by slit-lamp methods, particularly Scheimpflug photography, which optimizes clarity of lens images. The potential contribution of the zones of discontinuity to diminution of image quality on the retina remains to be analyzed, but it is clear from the mechanism by which slit-lamp photos are generated that they represent specific areas within the lens where more light is scattered than in adjacent areas.

2.3.3 Sutures

Studies of both non-primate and primate lens image formation have generally been based on a laser ray-tracing technique through isolated lenses suspended perpendicular to the direction of the impinging light "pencils". A set of parallel rays is traced through the lens and ocular medium to characterize the location of focus and various aberrations. One application for which this method has been especially useful is to analyze the effects of the sutures on image quality (Kuszak et al 1991; Kuszak et al 1994). In all eyes, the sutures reduce image quality through scatter and spherical aberration, but the degree of effect is much less in the primate eye than the non-primate. In non-primate lenses, the suture number is unchanged with lens growth, and addition of lens fiber cells to the lens surface is along suture planes that contain the polar axis. As a result, light that impinges on the suture planes is much more strongly scattered than light that passes between the planes, and its focal point is altered relative to unimpeded light; the effect is thus severe, but quite localized. In the primate lens, in contrast, the sutures are offset relative to each other (e.g., Fig. 2.2a), both within the same shell in comparing anterior to posterior patterns and between adjacent shells in the same lens region. As a result, the spherical aberration and scatter associated with the sutures are minimized for a given shell and spread evenly over the lens surface, again minimizing localized effects, when the sum of the shells is considered.

2.3.4 Central Sulcus

The central sulcus is a clear zone in the center of the lens nucleus that bisects it along the equatorial plane of the lens. It becomes more prominent in older adult eyes, where the surrounding lens nuclear region is beginning to scatter significant amounts of light relative to youthful lenses. It is not, however, a sulcus as the term is generally used. Like the zones of discontinuity, the central clear region may be mistakenly thought of as an artifact, but it is reproducible in a given lens and moves with the lens nucleus during accommodation. Modeling of lens nucleus ultrastructure at the cellular level provides the explanation – the lens fiber cells along the nuclear equator are oriented roughly parallel to incoming light compared to cells either anterior or posterior to this central region. This orientation maximizes the amount of light that goes through a single cell along its long axis, and minimizes the number of cells whose membranes cross the light path in this region (Fig. 2.4).

Figure 2.4. Schematic diagram of the central clear region's lens fiber cells in an unaccommodated (above) and accommodated (below) lens. Lens curvature is such that in this region light rays pass virtually parallel to the length of the cells, unlike the rest of the lens, where rays pass through cross-sections of lens fibers (e.g., Fig. 2.1).

2.3.5 Other Age-Related Changes

A variety of factors lead to the decreasing transparency of the normal crystalline lens with age. In addition to the ultrastructural factors at the cellular level that create apparent optical interfaces (e.g., the zones of discontinuity/suture shells), there are processes at the cellular, subcellular, and molecular levels that can cause increases in the amount of light scattered at high angles and distortion of the color spectrum of any light that does succeed in passing through the lens (Bettelheim & Paunovic 1979; Taylor et al 1996). It has been noted by a number of morphologists, for example, that the lens fiber cell membranes become more fragmented and convoluted the older (and thus deeper in the lens) they are. Combined with alterations in the distribution and orientation of integral membrane proteins and/or gap junctions, these changes may cause a reduction in lens transparency and/or increased scattering (as in the case of glare) with increasing lens age. Such changes are particularly noticeable in the older lens nucleus.

In addition to creating potential scattering centers, the changes in membrane and membrane protein organization is likely to affect transport and diffusion properties within the lens (Bron et al 2000; Moffat et al 1999). It has already been noted that the cellular organelles are lost with lens fiber cell differentiation; this means that there are few or no mechanisms supporting active transport to maintain optimum environmental conditions (e.g., pH, relative ion concentrations) or to repair lens proteins damaged or otherwise altered by these conditions. These proteins, which in the young lens avoid intermolecular interactions, may be modified with age to form intermolecular bonds. Such processes can lead to the formation of aggregates of proteins that are large enough to scatter light (e.g., Mie scattering) and, equally, lead to a "lumpy" cytoplasm. Certainly the relative amounts of high molecular weight soluble aggregates and insoluble aggregates increase in the lens with increasing age. It is possible that these aggregation mechanisms lead ultimately to the development of some of the various forms of cataract.

At the molecular level, there are a variety of biochemical processes whose by-products accumulate with age. These include mixed disulfide formation of certain proteins in the nucleus of the lens, yellowing of the lens, and brunescence (Lou & Dickerson 1992; Lutze & Bresnick 1991). These processes have been characterized as part of the aging of the normal lens, and they can be corrected for experimentally if the age of the individual whose lens is being studied is known. It is interesting to note that these biochemical changes exhibit an increase in rate with increasing age, with the major increase occurring at about the same time that accommodative ability is lost – near age 50 yr – and that the formation of larger scattering centers, causing increased problems with glare, also shows this bimodal pattern (Cook et al 1994). It may well be that the inability to change lens shape to focus on near objects also reduces the ability of the lens cells to deal with environmental inhomogeneities, causing irreversible chemical changes that accumulate at a higher rate in presbyopic eyes.

2.4 THE LENS IN THE ANTERIOR SEGMENT

2.4.1 Lens Placement Relative to the Cornea

The crystalline lens of the adult emmetropic human eye is positioned and oriented within the anterior segment to preserve specific refractive characteristics (Sorsby et al 1957; Sorsby et al 1961). With an average globe length of approximately 24 mm and a corneal refractive power of about 44 diopters, the distance from the cornea to the anterior lens surface and from the posterior surface of the lens to the retina become critical factors. There are strong indications from both A-scan ultrasonography and Scheimpflug slit-lamp photography that anterior segment length – and therefore vitreous chamber length, since most of the growth of the globe's axial length is complete by age 3 yr – are not altered with age. The vitreous chamber for a 24 mm globe is about 16 mm long, with the remaining length representing anterior segment length (the distance from the anterior cornea to posterior lens surface). Although there is a sizeable variation in these values within the emmetropic human population, the age dependence of each value has been found to be not statistically significant (Cook et al 1994; Koretz et al 1989a; Koretz et al 1989b).

Since the distance from the posterior lens surface to the retina is maintained, the refractive power of the anterior segment as a whole can be treated as a single unit. Although this simplified representation is not very useful for optical modeling of the detailed characteristics of the lens, it is nevertheless illustrative of the compensatory mechanisms operating in the eye to correlate corneal and lenticular refractive power, i.e. emmetropization. As Sorsby and coworkers showed several decades ago, an "inflatable" globe can exhibit emmetropia outside the average values; a larger globe will be compensated by a shallower cornea, and a smaller globe will have a more sharply curved cornea. Equally, there is an "inflatable" anterior segment, wherein variations in lens placement relative to the cornea lead to an emmetropic eye as a whole (Brown et al 1999; Koretz et al 1995). Breakdowns in these compensatory relationships appear to be correlated with ametropia (see Chapter 18 in this volume); while the major contribution to refractive error (Fig. 2.5) is globe asymmetry (prolate for myopia and oblate for hyperopia), there is also a breakdown in the correlations that allow the cornea and lens to act in a compensatory manner. It does not appear, however, that crystalline lens refractive power per se shows any variation for a given age, but rather that its placement alters its refractive contribution to image formation.

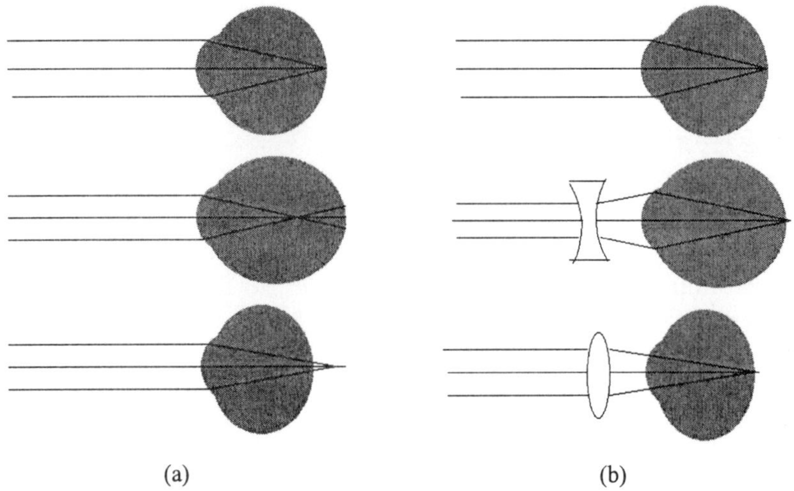

Figure 2.5. The role of globe length in generating refractive error. (a) When the effective focus of the anterior segment equals globe length (top), the eye is considered emmetropic. A globe that is too long (middle) for the refractive power of the anterior segment is called myopic, or near-sighted, because the focus is in front of the retina, while a globe that is too short (bottom) is called hyperopic, or far-sighted, because the focus is behind the retina. (b) Myopia is corrected using diverging lenses (middle) to reduce the refractive power of the anterior segment, while hyperopia is corrected with converging lenses (bottom) to increase its refractive power.

2.4.2 Aging of the Lens

As noted earlier, the youthful human lens appears to exhibit a constant sagittal thickness of about 3 mm up to the end of the second decade, whereupon nuclear compaction presumably ceases, and changes in lens thickness with time achieve a steady state. At about the same point in life, the equatorial dimensions of the lens cease to growth, so that its 6.5 mm diameter at birth expands to no more than about 9-10 mm at maturity. Addition of new lens fiber cells to the lens surface beyond this point, then, implies that lens thickness in the sagittal direction will increase with age and that lens shape will be altered due to geometrical constraints (Cook et al

1994; Koretz et al 2001b; Koretz & Handelman 1988; Koretz et al 1989a). However, since anterior segment length appears to remain essentially unchanged with increasing age, this indicates that the distance from the cornea to the anterior lens surface – anterior chamber depth – decreases with age at about the same rate that lens thickness increases. This also means that the lens center of mass changes in relation to its support structures, moving more anteriorly with age, and that the lens nucleus – presumably a region of higher refractive power than the surrounding lens cortex – is also translated anteriorly. Both effects will serve to increase overall refractive power if all other factors remain unchanged.

The sagittal thicknesses of internal lens regions change with age in an asymmetric fashion (Cook et al 1994). In the youthful lens, as previously noted, nuclear thickness decreases with compaction at a rate that falls to zero by the end of the second decade, while cortical thickness increases by the same amount. At this point, nuclear thickness remains unchanged for the adult lens, while both anterior and posterior cortical regions continue to increase in thickness at a steady rate; the sum of these increases is identical to the overall increase in sagittal lens thickness. As noted above, the growth of the posterior cortical region acts to translate the nucleus anteriorly. Since anterior segment length is preserved, the growth of the anterior cortical region brings the anterior lens surface closer to the posterior corneal surface. It should be noted, however, that the anterior and posterior cortical regions do not increase in thickness at the same rate; although the number of lens fiber cells is the same on both surfaces, the significant difference in the curvatures of these surfaces results in a difference in the cross-sectional shape of the lens fiber cells that cover the two surfaces. The asymmetry between the two cross-sectional shapes increases with age.

2.4.3 Lens Shape

The changes in lens shape with age have been difficult to characterize in a quantitative manner. Scheimpflug and other forms of slit-lamp photography can provide high-resolution, high-contrast images of the anterior segment and the internal organization of the lens. In these methods, a narrow, intense slit beam is used to illuminate a sagittal section of the anterior segment that includes the visual axis. The image of this section is captured by a special camera located so that the captured image is collected roughly perpendicular to the cornea; additionally, the image capture mechanism (e.g., film, CCD camera) is tilted relative to the plane of the slit beam so that compression of the image caused by looking at it at an angle is partially or completely neutralized. These images suffer, however, from non-linear distortions that are intrinsic to the visual system (e.g., the cornea) and to the optical system capturing the image (e.g., the tilted image plane, oriented specifically to increase image resolution).

An initial and highly influential attempt to compensate for these distortions was presented in the 1970s by Nicholas A. Phelps Brown (1972; 1973b), who is arguably responsible for the resurrection of the Scheimpflug principle for anterior segment studies. He used a set of nested curves to match as best as he could the central and peripheral radii of curvature of the anterior and posterior lens surfaces, and a magnification-matched image of a grid to compensate for anterior segment image distortions in estimation of sagittal spacings. The resultant analyses provided the first age-related quantitative assessment of this region. The age-dependent and -independent data listed in the Tables below come, in large part, from a subsequent generation of Scheimpflug images that have been analytically corrected for distortions.

An extremely important observation by Brown, one that angered and confused many vision scientists of the time, was the way human crystalline lens shape changes with age (Brown 1973a, 1974). He showed unequivocally that the surface curvature of the lens, and particularly anterior surface curvature, becomes more highly curved with increasing age. This is in apparent contradiction to the direction of loss of accommodative amplitude with age, where distance vision is preserved at the expense of near vision. All other factors being equal, a lens with increasingly highly curved surfaces should become a more powerful refractor rather than a less powerful one. This is Brown's lens paradox (Fig. 2.6). As has already been noted, other factors contributing to image formation, such as distance between the anterior segment and the retina, are independent of age, indicating that the only way to "solve" this paradox is to assume that the refractive index gradient of the lens is itself age-dependent in a fashion that compensates for increased lens curvature (Koretz & Handelman 1986). Models of the refractive index gradient will be discussed in detail later.

Re-examination of Brown's raw Scheimpflug data for the non-accommodated human eye, as well as collection and analysis of additional data using the same optical principle, has provided important quantitative information about changes in the lens with age (Koretz et al 2001a; Koretz et al 2001b; Koretz et al 1984). At the age where nuclear compaction is essentially complete and the equatorial diameter has attained its mature dimensions, lenticular curvatures on the anterior and posterior surfaces begin to become more rounded (smaller central radius of curvature). The anterior lens surface, which is much flatter than the posterior, also exhibits a much higher rate of change with age. In contrast, the much more rounded internal curvatures delineating the lens nucleus are approximately the same for anterior and posterior, and they show little age dependence; this would be expected, since overall morphological changes in the lens occur primarily in the cortical region. Internal boundaries defining the zones of discontinuity exhibit curvatures intermediate between nuclear and surface curvatures for each half, with the specific value linearly related to its location. It is interesting to note that all of these curves, whether surface or internal and

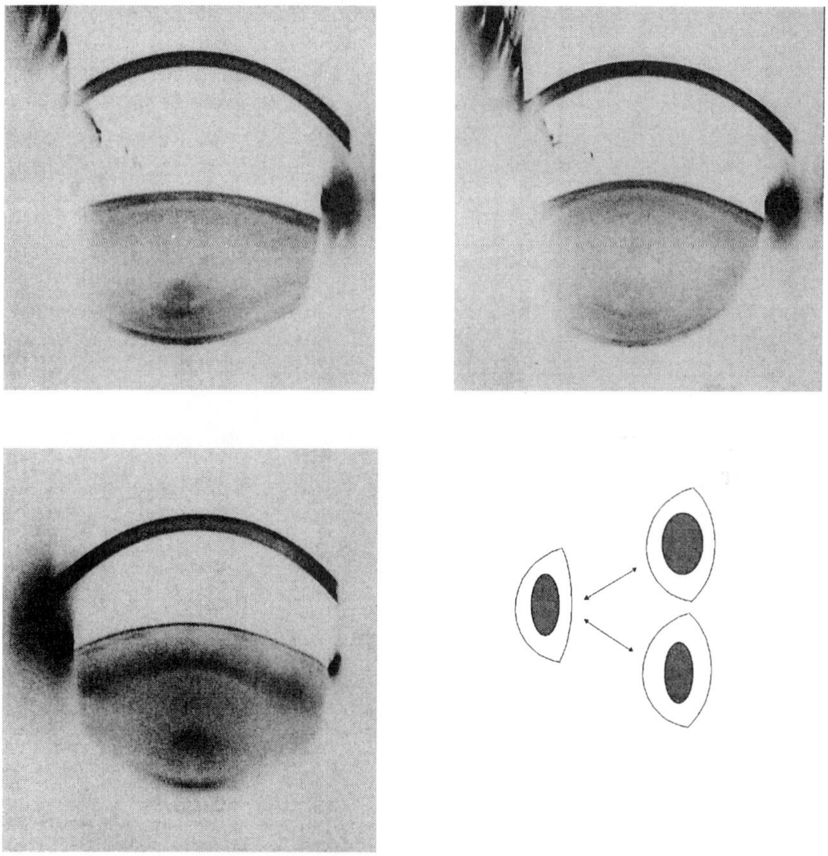

Figure 2.6. Scheimpflug photographs illustrating changes in lens shape with accommodation and age. The lens of an 18 yr old subject focused on infinity (above left) and at 10 diopters accommodation (above right). Note the change in lens curvature on the anterior surface particularly, as well as the reduction in anterior chamber depth and the massive change in shape of the lens nucleus. The lens of a 67-yr old focused on infinity (below left) appears to resemble the fully accommodated young lens in overall shape and anterior chamber depth; however, the two differ in nucleus shape and sagittal thickness, as well as in the overall thickness of the cortex. Note as well that the posterior surface of the lens is difficult to discern because the lens anterior scatters so much light. These differences are illustrated schematically at below right.

whatever the age, are best represented by paraboloids, where the coefficient of the square term is inversely proportional to the central radius of curvature (Cook & Koretz 1991; Cook & Koretz 1998). This parabolic curvature reduces any lens contribution to overall system spherical aberration, where peripheral light rays come to focus at a different point than central light rays (Fig. 2.7). Additionally, a paraboloid exhibits an increasing radius of curvature with distance from the optical axis, changing in the opposite manner from the ellipsoidal approximations many have used in the past to represent the lens surfaces.

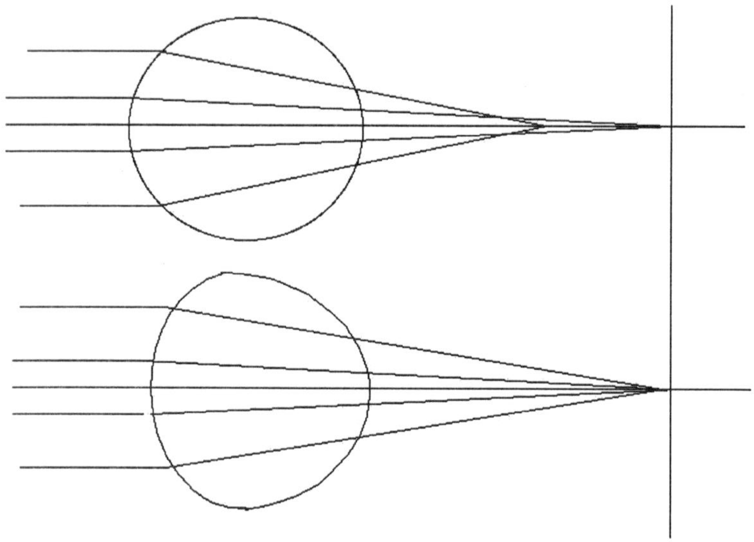

Figure 2.7. Spherical aberration arises from a difference in the refractive power of a lens in the center and at its periphery (above), such that focus is represented as a curved surface rather than a point. This form of aberration can be either positive or negative, depending on the shape of the curve, and can be reduced using a parabolic surface (below) rather than a spherical one.

Recent advances in imaging technologies and analysis protocols have increased the mechanisms by which lenticular shape can be analyzed. Magnetic resonance imaging (MRI) using a specially designed coil, for example, has enabled the visualization of the entire lens, albeit at a much lower resolution than photography (Strenk et al 1999). A description of lens shape in toto requires a fourth order polynomial; this reduces, however, to the anterior and posterior paraboloids (second order polynomials) obtained from Scheimpflug photography in the regions of interest for optical modeling. Unfortunately, it is not currently possible to use MRI to study the internal structure of the lens, but a combination of the two methods will allow a complete description. Additionally, slit-lamps designed with a tilted image plane, but not at the Scheimpflug angle, can nevertheless provide qualitative and semi-quantitative data about changes in lens shape and internal organization, and have been used productively in the study of cataractogenesis. A recent study (Dubbelman & Van der Heijde 2001) has also indicated that these images might be used for characterization of lens shape with the appropriate mathematical corrections for corneal refractive power and the lens's own refractive index gradient; if so, this will provide a more accurate analysis of lens shape in ametropic eyes.

2.5 OPTICAL MODELS OF THE LENS

2.5.1 Simple Geometric Models

Historically, the visual system has been simplified to resemble a set of lenses that can be characterized by geometric optics. To a rough approximation, the cornea and lens can be taken as two elements of a compound system, something like a camera, where the first refractor or cornea – the objective lens - provides the majority of the refractive power and the second refractor or lens – the focusing lens – the additional power needed for final focus at a point. For both the actual visual system and its geometric approximation, the size of the image is much smaller than the actual object, the final image is inverted, and the system's posterior focal point coincides with the retina for emmetropic eyes (Atchison & Smith 2000). Note, however, that these simple models can be reasonable approximations to the visual system only within central optical paraxial limits (i.e., where the angle and its sine are roughly identical).

Using this analogy between geometric optics and the visual system, limited though it might be, two optical models are generally presented in optometry and introductory physics textbooks. The simplest model of all is the Reduced Eye, where the refraction contributed by the elements and geometry of the anterior segment is represented by a single thin-lens equivalent, or a single refracting surface. This model generates two foci, plus the equivalent of one principal plane and one nodal point for the visual system. The principal planes of an optical system are those paired locations where an image located at one is seen as erect and unmagnified at the other. The nodal points are locations where, when an off-axis ray passes through one nodal point, it appears to pass through the one on the other side of the system, and vice versa; the angle that these rays make with the axis is the same for both sides of the system. For a model with essentially only one refracting surface, however, the location of the two principal planes is coincident, as is the location of the two nodal points (Davson 1990). Thus, the principal plane is located about 1.5 mm behind (posterior to) the anterior corneal surface, while the nodal point is 7.2 mm posterior. Then, F_a, the anterior principal focus, is 17.2 mm distant, and F_p, the posterior principal focus, is 22.9 mm distant from the principal plane. Of all these geometric locations, only the location of the second principal focus represents a real part of the emmetropic eye – the retina.

Another set of models is the Schematic Eye, where the cornea and crystalline lens are represented as separate refracting elements sharing a common axis (Howland 1994). This generates two principal foci, two nodal points, and two principal planes. For most purposes, the Gullstrand Schematic Eye is used, where the cornea is represented by two refracting surfaces and the crystalline lens by four, normally two for the lens surfaces and two for the boundaries of the nucleus within the lens (Fig. 2.8). Values for distances along the optic axis, radii of curvature, and refractive indices for the Gullstrand model are listed in Table 2.1. They are clearly different in

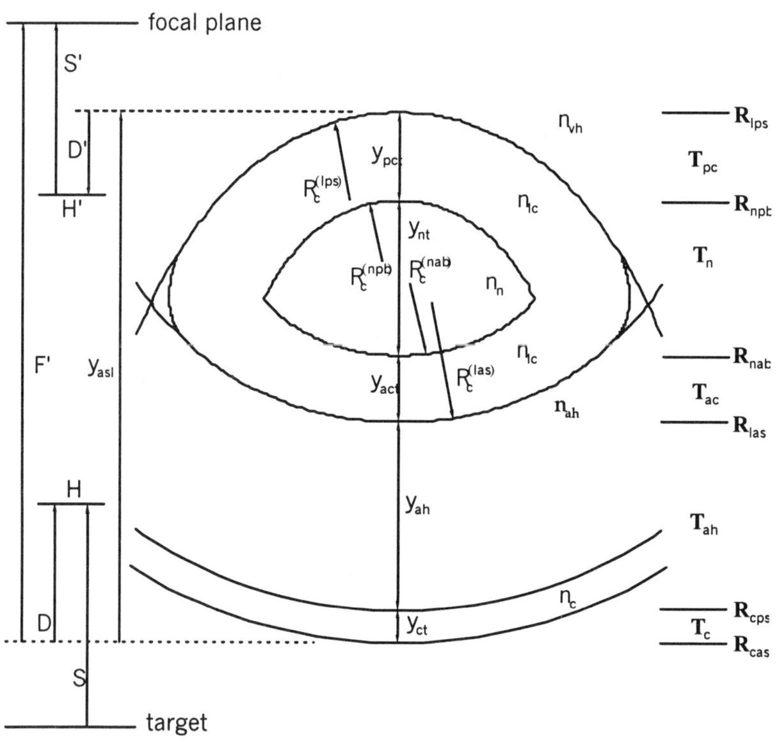

Figure 2.8. Schematic representation of a Gullstrand-type model of the anterior segment. Notation used in the paraxial modeling of this and other models is illustrated.

large part from the equivalent values measured experimentally in adult emmetropic eyes and summarized in Tables 2.2-2.4. However, it would be a straightforward task to create Gullstrand-type Schematic Eyes using the experimental data, adjusting the values of the nuclear and cortical refractive indices of the lens to obtain focus on the retina and following the effects of changes in the lens and anterior chamber with age.

Table 2.1. Values for parameters defining the Gullstrand schematic eye

Radii of Curvature (in mm)

Anterior cornea	7.7
Posterior cornea	6.8
Anterior lens	10.0
Anterior nucleus	7.91
Posterior nucleus	-5.76
Posterior lens	-6.0
Posterior of globe	-12.0

Other data

Region	Distance from anterior Corneal surface (in mm)	Refractive Index
Cornea	0.5	1.376
Anterior chamber	3.6	1.336
Anterior lens cortex	4.146	1.386
Nucleus	6.565	1.406
Posterior lens cortex	7.2	1.386
Vitreous (to retina)	20.0	1.336

The increased complexity of Gullstrand's crystalline lens representation over a pared-down Schematic Eye indicates the critical importance of the lens as a variable refractive element in an otherwise fixed refractive system. The very high lens refractive index required in a simple Schematic Eye, and the step-wise refractive index transition between it and the surrounding media, will generate wavefront aberrations that are definitely not physiologically realistic. The Gullstrand two-region lens is a more realistic representation of the crystalline lens, which provides an additional variable for fine-tuning of any model and reduces aberration by reducing the size of transitions between adjacent regions of differing refractive index. More complex representations of the crystalline lens are discussed below.

2.5.2 The Gradient Refractive Index

Characterization of crystalline lens shape as a function of age, initiated through the studies of Brown using Scheimpflug slit-lamp photography, indicates that the central radii of curvature on the anterior and posterior surfaces decrease with age. It would be expected that this change of curvature would lead to an increase in the refractive power of the lens, assuming all other factors being held constant. However, since the human visual system loses focusing ability at near with aging rather than distance focus - Brown's lens paradox - it is clear that one or more factors must also have changed to compensate for the age-related changes in lens shape. As summarized in Table 2.2, corneal refractive power, anterior and posterior segment lengths, and globe length are all independent of age in the adult eye. Decreases in anterior chamber depth are linked to increases in lenticular thickness; the former factor would contribute to a slight increase in refractive power by bringing refracting surfaces closer together, while the latter would contribute a slight decrease through the increased distance between the anterior and posterior lens surfaces. Taken together, there are no changes in the factors external to the lens that can effectively account for Brown's lens paradox.

Internally, however, it is likely that the aging of the lens material is associated with a reduction in refractive indices and/or the refractive index gradients that are presumed to affect the lens refractive contribution to the visual system. It has already been noted that significant differences in lens morphology, ultrastructure, and biochemical composition exist in different regions of the lens, and that some of these differences are also age-dependent. The capability of distinguishing the boundary between the nucleus and cortex, for example, indicates an optical difference between the two regions irrespective of its molecular or cellular basis. Since, however, the concentration gradient of the proteins within the lens is smooth and does not appear to change for a given region with age, and the lens hydration is also constant, the underlying cause of changes in refractive index must be

Table 2.2. Age-independent characteristics of emmetropic adult human eyes (from Koretz et al., 1989a). Asterisks (*) indicate age-dependent increase, $p < 0.05$.

Parameter	Women (n=68)	σ	Men (n=32)	σ	All (n=100)	σ
Age (yr)	43.7	14.1	45.2	15.6	44.2	14.6
Height (in)	64.7	3.0	69.7	3.5	66.3	4.0
Weight (lb)	*137.0	18.8	177.8	19.0	150.0	26.8
Corneal Thickness (mm)						
OD	0.47	0.04	*0.46	0.04	*0.47	0.04
OS	0.47	0.04	*0.46	0.04	*0.47	0.04
Ant. corneal Power (D)						
OD-vertical	44.1	1.2	43.6	1.3	43.9	1.2
OD-horizontal	43.6	1.3	43.1	1.3	43.4	1.3
OS-vertical	44.2	1.3	43.7	1.4	44.0	1.4
OS-horizontal	43.7	1.3	43.2	1.3	43.6	1.4
IOP (mm Hg)						
OD	*16.0	2.3	15.2	2.4	*15.7	2.4
OS	*15.6	2.3	15.1	2.4	*15.5	2.3
Ant. segment (mm)						
Ultrasound	*7.60	0.29	*7.71	0.26	*7.63	0.29
Corrected US	7.61	0.26	7.72	0.21	7.64	0.25
Slit-lamp photo.	7.70	0.34	7.75	0.26	7.71	0.32
Vitreous length (mm)	15.81	0.68	16.34	0.75	15.98	0.75
Globe length (mm)	23.43	0.76	24.08	0.81	23.64	0.83

more subtle (Fagerholm et al 1981; Siebinga et al 1991; Siebinga et al 1992). Formation of soluble and insoluble protein aggregates of increasing size with time means that the total number of aggregates per unit volume is decreasing. This has two consequences. One is that the lens fiber cell cytoplasm becomes more grainy as the protein is increasingly concentrated in clumps rather than evenly distributed. This effect will also contribute to decreasing lens transparency. Another consequence is that the relative proportion of water that is fully mobile increases, because water molecules that had been immobilized at protein surfaces or had interacted with surface polar groups are released when the protein surfaces bind to each other (and thus become inaccessible to water molecules). These age-related changes may also have implications for biomechanical models of lens accommodation and presbyopia.

Whatever the mechanism of lens aging at the molecular level, however, it is clear that there is some form of refractive index gradient across the lens, and that it changes with age so as to compensate optically for the increased sharpness of lens curvature. The comparatively small degree of spherical aberration generated by the lens is obviously due in large part to the paraboloid curvatures of the surfaces and internal boundaries, but smooth refractive index gradients rather than abrupt step-wise changes in refractive index will also contribute to diminution of aberration. The question then becomes what shape this gradient of refractive indices (GRIN) takes, and how it is altered with age. There have been a number of different models presented in the literature over the years – modified Gullstrand-type, linear, and non-linear – any one of which can be modified sufficiently to provide the appropriate refractive power to the lens. Testing three of these GRINs with the experimental aging data provided in Tables 2-4 and with accommodation data over the adult age range using a paraxial approach, Cook and Koretz (1995) were able to optimize each of them to provide focus on the retina. It is unfortunate but true that factors defining the human crystalline lens GRIN are underdetermined, leading to any of a broad family of solution types.

If there is no unique solution to the problem of the shape of the crystalline lens GRIN based on the optics of the visual system, additional factors must be considered in attempts to narrow the range of possibilities. Several GRIN models follow the nested nature of the lens itself, defining a set of isoindicial surfaces within the lens where the refractive index is incrementally increased with increasing distance from the lens surface. The linear or non-linear nature of this gradient, the initial refractive index value upon entering the lens and the final one upon reaching the lens center, and the extent, if any, of a constant refractive indexed central region are all variables to be determined.

An attempt was made to measure the GRIN directly using laser ray-tracing (Pierscionek & Chan 1989). This approach suffered from several shortcomings, including the assumption that equatorially, where the ray-tracing was done, lens GRIN would be equivalent to the sagittal GRIN. The protein concentration profiles of the human lens derived from the experimental work of Fagerholm et al. (1981), which themselves could serve as the basis for a GRIN model using the Gladstone-Dale relationship, were used to justify this approach and to validate the results; however, this experimental protocol , as well as the similar approach of Bours et al. (1990), was seriously flawed as well because of the use of freeze-drying for specimen preparation. Later versions of this model, as well as further attempts to measure the lens GRIN in the sagittal direction directly, appeared to indicate that central (nuclear) refractive index is independent of age, although very recent studies using MRI seem to show the contrary (Atchison, 2001). But for this model, any reduction in the refractive power of the lens material to compensate for increased lenticular sharpness of curvature must rely on defining a gradient shape outside of the nucleus that can be adjusted. If the nuclear region's refractive index is constant, this can only be achieved by increasing the initial value of the refractive index at the lens margin (making any gradient more shallow), by going to a non-linear gradient where the shape can be adjusted, or both in combination. When such a set of approaches was used, the GRIN shape became increasingly complex with age, requiring the contribution of higher order integer powers as the lens shape became increasingly sharply curved. Thus, while models like this can always be made to work, their increasing separation from the physiological and experimental reality makes their overall validity as models of the lens problematic.

Although direct measurement of the GRIN along the sagittal axis remains technically very difficult, there are nevertheless associated factors that can be followed more accurately as a function of depth within the lens and of lens age. The use of Raman microspectroscopy, for example, enables the direct measurement of free water concentrations without dehydration of the specimen, as well as protein concentrations (Siebinga et al 1991; Siebinga et al 1992). With this approach, the investigators found distinct differences in the shape of the protein concentration gradient and water context along the sagittal and equatorial axes of the lens, in direct contradistinction to the ray-tracing model. They additionally found that the free water content of the lens nucleus increases with increasing age, leading to an alternative GRIN formulation where nuclear refractive index is the variable, decreasing with increasing age, and with initial cortical refractive index held constant and the (linear) gradient between lens surface and lens nucleus thus becoming shallower with increasing age.

This model also can be made to work effectively, but the assumptions are much more solidly grounded in lens physiology than the power model. It is likely, for example, that differentiating and newly formed lens fiber cells will have about the same protein content irrespective of overall lens age, implying a constant value for the refractive index of the lens material at the surface interface. It is also likely that, with increasing lens fiber cell age (and therefore a location increasingly far from the lens surface), an increasing percentage of the protein will be involved in the formation of larger soluble and insoluble aggregates with the associated release of bound water. Thus, while neither protein concentration nor total water content changes significantly with time, their overall state – and therefore their effect on local refractive index – will be age-dependent.

The two isoindicial models described above can be considered representative of two classes of GRIN formulations. The first one holds the lens nuclear refractive index constant, and therefore requires that any solution to the lens paradox requires an age-dependent alteration in the cortical gradient, while the second essentially requires increasing changes in refractive index with increasing lens depth/fiber cell age, leading to a decreasing lens nuclear refractive index over time. Both can be made to work effectively with the lens curvature and spacing data presented here in a paraxial optical representation. A third class of models is derived independently from these considerations, relying instead on a wide-angle (non-paraxial) approach. When the iris is dilated far beyond the normal physiological values, it is possible to characterize the spherical aberrations associated with the visual system. These aberrations can then be used to test the validity of any lens model, since they provide an additional set of criteria for defining physiological function. This approach led to the development of what is arguably one of the first modern isoindicial models of the lens, where 98 nested isoindicial, aspheric layers surround a biconvex homogeneous nucleus (Pomerantzeff et al 1971; Pomerantzeff et al 1983). Both the asphericity and the refractive index of each layer vary linearly, with ratios between nuclear and cortical refractive indices, layer thickness, and change of asphericity used as parametric constraints. While details of the model are not provided, it is interesting to note that the resultant GRIN strongly resembles that derived from Raman microspectroscopic results.

2.5.3 Paraxial Models

At this point, it is possible to create a paraxial model of the visual system where all relevant parameters except for the crystalline lens GRIN are defined. For an emmetropic eye, the age-independent values for globe length and geometry are found in Table 2.2. The values for a -4 diopter myope would be roughly the same as for the average emmetrope, except that globe length differs significantly; this difference is taken up almost entirely by a difference in vitreous chamber length with other distances remaining unchanged (Δthickness] = 0.3mm/diopter). For all eyes, the refractive index of the aqueous and vitreous humor is 1.336, with the corneal refractive index generally taken as 1.372.

The age dependencies of lens surface and lens nuclear shapes, as well as age-dependent spacings between surfaces, have also been presented (Tables 2.3 and 2.4), where age in years is the independent variable in these simple linear equations of the form *y = intercept + slope x age (in yr)*.. Note that these data were derived from a group of subjects ranging from about 18 to 70 yr in age. It is likely that the lenses of many individuals whose age is around the upper limit or above will not be well described from these data, since they will be pre-cataractous and thus becoming more myopic. Data for visual systems younger than age 18 yr are limited, although it is likely that, for most purposes, the ultrasound data of Mutti et al (1998) can be combined with the cortical thickness data in Koretz et al (1994) for determination of nuclear thickness and nuclear placement within the lens.

Because paraxial models use only a very small cylindrical volume around the optical axis and are assumed to focus at a single point, it is possible to make a number of approximations. Corneal shape and thickness can be modeled in a manner similar to that of the Gullstrand model, using two nested spherical surfaces separated by about 0.5 mm, while lenticular surfaces and nuclear boundaries can also be approximated by the central radii of curvature for the nested parabolic surfaces that describe their true overall shape. For isoindicial models, where a set of nested curves define the anterior and posterior cortical region, these additional radii of curvature should decrease monotonically from the surface to the nuclear boundaries. Creation of a model of the lens that brings the system into focus can then be performed in one of three ways: direct calculation, following rays surface-by-surface through the globe analytically and using the approximation that, with a small enough angle, the angle and its sine are equal (e.g., Howland 1994); the matrix method of paraxial ray-tracing, where the transformation of a ray through a complex optical system is defined as a set of matrices which can return the locations of system power, system principal planes,

Table 2.3. Age-independent and -dependent changes in spacings along the adult emmetropic lens sagittal axis (in mm) . The independent variable is age in years , and the intercept and slope are provided for each distance (from Cook et al., 1994).

	Intercept	σ_I	Slope	σ_S	p
Anterior chamber depth	4.150	0.101	-0.012	0.002	<<0.005
Lens thickness	3.232	0.082	0.014	0.002	<<0.005
Anterior segment Length	7.156	0.087	0.000	0.0002	>0.05
Lens cortex					
Anterior cortex	0.571	0.042	0.012	0.001	<<0.005
Posterior cortex	0.593	0.031	0.005	0.001	<<0.005
Total cortex	1.178	0.060	0.017	0.001	<<0.005
Lens nucleus					
Anterior nucleus	0.966	0.031	-0.002	0.001	0.0089
Posterior nucleus	0.943	0.032	-0.001	0.001	>0.05
Central clear zone	0.190	0.040	0.000	0.001	>0.05
Total nucleus	2.096	0.069	-0.003	0.001	0.0256

Table 2.4 Age-dependent changes in lens curvatures at 0.0 D accommodation along the adult emmetropic lens sagittal axis (in mm). The independent variable is age in years, and intercept and slope are provided (from Koretz et al., 2001b).

	Intercept	σ_I	Slope	σ_S	p
Lens anterior surface	11.155	1.729×10^{-2}	-2.004×10^{-2}	3.806×10^{-4}	0.046
Lens posterior surface	-8.267	4.336×10^{-2}	2.025×10^{-2}	1.148×10^{-3}	0.042
Nucleus anterior boundary	3.782	1.376×10^{-2}	-5.996×10^{-3}	2.798×10^{-4}	0.231
Nucleus post. boundary	-3.500	1.947×10^{-2}	5.537×10^{-3}	4.126×10^{-4}	0.199

and/or lens principal planes (Blaker 1971; Klein & Furtak 1986); and computer-based optics packages such as ZEMAX, where the special characteristics of a lenticular model must be described in a macro that the program can use. The first method provides the least amount of information while the third provides the most, assuming that the system has been properly modeled.

Overall, the matrix approach is the most useful for paraxial modeling, since its formulation is straightforward and does not require special software other than a general utility package such as Mathematica or MatLab. Additionally, this method is standard in analysis of geometric optics, of which physiological optics is – or should be – a special case (Fig. 2.8). All the refracting surfaces are represented by their spherical approximations, which are centered along the symmetry axis. Thus, the paraxial translation of a light ray across a refractive interface α, represented by the two vectors $\mathbf{r}_{in}$ and $\mathbf{r}_{out}$:

$$\mathbf{r}_{in} = \begin{bmatrix} n\alpha_{in} \\ x_{in} \end{bmatrix} \rightarrow \mathbf{r}_{out} = \begin{bmatrix} n'\alpha_{out} \\ x_{out} \end{bmatrix} \tag{2.1}$$

can be described by the matrix transformation

$$\mathbf{r}_{out} = \mathbf{R}_{\alpha}\mathbf{r}_{in} \tag{2.2}$$

where the 2 x 2 transformation matrix **R** is given by

$$\mathbf{R}_{\alpha} = \begin{bmatrix} 1 & -P_a \\ 0 & 1 \end{bmatrix} \tag{2.3}$$

and where $-P_a$ is the power of the refracting surface given with respect to the indices of refraction of the incident and outgoing media and the spherical curvature of the surface **R** by

$$P_{\alpha} = \frac{n' - n}{R_{\alpha}} \tag{2.4}$$

The translation of a ray across a region is defined by the transformation matrix **T** , given by,

$$\mathbf{T}_{\alpha} = \begin{bmatrix} 1 & 0 \\ D_{\alpha} & 1 \end{bmatrix} \tag{2.5}$$

where the dilated distance D is defined in terms of the refractive index of the region and its thickness by,

$$D_\alpha = \frac{y_\alpha}{n_\alpha} \tag{2.6}$$

These transformation matrices are combined as a successive product to give the overall transformation through a compound optical element. Since matrix multiplication is non-commutative in this case, they must be applied in order with the convention being from left to right.

Consider first a matrix representation of the Gullstrand Schematic Eye (Fig. 2.8), which is composed of six refracting surfaces, two for the anterior and posterior cornea, two for the anterior and posterior lens surface, and two for the boundaries of the lens nucleus. The refractive powers of the anterior and posterior corneal surfaces (subscripted *cas* and *cps* respectively) are:

$$P_{cas} = \frac{n_c - n_{air}}{|R_{cas}|}, \; P_{cps} = \frac{n_{ah} - n_c}{|R_{cps}|} \tag{2.7}$$

and the subscripts c and ah refer to the cornea and aqueous humor. For cornea and anterior chamber displacements D_c and D_{ah} respectively, the relationships are:

$$D_c = \frac{y_{ct}}{n_c}, \; D_{ah} = \frac{y_{acd}}{n_{ah}} \tag{2.8}$$

The refractive powers of the lens anterior surface, nuclear anterior boundary, nuclear posterior boundary, and lens posterior surface (*las, nab, npb*, and *lps* respectively), with refractive indices for lens cortex, nucleus, and vitreous humor subscripted as *lc, n,* and *vh*, are:

$$P_{las} = \frac{n_{lc} - n_{ah}}{|R_{las}|}, P_{nab} = \frac{n_n - n_{lc}}{|R_{nab}|}, P_{np}b = \frac{n_n - n_{lc}}{|R_{npb}|}, P_{lps} = \frac{n_{lc} - n_{vh}}{|R_{lps}|}. \tag{2.9}$$

Finally, the displacements for the three lens regions – anterior cortex, nucleus, and posterior cortex – are:

$$D_{ac} = \frac{y_{act}}{n_{lc}}, \; D_n = \frac{y_{nt}}{n_n}, \; D_{pc} = \frac{y_{pct}}{n_{lc}}. \tag{2.10}$$

The system matrix for the crystalline lens of the Gullstrand Schematic Eye is then constructed by taking the product of the refraction and translation matrices as defined for each region, so that

$$\mathbf{M_L} = \mathbf{R_{lps}}\mathbf{T_{pc}}\mathbf{R_{npb}}\mathbf{T_n}\mathbf{R_{nab}}\mathbf{T_{ac}}\mathbf{R_{las}}. \tag{2.11}$$

The overall system matrix for the anterior segment of the eye is constructed in the same way, such that

$$\mathbf{M} = \mathbf{M_L}(\mathbf{T_{ah}}\mathbf{R_{cps}}\mathbf{T_c}\mathbf{R_{cas}}). \tag{2.12}$$

Other models can be developed and studied in the paraxial limit using this general approach, with the specific formulation of the lens matrix $\mathbf{M_L}$ being used to test various GRINs. Two such representations, one based on the Smith et al. (1991; 1992) power series (Model #1) and the other derived from the experimental data of Siebinga et al. (1991, 1992) (Model #2), are developed below as examples. For both models, the GRIN is generated in nested isoindicial laminae, where the curvatures are approximated by spherical shells. However, while the shape of the GRIN and the underlying assumptions about refractive index distributions differ considerably between the two, and both are fundamentally different from the very simple Gullstrand model, all three forms can be shown to satisfy the requirements of focus by these models in the paraxial limit (Cook & Koretz 1995).

2.5.3.1 Model #1

In this model, most of the change in refractive index occurs superficially at and near the lens boundaries with the surrounding medium, while the central region's refractive index remains unchanged with age. The anterior and posterior boundaries themselves, originally defined as ellipsoidal surfaces, intersect equatorially at a distance from the center of symmetry defined by the equator of the model. The same holds true for each anterior and posterior pair of nested curves, and the radius of curvature of the boundaries for each half decreases from that of the lens surface as linear function of depth down to zero at the plane of intersection. For the anterior

$$R_a = R_{las}\left\{1 - \frac{y}{y_e}\right\}, \qquad 0 \le y \le y_e \tag{2.13}$$

$$R_p = R_{lps}\left\{1 - \frac{y}{y_{lt} - y_e}\right\}, \quad 0 \le y \le y_{lt} - y_e, \tag{2.14}$$

and posterior halves of the lens, this is respectively, with y_e – the location of the equatorial intersection – measured relative to the anterior lens surface. In either lens half, the region between the lens surface and the boundary plane is divided into m steps of equal width, so that for m laminae there are $m+1$ surfaces for each lens half,

$$R_i^{(a)}(y) = R_{las}\left\{1 - \frac{i}{m}\right\}, \qquad i = 1, \ldots, m \tag{2.15}$$

$$R_i^{(p)}(y) = Rlps\left\{1 - \frac{i}{m}\right\}, \qquad i = 1, \ldots, m \tag{2.16}$$

with $i = 0$ for either lens surface. Note that the thickness of each lamina, *Dy,* will be the thickness of each lens half divided by m, and may differ for the anterior and posterior.

The refractive index value for each lamina will be the same in the anterior and posterior, and, for this model, is assigned based on where it intersects the equatorial plane, or,

$$n(r) = c_0 + c_1 r^P, \qquad 0 \le r \le r_e, \tag{2.17}$$

where r is the radial distance from the axis of symmetry normalized to unit dimension. This can be modified so that the same functional form is incorporated as a normalized distance along the symmetry axis from the plane of intersection to each lens surface. Then,

$$n_a(y) = n_n - \kappa\left\{1 - \frac{y}{y_e}\right\}^P, \qquad 0 \le y \le y_e, \tag{2.18}$$

$$n_p(y) = n_n - \kappa\left\{1 - \frac{y}{y_{lt} - y_e}\right\}^P, \qquad 0 \le y \le y_{lt} - y_e \tag{2.19}$$

and the power P of the model is a variable for optimization. Smith et al. used values of $n_n = 1.406$ and $\kappa = 0.066$ for all values of P. Then the refractive index of the ith lamina for either lens half is,

$$n_i = n_n - \kappa\left\{1 - \frac{i-1}{m}\right\}^P, \qquad i = 1, \ldots, m, \tag{2.20}$$

and i increases from either surface. Convergence of this model is not a problem with the constants provided by Smith et al. (1992), a value for the exponent greater than 2, and $m > 200$. For other conditions, however, convergence does not occur for $m < 500$.

2.5.3.2 Model #2

For this model, the refractive index of the cortical regions of the lens decreases linearly from the lens nucleus to the lens boundaries, with the refractive index of the nucleus being the variable for optimization. This is based on the studies of Siebinga et al. (1991, 1992) who showed that there was an age-related change in the ratio of bound to free water in the nuclear region; since change in this ratio is associated with change in the aggregation state of solutes (e.g., proteins), it is possible for a change in refractive index to occur without a change in total water and protein. The resultant GRIN representation would incorporate a gradual loss of refractive power in the older regions of the lens with age.

The geometry of this model is based on the Scheimpflug slit-lamp studies of the aging of the accommodative mechanism by Koretz and coworkers (Cook et al 1994; Koretz et al 1997; Koretz et al 2001a; Koretz et al 2001b;

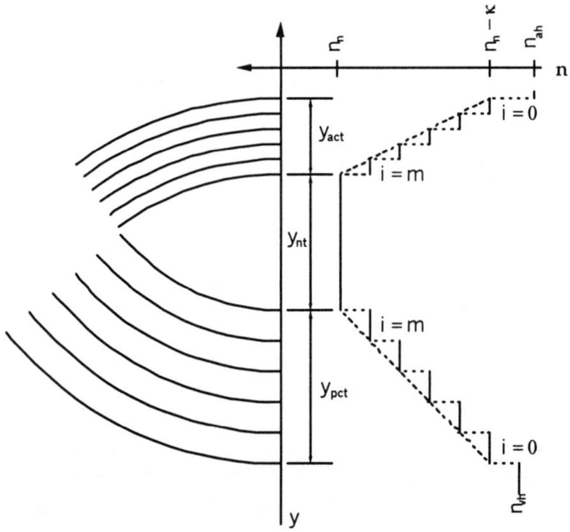

Figure 2.9. Schematic representation of Model #2, with a gradient based on the experimental data of Siebinga et al. (1991) relating to the relative amounts of free and bound water.

Koretz et al 1989a), from which the data in most of the tables in this chapter are derived. All of the lens curves, whether surfaces or internal boundaries, could be described as paraboloids, and there was a linear relationship between the radii of curvature of these paraboloids and their distance from the surface for each lens half. Thus, these curves would nest in the same manner as spherical shells. A linear model for the radius of curvature of a boundary as a function of distance from the surface for either lens half is then given by

$$R(y) = R_s + ay. \tag{2.21}$$

For the lens anterior, the relation is subject to the boundary conditions

$$R(0) = R_{las}, R(y_{nab} - y_{las}) = R_{nab}, \tag{2.22}$$

so that a_{ant} for this half is

$$a_{ant} = \frac{R_{nab} - R_{las}}{y_{act}}. \tag{2.23}$$

Similarly, for the posterior half, the boundary conditions

$$R(0) = R_{lps}, R(y_{npbb} - y_{lps}) = R_{npb}, \tag{2.24}$$

generate a_{post}, given by

$$a_{post} = \frac{R_{npb} - R_{lps}}{y_{pctt}}. \tag{2.25}$$

For each lens half, the region between the lens surface and the nuclear boundary is divided into m steps of equal width, as with Model #1. The radius of curvature of the ith boundary for the anterior and posterior halves respectively, is

$$R_i^{(a)}(y) = R_{las} + i\frac{R_{nab} - R_{las}}{m} \qquad i = 1, \ldots, m, \tag{2.26}$$

$$R_i^{(p)}(y) = R_{lps} + i\frac{R_{npb} - R_{lps}}{m} \qquad i = 1, \ldots, m. \tag{2.27}$$

The refractive index of the nucleus is, of course, the variable to be optimized in this model. The gradient refractive index in the cortical regions is represented by a linear decrease in nuclear value (Fig. 2.9) as a function of distance from the surface,

$$n(y) = n_n - \kappa\frac{y_t - y}{y_t}, \qquad 0 \le y \le y_t. \tag{2.28}$$

The value of κ is again 0.066 and the distance from the lens surface to the nucleus in either lens half is y_t. For either half, the refractive index of the ith region is given by

$$n_i = (n_n - \kappa) + \kappa\frac{i-1}{m}, \qquad i = 1, \ldots, m. \tag{2.29}$$

Again the convention used gives the region closest to either surface the lowest value of i. Convergence for Model #2 appears to occur rapidly, so values of m around 250 are more than sufficient.

2.5.3.3 Principal Planes

A major advantage to the matrix method of paraxial ray tracing is that the location of an optical system's principal planes, as well as the system's overall refractive power, can be derived from specific matrix elements. Lens refractive power p and system refractive power P are given by the upper right-hand element of the lens and system matrices respectively, where ***(L)*** refers to the lens,

$$p = -M_{12}^{(L)}, \tag{2.30}$$

$$P = -M_{12}. \tag{2.31}$$

As shown in Fig. 2.8, the principal planes for the system (H and H')are located relative to the centers of the anterior corneal surface and the posterior lens surface by the distances D and D' respectively. These are found from the system matrix through the equations,

$$D = \frac{1}{M_{12}}(1 - M_{11}), \tag{2.32}$$

$$D' = \frac{n_{vh}}{M_{12}}(1 - M_{22}), \tag{2.33}$$

with the convention that if D > 0, then H is located anterior to the cornea and if D' > 0, then H' is located posterior to the lens. The object distance S, the distance from the target to the anterior principal plane, and the image distance S', the distance from the posterior principal plane to the focal plane, are related by,

$$\frac{S}{n_{air}} + \frac{S'}{n_{vh}} - \frac{PSS'}{n_{air} n_{vh}} = 0, \tag{2.34}$$

where P is the system power. For the special case of focus on infinity,

$$S' = \frac{n_{vh}}{P}, \tag{2.35}$$

otherwise, S' is found from,

$$S' = \frac{n_{vh} S}{SP - 1}. \tag{2.36}$$

Then, the anterior and posterior principal plane are located relative to the anterior corneal surface and lens posterior surface respectively, and given by

$$H = -D, \tag{2.37}$$

$$H' = y_{asl} + D'. \tag{2.38}$$

Analogous equations are used for characterization of the anterior and posterior lens principal planes, such that

$$d = \frac{n_{ah}}{M_{12}{}^{(L)}}(1 - M_{11}{}^{(L)}), \tag{2.39}$$

$$d' = \frac{nvh}{M_{12}{}^{(L)}}(1 - M_{22}{}^{(L)}). \tag{2.40}$$

These distances are given relative to the anterior corneal surface by

$$h = y_{acd} - d, \tag{2.41}$$

$$h' = y_{asl} + d'; \tag{2.42}$$

where the sign convention for d and d' are as defined for D and D'.

The utility of this approach in assessing the optics of human eyes with age is illustrated in a recent publication that also evaluates the three models used as examples herein (Koretz and Cook, 2001c).

2.5.4 Wide-Angle Models

The availability of a description of lenticular shape that extends significantly far from the optical axis, as well as descriptions of the locations and shape of the lens nuclear boundaries, means that modeling of the lens is not restricted to paraxial approximations. The parabolic surfaces used to describe lens shape as a function of age for the non-accommodated emmetropic eye were derived from Scheimpflug photographs where the pupil was dilated to 8-10 mm, and they appear to provide a good fit over the entire field of view. As with the paraxial models, internal layers between the lens surface and the nucleus can be defined using a set of nested paraboloids, where the radii of curvature along each surface are inverse functions of the coefficient of the x^2 term. Of course, since the nested surfaces can no longer be treated as spherical, the distance between surfaces will vary according to their radius of curvature and distance from the optical axis.

A point that was overlooked or forgotten by Pomerantzeff and colleagues (1971, 1983), in their creation of a wide-angle model of the eye, is that NONE of the assumptions underlying a paraxial model will hold for a non-paraxial one. The cornea can no longer be approximated by two nested spherical surfaces, and in fact a cornea of about 0.5 mm central thickness exhibiting a conic constant between about –0.7 and -0.8 appears to be a more appropriate model (Applegate,1998). Equally important, the curvature of the posterior of the globe becomes significant in evaluating optical aberrations, since the focus is no longer a single point. To implement a wide-angle model will then require additional information about the posterior curvature of the globe and corneal shape, but will provide a mechanism for evaluating lenticular GRIN models using experimentally derived data about aberration as well as refractive power. For such model-building, the use of a computer-based package such as ZEMAX is almost a necessity. However, an appropriately constructed macro for lens shape and internal organization is essential for success. Double-pass experiments can also be simulated, or the results of such experiments used, like Pomerantzeff's spherical aberration data, to narrow the range of possibility for the variables involved in creating an optical model of the lens.

2.5.5 Accommodation and Presbyopia

Modeling of the lens, whether paraxially or in wide angle, has been described here primarily in terms of a single refractive state, where the eye is focused on infinity and any changes in parameter values such as lens shape or refractive index gradient are associated with aging (Koretz, 2000). As described previously and illustrated in Figure 2.6, however, change in lens shape is the mechanism by which the eye focuses on objects closer than infinity. This is a capability that is gradually eroded with age, so that the range over which focus can occur is reduced from about 15 diopters in youth to less than 2 diopters by age 50 (presbyopia). Most of this residual in middle age can be attributed to an increased depth of focus due to a contracted pupil (the "pinhole" effect). The changes in the anterior segment with age involve reduced anterior chamber depth, increased lens thickness and more sharply curved surfaces, especially on the anterior. The volume, thickness, and curvature of the nucleus are quite constant, so increased lens thickness along the sagittal axis is due almost entirely to increased cortical thickness. The accommodated anterior segment illustrated in Figure 2.6b also exhibits reduced anterior chamber depth, increased lens thickness and more sharply curved surfaces, especially on the anterior. Unlike the aged anterior segment of Fig. 2.6c, however, the change in lens shape in accommodation is due almost entirely to a change in the shape of the lens nucleus, which becomes much thicker and more sharply curved. The thickness of the cortical regions along the sagittal axis remains constant over the entire accommodative range, although of course this is not true off-axis.

Any optical model of the aging eye described in this chapter can be immediately and directly applied to the accommodated one simply by changing some of the values of parameters that have already been incorporated into the mathematical formulation. Corneal thickness, asymmetry, and curvatures will not change, and globe length is also constant. But most distances between boundaries along the sagittal axis will change as a function of accommodation. These changes are listed in Table

2.5 for a pre-presbyopic population and for a population that also includes presbyopes who can nevertheless summon a small degree of accommodation. It is interesting that these changes appear to be independent of age or accommodative amplitude, with only the size of the standard deviation increased in the larger population. Table 2.5 should be used in association with Table 2.3, which provides the intercept and slope for the linear relationship between a given spacing and subject age; set up the unaccommodated (0 diopter) distances for the eye of the selected subject age using the intercepts and slopes of Table 2.3, and then modify each value as a function of accommodative state for that age using Table 2.5. Tables 2.6 and 2.7 need to be used in association with Table 2.4, and therefore require

Table 2.5. Change in variable length per diopter of accommodation for a subset of pre-presbyopes (n=42) and all subjects who showed a change in accommodation (n=82). The standard deviation for each data set is also provided. Note that these rates are independent of age and total accommodative amplitude. (From Koretz et al., 1997).

Variable	Rate/diopter 18-45yr (n=42)	σ_{subset}	Rate/diopter 18-70 yr (n=82)	σ_{all}
Anterior chamber depth	-0.037	0.026	-0.038	0.139
Anterior cortex	0.002	0.013	0.002	0.034
Ant. lens to clear zone	0.025	0.019	0.025	0.088
Anterior nucleus plus clear zone	0.023	0.026	0.023	0.087
Posterior nucleus	0.018	0.023	0.018	0.069
Posterior cortex	0.000	0.031	0.000	0.076
Post. lens to clear zone	0.018	0.021	0.017	0.108
Nuclear thickness	0.041	0.029	0.041	0.120
Total lens thickness	0.043	0.027	0.043	0.145
Anterior segment length	0.003	0.035	0.003	0.174

some explanation. Table 2.4 provides the intercept and slope for the linear relationship between central radius of curvature and age in the unaccommodated (0 diopter) eye for each of the four lens boundaries. There is also a linear relationship for a given age between central radius of curvature of each of the four boundaries and accommodative amplitude, the slope of which changes with subject age. Table 2.6 provides the parameters that define the slope of that relationship (in mm/diopter) as a function of subject age. Similarly, for a given accommodative state at a given age, there is a linear relationship between the surface and internal lens curvatures for each lens half and their location within the lens; this slope changes with accommodative state for a given age and with age for a given accommodative state. In Table 2.7, the first Sub-Table provides the parameters to calculate the slope for the 0 diopter eye as a function of lens age, while the second and third Sub-Tables can be used to calculate the change in slope per diopter of accommodation for a given age.

Table 2.6. Linear intercept and slope of the rate of change of the central radius of curvature as a function of accommodation and age (in 1/ΔA yr) for the four principal lens boundaries (from Koretz et al., 2001a).

	Intercept	σ_I	Slope	σ_S	p
LAS	-4.736×10^{-1}	7.708×10^{-3}	4.705×10^{-3}	2.802×10^{-4}	0.062
LPS	2.778×10^{-1}	1.591×10^{-2}	-4.375×10^{-3}	5.648×10^{-4}	0.924
NAB	-7.760×10^{-2}	6.492×10^{-3}	-5.845×10^{-4}	2.355×10^{-4}	0.571
NPB	1.092×10^{-1}	1.192×10^{-2}	-1.010×10^{-3}	4.214×10^{-4}	0.719

Age dependence to age 45 years.

	Intercept	σ_I	Slope	σ_S	p
LAS	-5.898×10^{-1}	7.260×10^{-3}	9.230×10^{-3}	2.613×10^{-4}	0.058
LPS	2.552×10^{-1}	1.565×10^{-2}	-3.489×10^{-3}	5.535×10^{-4}	0.659
NAB	-6.949×10^{-2}	5.831×10^{-3}	-8.769×10^{-4}	2.066×10^{-4}	0.612
NPB	9.230×10^{-2}	1.022×10^{-2}	-4.023×10^{-4}	3.535×10^{-4}	0.406

Age dependence using the complete age range for data.

Table 2.7. Age dependence of the linear relationship between radii of curvature and curve locations for each lens half. (From Koretz et al., 2001a).

	Intercept	σ_I	Slope	σ_S	p
ant.	-10.189	1.592×10^{-2}	3.686×10^{-2}	3.090×10^{-4}	0.014
pos.	6.670	2.496×10^{-2}	-3.451×10^{-2}	5.681×10^{-4}	0.097

Age dependence of for anterior and posterior lens regions, with age in yr as the independent variable.

	Intercept	σ_I	Slope	σ_S	p
ant.	7.354×10^{-1}	8.959×10^{-3}	-1.219×10^{-2}	3.183×10^{-4}	0.170
pos.	-3.175×10^{-1}	1.656×10^{-2}	3.052×10^{-3}	5.938×10^{-4}	0.947

Dependence to age 45 years of the slope for anterior and posterior lens regions as a function of accommodation.

	Intercept	σ_I	Slope	σ_S	p
ant.	7.164×10^{-1}	7.788×10^{-3}	-1.134×10^{-2}	2.690×10^{-4}	0.110
pos.	-2.994×10^{-1}	1.565×10^{-2}	2.341×10^{-3}	5.566×10^{-4}	0.683

As above, but using the full age range for data.

Finally, the limit of accommodative range – the maximum average accommodation that can be attained - for a given age needs to be determined and used as a cut-off. The value of this as a function of age varies slightly between published data sets, and should be selected with this in mind. An empirical function that summarizes the change in maximum accommodative amplitude A_{max} with age (in years) for the data tabulated here is

$$\Delta A_{max} = 0.75 + \frac{13.135}{1.0 + 1.457\times10^{-2} e^{0.1476T}} \quad (2.43)$$

It should also be remembered that Eq. 2.43 and the data in Tables 2-7 were derived from an adult population whose age ranged from 18 to 70 yr, and can be extrapolated only with great care outside of these age and accommodative limits.

2.6 SHACK-HARTMANN ANALYSIS OF VISUAL ABERRATION

Each component of the visual system has the potential of inducing forms of aberration instead of contributing to a clear and focused image. As already noted, a globe length different from the location of focus leads to refractive error, while asymmetric deviations from a spherical surface for the cornea can lead to astigmatism. Astigmatism may also arise from the cornea as the result of refractive surgical procedures (e.g., photorefractive keratotectomy – PRK – or LASIK – laser assisted in-situ keratomileusis), although in these cases the aberration may be much more complex than simple astigmatism. The crystalline lens may also contribute to the reduction of image quality, due to subtle developmental asymmetries, the effects of certain forms of cataract, or other factors that can affect the orientation of the lens relative to the rest of the visual system.

Factors affecting image quality are often difficult to identify, and it is also often difficult to assess the effect of each on the entire visual system. There are algorithms which can be used to characterize overall visual quality (e.g., the modulation transfer function, or MTF), but they do not provide information about the source or type of aberration leading to its reduction. This lens chapter and other chapters in this book provide information about the nature of some of the aberrations that each portion of the eye might contribute, but may not provide a complete picture of the impact of those which are significantly asymmetric (e.g., cortical cataracts), nor a comprehensive protocol for modeling them. A technique originally developed in the 1970's to improve the quality of telescope images is the Shack-Hartmann method (also mentioned in the literature as the Hartmann-Shack method), which has recently been adapted for analysis of aspects of the visual system, including the visual wavefront (Liang et al 1994; Miller et al 1996). It provides a foundation for more complete assessment of aberration, as well as suggesting the potential for correlation, at least in part, between aberration and its sources (Guirao et al 2001; Hong & Thibos 2000; Thibos & Hong 1999).

The principle behind the Shack-Hartmann approach as applied to the visual system is straightforward: if light could radiate from a single point in the center of the retina outward through all of the elements of the eye and emerge from the cornea, the shape of the resultant wavefront would represent the sum of refractive and aberrational effects generated by a single pass through the eye's visual elements. Rather than being initiated external to the eye and leading to focus on the retina, this "backward" generation allows the shape of the wavefront to be detected, analyzed, and broken down into components related to different forms of aberration.

Characterization of the wavefront shape is accomplished experimentally using a two-dimensional array of miniature lenses, or lenslets, each of which focuses that portion of the wavefront falling upon it (Fig. 2.10). In the ideal case, where the radiating light is refracted without aberration, the wavefront would be flat and uniform across its breadth; the resultant intensity of light focused by each lenslet would be the same, and the location of focus of each lenslet would also be identical. Aberrations in the wavefront, which conceptually may be considered uneven bundling of light rays and an undulating surface, would lead to uneven intensity across the lenslet array and displacement of the image of some lenslet elements from their ideal grid location (Fig. 2.11 a and b). Analysis of these deviations in amplitude and location as a function of the total lenslet array can lead to a representation of the wavefront as the sum of a set of Zernicke polynomials (Fig. 2.11c) or a

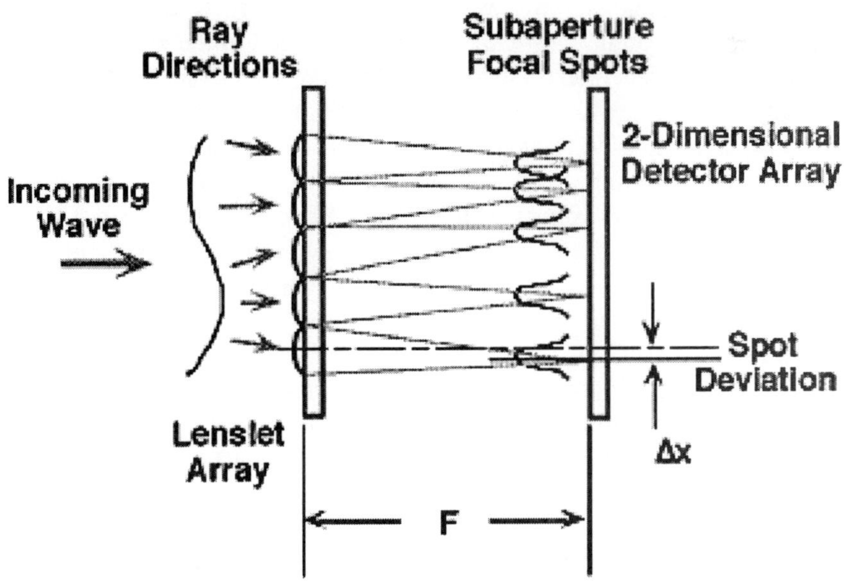

Figure 2.10. Cross-sectional view of a Shack-Hartmann lenslet array, illustrating the relationship between wavefront surface shape and spot displacement for each lens focal point. (Figure courtesy of Dr. Larry Schmutz, Design-X, Inc.).

three-dimensional representation of the wavefront surface (Fig. 2.11d). Although there is not a direct relationship between elements of the polynomial series and specific aberrations, the former can nevertheless be used in combination to assess the type and magnitude of the latter. They can also be used to generate more familiar metrics, such as the modulation transfer function.

Although this approach can lead to a much more comprehensive and quantitative evaluation of image formation and quality, there is a major drawback: the factors contributing to non-ideality cannot be directly correlated with elements of the visual system. Refractive error, for example, can arise from several different portions of the eye, but what is assessed by the Shack-Hartmann approach is the sum of all their contributions. Thus, it can arise from a single source or a multiplicity of sources, some of which may even cancel out the effects of others in a variety of different ways. As quantitative characterization of the anterior corneal surface becomes more and more accurate and reliable, it should theoretically be possible both to determine its contribution to the wavefront and to subtract its contribution in order to assess the rest of the visual system. Development of equally comprehensive quantitative descriptions of other elements of the visual system should eventually enable the dissection of the overall wavefront into individual contributions, but that must await advances in the future. In the meantime, a wide-angle model of the type described in the previous section can be used to generate Shack-Hartmann wavefronts in ZEMAX. The optical system can easily be oriented so that a point source of light located at the fovea generates a wavefront emerging from the cornea, and the software can be used to determine the Zernike polynomial description of this surface (Fig. 2.11c). A perfect optical system would, of course, lead to a flat, evenly illuminated surface of parallel rays, while the shape and characteristics of the model elements, most notably the lens and cornea, can be modified to determine their effects on overall image quality or to simulate actual experimental data.

It has been argued that attribution of aberrational contributions to the various components of the visual system is not necessary for their correction. The human eye is theoretically capable of a refraction of 20/8 – so-called super-vision - in the absence of confounding factors, and knowing the shape of an individual eye's wavefront defines the compensatory aberrations necessary for ideality. It has also been argued that a reshaping of the anterior corneal surface to provide such compensatory aberrations is all that is empirically necessary for super-vision, and it is certainly true that refractive surgical algorithms are becoming increasingly sophisticated (e.g., Arbelaez 2001). It should therefore be possible to develop individualized programs for this purpose. However the success of super-vision, even if it were possible, implicitly depends on the long-term stability of both the non-corneal aberrations and the corneal corrections. Since the crystalline lens continues to grow and change throughout life, super-vision is likely to be an ephemeral goal unless the lens is replaced as well.

On the other hand, it may be a reasonable goal to reshape the cornea to compensate for the changes in system aberration arising from an intraocular lens implant, although not necessarily to create 20/8 visual systems in the absence of any accommodative ability. Of course, there are also contact lenses, and a more reasonable goal, at least in the short term, might be their individualized design to compensate, in whole or part, for a given eye's aberration profile.

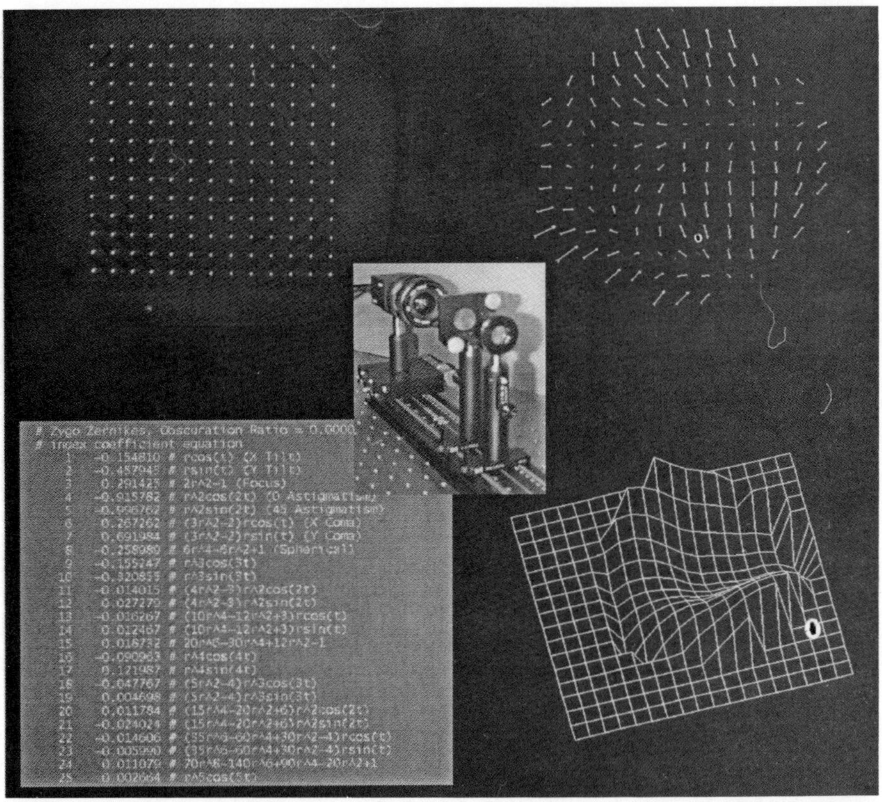

Figure 2.11. Shack-Hartman data collection and analysis. (a) image array from the lenslet output of a specific wavefront; (b) graphical representation as arrows of wavefront tilt vectors; (c) Zernike poynomial terms fitted to the experimental data; and (d) three-dimensional representation of the calculated fit to the experimental wavefront. (Figure courtesy of Dr. Larry Schmutz, Design-X, Inc.).

2.7 VISUALIZING THE RETINA

Optical models of the lens should not be confined to the process of image formation or the characterization of the age-dependent changes in lenticular GRIN. Just as a major concern for refractive surgeons is how best to improve refraction and image quality, a major concern of investigators who study the posterior of the globe is how to correct for the contributions of the anterior segment elements. In this sense, then, interest in the lens per se is not so much in its optical contribution which, with the cornea, can theoretically be compensated for, as in the ways it can interfere with collection of data posterior to it. To a rough approximation, there are three ways in which the lens can interfere: lens color, lens opacification, and lens aberrations. Any of these factors can lead to an age-dependent attentuation of the signal.

Lens color is the factor which is of primary concern in the literature (Gaillard et al 2000; Lou et al 1999; Lutze & Bresnick 1991; Pokorny et al 1987; Sparrow et al 1988; Teesalu et al 1997). As part of the normal process of aging, the lens becomes yellower, and this will affect the passage of short-wavelength light through it in either direction. For light whose wavelength is about 550 nm or longer, there is little effect. But for tasks like short wavelength automated perimetry (SWAP), this differential filtering is of major concern. There are several data sets used for characterization of age-related color effects, primarily derived from color-matching experiments, that exhibit similar distribution shapes – exponential or bimodal, depending on the method of analysis. The way these data should be used for compensation of lens color effects, the necessity of color correction for certain classes of experiments, and even the model of the lens used in the compensatory process - all remain the subject of discussion.

The same wavelength range that shows an age-related alteration in amplitude by passage through the lens will also be most strongly affected by scattering centers that are within 10-20 times the size of the wavelength (Mie scattering). As a result, there is some controversy as to whether age-related color effects should be separated from age-related lens opacification (density), or whether they are part of the same process of cataractogenesis (Chylack et al 1993; Johnson et al 1993; Occhipinti et al 1986; Xu et al 1997). It is interesting in this context to note that, using unrelated techniques and white light, Cook et al. (1994) showed that the increase in light scattered at high angles from Scheimpflug images of the lens was both age-related and exponential/bimodal, and Lou and Dickerson (1992) demonstrated the bimodal increase in mixed disulfide formation within the lens nucleus. For these results, as well as for the yellowing data already mentioned, the change in slope or inflection occurs at about the same age –

around 50 yr, or the cessation of accommodation - with the processes accelerating at older ages. These results strongly suggest that lens aging may be a unitary and universal process. However, if lens color and lens density are indeed different facets of the same aging phenomenon, then it is important to distinguish the influence of the lens substance centrally (paraxially) and peripherally, since different lens regions would be affected to differing extents.

Lens aberrations have already been briefly discussed in terms of whole-lens effects. That is, it was assumed for the purposes of modeling that the lens was centro-symmetric, which would generate centro-symmetric aberrations. The aging of the lens appears to involve differential changes in refractive index in different lens regions, at least in the early stages of cataractogenesis, that could result in a more complex effect. Certainly the development of nuclear cataract is associated both with increased myopia (suggesting an increase in nuclear refractive index) and with a thinner lens or lens nucleus, and it is not clear at this point whether that process is centro-symmetric. Other cataract types, such as cortical and posterior subcapsular, are asymmetric. The development of compensatory methods for these effects is still in its infancy. The use of a Shack-Hartmann approach to characterize lens contributions will be effective only if the cornea's optical properties can be separated from those of the lens. Other approaches are currently being investigated.

2.8 CONCLUSIONS

The crystalline lens of the human visual system is uniquely fascinating. Of all the bodily tissues, only the lens exhibits clear stages of growth, development, and aging that can be directly and non-invasively correlated to systemic aging. The lenticular compensatory mechanisms involved in preserving far focus at the expense of accommodation are elegant and effective, and it is only with the development of cataracts that this balance begins to erode. At the cellular level, the lens provides a model system for both cell differentiation and transport mechanisms, and at the molecular level, the range in age of the lens proteins from the nucleus to the periphery provides a well-defined model system for the study of functional genomics and proteomics. Understanding this integration of structure and function at all levels of organization will, in the future, illuminate both the life cycle of visual function and the life cycle itself.

2.9 ACKNOWLEDGMENTS

This work was supported in part by NIH grant EY02195. The author is deeply indebted to Prof. Jer Kuszak of Rush-Presbyterian-St. Luke's for the figures illustrating lens morphology and development, and for a critical (and very constructive) review of the relevant sections of this chapter. Dr. Ann E. Elsner of the Schepens Eye Research Institute was wonderfully informative about the problems the crystalline lens poses for retina and psychophysical studies and the current status of strategies to deal with them, and Dr. Stephen A. Burns provided supplementary information. Dr. Ben Platt, VP, Optics Technology Development, Calhoun Vision, Inc., helped immensely with the Shack-Hartmann section (and indeed with the development of the technique in the 1970s!), although any errors of fact are mine. Dr. Larry Schmutz, head of Design-X, Inc., was gracious in providing Figures 2.10 and 2.11, and especially in re-designing the latter. Last, but certainly not least, a massive debt of thanks is owed to Dr. Christopher A. Cook of Cytometrics, Inc., for having addressed many of the issues arising in this chapter as part of his doctoral thesis.

2.10 REFERENCES

Al-Ghoul, K. J., Nordgren, R. K., Kuszak, A. J., Freel, C. D., Costello, M. J., Kuszak, J. R., 2001, Structural evidence of human nuclear fiber compaction as a function of ageing and cataractogenesis, *Exp. Eye Res.* **72**: 199-214.

Applegate, R. A., 1998, Personal communication.

Arbelaez, M. C., 2001, Super vision: dream or reality, *J. Refract. Surg.* **17**: S211-218.

Atchison, D. A., 2001, Personal communication.

Atchison, D. A., Smith, G. 2000, *Optics of the Human Eye,* Butterworth-Heinemann, Boston.

Bessems, G. J., De Man, B. M., Bours, J., Hoenders, H. J., 1986, Age-related variations in the distribution of crystallins within the bovine lens, *Exp. Eye Res.* **43**: 1019-1030.

Bessems, G. J., Hoenders, H. J., Wollensak, J., 1983, Variation in proportion and molecular weight of native crystallins from single human lenses upon aging and formation of nuclear cataract, *Exp. Eye Res.* **37**: 627-637.

Bettelheim, F. A., Paunovic, M., 1979, Light scattering of normal human lens I. Application of random density and orientation fluctuation theory, *Biophys. J.* **26**: 85-99.

Blaker, J. W., 1971, *Geometric Optics - The Matrix Method,* New York: Marcel Dekker.

Bours, J., Wegener, A., Hofmann, D., Fodisch, H. J., Hockwin, O., 1990, Protein profiles of microsections of the fetal and adult human lens during development and ageing, *Mech. Ageing Dev.* **54**: 13-27.

Bron, A. J., Vrensen, G. F., Koretz, J., Maraini, G., Harding, J. J., 2000, The ageing lens. *Ophthalmologica.* **214**: 86-104.

Brown, N., 1972, An advanced slit-image camera, *Br. J. Ophthalmol.* **56**: 624-631.

Brown, N., 1973a, The change in shape and internal form of the lens of the eye on accommodation, *Exp. Eye Res.* **15**: 441-459.

Brown, N., 1973b, Quantitative slit-image photography of the anterior chamber, *Trans. Ophthalmol. Soc. U. K.* **93**: 277-86.

Brown, N., 1974, The change in lens curvature with age, *Exp. Eye Res.* **19**: 175-183.

Brown, N. A., Sparrow, J. M., Bron, A. J., 1988, Central compaction in the process of lens growth as indicated by lamellar cataract, *Br. J. Ophthalmol.* **72**: 538-544

Brown, N. P., Koretz, J. F., Bron, A. J., 1999, The development and maintenance of emmetropia, *Eye.* **13**: 83-92.

Chylack, L. T., Jr., Wolfe, J. K., Friend, J., Khu, P. M., Singer, D. M., et al., 1993, Quantitating cataract and nuclear brunescence, the Harvard and LOCS systems, *Optom. Vis. Sci.* **70**: 886-895.

Cook, C. A., Koretz, J. F., 1991, Acquisition of the curves of the human crystalline lens from slit lamp images: an application of the Hough transform, *Applied Optics.* **30**: 2088-2099

Cook, C. A., Koretz, J. F., 1995, Modeling the optical properties of the aging human crystalline lens from computer processed Scheimpflug images in relation to the lens paradox, *Technical Series on Vision Science and Its Applications (Op. Soc. Am.)*, pp. 138-141.

Cook, C. A., Koretz, J. F., 1998, Methods to obtain quantitative parametric descriptions of the optical surfaces of the human crystalline lens from Scheimpflug slit-lamp images. I. Image processing methods, *J. Opt. Soc. Am. A Opt. Image Sci. Vis.* **15**: 1473-1485.

Cook, C. A., Koretz, J. F., Pfahnl, A., Hyun, J., Kaufman, P. L., 1994, Aging of the human crystalline lens and anterior segment, *Vis. Res.* **34**: 2945-2954.

Davson, H., 1990, *Davson's Physiology of the Eye,* 5th Ed. ed. New York: Pergamon Press

Dubbelman, M., Van der Heijde, G. L., 2001, The shape of the aging human lens: curvature, equivalent refractive index and the lens paradox, *Vis. Res.* **41**: 1867-1877.

Fagerholm, P. P., Philipson, B. T., Lidstrom, B., 1981, Normal human lenses - the distribution of protein, *Exp. Eye Res.* **33**: 615-620.

Fu, S. C., Su, S. W., Wagner, B. J., Hart, R., 1984, Characterization of lens proteins. IV. Analysis of soluble high molecular weight protein aggregates in human lenses, *Exp. Eye Res.* **38**: 485-495.

Gaillard, E. R., Zheng, L., Merriam, J. C., Dillon, J., 2000, Age-related changes in the absorption characteristics of the primate lens, *Invest. Ophthalmol. Vis. Sci.* **41**: 1454-1459.

Garland, D. L., Duglas-Tabor, Y., Jimenez-Asensio, J., Datiles, M. B., Magno, B., 1996, The nucleus of the human lens: demonstration of a highly characteristic protein pattern by two-dimensional electrophoresis and introduction of a new method of lens dissection, *Exp. Eye Res.* **62**: 285-291.

Guirao, A., Williams, D. R., Cox, I. G., 2001, Effect of rotation and translation on the expected benefit of an ideal method to correct the eye's higher-order aberrations, *J. Opt. Soc. Am. A Opt. Image Sci. Vis.* **18**: 1003-1015.

Hong, X., Thibos, L. N., 2000, Longitudinal evaluation of optical aberrations following laser in situ keratomileusis surgery, *J. Refract. Surg.* **16**: S647-650.

Howland, H. C., 1994, Physiological Optics, in: *Principles and Practice of Ophthalmology: Basic Sciences*, ed. D. M. Albert and F. A. Jakobiec, W. B. Saunders Company, Philadelphia, PA.

Johnson, C. A., Howard, D. L., Marshall, D., Shu, H., 1993, A noninvasive video-based method for measuring lens transmission properties of the human eye, *Optom. Vis. Sci.* **70**: 944-955.

Klein, M. V., Furtak, T. E., 1986, *Optics,* John Wiley & Sons, New York, N.Y.

Koretz, J. F., 2000, Development and aging of human visual focusing mechanisms, in: *Trends in Optonics and Photonics: Vision Science and Its Applications, ed. V.* Lakshminarayanan, Optical Society of America, Washington, D.C. **35**: 246-258.

Koretz, J. F., Cook, C. A., Kaufman, P. L., 1997, Accommodation and presbyopia in the human eye. Changes in the anterior segment and crystalline lens with focus, *Invest. Ophthalmol. Vis. Sci.* **38**: 569-578.

Koretz, J. F., Cook, C. A., Kaufman, P. L., 2001a, Aging of the human lens: Changes in lens shape upon accommodation and with accommodative loss. *J. Opt. Soc. Am. A Opt. Image Sci. Vis., in press.*

Koretz, J. F., Cook, C. A., Kaufman, P. L., 2001b, Aging of the human lens: changes in lens shape at zero-diopter accommodation, *J. Opt. Soc. Am. A Opt. Image Sci. Vis.* **18**: 265-272.

Koretz, J. F., Cook, C. A., 2001c, Aging of the optics of the human eye: Lens refraction models and principal plane locations, *Opt. Vis. Sci.,* in press

Koretz, J. F., Cook, C. A., Kuszak, J. R., 1994, The zones of discontinuity in the human lens: development and distribution with age, *Vis. Res.* **34**: 2955-2962.

Koretz, J. F., Handelman, G. H., 1986, The "lens paradox" and image formation in accommodating human eyes, in: *Topics in Aging Research in Europe, vol.* 6, *The Lens: Transparency and Cataract*, pp. 57-64.

Koretz, J. F., Handelman, G. H., 1988, How the human eye focuses, *Sci. Am.* **259**: 92-99.

Koretz, J. F., Handelman, G. H., Brown, N. P., 1984, Analysis of human crystalline lens curvature as a function of accommodative state and age. *Vis. Res.* **24**: 1141-1151.

Koretz, J. F., Kaufman, P. L., Neider, M. W., Goeckner, P. A., 1989a, Accommodation and presbyopia in the human eye--aging of the anterior segment, *Vis. Res.* **29**: 1685-1692.

Koretz, J. F., Kaufman, P. L., Neider, M. W., Goeckner, P. A., 1989b, Accommodation and presbyopia in the human eye, 1: Evaluation of *in vivo* measurement techniques, *Applied Optics,* **28**: 1097-1102.

Koretz, J. F., Rogot, A., Kaufman, P. L., 1995, Physiological strategies for emmetropia, *Trans. Am. Ophthalmol. Soc.* **93**: 105-118; discussion 118-122.

Kuszak, J. R., 1995, The ultrastructure of epithelial and fiber cells in the crystalline lens, *Int. Rev. Cytol.* **163**: 305-350.

Kuszak, J. R., Peterson, K. L., Sivak, J. G., Herbert, K. L., 1994, The interrelationship of lens anatomy and optical quality, II. Primate lenses, *Exp. Eye Res.* **59**: 521-535.

Kuszak, J. R., Sivak, J. G., Weerheim, J. A., 1991, Lens optical quality is a direct function of lens sutural architecture [published erratum appears in Invest Ophthalmol Vis Sci 1992 May;33(6):2076-7], *Invest. Ophthalmol. Vis. Sci.* **32**: 2119-2129.

Liang, J., Grimm, B., Goelz, S., Bille, J. F., 1994, Objective measurement of wave aberrations of the human eye with the use of a Hartmann-Shack wave-front sensor, *J. Opt. Soc. Am. A.* **11**: 1949-1957.

Lou, M. F., Dickerson, J. E., Jr., 1992, Protein-thiol mixed disulfides in human lens, *Exp. Eye Res.* **55**: 889-896

Lou, M. F., Dickerson, J. E., Jr., Tung, W. H., Wolfe, J. K., Chylack, L. T., Jr., 1999, Correlation of nuclear color and opalescence with protein S-thiolation in human lenses, *Exp. Eye Res.* **68**: 547-552.

Lutze, M., Bresnick, G. H., 1991, Lenses of diabetic patients "yellow" at an accelerated rate similar to older normals. *Invest. Ophthalmol. Vis. Sci.* **32**: 194-199.

Miller, D. T., Williams, D. R., Morris, G. M., Liang, J., 1996, Images of cone photoreceptors in the living human eye, *Vis. Res.* **36**: 1067-1079.

Moffat, B. A., Landman, K. A., Truscott, R. J., Sweeney, M. H., Pope, J. M., 1999, Age-related changes in the kinetics of water transport in normal human lenses, *Exp. Eye Res.* **69**: 663-669.

Mutti, D. O., Zadnik, K., Fusaro, R. E., Friedman, N. E., Sholtz, R. I., Adams, A. J., 1998, Optical and structural development of the crystalline lens in childhood, *Invest. Ophthalmol. Vis. Sci.* **39**: 120-133.

Occhipinti, J. R., Mosier, M. A., Burstein, N. L., 1986, Autofluorescence and light transmission in the aging crystalline lens, *Ophthalmologica.* **192**: 203-209.

Pierscionek, B. K., Chan, D. Y., 1989, Refractive index gradient of human lenses, *Optom. Vis. Sci.* **66**: 822-829.

Pokorny, J., Smith, V. C., Lutze, M., 1987, Aging of the human lens, *Appl. Opt.* **26**: 1437-1440.

Pomerantzeff, O., Dufault, P., Goldstein, R., 1983, Wide-angle optical model of the eye, in: *Advances in Diagnostic Visual Optics,* Springer-Verlag, Berlin, pp. 12-21.

Pomerantzeff, O., Fish, H., Govignon, J., Schepens, C. L., 1971, Wide angle optical model of the human eye, *Ann. Ophthalmol.* **3**: 815-819.

Siebinga, I., Vrensen, G. F., De Mul, F. F., Greve, J., 1991, Age-related changes in local water and protein content of human eye lenses measured by Raman microspectroscopy, *Exp. Eye Res.* **53**: 233-239.

Siebinga, I., Vrensen, G. F., Otto, K., Puppels, G. J., De Mul, F. F., Greve, J., 1992, Ageing and changes in protein conformation in the human lens: a Raman microspectroscopic study. *Exp. Eye Res.* **54**: 759-767.

Smith, G., Atchison, D. A., Pierscionek, B. K., 1992, Modeling the power of the aging human eye, *J. Opt. Soc. Am. A.* **9**: 2111-2117.

Smith, G., Pierscionek, B. K., Atchison, D. A., 1991, The optical modelling of the human lens, *Ophthal. Physiol. Opt.* **11**: 359-369.

Sorsby, A., Benjamin, B., Davey, J. B., et al., 1957, *Emmetropia and its aberrations*, vol. 293, Her Majesty's Stationery Office, London.

Sorsby, A., Benjamin, B., Sheridan, M., 1961, *Refraction and its components during the growth of the eye from the age of three,* vol. 301, Her Majesty's Stationery Office, London.

Sparrow, J. M., Hill, A. R., Ayliffe, W., Bron, A. J., Brown, N. P., 1988, Human lens nuclear colour matching and brunescence grading in vivo, *Int. Ophthalmol.* **11**: 139-149.

Strenk, S. A., Semmlow, J. L., Strenk, L. M., Munoz, P., Gronlund-Jacob, J., DeMarco, J. K., 1999, Age-related changes in human ciliary muscle and lens: a magnetic resonance imaging study *Invest. Ophthalmol. Vis. Sci.* **40**: 1162-1169.

Taylor, V. L., al-Ghoul, K. J., Lane, C. W., Davis, V. A., Kuszak, J. R., Costello, M. J., 1996, Morphology of the normal human lens, *Invest. Ophthalmol. Vis. Sci.* **37**: 1396-1410.

Teesalu, P., Airaksinen, P. J., Tuulonen, A., Nieminen, H., Alanko, H., 1997, Fluorometry of the crystalline lens for correcting blue-on-yellow perimetry results, *Invest. Ophthalmol. Vis. Sci.* **38**: 697-703.

Thibos, L. N., Hong, X., 1999, Clinical applications of the Shack-Hartmann aberrometer, *Optom. Vis. Sci.* **76**: 817-825.

Willekens, B., Vrensen, G., 1981, The three-dimensional organization of lens fibers in the rabbit. A scanning electron microscopic reinvestigation, *Albrecht Von Graefes Arch. Klin. Exp. Ophthalmol.* **216**: 275-289.

Xu, J., Pokorny, J., Smith, V. C., 1997, Optical density of the human lens, *J. Opt. Soc. Am. A.* **14**: 953-960.

II

NEUROSENSORY SYSTEM MODELS

Chapter 3

Anatomy and Physiology of the Retina

Richard A. Normann,[1] Ph.D.; K. Shane Guillory,[1] Ph.D.

[1]Dept. of Bioengineering, University of Utah, 20 South 2030 East, Room 506, Salt Lake City, UT 84112, PH: (801) 581-7645, FX: (801) 581-8966, EM: normann@m.cc.utah.edu

3.1 INTRODUCTION

Because sight is such a highly valued sensory modality, the mechanisms by which spatial, intensity, chromatic and temporal patterns of retinal illumination evoke perceptions which are fully congruent with the physical world around us has been a subject which has intrigued scientists and natural philosophers for centuries. While the answers to this question are still not quite within our grasp, the work of many researchers has provided us with a fairly good understanding of the retinal contributions to this fascinating question. As we move deeper into the visual pathways, our understanding becomes less precise, and the problems posed become more complex.

The goal of this chapter is to provide the reader with a brief summary of what is known about how the vertebrate retina contributes to the first part of the process. We will start with an anatomical perspective to understand the gross structure of the retina, and what the neurons of the retina look like, for it will be difficult to understand the function of the retinal pathways without acquiring an understanding of its structure.

We will then review the mechanisms by which optical patterns of retinal illumination get converted into an electrical image by the rods and cones of the retina, and how this image is spatially and temporally modulated as the neural image is passed from cell layer to cell layer. We will briefly consider the mechanisms by which the retina can respond to only a few photons in its most sensitive state, yet can still produce high acuity image encoding in very bright environments.

The last part of this chapter will begin to look at how groups of ganglion cells work together to encode patterns of retinal illumination. Because individual ganglion cells respond to many features of a visual stimulus, it is only by looking at the simultaneous responses of large numbers of ganglion cells that the higher visual centers can begin to reassemble the response patterns of the retinal ganglion cells into a meaningful perceptual image.

The subject of this chapter, the anatomy and physiology of the retina, is a vast topic. Because many books have been written on the subject (Dowling, 1987; Rodieck, 1998; Kolb et al., 2000), it will be impossible to provide a comprehensive view of retinal information processing. Rather, the authors will try to present a summary of some of the work that has gone before. Our goal is to provide an anatomical and physiological framework upon which the subsequent chapters in this book can build. It is hoped that the reader will find that this chapter has enhanced his/her understanding of the retinal mechanisms that start the processes that end in visual perception.

3.2 ANATOMY OF THE RETINA

3.2.1 Gross Anatomical Features

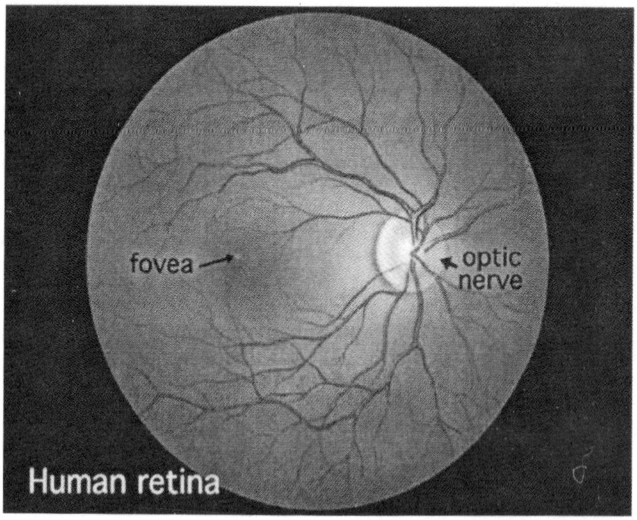

Figure 3.1. The human retina as it would be seen with an ophthalmoscope. Features readily seen are the retinal circulation, the optic nerve head, the macula lutea that surrounds the fovea. Reprinted from Kolb et al. (2000) with permission of WebVision.

The human retina is a thin network of neurons and glial cells (called 'Muller cells'). It is about 1/3 of a mm thick and has a surface area of about 1100 square mm (Kolb et al., 2000). Fig. 3.1 shows the three distinct morphological structures that are seen when the retina is viewed with an ophthalmoscope: the optic nerve head where the optic nerve exits the eye, the 'macula lutea' or the yellowish colored region that contains the fovea, and the retinal vasculature. The cells of the retina are nourished by blood that is delivered to the retina's two sides. The choroidal circulation at the back of the eye nourishes the cells of the outer retina. The retinal circulation emerges from the optic nerve head and nourishes the cells on the retina's vitreal side.

3.2.2 Retinal Neurons

The vertebrate retina contains six general classes of neurons organized into discrete layers or lamina. The main five of these classes are illustrated in Fig. 3.2.

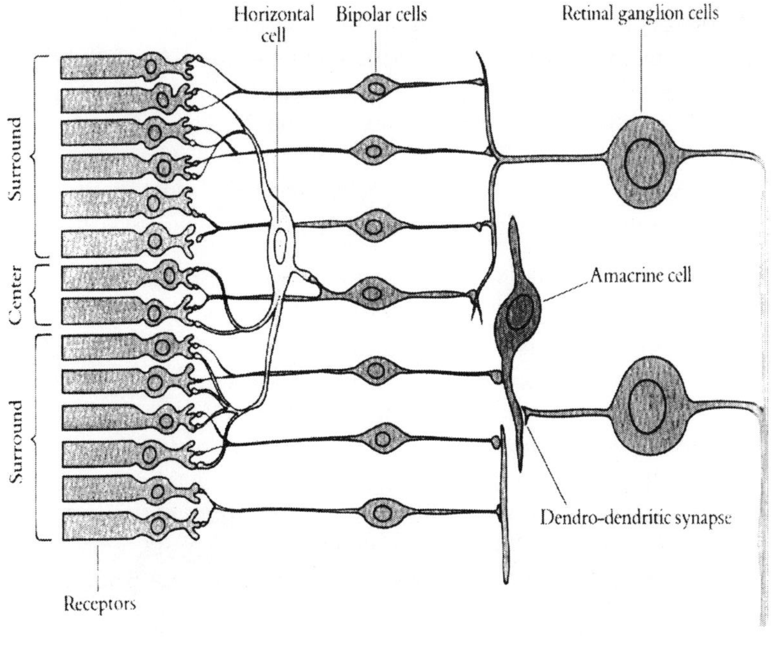

Figure 3.2. Cartoon illustrating the major neuronal classes in the vertebrate retina. Reprinted from Hubel (1988) with permission of W. H. Freeman and Co.

The neuronal classes and the functions the cells perform are described below. The anatomical classifications date from the early work of Cajal (translation, 1972).

- Rod and cone photoreceptors – Cone photoreceptors are functional in most viewing conditions, and they mediate color vision. Humans have three classes of cone photoreceptors which are named based on the nature of the photopigments that are located in their outer segments: 'l' or 'long-wavelength sensitive' or 'red' cones; 'm' or 'medium-wavelength sensitive' or 'green' cones; and 's' or 'short-wavelength sensitive' or 'blue' cones. Color blindness results from genetic defects that produce one or more classes of cones with either reduced or an absence of visual pigment. Rod photoreceptors are functional only in very dim viewing conditions, and therefore mediate 'night vision'. In contrast to cones, there is only one class of rod photoreceptor.

- Horizontal cells – The horizontal cells are laterally-oriented cells that contact photoreceptors at the first synaptic layer (the outer plexiform layer). There are two broad functional classes of horizontal cells. Luminosity-type cells respond to light stimulation with a response that is not strongly dependent upon the color of the stimulus. Luminosity-type cells are generally sensitive to the amount of light striking the photoreceptors that provide its input. Chromaticity-type horizontal cells have light evoked responses with kinetics that are strongly dependent upon the color of the light stimuli. There are also horizontal cell sub-classifications that are based upon morphological features of the cells. Horizontal cells are large, with extensive dendritic arbors that extend horizontally across the retina, and they are electrically coupled to neighboring horizontal cells of the same type.

- Bipolar cells – Bipolar cells are radially directed neurons that convey visual information from the photoreceptors in the outer plexiform layer to the ganglion cells in the second synaptic layer (the inner plexiform layer). These cells fall into one of two broad functional classes: 'on' and 'off' bipolar cells. These classifications result from the nature of the bipolar cell's response to small spots of light positioned over the center of the cell's receptive field. 'On' bipolar cells depolarize, whereas 'off' bipolar cells hyperpolarize to increases in illumination. Bipolar cells also have anatomical sub-classifications. The bipolar cell gets its name from the fact that it has two processes that extend out from opposite ends of the cell body.

- Amacrine cells – Amacrine cells form lateral and radial interconnections in the inner plexiform layer. Classification of these cells is more complex, as there are many different anatomical classes of cell and their functional counterparts have yet to be fully explored. The amacrine cell gets its name from its anatomy: it is distinguished by an axonless cell body.
- Ganglion cells – The ganglion cells are the output neurons of the retina. They have very long axons that leave the cell bodies, course across the retina and exit the eye through the optic nerve head. The axons are unmyelinated as they cross the retina, but they become myelinated as they exit the optic nerve head. These myelinated fibers comprise the optic nerve that transmits visual information from the retina to the lateral geniculate nucleus in the thalamus. Ganglion cells are also composed of a wide variety of anatomical subclasses whose precise functions have yet to be fully elucidated.
- Interplexiform cells – Not shown in Fig. 3.2 are the interplexiform cells. These cells provide a feedback pathway from the inner to the outer plexiform layers.

The number of subclasses in these cells is a function of the species and whether the subclasses are based on anatomical or physiological criteria. Some of these subclasses will be highlighted in the sections below.

3.2.3 Retinal Lamina

The human retina rests against the pigment epithelium, a layer of cells filled with black melanin pigment which is involved directly with the regeneration of photopigments that have been isomerized by incident photons. Because of its role in photopigment regeneration, the outer segments of the photoreceptors must abut the pigment epithelium, which causes the remaining retinal neurons to be interposed between the front of the eye and the photoreceptors. As a result, light must travel through all of the retinal neurons before it is absorbed by the photopigments located in the photoreceptors. However, as these retinal neurons are virtually transparent, the retinal image is not significantly degraded.

A histological section of a human retina normal to the plane of the retina is shown in Fig. 3.3. Because these neurons are virtually transparent, this section has been stained with a dye that colors cell nuclei dark. One feature that is immediately noted is the layered appearance of the section. This results from the orderly laminar arrangement of neurons and dendritic arbors that make up the retina. The cell bodies of the retinal neurons are located in 'nuclear layers', while the synaptic interconnections that mediate retinal

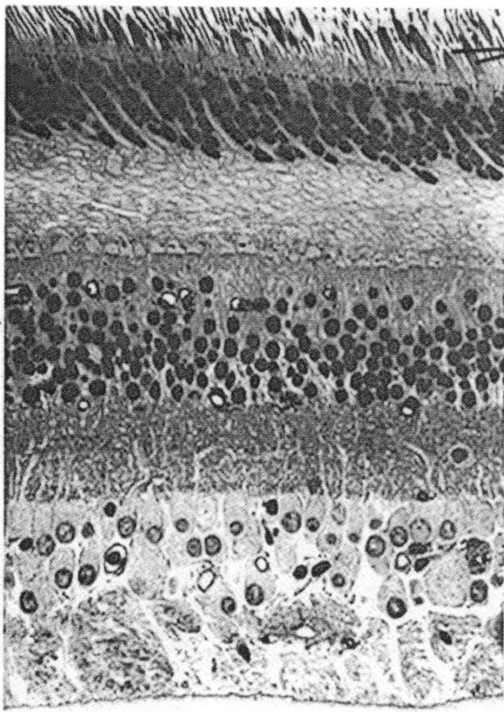

Figure 3.3. Radial light micrographic section of the human central retina showing the laminar arrangement of the neurons and the inner- and outer-plexiform layers; the two regions where synaptic interaction occurs between neurons. The regions are: OLM – outer limiting membrane; ONL-outer nuclear layer; INL – inner nuclear layer, GCL – ganglion cell layer; and ILM – inner limiting membrane. The nuclear layers contain cell bodies of the neurons Reprinted from Kolb et al. (2000) with permission of WebVision.

information processing occur in the 'plexiform layers'. The laminae, or layers, that make up the retina are:

- Photoreceptor outer segments (Ph) - the regions of the photoreceptors that contain light sensitive photopigments.
- Outer nuclear layer (ONL) - contains photoreceptor cell bodies.
- Outer plexiform layer (OPL) - the site of synaptic interactions between photoreceptors, horizontal cells and bipolar cells.
- Inner nuclear layer (INL) – contains horizontal cell, bipolar cell, amacrine cell, and innerplexiform cell bodies.
- Inner plexiform Layer(IPL) , the site of synaptic interactions between the bipolar, amacrine and ganglion cells.

- Ganglion cell layer (GCL) – composed of ganglion cell bodies
- Optic nerve fiber layer (not labeled in Fig. 3.3)– this region is composed of ganglion cell axons that course across the retina, exit the optic nerve head, and project to the lateral geniculate nucleus in the thalamus.

The thickness of the retina is the sum of the thicknesses of each of these lamina, and also depends upon where the retinal thickness is measured (ie., in the fovea, or peripheral retina).

3.2.4 Retinal Eccentricity

The retina is not an isotropic network of neurons. The size and density of the retinal neurons is a function of their location within the retina. This dependence on retinal location uses as a reference point the fovea (the visual axis), which is the region of the retina that subserves highest visual acuity. An imaginary line joining the fovea and the fixated target comprises the visual axis. Concentric rings of ever increasing extent, or eccentricity, about the fovea form the retinal parafovea and periphery. As one moves from the fovea to the peripheral regions of the retina, the signal processing that occurs in the retina also changes. Thus, there is a dependence on retinal eccentricity of both neural structure and signal processing function. Fig. 3.4 shows light micrographs of the fovea in a radial section, and the peripheral photoreceptor layer in a tangential section.

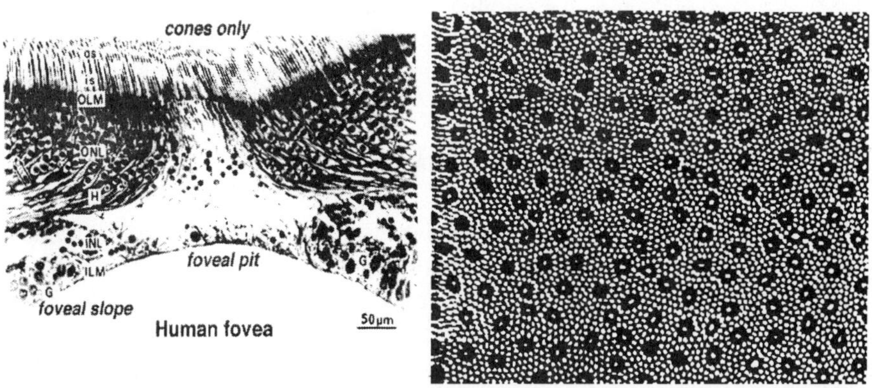

Figure 3.4. Retinal anatomical features. Left - Radial might micrographic section through the human fovea showing the foveal pit, the region of highest acuity vision. The fovea contains only cone type photoreceptors (taken from Kolb, 1994). Right - Tangential section through the photoreceptor outer segments in the peripheral retina shows a regular distribution of cone photoreceptor outer segments, each surrounded by a field of rod photoreceptor outer segments Reprinted from Hubel (1988) with permission of W. H. Freeman and Co.

If one were to design a visual encoder, there would be design compromises that must be made in order to achieve high acuity encoding. One can make an array of very small photosensors, but as the size of the photosensors is decreased, their number increases, which places considerable computational overhead on the post-photosensor processing. Alternatively, one can design an encoder that has a central region of very small, high-density photosensors, and surround this with larger, lower density photosensors. Such an encoder, however, would require active scanning of the visual world; objects of potential interest are detected by the peripheral sensors, and the encoder is then moved so the high acuity region is encoding the object of interest. This is the strategy used in the human visual system. The foveal region is about 6mm in diameter, and in its center is the region of ultra-high visual acuity known as the foveal pit. The outer segments (the region of the cones that contain the light absorbing photopigments) in this region of the human retina have diameters of about 1.5 microns. Individual photoreceptors abut one another and are packed into a high-density hexagonal array. In contrast, the cone photoreceptors in the far peripheral parts of the human retina are about 6 microns in diameter and are spaced 20 microns apart. The region between the cones in the peripheral retina is filled with rod type photoreceptors.

The foveal pit is located in the center of the fovea and is about 0.5 mm in diameter. It has a number of distinguishing anatomical features. In contrast to the remaining regions of the retina (parafoveal and peripheral retina), the foveal pit only contains cone photoreceptor outer segments. The inner segments and the other retinal neurons that make up the foveal retinal circuitry are displaced into the annular region that surrounds the foveal pit. This annular region is called the parafovea or the foveal slope. It contains a very large number of cells and is the thickest part of the entire retina.

3.2.5 Radial and Tangential Pathways

The morphological arrangements of the retinal neurons and their laminar organization provide a substrate that mediates the flow of information from the retinal input to its output. This framework is shown in Fig. 3.2. The photoreceptors, bipolar cells and ganglion cells constitute a simple coaxial neural network where photoreceptors provide synaptic input to bipolar cells that, in turn, provide input onto ganglion cells. The distribution of this triplet of cells across the entire retina makes up what is called the radial retinal pathways.

Information flow along each radial pathway is modulated by the flow of information in neighboring radial pathways. The cells that mediate this modulation are the horizontal cells and the amacrine cells. These neurons have dendritic arbors that spread laterally (not radially) across the retina. Further, each of these classes of neurons is strongly coupled to neighboring neurons of the same class. These features allow information in one part of the retina to flow considerable distances in the plane of the retina.

Retinal information processing (as is information processing in other sensory and motor modalities) is contextual or relativistic. The retinal output at each ganglion cell is not 'absolute'. It does not solely depend upon the neural activity in the radial pathway that provides its direct input, but is a function of the level of activity in neighboring radial pathways. Because the tangential pathways generally provide an inhibitory influence on the flow of information in the radial pathways, this mechanism provides the foundation for 'lateral inhibition'. An example of the contextual or relativisitic signal processing that takes place in our visual system is shown in the illusion of Fig. 3.5. This shows two equiluminant bars of light that are positioned against a background gradient of luminance. Our subjective judgment is that the two bars are of different intensities, but this is not the case. We make our subjective judgments contextually: we perceive that the right bar is darker because its local context is bright, while we perceive that the left bar is brighter because the local context around the bar is dark. Retinal mechanisms which underly this relativisitic signal processing are described more fully in Sections 3.3.4 and 3.3.5.

Figure 3.5. Illusion demonstrating the 'relativistic or contextual signal processing of the visual pathways. The bar on the left appears brighter than that on the right, even though they are equiluminant. The neurons processing the parts of the image surrounding the left bar are not well stimulated by the dark regions in the image, and the bar appears bright in the surrounding dark context. The opposite situation maintains for the right bar. Here the neurons that are processing the parts of the image surrounding the right bar are well stimulated by the image and the right bar appears dark in the context of the brighter surrounding regions.

3.2.6 The 'On' and 'Off' Pathways

The photoreceptors of the vertebrate retina transduce spatial, temporal, intensity, and chromatic patterns of light striking the retina into a spatio-temporal pattern of electrical activity, forming a 'neural image'. This neural image is further transformed by the outer and inner plexiform layers into a neural 'output image' that is sent by the optic nerve fibers of the ganglion cells to the higher visual centers. This parallel information processing results in an efficient neural code that compresses the visual information that is transmitted to the higher visual centers. The need for this neural information processing can be appreciated when one considers that there are about 150 million rod and cone photoreceptors in each eye that transduce the incident optical image into an electrical image, but there are only 1 million optic nerve fibers that can be used to send this image to the higher visual centers. The mechanisms by which this 150:1 compression of visual information takes place has motivated a great deal of research over many decades.

An important concept in retinal information processing is that the neural image is composed of 'on' and 'off' pathways. Specifically, the optical image that strikes the retina has a mean luminance. There are certain regions of the optical image that are brighter than this mean luminance, and other regions which are darker that the mean luminance. The retina must inform the higher visual centers of both the brighter and the darker regions of the optical image. This is accomplished by the 'on' and 'off' retinal neural pathways. The 'on' pathways increase their activity in response to parts of the image that are brighter than the mean luminance levels, and the 'off' pathways increase their activities for parts of the optical image that are dimmer.

3.2.7 Foveal Pathways

As we will see in subsequent sections, a striking anatomical and functional feature of the retinal circuitry is convergence and divergence of neural pathways and information as the retinal neural images are processed by the retinal neurons. However, these mechanisms are not seen to any significant degree in the fovea, where the neural circuitry has been specialized to achieve the high spatial acuity that characterizes this retinal region.

The foveal photoreceptors consist exclusively of very small cones that transform the light images incident onto the retina into a high-resolution neural image. Each of the three classes of cones is represented in this specialized retinal region, but not in equal numbers. The ratios of red to green to blue cone photoreceptors are: 45:45:10 (Kolb et al., 2000). The synaptic terminals of the cones are located on the foveal slope where each cone makes a synaptic contact with horizontal cells, and with two special classes of bipolar cells, called midget bipolar cells (Polyak, 1941). Each cone contacts one depolarizing midget bipolar cell (an 'on' bipolar cell) and one hyperpolarizing midget bipolar cell (an 'off' bipolar cell) (Kolb, 1970; Gouras, 1971). Thus, there is some divergence of each cone signal into two bipolar cell signals, but these diverge no further. Each midget bipolar cell contacts a single ganglion cell: the 'on-bipolar' cell contacts one 'on-ganglion cell', and the 'off-bipolar cell' contacts a single 'off-ganglion cell'. Local tangential pathways composed of the horizontal and amacrine cells also influence signal processing through these radial pathways.

3.2.8 Peripheral Pathways

The peripheral parts of our visual system provide a low acuity view of the visual world. The system has been designed to make us aware when objects of potential interest enter our peripheral field of view, so that we can move our gaze and examine these objects with our high acuity foveal vision. It is not surprising, therefore, that the peripheral retina is composed of larger cells, and that there is much greater convergence and divergence of information in this region compared to the fovea. The maximum cone density in the fovea is about 250,000/sq mm, while in the periphery, it is over 100 times less dense. Further, each bipolar cell in the peripheral radial pathways receives input not from a single cone, but from many cones, and each peripheral ganglion cell receives input from dozens of bipolar cells. While this data has not been obtained in the human retina, studies in the cat retina (Kolb, 1994) show that a single ganglion cell can receive input from upwards of 15,000 cone photoreceptors. Convergence and divergence of the radial pathways is clearly much more pronounced in the peripheral parts of the retina. A remarkable feature of the human visual system is that we are generally unaware of these vast differences in the acuity and convergence of foveal and peripheral vision. The perceptual transition from fovea to parafovea to peripheral vision is generally seamless.

3.3 PHYSIOLOGY OF THE RETINA

The laminar structure of the retina, and its accessibility have caused it to be one of the most intensely studied tissues in the nervous system. It has been studied by anatomists, biochemists, molecular biologists, and electrophysiologists. A great deal has been learned about the genetics and genetic defects of the phototransduction mechanisms, the biochemical pathways underlying phototransduction and synaptic transmission, and the biophysics of the six classes of cells that make up the retina. One of the earliest observations made about the biophysics of the retinal neurons was that the neurons of the outer and inner nuclear layers (the photoreceptors, horizontal cells, bipolar cells and certain amacrine cells) did not produce action potentials in response to light stimulation (Werblin and Dowling, 1969). Action potentials, the typical mode of response of most neurons in the central and peripheral nervous system, were only observed in the ganglion cells and in certain classes of amacrine cells. As the distances that ganglion cells must transmit information are large, it is not surprising that they use 'all-or-none' action potentials to reliably send information to the lateral geniculate nucleus. However, as the distances that the remaining retinal neurons communicate with their neighboring neurons are very short, one might expect these cells use a different form of communication; a so called 'low pass' form of communication that will be described in section 3.2.

3.3.1 Phototransduction

The molecular mechanisms underlying the transformation of incident photons into ionic currents flowing into and out of rod and cone photoreceptors have been well identified and are described in the literature. These mechanisms will be summarized below with reference to Figs. 3.6 and 3.7. Fig. 3.6 shows a cartoon of the rod photoreceptor. The rod outer segment contains the photosensitive pigment, rhodopsin. The inner segment contains a high concentration of mitochondria, the cell body and the synaptic terminal (called a spherule). The outer segment in the frog rod is composed of a stack of 1000 'discs' that is surrounded by the rod plasma membrane. Each disc resembles a red blood cell. The disc membranes are not part of the rod's plasma membrane (except at the base of the outer segment where new discs are synthesized on a daily basis). Each of the thousand discs contains about 1 million rhodopsin molecules. The rhodopsin molecules are composed of a large protein molecule called opsin, and a smaller chromophore called retinal.

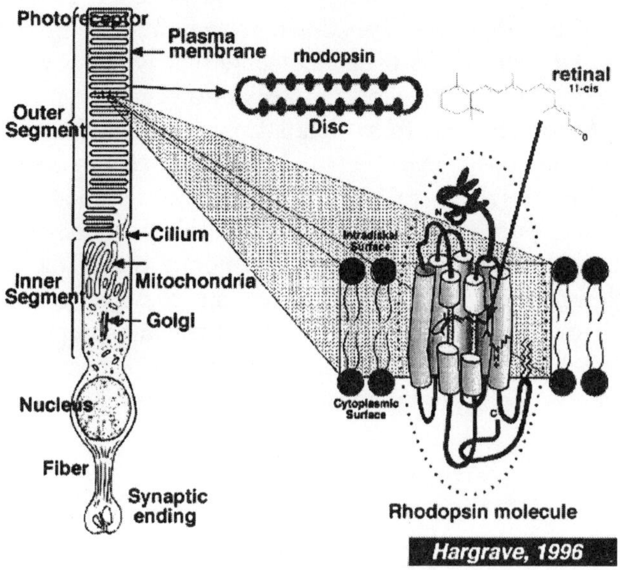

Figure 3.6. Cartoon showing the structure of the rod photoreceptor, with its outer segment that contains the light absorbing rhodopsin, the inner segment that contains high concentrations of mitochondria, the nucleus, and the synaptic ending (called the 'rod spherule'). The rhodopsin molecules are located on the membranes of 'discs' that fill the rod's outer segment (drawing courtesy of Paul Hargrave).

The ultimate consequence of the phototransduction process is the modulation of the release of neurotransmitter from the rod to the bipolar cell. This process is directly controlled by the membrane potential of the rod photoreceptor, which, in turn, is directly controlled by the sodium permeability of the rod photoreceptor plasma membrane. In the dark, the rod's sodium permeability is high, and the rod plasma membrane is depolarized to about –30 mV. In the light, the rod plasma membrane's sodium permeability is low, and the rod membrane potential hyperpolarizes to –65mV. So, how does the absorption of a photon by a rhodopsin molecule lead to a decrease in the sodium permeability of the rod's plasma membrane?

The biochemical steps by which this occurs are illustrated in Fig. 3.7, and involve a second messenger cascade similar to that found in other sensory and neuronal transmission systems (see Polans, 1996 for a review). The sodium permeability of the rod is directly increased by the internal second messenger, cyclic GMP. When the rod's cytoplasmic cGMP concentration is high, sodium channels in the plasma membrane are open, and the membrane is depolarized. When cGMP falls, sodium channels close and the membrane hyperpolarizes. The cGMP concentration is regulated by the molecule phosphodiesterase, which is controlled by a molecule called transducin. The cytoplasmic transducin concentration is directly controlled by a short-lived state of the opsin molecule that is created when a photon causes the retinal on the rhodopsin molecule to become isomerized from its 11-cis form to the all-trans form. The adsorption of a photon by rhodopsin and the subsequent isomerization of retinal are accompanied by a color change in the molecule called 'bleaching': a solution of unbleached rhodopsin has a purple color that becomes yellow when it is exposed to bright lights. The bleached retinal is transported out of the outer segment and into the pigment epithelium where an enzyme, retinal isomerase, converts it back into the 11-cis form. It is then transported back from the pigment epithelium into the outer segment, where it combines with a free opsin molecule, and the phototransduction process is ready to begin again.

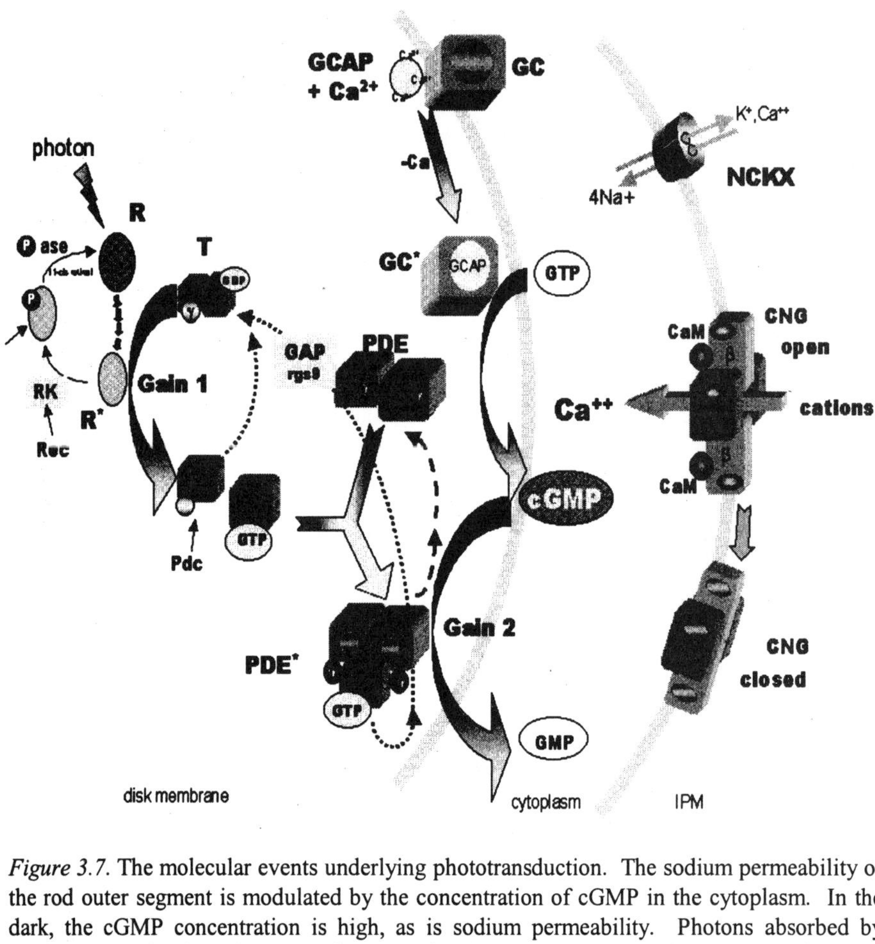

Figure 3.7. The molecular events underlying phototransduction. The sodium permeability of the rod outer segment is modulated by the concentration of cGMP in the cytoplasm. In the dark, the cGMP concentration is high, as is sodium permeability. Photons absorbed by rhodopsin molecules decrease the cytoplasmic concentration of cGMP, and, as a consequence, the sodium permability (drawing courtesy of Wolfgang Baehr).

The rate constants of the biochemical cascade illustrated in Fig. 3.7 are influenced by a number of ions and molecules. One of the most potent influences is from cytoplasmic calcium concentration. Calcium also enters the rod photoreceptor via the same channels as those for sodium ions. So, in the dark, the sodium and calcium membrane permeabilities are both relatively high, and the cytoplasmic calcium concentration is relatively high. Calcium is a co-factor in the production of cGMP, so the high cytoplasmic calcium concentration helps keep the cGMP concentration high, which keeps the sodium and calcium permeabilities high. When the cGMP concentration falls, the membrane permeability decreases, and therefore, the cytoplasmic calcium concentration also falls. This, in turn, reduces the cGMP production. Thus, cytoplasmic calcium acts as a powerful gain controlling mechanism in the rod and cone photoreceptors (see Polans, 1996, for a review).

3.3.2 The Photoreceptor's Photoresponse

Now that we have described the mechanisms that transduce light into sodium permeability changes, we will next consider how these changes are manifested in terms of the rod membrane potentials. The biophysics of photoreceptors has been studied in numerous ways, but most of our understanding has resulted from single microelectrode intracellular recording techniques.

One of the most fundamental questions of retinal function involves the ultimate sensitivity of the rods: 'Can a rod photoreceptor respond to a single photoisomerization in its outer segment?' This question was answered affirmatively in the elegant psychophysical experiments of Hecht, Schlaer, and Pirenne (1942) who showed that the perception of light was evoked by a brief incident photon flux of a few hundred photons. By the time these photons reached the rod outer segments, it was determined that the likelihood that any individual rod had absorbed more than one photon was very low. They concluded therefore that individual rods can detect single photons.

Baylor and his colleagues (Baylor et al., 1979), and others, have used whole cell patch clamp techniques to actually record the responses of rods to single photon absorption. This technique uses glass micropipettes as attachments to rod outer segments that have been mechanically isolated from their inner segments. Because the micropipette often completely seals around the base of the outer segment, the micropipette can be used to measure the change in outer segment conductances produced by photon absorption in the outer segment. It was shown that the absorption of a single photon produced a 1 picoamp (10^{-12}amps) peak decrease in membrane current that persisted for about 0.25 seconds. While this seems like a very small change in membrane current, this single photon event reduces the flux of sodium ions across the rod's plasma membrane by about one million ions. This is a very high gain phototransduction mechanism!

How does the photoreceptor membrane potential change when larger numbers of photons are absorbed by its photopigment? As it is changes in the membrane potential of the photoreceptors that modulates the synaptic transmission to the bipolar cells, this question is particularly relevant. In the left panel of Fig. 3.8 we show how the membrane potential of a turtle cone photoreceptor is modulated by a series of 0.5-second flashes of increasing intensities. This series of responses illustrates the 'low pass' nature of the responses of the neurons of the outer retina to flashes of light (the term 'low pass' is a systems theory term which describes the fact that the cell responds to low frequency stimuli and to constant stimuli). The responses are sustained (or 'tonic'): as long as the light is on, the cell remains hyperpolarized. Much of this low-pass temporal filtering occurs because of the relatively slow responses of the photoreceptors (slow relative to the incredibly fast responses of human engineered photodiodes built from silicon). Unlike the near instantaneous production of hole-electron pairs by quanta that strike a photodiode, the photochemical mechanisms in the rod and cone photoreceptors activated by photon absorption take time to develop and to disappear. The time course of the response of rod and cone photoreceptors to very brief flashes of light are the limiting factors on the 'speed' of the visual system; they are the slowest elements in the string of neurons that constitute the visual pathway (see Daly and Normann, 1985). The kinetics of photoreceptors are dependent on species, and on the nature of the photoreceptor: rods or cones.

The right panel of Fig. 3.8 is a plot of the amplitude of the cone photoresponse versus the logarithm of the light intensity that evoked the response. The amplitude of the response is graded with respect to the intensity of the stimulation. The cone manifests a threshold and a saturation: light stimuli that are very dim evoke no apparent response, while light stimuli that are brighter than some saturating value evoke responses whose amplitudes are constant (but whose durations begin to increase). Between threshold and saturation, the cone response is approximately proportional to the logarithm of the light intensity striking the cone. The cone has a dynamic range from threshold to saturating intensities of a factor of about one thousand (3 log units). Since the cones and rods synapse upon horizontal and bipolar cells, it is not surprising that these cells also exhibit 'low pass' kinetics in their response similar to those seen in the photoreceptors.

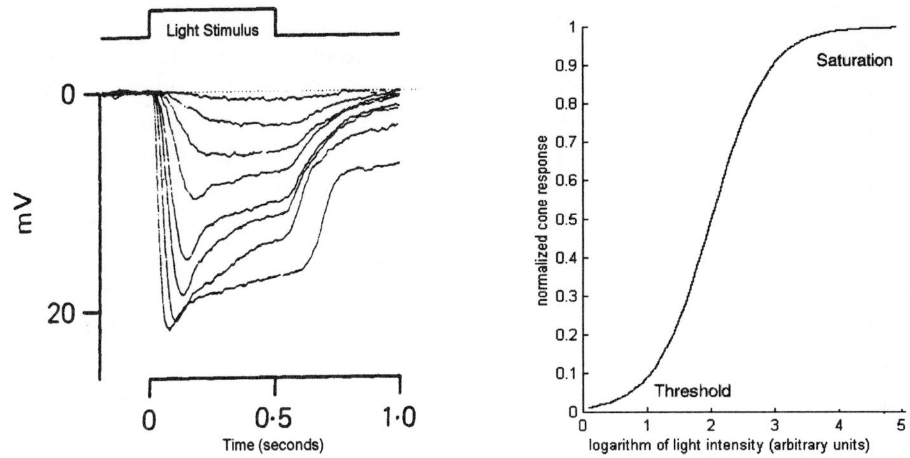

Figure 3.8. Left: Superimposed cone photoresponses to eight light stimuli, presented for 0.5 seconds. Each response is to a light intensity approximately three times brighter than the previous stimulus. The membrane potential of the cone photoreceptor is relatively depolarized in darkness, and hyperpolarizes when photons are absorbed by its rhodopsin. Right: Cone dark-adapted intensity response curve. The extent of the hyperpolarization is graded with light intensity over an approximate thousand-fold range of light intensities from threshold to saturation. Adapted from Normann and Perlman (1979) with permission of The Physiological Society.

3.3.3 Light- and Dark-Adaptation of the Vertebrate Photoreceptors

One of the most interesting features of the cones of the vertebrate retina is their intrinsic ability to adjust their sensitivity to the level of ambient illumination to which they are exposed. This capability is called light-adaptation, and it allows the photoreceptors to respond to local changes in retinal illumination over a thousand-fold range of ambient intensities. As mentioned in the previous section, the dark-adapted cone photoreceptor has a dynamic range from threshold to saturation of three log units. This range of intensities would be experienced by a human observer in a dimly lit restaurant, or at twilight. However, these same cones must be able to generate photoresponses when the ambient illumination is much greater, as one would experience at the beach or on the ski slopes on a sunny afternoon. The ranges of intensities over which the cones must respond in this situation are at least a factor of one thousand to ten thousand times greater than those encountered in the twilight situation. The mechanisms whereby the cone is able to achieve this remarkable capability were have been extensively studied by Normann and Perlman (1979).

The solid curve in Fig. 3.9 shows the dark-adapted cone's dynamic range redrawn from Fig. 3.8. If the cone is now illuminated with a steady bright background intensity greater than the saturation intensity, the cone initially produces a saturated, maximal response. The vertical line that intersects the right-most curve in Fig. 3.9 represents such an intensity (6 log units of intensity). As time progresses, the potential produced by the bright background intensity decreases and eventually reaches a steady-state potential about one half of the maximal potential evoked by the background light (indicated by the short horizontal line that intersects the right-most intensity response curve). If the intensity of the background light is now modulated, the cone is once again able to respond to these modulations. Specifically, briefly increasing or decreasing the background intensity evokes graded increment or decrement responses in the cone. If one plots the amplitude of the increment and decrement responses versus the intensity of the light that evoked the responses, one obtains the right-most intensity response curve in Fig. 3.9. It is interesting to note that the dark-adapted intensity-response curve and the light adapted intensity-response curves are virtually identical in shape but the light-adapted curve has been displaced along the log light intensity axis by the background illumination.

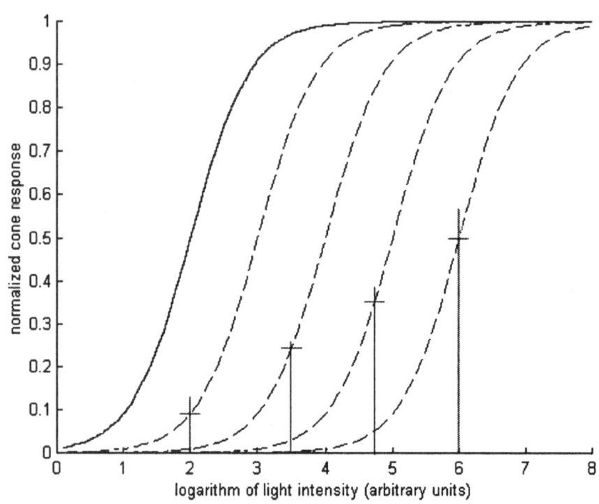

Figure 3.9. The effects of background illumination on the intensity response curves of cone photoreceptors. The solid curve was replotted from Fig. 3.8, and shows how the amplitude of the dark-adapted cone photoresponse varies with light intensity. The four right-hand curves are intensity response curves that would be measured in a cone around four different background intensities (2, 3.5, 4.7, and 6 log units of intensity). The effect of the background illumination is simply to shift the entire 3-log unit dynamic range of the cone such that it spans higher ranges of intensities. Note that the light adapted cones are able to respond both to increases and decreases in light striking the retina (see Normann and Perlman, 1979).

The light-adapted intensity response curves of Fig. 3.9 were generated with four different background intensities. Different background intensities (either brighter or darker) cause differing degrees of displacement of the intensity-response curves. A whole range of background intensities evokes a whole set of set of identical intensity-response curves. All curves, however, span a three-log unit range of light intensities. In this sense, the vertebrate cone photoreceptor has an automatic gain control mechanism that accomplishes the same goal as the shutter speed and lens aperture do in an automatic camera. These mechanical elements are used to match the film's dynamic range to the ranges of intensities the photographer is shooting in. Photographic film has a dynamic range of about two log units, but this fixed range can be used to produce perfectly exposed images in dim moonlight and in bright sunlight. Each cone photoreceptor has its own automatic gain controlling mechanism that accomplishes the same function. We are only aware of this mechanism when the automatic gain control is challenged by rapid and large changes in ambient illumination (entering a dark tunnel on a bright afternoon, or emerging from the tunnel into the bright light). So, we have seen that the very first stage of the retina, the cone photoreceptors, achieves considerable processing of the incident optical images: they transduce these signals into graded membrane potentials, they adjust their own gains, so that they are maximally sensitive to the light that strikes them, and they perform a tri-color separation of the incident light pattern.

3.3.4 The Cone Triad Synapse and Spatial Opponency

The vertebrate retina does a great deal more than simply transduce incident patterns of retinal illumination into patterns of electrical activity, although this is still one of its most important functions. Another important function of the retina is spatial filtering. While one might have expected that the retina acts as a spatial filter that has a low-pass nature due to the quality of the optical elements of the eye: the aperture of the pupil, and the finite size of the photoreceptors. In fact, the retina acts more like a band-pass spatial filter. Very low and very high spatial frequencies are attenuated relative to a pass band whose center spatial frequency depends upon retinal eccentricity. The band-pass spatial filtering of the retina results from an antagonistic interaction between the radial and the tangential pathways. The radial pathway, composed of the cones and bipolar cells, is basically spatially low-pass in character. The tangential pathway in the outer plexiform layer is composed of the horizontal cells, and it also is spatially low-pass in nature. However, this pathway has a much lower spatial corner, or cut-off, frequency due to the extensive lateral interconnections that exist between neighboring horizontal cells. The band-pass character of the outer plexiform spatial filter results from the antagonistic interaction between these radial and horizontal pathways. It is mediated by the feed-forward synapse between the cones and the bipolar cells, which is modulated by the horizontal cells at the cone 'triad synapse'.

The spatial band pass nature produced by these pathways is referred to as 'lateral inhibition', and it acts to augment neural activity across light-dark boundaries. The phenomenon of lateral inhibition was first discovered in the horseshoe crab by H. K. Hartline (1940), but the phenomenon is not limited to the visual system. Lateral inhibition is found in virtually all sensory modalities, and, just as for vision, the mechanism sharpens the effects of spatial discontinuities in each sensory modality. Thus, in the auditory system where spatial position along the basilar membrane is related to transduction of differing auditory frequencies (known as 'place coding'), lateral inhibition between neighboring groups of sensory hair cells on the basilar membrane of the cochlea have sharper tuning curves (they respond more selectively to specific frequencies of auditory stimulation) than they would if this mechanism were not operational. The mechanisms of lateral inhibition operate at all levels in the visual pathways, and give rise to a number of simple illusions such as those shown in Figs. 3.5 and 3.10. In Fig. 3.10, the scalloping seen in uniform light steps and the dark dots seen in the middle of the white panes in the 'Hermann grid' are due, in part to lateral inhibition at the outer plexiform layer.

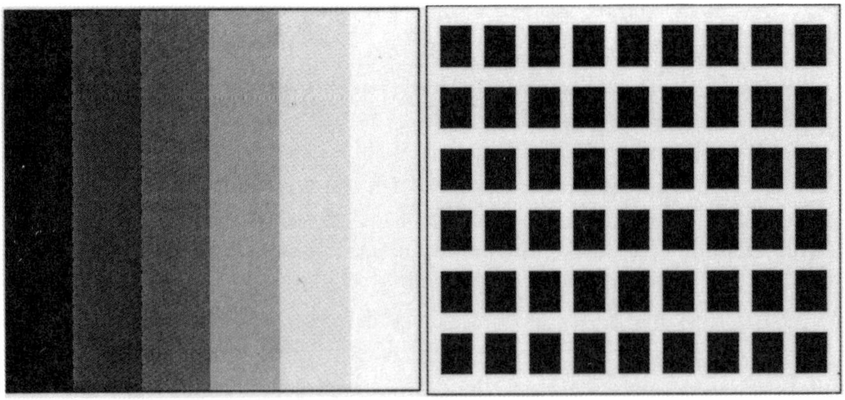

Figure 3.10. Visual illusions that are produced by the mechanisms of lateral inhibition. Left panel - The uniform steps of illumination appear scalloped at the intensity discontinuities. Right panel - The white horizontal and vertical stripes appear to have black fuzzy spots at their intersection.

3.3.5 A Simple Model of Lateral Inhibition

The visual signal processing produced by the mechanisms of lateral inhibition are generally transparent to us in our normal viewing of the visual world. However, certain patterns of retinal illumination can make us aware that these mechanisms exist, and these patterns produce percepts that fall into the class of visual illusions. Two examples of such illusions are shown in Fig. 3.10. While the left panel in Fig. 3.10 is a sequence of uniform intensity steps, the patterns appeared scalloped at the intensity discontinuities. In the right panel in Fig. 3.10, there appears to be black fuzzy dots at the intersection of the horizontal and vertical white stripes.

To illustrate how the mechanisms of lateral inhibition can alter our perceptions of simple light distributions, a simple model can be constructed using an Excel™ spreadsheet to predict the output of a network of interconnected cells. To further simplify the model, one can convert the two-dimensional retinal network into a one-dimensional problem by using spatial stimuli that vary only in one dimension like the example shown in the left panel of Fig. 3.10 and modeled in Fig. 3.11. We can represent a light distribution, incident on the retina, as a set of values entered into the spreadsheet's cells on the top row of a spreadsheet (cells $A_1 \ldots A_q$, in the spreadsheet's nomenclature). The gray squares in Fig. 3.11 represent the light intensities striking the retina. The illumination pattern is dark on the left (cells 1-11), bright in the middle (cells 24-35), and even brighter on the right (cells 43-50). If each of the cells directly under the top row (cells $B_1 \ldots B_q$) receives excitation from the cell directly above it, and inhibitory input from its neighboring cells, the iterated solutions of this network mimic the illusions shown above. Specifically, we represent the pattern of retinal illumination (a one dimensional pattern to simplify this demonstration), in cells A_1 through A_q, and compute the potential in each B cell as $B_n = A_n - 0.5*(B_{n+1}+B_{n-1})-0.3*(B_{n+2}+B_{n-2})-0.1*(B_{n+3}+B_{n-3})$. The response of this simple lateral inhibitory mechanism to the light distribution is shown in Fig. 3.11 as the black squares. The model predicts the subjective impression that the light distribution is not built from uniform intensity steps, but that it appears to be scalloped. It is clear from this simple simulation that the perceived contrast of the staircase will appear greater at the edge of the step. At the edge of the step, the darker side of the step will seem darker than the rest of the step, and the brighter side of the step will seem even brighter than the rest of the bright step. For the bar of light, its edges will seem brighter than the center of the bar.

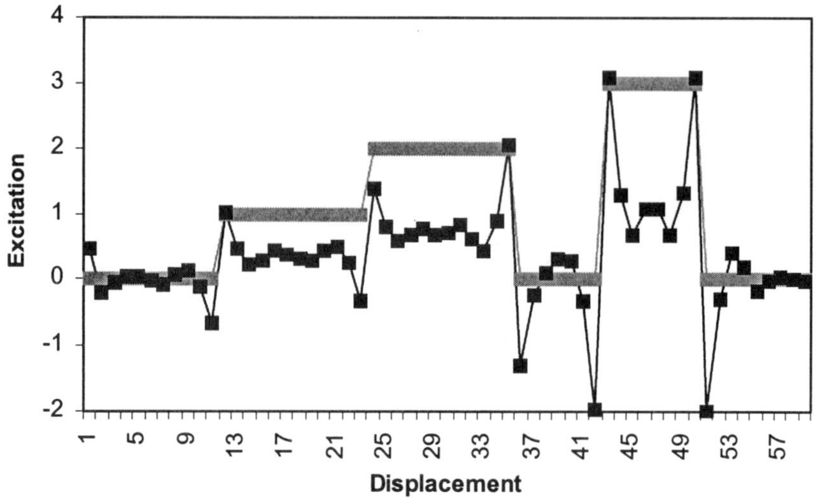

Figure 3.11. Results of a simple Excel™ model of the *consequences* of lateral inhibition on our perceptions of a set of intensity steps. The gray squares represent the pattern of light striking a retina, and the black squares model the results of lateral inhibition on this pattern of excitation.

3.3.7 Color Opponency and Horizontal Cells

Just as we have seen that spatial antagonism between the radial and tangential pathways leads to lateral inhibition, similar mechanisms operate in the color domain. In fact, this mechanism begins in the outer plexiform layer, where chromaticity-type horizontal cells have responses whose polarity is color dependent. Horizontal cells fall into two broad categories: luminosity type cells and chromaticity type cells. Fig. 3.12 shows examples of both luminosity type (Fig. 3.12a) and chromaticity type (Figs. 12b and c) horizontal cell responses evoked by a sequence of brief flashes of light of various colors. Just as was seen for the cone photoreceptors, horizontal cells also respond to a series of light flashes of increasing intensities with graded changes in membrane polarization. The polarity of the responses in the luminosity type cell are all hyperpolarizing to all colors. However, the polarity of the responses in the two chromaticity type horizontal cells is color dependent. The red light flashes (700 nm) evoke responses in the chromaticity type horizontal cell that are depolarizing, while the blue light flashes (450 nm) evoke responses in the same horizontal cell that are hyperpolarizing. Thus, these horizontal cell is said to manifest color opponency. Synaptic circuitry that could mediate this color opponency has been proposed by Lipetz, Stell, and by Simon and Fuortes (see Kolb and Lipetz, 1991, for a review).

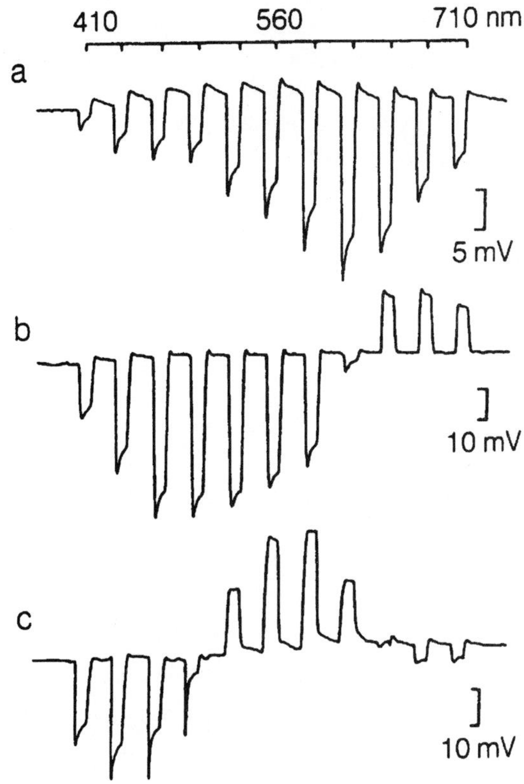

Figure 3.12. Horizontal cell responses to successive flashes of increasing wavelengths. a) Luminosity type cell responds with hyperpolarizing responses to flashes of all colors. b) Biphasic chromaticity type horizontal cell responds to long wavelength flashes with depolarizations and to short wavelength flashes with hyperpolarizations. c) Triphasic chromaticity type horizontal cell responds with hyperpolarizations to long and short wavelength flashes, but with depolarizations to medium wavelength flashes. Adapted from Toyoda, Kujiraoka et al. (1982) with permission of Alan Liss Publishers.

3.4 THE RETINAL OUTPUT

3.4.1 'On', 'Off', and 'On/Off' Ganglion Cell Responses to Light

The output of the vertebrate retina are the ganglion cells whose axons make up the optic nerve. When the retina is stimulated with light, the ganglion cells respond to light stimulation with action potentials. There are a large variety of functional and anatomical ganglion cell classifications, but every ganglion cell can generally be characterized as being either an 'Off' center cell, an 'On' center cell, or an 'On/Off' center cell. An 'On' center cell responds to a small spot of light in the center of its receptive field with an increase in its firing. The response can be either sustained (tonic) or transient (phasic). In a similar fashion, an 'Off' center ganglion cell will generally have a non-zero firing rate in the absence of visual stimulation that is decreased by light stimulation. When the light stimulus is extinguished, the 'Off' center ganglion cell will respond with a transient burst of action potentials. 'On/Off' center ganglion cells are generally phasic, and generate a burst of action potentials at both the onset or offset of the stimulation. Ganglion cells generally manifest spatial center-surround antagonism (lateral inhibition). Examples of 'On' and 'Off' center ganglion cell responses are shown in Fig. 3.13.

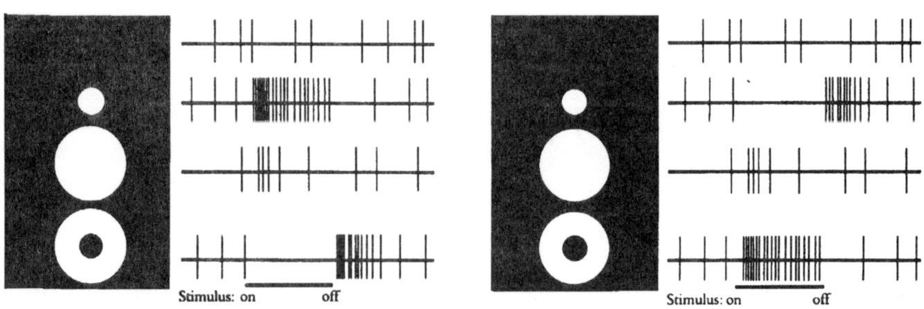

Figure 3.13. Examples of 'On' and 'Off' center ganglion cell responses to small spot, large spot, and annular illumination. The cell on the left is an 'On' center ganglion cell and produces spikes when a small spot of light is positioned over its center. A larger spot of light evokes lateral inhibitory mechanisms that result in a less robust response. Illumination of the surround region without illuminating the center causes the cell to become silent. The cell on the right is an 'Off' center ganglion cell and it is silenced by small spots of light positioned over its receptive field center. However, the 'Off' center ganglion cell responds robustly to annular illumination. Reprinted from Hubel (1988) with permission of W. H. Freeman and Co.

3.4.2 Classifications of Retinal Ganglion Cells

The classification of ganglion cells in different species has been the subject of many investigations. One of the most fundamental studies of ganglion cell types was conducted by Lettvin, Maturana, Pitts, and McCulloch (1959), who attempted to classify the types of ganglion cells in the frog retina based upon the types of stimuli the frog would be expected to experience naturally. Since frogs are particularly adept at capturing moving crickets, Lettvin et al postulated that there may be a class of ganglion cells in the frog retina that are particularly sensitive to small moving spots. Similarly, because a frog will produce an evasive maneuver if a large animal approaches it, Lettvin et al thought they might find ganglion cells with very large receptive fields. As expected, these researchers found examples of ganglion cells with just such receptive field properties, and they named them 'bug detectors', 'net convexity detectors', 'dimming detectors', etc. This type of behaviorally-relevant classification scheme, however, has not, in general, been championed by other retinal or cortical physiologists (see Ammermuller and Kolb, 1996).

The cat retina has been studied extensively, and its ganglion cells have been classified both anatomically and functionally. Enroth-Cugell and Robson (1966) based their classification scheme from an engineering perspective of the retina. They discovered that certain ganglion cells had a receptive field where the effects of light summed linearly in its center ('X cells'), another class of ganglion cells were decidedly non-linear in their summation properties ('Y cells'), and yet another class of cells had receptive field properties that did not fit well into either classification scheme ('W cells').

As much of the interest in visual system information processing has been motivated by understanding human vision, the monkey has been used as a more appropriate model. The primate retina has two broad classes of ganglion cells; 'M-type' and 'P-type' (see Shapley and Perry, 1986, for a review). This classification scheme actually satisfies both anatomical and physiological criteria, and the classification scheme has been applied also in the higher visual centers.

- **M-type.** These cells are called 'magno' type cells because of their large size. At every retinal eccentricity, the M-type cells have larger receptive fields, are more sensitive to low contrast stimuli, and are more phasic than their P-type counterparts. M-type cells project to the magnocellular layers of the lateral geniculate nucleus.

- **P-type.** These are called 'parvo' type cells because of their small size. They are generally more tonic in their response, have small receptive fields, and are color and spatially opponent. P-type cells project to the parvocellular layers of the lateral geniculate nucleus.

3.4.3 Color Sensitivity of Ganglion Cells

The P-type ganglion cells in the monkey retina have color-opponent responses that fall into two broad categories. Some have a red/green opponency, and others have a blue/yellow opponency. The red/green opponent cells have spatial opponency as well (such cells are said to be doubly opponent), where surround illumination inhibits the response of the ganglion cell to center illumination, but the degree of the inhibition is augmented by surround illumination with the opponent color. Specifically, a red/green ganglion cell will be excited by a small spot of red light and inhibited by a small spot of green light. A large spot of red light will evoke a smaller response because of the spatial opponency described above. However, if the retina is illuminated with a small spot of red light, surrounded by an annulus of green light, the cell's response will be virtually eliminated. This color opponency is illustrated in Fig. 3.14. Blue/yellow ganglion cells, however, do not have this form of spatial opponency; the cell's response to blue stimuli is inhibited by yellow light delivered anywhere in its receptive field (Dacy and Lee, 1994).

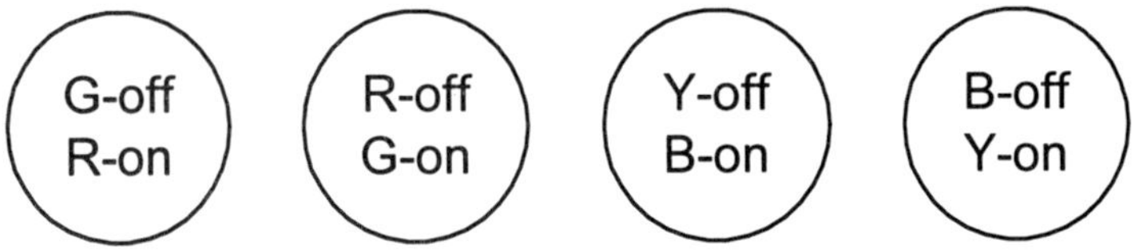

Figure 3.14. Four main classes of color opponent ganglion cells.

3.4.4 Temporal Response of Retinal Ganglion Cells

It is important to remember that the retina is not a simple video encoder. Instead, it is used within visual systems that scan the visual world with quick saccadic motions and smooth target pursuits. These eye motions partition the retinal input and response into discrete episodes that are often simulated by researchers through flashed presentations to the retina. These types of studies reveal that the temporal structure of the response can represent certain aspects of the stimulus.

In Fig. 3.15, the responses are shown for an On-Off cell in an isolated turtle retina presented with full-field stimuli at three different intensities. In this example, the latency of the spike response after the onset and offset of the stimulus is clearly modulated by the stimulus intensity. Another striking aspect of the figure is the variability of the responses for the same stimulus. These responses can be averaged by computing a Peri-Stimulus Time Histogram (PSTH) for each stimulus (shown at the bottom of the Fig. 3.15). The PSTH is simply a histogram of the spike counts at different times throughout the stimulation trial used to reveal the underlying repeatable characteristics of the response. However, recent work by Berry and Meister (1998) has suggested that the retinal response can be quite reproducible under certain stimulus conditions.

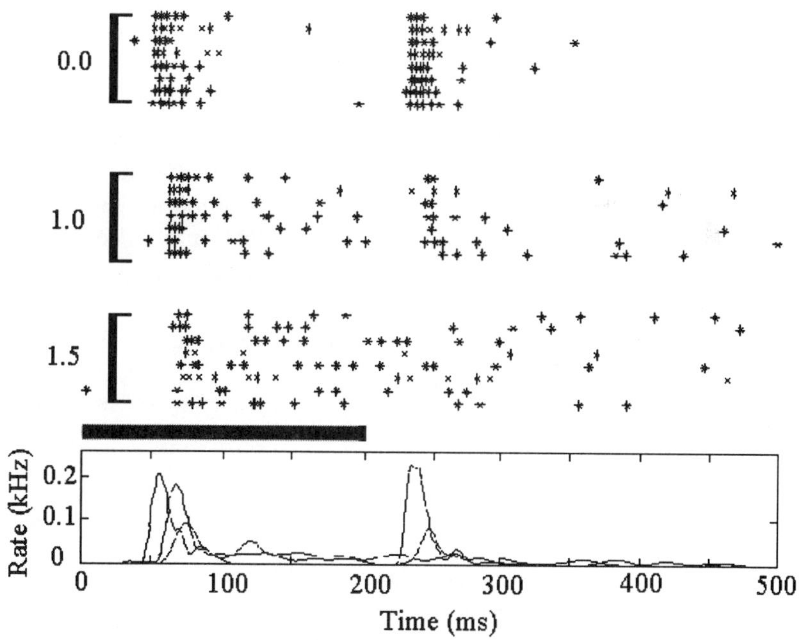

Figure 3.15. Variability of ganglion cell responses to eight repeated presentations of stimuli of three different intensities. The set of responses to 1.0 intensity were evoked by flashes ten times dimmer than the responses evoked by the 0.0 stimuli, and three times brighter than those used to evoke responses with the 1.5 stimuli. The sets of raster plots have been collapsed into three smoothed Peri-Stimulus Time Histograms (1ms resolution) at the bottom of this figure. Adapted from Normann et al. (2000) with permission of Elsevier Science.

Another notable feature of the responses in Fig. 3.15 is the nonlinear nature of the processing that allows both On and Off stimuli to generate responses. These nonlinearites have encouraged the use of blind, white noise model building methods (Naka and Sakai, 1991), as well as a variety of other linear-nonlinear (LN) cascade models. An excellent review of many of these quantitative methods was recently published by Meister and Berry (1999).

3.5 GANGLION CELL ENSEMBLES

Each ganglion cell in the retina provides visual information for only a small portion of the visual field, and no two ganglion cells code exactly the same portion of the visual field in exactly the same way. Each of these ganglion cells is also somewhat noisy in that the same stimulus in its receptive field can evoke different responses. The responses are also somewhat ambiguous in that many different stimuli can evoke similar responses. Consequently, the collective signaling and interpretation of all of the ganglion cells in the retina is necessary for an entire visual scene to be accurately perceived. Unfortunately, the precise mechanisms by which retinal ganglion cells and higher brain centers cooperate to effectively accomplish this task are not well understood. The central question to this understanding is why ganglion cells divide the labor of representing the visual field in the way that they do.

Some researchers have approached this question from the top-down direction by theoretically exploring effective strategies for vision, and then comparing these strategies to observations in living systems. For example, one can build vision-coding models under certain coding sparseness constraints that have responses similar to cells in the vision system (Olshausen, 1996; Bell, 1997; and Haft, 1998). Moreover, the types of training rules required for these models are often available in biological systems (Hyvarinen and Oja, 2000). Historically, however, the neuroscience community has largely opted for the bottom-up approach of studying living retinal systems and building empirical models to describe the observed transfer functions of various ganglion cell types. The assumption is that the rationale behind these models will become apparent as more of them are known. The contributions of both of these approaches, however, will likely be required in order to fully understand the coding mechanisms by the retina.

Before useful multi-electrode technologies were applied to the retina, researchers were limited in the number of cells from which they could simultaneously record. This problem is also faced by many researchers studying information coding in other neural areas. One standard approach that arose from these limitations was to re-play stimulation sequences, and individually record the responses of different neurons. Conversely, one can also simulate an array of neurons in the retina by recording from a single ganglion cell, while scanning a single stimulus over the retina (Jacobs and Werblin, 1998). An example of this technique is shown in Fig. 3.16 where a single ganglion cell's response to the presentation of a flashed square was recorded for many locations of the flashed square.

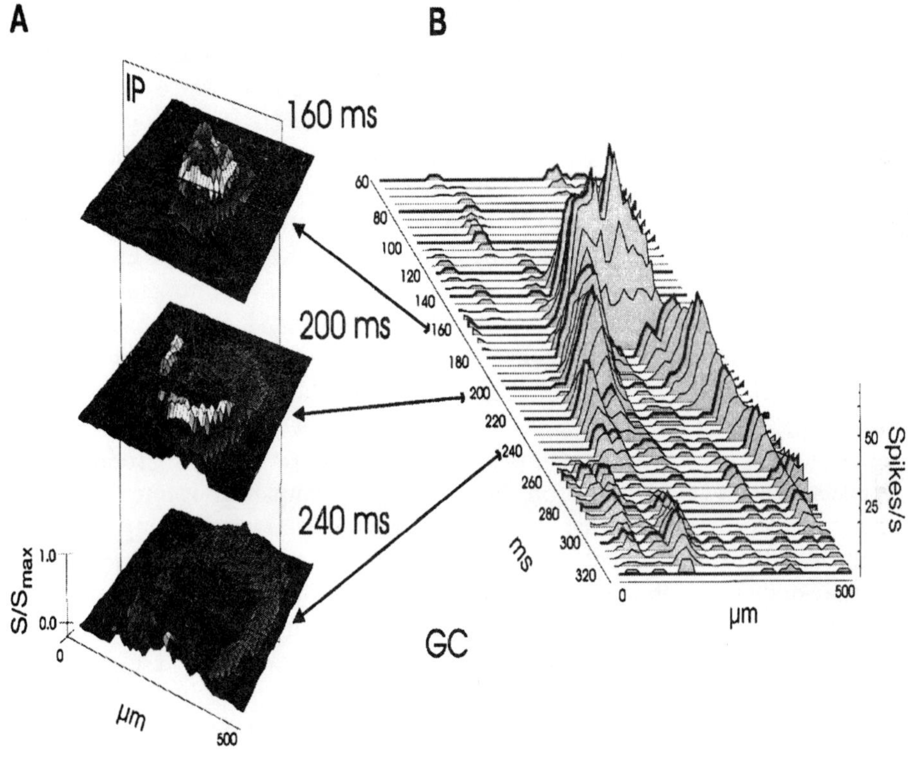

Figure 3.16. Spike activity of a simulated array of ganglion created with a single square stimulus flash that was scanned over a single cell in the salamander retina. (A) 2-D plots of the activity at 160ms, 200ms, and 240ms following stimulation (B) 1-D slice of activity across the center of the simulated cell array. Modified from Jacobs and Werblin (1998) with permission of The Physiological Society.

Assuming that there is a field of identical, spatially-distributed cells with similar properties to the recorded cell, this method can be used to approximate the ensemble response of these cells to a single flashed square. The technique also allows individual ganglion cell responses to be visualized in a more intuitive form for the understanding of how groups of ganglion cells could process information. For example, it is well known that the latency of a ganglion cell response increases as visual stimuli are displaced from the receptive field center (Shapely and Perry, 1986). Indeed, activity plots of simulated ganglion cell ensembles provide an elegant demonstration that this effect leads to waves of activity that propagate away from areas of visual stimulation. Similar types of moving waves along the retina were later reported by Berry et al. (1999) along with speculation of the role of these waves in motion processing.

The two main assumptions of single cell techniques are that the responses are somewhat repeatable, and that cells can be characterized independently of the activity of surrounding cells. It is also possible, however, that the retina utilizes synchronous or other types of synergistic coding strategies between groups of ganglion cell spikes. The potential utility of these types of codes in neural systems has been explored by a variety of neuroscientists (Abeles, 1991; Shadlen and Newsome, 1994; and Softky, 1995). To investigate these types of coding schemes in the retina, simultaneous recordings from many ganglion cells is necessary. Fortunately, recent innovations in microelectrode array technologies and multi-channel data acquisition systems (Meister et al., 1994; Nicolelis et al., 1997; Guillory and Normann, 1999; and Normann et al., 2000) have begun to make these types of experiments possible.

An early model for synchrony codes in the retina was proposed by Meister et al. (1995) based on data from the salamander retina. In their experiments, they found that synchronous spikes from neighboring ganglion cells signaled the presence of stimuli that were not simply the sum of the stimuli signaled by the two individual cells. Instead, the cells appeared to be responding to light in a separate receptive field and communicating this information through synchrony. Data from their experiments are shown in Fig. 3.17.

Meister (1996) also presented a convincing explanation for the potential utility of this signaling mechanism in the salamander, including improved data capacity and reduced size for the optic nerve. The results of this work provided evidence of synchrony codes in even the initial levels of vision processing. However, Masland (1996) noted that this phenomenon might be specific to certain species and not a generalized feature of vertebrate retinas. Chichilnisky and Baylor (1999) have also recently presented work indicating that these synchrony codes do not appear in primates. Regardless, these experiments demonstrate that, in at least some species, individual ganglion cells might not act as independent coders of visual information, but might interact with each other in higher-order ways that encode additional information. Complete models of retinal function in these species would need to include these interactions.

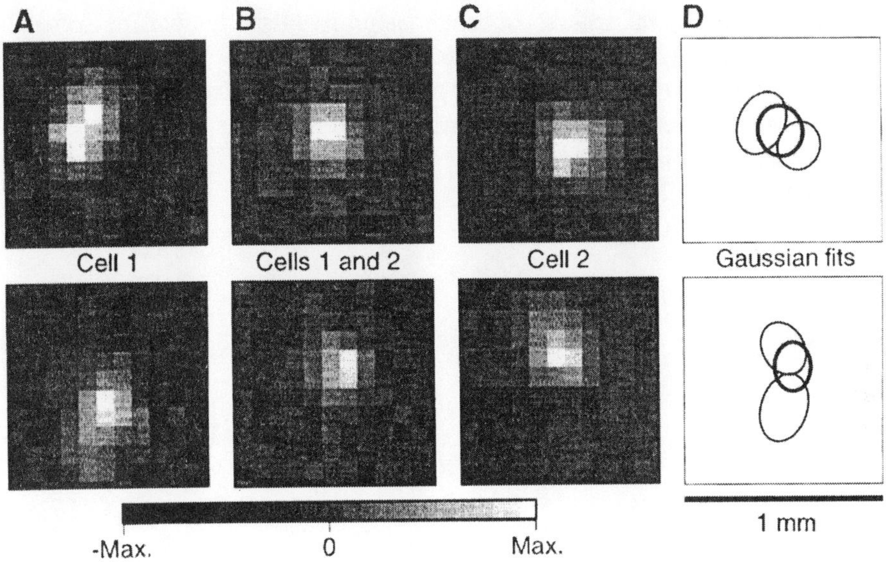

Figure 3.17. Receptive fields generated by reverse correlation between the stimulus and spike times for 4 cells in the retina (A and C), receptive fields from synchronous spikes between them (B) and Gaussian fits to these receptive fields. Reprinted from Meister et al. (1995) with permission of Science. If the cells operated independently, the synchronous spikes should reverse correlate to the sum of the two receptive fields, not the intersection.

3.5.1 Decoding the Retinal Output

Armed with experimental data, one can begin to make models that produce appropriate spike patterns in response to light stimuli. However, it is difficult to assess the accuracy of these models because it is not clear which features of spike trains are relevant to the subsequent visual pathways. It is, therefore, somewhat attractive to investigate methods for decoding the spike trains and reconstructing the visual stimulus. In this case, algorithm performance can be assessed by comparing the original stimulus image and the reconstructed stimulus. The comparison of time varying signals and images is somewhat more natural and intuitive for humans, and there are a variety of mathematical and information metrics that can be used.

One of the more straightforward approaches for decoding multi-neuron spike trains from sensory systems was the optimal linear estimation model proposed by Bialek et al. (1991) and Warland et al. (1996). In this method, a linear contribution to the reconstruction is computed for the spikes of each neuron in the data record in response to a training stimulation sequence.

These contributions are normalized versions of the average stimulus waveform preceding spikes of each neuron (also known as reverse correlation). The normalizations are based upon the histogram of times between spikes in the cell (also known as the inter-spike interval histogram) and the response characteristics of the other cells in the population. This technique was applied to the retina by Warland et al. (1997), and the method and results for 2 cells are shown in Fig. 3.18.

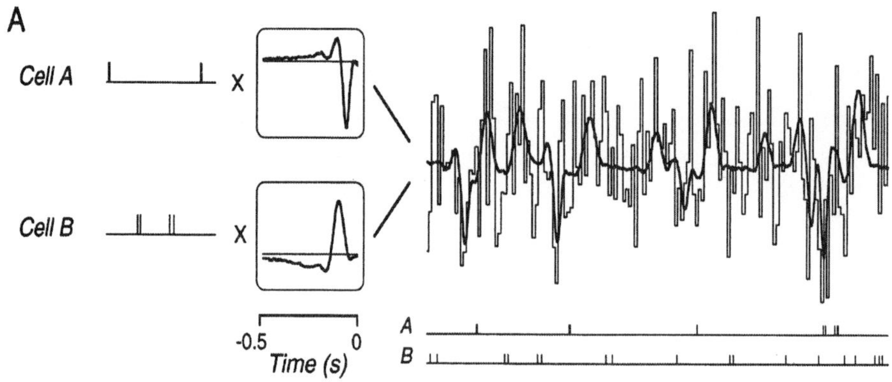

Figure 3.18. Optimal Linear Estimator method and results for estimating the intensity of a full-field, uniform, randomly-modulated visual stimulus with an 'Off' cell (A) and an 'On' cell (B). Reprinted from Warland et al. (1997) with permission of The Physiological Society.

One can also use blind, non-linear estimation methods such as perceptron networks for estimation of stimuli based upon the retinal response (Warland et al., 1997; and Normann et al., 2000). Although these methods can be used to determine minimum limits on information present in the spike trains of the cells, it is often difficult to extract the forward-coding models of the ganglion cells that are used by these methods.

One approach that is gaining popularity in the neuroscience community is the application of stochastic modeling and decoding for neural systems (Reike et al., 1997). In this approach, neurons are viewed as modulating their probability of generating spikes according to a stimulus or represented value, but the spike generation mechanisms are assumed to be inherently random. A stochastic model of retinal ganglion cell coding, for example, would respond to repeated stimulus presentations with a constant distribution for the evoked spikes, but each trial would contain random variations within the distribution. These types of models have been applied to the auditory system (Johnson, 1996), motor cortex (Sanger, 1996), and hippocampal place cells (Brown et al., 1998).

One advantage of stochastic modeling of neural encoding is that the encoding models can be inverted with Baye's rule (Reike, et al., 1997) to determine the probabilities of different stimuli given an observed neural response. In these approaches, a neural response does not precisely specify the stimulus in the receptive field of the cell, but it narrows the space of possible stimuli that could be present. By combining information from additional cells and adding prior knowledge about what stimuli are likely, the likelihoods for possible stimuli are further limited. This stimulus likelihood information can then be used for decision processes, or the most likely stimulus can be chosen for maximum likelihood estimation.

The challenge is to build appropriate stochastic descriptions of the spike activity of the cells. One of the commonly employed models is the inhomogeneous Poisson process in which the probability of generating a spike is completely specified by an instantaneous rate value that can be modulated in time. In neurons of sensory systems, this rate is usually a nonlinear function of the sensory input with modulations for sensory adaptation. The Poisson model also assumes that the spikes from a particular neuron are generated independently of each other. Real neurons, however, also include refractory processes that limit their maximum firing rates. Interestingly, these limitations can be shown to increase the reliability of retinal responses to continuous stimuli during periods of high firing rates (Berry and Meister, 1998).

To evaluate the importance of the spike patterns within discrete stimulus presentations, Normann et al. (2000) have conducted experiments in the turtle retina where they have simultaneously recorded responses from 15 ganglion cells to random presentations of 8 different colored stimuli. The spike responses were quantified by 1) rate counts across the entire 500ms trials, 2) separate rate counts for the 200 ms 'On' and 300 ms 'Off' responses, and 3) inhomogeneous Poisson rate functions that could vary on 1 ms resolution within the 500 ms trial. In the first two cases, probabilities for the different counts for each cell and stimulus were compiled across 8000 stimulus trials. In the third case, Poisson rate functions were estimated for each cell and stimulus by smoothing the PSTH of the responses. With these stochastic models, maximum likelihood analysis was performed to estimate the stimulus based on the neural response for test data sets. Performance of the three different techniques are shown in Fig. 3.19.

As one would expect, the accuracy of the estimation process increases with the number of ganglion cells used for estimation. The best performance for this stimulus class came from the Poisson model and likelihood methods that reached an accuracy of 75% correct with 10 cells (as opposed to chance performance of 12.5%). This work suggests that information about the visual field is encoded within the temporal structure of the response for flashed stimuli in addition to information gained from the total number of spikes across a trial or saccade.

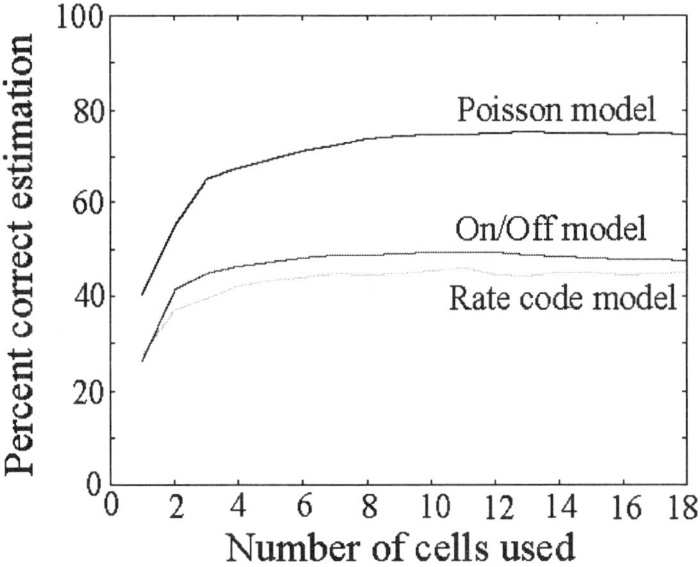

Figure 3.19. Comparison of estimation performance for estimators built on three different stochastic measures of the retinal response. The 'Rate code' model used the spike counts for each cell across the entire stimulation trial. The 'On/Off code' model used separate counts for the 'On' and 'Off' portions of the stimulus. The 'Poisson' model used a smoothed model of the PSTH as an inhomogeneous Poisson rate function (sampled at 1 ms). Estimation was performed by likelihood analysis that the observed spike train came from the rate function for each stimulus. The results suggest that information about the stimulus is contained within the fine temporal structure of the response. Adapted from Normann et al. (2000) with permission of Elsevier Science.

3.6 SUMMARY

In this chapter, we have discussed some of the anatomical and physiological features of the retina, along with a brief overview of some of the ensemble techniques that have been applied to study retinal coding. There are many challenges remaining for modeling the neural processing of the retina. One aspect that should not be overlooked in future efforts should be the correlation between model functions and physiological and anatomical features of real living systems. In addition to models of the light-adapted retina, future work should also include better models of adaptation mechanisms. These efforts will provide better understandings of the physiology of the retina, better models for machine vision and neuroprosthetic applications (Humayun et al., 1995; Normann et al., 1996; Wyatt and Rizzo, 1996; Eckmiller, 1997; and Zrenner et al., 1997), and better insight into the strategies employed by the higher visual centers for processing the retinal output.

3.7 REFERENCES

Abeles, M. 1991. *Corticonics*. Cambridge University Press, Cambridge, MA.

Baylor, D. A., Lamb, T. D. Yau, K.W., 1979, The membrane current of single rod outer segments, *J. Physiol.* **288**: 589-611.

Bell, A. J. and Sejnowski, T. J., 1997, The independent components of natural scenes are edge filters, *Vis. Res.* **37**: 3327-3338.

Berry, M. J., Brivanlou, I. H., Jordan, T.A. and Meister, M., 1999, Anticipation of moving stimuli by the retina, *Nature.* **398**: 334-338.

Berry, M. J. and Meister, M., 1998, Refractoriness and neural precision, *J. Neurosci.* **186**: 2200-2211.

Bialek, W., Rieke, F., de Ruyter van Steveninck, R.R., Warland, D., 1991, Reading a neural code, *Science.* **252**: 1854-1857.

Brown, E. N., Frank, L. M., Tang, D., Quirk, M.C. and Wilson, M.A., 1998, A statistical paradigm for neural spike train decoding applied to position prediction from ensemble firing patterns of rat hippocampal place cells, *J. Neurosci.* **18**: 7411-7425.

Cajal, S. R., 1972, *The Structure of the Retina*, Thomas, Springfield Il,

Chichilnisky, E. and Baylor, D., 1999, Synchronized firing by ganglion cells in monkey retina, *Society for Neuroscience* annual meeting, Miami, FL.

Dacey, D. M. and Lee, B. B., 1994, The 'blue-on' opponent pathway in primate retina originates from a distinct bistratified ganglion cell type, *Nature.* **367**: 731-735.

Daly, S. J. and Normann, R. A., 1985, Temporal information processing in cones: effects of light adaptation on temporal summation and modulation, *Vis. Res.* **25**: 1197-1206.

Dowling, J. E., 1987, *The Retina: An Approachable Part of the Brain*. Harvard University Press, Cambridge, MA.

Eckmiller, R., 1997, Learning retina implants with epiretinal contacts, *Ophthalmic Res.* **29**: 281-289.

Enroth-Cugell, C. and Robson, J. G., 1966, The contrast sensitivity of retinal ganglion cells of the cat, *J. Physiol.* **187**: 517-552.

Gouras, P., 1971, The function of the midget cell system in primate color vision, *Vis. Res.* **Suppl 3**: 397-410.

Guillory, K. S. and Normann, R. A., 1999, A 100-Channel System for Real Time Detection and Storage of Extracellular Spike Waveforms, *J. Neurosci. Meth.* **91**: 21-29.

Haft, M. and v. Hemmen, J. L.,1998, Theory and implementation of infomax filters for the retina, *Network.* **9**: 39-71.

Hartline, H. K., 1940, The nerve messages in the fibers of the visual pathway, *J. Opt. Soc. Am.* **30**: 239-247.

Hecht, S, Schlaer, S, and Pirenne, M.H., 1942, Energy, quanta and vision, *J. Gen. Physiol.* **25**, 819-840.

Hubel, D., 1988, *Eye, Brain and Vision*, W. H. Freeman and Co., New York, NY.

Humayun, M., Sato, Y., Propst, R., and de Juan, E., Jr.. 1995, Can potentials from the visual cortex be elicited electrically despite severe retinal degeneration and a markedly reduced electroretinogram? *Gen. J. Ophthalmol.* **4**: 57-64.

Hyvarinen, A. and Oja, E., 2000, Independent component analysis: algorithms and applications, *Neural Netw.* **13**: 411-430.

Jacobs, A. L. and Werblin, F.S., 1998, Spatiotemporal patterns at the retinal output. *J. Neurophysiol.* **80**: 447-451.

Johnson, D. H., 1996, Point Process Models of Single-Neuron Discharges, *J. Computational Neuroscience.* **3**: 275-299.

Kolb, H., 1970, Organization of the outer plexiform layer of the primate retina: electron microscopy of Golgi-impregnated cells, *Phil. Trans. Roy. Soc. B.* **258**: 261-283.

Kolb, H., 1994, The architecture of functional neural circuits in the vertebrate retina. The Proctor Lecture, *Invest. Ophthalmol. Vis. Sci.* **35**: 2385-2404.

Kolb, H. and Lipetz, L., 1991, The anatomical basis for colour vision in the vertebrate retina, in: *The Perception of Colour,* P. Gouras, ed., Macmillan Press Ltd., London, pp. 128-145.

Kolb, H., Nelson, R. Fernandezs, E., 2000, Webvision, 'www.webvision.med.utah.edu'.

Lettvin, J.Y., Maturana, H.R. McCulloch, W.S., and Pitts, W.H., 1959, What the frog's eye tells the frog's brain. *Proc. Insts. Radio Engineers.* **47**: 1940-1951.

Masland, R. H., 1996, Processing and encoding of visual information in the retina, *Curr. Opin. Neurobiol.* **6**: 467-474.

Meister, M., 1996, Multineuronal codes in retinal signaling, *Proc. Natl. Acad. Sci. USA.* **93**: 609-614.

Meister, M. and Berry, M. J., 1999, The neural code of the retina, *Neuron.* **22**: 435-450.

Meister, M., Lagnado, L. and Baylor, D.A., 1995, Concerted signaling by retinal ganglion cells, *Science.* **270**: 1207-1210.

Meister, M., Pine, J. and Baylor, D.A., 1994, Multi-neuronal signals from the retina: acquisition and analysis, *J. Neurosci. Meth.* **51**: 95-106.

Naka, K. and Sakai, H. M., 1991, The messages in optic nerve fibers and their interpretation, *Brain Res. Rev.* **16**: 135-149.

Nicolelis, M. A., Ghazanfar, A. A., Faggin, B.M., Votaw, S. and Olivera, L.M., 1997, Reconstructing the Engram: Simultaneous, Multisite, Many Single Neuron Recordings, *Neuron.* **18**: 529-537.

Normann, R. A., Maynard, E. M. Guillory, K.S. and Warren, D.J., 1996, Cortical Implants for the Blind, *IEEE Spectrum.* **May**: 54-59.

Normann, R. A. and Perlman, I., 1979, The effects of background illumination on the photoresponses of red and green cones, *J. Physiol.* **286**: 491-507.

Normann, R. A., Warren, D. J., Ammermuller, J., Fernandez, E. and Guillory, K.S., 2000, High-Resolution Spatio-Temporal Mapping Of Visual Pathways Using Multi-Electrode Arrays, *Vis. Res.* **41,** in press.

Olshausen, B. A. and Field, D. J., 1996, Emergence of simple-cell receptive field properties by learning a sparse code for natural images, *Nature.* **381** 6583: 607-609.

Polans, A., Baehr, W. and Palczewski, K., 1996, Turned on by calcium! The physiology and pathology of calcium binding proteins in the retina, *Trends in Neuroscience.* **19**, 547-554.

Polyak, S., 1941, *The Retina,* Univ. of Chicago Press, Chicago, IL.

Reike, F., Warland, D. deRuyter vonSteveninck, R. and Bialek, W., 1997, *Spikes: Exploring the Neural Code*, The MIT Press, Cambridge, MA.

Rodieck, R. W., 1998, *The First Steps in Seeing*, Sunderland, MA, Sinauer Associates.

Sanger, T. D., 1996, Probability density estimation for the interpretation of neural population codes, *J. Neurophysiol.* **76**: 2790-2793.

Shadlen, M. N. and Newsome, W. T., 1994, Noise, neural codes and cortical organization, *Curr. Opin. Neurobiol.* **4**: 569-579.

Shapley, R. and Perry, V. H., 1986, Cat and monkey retinal ganglion cells and their visual functional roles, *Trends in Neurosciences.* **9**: 229-235.

Softky, W. R., 1995, Simple codes versus efficient codes, Curr. Opin. Neurobiol. **5**: 239-247.

Toyoda, J. I., Kujiraoka, T. and Fujimoto, M., 1982, The opponent color process and interaction of horizontal cells, in: *The S-Potential*, B. D. Drujan and M. Laufer, eds., Alan Liss, New York, pp. 151-160.

Warland, D. K., Reinagel, P. and Meister, M., 1997, Decoding visual information from a population of retinal ganglion cells, *J. Neurophysiol.* **78**: 2336-2350.

Werblin, F. S. and Dowling, J. E., 1969, Organization of the retina of the mudpuppy, Necturus maculosus. II. Intracellular recording, *J. Neurophysiol.* **32**: 339-355.

Wyatt, J. and Rizzo, J., 1996, Ocular implants for the blind. *IEEE Spectrum*, **May**: 47-53.

Zrenner, E., Miliczek, K. D., Gabel, V. P., Graf, H. G., Guenther, E., Haemmerle, H., Hoefflinger, B., Kohler, K., Nisch, W., Schubert, M., Stett, A. and Weiss, S., 1997, The development of subretinal microphotodiodes for replacement of degenerated photoreceptors, *Ophthalmic Res.* **29**: 269-280.

Chapter 4

Intraocular Retinal Prostheses and Related Signal Processing

Dean Scribner[1], Eyal Margalit[2a], Kah-Guan Au Eong[3], James Weiland[2b], E. de Juan, Jr. [4], and Mark S. Humayun[5]

[1] U.S. Naval Research Laboratory, Code 5636, Washington, DC 20375, PH: (202) 767-2225, FX: (202) 767-9203, EM: scribner@nrl.navy.mil

[2] Johns Hopkins University, Wilmer Eye Institute, 600 N. Wolfe Street, Building B28, Baltimore, MD 21287, PH: (410) 614-3816, (410) 614-0372, FX: (410) 614-2637, (410) 614-7555, EM: (a) eymargalit@jhmi.edu; (b) jweiland@jhmi.edu

[3] Department of Ophthalmology, Tan Tock Seng Hospital, 11 Jalan Tan Tock Seng Hospital, Singapore 308433, EM: Kah_Guan_Au_Eong@notes.ttsh.gov.sg

[4] Johns Hopkins University, Wilmer Eye Institute, 600 N. Wolfe Street, 738 Maumenee, Baltimore, MD 21287, PH: (410) 502-5383, FX: (410) 614-9315, EM: edejuan@welchlink.welch.jhu.edu

[5] Johns Hopkins University, Wilmer Eye Institute, 600 N. Wolfe Street, 738 Maumenee, Baltimore, MD 21287, PH: (410) 955-5108, FX: (410) 614-7555, EM: mhumayun@jhmi.edu

4.1 INTRODUCTION

For millennia, restoring sight to the blind has been viewed as being nothing less than miraculous. It has only been in the last few years that the fields of electronic microfabrication, neurophysiology, and retinal surgery have advanced to the point where an implantable visual prosthesis system, based on electrical stimulation, is considered feasible.

This was preceded by efforts to develop implantable bioelectronic devices for other types of neural prostheses, for example, thousands of cochlear prostheses for deaf patients have been successfully implanted (Agnew and McCrerry, 1990; Heiduschka and Thanos, 1998). Although there are a number of approaches to visual prosthesis, this chapter focuses primarily on the development of an intraocular electronic stimulator array. Many technical issues need to be resolved before successful implants become practical, not the least of which is the necessary signal processing of images before electrical stimulation of the retina.

The first section of this chapter, Section 4.1, gives a brief overview of the retina and presents the intraocular retinal prosthesis (IRP) concept. Section 4.2 gives a brief history of visual prosthesis. Signal processing for retinal prosthesis is discussed in Section 4.3, followed by neurophysiology and interface issues in Section 4.4. Current efforts to design and fabricate an advanced IRP device are described in Section 4.5.

4.1.1 Neuroanatomy of the Retina

The retina is the innermost layer of the eye. It is basically composed of two layers, the outer retinal pigment epithelium (RPE) and the inner neural (sensory) retina (Figs. 4.1, 4.2). The sensory retina is a delicate sheet of transparent tissue varying in thickness from 0.4-0.15 mm). The anatomical site for detailed fine vision, called the fovea, is in the center of the macula. The outermost layer of the sensory retina consists of photoreceptors; in the

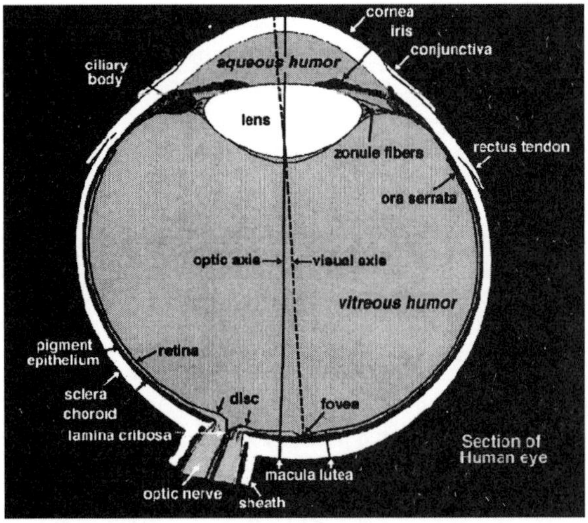

Figure 4.1. Sagittal section of adult human eye. Adapted from Ogden (1989) with permission of Harcourt Health Sciences, as modified by Kolb (2001).

macular region, the photoreceptors are mostly cones (color-sensitive). Other more inner layers of the sensory retina are the inner nuclear layer with bipolar, amacrine, and horizontal cells, and the ganglion cell layer. The axons of the ganglion cells form the optic nerve after traversing the nerve fiber layer.

Photoreceptor loss from diseases such as retinitis pigmentosa (RP) and age-related macular degeneration (AMD) are the leading cause of legal blindness. Despite near-total loss of photoreceptors, there is relative preservation of the other retinal neurons. By stimulating the remaining functional retinal layers, it may be possible to restore visual perception. In other diseases, this approach may not be practical. For example, in glaucoma (high intraocular pressure with optic nerve damage), the ganglion cells are primarily damaged. In diseases such as retinopathy of prematurity, diabetic retinopathy, and vascular diseases of the retina, all the layers are affected. In these diseases, it is highly unlikely that electrical stimulation of the retina can restore visual function, and other approaches such as retinal transplantation or electrical stimulation of the visual cortex should be investigated (see Section 4.2).

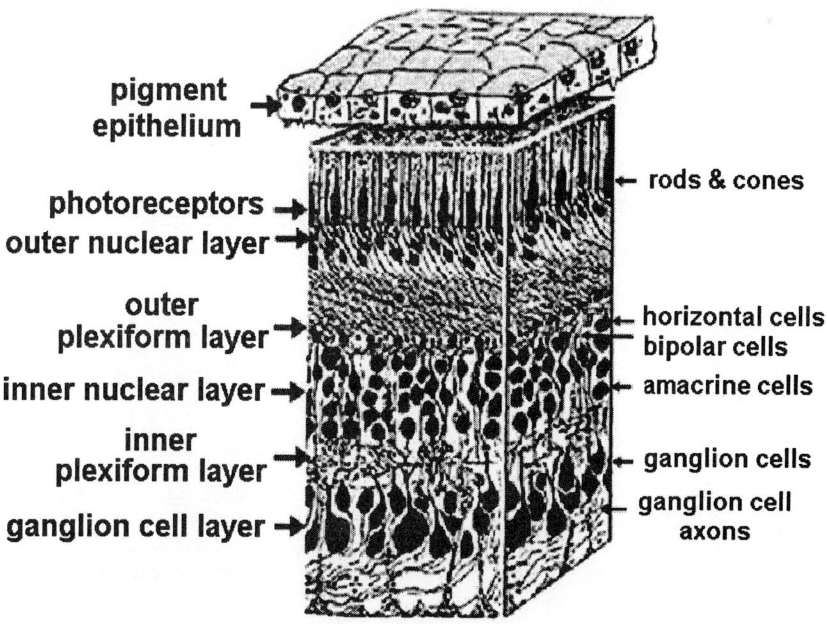

Figure 4.2. Three-dimensional section of human retina. Adapted from Polyak, (1941) with permission of The University of Chicago Press, as modified by Kolb (2001).

4.1.2 Intraocular Retinal Prosthesis

The concept of an IRP for blind patients has been hypothesized by a number of researchers and is currently an active area of medical research (Humayun, 1994; Humayun et. al., 1996). The basic concept is straightforward: visual images can be produced in the brain by electrical stimulation of retinal cells. A layer of retinal cells, such as a ganglion cell layer, can be stimulated by using an adjacent array that inputs electrical impulses to create the perception of an image. The axons of the stimulated ganglion cells then transmit the image through the optic nerve to cells in the visual cortex. This is in place of the normal photo-transduction process that occurs in a healthy retina. In a large percentage of blind patients, the photoreceptors are diseased, but the other retinal layers are still responsive to electrical stimulation (de Juan et al., 1989).

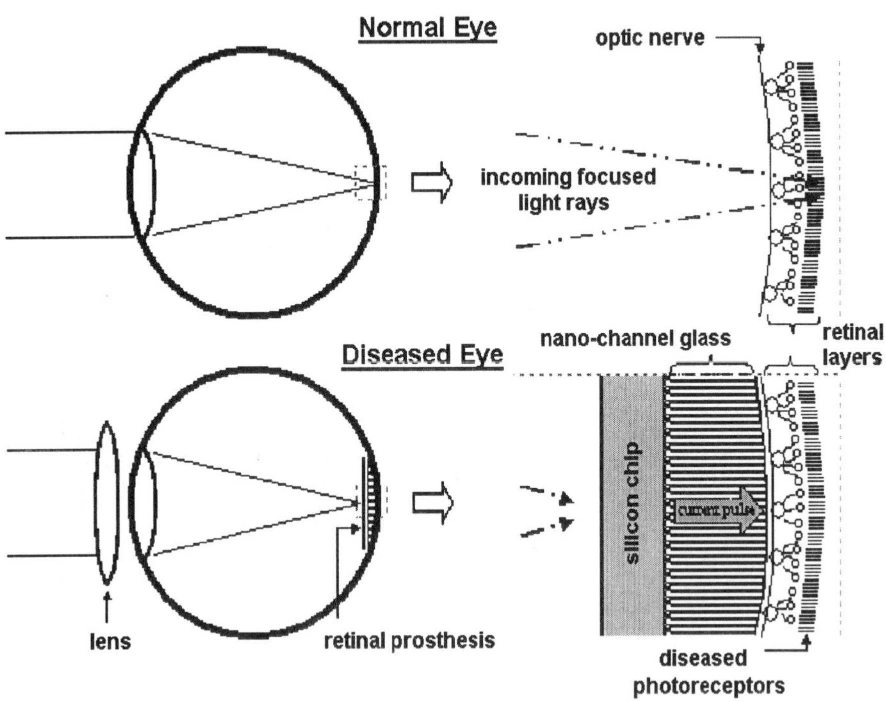

Figure 4.3. Basic concept for a retinal prosthesis devices. A ray trace of photons incident on a retina without a prosthesis device is shown in top half of the figure. Incoming photons pass through several layers of clear retinal cells before being absorbed by the photoreceptors. In the bottom half of the figure, a retinal prosthesis device is shown positioned against the retina.

One concept for a high-resolution retinal prosthesis device is shown in Fig. 4.3. A ray trace of photons incident on a retina without a prosthesis device is shown in top half of Fig. 4.3. Note that the incoming photons pass through several layers of transparent retinal cells before being absorbed by the photoreceptors. In the bottom half of Fig. 4.3, a retinal prosthesis device is shown positioned against the retina. In this case, the photons are absorbed by a microelectronic imaging array that is hybridized to a glass disk containing an imbedded array of microwires. The glass disk has one flat side, while the other side has a curved surface that conforms to the inner radius of the retina. The microelectronic imaging array is made of thin silicon containing very-large-scale integrated (VLSI) circuitry and photon detectors that convert the incident photons to electronic charge. The charge is then converted to a proportional amount of electronic current that is input into the retinal cells. The cells fire and signal is transmitted on through the optic nerve.

A number of technical issues must be addressed in designing and fabricating a retinal prosthesis device that will generate a high-resolution image. First, there is the issue of creating an electrical interface between the high-density electrode array and the curved surface of the retina. The electrode array must have a spherical, convexed shape to conform to the spherical concaved surface of the retina. The electrode array must be biocompatible and safe for permanent implantation. Second, the electrical stimulation pulse shapes and repetition rates need to be determined in general and may need to be optimized for each patient. Third, direct electrical stimulation of the ganglion cells precludes certain image processing functions that normally would have occurred in earlier layers of the retina. Therefore, image preprocessing operations may need to be performed on the image before stimulation of the retina. Fourth, supplying power to a permanent implant will need to be engineered in a manner such that there are no wires or cables through the eye wall. Fifth, because a normal retina processes image information created by the photoreceptors in a simultaneous manner, it is assumed that a prosthesis device should similarly excite retinal cells in a simultaneous manner (as opposed to a sequential raster scan like that used in video displays).

All of these design issues will be discussed in the following sections. However, to put the discussion in its proper context, the next section gives a brief historical overview of visual prosthesis.

4.2 HISTORY OF VISUAL PROSTHESIS

During the 18th century, scientists began to understand that electricity could stimulate biological tissues. Galvani showed that electrical stimulation could cause contraction in muscle preparation (Galvani, 1791). The electrical excitability of the cerebral cortex was demonstrated in a dog by Fritsch and Hitzig (1870). They were able to use this finding to localize electrophysiological functions of the brain. Glenn and Mauro (1959) developed a totally implanted heart pacemaker using radio-frequency waves to transfer information. This technological breakthrough overcame the problem of stimulating deep neural structures without the danger of infection that can accompany percutaneous leads. Electrical stimulation of the acoustic nerve in a totally deaf human by direct application of an electrode in the inner ear was reported by Djourno and Eyries (1957).

Today, a number of research projects around the world are aimed at developing prosthetic vision systems. The various approaches can be categorized most simply by where the actual stimulation occurs. The device discussed in Section 4.1 addresses the technical problem of positioning a high-density electrode array against the retina to achieve very high-resolution imagery. Other efforts are proceeding in the United States, Germany, and Japan that build on the basic idea of stimulating retinal cells with a small number of electrodes on a microelectronic chip.

In the past, another approach has been to skip the retina altogether and stimulate the visual cortex of the brain. In this approach, an array, with penetrating microelectrodes is positioned against the visual cortex. This involves invasive brain surgery through the cranium. Both of these approaches are discussed individually in the two subsections below.

4.2.1 Cortical Prosthesis

A visual prosthesis based on electrical stimulation had its origins in the studies of a German neurosurgeon, Foerester (1929). He was the first to expose and electrically stimulate the human occipital pole under local anesthesia. He noted that stimulation caused his subject to see a spot of light (phosphene) and that the position of the light was dependent on where the stimulation probe was placed. Krause and Schum (1931) noted similar results in a blind patient. These results were later confirmed by others (Penfield and Rasussen, 1952; Penfield and Jasper, 1954).

Brindley and Lewin (1968a; 1968b) implanted a device consisting of 80 electrodes on the visual cortex of a blind nurse. This experiment represented a large advance in the amount of information that could be transferred into the central nervous system. After the late sixties, it was clear that blind humans can perceive electrically elicited phosphenes in response to either retinal stimulation (Potts et al., 1968; 1969; 1970) or cortical stimulation (Sato et al., 1982; Dobelle and Mladejovsky, 1974; Pollen, 1977). The artificial visual system through direct stimulation of the visual cortex was reviewed in 1975 (Karny, 1975).

Since then, attempts to develop visual prostheses have taken different approaches. The main two are a visual cortical prosthesis that will allow bypassing all malfunctioning visual pathways except diseased primary and secondary visual cortical centers. The other approach is to implant a retinal prosthesis to bypass only diseased photoreceptors in retinitis pigmentosa (RP) or age-related macular degeneration (AMD) patients and stimulate healthy inner retinal layers that include viable ganglion and bipolar cells.

Currently there are two concepts to cortical electrical stimulation. The first is surface electrical stimulation. Early studies used large surface electrodes for cortical stimulation (Dobelle, 1974; Polen, 1977; Brindley and Lewin, 1968a, 1968b). The results indicated that in spite of the possibility of inducing phosphenes, there were many difficulties to overcome: high currents were used to induce phosphenes, interactions between phosphenes were noted when electrodes were placed less than 2.4 mm apart. The same stimulating electrode induced multiple phosphenes, and the phosphenes were not consistently produced. Pain was induced occasionally due to meningeal stimulation. There were other limiting factors such as the question of two-point discrimination or resolution, local heating, and electrode electrolysis (Karny, 1975). Phosphenes persisted following cessation of electrical stimulation (Bak et al., 1990), and most studies (Dobelle and Mladejovsky, 1974; Brindley and Lewin, 1968; Dobelle et al., 1976) reported flickering during surface stimulation.

The persistence of phosphenes following cessation of the stimulus and flicker phenomenon could possibly reflect neuronal epileptic activity (Pollen, 1977). In spite of all the difficulties, Dobelle et al. (1976) reported a chronic implantation of a 64-channel platinum disc electrode array at the occipital cortex of a blind patient. After practicing, the prosthesis allowed the patient to read "phosphene Braille" two to three times faster than tactile Braille.

The cortical stimulation system is an electronic camera (light and ultrasonic sensors) linked to a processing computer, which transmits stimulated signals to electrodes via a connecting "pedestal" that perforates the scalp. This pedestal allows the passage of wires through the skin without infection and permits easy upgrade of the electronics over the years (Dobelle, 1994).

Another cortical prosthesis approach is the intracortical microstimulation. Schmidt et al. (1995) developed a system with multiple (38) microelectrodes that were implanted in an area of 40.8 mm by 19.2 mm into the visual cortex of a human volunteer. They demonstrated that phosphenes could be evoked with currents that were in orders of magnitude (10-100 times) lower then those used with surface stimulation of the cortex, and that simple patterned perceptions could be evoked by electrical stimulation via small groups of these microelectrodes. Phosphenes could be resolved with simultaneous stimulation electrodes as close as 500 μm, and they have simple, stable, and predictable forms. The phosphenes could have the perception of color. One theory to try to explain these lower threshold colored phosphenes generated due to intracortical microstimulation suggest that patients can be starting to see the percepts encoded by single ocular dominance columns or color blobs in the absence or reduction of surround inhibition (Bak et al., 1990).

The intracortical microelectrode stimulation allows for reducing phosphene interactions, increasing the number of electrodes, improving the power requirements, and decreasing the current per microelectrode (Bak et al., 1990; Schmidt et al., 1996). Another intracortical microelectrodes stimulation approach was developed by Normann et al. at the University of Utah (Normann, 1999; Jones and Normann, 1997; Maynard et al., 1997; Nordhausen et al., 1996). The Utah electrode array has a large number of electrodes (typically 100) concentrated on a 4 × 4 mm area. Undoubtedly, the lower threshold of the intracortical microstimulation, the predictable forms of the generated phosphenes, and the absence of flicker phenomenon are the main advantages of the intracortical microstimulation approach.

Since the early surface and intracortical stimulation experiments, different electronic devices have been proposed to convey visual information to blind patients. Some convert visual information and present it as an auditory or tactile input (i.e., sensory substitution devices (Brabyn, 1982; Kaczmarek, 1998)).

There are two major advantages of the cortical stimulation approach (Normann, 1999). First, the skull is a stable stimulation site and will protect the electronics and the electrode array. Second, the approach bypasses all distal visual pathway pathologies. However, a number of disadvantages are associated with this approach. The retinotopic mapping on the cortical surface is poorly understood, so patterned stimulation may not produce patterned perception. Furthermore, it is unclear what visual perceptions will be evoked by stimulation of cortical neurons. Also, the complex topography of the cortical anatomy makes it difficult for implantation. Finally, surgical complications can lead to very severe consequences.

4.2.2 Retinal Prosthesis

Other groups are attempting to develop retinal, or optic nerve, prostheses that will cause visual perception by electrical stimulation of the healthy inner layers of the retina in patients who suffer from diseases such as retinitis pigmentosa (RP) and age-related macular degeneration (AMD). Progress in the field of neural prostheses has converged with advances in retinal surgery to enable the development of an implantable retinal prosthesis. Initial experiments with intraocular stimulation, were performed by de Juan and Humayun several years ago (de Juan and Humanun, 1994). Since that time, several research groups have begun the development of retinal prostheses (Zrenner et al., 1999; Humayun et al., 1999; Chow and Peachey, 1998; Eckmiller, 1997; Wyatt and Rizzo, 1996; Veraart et al., 1998; Yagi and Hayashida, 1999). These groups can be classified according to where their device will be positioned–on the retinal surface (epiretinal) or in the subretinal space (subretinal).

Epiretinal implantation has the advantage of leaving the retina intact by placing the implant in the vitreous cavity, a naturally existing and fluid-filled space. Studies at John Hopkins University Hospital have demonstrated that this array position is biocompatible (Majji et al., 1999). Other groups are examining this approach as well (Eckmiller, 1997; Rizzo and Wyatt, 1997). The basic concept that has been described in the past is to mount a miniature video camera (e.g., charge-coupled device - CCD) on a pair of glasses. The video signal and power of the output will be processed by a data processor, and the information will be transferred to intraocular electronics by either 820-nm wavelength laser (Rizzo and Wyatt, 1997) or by radio frequency transmission from an external metal coil to an intraocular coil (Troyk and Schwan, 1992; Heetderks, 1988). The power and data transmitted from the laser or the coil is converted to electrical current on a stimulating chip that then controls the distribution of current to the epiretinal electrode array. Section 4.5 discusses a means of naturally imaging light onto an epiretinal prosthesis.

Subretinal implantation of a retinal prosthesis is being developed by Zrenner (Zrenner et al., 1999; Guenther et al., 1999) and Chow (Chow and Peachey, 1998; Chow and Chow, 1997; Peyman et al., 1998) . This approach essentially replaces the diseased photoreceptors with a microelectronic stimulator device. However, the surgical implantation requires detaching the retina, and the location of the device may be disruptive to the health of the retina (Zrenner et al., 1999). The histology of the retina after chronic device implantation showed declining of inner nuclear and ganglion cell layer densities (Peyman and Chow, 1998). The nourishment of the outer retinal layers of the retina comes from the choroid. For this reason, Zrenner's group implemented nutrition openings in each unit of their device. These issues are being examined in recently announced phase I clinical trials of a subretinal implantation by Chow and colleagues in Chicago. A disadvantage of this approach is that it is not applicable to patients with AMD, because the retina is no longer transparent.

Another retinal prosthesis approach was proposed by Yagi from the Kyushu Institute of Technology, Japan (Yagi and Hayashida, 1999; Yagi and Watanabe, 1998). He proposed to develop a device called the "hybrid retinal implant." This device will be an integrated circuit and will include both electronic and cellular components. The neurons on the device will extend their axons to the central nervous system (CNS) and thus create a natural device-CNS interface.

Veraart chronically implanted a self-sizing spiral cuff electrode with four contacts around the optic nerve of a 59-year old blind RP patient (Veraart et al., 1998). Charge-balanced electrical stimuli applied to the nerve produced localized, often colored, visual sensations that were broadly distributed throughout the visual field and could be varied by changing the stimulating conditions (pulse duration, pulse train frequency), even when using the same contact on the cuff electrodes.

The epiretinal and subretinal approaches have several advantages over the cortical approach discussed above. They both have the ability to use existing physiological optics of the eye, less severe consequences in case of infection, obvious spatial mapping or retinotopic organization, and natural processing of the electrically stimulated images along the proximal visual pathways. However, the retina encodes many properties of the image properties that are passed on to the higher visual centers (color, intensity, motion, etc.). Therefore, it may be necessary to integrate some image processing functions into a retinal prosthesis device. This issue is the subject of the next section.

4.3 SIGNAL PROCESSING FOR IRP

The development of successful signal processing methods for an IRP device should logically begin with a complete understanding of the visual system. This would include the visual pathways, of which there are several, that branch apart in the retina and proceed through the lateral geniculate nucleus and then to the visual cortex. To fully understand these pathways, one must learn the hierarchy of image processing functions performed as information proceeds from the photoreceptors through the retina and into the higher processing levels of the brain.

Unfortunately, a complete understanding of the visual pathways and their image processing functions does not exist at this time. Nevertheless, many interesting research results can be used collectively to construct a general representation of information flow and processing in the human visual system.

Knowing where along the pathway an IRP device stimulates neural cells, one could then assume what processing functions need to be performed on the incoming image. For example, assume an IRP device stimulates bipolar cells, then it may not be desirable to stimulate these cells with the raw image. Instead, the stimulating image should be like that processed and transmitted by the photoreceptors and horizontal cells that occur earlier in the visual pathway. Regardless of our knowledge of the visual system, it will not be easy to develop artificial stimulation for the blind that completely replicates those natural processes that occur in healthy, functioning, retina. It needs to be clear from the outset that our ability to understand this processing and substitute for it in blind subjects will be quite limited. However, as has been the case with cochlear prosthesis devices, even a less than optimal stimulation capability can be successful since the human brain is adaptable.

In an effort to define the minimum acceptable resolution for useful vision, several psychophysical experiments have been performed. As early as 1965, it was suggested that 600 channels or points of stimulation would be sufficient for reading ordinary print (10 letters per presentation and 120 presentations per minute) (Cha et al., 1992). Others suggested that 80-120 points would be adequate for easy reading, while 200 points could allow recognition of simple obstacles (Thompson et al., 2000).

More recent studies confirmed these estimated numbers (Cha et al., 1992a; 1992b; 1992c). Therefore, even with a low-resolution IRP device, the brain might be able to extract a very useful amount of information. Nevertheless, it is expected that the better the IRP resolution and the better the associated preprocessing, the more the brain will be able to perceive. In the subsection below, a brief overview is given of some the salient features of the retina and related signal processing.

4.3.1 Physiological and Pychophysical Models

A general understanding of the visual system has emerged from several different fields of study. Detailed physiological studies of the retina and later stages of the visual system reveal the anatomy, connectivity, and pharmacology of the neural cells. Major advances in this area have been made over the past few decades with the advent of patch clamp measurements. These are direct measurements of electrical activity at the individual cell level due to light incident on the retina. The actual clamp onto the cell is done with a fine glass tube containing an internal wire that allows the measurement of changes in the cell potential. Knowing the response of individual neuron layers to simple spatio-temporal stimuli provides enough basic information to construct the *receptive* field (described by a plot of the response of a retinal neuron to light in a nearby area) for each neuron layer in the retina. Knowing these relations at each layer allows the construction of retinal processing models for signals moving through the retina. The cell measurements are made using a wide variety of animal species. This is highly relevant because, in all vertebra species, the retina consists of the same five basic layers and cell types. Historically, there has been particular interest in those species with large cells that make cellular recordings the easiest. With the data obtained and the availability of modern, powerful computers, it is possible to perform detailed retinal simulations.

In contrast to single-cell recording studies, psychophysical studies tend to treat the visual system as a *black box* in which various spatio-temporal shapes and forms are presented to a human viewer, and the perceived sensations are described and analyzed. These results are then used in an inverse manner to identify the processing that was performed by the system as a whole. In essence, this is similar to the subject of *systems identification* used in electrical engineering. A special topic within the field of psychophysical studies is that of illusions, where psychophysical phenomenon provide insight into some of the salient processing features of the retina. Although the reduction of the retina to a simple set of analytical expressions would be extremely useful for our purposes, some difficulties exist. For example, retinal functions are, to a degree, nonlinear processes. Therefore, analytic solutions to the systems identification problem may require many simplifying approximations, thereby limiting the usefulness of the results.

4.3.2 Connectivity and Cellular Interaction

The retina is a highly complex biological system that has been extensively studied over the past century. Nevertheless, there are still many unknowns regarding the anatomy, pharmacology, and flow of information. The field has attracted many renowned biologists, and a full description of this topic is beyond the scope of this paper. However, as background for further discussion, consider the highly simplified cross-sectional diagram of the retina shown in Fig. 4.2. Image information is converted from photons into neural signals by the photoreceptors (either rods or three types of cones). The image information flows through the retina in a synchronous, parallel manner through as many as 10 types of bipolar cells and finally to the ganglion cells, of which there are 20 to 25 types. The signals from the ganglion cell bodies flow through their output axons, which converge to form the optic nerve that channels the signals on to higher processing levels in the brain.

The image information goes through a number of transformations before leaving the retina. Some of this processing is spatial in nature and is performed predominantly by the intermediate cell layers containing the horizontal cells (between the photoreceptor and the bipolar cells) and the amacrine cells (between the bipolar and the ganglion cells). Mathematically, one of the processing functions performed by these lateral layers of cells can generally be represented as a spatial convolution. Another form of processing that occurs is temporal in nature. Specifically, as information flows directly from layer to layer, the individual cells perform adaptive processing functions. For example, the photoreceptors adapt their response to the ambient light levels, and the bipolar cells may perform temporal filtering. Ganglion cells receive input processed by the previous layers and generate coded pulses that are transmitted through the optic nerve to the higher visual processing centers.

The construction of a simplified retinal wiring diagram on which adequate mathematical model could be based is research that is still in progress. One complication is that the retina is not a simple feed-forward network. Feedback to earlier layers is a known feature of the retina. Furthermore, as mentioned above, the signals branching into many parallel pathways make for a highly complex system.

It is a generally accepted view that the outer plexiform layer (OPL), consisting of the photoreceptors, horizontal cells, and bipolar cells, splits the visual pathway into two channels known as the ON and OFF pathways. Other pathways that originate in the OPL are color vision, based on cone outputs, and scotopic vision (night vision), based on the output of the rod photoreceptors. In the central region of the retina known as the fovea, there are no rods, only cones. However, moving outward from the central region of the retina, the situation is reversed. Beyond 10 degrees in eccentricity from the central fovea, the rod density peaks, and there are very few cones, as shown in Fig. 4.4. Overall, there are typically more than 100 million rods in the human retina and only about 6 million cones.

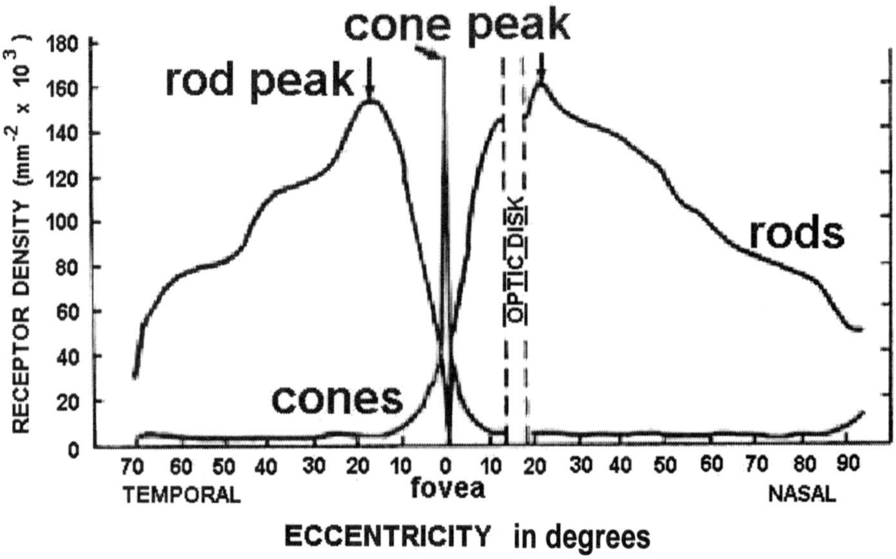

Figure 4.4. Density plot of rods and cones on the horizonal meridian of the human retinal. Adapted from Østerberg (1935) with permission of Acta Ophthalmologica Scandinavica.

The situation is much different in the inner plexiform layer (IPL), consisting of the bipolar cell outputs, the amacrine cells, and the ganglion cells, where the visual pathway appears to split into many channels. It is here that the bipolar cell outputs connect to a wide variety of amacrine and ganglion cell types. From cell-staining studies, it has been suggested that there are at least 25 different amacrine cell types in the human retina. A similar number of ganglion cells have also been identified (Kolb et al., 1992). Aside from their differences in size, shape, and dendrite spread, their synapses occur at five different strata levels within the IPL. It is fairly clear that these differences again relate to more branching of the visual pathways into the brain. These additional pathways may include preferential motion filtering.

It is believed that the various visual pathways branch apart early in the retina. Therefore, it is not surprising that there are a wide variety of retinal cell types. From the perspective of IRP design, it would be desirable to stimulate the retina at as deep a layer as possible. Or put another way, as early in the branching process as possible. One reasonable strategy is to preferentially stimulate the bipolar cell layer. Stimulation of the photoreceptors is not practical because, in retinitis pigmentosa, it is the photoreceptors that degenerate and fail. Stimulation of the ganglion cells might be thought of as easiest because they are large and closest to an IRP device, but this may cause some complications. As stated above, there are many different types of ganglion cells, indicating visual pathways that have already branched apart. One strategy is to stimulate the bipolar cells before some of this branching occurs, thus removing the need to replicate the missing signal processing that would be lost as well.

4.3.3 Retinal Image Preprocessing Functions

As mentioned above, by directly stimulating ganglion cells, certain image processing functions that normally would have occurred in earlier layers of the retina are skipped. Therefore, computationally based image preprocessing operations may need to be performed on the image before stimulation of the ganglion cell layer. Some of the more important preprocessing functions believed to occur in the retina are described below and discussed with respect to IRP designs. It should be noted that the discussions below are meant only as a simple overview. The complexity of retinal processing and the ability to stimulate the retina in a coherent manner are both areas of study that are in their formative stages.

Adaptive photoreceptor gain control: By adapting gain of the photoreceptors to varying light levels, the sensitivity of the retinal system can be self-adjusted to match the photon flux levels incident on the retina. The adjustment of gain is an adaptive process that is dependent on the very broad range of ambient light levels in which humans retain visual function. Laboratory demonstrations have shown that humans can detect a single photon in spite of inherent noise (Zuidema et al., 1983). The same retinal processing must also be capable of imaging in bright sunlight while avoiding saturation. In the latter case, the signal is so high that the gain mechanism is used to attenuate the incoming signal. By continuously adapting, the retinal response remains approximately the same to each scene in spite of changes in illumination. Thus the brain is free to interpret images without having to constantly adjust for changes in illuminant due to the solar cycle, shading, or artificial light sources.

Although gain control occurs in various forms throughout the retinal system, the greatest contribution occurs at the photoreceptor level. The useful range of luminance that photoreceptors must image (from brightest to darkest) is approximately 10^{10} to 1, or 10 orders-of-magnitude. The lower three orders of magnitude are processed by the rods. The cones handle the upper six orders of magnitude. Actually, the response of the rods and cones may overlap by one order of magnitude, but the overall range of the visual system is extended by dilation of the pupil and other processing mechanisms. The major processing required with this broad range of photon flux is for the photoreceptors to adaptively control their gain, because the next layer of retinal processing has a dynamic range of approximately 100.

To accomplish this task, the photoreceptors have a nonlinear response function that adapts to the ambient light level and performs a transform described by the Naka-Rushton equation (Eq. 4.1; Shapley and Enroth-Cugell, 1984):

$$R = R_{max}\left(\frac{I}{I + I_S}\right) \tag{4.1}$$

where R is the response of the neuron measured as the change in membrane potential from its totally dark adapted level; I is the illumination of the simulus; I_s is an adjustable parameter that causes a *curve-shifting* effect; and R_{max} is the saturation maximum. Naka-Rushton response functions are plotted in Fig. 4.5 as a family of curves. Notice that the curves are approximately logarithmic in the center of their output domain. The value of I_s is dependent on the value of I and its recent history. One beautiful property of this process is, regardless of the background illumination, after adaptation the output tends to always be centered in a common region. Restated, the curve-shifting feature defines a response that allows the gain to be adjusted to match the range of signal in a very elegant manner. Note that using fixed logarithmic response functions would allow a very large range to be covered (without saturation), but would entail the use of a response function with a very low gain over most of the operating range.

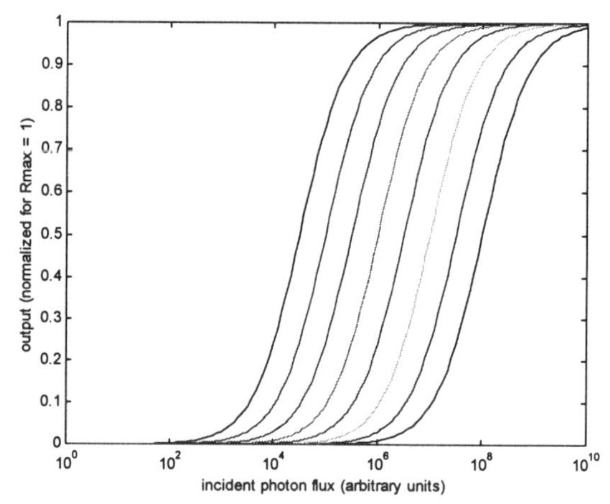

Figure 4.5. Adaptive response function of photoreceptor cells. The family of curves represents the curve shifting property as described by the Naka-Rushton equation (Eq. 4.1).

In terms of incorporating this feature on an IRP device, a number of electronic imaging arrays have addressed this problem (Delbruck and Mean, 1994). Recently, Schwarz (1999) has customized this concept specifically for retinal prosthesis applications.

Contrast control: This processing function involves an image renormalization that is thought to involve the horizontal and bipolar cells. Essentially, the horizontal cells, through local-neighbor interconnects, create a blurred image that is the result of a spatial convolution operation. The resulting blurred image is then differenced with the photoreceptor image. This operation is similar to masking techniques in image processing and photography. By comparison, these techniques are slow and cumbersome to execute in conventional electronics, whereas in the retina, the processing is very fast because of the inherent massively parallel structure. The resulting image is essentially high-pass spatial filtered. That is, most of the low spatial frequencies have been removed, leaving the high spatial frequencies. The resulting image retains edges clearly but removes the softer changes across the scene, thus increasing the contrast between sharp objects and the background. Furthermore, it reduces the overall dynamic range required to pass relevant image information on through the retina. Another form of high-pass filtering is believed to occur within the bipolar cells. In essence, the bipolar cells operate basically as an array of temporal low-pass filters, one for each channel. The process has been modelled by Curlander and Marmarellis (1983) as a two-pole filter (Eq. 4.2):

$$h_b = k_b \left(e^{-\gamma_1 t} + e^{-\gamma_2 t} \right) \tag{4.2}$$

where the filter parameters are defined to be $k_b = 1.0$, $\gamma_1 = 0.008$, and $\gamma_2 = 0.032$. Differencing the temporal low-pass filtered image with the original input image into the bipolar cells results in another high-pass filter. Although this process is a temporal operation performed in each bipolar cell, it also acts to remove low spatial frequencies from the image because of eye motion. It may also have the effect of performing an offset correction and is therefore a form of nonuniformity correction.

Color processing: The origin of color signals in the retina is at the photoreceptor layer–namely in the three cone types (nominally named red, green, and blue, with respect to their approximate spectral sensitivity). For an IRP device, this poses a major problem with respect to creating the perception of a meaningful color image; the stimulating electrodes would need to distinguish between the various ganglion (or bipolar) cells that are transmitting the different types of chromatic information. Even if the cell types could be distinguished, the actual retinal processing of color information would need to be understood. It is generally accepted that signals from the three cone types combine in an antagonistic or opponent manner (Dacey, 1996). In essence, the retina transforms the three images (form the red, green, and blue cone types) into two opponent pathways.

One is the red-green opponency pathway, in which the red cone image information is opposed to that from the green cones. The second is the blue-yellow pathway, in which the blue cone information is opposed to that from the combined image information from the red and green cones. A brightness pathway also exists that combines the image information from all three cone types. This information gives sensations as a mixture of black and white–intermediate values being shades of gray. In a near-term retinal prosthesis, the brightness pathway would be a primary target for stimulation. Stimulating color pathways will be a much more difficult challenge.

Other processing: Two other retinal functions that are not well understood but may have importance in a prosthesis design are hyperacuity and kinematic movement processing. Hyperacuity is the ability to resolve objects with an acuity better than that predicted by the optical limitation of the eye's lens or the grain of the retinal mosaic. It is typically measured by testing the human ability to align a vernier. Hyperacuity does not violate any laws of physics or information theory (such as the Nyquist theorem). It is believed to be achieved by a combination of eye movement and sampling over time. The actual processing starts in the retina but results are accumulated in the visual cortex. A complete discussion of hyperacuity is beyond the scope of this paper, but it may have important implications for designing a retinal prosthesis. Kinematic movement processing is demonstrated by the effectiveness of the visual system to recreate a moving image when viewing cinematography–which, in essence, is a succession of frozen images.

4.3.4 Modeling and Simulation Results

The preceding discussion of the connectivity and function of the retina has been qualitative in nature. To develop successful signal processing methods for a retinal prosthesis device, a more quantitative description of retinal function will be required. Ideally, what is needed is a full mathematical model that can be used to numerically represent the input and subsequent output of the retina. By knowing what image processing occurs within each layer of the retina, the ganglion cell output can be predicted as a function of the input image. Conversely, if a prosthesis is stimulating the bipolar cells, for example, then the model can be inverted to determine what the corresponding input stimulation pattern should be. One of the most comprehensive models that has been developed to date is that by the Werblin

Laboratory at the University of California at Berkeley (Teeters et al, 1997). The work initially focused on the tiger salamander, but current work involves study of the mammalian retina. Fig. 4.6 is a simplified diagram of the model.

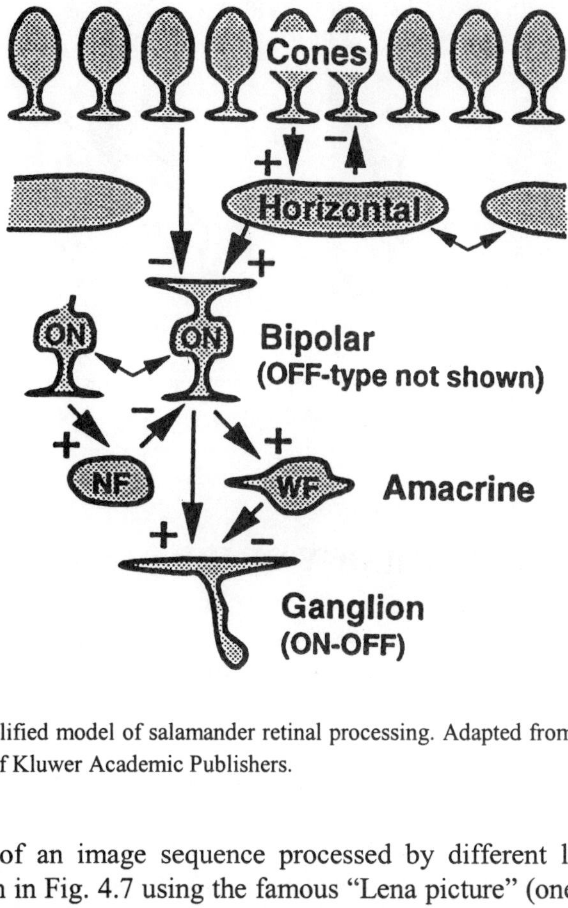

Figure 4.6. Simplified model of salamander retinal processing. Adapted from Teeters (1997) with permission of Kluwer Academic Publishers.

A sample of an image sequence processed by different layers of the retina is shown in Fig. 4.7 using the famous "Lena picture" (one of the most widely used standard test images for compression algorithms). Note that the image as it exists in the ganglion cell layer is dramatically different than the input image. Although not evident from the static images shown in Fig. 4.7, the temporal characteristics are dramatically different as well. It is expected that future IRP devices will require processing like that shown in this model. Further discussion of the model and its implementation into future IRP devices is beyond the scope of this paper, but it is clear that there is much to learn and many challenges involved.

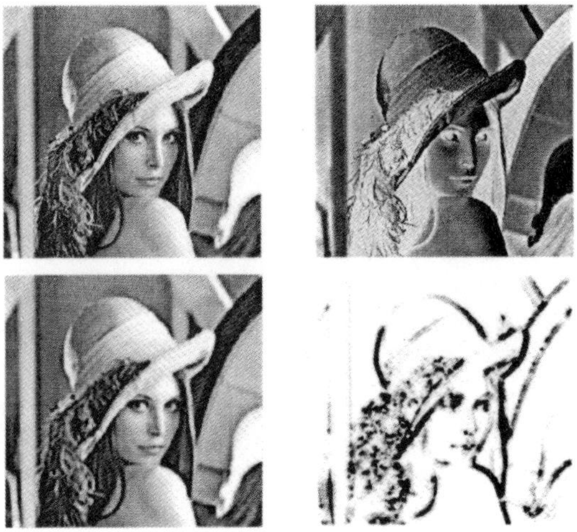

Figure 4.7. Results from the simulation model showing the processing changes that occur layer-by-layer as the "Lena test image" passes through the retina.

4.4 NEUROPHYSIOLOGY OF IRP

Many questions and concerns arise when interfacing an electronic device to neural tissue. A number of these questions are addressed individually below.

4.4.1 Preferential Stimulation of Retinal Cell Layers

The advantage of stimulating retinal cells other than ganglions was mentioned in Section 4.1.1 and discussed in Section 4.3. Histological analysis of postmortem eyes of RP (Humayun et al., 1999; Santos et al., 1997; Stone et al., 1992) and AMD patients reveals apparently healthy ganglion and bipolar cells in the macular region. Experimentally, it has been shown that phosphenes could be elicited in patients with advanced outer retinal degeneration via electrical stimulation (Potts and Inoue, 1970; Weiland et al., 1999; Humayun et al., 1996; Rizzo et al., 2000). These electrically elicited responses require and indicate the presence of functioning retinal cells.

As mentioned in Section 4.1.1, retinal ganglion cells (RGC) lie close to the surface of the retina facing the vitreous cavity and send mostly unmyelinated axons in a more superficial layer toward the optic disc. As the human RGC axons exit the eye, they become myelinated and form the optic nerve. The cell bodies (somas) of these ganglion cells are mapped over the surface of the retina in a manner that approximates the projection of the visual world onto the surface of the retina. However, at any particular location on the surface of the retina, axons from distant sites overlay the individual ganglion cell bodies. If these superficial passing axons were preferentially stimulated, groups of ganglion cells from large areas of the retina would be excited. One might expect the visual perception of such a stimulus to appear as a wedge. On the other hand, if the ganglion cell bodies or deeper retinal cells were stimulated, one would expect the visual perceptions to be focal spots. RP Patients that were stimulated with 50-200 μm diameter platinum disk electrodes reported seeing spots, not wedges, of light (Humayun, 1999; Weiland and Humayun, 1999; Humayun et al., 1996).

To explore the possibilities retinal electrical stimulation, a computational model of extracellular field stimulation of the RGC has been constructed (Greenberg, 1999). The model predicted that the stimulation threshold of the RGC soma is 58-73% lower than a passing axon, even though the axon was closer to the electrode. Nevertheless, a factor of less then two does not explain the source of visual perceptions observed during previous intraocular patient experiments.

Another possibility was that deeper retinal cells were stimulated. Postmortem morphometric analysis of the retina of RP patients revealed that many more inner nuclear layer cells retain functionality (e.g., bipolar cells and others, with 78.4%) compared to the outer nuclear layer (photoreceptors-4.9%) and the ganglion cell layer (29.7%) (Santos et al., 1997). Early electrophysiological experiments showed that cathodic stimulation on the vitreous side of the retina depolarizes presynaptic end-terminals of the photoreceptors (Knighton, 1975a; 1975b) and bipolar cells (Toyoda and Fujimoto, 1984). Recently, latency experiments in frog retinas showed that higher currents stimulate RGC directly, while lower currents activate other cells (photoreceptors, bipolar cells) (Greenberg, 1998).

Another finding in those experiments was that shorter stimulating pulses (<0.5 msec) have an effect different than longer stimulating pulses (>0.5 msec). There are well-defined relationships between the threshold current and the stimulus pulse duration required for neuronal activation (West and Wolstencroft, 1983). As the stimulus pulse duration decreases, the threshold increases exponentially. Also, as the pulse duration increases, the threshold current approaches a minimum value, called the rheobase. A chronaxie is the pulse width for which the threshold current is twice the rheobase current. Greenberg (1998) showed that deeper retinal cells have unusually long chronaxies compared to RGC. In human experiments, short stimulation duration (<0.5 msec) created elongated phosphene percepts, while longer stimulation (1-8 msec) created rounded percepts (Greenberg, 1998). It can be speculated from these results that there is a preferential stimulation of RGC cells/axons for short pulses and deeper cellular elements for long pulses.

Thus, several factors give interest to preferential electrical stimulation of bipolar cells. First, intraocular stimulation trials with 1-8 msec pulses have shown that human RP patients perceived focal, not wedge-shaped, phosphenes. Second, in severe RP patients there are more viable bipolar cells then RGC. Third, preferential stimulation of deeper retinal layers has been demonstrated in a frog given long stimulation pulses.

4.4.2 Interfacing IRP Electrodes to Retinal Tissue

Several specific physiological questions and concerns arise when interfacing an electronic device to neural tissue. A number of these questions are addressed individually below.

What is the minimum current for neuron activation? In 1939, Cole and Curtis found that during the action potential propagation in the axon of the giant squid, the conductance of the membrane to ions increases dramatically (Cole and Curtis, 1939). In 1949, Cole designed an apparatus known as the voltage clamp to overcome the problems of experimentally measuring the Na+ and K+ currents through the axon's membrane. The amount of current that must be generated by the voltage clamp to keep the membrane potential from changing provides a direct measure of the current flowing across the membrane. The voltage clamp technique and the squid axon was used by Hodgkin and Huxley (1952a; 1952b) to give the first complete description of the ionic mechanisms underlying the action potential. According to the Hodgkin-Huxley model, an action potential involves the following sequence of events: a depolarization of the membrane causes Na+ channels to open rapidly resulting in an inward Na+ current (because of a higher concentration of this ion *outside* the cell membrane). This current causes further depolarization, thereby opening more Na+ channels and results in increased inward current; the regenerative process causes the action potential.

Electrical stimulation elicits a neural response by "turning on" the voltage sensitive ion channels, bypassing the chemically gated channels in the stimulated cell. There are different methods by which neurons can be activated. The first is cathodic threshold activation. This is the minimum stimulus amplitude and duration required to initiate an action potential. Once the membrane reaches a certain potential, a trigger mechanism is released and an action potential results (all-or-nothing mechanism). Other methods to stimulate neurons are anodic pulses and biphasic pulses.

There are well-defined relationships between the threshold charge and pulse duration (West and Wolstencroft, 1983). Charge and threshold have different minimum requirements during neuronal stimulation. A minimum charge is required for shorter pulse duration in contrast to threshold current, which is minimized at long pulse duration. Experiments were performed at Johns Hopkins University to define threshold currents for retinal electrical stimulation. One study assessed the effect of changing parameters of the stimulating electrode and the stimulus pulse by recording electrically elicited action potential responses from retinal ganglion cells in isolated rabbit retina (Shyu et al., 2000). It was concluded that the threshold for stimulation from the ganglion side is lower then from the photoreceptor side, especially when using microelectrodes (19.05 μA versus 48.89 μA, with pulse duration of 0.5 msec). Recently, similar experiments with very small electrodes (10 μm diameter) demonstrated successful stimulations with currents as low as 0.14 – 0.29 μA (Grumet et al, 1999; 2000).

A second type of experiment compared the electrical stimulation threshold in normal mouse retina versus different aged retinal degenerate (rd) mouse retina (Suzuki et al., 1999). Retinal ganglion cell recordings were obtained from anesthetized 8- and 16-week old rd mice, and 8-week old normal mice in response to a constant current electrical stimulus delivered via a platinum wire electrode on the retinal surface. The excitation thresholds were significantly higher in the 16-week old rd mouse (0.075 μC for 0.08 msec square pulse) versus the 8-week old rd (0.048 μC for 0.08 msec square pulse) ($P < 0.05$) and versus normal mouse rd (0.055 μC for 0.08 msec square pulse) ($P < 0.05$). In all groups, short-duration pulses were more efficient than longer pulses (lower total charge) ($P < 0.05$). A related experiment, involved the electrical stimulation of normal and rd mouse retina and the elicited visual cortical responses (Chen et al., 1999). A square-wave stimulus (240 ± 58 μA) was more efficient than the sine waveform (533 ± 150 μA) or pulse-train (1000 ± 565 μA) waveform (p = 0.002).

In human experiments at Johns Hopkins University Hospital, typical thresholds observed for retinal stimulation of RP patients was 500 μA with a 2 msec half-pulse stimulus duration (1 μC/phase) using electrode sizes from 50 μm to 200 μm diameter disks that were very near, but not touching the retina (Humayun, 1996). The quantity *charge-per-phase* is defined as the integral of the stimulus current over one half-cycle of the stimulus duration. In summary, the measurements that have been made to date serve as useful guides for threshold levels needed to stimulate retinal neurons, however, a quantitative relationship between minimum currents, electrode size, proximity, and pulse shape is still incomplete.

What is the maximum current that can be used before impairing physiological function of retinal cells? Among the early studies that have addressed this issue are the histopathological studies of long-term stimulation by Pudenz et al. (1975a; 1975b; 1975c) as well as the electrochemical studies of the electrode-electrolyte interface by Brummer and Turner (1975). Lilly (1961) demonstrated the relative safety of biphasic, charge-balanced waveforms compared to monophasic waveforms. McCreery et al. (1969) showed that stimulation-induced neural damage derives from processes associated with passage of stimulus current through tissue, rather than from electrochemical reactions at the electrode-tissue interface.

They also showed that the threshold of tissue damage from electrical stimulation is dependent on current amplitude and pulse frequency, but more importantly, on charge density and charge per phase (McCreery et al., 1988; 1990). Charge density is defined as *charge-per-phase* divided by the electrochemically active electrode surface area. Since total charge density is responsible for the damage of tissue and electrodes, there is a theoretical limit to how small the electrodes can be (Brown et al., 1977; Tehovnik, 1996). Using simple waveforms, conservative charge density limits for chronic stimulation with platinum are 100 $\mu C/cm^2$ and 1 μC/phase. For activated iridium oxide electrodes, the limit is 1 mC/cm^2 and 16 nC/phase. Most of the studies that were done to determine these limits were performed with superficial cortical electrodes (McCrerry et al., 1988; McCreery et al., 1990), or intracortical microstimulation (Bullara et al., 1983). Chronic *in vivo* retinal stimulation tests still need to be performed to define tissue damage thresholds.

What are the optimum conditions for a stimulating retinal neurons and what is the desired response? One of the conditions for safe electrical stimulation of neural tissue is a reversible faradaic process. These reactions involve electron transfer across the electrode tissue-interface. Some chemicals are either oxidized or reduced during these reactions. These chemicals remain bound to the electrode surface and do not mix with the surrounding solution. It is also necessary to know the chemical reversibility of electrode materials and stimulation protocols. Chemical reversibility requires that all processes occurring at an electrode due to an electrical pulse, including H_2 and O_2 evolution, will be chemically reversed by a pulse of opposite polarity.

There are two basic waveforms for neural stimulation: sinusoidal and pulsatile. The sinusoidal waveform is completely described by the amplitude and the frequency. Pulsatile waveform is completely described by the pulse amplitude, its duration, polarity, and repetition frequency (Gorman and Mortimer, 1983).

Over time, any net DC current can lead to charge accumulation and irreversible electrolyte reactions. A biphasic current waveform consisting of two consecutive pulses of equal charge but opposite polarity avoids these problems. A simple monophasic waveform is similarly unacceptable. Studies with both isolated rabbit retina, normal and rd mice showed that the electrophysiological response has the lowest threshold when using a cathodic wave first. These studies also showed that response threshold was lower when using square-wave electrical stimulus (Shyu et al., 2000; Suzuki et al., 1999; Chen et al., 1999).

4.4.3 Electrode Biocompatibility

Because any future implantable device could be positioned against neural tissue for very long periods of time, potentially decades, a number of biocompatibility issues need to be addressed.

What kind of electrode array and attachment methods should be used for minimizing any possible damage to neural tissue? The biocompatibility between an implanted medical device and the host tissue is as important as its mechanical durability and functional characteristics. This includes the effects of the implant on the host and vice versa. Effects of the implant on the tissue include inflammation, sensitivity reactions, infections, and carcinogenicity. Effects of the tissue on the implant are corrosion and other types of degradation. Sources of toxic substances are antioxidants, catalysts, and contaminants from fabrication equipment.

Microfabricated electrodes were initially conceived in the early 1970s (Wise et al., 1970). In subsequent years, the dimensions of these electrodes have been decreased, using concurrent advances in the microelectronics industry. Today, micromachined silicon electrodes with conducting lines of 2 μm are standard (Hetke et al., 1994; BeMent et al., 1986; Kovacs et al., 1992; Turner et al., 1999). Methods for depositing thin-film metal electrodes have been established. Chronic implantation and *in vitro* testing have demonstrated the ability of silicon devices to maintain electrical characteristics during long-term implantation (Weiland and Anderson, 2000).

Even the "noble" metals (platinum, iridium, rhodium, gold, and palladium) corrode under conditions of electrical stimulation (McHardy et al., 1980; Laing et al., 1967). Platinum and its alloys with iridium are the most widely used. Using simple waveforms, conservative charge density limits for chronic stimulation with platinum are 100 $\mu C/cm^2$. For activated iridium oxide electrodes, the limit is 1 mC/cm^2 (Beebe and Rose, 1988). Platinum-iridium alloys are mechanically stronger then platinum alone.

Iridium oxide electrodes belong to a new category termed "valence change oxides." Iridium oxide layers can be formed by electrochemical activation of iridium metal, by thermal decomposition of an iridium salt on a metal substrate, or by reactive sputtering from an iridium target. Activated iridium is exceptionally resistant to corrosion. It appears to be a very promising electrode material. Most neural prostheses use platinum stimulating electrodes, the exception being the BION microstimulator (Advanced Bionics, Sylmar, California), which uses iridium oxide. Iridium oxide *in vitro* has been shown to have a safe stimulation limit of 3 mC/cm^2 (Beebe and Rose, 1988). Recently, a titanium nitride, thin-film electrode has demonstrated charge injection limits higher than both platinum and iridium oxide (Janders et al., 1996). Titanium nitride has an *in vitro* limit of 22 mC/cm^2 (Janders et al., 1996).

Stabilizing the electrode array on the surface of the retina is an especially formidable problem. The biocompatibility and the feasibility of surgically implanting an electrode array onto the retinal surface has been examined Johns Hopkins University Hospital. In one experiment, a 5 × 5 electrode array (25 disc-shaped platinum electrodes in a silicone matrix) was implanted on to the retinal surface using retinal tacks in each of four mixed-breed sighted dogs for a maximum period of time of one year. No retinal detachment, infection, or uncontrolled intraocular bleeding occurred in any of the animals. Retinal tacks and the retinal array remained firmly affixed to the retina throughout the follow-up period. It was concluded that implantation of an electrode array on the epiretinal side (i.e., side closest to the ganglion cell layer) is surgically feasible, with little if any significant damage to the underlying retina, and that platinum and silicone arrays as well as the metal tacks are biocompatible in the eye (Majji et al., 1990).

Another method for attaching electrode arrays is by biocompatible adhesives. Nine commercially available compounds were examined for their suitability as intraocular adhesives: commercial fibrin sealant, autologous fibrin, Cell-Tak®, three photocurable glues, and three different polyethylene glycol hydrogels. One type of hydrogel (SS-PEG, Shearwater Polymers, Inc.) proved to be nontoxic to the retina (Margalit et al., 2000). Hydrogels proved superior for intraocular use in terms of consistency, adhesiveness, stability, impermeability, and safety.

4.4.4 IRP Experiments

A number of *in vivo* and *in vitro* retinal stimulation studies have been performed in animals and humans at Johns Hopkins University Hospital.

The next major step will be chronic implantation of active devices in animal models to examine the efficacy and safety of such devices.

Specific parameters that will be examined during these experiments are the clinical appearance of the retina during the time period of the implantation, and electrophysiological responses–electroretinogram (ERG) and visual evoked potentials (VEP). The VEP can be examined by surface scalp electrodes, subdural surface electrodes, or intracortical recording of single neurons. These experiments will be examined in both normal and retinal degenerate animals. After chronic implantation in animal models, the retina and cortex will be examined using optical and electron microscopy for any histological damage. If animal model experiments prove successful, chronic human experiments would follow with blind volunteer patients.

Future devices will contain more electrodes, more advanced electronics, and radio frequency or other wireless communication links. Proper hermetic sealing and the use of advanced biocompatible materials will improve the host response and ensure the long-term integrity of the device. After successful demonstration of prototype devices, the issues of biocompatibilty will become the most challenging aspect of this technology.

Research programs to develop retinal and cortical visual prosthesis are progressing in parallel tracks, and it is too early to say if either will provide therapeutic benefit. Chronic experiments in animal models and humans will give some idea regarding the future of these projects in 5 years. A successful retinal prosthesis device will aid only blind patients affected with outer retinal degeneration diseases such RP and AMD. Other blind patients would benefit from a cortical visual prosthesis. It is not clear whether humans who became blind early in life will benefit from either prosthesis type. The visual cortex is desensitized to visual information if the eyes are deprived of visual stimuli during a critical period during early childhood. This process is called amblyopia. Patients who have been blind from early childhood may have lost their ability to process visual stimuli in the visual centers of their cortex and therefore lose their ability to benefit from any type of visual prosthesis.

The cortical visual prosthesis, if and when proven successful, can provide very important information on the mechanisms and development of amblyopia. Patients whose blindness is caused by ischemic events, trauma, tumor, or other destructive processes of the occipital cortex may also have lost their ability to benefit from the visual prosthesis as well. These questions will be answered in future experiments.

4.5 IRP DEVICE DEVELOPMENT

A microelectronic stimulator array is described below that addresses many of the technical issues discussed above. The current joint effort between the Naval Research Laboratory and Johns Hopkins University is aimed at developing a microelectronic IRP stimulator array that will be used in preliminary short-term tests in an operating room environment. The IRP test device will receive input images from an external camera connected via a microcable. These tests will determine requirements for a permanent IRP implant device that images incident photons, as shown in the bottom half of Fig. 4.3.

4.5.1 IRP Test Device

The IRP test device will enable short-term human experiments (less than an hour) to study basic issues involved with interfacing a massively parallel electrode array to retinal tissue. The design combines two technologies: (1) electrode arrays fabricated form nanochannel glass (NCG) (Tonucci et al., 1992), and (2) infrared focal plane array (IRFPA) multiplexers (Scribner et al., 1991).

NGC is a technology that uses fiber optic fabrication techniques to produce thin wafers of glass with very small channels perpendicular to the plane of the wafer (Tonucci and Justus, 1993a; 1993b). Typical NCG wafers that will be required for retinal prosthesis devices are several millimeters in diameter and contain millions of channels, with channel diameters on the order of one micron. A sample of NCG is shown in Fig. 4.8. The channels are filled with a good electrical conductor, and one surface of the glass is ground to a spherical shape consistent with the radius of curvature of the inside of the retina. The electrical conductors on the curved surface should protrude slightly to form efficient electrodes.

For the IRP test device, a microelectronic multiplexer is required. The IRFPA community has been developing a similar multiplexer technology over the past decade. IRFPAs use microelectronic multiplexers that are fabricated at silicon foundries. The multiplexer is a two-dimensional array that reads-out the infrared images captured by a complementary detector array that converts photons into electrical charge. The charge is integrated and stored in each pixel (sometimes referred to as a unit cell) for a few milliseconds. The full image is then multiplexed off the array at frame rates compatible with commercial video. For an IRP test device, the process is essentially reversed, and the device acts as demultiplexer. That is, an image is read onto the stimulator array. Although devices discussed here for IRP will perform demultiplexing operations, they are simply referred to as multiplexers.

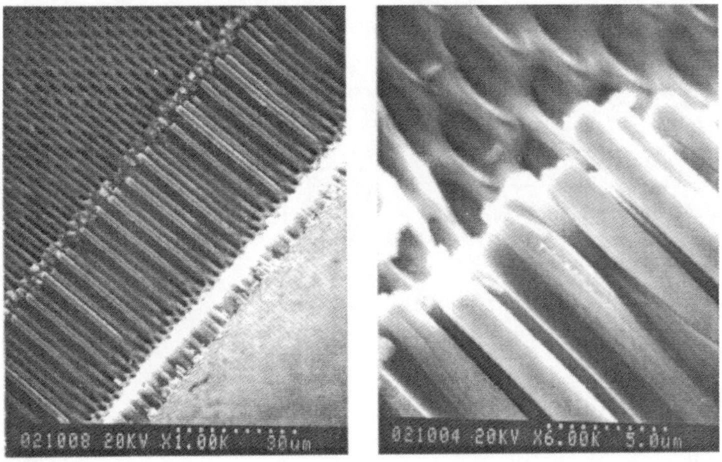

Figure 4.8. Scanning electron microscope images of platinum micowires grown in nano-channel glass, shown at two different scales. The glass piece has been cleaved revealing the wires in side view running through the glass. The wires have a diameter of 1.6 microns and a length of 30 microns.

Fig. 4.9 shows an IRP test device positioned against the retina as it would be in an acute human experiment performed by an ophthalmologist. The experimental procedure uses standard retinal surgical techniques identical to those in an operating room environment. It is necessary that the patient be administered local (rather than general) anaesthesia so that he is conscious during the procedure.

Fig. 4.10 shows a side view of the fully packaged IRP test device. The NCG is hybridized to the multiplexer using indium bump bonds–again, this is similar to hybridization techniques used in IRFPAs. The image is serially input onto the multiplexer via a very narrow, flexible micro-cable. The ceramic carrier with gold *via-holes* (conducting wires penetrating from the front to the back) provides a mechanically convenient means of routing interconnects from the top-side of the device to the back-side. By designing the ceramic carrier such that the via-holes are in close proximity to the bond pads on the silicon multiplexer, the interconnection can be made with conventional *tab-bonds* (thin gold ribbons fused to interconnects with mechanical pressure). This keeps all the interconnects from protruding above the spherical curved envelope defined by the polished NCG surface and therefore protects the retina from damage as well as reduces the risk of breaking a tab bond.

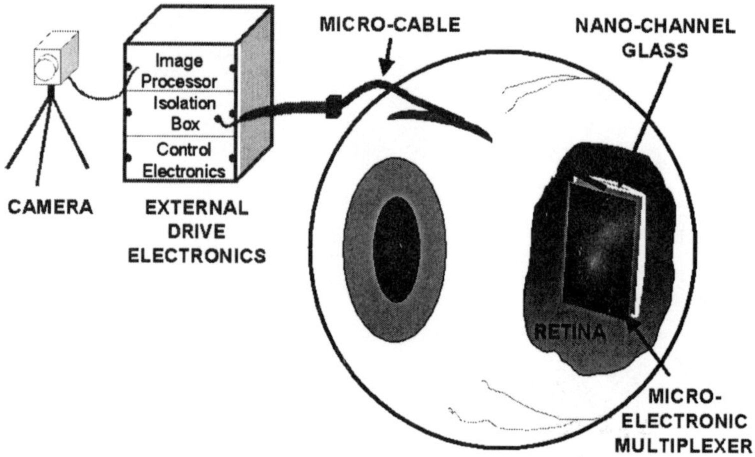

Figure 4.9. An IRP test device positioned against the retina as it would be in an acute human experiment performed by an ophthalmologist. External drive electronics are needed to control the device and interface it with a standard video camera.

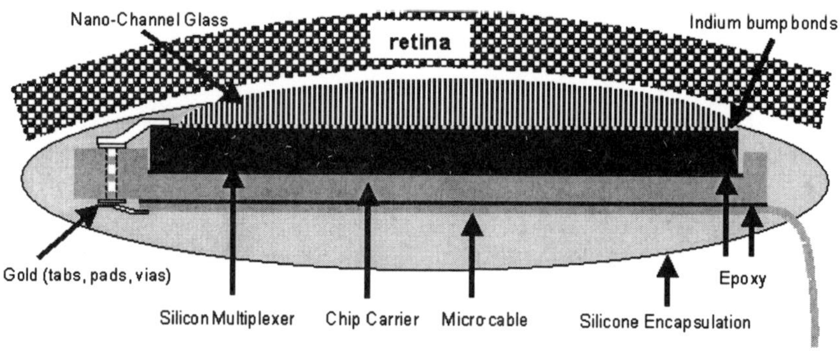

Figure 4.10. Side view of the fully packaged *IRP test device.* The NCG is hybridised to the multixplexer using indium bump bonds similar to hybridisation techniques used in IRFPAs.

As discussed in Section 4.4, a critical issue for any neural prosthesis device is biocompatiblilty and safety. Because the duration of any tests with the IRP test device are very short (less than an hour), biocompatibility issues are primarily reduced to acute effects and need not address the more difficult chronic issues that arise with permanent implants. Note that the surface of the packaging shown in Fig. 4.10 consists only of glass, platinum electrodes, and silicone encapsulation. However, as with any electronic medical instrumentation, a major safety issue is electrical shock hazard. The purpose of the device is to provide minimal electrical stimulation of retinal tissue using very low voltages and the smallest currents possible. During this procedure the patient must be coupled to the external instrumentation. To protect the patient from any electrical shock, the patient is isolated from high voltages using optocouplers that are powered by low voltage batteries.

NCG features: Specific requirements for the NCG are that the channels be small enough so that many microwires can be connected to each unit cell. This provides redundancy, but more importantly helps simplify the hybridization alignment. If the NCG microwires were to approach the size of the unit cells, then a one-to-one alignment and precise hybridization would be required. This would be very problematic because of irregularities in the channel periodicity and the possibly of shorting nearest-neighbor cells. On the other hand, very narrow channels imply very high length-to-width aspect ratios for the channel geometry. This makes it difficult to fabricate large-area NCG samples with the proper thickness. Therefore, a reasonable design goal for the channel width is about a one micron diameter.

The NCG channels must be filled with a high conductivity material to create microwires. The microwires can be fabricated by using electrodeposition or infusion of molten metal under pressure. After the channels have been filled with a conductive material and the continuity of the microwires has been confirmed, one side of the glass must be curved to create a spherical surface. Grinding and polishing techniques similar to those used in lens fabrication can be applied to the NCG pieces. The radius of curvature is nominally half an inch to provide a conformal fit against the inside of the retina. This is critically important as it allows positioning of the high-density electrodes in direct contact with the retinal tissue. The polishing process will create slightly recessed microwires with respect to the curved NCG surface. This is because the metal is softer than the glass. Therefore further processing is necessary to create electrodes that protrude slightly above the curved surface. This can be accomplished by applying a chemical etch to the surface that removes a few microns of glass.

In preparation for hybridizng the NCG to the multiplexer, indium bumps can be deposited on the flat side of the NCG. Alternatively, the microwires can be hybridised directly to the indium bumps on the multiplexer if they are formed to protrude slightly from the NCG. Getting the microwires to protrude could again be accomplished by chemical etching like that described above for forming protruding electrodes on the curved side of the NCG.

Multiplexer features: The silicon multiplexer performs several operations in a sequential order. During the first step, an image frame is read onto the multiplexer, pixel-by-pixel, to each unit cell. Row-by-row, each unit cell samples the analog video input and stores the pixel value as charge on a MOS capacitor. A full frame is completed every 60th of a second in a manner compatible with the RS-170 television format; this allows the use of the test prosthesis with standard video equipment. Fig. 4.11 is a highly simplified layout of the multiplexer. The digital electronics block is of major importance because it generates the switching pulses that route image data into the unit cells. Without the on-chip digital electronics, a dozen or more clocks would need to be input to the device. That would make the cable through the eye wall much larger and more cumbersome. The use of IRFPA multiplexer technology greatly simplifies the cable problems through the eye wall.

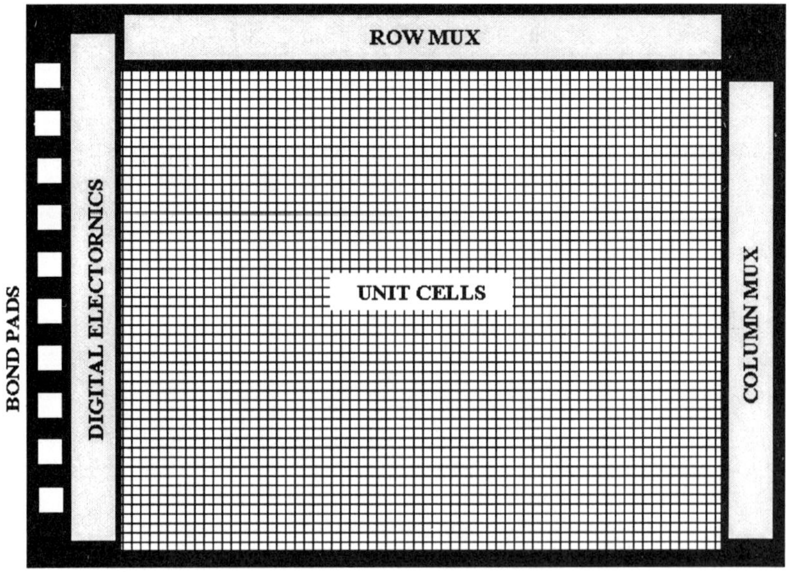

Figure 4.11. The conceptual layout of the Si chip for *IRP test device.*

After all the unit cells have been loaded with the pixel values for the current frame, the next step is to send a biphasic pulse to each unit cell, which in turn is modulated in proportion to the pixel value stored in each unit cell. The biphasic pulse flows from an external source, through each unit cell, thus stimulating retinal neurons in a simultaneous manner. This is an important feature of the design because it is a synchronistic action analogous to imaged photons stimulating photoreceptors in a normal retina. Finally, the electrodes are all connected to ground to prevent any possible charge buildup at the electrode-neuron interface.

There are several important considerations in designing a device that performs all these operations successfully. First, the multiplexer operation should be designed with many of the requirements that exist for an IRFPA, for example, good uniformity, low noise, and high dynamic range. Of course, the retinal prosthesis test device moves image data in the opposite direction than an IRFPA multiplexer, that is, the image moves onto the device rather than off the device, but otherwise the specifications are analogous. Another consideration is that the unit cell stores the pixel value and then uses it to modulate the biphasic pulse that is input to the retinal tissue through the NCG. Fig. 4.12 shows a simplified circuit design for such a unit cell. Note that the biphasic pulse and the image data are both

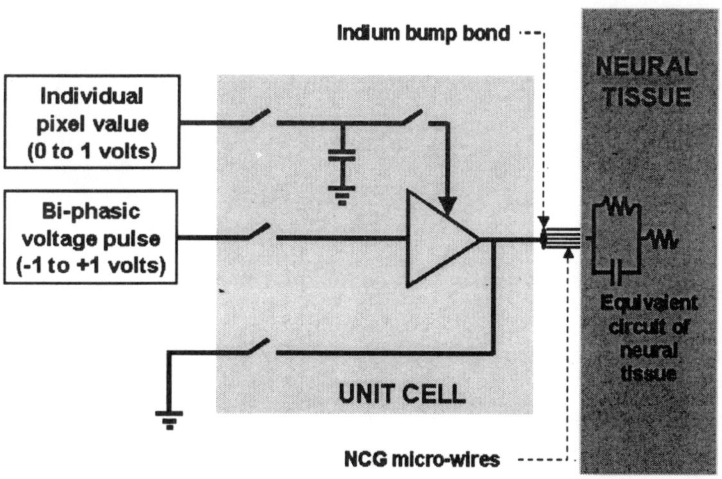

Figure 4.12. Conceptual design of the unit cell for an *IRP test device* showing the external inputs from off-chip. The pixel values are acquired from a camera (or any other video system that generates RS-170 signals) and are routed to each unit cell via a pixel-by-pixel raster scan through the on-chip multiplexer. The biphasic pulse is generated off-chip and delivered to each unit cell via a global connection.

generated off-chip. This allows for greater flexibility during human testing as any image sequence can be input and combined with any shape of biphasic pulse. The switch at the bottom of Fig. 4.12 provides the capability to connect the retinal tissue to ground to avoid any possibility of charge build-up.

Ancillary electronics: The operation of the IRP test device during acute experiments is controlled and powered by external ancillary electronics. Fig. 4.13 is a block diagram of the ancillary electronics. The input signal is an image sequence at data rates fast enough to achieve 60 frames per second. As mentioned above, the multiplexer can be designed to sample the multiplexed input signal in a manner compatible with the RS-170 format.

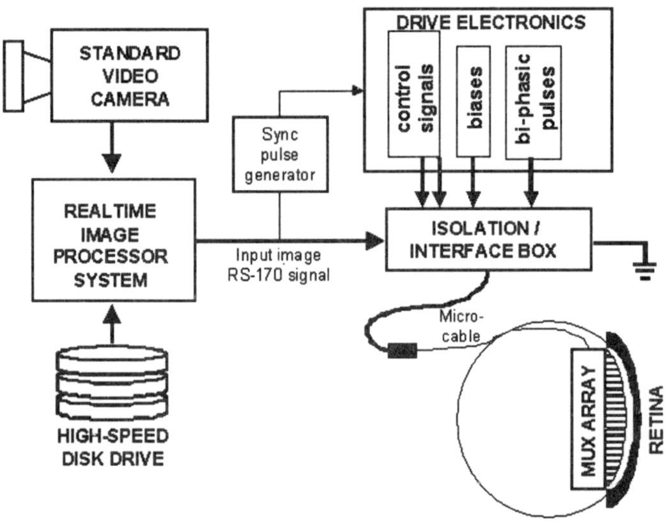

Figure 4.13. Block diagram of the ancillary electronics for an *IRP test device.*

This allows the IRP test device to be interfaced directly with any standard video camera. This includes the use of a personal computer that stores digital imagery and can display sequential fields at a 60 Hz rate (RS-170 interlaces two fields per frame at a rate of 30 frames per second). The actual control of the microelectronic multiplexer is done with precisely timed pulses generated by a set of signal clocking boards in a manner similar to that used in typical IRFPAs. The sync pulse generator is used to synchronize the RS-170 signal with the clocking pulses. Briefly, the sync pulse generator detects the beginning of each RS-170 field and then sends a corresponding pulse to the drive electronics box that triggers the clocking signals that control each field of image data input to the multiplexer. The isolated breakout box is used to isolate the human subject from any high voltage power supplies. The box contains opto-couplers that isolate the clocks and biphasic pulse signals and low voltage batteries to for supplying bias potentials.

The biphasic pulses used to stimulate the retinal tissue can be programmable such that any pulse shapes can be tested. This has several important implications for the development process. First, because the input impedance to the retinal tissue has both a resistive and capacitive reactance associated with it, a square-wave voltage pulse will not produce the desired square-wave current pulse. Neurobiologists have found that square-wave current pulses are best to achieve effective neural stimulation. With knowledge of the output impedance at the electrode-retina interface, a voltage shape can be computed that will provide a square-wave current pulse, thus providing efficient stimulation. Second, there is evidence that different shaped pulses will stimulate different layers of the retina– probably because of their differing frequency responses. Specifically, it is expected that either the ganglion or bipolar cells can be selectivity stimulated. Stimulating the bipolar cells instead of the ganglion cells has the advantage of reaching more deeply in the retina, allowing more use of the natural signal processing.

Algorithms: As mentioned above, direct electrical stimulation of the ganglion cells precludes certain processing functions that normally would have occurred in the earlier layers of the retina. Therefore, it may be necessary to perform certain functions on the incoming imagery before stimulation to compensate for the missing processing. Unfortunately, a detailed model of human retinal functions has never been confirmed. Nevertheless, as discussed in Section 4.3.4, animal models of retinal processing exist and are suitable for use in defining processing algorithms.

4.5.2 IRP Permanent Implant

As a future concept, a permanent IRP implant device that responds to incident photons naturally imaged through the lens of the eye is shown in the bottom half of Fig. 4.3. It would be surgically implanted, with no external connections passing through the eye wall. The basic design for the device is based extensively on the IRP test device described above. Specifically, the permanent IRP implant device would use an NCG array hybridized to a silicon chip in an identical manner to the IRP test device. However, the unit cell circuitry would need to be redesigned in that the image is no longer being multiplexed onto the chip through an electrical lead from an external camera. Instead, the image is simultaneously generated within each unit cell through a photon-to-electron conversion using a silicon photodiode. The unit cell design is shown conceptually in Fig. 4.14. The photons can propagate directly into the each unit cell because the silicon chip can be used in a back-side illuminated configuration–essentially the photons enter through the backside of the silicon chip. To improve the quantum efficiency, the silicon

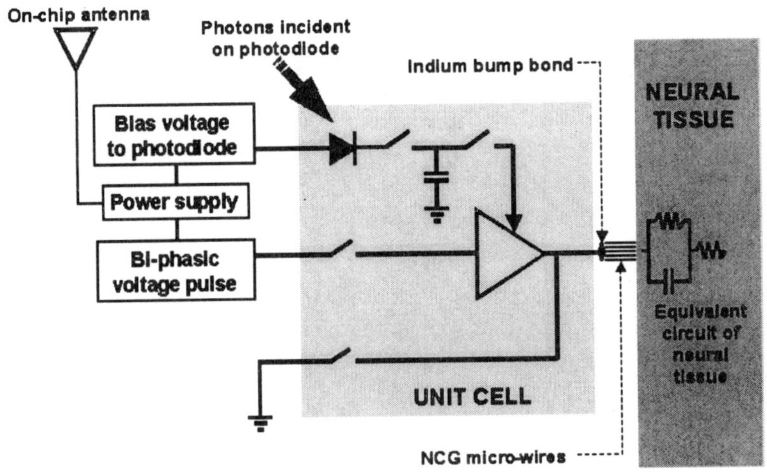

Figure 4.14. Conceptual design of the unit cell for a *permanent IRP implant device* – note that the only connection to outside the eye is wireless. The biphasic pulse is generated off-chip and delivered to each unit cell via a global connection.

chip can be thinned.

The packaging of the device would be different than that shown previously for the IRP test device. An implanted device would need to allow photons to pass through the backside of the device. This is a simple matter of eliminating the ceramic carrier. Fig. 4.15 shows the packaging scheme. The silicon chip can be thinned to a few tens of microns so that the overall mass of the object is primarily that of the NCG array, making it more amenable for surgical attachment to the retina. Note that because there is no need for any multiplexing functions, i.e., input of imagery onto the chip from an external camera, the design of the silicon chip becomes much simpler. There are no ancillary electronics as was needed in the case of the IRP test device; however, the silicon chip is significantly different, as shown in Fig. 4.16. Although there are no multiplexing requirements, there are two new requirements. Specifically, these are external power and a command link to adjust the operation of the device. Transmitting power and signals onto the device can be implemented with an inductively driven coil or antenna (Liu et

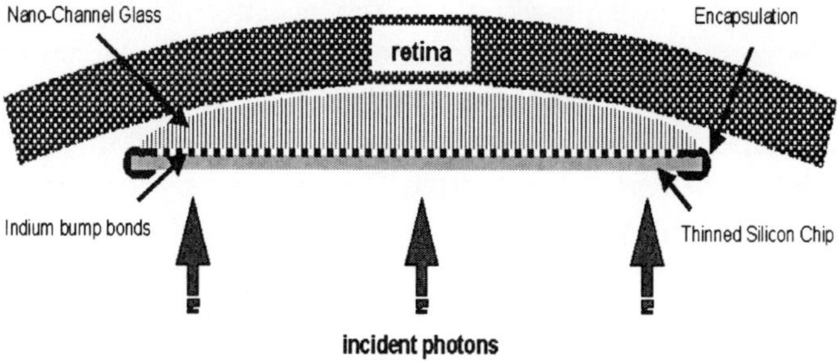

Figure 4.15. Side view of the fully packaged permanent *IR implant device.*

al., 1999). The major on-chip electronic controls needed are adjustments of bias supplies and the biphasic pulse generator plus the standard digital electronics that supply timing for simultaneous operation of the unit cell sequences. Again, the operation of the device is to collect charge in the storage capacitors via the photon-to-electron conversion process, stimulate the neural tissue with biphasic pulses in proportion to the stored charge, and reset the storage capacitors to repeat the process. The update rate would nominally be 60 frames per second but because there is no longer a need to be compatible with the RS-170 format, the frame rate can be chosen as determined by future research.

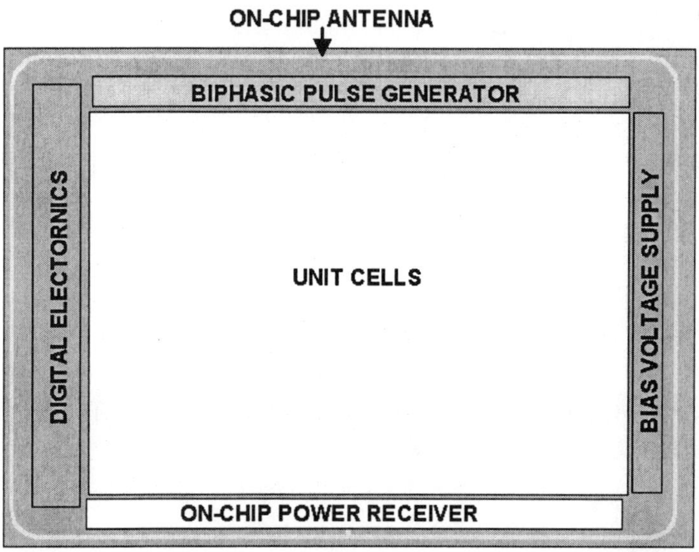

Figure 4.16. The conceptual layout of the Si chip for a permanent IRP implant device.

The packaging of the permanent IRP implant device is very demanding. Along with issues of biocompatiblilty is the question of device lifetime. Permanent implants might need to operate for several decades. Similar requirements exist for other electronic implants such as cardiac pacemakers and cochlear prosthetics. Encapsulation of the permanent IRP implant device is easier in some respects than that of the IRP test device because there are no connecting cables to the device. In the case of the latter, encapsulation was not as critical because the duration of the experiments are typically less than one hour. Cables connected to any neural prosthesis are subject to mechanical forces that over time can damage seals and ultimately cause failures. Because the implant device will be completely wireless–no cable connections–a simple encapsulation should be achievable with very high integrity.

4.6 SUMMARY

The hope of restoring vision to the blind is now believed to be a real possibility using neural prosthetics. However, many technical problems remain and many engineering issues must be resolved before complete clinical success is achieved. Not the least of these problems will be obtaining a better understanding of the visual pathways and the plasticity of the nervous system with respect to artificial electrical stimulation.

Other issues that will require great attention are biocompatibility and the reliability of a device that will be implanted and expected to function without degradation for decades. Ultimately, the true measure of success will be the acceptance of this approach by the blind community. Hopefully this success will parallel that of the cochlear implant that, although initially slow, continues to grow exponentially each year and is now a fully commercialized medical product.

4.7 ACKNOWLEDGMENTS

Work on an IRP test device is being sponsored by the DARPA Tissue Based Biosensors Program managed by Dr. Alan Rudolph.

4.8 REFERENCES

Agnew, W. F. and McCreery, D. B. (eds), 1990, *Neural Prosthesis*, Prentice-Hall, New York.

Bak, M., Girvin, J. P., Hambrecht, F. T., et al., 1990, Visual sensations produced by intracortical microstimulation of the human occipital cortex, *Med. Biol. Eng. Comput.* **28**: 257-259.

Beaudot, W., 1996, Adaptive spatiotemporal filtering by a neuromorphic model of the vertebrate retina, *Proc. IEEE Int. Conf. Image*, Vol. **1**, pp. 427-430.

Beebe, X. and Rose, T. L., 1988, Charge injection limits of activated iridium oxide electrodes with 0.2 ms pulses in bicarbonate buffered saline, *IEEE Trans. Biomed. Eng.* **135**: 494-495.

BeMent, S. L., Wise, K. D., Anderson, D. J., et al., 1986, Solid-state electrodes for multichannel multiplexed intracortical neuronal recording, *IEEE Trans. Biomed. Eng.* **33**: 230-241.

Bostock, H., 1983, The strength-duration relationship for excitation of myelinated nerve: computed dependence on membrane parameters, *J. Physiol. (London)*. **341**: 59-74.

Brabyn, J. A., 1982, New developments in mobility and orientation aids for the blind, *IEEE Trans. Biomed. Eng.* **29**: 285-289.

Brindley, G. S. and Lewin, W. S., 1968a, The sensations produced by electrical stimulation of the visual cortex, *J. Physiol. (London)*. **196**: 479-493.

Brindley, G. S. and Lewin, W. S., 1968b, The visual sensations produced by electrical stimulation of the medial occipital cortex, *J. Physiol. (London)*. **194**: 54-55P.

Brindley, G. S., 1965, The number of information channels needed for efficient reading, *J. Physiol.* **177**: 44P.

Brown, W. J., Babb, T. L., Soper, H.V., et al., 1977, Tissue reactions to long-term electrical stimulation of the cerebellum in monkeys, *J. Neurosurg.* **47**: 366-379.

Brummer, S. B. and Turner, M. J., 1975, Electrical stimulation of the nervous system: The principle of safe charge injection with noble metal electrodes, *Bioelectrochem. Bioenerg.* **2**: 13-25.

Bullara, L. A., McCreery, D. B., Yuen, T. G., and Agnew, W. F., 1983, A microelectrode for delivery of defined charge densities, *J. Neurosci. Methods.* **9**:15-21.

Cha, K., Horch, K. W., and Normann, R. A., 1992a, Mobility performance with a pixelized vision system, *Vision Res.* **32**: 1367-1372.

Cha, K., Horch, K. W., Normann, R. A., Boman, D. K., 1992b, Reading speed with a pixelized vision system, *J. Opt. Soc. Am.* **9**: 673-677.

Cha, K., Horch, K., Normann, R. A., 1992c, Simulation of a phosphene-based visual field: visual acuity in a pixelized vision system, *Ann. Biomed. Eng.* **20**: 439-449.

Chen, S. J., Humayun, M. S, Weiland, J. D., et al., 2000, Electrical stimulation of the mouse retina: A study of electrically elicited visual cortical responses, *Invest. Ophthal. Vis. Sci.* **40**: S889.

Chow, A. Y. and Chow, V. Y., 19997, Subretinal electrical stimulation of the rabbit retina, *Neurosci. Lett.* **225**: 13-16.

Chow, A. Y. and Peachey, N. S., 1998, The subretinal microphotodiode array retinal prosthesis [letter; comment], *Ophthalmic Res.* **30**: 195-198.

Cole, J. and Curtis, H., 1939, Electric impedance of the squid giant axon during activity, *J. Gen. Physiol.* **22**: 649-670.

Curlander, J. C. and Marmarrelis, V. Z., 1983, Processing of visual information in the distal neurons of the vertebra retina, *IEEE,* SMC-**13**: 934-943.

Dacey, D. M., 1996, Circuitry for color coding in the primate retina, *Proc. Nat. Acad. Sci.*, **93**: 582-588.

de Juan, E., Humayun, M. S., Hatchell, D., and Wilson, D., 1989, Histopathology of experimental retinal neovascularization, *Invest. Ophthal. Vis. Sci.* **30**: 1495.

Delbruck, T. and Mead, C.A., 1994, Adaptive photoreceptor with wide dynamic range, *Proc. IEEE Int. Symp. on Circuits and Systems, ISCAS '94*, Vol. **4**, pp.339-342.

Djourno, A. and Eyries, C., 1957, Prothese auditive par excitation electrique a distance du nerf sensorial a l'aide d'un bobinage inclus a demeure, *Presse Med.* **35**: 14-17.

Dobelle, W. H. and Mladejovsky, M. G., 1974, Phosphenes produced by electrical stimulation of human occipital cortex, and their application to the development of a prosthesis for the blind, *J. Physiol. (London).* **243**: 553-576.

Dobelle, W. H., 1994, Artificial vision for the blind. The summit may be closer than you think, *ASAIO J.* **40**: 919-922.

Dobelle, W. H., Mladejovsky, M.G, Evans J. R., et al., 1976, "Braille" reading by a blind volunteer by visual cortex stimulation, *Nature.* **259**: 111-112.

Dowling, J. E., 1987, *The Retina: An Approachable Part of the Brain*, Belknap Press, Cambridge.

Eckmiller, R., 1997, Learning retina implants with epiretinal contacts, *Ophthalmic Res.* **29**: 281-289.

Foerster, O., 1929, Beitrage zur pathophysiologie der sehbahn und der spehsphare, *J. Psychol. Neurol. (Lpz).* **39**: 435-463.

Fritsch, G. and Hitzig J., 1870, Ueber die elecktrische erregbarkeit des grosshirns. *Arch. Anat. Physiol.* **37**: 300-332.

Galvani L., 1791, De viribus electricitatis in motu musculary, commentarius. *De Bononiensi Scientiarum et Artium Instituto atque Academia.* **7**: 363-418.

Glenn ,W., Mauro, E., Longo, P. , et al., 1959, Remote stimulation of the heart by radio frequency transmission, *New Eng. J. Med.* **261**: 948.

Gorman, P. H. and Mortimer, J. T., 1983, The effect of stimulus parameters on the recruitment characteristics of direct nerve stimulation, *IEEE Trans. Biomed. Eng.* **30**: 407-414.

Greenberg, R. J., 1998, *Analysis of Electrical Stimulation of the Vertebrate Retina–Work Towards a Retinal Prosthesis*, Ph.D. Dissertation, The Johns Hopkins University, Baltimore, MD.

Greenberg, R. J., Velte, T.J., Humayun, M. S., et al., 1999, A computational model of electrical stimulation of the retinal ganglion cell, *IEEE Trans. Biomed. Eng.* **46**: 505-514.

Grumet, A. E., Rizzo, J. F., and Wyatt, J. L., 1999, Ten Micron Diameter Electrodes Directly Stimulate Rabbit Retinal Ganglion Cell Axons, *Invest. Ophthal. Vis. Sci.* **40**: S734.

Grumet, A. E., Rizzo, J. F., and Wyatt, J., 2000, In-vitro electrical stimulation of human retinal ganglion cell axons, *Invest. Ophthal. Vis. Sci.* **41**: S10.

Guenther, E., Troger, B., Schlosshauer, B., and Zrenner, E., 1999, Long-term survival of retinal cell cultures on retinal implant materials, V*ision Res.* **39**: 3988-3994.

Heetderks, W. J., 1988, RF powering of millimeter and submillimeter sized neural prosthetic implants, *IEEE Trans. Biomed.* Eng. **35**: 323-326.

Heiduschka, P. and Thanos, S., 1998, Implantable bioelectronic interfaces for lost nerve functions, *Prog. Neurobiol.* **55**: 433.

Hetke, J. F., Lund, J. L., Najafi, K., et al., 1994, Silicon ribbon cables for chronically implantable microelectrode arrays, *IEEE Trans. Biomed. Eng.* **41**: 314-321.

Hodgkin, A. and Huxley, A., 1952a, Currents carried by sodium and potassium ions through the membrane of the giant axon of loligo, *J. Physiol.* **116**: 472-49.

Hodgkin, A. and Huxley, A., 1952b, A quantitative description of membrane current and its application to conduction and excitation in nerve, *J. Physiol.* **116**: 500-544.

Humayun, M., 1994, *Is Surface Electrical Stimulation of the Retina a Feasible Approach Towards the Development of a Visual Prosthesis?* Ph.D. Dissertation, Johns Hopkins University School of Medicine, Baltimore, MD.

Humayun, M., de Juan, E., Jr., Dagnelie,G., Greenberg, R., Propst, R., and Phillips, H., 1996, Visual perception elicited by electrical stimulation of retina in blind humans, *Arch. Ophthalmol.* **114**: 40-46.

Humayun, M. S., de Juan, E. J, Weiland, J.D., et al., 1999, Pattern electrical stimulation of the human retina, *Vision Res.* **39**: 2569-2576.

Humayun, M. S., de Juan, E. J., Dagnelie, G., et al., 1996, Visual perception elicited by electrical stimulation of retina in blind humans, *Arch. Ophthalmol.* **114**: 40-46.

Humayun, M. S., Prince, M., de Juan, E. J., et al., 1999, Morphometric analysis of the extramacular retina from postmortem eyes with retinitis pigmentosa, *Invest. Ophthal. Vis. Sci.* **40**: 143-148.

Humayun, M. S., Propst, R., de Juan, E. J., et al., 1994, Bipolar surface electrical stimulation of the vertebrate retina, *Arch. Ophthalmol.* **112**: 110-116.

Janders, M., Egert, U., Stelze, M., and Nisch, W., 1996, Novel thin-film titanium nitride micro-electrodes with excellent charge transfer capability for cell stimulation and sensing applications, *Proc. 19th Int. Conf. IEEE/EMBS*, pp. 1191-1193.

Jones, K. E. and Normann, R. A., 1997, An advanced demultiplexing system for physiological stimulation, *IEEE Trans. Biomed. Eng.* **44**:1210-1220.

Karny, H., 1975, Clinical and physiological aspects of the cortical visual prosthesis, *Surv. Ophthalmol.* **20**: 47-58.

Knighton, R. W., 1975a, An electrically evoked slow potential of the frog's retina. I. Properties of response, *J. Neurophysiol.* **38**: 185-197.

Knighton, R. W., 1975b, An electrically evoked slow potential of the frog's retina. II. Identification with PII component of electroretinogram, *J. Neurophysiol.* **38**: 198-209.

Kolb, H., Fernandez, E., and R. Nelson, *Web Vision* an internet resource, located at http://webvision.med.utah.edu

Kolb, H., Linberg, K. A., and Fisher, S. K., 1992, The neurons of the human retina: a Glogi study, *J. Comp. Neurol.* **318**: 147-187.

Kovacs, G. T., Storment, C. W., Rosen, J. M., 1992, Regeneration microelectrode array for peripheral nerve recording and stimulation, *IEEE Trans. Biomed. Eng.* **39**: 893-902.

Krause, F. and Schum, H., 1931, Die epiliptischen erkankungen, in: *Neue Deutsche Shirurgie*, H. Kunter, ed., Stuttgart, Chap. 49a, pp 482-486.

Laing, P. G., Ferguson, A. B., Jr., and Hodge, E. S., 1967, Tissue reaction in rabbit muscle exposed to metallic implants, *J. Biomed. Mater. Res.* **1**: 135-149.

Lilly, J. C., 1961, Injury and excitation by electric currents: The balanced pulse-pair waveform, in: *Electrical Stimulation of the Brain,* D. E. Sheer, ed., Hogg Foundation for Mental Health, pp. 60-64.

Liu W, Vichienchom K, Clements M, Demarco C, Hughes C, McGucken E, Humayun MS, de Juan E. Jr., Weiland J. D., 2000, A neuro-stimulus chip with telemetry unit for retinal prosthesis device. *IEEE Solid-State Circuits.* **35**: 1487-1497.

Majji, A. B, Humayun, M. S, Weiland, J. D., et al., 1999, Long-term histological and electrophysiological results of an inactive epiretinal electrode array implantation in dog, *Invest. Ophthal. Vis. Sci.* **40**: 2073-2081.

Margalit,, E., Fujii, G., Lai J, et al., 2000, Bioadhesives for intraocular use, *Retina*, **20**: 469-477.

Maynard, E. M., Nordhausen, C. T., and Normann, R. A., 1997, The Utah intracortical electrode array: a recording structure for potential brain-computer interfaces, *Electroencephalogr. Clin. Neurophysiol.* **102**: 228-239.

McCreery, D. B., Agnew, W. F., Yuen, T. G., and Bullara, L. A., 1988, Comparison of neural damage induced by electrical stimulation with faradaic and capacitor electrodes, *Ann. Biomed. Eng.* **16**: 463-481.

McCreery, D. B., Agnew, W. F., Yuen, T. G.H., and Bullara, L., 1990, Charge density and charge per phase as cofactors in neural injury induced by electrical stimulation, *IEEE Trans. Biomed. Eng.* **37**: 996-1001.

McHardy, J., Robblee, L. S., Marston, J. M., and Brummer, S. B., 1980, Electrical stimulation with pt electrodes. IV. Factors influencing Pt dissolution in inorganic saline, *Biomater.* **1**: 129-134.

Nordhausen, C. T., Maynard, E. M., and Normann, R. A.., 1996, Single unit recording capabilities of a 100 microelectrode array, *Brain Res.* **726**: 129-140.

Normann, R. A., 1999, MERPWD. A neural interface for a cortical vision prothesis, *Vision Res.* **39**: 2577-2587.

Østerberg, G. (1935) Topography of the layer of rods and cones in the human retina, Acta Ophthal. supp. **6**: 1-103.

Ogden, T.E. (1989) *Retina: Basic Science and Inherited Retinal Disease, Vol 1.* The CV Mosby Co., St. Louis.

Penfield, W. and Jasper, H., 1954, *Epilepsy and the Functional Anatomy of the Human Brain*, Churchill, London.

Penfield, W. and Rasussen, T., 1952, *The Cerebral Cortex of Man*, Macmillan, New York, pp. 135-147.

Peyman, G., Chow, A. Y, Liang, C., et al., 1998, Subretinal semiconductor microphotodiode array, *Ophthalmic Surg. Lasers.* **29**: 234-241.

Pollen, D. A., 1977, Responses of single neurons to electrical stimulation of the surface of the visual cortex, *Brain Behav. Evol.* **14**: 67-86.

Polyak, S.L. (1941) *The Retina.* University of Chicago Press, Chicago.

Potts, A. M. and Inoue J., 1970, The electrically evoked response of the visual system (EER) III. Further consideration to the origin of the EER, *Invest. Ophthal Vis. Sci..* **9**: 814-819.

Potts, A. M. and Inoue, J., 1969, The electrically evoked response (EER) of the visual system II. Effect of adaptation and retinitis pigmentosa, *Invest. Ophthal. Vis. Sci.* **8**: 605-612.

Potts, A. M., Inoue, J., and Buffum, D., 1968, The electrically evoked response of the visual system (EER), *Invest. Ophthal. Vis. Sci.* **7**: 269-278.

Pudenz, R. H., Bullara, L. A., Dru, D., and Talalla, A., 1975a, Electrical stimulation of the brain. II. Effects on the blood-brain barrier, *Surg. Neurol.* **4**: 265-270.

Pudenz, R. H., Bullara, L. A., Jacques, S., and Hambrecht, F. T., 1975b, Electrical stimulation of the brain. III. The neural damage model, *Surg. Neurol.* **4**: 389-400.

Pudenz, R.H., Bullara, L.A., and Talalla, A., 1975c, Electrical stimulation of the brain. I. Electrodes and electrode arrays, *Surg. Neurol.* **4**: 37-42.

Rita, P., Kaczmarek, K A, Tyler, M. E., and Garcia-Lara, J., 1998, Form perception with a 49-point electrotactile stimulus array on the tongue: a technical note, *J. Rehabil. Res. Dev.* **35**: 427-430.

Rizzo, J. and Wyatt, J., 1997, Prospects for a visual prosthesis, *Neuroscientist*, **3**: 251-262.

Rizzo, J., Wyatt, J., Loewenstein, J., and Kelly, S., 2000, Acute intraocular retinal stimulation in normal and blind humans, *Invest. Ophthal. Vis. Sci.* **41**: S102.

Robblee, L. S., Mangaudis, M., Lasinski, E., et al., 1986, Charge injection properties of thermally-prepared iridium oxide films, *Mat. Res. Soc. Symp. Proc.* **55**: 303-310.

Santos, A., Humayun, M. S., de Juan, E. J., et al., 1997, Preservation of the inner retina in retinitis pigmentosa. A morphometric analysis, *Arch. Ophthalmol.* **115**: 511-515.

Sato, S., Sugimoto, S., and Chiba, S., 1982, A procedure for recording electroretinogram and visual evoked potential in conscious dog, *J. Pharmacol. Methods.* **8**: 173-181.

Schmidt, E. M, Bak, M. J, Hambrecht, F. T, et al., 1996, Feasibility of a visual prosthesis for the blind based on intracortical microstimulation of the visual cortex, *Brain.* **119** (Pt 2): 507-522.

Schmidt, E. M, Bak, M. J., and Christensen, P., 1995, Laser exposure of Parylene-C insulated microelectrodes, *J. Neurosci. Methods* **62**: 89-92.

Schwarz, M. et al., 1999, Single-chip CMOS image sensors for a retina implant system, *IEEE Trans. Circuits Syst.-II: Analog Dig. Signal Proc.* **46**: 870-877.

Scribner, D. A., Kruer, M. R., and Killiany, J. M., 1991, Infrared focal plane array technology, *Proc. IEEE.* **79**: 65-85.

Shapley, R. and Enroth-Cugell, C., 1984, Visual adaptation and retina gain controls, in *Progress in Retinal Research*, Vol. **3**, N. N. Osborne and G. J. Chader, eds. , Pergamon, New York.

Shyu, J., Maia, M., Weiland, J., et al., 2000, Electrical Stimulation of Isolated Rabbit Retina, Biomedical Engineering Society Annual Meeting, Seattle, WA. **28**: S115.

Sterling, T. D. and Vaughn, H. G., Jr., 1971, Feasibility of electrocortical prosthesis, in: *Visual Prosthesis: The Interdisciplinary Dialogue*, T.D. Sterling et al., ed., Academic Press, New York.

Stone, J. L., Barlow, W. E., Humayun, M. S., et al., 1992, Morphometric analysis of macular photoreceptors and ganglion cells in retinas with retinitis pigmentosa, *Arch. Ophthalmol.* **110**: 1634-1639.

Suzuki, S., Humayun, M., de Juan, E., et al., 1999, A comparison of electrical stimulation threshold in normal mouse retina vs. different aged retinal degenerate (rd) mouse retina, *Invest. Ophthal. Vis. Sci.* **40**: S735.

Teeters, J., Jacobs, A., and Werblin, F., 1997, How neural interactions form neural responses in the Salamander retina, *J. Comp. Neurosci.* **4**: 5.

Tehovnik, E., 1996, Electrical stimulation of neural tissue to evoke behavioral responses, *J. Neurosci. Methods.* **65**: 1-17.

Thompson, R., Barnett, D., Humayun, M., and Dagnelie, G., 2000, Reading speed and facial recognition using simulated prosthetic vision, *Invest. Ophthal. Vis. Sci.* **41**: S860.

Tonucci, R.J. and Justus, B.L., 1993a, *Nanocchannel Glass Matrix Used in Making Mesoscopic Structures*, U.S. Patent 5,264,722, issued November 1993.

Tonucci, R.J. and Justus, B.L., 1993b, *Nanochannel Filter*, U.S. Patent 5,234,594, issued August 1993.

Tonucci, R. J., Justus, B. L., Campillo, A. J., and Ford, C. E., 1992, Nanochannel array glass, *Science.* **258**: 783-785.

Toyoda, J. and Fujimoto, M., 1984, Application of transretinal current stimulation for the study of bipolar-amacrine transmission, *J. Gen. Physiol.* **84**: 915-925.

Troyk, P. and Schwan, M., 1992, Closed-loop class E transcutaneous power and data link for microimplants, *IEEE Trans. Biomed. Eng.* **39**: 589-599.

Turner, J. N., Shain, W., Szarowski, D. H., et al., 1999, Cerebral astrocyte response to micromachined silicon implants, *Exp. Neurol.* **156**: 33-49.

Veraart, C., Raftopoulos, C., Mortimer, J. T., et al., 1998, Visual sensations produced by optic nerve stimulation using an implanted self-sizing spiral cuff electrode, *Brain Res.* **813**: 181-186.

Weiland, J. D. and Anderson, D. J., 2000, Chronic neural stimulation with thin-film, iridium oxide stimulating electrodes, *IEEE Trans. Biomed. Eng.* **47**: 911-918.

Weiland, J. D., Humayun, M. S., Dagnelie, G., et al., 1999, Understanding the origin of visual percepts elicited by electrical stimulation of the human retina, *Graefes Arch. Clin. Exp. Ophthalmol.* **237**: 1007-1013.

West, D. C. and, Wolstencroft, J. H., 1983, Strength-duration characteristics of myelinated and non-myelinated bulbospinal axons in the cat spinal cord, *J. Physiol. (London).* **337**: 37-50.

Wiley, J. D. and Webster, J. G., 1982, Analysis and control of the current distribution under circular dispersive electrodes, *IEEE Trans. Biomed. Eng.* **29**: 381-385.

Wise, K. D., Angell, J., and Starr, A., 1970, An integrated-circuit approach to extracellular microelectrodes, *IEEE Trans. Biomed. Eng.* BME-**17**:238-247.

Wyatt, J. and Rizzo, J. F., 1996, Ocular implants for the blind, *IEEE Spectrum.* **112**: 47-53.

Yagi, T. and Hayashida, Y., 1999, Implantation of the artificial retina, *Nippon Rinsho.* **57**: 1208-1215.

Yagi, T. and Watanabe, M. A., 1998, A computational study on an electrode array in a hybrid retinal implant, *Proc. of 1998 IEEE Int. Joint Conf. on Neural Networks*, pp. 780-783.

Zrenner, E., Stett, A., Weiss, S., et al., 1999, Can subretinal microphotodiodes successfully replace degenerated photoreceptors? *Vis. Res.* **39**: 2555-2567.

Zuidema, P., Koenderink, J. J., and Bouman, M. A., 1983, A mechanistic approach to threshold behavior of the visual system, *IEEE Trans. Syst. Man Cybernet.* **SMC-13**: 923.

Chapter 5

Push-Pull Model of Dopamine's Action in the Retina

Iván Bódis -Wollner[1] and Areti Tzelepi[1, 2]

[1]Department of Neurology, State University of N.York, Health Science Center at Brooklyn, Box 1213, 450 Clark Ave., Brooklyn, NY 11203-2098, PH: (718) 270-1482, FX: (718) 270-3840, EM: bodisi01@bmec.hscbklyn.edu

[2]Dept. of Computer Science, Hellenic Naval Academy, Terma Hatzikiriakou St., Piraeus 18539, Greece, PH: +30-1-4225292, EM: ikoyl@tee.gr

5.1 RETINAL CIRCUITRY

Although our knowledge of the interactive connections in the mammalian retina has greatly expanded over the last 20 years, the exact wiring and functional role of this complex network has not yet been elucidated. Ganglion cells receive input from the vertical pathways (photoreceptors – bipolar cells – ganglion cells) and lateral pathways (photoreceptors – horizontal cells – bipolar cells – amacrine cells) of retinal interneurons (Werblin and Dowling, 1969). Contrary to the ganglion cell responses which can be recorded from the optic nerve, the three classes of retinal interneurons (horizontal, amacrine and bipolar cells) are not easily accessible and recordings are further limited by their small size, especially in higher vertebrates. Together with photoreceptors and the ganglion cells, they make up the complex network of the mammalian retina. Ganglion cells have a characteristic center-surround antagonistic organization in their receptive fields (Kuffler, 1953; Rodieck and Stone, 1966; Enroth-Cugell and Robson, 1966) (Fig. 5.1). It has been shown, however, that bipolar neurons of the primate, which precede ganglion cells in the vertical pathway, also have surrounds (Dacey et al, 2000).

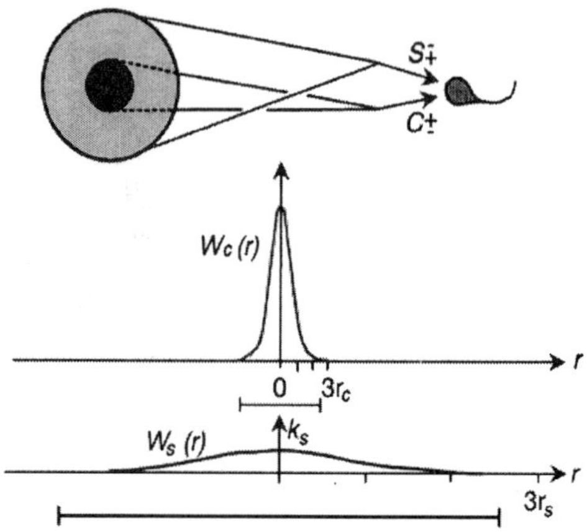

Figure 5.1. The receptive field model representing signal summation over a retinal ganglion cell receptive field, as described by Enroth-Cugell and Robson (1966), with permission of J. Physiology. Upper diagram illustrates the concentric center and surround region. Signals from the center (C) and surround (S) have an antagonistic effect on the ganglion cell, expressed by the opposite sign of the C and S signal, either an on center (+C) or an off-center (-C). Lower diagrams show the gaussian profiles, assumed to describe the sensitivity of the center and surround. The bars drawn below the center and the surround weighting functions are $5r_c$ and $5r_s$ respectively.

Based on studies on lower vertebrates, horizontal cells are thought to dominate in the surround organization, while their direct influence on the center is negligible (Mangel and Dowling, 1985). Horizontal cells can form local circuits that sum information from a wide spatial area in the outer plexiform layer through feedback connections. A feedback connection from horizontal cells to the photoreceptors has also been described in lower vertebrates (Baylor et al., 1971; Djamgoz and Kolb, 1993). The bipolar cells, driven by the photoreceptor input, include an inhibitory response from a large area in the retina, resulting in their surround organization.

Bipolar cells represent the more direct pathway from photoreceptors to ganglion cells, carrying information from the outer to inner plexiform layer. Anatomical studies using various techniques revealed at least nine different types of bipolar cells (Boycott and Wassle, 1991; Kolb et al., 1992) in the human retina, eight related to cones and one related to rods. Five of them receive convergence of information from many cones and are known as diffuse cone bipolars. Three cone bipolar types appear to have single cone contacts in a one-to-one relationship and, are known as midget bipolar, or blue cone of a specific type. The center of their receptive field appears to be directly connected to the cones. Receptive field surrounds at the bipolar cell level have also been described in the non-mammalian retina (Werblin and Dowling, 1969; Matsumoto and Naka, 1972; Schwartz, 1974), originating from horizontal cell feedback connections (Kamermans and Spekreijse, 1999). Recently, Dacey et al (2000) demonstrated evidence of surround organization of bipolars in the monkey. Interestingly, they showed that although both midget and diffuse bipolar types are characterized by center-surround organization, small "midget" bipolars are more center dominated, while "large" diffuse bipolars are more surround dominated.

Amacrine cells in the inner plexiform layer belong to the lateral interconnecting network in the pathway from photoreceptors to ganglion cells. (Mariani, 1990; Kolb et al. 1992). Similar to horizontal cell feedback connections in the outer plexiform layer, amacrine cells support feedback mechanisms in the innerplexiform layer. Consequently, amacrine cells also participate in the formation of the surround receptive field in bipolars. From neuroanatomical studies, at least 25 morphologically different amacrine cell types have been identified in the primate retina.

Dopaminergic (DA) cells are found in all vertebrates, including primates, mostly in amacrine cells. In fish (Lasater and Dowling, 1985) and turtle (Piccolino et al, 1989) retina, DA cells are found to influence significantly horizontal cell activity by uncoupling horizontal cell junctions. In the mammalian retina, a similar but weaker effect has been reported (Xin and Bloomfield, 1999). However, in mammals, amacrine cells are dopaminergic and their effects are expected to be more pronounced in the inner plexiform layer. They may influence other amacrines in a lateral interconnecting circuit in the inner plexiform layer, or connect other neurons in a feedback manner via paracrine release of DA. Additionally, they may electrotonically interact with each other, as seen with electrotonic cell junctions of horizontal cells in the outer plexiform layer. Some amacrines may uncouple interconnected amacrine cells, as has been observed with AII cell type (Vaney, 1990; 1994). In the mammalian retina, various DA receptors have been identified in several levels, falling in two major classes: D1 and D2 receptors (McGonigle et al, 1988; Denis et al, 1990; Schorderet and Nowak, 1990; Strormann et al, 1990).

Ganglion cells are larger than the preceding interneurons and through their axons which form the optic nerve, the visual information is passed to higher brain centers. A great deal of our knowledge of ganglion cells came from studies on the cat retina, which is probably the most extensively studied retina in mammals. There are different types of ganglion cells in the cat retina. A successful morphological classification scheme was proposed by Boycott and Wassle (1974), describing the most common ganglion cells into four different cell types, alpha, beta, gamma and delta types. This morphological classification has been related to the physiological classification in X, Y and W cells (Boycott and Wassle, 1974, Enroth-Cugell and Robson, 1966; Cleland and Levick, 1974). The correspondence between morphology and physiology in the two independent pathways hints to a parallel organization in the retina.

Enroth-Cugell and Robson (1966) described the existence of two ganglion cell types, X and Y, with different spatiotemporal characteristics. Their "null" test, the introduction and withdrawal of a sinusoidal grating at $\pm 90^\circ$ relative to the receptive field center, left no doubt about X ganglion cells linear response versus Y ganglion cells non-linear response. Subsequent studies further supported a correspondence between alpha and Y cells, and beta and X cells. W cells revealed various morphological types including gamma and delta cells (Fukuda et al, 1984; Fukuda et al, 1985). Another classification scheme was proposed by Kolb et al (1981), who described morphological types of ganglion cells including and going beyond the gamma and delta cells, named from G4-G23 (from smallest types, G4 to largest types at G23).

In the human retina, neuroanatomical, but not neurophysiological, studies have revealed at least 16 morphological types of ganglion cells. Almost all of them correspond to ganglion cells in the cat retina. The most common ganglion cell types in human are the parasol and the midget ganglion cells (Rodieck et al., 1985; Kolb et al., 1992; Dacey, 1993), which correspond to the cat a and b cells, respectively. They are also known as P cells (midget ganglion cells) because they project to the parvocellular layers of the LGN, and M cells (parasol ganglion cells) because they project to the magnocellular layers of the LGN (Shapley and Perry, 1986; Kaplan and Shapley, 1986). P and M systems mediate different signal properties. M cells have larger receptive fields and respond to large stimuli concerned more about gross features and movement. P cells have smaller receptive fields. They respond to smaller stimuli, and are thought to process fine detail and color. The most common ganglion cell types in the cat and the primate retina and their correspondence are summarized in Table 5.1.

Table 5.1. Morphological and physiological correlates of the most common ganglion cell types in the cat and the primate retina.

Cat	α	β	γ	δ	morphology
	Y	*X*	*W*		physiology
Primate	*Parasol*	*midget*			morphology
	M	*P*			physiology

The spatial arrangement of the two ganglion cell types is not random. In the cat, small beta cells show high concentration in the central part of the retina, converging information from only a few cones. In humans, this concentration of P or midget cells in the fovea can reach a one-to-one relationship with cones. Large alpha cells dominate mostly in the periphery, but with a much lower convergence ratio.

5.2 RECEPTIVE FIELDS OF GANGLION CELLS

The output of ganglion cells is based on the antagonism between the center and the surround of their circular receptive field. Kuffler (1956), using small spots of light, showed that these excitatory and inhibitory areas are organized concentrically and were best described as a circular center surrounded by a ring of opposite polarity.

This mechanism has been studied extensively using small spots of light, as well as, sinusoidal grating patterns. Enroth-Cugell and Robson (1966), and Rodieck (1965), proposed a model of the experimental results based on the difference of two Gaussian functions (Fig. 5.1). The center and the surround of the receptive field can be described by one Gaussian, as defined by its radius r and strength w, respectively. Each Gaussian peaks at the receptive field center, and their interaction is subtractive. As an extension of Kuffler's concept, it was shown that "center" and "surround" are co-extensive. While this model cannot account for all ganglion cells responses, and it mainly explains the behavior of the X-cells of the cat retina, it does give a satisfactory explanation for the most characteristic property of retinal ganglion cells: spatial tuning. Different ganglion cells respond optimally to different stimulus sizes. Spatial tuning of individual ganglion cells is reflected in the contrast sensitivity function that has an inverted "U" shape. The contrast sensitivity function expresses the minimum contrast required to detect a stimulus as a function of its size under equiluminant conditions. In other words, it is a measure of retinal temporal or spatial modulation of the local energy distribution. The overall contrast sensitivity curve, obtained with psychophysical methods in either human or monkey, initially increases with spatial frequency, reaches a peak around 5 cycles per degree (cpd), and then it falls off sharply with increasing spatial frequency. The difference-of-Gaussians model can successfully describe the contrast sensitivity function. The striking similarity of the contrast sensitivity curve to the X ganglion cell response raises the question of whether or not contrast sensitivity simply reflects the profile of a single group of cells. Numerous clinical studies have revealed abnormal contrast sensitivity functions (Bódis-Wollner, 1972; Bódis-Wollner and Camisa, 1980), suggesting selective damage of different pathways. It is unknown, however, whether selective spatial frequency losses, such as shown in multiple sclerosis (Regan et al, 1991) represent optic nerve deficits, i.e. retinal ganglion cell properties.

Not all mammalian ganglion cells code contrast exclusively. There are certain types that exhibit directional selectivity. Directional selective cells respond to stimuli moving in a preferred direction and are inhibited by stimuli moving in the opposite direction, with this depending also on how fast or slow the stimulus moves. Most of the studies on directional selectivity come from the rabbit retina, which has a different organization from the cat and the primate. Directional selective cells have also been observed in the monkey (DeMonasterio and Gouras, 1975). The exact function and wiring of these cells is still not known. The challenge of a model of directionally-selective ganglion cells in the rabbit has been attempted by Grzywacz et al (1998), and also He and Masland (1997).

5.3 RETINA PROCESSING AND DA ACTION

Massed retinal ganglion cell processes in vivo in humans can be studied by recording the pattern electroretinogram (PERG). This electrophysiological measurement most likely represents the activity of diverse ganglion cells (Maffei and Fiorentini, 1986) and indirectly the preganglionic processing in the retina. In monkeys, optic nerve section results in the loss of the PERG, while diffuse light still elicits an ERG response (Maffei et al, 1985). PERG studies using sinusoidal gratings stimuli have clearly shown that the PERG does depend on the spatial frequency of the eliciting stimulus, and furthermore that it actually responds as a spatial transfer tuning function, similarly to the contrast sensitivity curve in monkey and human.

The peak of the spatial transfer function and its descending limbs are differentially vulnerable to neuronal pathology (Marx et al, 1988; Ghilardi et al, 1988; Bódis-Wollner, 1996; Tagliati et al, 1996). The decrease at low spatial frequencies can be attributed to decreasing contribution of low frequency selective neurons in the central retina, either because of their scarcity or the dominating surround mechanisms, or both. The sharp decrease at higher spatial frequencies can be related to the relatively increasing contribution from center mechanisms.

Dopamine has a differential effect on the two limbs of the PERG spatial transfer function. The two major classes of dopamine receptors, D1 and D2, although acting antagonistically at the cellular level (Picollino et al, 1987), act synergistically at the overall-response level to establish the tuned spatial transfer function in the retina. More specifically, we suggested that dopamine modulates spatial contrast by a "push-pull" action on center and surround mechanisms of different ganglion cells in the primate retina (Bódis -Wollner and Tzelepi, 1998). In the following pages, we will summarize previous experimental results using different DA manipulations, on which we will base a model for dopamine action in retinal spatial processing.

5.4 DOPAMINERGIC EFFECTS ON THE PERG IN THE MONKEY

5.4.1 Retinal spatial tuning in the MPTP primate model

One method for dopamine depletion in the macaque retina is by 1-methyl 4-phenyl 1,2,3,6-tetrahydropyridine (MPTP) treatment. It has been shown that MPTP causes a Parkinsonian syndrome in monkey and man via its oxidation product, MPP+ (Nicklas et al, 1987). Histological and neurochemical findings suggest that MPTP decreases the release of dopamine, and that it is selectively toxic to the pigmented dopaminergic neurons of the pars compacta of the substantia nigra in monkeys (Burns et al, 1983; Ghilardi et al, 1988), similarly to the MPTP intoxication observed in humans (Davis et al, 1979). The monkey model is behaviorally and pharmacologically nearly identical to the human Parkinson disease (Burns et al, 1983). MPTP was first shown to cause DA cell destruction in the rabbit retina (Wong et al, 1985). Using systemic MPTP treatment (5mg/kg), the monkeys developed a bilateral Parkinsonian syndrome. Simultaneous recordings of VEPs and PERG revealed spatial frequency dependent losses in monkeys treated with MPTP (Ghilardi et al, 1988). Administration of L-Dopa with carbidopa improved significantly the VEP and PERG (Fig. 5.2). The tuning ratio, which is defined as the ratio of the amplitude to the peak response (around 3-4cpd) divided by the response amplitude to the low spatial frequency (0.5cpd), provides a good estimate of retinal spatial tuning. The tuning ratio, after MPTP treatment, decreased on the average by a factor of 2. PERG responses to 2.5 and 3.5cpd stimuli became abnormal, while responses at 0.5 and 1.2cpd were less impaired. The average tuning ratio of five monkeys was 0.88 before (baseline condition) and 0.38 after MPTP treatment.

5.4.2 The Effects of Intravitreal 6-OHDA on Spatial Tuning

Another method for dopamine depletion in the macaque retina is based on injection of the neurotoxin 6-hydroxydopamine (6-OHDA) into the vitreous of the eye (Ghilardi et al, 1989). It has been proposed that this drug destroys dopaminergic neurons in the nigrostriatal system and decreases dopaminergic neurotransmission (Ungerstedt and Arbuthnott, 1970). The retinal effect of 6-OHDA was demonstrated in the carp (Cohen and Dowling, 1983), in the turtle (Witkovsky et al, 1987), and in the rabbit (Oliver et al, 1986). Stimuli of the same spatial frequency as in the MPTP study were used to compare results. The same effect was observed: PERG and VEP responses were significantly attenuated in response to 2.5 and 3.5cpd stimuli, while responses at lower spatial frequencies (0.5cpd and 1.2cpd) were less impaired after repeated treatments. The final results show a profound effect of 6-OHDA in three monkeys on the responses at 3.5cpd patterns. The spatial tuning ratio became less than one in all three monkeys. The tuning ratio was changed by a factor of 2, 5 and 3, respectively, in the three monkeys. Thus the effects of systemically administered MPTP on the PERG and the VEP are similar to the effects of intravitreal injections of 6-OHDA. However repeated administration of 6-OHDA has a more profound effect on spatial tuning than MPTP does.

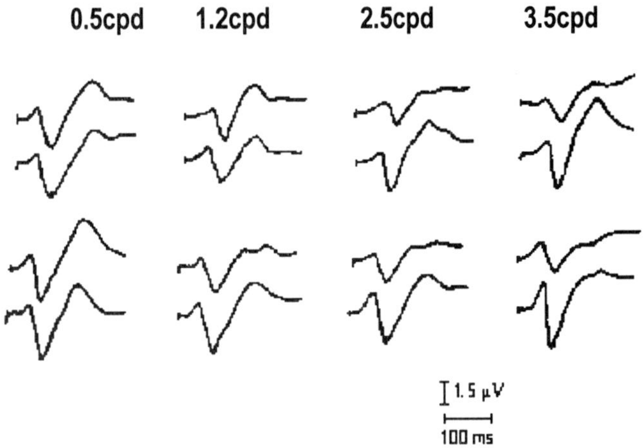

Figure 5.2. Spatial frequency dependent effect of dopamine in the transient PERG of the macaque retina. Traces from two runs for the same condition, are superimposed. For each row, top represents the PERG in the dopamine deficient monkey model of Parkinson disease (PD), induced by MPTP; bottom represents the transient PERG after administration of L-Dopa with carbidopa. A. Responses 20 days following the initial MPTP treatment. B. Responses 40 days following the initial MPTP treatment. Notice the transient and spatial frequency dependent effect of levodopa: it increased PERG amplitude (and slightly increased latency) for "peak" spatial frequencies without much effect at low spatial frequencies (after Ghilardi et al, 1988 with permission of Elsevier Science).

5.4.3 The Effect of the D2 Receptor Blocker L-Sulpiride on Spatial Frequency Tuning

Besides dopaminergic neuronal toxins, three different types of dopamine receptor-ligands in the monkey were used. L-Sulpiride, a D2 antagonist (Tagliati et al, 1994), was injected in three monkeys. The concentration of L-

Sulpiride was varied to evaluate whether there was a dose-dependent effect of the drug. For this reason, we separately explored the effects of 0.07 and 0.35 mg/kg, in separate experiments. The PERG was recorded before and after the administration of L-Sulpiride. The higher dose suppressed responses at all spatial frequencies. The lower dose attenuated responses only at the peak of the spatial frequency curve, resulting in a decrease of the tuning ratio (by a factor of 2.5, 1.7 and 3.2) in the three monkeys (Fig. 5.3).

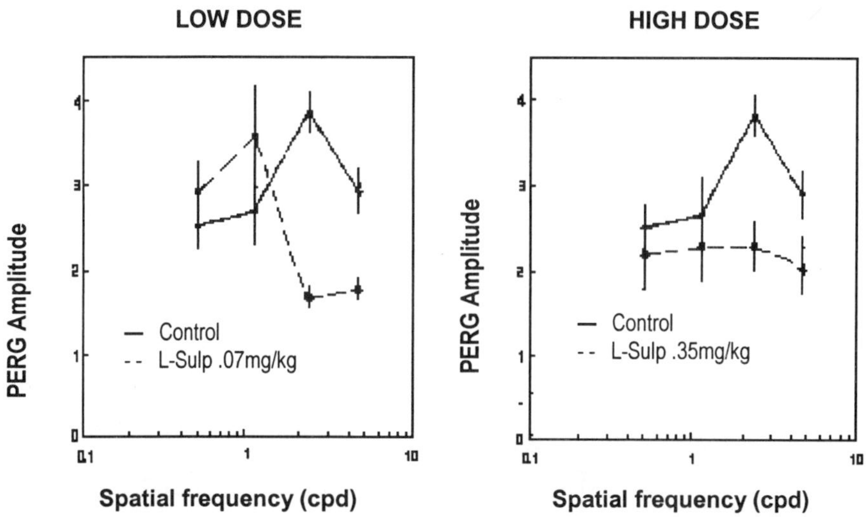

Figure 5.3. Effect of L-Sulpiride on the PERG spatial transfer function for one monkey. In lower dose, L-Sulpiride attenuates peak spatial frequency responses. In higher dose, lower spatial frequencies are also affected. The plot represents the average of 3 runs for each spatial frequency. (after Tagliati et al, 1994 with permission of Elsevier Science).

5.4.4 The Effect of CY 208-243, a D1 Agonist on Spatial Tuning

The PERG responses to a range of spatial frequencies between 0.5 and 6.9cpd were recorded before and after the administration of a D1 receptor agonist, CY 208-243. Following drug administration, the responses to the low spatial frequency stimuli were significantly decreased. As can be seen in Fig. 5.4, the amplitude of the responses at 0.5cpd after drug administration was at the noise level, while the responses to 2.3cpd remain almost intact. The suppression of low spatial frequency responses after using a D1 agonist is opposite to the effect of the low dose D2 antagonist L-Sulpiride, which attenuated middle spatial frequency responses.

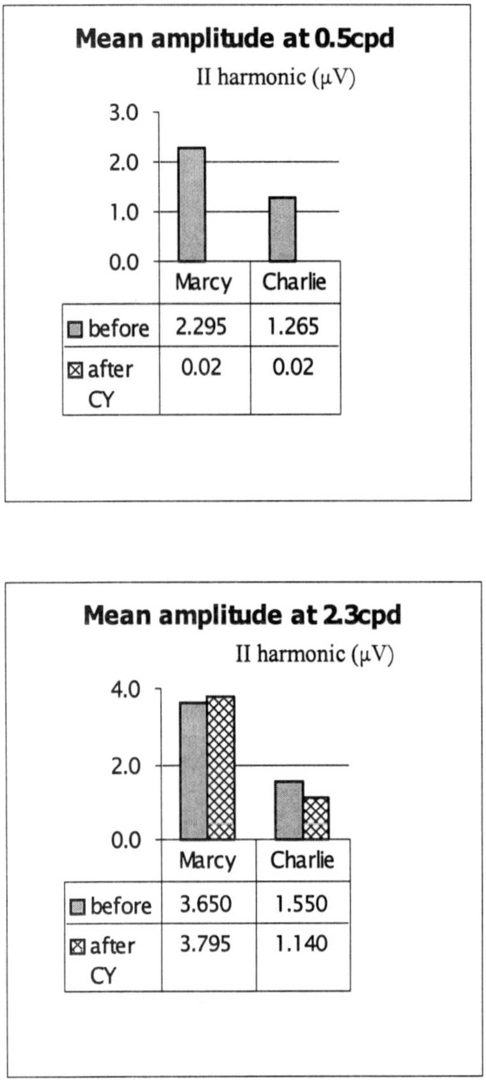

Figure 5.4. The effect of a D1 agonist, CY208-243 on the PERG responses on two monkeys, Marcy and Charlie (after Peppe et al, 1998 with permission of Elsevier Science). Notice that this D1 agonist suppressed selectively low spatial frequencies.

5.4.5 Synthesis of Experimental Results

From the experiements, we observed that different manipulation of DA mechanisms revealed three different kind of changes in the PERG spatial tuning function (Table 5.2).

First, we observed similar results from the PERG experiments with 6-OHDA and MPTP. They showed that dopamine depletion resulted in decreased amplitude at peak spatial frequencies. A slight increase in low spatial frequencies was also observed; however, it was not significant. The loss of the band-pass nature of the spatial transfer function was not identical in the two experiments. The use of 6-OHDA yielded a more profound loss than MPTP did. The same effect was observed with a D2 blocker in low dose L-Sulpiride. This similarity suggested that a deficiency related to the D2 receptors attenuated medium spatial frequency responses, thereby rendering the spatial transfer function effectively a low-pass filter. It also showed that the degree of loss varied, although the qualitative effect was the same. However, the similarity of pre-synaptic neuronal damage to the effects of low dose L-Sulpiride suggests that it is probable that D2 receptors are located post-synaptically in the preganglionic retina.

The opposite effect was observed with CY208-243, a D1 agonist. The PERG spatial transfer function showed a suppression of low spatial frequencies. It follows that a D1 deficiency would result in an increase of lower spatial frequency responses.

Finally, we observed a global loss in all spatial frequencies with a high dose of L-Sulpiride, a D2 antagonist. The fact that a D2 antagonist in high concentration has a similar effect to a D1 agonist and also reduces low spatial frequency responses argues for interaction of D1 and D2 pathways. This argument is also supported by the small, albeit not statistically significant, increase of low spatial frequencies observed with low dosage L-Sulpiride. If low affinity presynaptic D2 receptors are on the D1 pathway, dopamine and D2 agonists in high concentration could provide inhibitory pre-synaptic input to the D1 pathway. Consistent with this interpretation is that blocking D2 input results in enhanced D1 response.

Table 5.2. Summary of different experimental manipulations on D1 and D2 responses

	DA neuronal toxins		**Dopamine receptor ligands**		
	MPTP	**6-OHDA**	**D2 antagonist**		**D1 agonist**
			L-Sulpiride		**CY208-243**
Effect at: ***(spat.freqs)***			***Low Dose***	***High Dose***	
low	n.s. ↑	n.s. ↑	n.s. ↑	↓	↓
middle-high	↓	↓	↓	↓	no effect

↑ = increase, ↓ = decrease, n.s. = not significant

5.5 THE MODEL

First, we assume that a retinal ganglion cell response can be modeled, as described by Enroth-Cugell and Robson (1966), by the difference of two Gaussian functions. Then, we assume that the PERG spatial tuning function represents the envelope of the response profiles of all ganglion cells in the central retina which was stimulated. As explained earlier, this is a reasonable assumption. Also we postulate that the PERG function represents the response of the "equivalent" ganglion cell based on the difference of two Gaussians. We realize, however, that Enroth-Cugell and Robson based their model on the sensitivity profile of a cell, while our experimental data represents a response profile elicited by higher contrast stimuli. Finally, we do not include any other interactions of dopamine with different neurotrasmitter systems of the retina.

According to this model, signal summation over a retinal ganglion cell receptive field can be described by the difference of the center (C) and surround (S) response. Each of these response profiles can be represented with a one-dimensional Gaussian function:

$$C(r) = w_c * \frac{1}{2\sqrt{\pi}} * \frac{1}{r_c}\ e^{-\left(\frac{r}{r_c}\right)^2} \qquad S(r) = w_s * \frac{1}{2\sqrt{\pi}} * \frac{1}{r_s}\ e^{-\left(\frac{r}{r_s}\right)^2}$$

where r_c, r_s are the radii of the receptive field center and surround, respectively, and w_c, w_s their respective gains.

We then modeled the data from the PERG spatial transfer function as obtained with the L-Sulpiride experiment as the difference of two Gaussian curves. The curve which fitted the experimental results in the normal condition had a gain ratio $w_s/w_c = 0.19$ and radius ratio $r_s/r_c = 3$ (r_c=10, r_s=30, w_c=0.17, w_s=0.032).

5.5.1 Decreased Response to Low Spatial Frequency Stimuli

In the model, if the surround's role in the antagonistic relation is enhanced, then responses to low spatial frequencies are suppressed. A stronger surround can be obtained if its radius r_s and gain w_s are increased in the model. The introduction of these changes to the model yielded similar results to those obtained in the experiment with the D1 agonist, CY 208-243, as shown in Fig. 5.5A. The simulated curves in the two conditions (before and after CY), are shown in Fig. 5.5B. Low spatial frequencies are suppressed, while middle and high spatial frequencies are the same, with a slight increase around the peak of the curve. The response profile of the "equivalent" ganglion cell, before and after CY administration, are shown in Fig. 5.5C.

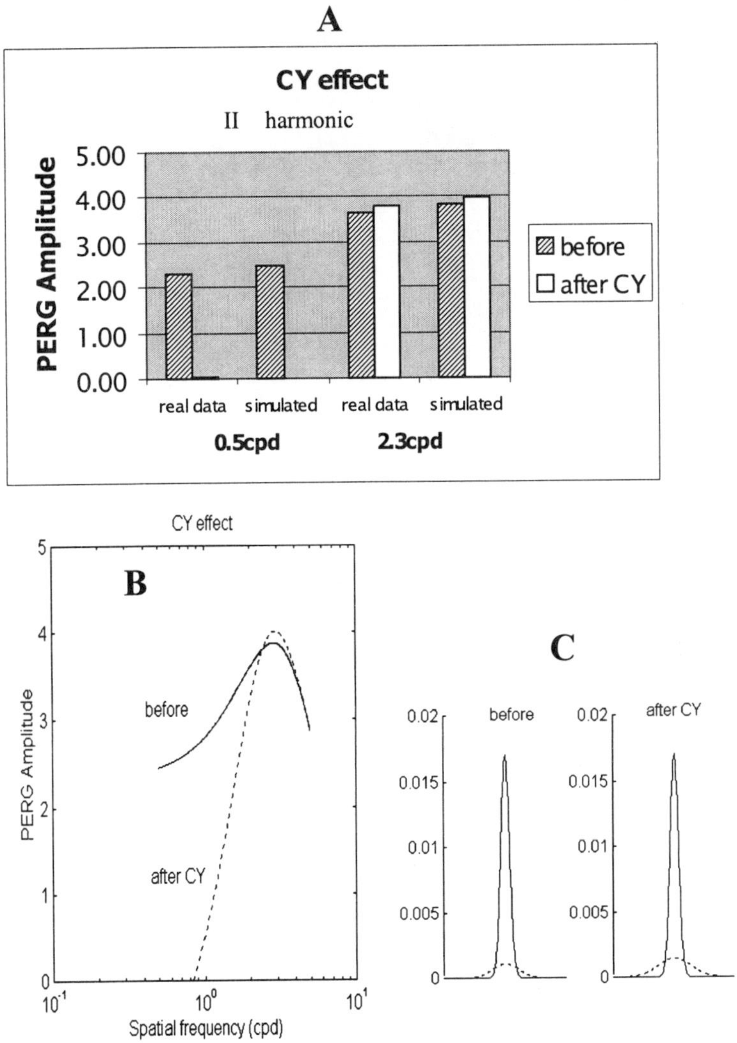

Figure 5.5. The effect of a stronger surround mechanism in the receptive field model of the difference of two Gaussians for the "equivalent ganglion cell". The PERG spatial transfer function before CY administration is decomposed to its center/surround parameters. **A**. Simulated results with a stronger surround contribution are similar to the experimental ones using a D1 agonist, CY 208-243. **B**. Simulated transfer functions before (solid line) and after (dashed line) the enhancement of the surround's role. **C**. The Gaussian profile of center and surround for the "equivalent ganglion cell", left normal retina, right CY effect.

Variation in the surround's radius and gain is reflected in the spatial transfer function (Fig. 5.6). Depending on experimental modifications, the amplitude of middle spatial frequency responses can be slightly higher or lower than normal. The results obtained in the CY experiment with monkey "Marcy" presented above are consistent with a slight increase, while the results with monkey "Charlie" are consistent with a slight decrease of the response at middle spatial frequencies.

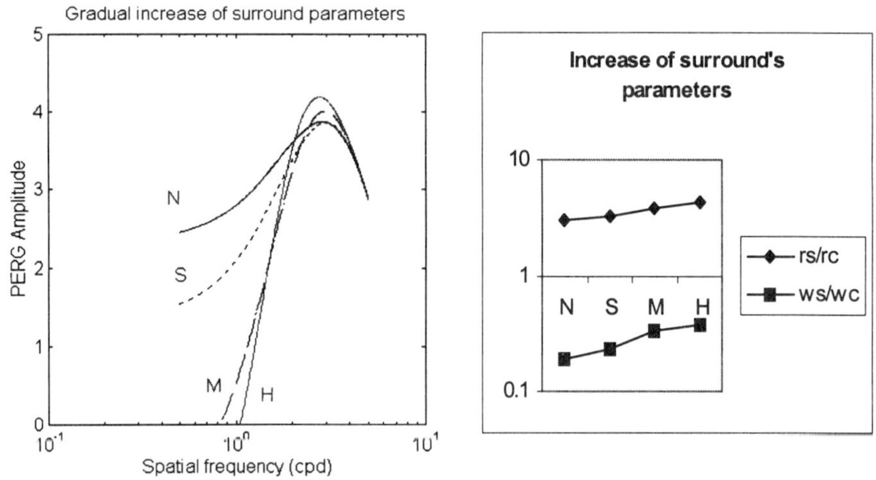

Figure 5.6. The effect of increasing surround's parameters in the model: a gradual increase of the radius r_s and gain w_s from **N**ormal values (corresponding to the simulation of experimental PERG spatial transfer function), to values with a **S**mall, **M**oderate and **H**igh increase (relatively to **N**ormal), as shown in the right panel. This gradual increase results in a gradual attenuation at low spatial frequencies as can be seen in the corresponding curves on the left panel. This is similar to the D1 agonist, CY 208-243, effect in the experiments. Moderate and high changes slightly increase responses around the peak.

5.5.2 Loss of spatial tuning

A different effect is observed if the center's contribution becomes weaker. A weaker center could result from a larger radius and lower gain, leading to loss of middle and high spatial frequency responses. The same effect was observed in the experiments with MPTP and 6-OHDA (Ghilardi et al, 1988; Ghilardi et al, 1989; Bódis-Wollner, 1996), which predominantly destroy neurons producing the D2 pathway (Mariani, unpublished data). The experimental results with a D2 receptor blocker, L-Sulpiride in high dosage further suggest as a possible explanation that D2 presynaptic receptors are on the D1 pathway that provide an inhibitory input to the pathway primarily affecting D1 postsynaptic receptors. As a consequence, blocking D2 receptors leads to an enhanced D1 response.

Following these observations, we included the changes in the model. Gradual increase of radius r_c and gradual decrease of gain w_c should be accompanied with D1 enhancement, i.e. increase of r_s and w_s. The results of these changes are shown in Fig. 5.7A, where the curve of L-Sulpiride experiment is simulated using the model. The corresponding changes to the response profile of the "equivalent" ganglion cell are shown in Fig. 5.7B. The more the center deteriorates, the greater is the enhancement of the surround. However, the center's decrease is more rapid as compared to the surround's increase (Fig. 5.7D), which is consistent with experimental results. The "synergistic" antagonism between center and surround is illustrated in Fig. 5.7C. While the center gradually looses its gain, surround finds space to take over, while at the same time they cooperate in keep the overall normalized response stable.

The simulated curves of Fig. 5.7A differ from the experimental ones at middle and high frequencies. Our model resulted in values at zero or close to zero level, while experimentally middle and high frequency responses are reduced significantly, but they do not drop to zero. One possible explanation is that a close to noise level measurement would be arithmetically equivalent to a zero value, and our model should be based on "sensitivity" and not on response profile (as the Enroth-Cugell and Robson model). A different manipulation of parameters would not result in the characteristic slope at low spatial frequencies.

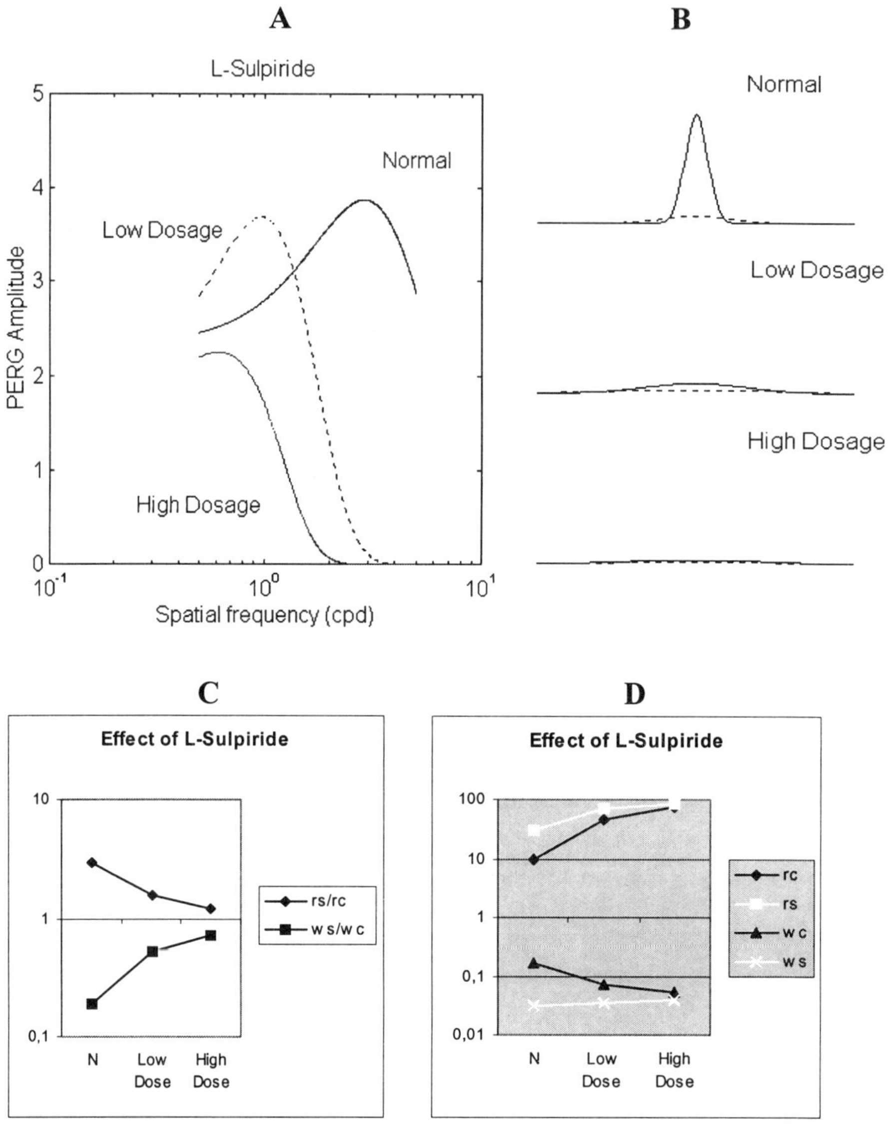

Figure 5.7. The effect of a "weaker" center mechanism in the receptive field model of the difference of two Gaussians for the "equivalent ganglion cell". The PERG spatial transfer function before administration of a D2 antagonist, L-Sulpiride, is decomposed to its center/surround parameters. A weaker center, which provides inhibitory feedback to the D1 pathway was introduced to the model. The results simulated the L-Sulpiride effect. **A**. Simulated transfer functions for normal retina (solid line), low dose L-Sulpiride effect (dashed line) and high dose L-Sulpiride effect (grey line). **B**. The corresponding gaussian profile of center and surround for the "equivalent ganglion cell", in the three conditions. **C**. Changes of the model parameters. As the center becomes weaker, the surround's role is enhanced. **D**. The center (black parameters) deteriorates faster than the surround (white parameters) is enhanced, induced by the inhibitory feedback loop.

5.6 DOPAMINE'S ROLE IN RETINAL MECHANISMS

It is well known that there are two major classes of dopamine receptors in all species, D1 and D2. Their action is combined in forming a paradoxical collaboration: they act antagonistically at the cellular level, but it is this antagonism which regulates their action resulting in a behavioral synergy. Pharmacological studies in the turtle retina (Piccolino et al, 1987) have clearly shown that at the horizontal cell level, D1 and D2 produce opposite effects. Electrophysiological studies (PERG) in the monkey, using different DA manipulations, demonstrated that D1 and D2 had a differential effect on the spatial transfer function. Based on the selective action of D1 and D2 on the spatial transfer function, it is tempting to assume that receptive field properties of retinal ganglion cells in the primate are altered in DA deficiency, with D1 and D2 playing different roles in the modulation of receptive field properties.

The responses of most retinal ganglion cells are based on center-surround antagonism of the receptive field. This function is quantitatively described by the difference of two Gaussians functions as described by Enroth-Cugell and Robson (1966). In this way, it is possible that different manipulation of the model parameters, namely center's radius and gain and surround's radius and gain, can simulate the experimental results (Table 5.3).

Table 5.3. Variation of parameters in the difference of Gaussians model and overall effect in the receptive field's center/surround, which simulate the experimental D1 and D2 responses.

Receptive field parameters		**Simulation**			
		D1 agonist		**D2 antagonist**	
center	r_c	-	*no change*	↑	*"weaker"*
	w_c	-		↓	
surround	r_s	↑	*"stronger"*	↑	*"stronger"*
	w_s	↑		↓	

↑ = increase, ↓ = decrease.

We found that the suppression at low spatial frequencies caused by a D1 agonist is consistent with a "stronger" surround with larger radius and higher gain. Conversely, D1 deficiency would weaken the surround response and enhance low spatial frequencies. It has been shown that dopamine modulates horizontal cell coupling via D1 receptors (Piccolino et al, 1985), and dopamine in this pathway reduces coupling among horizontal cells, forming smaller summation units (Gershenfeld et al; Teranishi et al, 1983; Teranishi et al, 1984; Lasater and Dowling, 1985; Hankins and Ikeda, 1991). One possible effect of smaller horizontal cell units is a dissipation of signal spread, i.e. the surround of each ganglion cell becomes stronger. In this way, D1 action leads to a stronger surround of neurons with large surrounds. This interpetation is consistent with the results of D1 and D2 manipulation on the horizontal cells in the turtle. Piccolino et al (1985) have shown that under increased levels of DA in the turtle retina, D2 agonists reverse horizontal cell narrowing, thereby resulting in wider and larger surround response. The results of their experiments are consistent with a presynaptic D2 pathway reducing D1 responses.

We postulate that a different dopamine dependent mechanism produces the effect observed with MPTP and 6-OHDA. It is likely that they both predominantly destroy the type of DA neurons which affect D2 receptors. Damage to these neurons and/or D2 receptor blockade at low concentrations causes attenuated peak spatial frequencies responses. In this case, we found that introduction into the model of a "weaker" center resulted in a loss at middle and high spatial frequencies. The higher the gain and the smaller its summating area, the more efficiently and selectively a center mechanism can perform. A lower gain and a higher radius of the center's response in the model resulted in a decrease at middle and high spatial frequencies. It is thought that dopamine uncouples AII amacrine cells in the mammalian retina (Hampson et al., 1992; Nguyen-Legros et al, 1999). It is interesting that in a model simulation of AII amacrine cells, increased coupling in the AII network resulted in expanded receptive field centers of amacrine and ganglion cells (Vardi and Smith, 1996).

Furthermore, we suggest that D2 autoreceptors are involved in the "surround" D1 dopamine pathway in the primate; blocking them allows a greater DA effect on D1 receptors and enhances "surround" signals leading to attenuated low spatial frequency responses, as it is observed with high dose L-Sulpiride. This is also consistent with the small increase at low spatial frequencies with low dose L-Sulpiride. Through this negative feedback circuit, D1 and D2 can regulate the cell's response. The combined effect of D1 and D2 action is illustrated in Fig. 5.8.

Based on the results, the spatial contrast curve cannot represent a single type of ganglion cell. Rather the results compel us to assume that D1 action reduces the response at low spatial frequencies alone, because D1 has little effect on the center. D1 increases the surround strength and only for low spatial frequencies. This means that D1 is important for ganglion cells with large surrounds.

Conversely, the use of a D2 antagonist reduced the peak spatial frequency response, but not the low spatial frequency response. It is known that neurons which respond to high spatial frequencies are center dominated and have little surround. This can be also seen in the spatial contrast transfer function. An exponential decay is observed at high spatial frequencies which

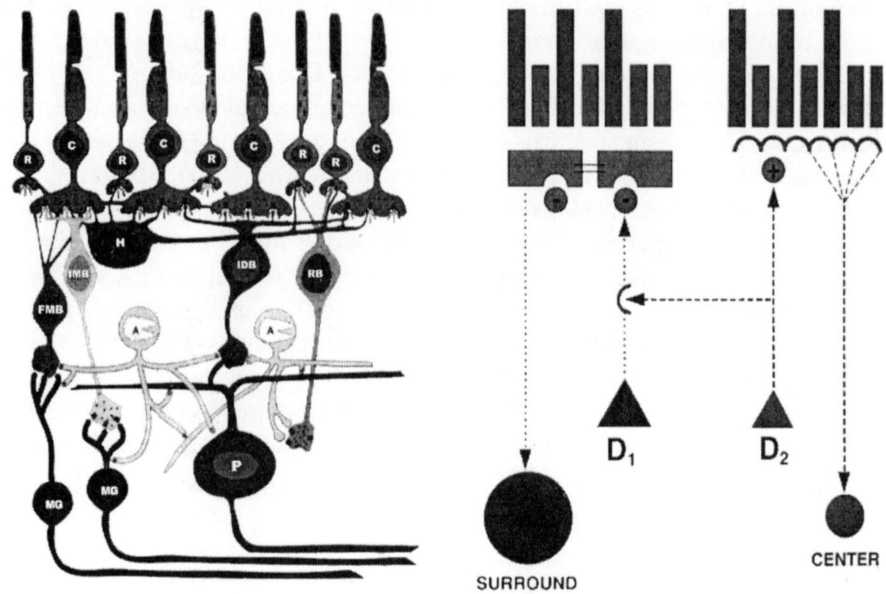

Figure 5.8. **Left**. The network of the mammalian retina consists of five different classes of neurons arranged in different layers: photoreceptors (rods R and cones C), horizontal cells (H), bipolar cells of different classes (invaginating midget bipolars IMB, flat midget bipolars FMB, invaginating diffuse bipolars DB, and rod bipolars RB), amacrine cells (A), and ganglion cells (falling into two main classes, midget ganglion cells MG, and parasol ganglion cells P). **Right**. Simplified schema of the D1-D2 interaction in the retina. D1 dopamine pathway enhances the "surround" signal, while D2 pathway enhances the "center" signal. Experimental results suggest that these two dopamine pathways are not independent from each other: D2 is involved in the D1 pathway participating in a negative feedback loop, providing a greater D1 effect when D2 receptors are blocked.

is consistent with pure spatial summation without much inhibition from the surround's response. Thus, D2 is important for neurons with smaller, but dominant centers, compared to neurons which respond to low spatial frequencies.

Hence, the results of various dopamine manipulations on the massed response of ganglion cells in the primate fovea suggests that two types of ganglion cells exist. One with dominant surrounds (corresponding to D1) mediates the response to low spatial frequencies, while the other with dominant centers (corresponding to D2) mediates the response to middle and high spatial frequencies. The notion of two types of ganglion cells gains further support from the two types of bipolar cells described by Dacey et al (2000). They exhibit two types of center-surround organization: the smaller ones have stronger centers, and the larger ones have relatively stronger surrounds. Thus, it is possible that each type of ganglion cell could receive input from similar bipolar input.

Dopaminergic systems vary among species. The exact function of DA in humans is not clear. Most of our knowledge comes from non-primate studies, mostly from the cat (Maguire and Smith, 1985; Maguire and Hamasaki, 1994) and the rabbit (Jensen and Daw, 1986). However, the functional differences between non-primate and adult primate retinas do not allow a direct comparison. Yet, from PERG experiments in the monkey, we suggest that dopamine has a push-pull effect in the primate retina, similar to non-primate retina: it strengthens the response of neurons with small centers and the surround response of neurons with large surrounds (Fig. 5.9). D1 and D2 act antagonistically at the cellular level, while at the overall retina level they synergistically modulate the properties of ganglion receptive field. The net result of DA's push-pull action is a tuned spatial transfer function, without which the function is low-pass.

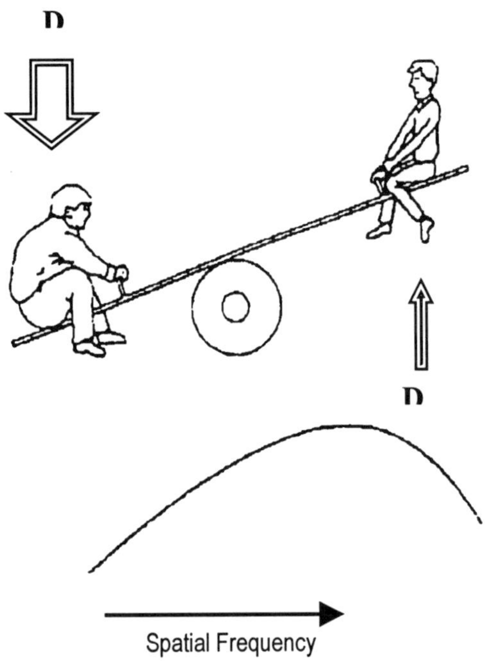

Figure 5.9. The antagonistic effect of D1 and D2 receptor activation acting on two different arms of the see-saw. As a consequence, the doubly opposite effects produce an overall synergistic action. The space underneath the curve represents the overall spatial frequency transfer function of the retina: low frequency decline occurs where D1 receptors are active. The peak of the curve is created by the see-saw pointing to the right (from Bódis-Wollner et al, 1993 with permission of Elsevier Science).

5.7 REFERENCES

Barlow, H.B., and Hill, R.M., 1963, Selective sensitivity to direction of movement in ganglion cells of the rabbit retina, *Science.* **139**: 412-414.

Baylor, D.A., Fuortes, M.G.F., and O'Bryan, P.M., 1971, Receptive fields of the cones in the retina of the turtle, *J. Physiol. (Lond.)* **214**: 265-294.

Bódis-Wollner, I., 1972, Visual acuity and contrast sensitivity in patients with cerebral lesions, *Science.* **178**: 769-771.

Bódis-Wollner, I., and Camisa, J.M., 1980, Conrast sensitivity in clinical diagnosis, in: *Neuro-opthalmology,* Lessell, S., Van Dalen, J.T.W., eds., Elsevier Science, Amsterdam, pp. 373-401.

Bódis-Wollner, I., Marx, M., and Ghilardi, M.F., 1989, Systematic Haloperidol administration increases the amplitude of the light and dark adapted flash ERG in the monkey, *Clin. Vis. Sci.* **4**: 19-26.

Bódis-Wollner, I., 1990, Visual deficits related to dopamine deficiency in experimental animals and Parkinson's disease patients. *Trends Neurosci.* **13**: 296-302.

Bódis-Wollner, I., Tagliati, M., Peppe, A., and Antal, A., 1993, Visual and visual perceptual disorders in neurodegenerative diseases, *Bailliere's Clinical Neurology.* **2**: 461-490.

Bódis Wollner, I., 1996, Electrophysiological assessment of retinal dopaminergic deficiency, *Funct. Neurosci.* **46**: 35-41.

Bódis -Wollner, I., and Tzelepi, A., 1998, The push-pull action of dopamine on spatial tuning of the monkey retina: the effects of dopaminergic deficiency and selective D1 and D2 receptor ligands on the pattern electroretinogram, *Vis. Res.* **38**: 1479-1487.

Boycott, B.B., and Wassle, H., 1974, The morphological types of ganglion cells of the domestic cat's retina, *J. Physiol. (Lond.)* **240**: 397-419.

Boycott, B.B., and Wassle, H., 1991, Morphological classification of bipolar cells of the primate retina, *Eur. J. Neurosci.* **3:** 1069-1088.

Burns, R.S., Chiuech, C.C., Markey, S., Ebert, M.H., Jacobowitz, D.M., and Kopin, J., 1983, A primate model of Parkinson's disease: selective destruction of substantia nigra pars compacta dopaminergic neurons by N-methyl-4-phenyl-1,2,3,6-tetrahydropyridine, *Proc. Natl. Acad. Sci. USA* **80**: 4546-4550.

Cleland, B.G., and Levick, W.R., 1974, Properties of rarely encountered types of ganglion cells in the cat's retina, *J. Physiol.(Lond.)* **240**: 457-492.

Cohen, J.L., and Dowling, J.E., 1983, The role of the retinal interplexiform cell: effects of 6-hydroxydopamine on the spatial properties of carp horizontal cells, *Brain Res.* **264**: 307-310.

Dacey, D., Packer, O., Diller, L., Brainard, D., Peterson, B., and Lee, B., 2000, Center surround receptive field structure of cone bipolar cells in the primate retina, *Vis. Res.* **40**: 1801-1811.

Dacey, D.M., 1993, The mosaic of midget ganglion cells in the human retina. *J. Neurosci.* **13**: 5334-5355.

Davis, G.C., Williams A.C., Markey, S.P., Ebert, M.H., Caine, E.D., Reichert, C.M., and Kopin I.J., 1979, Chronic parkinsonisms secondary to intravenous injections of meperidine analogues, *Psychiatry Res.* **1**: 249-254.

De Monasterio, F.M., and Gouras, P., 1975, Functional properties of ganglion cells of the rhesus monkey retina. *J. Physiol. (Lond.)* **251:** 167-195.

Djamgoz, M.B.A., and Kolb, H., 1993, Ultrastructural and functional connectivity of intracellularly stained neurons in the vertebrate retina: correlative analyses, *Microsc. Res. Tech.* **24**: 43-66.

Enroth-Cugell, C., and Robson, J.G., 1966, The contrast sensitivity of retinal ganglion cells of the cat, *J. Physiol.(Lond.)* **187**: 517-552.

Fukuda, Y., Hsiao, C.F., Watanabe, M., and Ito, H., 1984, Morphological correlates of physiologically identified Y-, X-, and W-cells in cat retina. *J. Neurophysiol.* **52:** 999-1013.

Fukuda, Y., Hsiao, C.F., and Watanabe, M., 1985, Morphological correlates of Y, X and W type ganglion cells in the cat's retina, *Vis. Res.* **25**: 319-27.

Gerschenfeld, H.M., Neyton, J., Piccolino, M., and Witkovsky, P., 1982, L-horizontal cells of the turtle: network organization and coupling modulation, *Biomed Res.* **3**: 21-32.

Ghilardi, M.F., Bódis-Wollner, I., Onofrj, M.C., Marx, M.S., and Glover AA., 1988, Spatial frequency-dependent abnormalities of the pattern electroretinogram and visual evoked potentials in a Parkinsonian monkey model, *Brain.* **11**: 131-184.

Ghilardi, M.F., Marx, M.X., Bódis-Wollner, I., Camras, C.B., and Glover, A., 1989, The effect of intraocular 6-hydroxydopamine on retinal processing of primates, *Ann. Neurol.* **25**: 359-364.

Grzywacz, N.M., Merwine, D.K., and Amthor, F.R., 1998, Complementary roles of two excitatory pathways in retinal directional selectivity, *Vis Neurosci.* **15**: 1119-1127.

Hampson, E.C., Vaney, D.I., and Weiler, R., 1992, Dopaminergic modulation of gap junction permeability between amacrine cells in mammalian retina. *J. Neurosci.* **12**: 4911-4922.

Hankins, M.W., and Ikeda, H., 1991, The role of dopaminergic pathways at the outer plexiform layer of the mammalian retina, *Clin. Vis. Sci.* **6**: 87-93.

He, S., and Masland, R.H., 1997, Retinal direction selectivity after targeted laser ablation of starburst amacrine cells, Nature. **389**: 378-382.

Jensen, R.J., and Daw, N.W., 1986, Effects of dopamine and its agonists and antagonists on the receptive field properties of ganglion cells in the rabbit retina, *Neurosci.* **17**: 837-855.

Kamermans, M., and Spekreijse, H., 1999, The feedback pathway from horizontal cells to cones – A mini review with a look ahead, *Vis. Res.* **39**: 2449-2468.

Kaplan, E., and Shapley, R.M., 1986, The primate retina contains two types of ganglion cells with high and low contrast sensitivity, *Proc. Natl. Acad. Sci. USA* **83**: 2755-2757.

Kolb, H., Nelson, R., and Mariani A., 1981, Amacrine cells, bipolar cells and ganglion cells of the cat retina: a Golgi study, *Vis. Res.* **21**: 1081-1114.

Kolb, H., Linberg, K.A., and Fisher, S.K., 1992, Neurons of the human retina: A Golgi study, *J. Comp. Neurol.* **31**: 147-187.

Kuffler, S.W., 1953, Discharge patterns and functional organization of mammalian retina, *J. Neurophysiol.* **16**: 37-68.

Lasater, E.M., and Dowling, J.E., 1985, Dopamine decreases conductance of the electrical junctions between cultured retinal horizontal cells, *Proc. Natl. Acad. Sci. USA* **82**: 3025-3029.

Maffei, L., and Fiorentini, A., 1981, Electroretinographic responses to alternating gratings before and after section of the optic nerve, *Science.* **211**: 953-955.

Maffei, L., Fiorentini, A., Bisti, S., and Hollander, H., 1985, Pattern ERG in the monkey after section of the optic nerve, *Exp. Brain. Res.* **59**: 423-425.

Maguire, G.W., and Hamasaki, D.I., 1994, The retinal dopamine network alters the adaptational properties of retinal ganglion cells in the cat, *J. Neurophysiol.* **72**: 730-741.

Maguire, G.W., and Smith, E.L.III., 1985, Cat retinal ganglion cell receptive-field alterations after 6-hydroxydopamine induced dopaminergic amacrine cell lesions, *J. Neurosphysiol.* **53:**1431-43.

Mangel, S.C., and Dowling, J.E., 1985, Responsiveness and receptive field size of carp horizontal cells are reduced by prolonged darkness and dopamine, *Science.* **229**: 1107-1109.

Mariani, A.P., 1990, Amacrine cells of the rhesus monkey retina, *J. Comp. Neurol.* **301**: 382-400.

Marx, M.S., Podos, S.M., Bódis-Woller, I., Lee, P.Y., Wang, R.F., and Severin, C., 1988, Signs of early damage in glaucomatous monkey eyes: low spatial frequency losses in the pattern ERG and VEP, *Exp. Eye Res.* **46**: 173-184.

Matsumoto, N., and Naka, K.I., 1972, Identification of intracellular responses in the frog retina, *Brain Res.* **42**: 59-71.

Nguyen-Legros, J., Versaux-Botteri, C., and Vernier, P., 1999, Dopamine receptor localization in the mammalian retina, *Mol. Neurobiol.* **19**: 181-204.

Nicklas, W.J., Youngster, S.L., Lindt, M.V., and Heikkila, R.E., 1987, Molecular mechanisms of MPTP induced toxicity. I.V. MPTP, MPP+ and mitochondrial function, *Life Sci.* **40**: 721-729.

Oliver, P., Jolicoeur, F.B., Lafond, B., Drumheller, A.T., and Brunette, J.R., 1986, Dose-related effects of 6-OHDA on rabbit retinal dopamine concentrations and ERG b-wave amplitudes, *Brain Res. Bull.* **16**: 751-753.

Peppe, A., Antal, A., Tagliati, M., Stanzione, P., and Bódis-Wollner, I., 1998, D1 agonist CY 208-243 attenuates the pattern electroretinogram to low spatial frequency stimuli in the monkey, *Neurosci Lett.* **242**: 1-4.

Piccolino, M., De Montis, G., Witkovsky, P., Bódis-Wollner, I., and Mirolli, M., 1987, D1 and D2 dopamine receptors involved in the control of electrical transmission between retinal horizontal cells, in: *Symposia in Neuroscience: Central and Peripheral Dopaminergic Receptors,* Biggio, G., Spano, P.F., Toffano, G. and Gessa, G.L. eds.,. Liviana Press, Padova, pp. 1-12.

] iccolino, M., Witkovsky, P., and Trimarchi C., 1987, Dopaminergic mechanisms underlying the reduction of electrical coupling between horizontal cells of the turtle retina induced by d-amphetamine, bicuculline, and veratridine, *J Neurosci.* **7**: 2273-2284.

Regan, D., Kothe, A.C., and Sharpe, J.A., 1991, Recognition of motion-defined shapes in patients with multiple sclerosis and optic neuritis. *Brain.* **114**: 1129-1155.

Rodieck, R.W., and Stone, J., 1965, Analysis of receptive fields of cat retinal ganglion cells, *J. Neurophysiol.* **28**: 833-849.

Rodieck, R.W., Binmoeller, K.F., and Dineen, J., 1985, Parasol and midget ganglion cells of the human retina, *J. Comp. Neurol.* **233**: 115-132.

Schwartz, E.A., 1974, Response of bipolar cells in the retina of the turtle. *J. Physiol. (Lond.)* **236**: 211-224.

Shapley, R., and Perry, V.H., 1986, Cat and monkey retinal ganglion cells and their visual functional roles, *Trends Neurosci.* **9**: 229-235.

Tagliati, M., Bódis-Wollner, I., Kovanesz, I., and Stanzione, P, 1994, Spatial frequency tuning in the monkey retina depends on D2 receptor-linked action of dopamine, *Vis. Res.* **34**: 2051-2057.

Tagliati, M., Bódis-Wollner, I., and Yahr, M.D., 1996, The pattern electroretinogram in Parkinson's disease reveals lack of retinal spatial tuning, *Electroenceph. Clinic. Neurophysiol.* **100**: 1-11.

Teranishi, T., Negishi, K., and Kato, S., 1983, Dopamine modulates S-potential amplitude and dye-coupling between external horizontal cells in carp retina, *Nature.* **301**: 243-246.

Teranishi, T., Negishi, K., and Kato, S., 1984, Regulatory effect of dopamine on spatial properties of horizontal cells in carp retina, *J. Neurosci.* **4:** 1271-1280.

Ungerstedt, U., and Arbuthnott, G.W., 1970, Quantitative recording of rotational behavior in rats after 6-hydroxydopamine lesions of the nigrostriatal dopamine system, *Brain Res.* **24**: 485-493.

Vaney, D.I., 1994, Patterns of neuronal coupling in the retina, *Prog. Ret. Eye Res.* **13**: 301-355.
Vaney, D.I., 1990, The mosaic of amacrine cells in the mammalian retina, *Prog. Ret. Res.* **9**: 49-100.
Vardi, N., and Smith, R.G., 1996, The AII amacrine network: coupling can increase correlated activity, *Vis. Res.* **36**: 3743-3757.
Werblin, F.S., and Dowling, J.E., 1969, Organization of the retina of the mudpuppy, Necturus maculosus II. Intracellular recording, *J. Neurophysiol.* **32**: 339-355.
Witkovsky, P., Alones, V., and Piccolino, M., 1987, Morphological changes induced in turtle retinal neurons by exposure to to 6-hydroxydopamine and 5,6-hydroxytryptamine, *J. Neurocytol.* **16**: 55-67.
Wong, C., Ishibashi, T., Tucker, G., and Hamasaki, D., 1984, Responses of the pigmented rabbit retina to NMPTP, a chemical inducer of Parkinsonism, *Exp. Eye Res.* **40**: 509-519.
Xin, D., and Bloomfield, S.A., 1999, Dark- and light-induced changes in coupling between horizontal cells in mammalian retina, *J. Comp. Neurol.* **405**: 75-87.

Chapter 6

Laminar Organization of the Visual Cortex:
A Unified View of Development, Learning, Attention, and Grouping

Stephen Grossberg

Dept. of Cognitive and Neural Systems, Boston University,677 Beacon St., Boston, MA 02215, PH: (617) 353-7858, FX: (617) 353-7755, EM: steve@bu.edu

6.1 INTRODUCTION

The cerebral cortex is the seat of the highest forms of biological intelligence in all sensory and cognitive modalities. The neocortex has an intricate design which exhibits a characteristic organization into six distinct cortical layers (Brodmann, 1909; Martin, 1989). Differences in the thickness of these layers and the sizes and shapes of neurons led the German anatomist Korbinian Brodmann to identify more than fifty divisions, or areas, of neocortex. This classification has been invaluable as a basis for classifying distinct functions of different parts of neocortex. The functional utility of such a laminar organization in the control of behavior has, however, remained a mystery until recently.

Why are all sensory and cognitive neocortex organized into layered circuits? How do these layers organize circuits that form functional columns in cortical maps? How do bottom-up, top-down, and horizontal interactions within the cortical layers generate adaptive behaviors. This chapter summarizes an evolving neural model, called the LAMINART model, which suggests how these interactions help the visual cortex to realize: (1) the binding process whereby cortex groups distributed data into coherent object representations; (2) the attentional process whereby cortex selectively processes important events; and (3) the developmental and learning processes whereby cortex shapes its circuits to match environmental constraints. It is suggested that the mechanisms which achieve property (3) imply properties of (1) and (2). New computational ideas about feedback systems suggest how neocortex develops and learns in a stable way, and why top-down attention requires converging bottom-up inputs to fully activate cortical cells, whereas perceptual groupings do not (Grossberg, 1999a; Grossberg, Mingolla, & Ross, 1997; Grossberg & Raizada, 2000; Grossberg & Williamson, 2000; Raizada & Grossberg, 2000; Ross, Grossberg, & Mingolla, 2000). These functional roles also appear to be generalizable to other forms of sensory and cognitive processing.

The present chapter also suggests how perceptual, attentional, developmental, and learning properties that are known to be carried out by visual cortex place severe, and seemingly contradictory, demands on cortical organization. When one tries to realize all of these constraints within a single, unified cortical architecture, a model of how this is accomplished can be identified, in which every component is supported by neurobiological data. This LAMINART model has provided a unified explanation of many behavioral and neurobiological data for which no alternative explanation has yet arisen, and moreover has also made a number of testable predictions. Some of the properties that the LAMINART model helps to clarify are called the stability-plasticity dilemma, the attention-preattention interface property of analog coherence, and the properties of top-down matching and attentional focusing.

6.2 A UNIFIED APPROACH TO PERCEPTUAL GROUPING AND ATTENTION

The main constraints on the model can be stated in terms of perceptual processes that are familiar to us all. During visual perception, the visual cortex can generate perceptual groupings and can focus attention upon objects of interest. *Perceptual grouping* is the process whereby the brain organizes image contrasts into emergent boundary structures that segregate objects and their backgrounds in response to texture, shading and depth cues in scenes and images (Julesz, 1971; Ramachandran & Nelson, 1976; Beck, Prazdny, & Rosenfeld, 1983; Polat & Sagi, 1994). Perceptual grouping is a basic step in solving the "binding problem", whereby spatially distributed features are bound into representations of objects and events in the world. Illusory contours are a particularly vivid form of perceptual grouping. Fig. 6.1 shows how an illusory contour can form over image positions that do not receive bottom-up inputs from image or scenic contrasts. Perceptual groupings can form *preattentively* and automatically, without any top-down influences (Moore & Egeth, 1997).

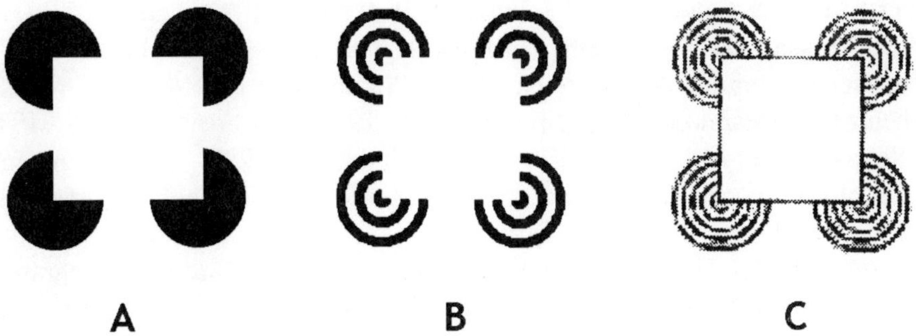

Figure 6.1. An illusory Kaniza square can be perceived (A) colinear to edge inducers and (B) perpendicular to line-end inducers. (C) Model simulation of the latter type of boundary grouping (reprinted from Grossberg, Mingolla, and Ross (1997), pg. 107, Fig. 1, with permission of Elsevier Science), down on-center off-surround network, as discussed in the text.

Attention enables humans and other animals to selectively process information that is of interest. In contrast to perceptual grouping, top-down attention does not form visible percepts over positions that do not receive bottom-up inputs. Attention can sensitize, modulate, or prime, an observer to expect an object to occur at a given location, or with particular stimulus properties (Posner, 1980; Duncan, 1984). But attention, by itself, is no substitute for the object's actual appearance. Were attention able to routinely generate fully formed perceptual representations, we could not tell the difference between external reality and internal fantasy, and would regularly hallucinate percepts of objects that were not really there. In fact, a breakdown in the modulatory property of attention *can* cause hallucinations during various mental disorders, including schizophrenia (Grossberg, 2000).

Given that perceptual grouping and attention make opposite requirements on bottom-up inputs, the question of how they are simultaneously realized within the same circuits of the visual cortex is challenging. One possible answer to this question is that these circuits are *not* simultaneously realized within the same cortical areas. This answer is not, however, supported by recent cortical data. For example, it has been shown that short-range perceptual groupings can occur within cortical area V1 (Polat *et al.*, 1998; Redies, Crook, & Creutzfeldt, 1986; Grosof, Shapley, & Hawken, 1993; Kapadia *et al.*, 1995; Sheth *et al.*, 1996) and that longer-range perceptual groupings can occur within cortical area V2 (Von der Heydt, Peterhans, & Baumgartner, 1984; Peterhans & von der Heydt, 1989). It has recently been shown that attentional focussing occurs from the earliest visual cortical area V1 and then top-down to the Lateral Geniculate Nucleus, or LGN (Sillito *et al.*, 1994) as well as in visual cortical areas V1, V2, and V4 (Motter, 1994a, 1994b; Beauchamp, Cox, & DeYoe, 1997; Hupé *et al.*, 1997; Ito,

Westheimer, & Gilbert, 1997; Johnson & Burkhalter, 1997; Lamme, Zipser, & Spekreijse, 1997; McAdams & Maunsell, 1997; Press & van Essen, 1997; Roelfsema, Lamme, & Spekreijse, 1998; Watanabe *et al.*, 1998; Reynolds, Chelazzi & Desimone 1999; Somers *et al.*, 1999) and areas MT and MST (O'Craven *et al.*, 1997; Treue & Maunsell, 1997). Many recent neurophysiological experiments show that attentional processing operates throughout the visual cortex.

Both perceptual grouping and attentional modulation are thus integrated within the visual cortical areas V1 and V2. How does this circuitry form perceptual groupings that *can* complete a boundary grouping over locations which receive no bottom-up visual inputs, whereas top-down attention *cannot* fully activate target cells that receive no bottom-up inputs? Why *should* attention be deployed throughout the visual cortex, including cortical areas which previously were thought to accomplish purely preattentive processing? An answer can be found by exploring the link between attention and learning, and using this link to further constrain the model.

6.3 THE LINK BETWEEN ATTENTION AND LEARNING

Earlier modeling work has suggested that top-down attention is a key mechanism whereby the brain solves the *stability-plasticity* dilemma (Grossberg, 1980, 1982; Grossberg & Stone, 1986; Carpenter & Grossberg, 1991, 1993; Grossberg, 1995, 1999b; Grossberg & Merrill, 1996). The stability-plasticity dilemma concerns that fact that our brains can rapidly learn enormous amounts of information throughout life, yet it does not revert in the opposite direction by just as rapidly forgetting what they already know. Brains are *plastic* and can rapidly learn new experiences, without losing the *stability* that prevents catastrophic forgetting.

The fact that multiple thalamic and cortical levels develop through experience-dependent learning (Hubel, Wiesel, & LeVay, 1977; Stryker & Harris, 1986; Calloway & Katz, 1990; Antonini & Stryker, 1993a, 1993b; DeAngelis, Ohzawa, & Freeman, 1993; Ghose, Freeman, & Ohzawa, 1994; Galuske & Singer, 1996) raises the question of how such attentive processes may be realized within neocortex in order to stabilize this learning through time. This question, in turn, leads to further constraints on cortical design, because at least some perceptual groupings can form preattentively, and provide the substrate upon which higher-level attentional processes can act. How can the preattentive grouping mechanisms develop in a stable way, before any higher-order attentional processes can develop with which to stabilize them? Why does not this problem lead to an infinite regress;

namely, attentional mechanisms cannot develop until preattentive mechanisms do, but preattentive mechanisms cannot develop stably in the absence of attention. I call this the *attention-preattention interface problem.* It is an *interface* problem because it is shown below how laminar cortical circuits enable preattentive grouping processes to use some of the same circuitry that attentive mechanisms use, even before attentive mechanisms may come into play, in order to stabilize their own cortical development and learning.

The solution proposed herein to the attention-preattention interface problem builds upon earlier efforts to solve the stability-plasticity dilemma. Adaptive Resonance Theory, or ART, has been proposed as a solution to how attention solves the stability-plasticity dilemma (Grossberg, 1976a, 1976b, 1978, 1980, 1982) by modeling how bottom-up signals activate top-down expectations whose signals are matched against bottom-up data (Fig. 6.2A). Both the bottom-up and top-down pathways contain adaptive weights, or long-term memory traces, that may be modified by experience. The learned top-down expectations "focus attention" upon information that matches them (Fig. 6.2B). Such a feedback network can select, synchronize, and amplify the activities of cells within the attentional focus, while suppressing the activities of irrelevant cells, which could otherwise be incorporated into previously learned memories and thereby destabilize them.

The cell activities which survive such top-down attentional focusing rapidly reactivate bottom-up pathways, thereby generating a type of feedback *resonance* between bottom-up and top-down signal exchanges (Fig. 6.2A). Such resonances rapidly bind distributed information at multiple levels of brain processing into context-sensitive representations of objects and events. These resonances are proposed to support slower processes of learning; hence the name *adaptive* resonance.

ART has been used to demonstrate, by means of using mathematical proofs and data explanations, how the learning of receptive field properties during early development, and the learning of perceptual and cognitive representations during adulthood, could easily suffer catastrophic forgetting in response to a changing world (Grossberg, 1976a, 1976b; Carpenter & Grossberg, 1987) On the other hand, it has also been shown how bottom-up activation and top-down attention can stabilize learning if they satisfy four properties (Carpenter & Grossberg, 1987, 1991), which together are called the ART Matching Rule:

Bottom-Up Automatic Activation: A cell, or cell population, can become active enough to generate output signals if it receives a large enough bottom-up input, other things being equal. Such an input can drive the cell to supraliminal levels of activation.

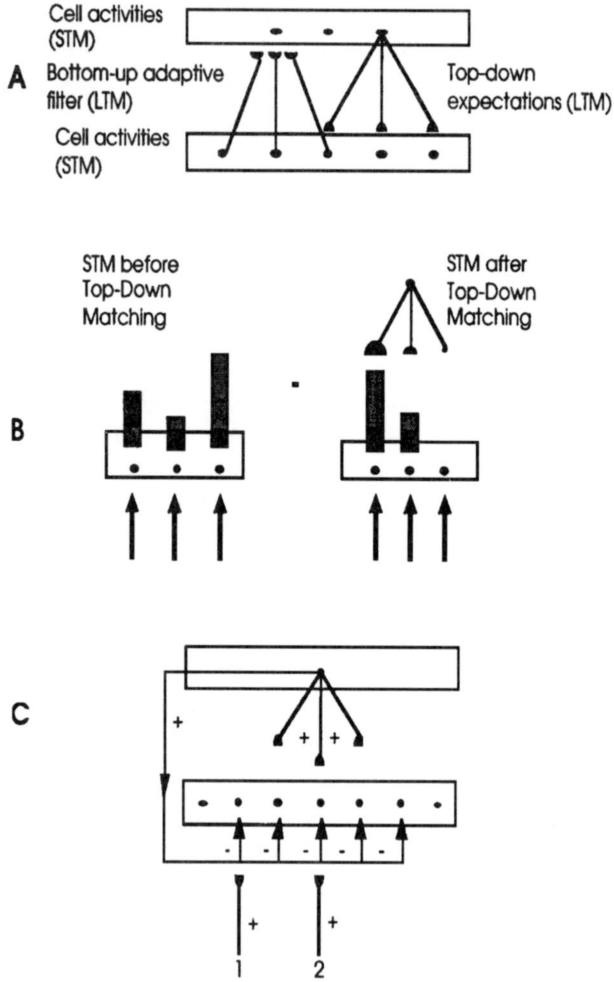

Figure 6.2. (A) Patterns of activation, or short-term memory (STM), on a lower processing level send bottom-up signals to a higher processing level. These signals are multiplied by adaptive weights, or learned long-term memory (LTM) traces (in the hemidisks), which influence the activation of the cells at the higher processing level. These latter cells, in turn, activate top-town expectation signals that are also multiplied by learned LTM traces (in the hemidisks). These top-down expectations are matched against the STM pattern that is active at the lower level. (B) This matching processes confirms and amplifies STM activations that are supported by large LTM traces in an active top-down expectation, and suppresses STM activations that do not get top-down support. The size of the hemidisks at the end of the top-down pathways represents the strength of the learned LTM trace that is stored in that pathway. The rightmost LTM trace is too small to support the corresponding bottom-up activation. (C) The ART Matching Rule may be realized by a top-down on-center off-surround network, as discussed in the text.

Top-Down Priming: A cell becomes subliminally active if it receives only a large top-down expectation input. Such a top-down priming signal can sensitize, or modulate, the cell, and thereby prepare it to react more quickly and vigorously to subsequent bottom-up inputs that match the top-down prime. The top-down prime by itself cannot, however, generate supraliminal output signals from the cell.

Match: A cell becomes active if it receives large convergent bottom-up and top-down inputs. Such a matching process can generate enhanced activation as resonance takes hold.

Mismatch: A cell's activity is suppressed, even if it receives a large bottom-up input, if it also receives only a small, or zero, top-down expectation input.

It was originally shown how four related circuits could all realize these properties (Carpenter & Grossberg, 1987) and thereby stabilize the learning process. More recent data analyses have suggested that variants of the simplest circuit, a top-down on-center off-surround network (Fig. 6.2C), is used by the brain (Grossberg, 1995, 1999b). Figure 6.2C clarifies how such a circuit can achieve all four properties. In particular, when only bottom-up signals are active, all cells that receive large enough inputs can fire. When only top-down attention is active, cells that receive inhibition but no excitation get inhibited, while cells that receive a combination of excitation and inhibition can get at most subliminally activated due to the balance between excitation and inhibition. When bottom-up and top-down inputs match, as in pathway 2 of Fig. 6.2C, the two excitatory sources of excitation that converge at the cell can overwhelm the one inhibitory source; it is a case of "two-against-one." When bottom-up and top-down inputs mismatch, as in pathway 1 of Fig. 6.2C, the top-down inhibition can neutralize the bottom-up excitation; it is a case of "one-against-one."

6.4 RECONCILING FAST FEEDFORWARD PROCESSING AND MODULATORY ATTENTION

Many scientists have resisted the concept that top-down attention plays an important role, despite the fact that there are massive top-down pathways throughout the cortex (Macchi & Rinvik, 1976; Tsumoto, Creutzfeldt, & Legendy, 1978; van Essen & Maunsell, 1983; Felleman & van Essen, 1991), due to the belief that feedback cannot operate quickly enough to influence cortical information processing. Two mathematical properties of ART systems show such concerns to be groundless: First, cortical interactions that obey ART properties can choose the correct cells on the first pass of bottom-up signalling, if the input pattern is unambiguous and familiar to the system;

second, resonance can stabilize within even a single processing cycle of attentional feedback (Carpenter & Grossberg, 1987, 1991; Grossberg & Somers, 1991).

Another roadblock to understanding derives from the properties, explained by ART, that top-down attention accomplishes modulatory, or subthreshold, *priming* and *matching*. By itself, it cannot activate cells enough to generate output signals, and thus seems to be too "weak" to significantly affect cortical processing. For example, Zeki and Shipp (1988, p. 316) wrote that "backward connections seem not to excite cells in lower areas, but instead influence the way they respond to stimuli". In a similar vein, the data of Sillito *et al.* (1994, pp. 479-482) on attentional feedback from V1 to LGN led them to conclude that "the cortico-thalamic input is only strong enough to exert an effect on those dLGN cells that are additionally polarized by their retinal input...the feedback circuit searches for correlations that support the 'hypothesis' represented by a particular pattern of cortical activity". Their experiments demonstrated all of the properties of the ART Matching Rule, since they found in addition that "cortically induced correlation of relay cell activity produces coherent firing in those groups of relay cells with receptive-field alignments appropriate to signal the particular orientation of the moving contour to the cortex...this increases the gain of the input for feature-linked events detected by the cortex". In summary, top-down priming, by itself, cannot fully activate LGN cells; it needs matched bottom-up retinal inputs to do so; and those LGN cells whose bottom-up signals support cortical activity are synchronized and amplified by this feedback. In addition, anatomical studies have shown that the top-down V1 to LGN pathway realizes a top-down on-center off-surround network (Dubin & Cleland, 1977; Weber, Kalil, & Behan, 1989), as in Fig. 6.2.

6.5 HOW TO STABILIZE PERCEPTUAL DEVELOPMENT AND LEARNING

Adaptive Resonance Theory suggests that top-down attentional mechanisms should be present in *every* cortical area wherein learning can occur, since without top-down learned expectations that focus attention via the ART Matching Rule, any such learned memories could easily be degraded due to catastrophic forgetting.

These analyses should, in particular, apply to the perceptual grouping process, because the cortical horizontal connections that support perceptual grouping in areas like V1 develop through a learning process that is influenced by visual experience (Luhmann *et al.*, 1986; Calloway & Katz,

1990; Gilbert & Wiesel, 1992; Lowel & Singer, 1992; Antonini & Stryker, 1993a; Galuske & Singer, 1996). It is also known that many developmental and learning processes, including those that control horizontal connections, are stabilized dynamically, and can be reactivated by lesions and other sources of cortical imbalance (Das & Gilbert, 1995; Gilbert & Wiesel, 1992), and that adult learning uses the same types of mechanisms as the infant developmental processes upon which it builds (Bailey *et al.*, 1992; Kandel & O'Dell, 1992; Mayford *et al.*, 1992). What cortical mechanisms ensure this type of dynamical stability?

Solving this problem for the case of perceptual groupings is challenging for two reasons which, on the surface, seem to be unrelated, but which are really intimately related on a mechanistic level. The first reason is that perceptual groupings can form *preattentively*. How, then, can attention control their stability during infant development and adult learning? This is the *attention-preattention interface problem* that was mentioned above. The second reason, also noted above, is that perceptual groupings can form over positions that do not receive bottom-up inputs, as in the case of illusory contours. They therefore seem to violate the ART Matching Rule. How, then, can the horizontal connections that generate perceptual groupings maintain themselves in a stable way? Why are they not destabilized whenever an illusory contour forms over positions that do not receive a bottom-up input?

My proposed answer to this question unifies two types of neural models which have been developed along separate paths for three decades: The attentive ART model, and the preattentive perceptual grouping model that is called the Boundary Contour System, or BCS (Grossberg & Mingolla, 1985; Grossberg, 1994; Gove, Grossberg, & Mingolla, 1995). Until the present work was done, it has not been possible to understand how attentive ART circuits are embodied within the laminar architecture of visual cortex. My proposed synthesis suggests how top-down attentional mechanisms that obey the ART matching law are integrated within processes of preattentive perceptual grouping, development, and perceptual learning within the laminar cortical circuits. Because it shows how the laminar structure of neocortex may realize basic ART properties, I have called the model LAMINART.

6.6 ANALOG COHERENCE OF PERCEPTUAL GROUPINGS

This refined BCS model proposes how the laminar circuitry of visual cortex enables perceptual groupings to maintain their analog and spatial context-sensitivity in response to changes in stimulus properties (Grossberg, Mingolla, & Ross, 1997). Analog sensitivity means that perceptual groupings can alter their form in response to graded changes in stimulus properties. For example, one boundary grouping, such as an illusory contour, may form if some signals are weak and others strong, whereas a different grouping may form if the reverse relative signals strengths occur. Spatial context-sensitivity means that the cortex can respond flexibly to spatial rearrangements of the stimuli that are to be grouped.

The grouping process actively selects and binds together the most salient groupings for conscious perception, while suppressing less salient groupings. This selection, or binding, process endows each grouping with an inner coherence, so that object representations are not merely the sum of their features. Unfortunately, processes that select winning groupings while inhibiting losing groupings tend to wipe out analog sensitivity; they tend to generate binary outcomes. Because analog values carry useful information about objects and events, it is important to understand how the cortex can bind distributed information into coherent representations without a loss of analog sensitivity. Remarkably, the laminar circuitry of visual cortex is capable of robustly realizing the key property of *analog coherence* whereby winning groupings can form coherently without losing analog or spatial context-sensitivity.

6.7 PREATTENTIVE CORTICAL GROUPING

Four circuit properties are proposed to demonstrate how the visual cortex, notably areas V1 and V2, uses its laminar design to generate perceptual groupings that preserve analog coherence (Fig. 6.3A-D). Each design principle will be described along with cortical data that it explains. Then four more circuit properties will be proposed whereby attention, development, and learning are integrated into this laminar design. Quantitative simulations of how the model develops into an adult laminar architecture that can group and attend to perceptual images in a manner that mimics adult neurophysiological, anatomical, and perceptual data are detailed in our recent publications; e.g., Grossberg, Mingolla, & Ross (1997), Grossberg & Raizada (2000), Grossberg & Williamson (2000), Raizada & Grossberg (2000), and Ross, Grossberg, & Mingolla (2000).

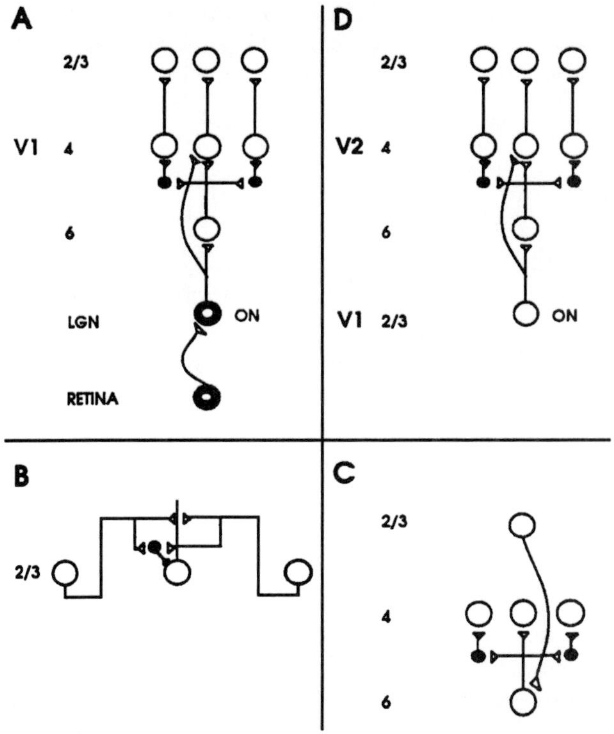

Figure 6.3. A model circuit of retinal, lateral geniculate nucleus (LGN), and cortical V1 interactions: Open symbols indicate excitatory interactions and closed symbols inhibitory interactions. (A) Feedforward circuit from retina to LGN to cortical layers 4 and 6: Retina: Retinal ON cells have an on-center off-surround organization. Retinal OFF cells have an off-center on-surround organization. LGN: The LGN ON and OFF cells receive feedforward ON and OFF cell inputs from the retina. Layer 4: Layer 4 cells receive feedforward inputs from LGN and layer 6. LGN ON and OFF cell excitatory inputs to layer 4 directly establish oriented simple cell receptive fields. Layer 6 cells excite layer 4 cells with a narrow on-center and inhibit them using inhibitory interneurons that span a broader off-surround. Like-oriented layer 4 simple cells with opposite contrast polarities compete (not shown) before generating half-wave rectified outputs that converge on layer 2/3 pyramidal (complex) cells. Layer 2/3: The converging simple cell outputs enable complex cells to respond to both polarities. As a result, complex cells full-wave rectify the image. (B) Horizontal grouping interactions in layer 2/3: After being activated by inputs from layer 4, layer 2/3 pyramidal (complex) cells excite each other monosynaptically via horizontal connections, primarily on their apical dendrites. They also inhibit one another via disynaptic inhibition that is mediated by model smooth stellate cells. Multiple horizontal connections share a common pool of stellate cells near each target pyramidal cell. This ensures that boundaries form inwardly between pairs or greater numbers of boundary inducers, but not outwardly from a single inducer. (C) Cortical feedback loop from Layer 2/3 to Layer 6: Layer 6 cells receive excitatory inputs from layer 2/3. The long-range cooperation hereby activates the feedforward layer 6-to-4 on-center off-surround network, which then influences the activity of layer 2/3 cells. This "folded feedback" loop can

Fig. 6.3 (con't) - select winning groupings without a loss of analog coherence. (D) Outputs from layer 2/3 to area V2 directly excite layer 4 cells and layer 6 cells, which indirectly influence layer 4 cells via an on-center off-surround network, as in area V1. The bottom-up circuits in Fig. 6.3A and 3D replace the more abstract bottom-up pathways in Figs. 6.2A–C.

6.7.1 Analog Sensitivity to Bottom-Up Sensory Inputs

Bottom-up inputs from the retina go through the Lateral Geniculate Nucleus (LGN) on their way to cortex. LGN outputs directly excite layer 4 (Hubel & Wiesel, 1962; Chapman, Zahs, & Stryker, 1991; Reid & Alonso, 1995). LGN inputs also excite layer 6, which then indirectly influences layer 4 via an on-center off-surround network of cells (Ferster & Lindstrom, 1985; Grieve & Sillito, 1991a, 1991b, 1995), as in Fig. 6.3A. A functional explanation of why this dual input pathway to cortex exists will be given below. The net effect of LGN inputs on layer 4 cells via this dual pathway is an on-center off-surround network. Such a feedforward on-center off-surround network of cells can normalize and preserve the analog sensitivity of the activities of target cells if these cells obey the membrane equations of neurophysiology. In particular, the off-surround inputs interact with the shunting terms in the membrane equation to divide the inputs that are processed by the on-center, thereby allowing cell activities to compute relative input size, even if the individual inputs are chosen very large. This was first proved in Grossberg (1973; see also Grossberg, 1980) and has since become a standard component of cortical models (Douglas *et al.*, 1995; Heeger, 1992). In the present case, such a network preserves the analog sensitivity of layer 4 cells in response to LGN inputs that may vary greatly in intensity.

6.7.2 Bipole Boundary Grouping

The active layer 4 cells input to pyramidal cells in layer 2/3. These cells initiate the formation of perceptual groupings. They generate excitatory signals among themselves using monosynaptic long-range horizontal connections, and inhibition using short-range disynaptic inhibitory connections (Hirsch & Gilbert, 1991; McGuire *et al.*, 1991), as in Fig. 6.3B. These interactions support inward perceptual groupings between two or more boundary inducers (von der Heydt, Peterhans, & Baumgartner, 1984; Peterhans & von der Heydt, 1989), but not outward groupings from a single inducer (Hirsch & Gilbert, 1991; Cannon & Fullenkamp, 1993; Knierim & van Essen, 1992; Somers, Nelson, & Sur, 1995; Stemmler, Usher, & Niebur, 1995), as in the case of illusory contours (Fig. 6.1). If a single inducer could generate groupings, our percepts would become crowded with webs of boundaries that spread out from every feature in a scene.

These grouping properties are obtained from the following intercellular interactions: When a single active pyramidal cell sends horizontal monosynaptic excitation to other pyramidal cells, this excitation is inhibited by the disynaptic inhibition that it also generates; this is another case of "one-against-one". This (approximate) balance between excitation and inhibition within layer 2/3 is one factor that helps to self-stabilize the development of these circuits in response to visual inputs in the infant and to dynamically maintain them throughout life (Grossberg & Williamson, 2000). A different result obtains when two or more pyramidal cells are activated at positions that are located at opposite sides of a target pyramidal cell, and all the cells are approximately colinear across space. Then the excitation from the active pyramidal cells summates at the target cell, thereby generating a larger total excitatory input than a single pyramidal cell could. In addition, the active cells all excite a single population of disynaptic inhibitory interneurons, which generates a saturating, or normalized, inhibitory output to the target cell. Total excitation is bigger than inhibition in this case, so that grouping can occur; it is another case of "two-against-one." This combination of constraints is called the *bipole* property. Layer 2/3 pyramidal cells may hereby become active either due to direct inputs from layer 4, or due to bipole boundary groupings that form in response to other active layer 2/3 cells.

Thus, when horizontal excitation is balanced against inhibition in a way that enables the cortex to develop stably in the infant, then the result in the adult is the bipole grouping property. If an input image or scene has complete object edges, or incomplete edges for which an unambiguous grouping exists, then this circuit can quickly complete a grouping within a single feedforward pass of information processing from layer 4 to layer 2/3, and then on to subsequent cortical areas; cf., Thorpe, Fize, and Marlot (1996). When, however, multiple groupings are possible, they competitively suppress one another via inhibitory interneurons in layers 6 and 4. Further processing is needed to select and amplify the statistically most favored grouping and to suppress less favored groupings. To achieve this selection, intracortical feedback of the following type may be necessary.

6.7.3 Folded Feedback and Analog Coherence

The active cells in layer 2/3 send excitatory feedback signals to layer 6 via layer 5 (Gilbert & Wiesel, 1979; Ferster & Lindstrom, 1983), as in Fig. 6.3C. Layer 6, in turn, once again activates the on-center off-surround network from layer 6 to 4. This feedback process is called *folded feedback*, because feedback signals from layer 2/3 to layer 6 get transmitted in a feedforward fashion back to layer 4. The feedback is hereby "folded" back into the feedforward flow of bottom-up information within the laminar cortical circuits.

Folded feedback turns the cortex into a feedback network that binds the cells throughout layers 2/3, 4, and 6 into functional columns (Mountcastle, 1957; Hubel & Wiesel, 1962, 1977). The on-center off-surround network helps to select the strongest groupings that are formed in layer 2/3 and to inhibit weaker groupings, while preserving the analog values of the selected groupings. In particular, the on-center signals from layer 6-to-4 support the activities of those pyramidal cells in layer 2/3 that are part of the strongest horizontal groupings. The off-surround signals of the strongest groupings can inhibit inputs to layer 4 that were supporting less active groupings in layer 2/3. In this way, signals from layer 4 to the less active groupings in layer 2/3 are removed, and thus these groupings collapse. The on-center off-surround network also helps to ensure the analog sensitivity of the selected grouping, and the feedback binds the selected cells into a coherent whole.

6.7.4 Self-Similar Hierarchical Boundary Processing

Converging evidence suggests that area V2 replicates aspects of the structure of area V1, but at a larger spatial scale (Kisvarday *et al*., 1995). In particular, layer 2/3 in area V1 sends bottom-up inputs to layers 4 and 6 of area V2, much as LGN sends bottom-up inputs to layers 4 and 6 of area V1 (van Essen & Maunsell, 1983; Felleman & van Essen, 1991); see Fig. 6.3D. This input pattern from V1 to V2 can preserve the analog sensitivity of layer 4 cells in V2 for the same reason that the LGN inputs to V1 can preserve the analog sensitivity of layer 4 cells in V1. The shorter perceptual groupings in layer 2/3 of area V1 (Redies, Crook, & Creutzfeldt, 1986; Grosof, Shapley, & Hawken, 1993) are proposed to group together, and thereby enhance the signal-to-noise ratio of, nearby V1 cells with similar orientation and disparity selectivity. The longer perceptual groupings in area V2 (van der Heydt, Peterhans, & Baumgartner, 1984; Peterhans & von der Heydt, 1989) are proposed to build long-range boundary segmentations that separate figure-from-background; generate 3-D groupings of the edges, textures, shading, and stereo information that go into object representations; and complete boundaries across gaps in bottom-up signals due to the retinal blind spot and veins (Grossberg, 1994, 1997; Grossberg & McLoughlin, 1997; Lamme, 1998). Both types of groupings achieve analog coherence by using the same type of laminar circuitry.

6.8 TOP-DOWN ATTENTION AND MATCHING

How does top-down attention fit into these layered circuits? Four attentive circuit properties are proposed to accomplish this. We also show how top-down signals (including the folded feedback mentioned in Section 6.7.3) can help cortical development and learning to proceed in a stable fashion. This synthesis of grouping and attention enables attention to selectively prime entire (preattentively formed) object boundaries, not just ungrouped bags of features (Grossberg & Raizada, 2000; Raizada & Grossberg, 2000; Roelfsema *et al.*, 1998).

6.8.1 Top-Down Feedback from V1 to LGN

As noted above, layer 6 of area V1 sends a top-down on-center off-surround network to the LGN (Murphy & Sillito, 1987; Weber, Kalil, & Behan, 1989; Murphy & Sillito, 1996), as in Fig. 6.4A. This top-down pathway carries out a type of automatic attentional focusing and gain control on those LGN cells whose activities succeed in activating V1 cells. Data of Sillito *et al.* (1994) show that this feedback obeys the ART Matching Rule, and thus can only subliminally activate, or modulate, LGN cells; matched bottom-up inputs are needed to fully activate LGN cells while top-down signals are active.

Such a top-down matching process was predicted to help stabilize the development of cell receptive fields in V1 during the visual critical period (Grossberg, 1976b, 1978, 1980). Although the matching portion of the prediction has been supported, the possible role of top-down attentive matching from V1 to LGN in stabilizing the experience-dependent development of bottom-up and top-down connections between LGN to V1 still needs to be tested.

6.8.2 Folded Feedback from Layer 6 of V2 to Layer 4 of V1

Such a top-down attentive process seems to occur at all stages of visual cortex, and probably beyond. Layer 6 in a given cortical area, such as V2, generates top-down cortical signals to layer 6 of lower cortical areas, such as V1, where they activate the layer 6-to-4 folded feedback network in the lower area (Fig. 6.4B). One such known top-down pathway exits layer 6 in V2 and activates V1 via layer 1 (Pandya & Yeterian, 1985). This pathway activates layer 1 apical dendrites of layer 5 cells, which relay them to layer 6 cells in V1 (Cauller & Connors, 1994; Rockland, 1994), as in Fig. 6.4C. Top-down feedback hereby activates a top-down on-center off-surround circuit, much like the ART circuit in Fig. 6.2. I propose that it is realized in the cortex using outputs from layer 6 of a given cortical area to activate layer 4 of a lower cortical area via layer 6-to-4 folded feedback. This proposal is supported by neurophysiological data showing that top-down signals activate the center and inhibit the surround of area V1 cells (Bullier *et al.*, 1996). Attention may also modulate activation of layer 2/3 by contacting apical dendrites of both pyramidal cells and disynaptic inhibitory interneurons; see Raizada & Grossberg (2000) for further discussion of this pathway.

6.8.3 Layer 6-to-4 Signals are Modulatory

The ART Matching Rule predicts that this top-down pathway *subliminally* activates, or modulates, cells in layer 4. I propose that this modulatory property is due to the fact that the excitatory and inhibitory connections in the on-center from layer 6-to-4 are balanced so that at most a weak excitatory effect occurs after activating the circuit via top-down feedback. This hypothesis is consistent with neurophysiological data from

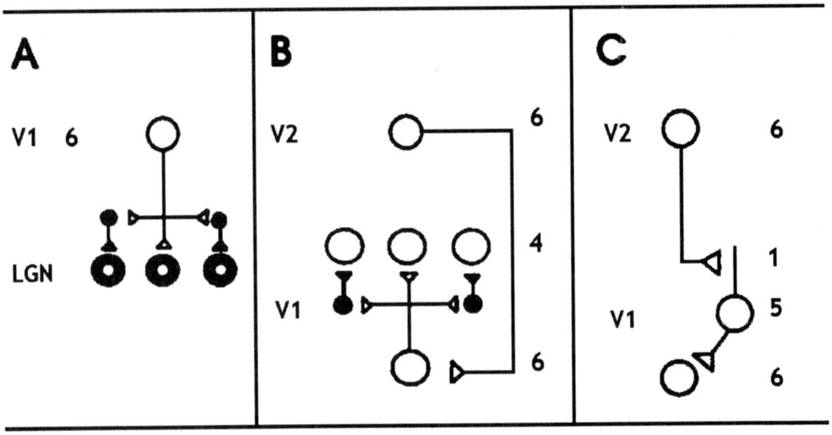

Figure 6.4. (A) Top-down corticogeniculate feedback from Layer 6: LGN ON and OFF cells receive topographic excitatory feedback from layer 6 in V1, and more broadly distributed inhibitory feedback via LGN inhibitory interneurons that are excited by layer 6 signals. The feedback signals pool outputs over all cortical orientations and are delivered equally to ON and OFF cells. Cortiogeniculate feedback selects, gain-controls, and synchronizes LGN cells that are consistent with the cortical activation that they cause, thereby acting like a type of automatic attentional focus. (B) Attentional feedback from V2 to V1: Layer 6 in V2 activates layer 6 in V1, which then activates the layer 6-to-4 on-center off-surround network that attentionally primes layer 4 cells. (C) One feedback pathway arises from Layer 6 cells in V2 and activates apical dendrites in Layer 1 of V1. Cells in Layer 5 are activated through these apical dendrites and thereupon activate Layer 6 cells. The top-down circuits in Figs. 6.4A–C replace the more abstract top-down pathways in Figs. 6.2A–C.

ferret visual cortex showing that the layer 6-to-4 circuit is functionally weak (Wittmer, Dalva, & Katz, 1997). In addition, Hupé *et al.* (1997, p. 1031) note: "feedback connections from area V2 modulate but do not create center-surround interactions in V1 neurons." Thus top-down feedback from layer 6 of V2 is predicted to be able to supraliminally activate layer 6 of V1, but not layer 4 of V1. (It is possible that layer 4 may be weakly activated, but not enough to drive grouping within layer 2/3.)

Grossberg and Williamson (2000) present simulations showing that the (approximate) balance of excitation and inhibition within the on-center is necessary to achieve stable development of interlaminar cortical connections. If the balance is upset too greatly, then proliferation or suppression of connections can occur. This constraint on stable development in the infant becomes the constraint that attention is modulatory in the adult.

The modulatory nature of attention does not prevent it from having a major effect on cortical cell activations when the cortex is activated bottom-up by visual inputs. In particular, even though the on-center may be modulatory, strong inhibition can suppress the activities of layer 4 cells in the off-surround. In this way, top-down attention can dramatically reorganize the *balance* of activation across the cortex. This analysis predicts how attentional and grouping constraints from higher levels of cortex can feed back to selectively bias the groupings that arise at lower cortical levels. In particular, the "higher-order" boundary completion and figure-ground perception grouping properties of V2 can select cells in V1 which are consistent with them, and can let attention selectively prime whole object groupings (Grossberg & Raizada, 2000; Lamme, 1998; Roelfsema *et al.*, 1998).

6.8.4 Two Bottom-Up Input Sources to Layer 4

A simple functional explanation can now be provided of why there are direct bottom-up inputs to layer 4, as well as indirect bottom-up inputs to layer 4 via layer 6, in many cortical areas (e.g., Figs. 6.3A and D). Why are two separate input pathways needed? Why, in particular, is not the indirect layer 6-to-4 pathway sufficient to fully activate layer 4 cells *and* to maintain their analog sensitivity using its on-center off-surround network? The discussion above suggests that the indirect 6-to-4 pathway must be modulatory for cortical development and learning to proceed in a stable way. Thus a direct pathway is needed to fully activate layer 4 cells in response to bottom-up inputs.

Taken together, these eight cortical design principles lead to the circuit diagram for perceptual grouping and attention between LGN, V1, and V2 that is shown in Fig. 6.5. Although this figure shows connections only with respect to a single input channel, they are replicated in a regular spatial array across all input channels. In addition to these principles, one also needs to model how, for example, oriented simple cells may be realized through these interactions in layer 4, and so on. The technical modeling articles go into these related issues. I propose that the same types of laminar cortical circuits may explain data at multiple levels of cortical organization; for example, they have already proved sufficient to explain attentional data collected from macaque cortical areas V2 and V4 (Motter, 1994a, 1994b; Reynolds, Chelazzi & Desimone 1999), wherein top-down attention once again selectively primes features within the attentional focus while suppressing the effects of features that are not. See Grossberg & Raizada (2000) for model simulations of these data.

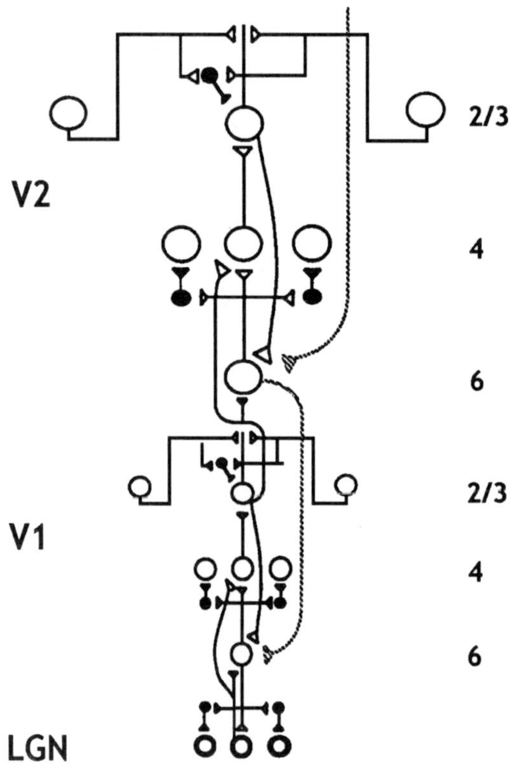

Figure 6.5. A LAMINART model synthesis of bottom-up, top-down, and horizontal interactions in LGN, V1, and V2. Open cells and connections denote preattentive excitatory mechanisms that are involved in perceptual grouping. Black cells and connections denote inhibitory mechanisms. Stippled connections denote top-down attentional pathways.

6.9 THE PREATTENTIVE PERCEPTUAL GROUPING IS ITS OWN ATTENTIONAL PRIME

Having outlined the circuit of Fig. 6.5, we can now supply an answer to the fundamental question of how the horizontal connections within cortical area V1 can *stably* develop through an experience-dependent learning process, even though they do not seem to satisfy the ART matching rule. The key fact is that preattentive perceptual groupings within V1 and attentive feedback from V2 to V1 both generate feedback signals to layer 6 of V1. Both types of feedback activate the folded feedback circuit from layer 6-to-4. Top-down attention uses this circuit to focus attention within V1 by inhibiting layer 4 cells that are not supported by excitatory 6-to-4 feedback. Perceptual grouping uses the folded feedback circuit *in the adult* to inhibit layer 4 cells that would otherwise form incorrect groupings. I claim that, *in the infant*, the same folded feedback circuit can also prevent the wrong combinations of cells in layers 4 and 2/3 from being active simultaneously. By the principle that "cells that fire together wire together," this selection mechanism enables the correct horizontal connections to be learned between these active cells.

The folded feedback circuit from layer 6-to-4 can be activated by bottom-up signals to layer 2/3 or by horizontal perceptual grouping signals within layer 2/3 via layer 2/3-to-6 connections. This activation can occur at *all* positions of a grouping, even those positions that do not receive bottom-up inputs. Layer 2/3 cells can hereby activate the modulatory on-center off-surround circuit at all positions of the grouping. The ART Matching Rule is thus satisfied at all such positions, and the source of the "top-down expectation" is the preattentively formed perceptual grouping itself. In summary, the *preattentive perceptual grouping is its own attentional prime* (Fig. 6.6). In this way, *intra*cortical feedback using the 2/3-to-6-to-4 circuit is able to control and stabilize the growth of connections within a cortical area even before *inter*cortical attentional feedback may be able to influence this development.

Model simulations of how cortical area V1 develops its horizontal connections in layer 2/3 and its interlaminar connections between layers 6 and 4 have demonstrated that stable development can be achieved if the excitation and inhibition in the on-center from layer 6-to-4 layer 4 cells is approximately balanced, and thus modulatory (Grossberg & Williamson, 2000). When this constraint is not realised, it is much harder to stabilize development in the model. By extension, these results suggest how perceptual learning in the adult can be stabilized using the same mechanisms.

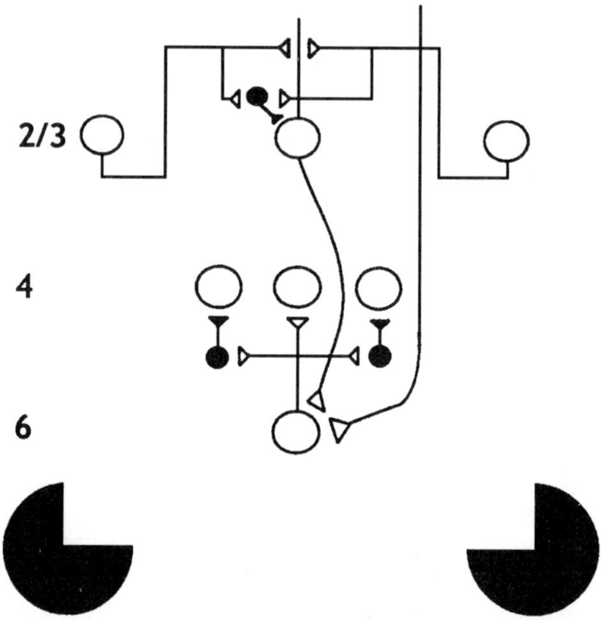

Figure 6.6. The preattentive perceptual grouping is its own attentional prime: An intracortical perceptual grouping, such as an illusory contour, uses the same layer 6-to-4 on-center off-surround network as does intercortical attentional priming. Its off-surround can hereby inhibit incorrect layer 4 cells that could otherwise enter into incorrect associations with active layer 2/3 cells. This mechanism works at all positions of the grouping, even those that do not receive bottom-up inputs.

6.10 AN ILLUSTRATIVE SIMULATION

Computer simulations with the LAMINART model have demonstrated the following data properties: sensitivity of illusory contour strength to different numbers, densities, and shapes of inducers, formation of illusory contours in both cortical areas V1 and V2, or just V2, depending upon the sizes and distances between image inducers, Gestalt grouping laws, and enhancement of camouflaged groupings while automatically suppressing noise (Grossberg, Mingolla, and Ross, 1997; Ross, Grossberg, and Mingolla, 2000); attentional enhancement of entire objects via propagation of attention along real and illusory object contours, colinear facilitation or suppression of targets by colinear flankers at different target and flanker contrasts, orientation-contrast and pop-out of targets surrounded by parallel or perpendicular image contrasts, protection of targets from flanker suppression by focal spatial attention, and great attentional enhancement of low contrast targets than high contrast targets (Grossberg and Raizada, 2000; Raizada and Grossberg, 2000); enhanced development of the projection range and orientation bias of horizontal cell connections in response to visual inputs, development of horizontal projection ranges that are much larger than the classical receptive fields of the respective cells, development of the measured cortical point spread functions in the adult, and development into a cortical architecture that performs like an adult observer in response to psychophysical displays, including properties of analog coherence of perceptual groupings (Grossberg and Williamson, 2000). These properties are in addition to the model's ability to provide functional explanations and new predictions concerning many known anatomical and neurophysiological properties of cells and circuits in areas V1, V2, and LGN.

Here one simulation is summarized for definiteness. This simulation illustrates how the model can explain interactions between attention and grouping, using the intimate connection between attentional and grouping circuits that enables the preattentive grouping to be its own attentional prime. In particular, it has recently been shown, through neurophysiological recordings from macaque monkey neurons in area V1, that spatial attention can propagate along a grouping to "light up" an entire object boundary. Attention enhancement spreads along line segments which are grouped together to form a smooth curve, but not into contiguous segments which are grouped as a different object. Roelfsema, Lamme, and Spekreijse (1998) showed this with monkeys who were trained to perform a visual curve-tracing task (Fig. 6.7). The monkey's task was to make a saccadic eye movement to the end of a target curve whose other end is at the fixation point. Recordings along the target curve showed attentional enhancement relative to recordings along a parallel distractor curve. Figure 6.6 shows how this can happen when attention enhances a grouping via the 6-to-4-to-2/3 pathway. By this route, attention can activate complex cells in layer 2/3, which can then horizontally propagate the attentional enhancement along the entire grouping of which they form a part. Raizada and Grossberg (2000) have also pointed out that attention may also directly modulate layer 2/3 via its effects on layer 1 apical dendrites of layer 2/3 complex cells and inhibitory interneurons. Attention remains modulatory along this route as well, due to the balance between excitation and inhibition which leads to the bipole property for perceptual grouping.

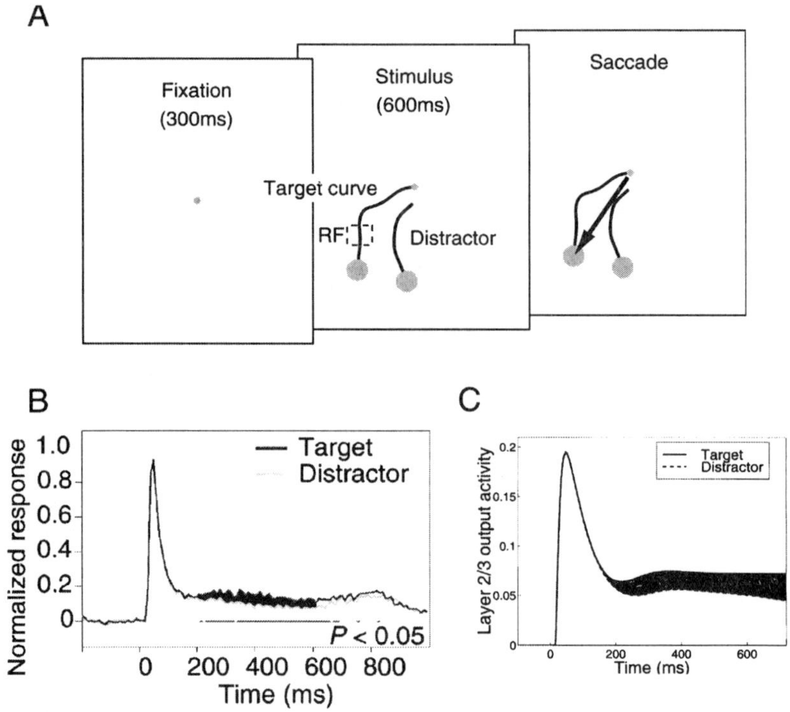

Figure 6.7. Spread of visual attention along an object boundary grouping, from an experiment by Roelfsema *et al.* (1998). (A) The experimental paradigm. Macaque monkeys performed a curve-tracing task, during which physiological recordings were made in V1. A fixation spot was presented for 300ms, followed by a target curve and a distractor curve presented simultaneously; the target was connected at one end to the fixation point. While maintaining fixation, the monkeys had to trace the target curve, then, after 600 ms, make a saccade to its endpoint. (B) Neurophysiological data showing attentional enhancement of the firing of a neuron when its receptive field (RF) lay on the target curve, as opposed to the distractor. Note that the enhancement starts later in distal curve segments, far from the fixation point, than it does in proximal segments, closer to fixation (Pieter Roelfsema, personal communication). This suggests that attentional signals propagate along the length of the target curve. Figures (A) and (B) adapted from Roelfsema et al. (1998) with permission of Macmillan Magazine Ltd. (C) Model simulation of the Roelfsema et al. Data. Figure (C) reprinted from Grossberg and Raizada (2000) with permission of Elsevier Science.

6.11 LEARNING, ATTENTION, AND GROUPING IN SENSORY AND COGNITIVE NEOCORTEX

The LAMINART model suggests how bottom-up, top-down, and horizontal interactions are organized within visual cortical areas V1 and V2. A key contribution of the model is to suggest how the cortex may use the folded feedback circuit from layer 6-to-4 to achieve both top-down attentional priming and analog coherence of preattentive perceptual groupings. The hypothesis that this priming circuit obeys the ART Matching Rule enables all previous results about how ART interactions stabilize development and learning to be applied to the case of cortical development and adult perceptual learning. This includes recent neurophysiological data which have supported the ART predictions that many developmental and learning processes are stabilized dynamically using top-down modulatory on-center off-surround signals, and that adult learning uses the same types of mechanisms as the infant developmental processes upon which it builds. These data include: shared molecular substrates of neonatal development and adult learning (Bailey *et al.*, 1992; Kandel & O'Dell, 1992; Mayford *et al.*, 1992); attentional modulation of cortical development (Singer, 1982); plasticity of adult cortical representations after lesions (Merzenich *et al.*, 1988); dynamical reorganization of long-range connections in the visual cortex (Gilbert & Wiesel, 1992; Zohary *et al.*, 1994); fast perceptual learning in the adult (Karni & Sagi, 1991; Poggio, Fahle, & Edelman, 1992); and fast cortical synchronization (Eckhorn *et al.*, 1988; Gray & Singer, 1989).

Because these issues are so general, and because cortical laminar circuitry is ubiquitous in all sensory and cognitive areas of neocortex, it is plausible that similar LAMINART circuits may be used throughout the neocortex. Such a hypothesis is consistent with the fact that ART models have already been able, albeit without a laminar cortical interpretation, to explain developmental, cognitive, and neurobiological data about normal and amnesic recognition learning, categorization, working memory, memory search, and hypothesis testing (Grossberg, 1980, 1982; Grossberg & Stone, 1986; Carpenter & Grossberg, 1991, 1993; Grossberg, Mingolla, & Ross, 1994; Grossberg, 1995; Grossberg & Merrill, 1996; Grossberg, Boardman, & Cohen, 1997; Grossberg, 1999b; Grossberg & Myers, 2000). In many of these cognitive examples, an orienting system, which has been proposed to be at least partly realized in the hippocampal system, interacts with the attentional thalamocortical circuits that do the learning. This interaction drives hypothesis testing, or memory search, for new, or more task-appropriate, recognition categories within the attentional system. Such an orienting system enables attention to be allocated in a more flexible way than can be achieved by the attentional mechanisms on their own, and permits an enormous expansion in the amount of information that can be learned. An orientive system is not shown in Fig. 6.5. How it can be integrated within a LAMINART model is an open research problem.

Most of these ART analyses have focused on interactions between bottom-up and top-down learning and binding. The LAMINART model suggests how to integrate horizontal associative learning and grouping into this picture. In this regard, long-range horizontal connections are known to occur in many areas of neocortex; for example, in the auditory and language areas of the human temporal cortex (Schmidt *et al.*, 1997). From this perspective, we can recognize that the type of perceptual grouping that has been discussed in this chapter is just one type of horizontal association and binding of distributed information. The LAMINART model opens the way towards integrating bottom-up, top-down, *and* horizontal learning and binding within the laminar circuits of a neocortical module. It remains to be seen whether such model circuits can generalize to show how other sensory and cognitive regions of the neocortex are functionally organized.

6.12 ACKNOWLEDGMENTS

Preparation of this chapter was supported in part by the Defense Advanced Research Projects Agency and the Office of Naval Research (ONR N00014-95-1-0409), the National Science Foundation (NSF IRI-97-20333), and the Office of Naval Research (ONR N00014-95-1-0657). The author wishes to thank Robin Amos for his valuable assistance in the preparation of the manuscript. Address correspondence to: Stephen Grossberg, Department of Cognitive and Neural Systems, Boston University, 677 Beacon St., Boston, MA 02215; or via e-mail at: steve@bu.edu.

6.13 REFERENCES

Antonini, A. and Stryker, M. P., 1993a, Functional mapping of horizontal connections in developing ferret visual cortex: Experiments and modeling, *J. Neurosci.* **14**: 7291-7305.

Antonini, A. and Stryker, M. P., 1993b, Rapid remodeling of axonal arbors in the visual cortex, *Science.* **260**: 1819-1821.

Bailey, C. H., Chen M., Keller, F., and Kandel, E. R., 1992, Serotonin-mediated endocytosis of a pCAM: An early step of learning-related synaptic growth in aptysia, *Science.* **256**: 645-649.

Beauchamp, M. S., Cox, R. W., and DeYoe, E. A., 1997, Gradients of attention in the human visual motion processing system, *Soc. Neurosci. Abstracts.* Abstract 179.3, **23**: 457.

Beck, J., Prazdny, K., and Rosenfeld, A., 1983, A theory of textural segmentation, in: *Human and Machine Vision*, J. Beck, B. Hope and A. Rosenfeld, eds., Academic Press, New York.

Brodmann, K., 1909, *Vergleichende Lokalisationslehre der Grosshirnrinde in ihren Prinzipien dargestellt auf Grund des Zellenbaues*. Barth, Leipzig.

Bullier , J., Hupé, J. M., James, A., and Girard, P., 1996, Functional interactions between areas V1 and V2 in the monkey, *J. Physiol. (Paris)*. **90**: 217-220.

Calloway, E. M. and Katz, L. C., 1990, Emergence and refinement of clustered horizontal connections in cat straite cortex, *J. Neurosci.* **10**: 1134-1153.

Cannon, M. W. and Fullenkamp, S. C., 1993, Spatial interactions in apparent contrast: Individual differences in enhancement and suppression effects, *Vis. Res.* **33**: 1685-1695.

Carpenter, G. and Grossberg, S., 1987, A massively parallel architecture for a self-organizing neural pattern recognition machine, *Comp. Vis., Graph., Image Proc.* **37**: 54-115.

Carpenter, G. and Grossberg, S., eds., 1991, *Pattern Recognition by Self-Organizing Neural Networks,* M.I.T. Press, Cambridge.

Carpenter, G. and Grossberg, S., 1993, Normal and amnesic learning, recognition, and memory by a neural model of cortico-hippocampal interactions, *Trends Neurosci.* **16**: 131-137.

Cauller, L. J. and Connors, B. W., 1994, Synaptic physiology of horizontal afferents to layer I in slices of rat SI cortex, *J. Neurosci.* **14**: 751-762.

Chapman, B., Zahs, K. R. and Stryker, M. P., 1991, Relation of cortical cell orientation selectivity to alignment of receptive fields of the geniculocortical afferents that arborize within a single orientation column in ferret visual cortex, *J. Neurosci.* **11**: 1347-1358.

Das, A. and Gilbert, C. D., 1995, Long-range horizontal connections and their role in cortical reorganization revealed by optical recording of cat primary visual cortex, *Nature.* **375**: 780-784.

DeAngelis, G. C., Ohzawa, I. and Freeman, R. D., 1993, Spatiotemporal organization of simple-cell receptive fields in the cat's striate cortex. 1. General characteristics and postnatal development, *J. Neurophysiol.* **69**: 1091-1117.

Douglas, R. J., Koch, C., Mahowald, M., Martin, K. A. C. and Suarez, H. H., 1995, Recurrent excitation in neocortical circuits, *Science.* **269**: 981-985.

Dubin, M. W. and Cleland, B. G., 1977, Organization of visual inputs to interneurons of lateral geniculate nucleus of the cat, *J. Neurophysiol.* **40**: 410-427.

Duncan, J., 1984, Selective attention and the organization of visual information, *J. Exp. Psychol.: Gen.* **113**: 501-517.

Eckhorn, R., Bauer, R., Jordan, W., Brosch, M., Kruse, W., Munk, M. and Reitbock, H. J., 1988, Coherent oscillations: A mechanism of feature linking in the visual cortex? *Biol. Cyber.* **60**: 121-130.

Felleman, D. J. and van Essen, D. C., 1991, Distributed hierarchical processing in the primate cerebral cortex, *Cereb. Cortex.* **1**: 1-47.

Ferster, D. and Lindström, S., 1983, An intracellular analysis of geniculo-cortical connectivity of area 17 of the cat, *J. Physiol.* **342**: 181-215.

Ferster, D. and Lindström, S., 1985, Synaptic excitation of neurones in area 17 of the cat by intracortical axon collaterals of cortico-geniculate cells, *J. Physiol.* **367**: 233-252.

Galuske, R. A. W. and Singer, W., 1996, The origin and topography of long-range intrinsic projections in cat visual cortex: A developmental study, *Cereb. Cortex.* **6**: 417-430.

Ghose, G. M., Freeman, R. D. and Ohzawa, I., 1994, Local intracortical connections in the cat's visual cortex: Postnatal development and plasticity, *J. Neurophysiol.* **72**: 1290-1303.

Gilbert, C. D. and Wiesel, T. N., 1979, Morphology and intracortical projections of functionally characterised neurones in the cat visual cortex, *Nature.* **280**: 120-125.

Gilbert, C. D. and Wiesel, T. N., 1992, Receptive field dynamics in adult primary visual cortex, *Nature.* **356**: 150-152.

Grieve, K.L . and Sillito, A. M., 1991, The length summation properties of layer VI cells in the visual cortex and hypercomplex cell end zone inhibition, *Exp. Brain Res.* **84**: 319-325.

Grieve, K. L. and Sillito, A. M., 1991, A re-appraisal of the role of layer VI of the visual cortex in the generation of cortical end inhibition, *Exp. Brain Res.* **87**: 521-529.

Grieve, K. L. and Sillito, A. M., 1995, Non-length-tuned cells in layers II/III and IV of the visual cortex: The effect of blockade of layer VI on responses to stimuli of different lengths, *Exp. Brain Res.* **104**: 12-20.

Gove, A., Grossberg, S. and Mingolla, E., 1995, Brightness, perception, illusory contours, and corticogeniculate feedback, *Vis. Neurosci.* **12**: 1027-1052.

Gray, C. M. and Singer, W., 1989, Stimulus-specific neuronal oscillations in orientation columns of cat visual cortex, *Proc. Nat. Acad. Sci.* **86**: 1698-1702.

Grosof, D. H., Shapley, R. M. and Hawken, M. J., 1993, Macaque V1 neurons can signal "illusory" contours, *Nature.* **365**: 550-552.

Grossberg, S., 1976a, Adaptive pattern classification and universal recoding, I: Parallel development and coding of neural feature detectors, *Biol. Cyber.* **21**: 117-158.

Grossberg, S., 1976b, Adaptive pattern classification and universal recoding, II: Feedback, expectation, olfaction, and illusions, *Biol. Cyber.* **23**: 187-202.

Grossberg, S., 1973, Contour enhancement, short-term memory, and constancies in reverberating neural networks, *Stud. Applied Math.* **52**: 217-257.

Grossberg, S., 1978, A theory of human memory: Self-organization and performance of sensory-motor codes, maps, and plans in: *Progress in Theoretical Biology (Vol. 5*), R. Rosen and F. Snell, eds., Academic Press, New York.

Grossberg, S., 1980, How does a brain build a cognitive code? *Psych. Rev.* **87**: 1-51.

Grossberg, S., 1982, *Studies of Mind and Brain*, Kluwer, Amsterdam.

Grossberg, S., 1994, 3-D vision and figure-ground separation by visual cortex. *Percept. Psychophy.* **55**: 48-120.

Grossberg, S., 1995, The attentive brain. *Am. Scientist.* **83**: 483-449.

Grossberg, S., 1997, Cortical dynamics of three-dimensional figure-ground perception of two-dimensional pictures, *Psych. Rev.* **104**: 618-658.

Grossberg, S., 1999a, How does the cerebral cortex working? Learning, attention, and grouping by the laminar circuits of visual cortex, *Spat. Vis.* **12**: 163-185.

Grossberg, S., 1999b, The link between brain learning, attention, and consciousness, *Consci. Cogn.* **8**: 1-44.

Grossberg, S., 2000, How hallucinations may arise from brain mechanisms of learning, attention, and volition, *J. Internat. Neuropsychol. Soc.* **6**: 583-592.

Grossberg, S., Boardman, I. and Cohen, M., 1997, Neural dynamics of variable-rate speech categorization, *J. Exp. Psych.* **2**: 481-503.

Grossberg, S. and Merrill, J. W. L., 1996, The hippocampus and cerebellum in adaptively timed learning, recognition, and movement, *J. Cog. Neurosci.* **8**: 257-277.

Grossberg, S. and Mingolla, E., 1985, Neural dynamics of perceptual grouping: Textures, boundaries, and emergent segmentations, *Percept. Psychophy.* **38**: 141-171.

Grossberg, S., Mingolla, E. and Ross, W. D., 1997, Visual brain and visual perception: How does the cortex do perceptual grouping? *Trends Neurosci.* **20**: 106-111.

Grossberg, S. and Myers, C., The resonant dynamics of speech perception: Interword integration and duration-dependent backwards effects, *Psych. Rev.*, in press.

Grossberg, S. and Raizada, R. D. S., 2000, Contrast-sensitive perceptual grouping and object-based attention in the laminar circuits of primary visual cortex, *Vis. Res.* **40**: 1413-1432.

Grossberg, S. and Somers, D., 1991, Synchronized oscillations during cooperative feature linking in a cortical model of visual perception, *Neural Net.* **4**: 453-466.

Grossberg, S. and Stone, G. O., 1986, Neural dynamics of word recognition and recall: Attentional priming, learning, and resonance, *Psych. Rev.* **93**: 46-74.

Grossberg, S. and Williamson, J. R., A neural model of how horizontal and interlaminar connections of visual cortex develop into adult circuits that carry out perceptual grouping and learning, *Cereb. Cortex,* in press.

Heeger,, D. J., 1992, Normalization of cell responses in cat striate cortex, *Vis. Neurosci.* **9**: 181-197.

Hirsch, J. A. and Gilbert, C. D., 1991, Synaptic physiology of horizontal connections in the cat visual cortex, *J. Neurosci.* **11**: 1800-1809.

Hubel, D. H. and Wiesel, T. N., 1962, Receptive fields, binocular interaction and functional architecture in the cat's visual cortex, *J. Physiol.* **160**: 106-154.

Hubel, D. H., Wiesel, T. N. and LeVay, S., 1977, Plasticity of ocular dominance columns in monkey striate cortex, *Philosoph. Trans. Royal Soc. London (B).* **278**: 377-409.

Hubel, D. H. and Wiesel, T. N., 1977, Functional architecture of macaque monkey visual cortex, *Proc. Royal Soc. London.* **198**: 1-59.

Hupé, J. M., James, A. C., Girard, P. and Bullier, J., 1997, Feedback connections from V2 modulate intrinsic connectivity within, *Soc. Neurosci. Abstracts.* Abstract 406.15, **23**: 1031.

Ito, M., Westheimer, G. and Gilbert, D. C., 1997, Attention modulates the influence of context on spatial integration in V1 of alert monkeys, *Soc. Neurosci. Abstracts.* Abstract 603.2, **23**: 1543.

Johnson, R. R. and Burkhalter, A., 1997, A circuit for amplification of excitatory feedback input from rat extrastriate cortex to primary visual cortex, *Soc. Neurosci. Abstracts.* Abstract 651.7, **23**: 1669.

Julesz, B., 1971, *Foundations of Cyclopean Perception*, University of Chicago Press, Chicago.

Kandel, E. R. and O'Dell, T. J., 1992, Are adult learning mechanisms also used for development? *Science.* **258**: 243-245.

Kapadia, M. K., Ito, M., Gilbert, C. D. and Westheimer, G., 1995, Improvement in visual sensitivity by changes in local context: Parallel studies in human observers and in V1 of alert monkeys, *Neuron.* **15**: 843-856.

Karni, A. and Sagi, D., 1991, Where practice makes perfect in textural discrimination: Evidence for primary visual cortex plasticity, *Proc. Nat. Acad. Sci.* **88**: 4966-4970.

Kisvárday, Z. F., Tóth, E., Rausch, M. and Eysel, U. T., 1995, Comparison of lateral excitatory and inhibitory connections in cortical orientation maps of the cat, *Soc. Neurosci. Abstracts.* Abstract. 360.11, **21**: 907.

Knierim, J. J. and van Essen, D. C., 1992, Neuronal responses to static texture patterns in area V1 of the alert macaque monkey, *J. Neurophysiol.* **67**: 961-980.

Lamme, V. A. F., Zipser, K. and Spekreijse, H., 1997, Figure-ground signals in V1 depend on consciousness and feedback from extra-striate areas, *Soc. Neurosci. Abstracts.* Abstract 603.1, **23**: 1543.

Lowel, S. and Singer, W., 1992, Selection of intrinsic horizontal connections in the visual cortex by correlated neuronal activity, *Science.* **255**: 209-212.

Luhmann, H. J., Martinez, Millan, L. and Singer, W., 1986, Development of horizontal intrinsic connections in cat striate cortex, *Exp. Brain Res.* **63**: 443-448.

Macchi, G. and Rinvik, E., 1976, Thalamo-telencephalic circuits: A neuroanatomical survey, in: *Handbook of Electroencephalography and Clinical Neurophysiology Vol. 2, Pt. A*, A. Remond, ed., Elsevier Press, Amsterdam.

Martin, J. H., 1989, *Neurroanatomy: Text and Atlas,* Appleton and Lange, Norwalk.

Mayford, M., Barzilai, A., Keller, F., Schacher, S. and Kandel, E. R., 1992, Modulation of an NCAM-related adhesion molecule with long-term synaptic plasticity in aplysia, *Science.* **256**: 638-644.

McAdams, C. J. and Maunsell, J. H. R., 1997, Spatial attention and feature-directed attention can both modulate neuronal responses in macaque area V4, *Soc. Neurosci. Abstracts.* Abstract 802.5, **23**: 2062.

McGuire, B. A., Gilbert, C. D., Riolin, P. A. and Wiesel, T. N., 1991, Target of horizontal connections in macaque primary visual cortex, *J. Comp. Neurology.* **305**: 370-392.

Merzenich , M. M., Recanzone, E. G., Jenkins, W. M., Allard, T. T. and Nudo, R. J., 1988, in: *Neurobiology of Neocortex*, P. Rakic and W. Singer, eds., Wiley, New York.

Moore, C. M. and Egeth, H., 1997, Perception without attention: Evidence of grouping under conditions of inattention, *J. Exp. Psych.: Human Percept. Perform.* **23**: 339-352.

Motter, B. C., 1994a, Neural correlates of attentive selection for color or luminance in extrastriate area V4, *J. Neurosci.* **14**: 2178-2189.

Motter, B. C., 1994b, Neural correlates of feature selective memory and pop-out in extrastriate area V4, *J. Neurosci.* **14**: 2190-2199.

Mountcastle, V. B., 1957, Modality and topographic properties of single neurons of cats somatic sensory cortex, *J. Neurophysiol.* **20**: 408-434.

Murphy, P. C. and Sillito, A. M., 1987, Cortico fugal feedback influences the generation of length tuning in the visual pathway, *Nature.* **329**: 727-729.

Murphy, P. C. and Sillito, A. M., 1996, Functional morphology of the feedback pathway from area 17 of the cat visual cortex to the lateral geniculate nucleus, *J. Neurosci.* **16**: 1180-1192.

Meyer, G., Lawson, R. and Cohen, W., 1975, The effects of orientation-specific adaptation on the duration of short-term visual storage, *Vis. Res.* **15**: 569-572.

O'Craven, K. M., Rosen, B. R., Kwong, K. K., Treisman, A. and Savoy R. L., 1997, Voluntary attention modulates fMRI activity in human MT-MST, *Neuron.* **18**: 591-598.

Pandya, D. N. and Yeterian, E. H., 1985, Architecture and connections of cortical association areas, in: *Cerebral Cortex 10*, A. Peters and E.G. Jones, eds., Plenum Press, New York.

Peterhans, E. and von der Heydt, R., 1989, Mechanisms of contour perception in monkey visual cortex. II. Contours bridging gaps, *J. Neurosci.* **9**: 1749-1763.

Poggio, T., Fahle, M. and Edelman, S., 1992, Fast perceptual learning in visual hyperacuity, *Science.* **256**: 1018-1021.

Polat, U., Mizobe, K., Pettet, M. W., Kasamatsu, T. and Norcia, A. M., 1998, Collinear stimuli regulate visual responses depending on cell's contrast threshold, *Nature.* **391**: 580-584.

Polat, U. and Sagi, D., 1994, The architecture of perceptual spatial interactions, *Vis. Res.* **34**: 73-78.

Posner, M. I., 1980, Orienting of attention, *Quarterly J. Exp. Psych.* **32**: 3-25.

Press, W. A. and van Essen, D. C., 1997, Attentional modulation of neuronal responses in macaque area V1, *Soc. Neurosci. Abstracts.* Abstract 405.3, **23**: 1026.

Raizada, R. D. S. and Grossberg, S., Context-sensitive binding by the laminar circuits of V1 and V2: A unified model of perceptual grouping, attention, and orientation contrast, *Spatial Cog.*, in press.

Ramachandran, V. S. and Nelson, J. I., 1976, Global grouping overrides point-to-point disparities, *Perception.* **5**: 125-128.

Redies, C., Crook, J. M. and Creutzfeldt, O. D., 1986, Neural responses to borders with and without luminance gradients in cat visual cortex and dLGN, *Exp. Brain Res.* **61**: 469-481.

Reid, R. C. and Alonso, J-M., 1995, Specificity of monosynaptic connections from thalamus to visual cortex, *Nature.* **378**: 281-284.

Reynolds, J., Chelazzi, L. and Desimone, R., 1999, Competitive mechanisms subserve attention in macaque areas V2 and V4, *J. Neurosci.* **19**: 1736-1753.

Rockland, K. S., 1994, The organization of feedback connections from area V1(18) to V1(17), in: *Cerebral Cortex* 4, A. Peters and K.S. Rockland, eds., Plenum Press, New York.

Roelfsema, P. R., Lamme, V. A. F. and Spekreijse, H., 1998, Object-based attention in the primary visual cortex of the macaque monkey, *Nature.* **395**: 376-381.

Ross, W. D., Grossberg, S. and Mingolla, E., 2000, Visual cortical mechanisms of perceptual grouping: Interacting layers, networks, columns, and maps, *Neural Net.* **13**: 571-588.

Schmidt, K. E., Schlote, W., Bratzke, H., Rauen, T., Singer, W. and Galuske, R. A. W., 1997, Patterns of long range intrinsic connectivity in auditory and language areas of the human temporal cortex, *Soc. Neurosci. Abstracts.* Abstract 415.13, **23**: 1058.

Sheth, B. R., Sharma, J., Rao, S. C. and Sur, M., 1996, Orientation maps of subjective contours in visual cortex, *Science.* **274**: 2110-2115.

Sillito, A. M., Jones, H. E., Gerstein, G. L. and West, D. C., 1994, Feature-linked synchronization of thalamic relay cell firing induced by feedback from the visual cortex, *Nature.* **369**: 479-482.

Singer, W., 1982, The role of attention in developmental plasticity, *Human Neurobiol.* **1**: 41-43.

Somers, D. C., Male, A. M., Seiffert, A. E. and Tootell, R. B., 1999, Functional MRI reveals spatially specific attentional modulation in human primary visual cortex, *Proc. Nat. Acad. Sci., USA* **96**: 1663-1668.

Somers, D. C., Nelson, S. B. and Sur, M., 1995, An emergent model of orientation selectivity in cat visual cortical simple cells, *J. Neurosci.* **15**: 5448-5465.

Stemmler, M., Usher, M. and Niebur, E., 1995, Lateral interactions in primary visual cortex: A model bridging physiology and psychophysics, *Science.* **269**: 1877-1880.

Stryker, M. P. and Harris, W., 1986, Binocular impulse blockade prevents the formation of ocular dominance columns in cat visual cortex, *J. Neurosci.* **6**: 2117-2133.

Thorpe, S., Fize, D. and Marlot, C., 1996, Speed of processing in the human visual system, *Nature.* **381**: 520-522.

Treue, S. and Maunsell, J. H. R., 1997, Attentional modulation of visual motion processing in cortical areas MT and MST, *Nature.* **382**: 539-541.

Tsumoto, T., Creutzfeldt, O. D. and Legendy, C. R., 1978, Functional organization of the corticofugal system from visual cortex to lateral geniculate nucleus in the cat, *Exp. Brain Res.* **25**: 291-306.

van Essen, D. C. and Maunsell, J. H. R., 1983, Hierarchical organization and functional streams in the visual cortex, *Trends in Neurosci.* **6**: 370-375.

von der Heydt, R., Peterhans, E. and Baumgartner, G., 1984, Illusory contours and cortical neuron responses, *Science.* **224**: 1260-1262.

Watanabe, T., Sasaki, T., Nielsen, M., Takino, R. and Miyakawa, S., 1998, Attention-regulated activity in human primary visual cortex, *J. Neurophysiol.* **79**: 2218-2221.

Weber, A. J., Kalil, R. E. and Behan, M., 1989, Synaptic connections between corticogeniculate axons and interneurons in the dorsal lateral geniculate nucleus of the cat, *J. Comp. Neurol.* **289**: 156-164.

Wittmer, L. L., Dalva, M. B. and Katz, L. C., 1997, Reciprocal interactions between layer 4 and layer 6 cells in ferret visual cortex, *Soc. Neurosci. Abstracts.* Abstract 651.5, **23**: 1668.

Zeki, S. and Shipp, S., 1988, The functional logic of cortical connections, *Nature.* **335**: 311-317.

Zohary, E., Cerebrini, S., Britten, K. H. and Newsome, W. T., 1994, Neuronal plasticity that underlies improvement in perceptual performance, *Science.* **263**: 1289-1292.

Chapter 7

Beyond Classical Retinotopy: Striate Cortical Mechanisms Associated with Voluntary Saccades and Attention

Iván Bódis-Wollner

Dept. of Neurology, State University of New York Health Center at Brooklyn, 450 Clarkson Avenue, Brooklyn, NY 11203, USA; PH: (718) 270-1482; FX: (718) 270-3840; EM: bodisi01@hscbklyn.edu

7.1 INTRODUCTION

Most traditional models of visual attention and eye movements consider the visual primary cortex (V1) simply as a way-station between the retina and the brain whose purpose is to provide information about the location and elementary feature of visual stimuli. Current evidence suggests, however, that visual information is modified in V1 by visual context, spatial attention and by eye movement contingent spatial rescaling.

Single cell neurophysiology studies in V1 has shown a correspondence between single cell receptive field location and retinotopy. Within a retinotopic map, individual V1 neurons could be classified as passive spatio-temporal filters that do not participate in perceptual decisions (Crick and Koch, 1995a and b). Many observations, however, go beyond this scheme. New techniques using multiple microelectrode recordings and imaging studies demonstrate neuronal ensemble properties not predictable from single cell properties. Indeed, the results are often inconsistent with the concept of simple retinotopy, because they reveal the existence of intracellular lateral interconnections and non point-to-point feedback connections between various visual cortices. The implication of these interconnections in V1 is that visual processing, visual attention and eye movement signals may influence each other.

The study of "attention" has been parcelled in several ways. Many subdivisions are driven by experimental paradigms within a prior framework of psychology. An explicit neurobiological model of visual attention has been proposed based only on forward (from V1) routing of visual information (Olshausen et al, 1993). In this chapter, attention is considered within the framework of electrophysiological studies that separate "contextual modulation" and "spotlight of attention," and also include feedback to V1.

In addition to these attention-related mechanism that occur in the absence of eye movements, saccade associated V1 activity have been demonstrated and will be summarized. Rapid saccadic eye shifts allow the eyes to foveate a new target, in the peripheral field of view. Originally, there was no evidence for feedback from the muscles of the eye concerning eye movements or eye position (Merton, 1964). More recent psychophysical (Velay et al 1994, 1995) and anatomical (Gentle and Rushell, 1997) studies, however, argue for the central influence of proprioceptive signals emanating either from the eye itself during saccades. There is evidence, however, that after proprioceptors are lesioned, monkeys can still accurately aim their eyes (Guthrie, Porter and Sparks, 1983). In any case, there is evidence from microelectrode recording (Weyand and Malpeli 1993) and MRI (Bódis-Wollner et al, 1997) studies that the striate cortex is actively possible associated with saccadic eye movements and blinks, in the absence of visual input. One role of V1 in association with saccades that it may serve to provide visual stability. It has been long known that during rapid eye movements, no sensation of object motion or self-motion occurs. The most influential concept to account for visual stability during saccades originates from H. von Helmholtz (1866). The concept, later named "corollary discharge" (Sperry, 1950) or "Reafferenzprinzip" (von Holst and Mittelstaedt, 1950), postulates that every intended (eye) movement is accompanied by signals from the motor command center of the brain to neurons which will match this pre-motor input with the sensory signals following the executed movement. In the case of saccades, corollary discharge can cancel any motion signals generated by relative object motion on the retina and suppress the visibility of targets so any motion sensation that might be induced is minimized. Recently, it has been suggested that a stable percept involves the reorganization of spatial scales in association with saccades (Ross et al, 1997). This concept has far reaching consequences, as I will discuss later on.

Blinks may occur conjointly with saccades. However, the saccadic metric is affected by blinks in a non-additive manner (Rottach et al, 1998), suggesting that the conjoint movement is centrally determined. Blinks require a different logic than saccades for maintaining visual stability. During blinks, the visual frame does not shift, or does so minimally. On the other hand, blinks interrupt vision for a considerable time and one is not aware of blanking out of vision during blinks (Volkman, 1962; Volkman et al, 1980). It is conceivable, however, that lack of awareness of blanking out of vision during, a blink is a necessary prerequisite of maintaining visual continuity in time, somewhat analogously to maintaining visual continuity in space during a saccade.

In this chapter, experimental evidence that the striate cortex is actively associated with oculomotor movements, even in the absence of visual input, will be reviewed. Additionally, an important role of the precuneus, previously not highlighted, will be included in discussing continuity in time and space. Results of electrophysiological as well as studies in humans using such diverse techniques as EEG, functional imaging, psychophysics, and transcranial magnetic stimulation (TMS), in behaving monkeys will be summarized.

7.2 ATTENTION AND THE PRIMARY VISUAL CORTEX

7.2.1 Contextual Modulation of V1 Neurons

Attention-related mechanisms of V1 neuronal groups are two kinds: a low level, sometimes called pre-attentive contextual modulation of the visual response, and secondly, attentional modulation of context- related primary visual responses. "Context" is not a precisely defined concept. In this chapter, "context" is used in a sense of having other stimuli or an array of stimuli present outside the receptive fields of the neurons that are being modulated. There is a dynamic interplay between modulated and modulating neurons. The psychological concept of stimulus-related attention may be traced to W. James as "taking possession by the mind, in clear and vivid form, of one out of what seem simultaneously possible objects"…. (Moray, 1969). The cellular neurophysiology of modulation of neurons by visual context originates with the studies of McIlwain (1964, 1966), who showed the so-called periphery effect in single ganglion cell responses of the cat retina. The traditional or classical receptive field (RF) boundary is defined as the area of the visual field from which stimuli can induce a neuron to respond, i.e. increase or decrease its firing rate. McIlwain showed that a stimulus well outside the RF boundary on its own cannot induce a response, but it can modulate the response of the neuron when it is responding to another stimulus inside the RF. Orientationally selective end-stopped cortical neurons exhibit a non-monotonic length function (inhibition or facilitation) (Bódis-Wollner et al, 1976; Rose, 1977; Orban et al, 1979). These findings were originally not interpreted as an effect of the stimulation of the non-classical receptive field. It was shown (Li and Li, 1994) in cat striate cortical neurons that the "non-classical" receptive field has similar orientation and tuning properties to the "classical" RF. These stimulus length dependent effects are consistent with those originally described by Maffei and Fiorentini (1976) in the cat cortex. Haenny and Schiller (1988) first reported that single cell responses in the striate cortex of alert rhesus monkeys are modulated by visual stimuli whose relevance was varied by a behavioral task. They hypothesized that V1 filter specific modulatory effects are caused by higher-order feedback systems. Gilbert and Wiesel (1990) reported the influence of contextual stimuli on orientation selectivity of V1 cells. Kapadia et al (1995) proposed facilitatory influences between neurons with similar stimulus selectivity. Kapadia et al (1995) showed improvement in visual sensitivity of monkey V1 complex cells when a second bar stimulus, placed outside the classical RF, was presented in the optimal orientation, colinear with the target. Using the same task in humans, it was shown that contrast thresholds were reduced. When stimuli flanking the target were made randomly non-colinear, responses attenuated in V1 cells and similarly, the human psychophysical threshold increased. Lee et al (1998) reported background (context) dependent enhancement of neuronal responses in V1. The enhancement was reflected in the latter part (80-200 ms) of the response. They suggested that feedback to V1 neurons, first hypothesized by Haenny and Schiller (1988), mediates this type (non-colinear) of contextual modulation. In behaving monkeys, Vidyasagar (1998) reported the importance of a time window from 50 to 200 milliseconds after stimulus onset for quantifying spatial attentional modulation of V1 cellular activity. A limited capacity attentional window from 50 msec following stimulus onset to about 200 msec was reported by Bódis-Wollner (1973) based on psychophysics. In that observers had to report foveal and parafoveal trigrams, without eye movements. It was found that detection of parafoveal trigrams began around 50 msec and became accurate around 200 msec. However, in the study by Lee et al (1998), the monkeys had to "report" the target by an appropriate saccade, possibly confounding the timing of attentional and saccade-associated V1 activity (see the section on saccades and V1). Studies in the monkey by Zipser et al (1996) Polat et al (1998) and Roelfsema et al (1998) have substantially confirmed contextual modulation of V1 neurons, albeit using stimulus/background configurations substantially different (i.e. non-colinear) from those employed by Gilbert and Wiesel (1990). Zipser et al (1996) demonstrated, using single and multiunit recordings of V1 cells of behaving monkeys, that extra-RF texture elements, up to 8-10 deg, have a strong modulatory effect on the later portion of the neurons firing rate. The anatomical basis of contextual modulation requires the existence of a plexus of lateral interconnecting axonal branches in the primary visual cortex (Gilbert and Wiesel, 1983). Kapadia et al (1999) recorded V1 activity in the superficial layers and studied contrast and surround-dependent length summation profiles of single cells. The results showed that V1 receptive fields are not static and that V1 neurons are dynamically modulated by context. Current models (see Chapter 6 of in this volume) of the common mechanism for attention and visual response postulate that the laminate circuits of the primary visual cortex (Grossberg and Raizada, 2000) play an essential role in contextual modulation.

Dynamic effects of extensive field integration in V1 neurons may not necessarily be obligatory and could represent a critical step in visual attention (see below). Based on psyhophysical and perceptual studies, it may well be inferred that V1 contextual modulation is likely to be important for perceptual grouping based on simple stimulus features (Lamme, 1995; Ito et al, 1998).

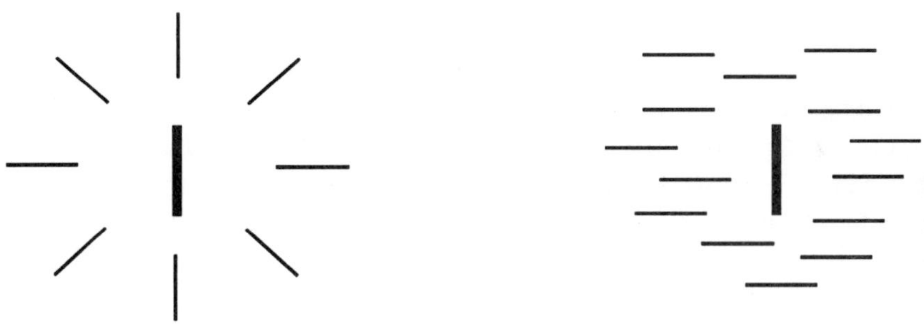

Figure 7.1. A sketch symbolizing contextual modulation of V1 neurons. The preferred stimulus is a vertical bar. On the left, co-linear texture elements, outside the "traditional" receptive field are shown. Neuronal firing rate in the initial 100 msec have been shown to increase in the presence of co-linear elements (Kapadia et al, 1995). On the right the texture elements are orthogonal. It appears that when this background is presented neuronal firing rate increases late: following the initial 150 msec (Zipser et al, 1996).

7.2.2 Spotlight of Attention in V1

As described above, many studies present strong arguments in favor of the existence of visual stimulus contingent contextual modulation of V1 responses. There is however, far more controversy concerning the "spotlight of attention" and V1. The disputed role of V1 in focal attention emanates from single cell studies of Wurtz and Mohler (1976a) and Mohler and Wurtz

(1977). These and other early negative studies, however, may not have used stimuli with strong activation of V1 neurons, and the paradigms may not be suitable to evaluate the effects of the spotlight of attention during contextual modulation. Attentional modulation of V1 neurons was shown in monkeys trained to direct attention to or away from a stimulus presented in an array of stimuli. The presence of competing stimuli enhanced the response to the attended stimulus in V1 neurons (Motter 1993). This enhancement was most evident for optimally oriented stimuli. Conversely, Luck et al (1997) demonstrated the effect of spatial selective attention in V2 and V4, but not in V1. In that study, monkeys were trained either in a spatial or temporal attention task. The study was designed to resolve whether a single neuron in the extra-striate visual cortices, having a large receptive field, can signal the attended of two stimuli, both falling into the receptive field. When the two targets were within the same receptive field ("inside" condition) the responses were enhanced to the attended stimulus, whether or not it was a target or non-target. Such a visuo-spatial focal attention could not be demonstrated in V1. One possible reason may be that, as the authors point out, RFs are small in V1, hence it is difficult to present two stimuli within the same RF. Indeed, no oscillatory neuronal responses (see later sections of this chapter) could be demonstrated for this special form of intracellular attention. Moreover stimulus differences between non-identical pairs presented to V1 neurons in this study may have de-activated the "like-with-like" interactions, and in turn the contextual modulation and consequent attentional modulation of these neurons (as proposed by Ito and Gilbert, 1999) could not occur.

Ito and Gilbert (1999) suggested that spotlight of attention is achieved by extrastriate signals modulating the intrinsic V1 organization which is responsible for contextual modulation in V1 neurons. Response properties of cells in V1 in alert animals trained in focal and distributed attention tasks were studied in three attentive conditions: toward the positions or away from the RF, or to four locations simultaneously. While attention had no effect on the target response, it did so when the response to the target had been amplified by the presence of a contextual (colinear) line. This attentional effect may be achieved by an active non-linear interaction between the sites surrounding the attended area. However, similar modulatory effects were explained as "filtering" between target and distractors (Moran and Desimone, 1985; Chelazzi et al, 1993, Treue and Maunsell, 1996; and Luck et al 1997). This explanation is anchored on some traditional concepts of attention, but has not received current neurophysiological support for V1 neurons.

Many earlier studies have conceptualized that attention (see for instance Grindley and Townsend, 1968) "filtered out" unwanted information and served to provide the low level peripheral gating. Gating may protect higher centers from sensory overload and limit information being processed (Broadbent, 1958). The "early selection" theory of selective attention (Treisman, 1966) implies a precortical automatic gating process. Ito and Gilbert (1999) showed that gating of information is not the major explanation for attention, since V1 is modulated from higher visual (or perhaps non-visual) cortices in a feedback manner to V1. Current evidence favors the importance of dynamic modulation in V1 (Kapadia et al, 1999) rather than passive filtering as a first step in visual attention.

7.2.3 Feedback to V1 from Higher-Order Cortices

Neuroanatomical and multi-unit neurophysiological studies (Sandell and Schiller, 1982; Kennedy and Bullier, 1985; Mignard and Malpeli, 1991; Salin and Bullier, 1995; Vanduffel et al, 1997) reveal multiple cortico-cortical connections. However, caution is required since much of the physiological information is obtained in anaesthetized and paralyzed monkeys (Haupé et a, 1998). Reversible de-activation of a relatively large (larger than 10 millimeter square) area of the superior temporal sulcus containing V5 has a selective, dynamic effect on a V1 neuron, depending on the background and the relationship of the background and the optimal stimulus. When V5 is deactivated, one third of V1 neurons decrease their response, thus suggesting a form of positive feedback on V1 neurons, exclusive of layer IV c. The feedback response modification depends on local contrast between the optimal stimulus and the background of "irrelevant" features. Simply stated, when the stimulus is weaker V5 deactivation is more evident then when the stimulus is stronger. This suggests that V5 feedback into V1 is local and serves to enhance the visibility of relevant but weak stimuli. These effects are even more pronounced on V3 neurons; however, studies have not yet addressed whether V1 and V3 are gated in parallel or sequentially by V5. The existence of a functionally relevant feedback loop is made likely by the studies of Zipser et al (1996) based on latency of responses in V1. They showed in behaving monkeys that the latency of contextual effects on V1 neurons (80 milliseconds or more) provides sufficient time for a signal to travel to and from higher cortices to V1. However several questions remain. For example, do these effects represent a simple automatic gain control process? Another question whether a V5 to V1 feedback loop is necessary for attention?

7.2.4 V1 Connections to Subcortical Nuclei

The concept of efferent connections controlling information flow at lower levels was proposed (for the auditory system) by Galambos (1955). The evidence for backward gating of visual information flow to thalamic structures from the cortex relies on anatomical and physiological data. Deep cortical layers send descending projections to the lateral geniculate nucleus (LGN) (Payne et al 1996). Thalamic relay cells show feature-linked synchronization under feedback from V4 (Sillito et al 1994). In the cat, properties of relay cells of the LGN change following directed saccades (Noda, 1975; Fischer et al, 1996). Surprisingly, however, cortical deactivation has a rather small effect on LGN neurons. Several studies (McClurkin et al, 1994; Sillito et al, 1994) suggest that cortical feedback is not focal, but rather it maps group activity to group activity in the following manner: a) feedback signals are determined by the global context rather than representing focal retinotopy and, b) the feedback itself targets groups of LGN neurons, synchronizing and improving their signal-to-noise ratio. These visual feedback ensemble properties are reminiscent of auditory cortical feedback onto geniculate neurons (Villa et al, 1991; Ghosh et al, 1994; Diamond et al, 1992).

At this point, the bulk of the data suggest that a cortical to LGN feedback system is non-focal, and may be distributed to groups of neurons. This feedback may be useful for gating or filtering and serve to guide global attention or vigilance, but it is a less likely candidate to enhance the "spotlight" of focal attention.

Pulvinar to V1 connections (Ogren and Hendricksen, 1977) have been proposed (LaBerge and Buchsbaum, 1990; Olshausen et al, 1993) to control attention in V1 and posterior parietal cortex (Petersen at al 1987). Substantial collicular information may also be directed to V1 via the pulvinar (Posner and Petersen 1990).

7.2.5 Non-Visual Feedback onto V1

Direct visual projections between prefrontal cortex and visual cortex (extrastriate and striate) have not been shown (Selemon and Goldman-Rakic, 1988). On the other hand, both on anatomical and physiological grounds, one can safely assume the existence of interconnections between posterior parietal, temporal (Rockland and Van Hoesen, 1994) and occipital cortices. Lateral inferior parietal (LIP) neurons have been shown to code for intended reaching and intended saccades for specific spatial locations of visual stimuli (Anderson 1989, 1995). These neurons must have access to precise retinotopical information. Given the response of LIP neurons prior to saccades, they are ideal candidates to mediate perceptual loops, signaling the coincidence visual input as a result of an eye movement and motor output in vectorial space. Contextual modulation and attention, performed in V1 neurons preceding the retinotopical input to LIP, may be a necessary part of this circuit. As summarized earlier, contextual attention in V1 may be modulated by feedback from higher cortices. This loop could select relevant or irrelevant retinotopical signals.

7.2.6 Imaging Studies of Attention and V1 in Humans

High resolution functional MRI (fMRI) takes advantage of the difference in magnetization properties of hemoglobin and deoxyhemoglobin; while metabolic PET methods reflect blood flow changes. Luck et al (1997) studied attentional effects in monkey visual cortex using PET. They proposed that increased blood flow might be caused by shifts in baseline neuronal firing rates rather than changes in the stimulus evoked activity. The significance of decreased blood flow over certain cortical areas, reported (Paus et al, 1995; Nobre et al, 1997; Law et al, 1997; Shulman et al, 1997) in some PET studies, may therefore represent a complementary change. However the neural correlates of these changes still need further evaluation. On the other hand, fMRI directly reflects oxygen utilization in capillary structures. While the best PET resolution is above 2-3mm, fMRI resolution is 1 millimeter/square or even slightly better. (Grinvald et al, 1994). Indeed fMRI studies reveal a high degree of retinotopy in the appropriate visual cortices, especially V1, but also beyond, for elementary visual stimuli (Engel et al, 1997; Tootell et al, 1998; Hadjikhani et al, 1998), but also for illusory and real contours (Mendola et al, 1999) and for attention (Somers et al, 1999). Although several studies reported negative fMRI results for attention in V1, this is possibly due to the method of signal evaluation and fixation requirements (see later). The relationship of fMRI activity and neuronal electrophysiological and metabolic properties has not been clarified. Clarifying this relationship would be of great value, in particular with respect to the neuronal mechanisms of the two kinds of attentional paradigms elicited in V1. Discussing this issue would go beyond the aims of this review. Suffice to say that fMRI reveals neuronal properties while electrophysiology may be strongly dominated by connected neurons.

Many imaging studies used stimulus feature attentional paradigms which would most likely engage contextual attention. In a large number of studies, however, the two types of attentional processes were not always clearly separated. Le et al, (1998) studied the straightforward visual stimulus feature attentional paradigm using fMRI. Attention was required to be either directed upon a single stimulus feature (such as shape or color) or rapidly shifted between attentional and non-attentional states but without a break in fixation. The difference between the two attention conditions was the variable for the fMRI. The results showed no activation of the striate (or other visual) cortices. The conclusions however, are weakened by the method of subtracting fMRI signals between attention and fixation condition. It may be possible that fixation may either excite or inhibit neurons. If the striate cortex is active in both conditions (as it should be, given the requirement to maintain fixation on the visual targets), subtraction will yield no or diminutive responses. In a pure stimulus feature paradigms applied by Le et al (1998), no V1 effect was seen. Büchel and colleagues (1998) used motion stimuli in a Gibsonian "optic flow" paradigm. These stimuli have a strong effect on MT neurons of the monkey (Dubner and Zeki, 1971), a homologue to V5 (Payne et al, 1993). The paradigm is relevant to objects moving to or from the observer and minimizes saccades but doesn't clearly separate "context" from "spotlight." Looking at the difference in optic flow patterns between attention versus simply gazing resulted in strong modulation of V5 and also V3, which is consistent with the concept of feedback from V5 to V3. Weaker but definite attentional modulation was observed on the V1/V2 border, thus suggesting feedback to V1 as well (Greenlee, 2000). This interpretation is consistent with the conclusions of Salin and Bullier (1995) in the monkey showing the effects of reversible cooling of V5 on V1. When V5 is cooled V1 neurons are also affected; i.e. their responses decrease, which is consistent with the concept of modulatory feedback effects of V5 on V1.

In many recent studies, observers maintained fixation while directing attention in a spatial paradigm (spotlight) to some particular stimulus feature. Studies by O'Craven et al (1997), Corbetta (1998) and Corbetta et al (1998) discerned no spotlight effect in V1. Corbetta (1998) and Corbetta et al (1998) used both a spotlight and a stimulus feature attention paradigm and found no V1 effect for either case. In a separate study, Corbetta et al (1999) again reported a cue/target paradigm fMRI activity only in extrastriate areas. Similarly, Kastner et al (1999) found only extrastriate activity in a directed attention paradigm. On the other hand, Perry and Zeki (1999) reported striate and extrastriate activity for covert spatial attention. Somers et al (1999) explored attention in a fovea – extrafovea paradigm. Attentional effects were robust in V1 and spatially specific, thus obeying retinotopy. However, when the observer directed attention to the periphery, fMRI responses were absent at the foveal cortical representation. Sasaki et al (1999) reported fMRI responses in the cortical representation of the fovea in a local stimulus feature attention paradigm, and anatomically more widespread activity for attention being directed to global features. In a spotlight paradigm Brefczynski and de Yoe (1999), showed clear responses in V1. Also Ghandi et al (1999) demonstrated strong spotlight attentional modulation of V1, which was separable from stimulus feature modulation. At this point, there is evidence that spotlight paradigms evoke V1 activity. However the evidence is not uniformly supported by all imaging studies.

7.3 VOLUNTARY SACCADES, BLINKS AND THE PRIMARY VISUAL CORTEX

Quick, darting eye movements, called saccades (S) and blinks (B) are executed constantly by awake humans. One is neither aware of the occurrence of saccades and blinks, nor of a shifting visual scene by S nor an interruption of vision by B. Saccades subserve various visual roles. (Leigh and Zee 1999). However, much current interest is devoted to analyzing saccades as keys to understanding detailed functions of cortical networks that are dedicated to attentive and memory functions and sensory-motor organization. While much has been learned recently on how frontal and parietal cortical mechanisms ensure visual stability and continuity associated with S and B, the role of the occipital cortex remains an enigma. Many current reviews (Paus, 1995; Carter and Zee, 1997; Everling and Fischer, 1998; Sharpe 1998; Gaymard et al, 1998; Schall and Thompson, 1999) of imaging studies and magnetic stimulation experiments on voluntary and reflex saccades may lead one to conclude that the occipital cortex is solely there to provide signals of the location of punctuate visual stimuli i.e. a retinotopical map. However, a considerable number of studies performed in humans with different techniques do not fit this simple concept, and suggest that intrinsic occipital activity is part of the network subserving S and B. Bódis-Wollner et al (1997) and (1999) have shown that the striate cortex, hitherto not considered to play a role in saccades and saccade-related visual stability, is actively in associated with S and B, even in the dark in the absence of visual input. More recently Law and colleagues (1997, 1998) and Wenzel (2000) reported V1 activity in association with saccades, even in the dark.

7.3.1 Intrinsic V1 Organizations

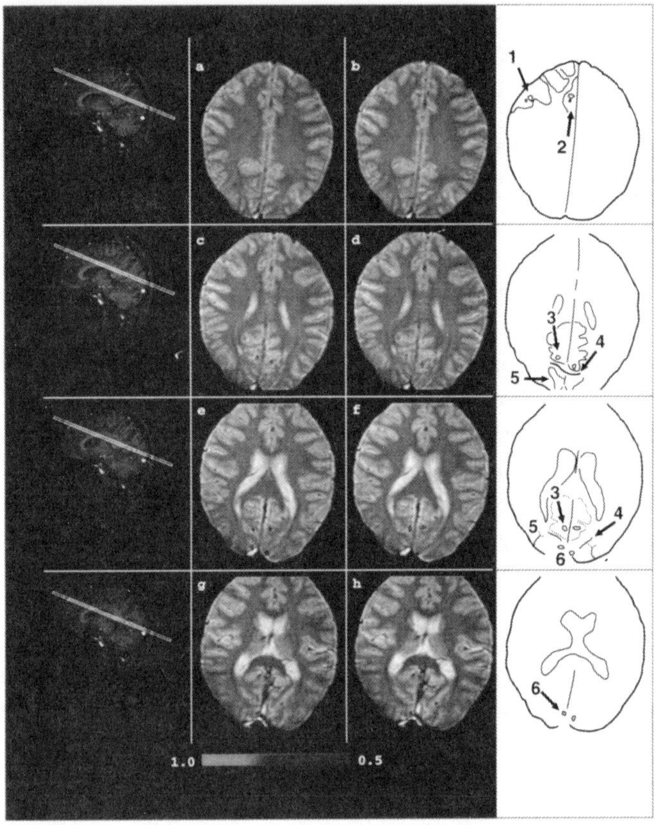

Figure 7.2. Functional MRI of Saccades and Blinks (Bódis-Wollner et al., 1997, 1999). *T2*-weighted oblique transverse MR images of four sections of a representative subject with superimposed activation maps (TR/TE=63-30 ms, a=10°0) associated with voluntary vertical eye movements (a, c, e, g), and voluntary blinks (b, d, f, h). The scout displaying the investigated sections is given in the left column. The gray-coded correlation coefficient scale ranges from 0.5 (dark) to a maximum of 1.0 (white). During voluntary vertical saccades, the activated areas included the posterior median frontal gyrus (a), the medial part of the superior frontal gyrus (a), the posterior (c,e), and the visual cortex (e,g). The parietal cortex was identified on successive horizontal sections by locating the intraparietal-intraoccipital sulcus and the transverse occipital sulci. As described by Valente et al (1998), they form a "striking parenthesis" around the parieto-occipital sulcus. This can be identified on sections c and d on our figures. Anterior to the parieto-occipital demarcation is the inferior part of the parietal lobe. In the right hand column are successive tracings of the axial slices. On each tracing the relevant features have been redrawn from the original.*

The numerical labels refer to: 1. Posterior medial part of frontal gyrus 2. Middle superior frontal gyrus 3. Parietal activation in the inferior lobule-intraparietal sulcus. 4. Intraparietal-intraoccipital sulcus 5. Transverse occipital sulcus 6. Striate cortex. Our identification of the posterior parietal, more precisely intraparietal sulcal area showing significant saccade and blink associated activity is some what tentative, since in general anatomical relationships are more constant in frontal than in parietal areas (Naidich and Brightbill, 1995).

In contrast to the number of multiunit studies showing attention in monkey V1, little is known at this point of the intrinsic organization of V1 in reference to saccades and blinks. Duffy and Burchfiel (1975) reported suppression of spontaneous neuronal activity in the striate cortex of encéphale isolé monkeys associated with spontaneous saccades. The time course was consistent with human physchophysical data on saccadic suppression. Fischer et al (1981) reported in behaving monkeys no suppression in V1 only in "area A19" during saccades. [An increase in neuronal response occurs in the lateral parietal area of the monkey preceding saccades (Colby et al, 1996)]. Recently, however, Martinez-Conde et al, (2000) reported that monkey V1 cells increased their firing rate to optimally oriented stimuli following microsaccades, which are of small amplitude (between 3 minutes and 2 degrees) and slower than "regular" saccades. They found no suppression of activity associated with the beginning of a saccade. One may infer that horizontal interconnections, which have been shown to be of primary importance for contextual attentional modulation in V1, may also be relevant to S and B. The targets of horizontal connections are 80% pyramidal cell dendrites, and the remaining are inhibitory (GABA-ergic) stellate neurons (McGuire et al, 1991). Horizontal interconnections may be particularly relevant to the concepts of Burr et al (1994) (see below) concerning the initiation of stimulus-dependent selective reorganization of spatial scales by saccades. Their studies suggest that veridical spatial metrics are non-linearly modified during saccades. The metric changes differently for the spatial regions ahead and behind the new fixation point. If this scale change takes place in V1, one would assume that the shortest latency neurons of layers 4c alpha and beta neurons undergo differential binding changes in association with S and B. It is more than conceivable that higher frequency oscillations of the visual cortex are relevant to spatial reorganization and changes in visual space perception associated with saccades. V1 neurons show stimulus locked oscillatory responses in the range between 50-100Hz (Maunsell and Gibson, 1992). Such high frequency, so-called gamma rhythms have been suggested as candidates for neuronal "binding" for visual perception and for attention (Eckhorn et al, 1988 and 1993; Gray et al 1989; Niebur et al, 1993; Singer and Gray, 1995). If visual space is reorganized in association with saccades and represent a collapse across space (Burr et al, 1994), high frequency EEG rhythms may be instrumental in changing spatial metrics in association with the saccades.

7.3.2 Perisaccadic Occipital EEG Changes

The neurophysiological interpretation of sustained presaccadic EEG negativity and positivity, spike potentials and lambda waves (ref) is still being debated. Frequency analysis of the EEG has become recently a valuable tool for investigating electrophysiological ensemble properties of the human cortex. Two high frequency bands from 16-34 Hz, called "upper beta" or lower gamma, and a higher frequency band above 34 Hz (gamma) have been shown to be differentially related to visual processing (Tzelepi et al, 2000). An attenuation of different frequency bands of the EEG both preceding and following visually-triggered saccades has been reported by Skrandies and Luschke (1997). In that study no change in perisaccadic higher frequency (gamma) rhythm was reported, but unfortunately the low sampling rates and relatively low-pass filtering applied in the study would have made any interpretation of frequencies above 25 Hz exceedingly difficult. Results by Mari et al (2000) revealed statistically significant changes both prior to and following saccades, as expressed differentially in low gamma (high beta) and high gamma range. Roughly 80 msec following a saccade, there is increased synchrony and power in the occipital gamma band.

Tzelepi, Bezerianos and Bódis-Wollner (2000) used discrete wavelet transform --DWT- (Mallat, 1989; Daubechies, 1990) to specify both time and frequency of high frequency EEG changes to visual stimulation. DWT as opposed to extended Fourier analysis decomposes several seconds long segments of the EEG into orthogonal, equal bandwidth components and specifies peak energy timing in a time window from 50 msec to 100 msec following stimulation. Lower range gamma (beta) lateralized with the hemifield presentation of the sinusoidal grating visual stimulus, while the higher range (near the classical 40 Hz range activity) also varied with stimulus spatial frequency. However, it did not exhibit retinotopical constraint. These differences could suggest that lower range gamma (beta) is the result of thalamo-cortical inputs while the higher frequency component is intrinsic to the cortex representing post-synaptic organization. Occipital upper range beta activity has been suggested as one mechanism by which neuronal linking across space is achieved as a first step in object recognition (Tallon-Baudry and Bertrand, 1999). However, consistent with the functional properties discerned by Tzelepi et al (2000) are the results from a recent MEG study (Sokolov et al, 1999). They suggested that low and high frequency gamma differ for coherent motion stimuli and that attention primarily modulates the higher frequency gamma band of the MEG.

7.3.3 Imaging Studies in Humans and Multi-Unit Recordings in Behaving Monkeys

7.3.3.1 Saccades

Voluntary and reflex saccades are associated with activity in distributed cortical areas (in addition to the known subcortical organization). PET and more recently fMRI and TMS have demonstrated (for a review, see Carter and Zee, 1997) in humans activity of the frontal eye field (FEF) and the supplementary eye field (SEF) in association with saccades. The parietal eye fields (PEF) are subdivided into five parts in the monkey (Colby and Goldberg, 1999). They are relevant to a dynamic updating of spatial representations. The lateral intraparietal area (LIP) transfers visual information from the entire visual field and performs dynamic re-mapping – presumably recalculating saccadic and occipital retinopic information (Colby et al, 1993 and 1996). Neurons in the medial intra- parietal area (MIP) of the monkey have been less well explored, but one suggested role is the maintenance of targets during memory guided tasks. Human PET (see for instance Sweeney et al, 1996), and more recent, fMRI studies (Bódis - Wollner et al, 1997; Law et al, 1998; Wenzel et al, 2000) reveal perisaccadic activity in the occipital cortex. The time resolution of fMRI and PET are low; hence the results may occur due to activity prior to, during or following saccades. Thus, it is appropriate to call this activity "perisaccadic". In the PET study by Sweeney et al (1996), striate cortex activity was well described only for visually evoked reflex saccades, but not for saccades without visual cues. On the other hand, fMRI studies demonstrate powerful striate cortical signals to voluntary saccades in the dark as well, suggesting that visual stimulation is not the only cause of V1 activity (Bódis-Wollner et al, 1997).

7.3.3.2 Blinks

Voluntary blinks have been shown to lead to the activation ($p < 0.05$) of the FEF, the SEF, the posterior parietal cortex ["parietal eye field" (PEF)] and the visual cortex. Voluntary blinking produces activity in the same cerebral structures as voluntary saccades. However, the number of activated voxels is smaller during voluntary blinking than during voluntary saccades in the visual cortex and in the FEF ($p < 0.01$). In contrast, the extent of activation is significantly higher ($p < 0.003$) in the SEF and in the PEF during voluntary blinking (Bódis-Wollner et al, 1999). Besides frontal and occipital activity, activity has been identified in the posterior bank of the intraparietal sulcus (Naidich and Brightbill, 1995; Valente et al, 1998). This deep inferior and midline area in the precuneus had not been described as part of the S and B system in man. A magneto-encephalographic topographic study localized blink-associated sources in the lateral posterior parietal cortex (Hari et al 1994).

In monkeys, an area in the caudal posterior medial parietal cortex has been found to be related to saccadic activity (Thier and Andersen, 1998) and in the maintenance of target location during memory guided tasks (Snyder et al, 1997 and 2000). Moreover, precuneus activation was demonstrated in memory-related imagery in humans (Fletcher et al, 1995) thus suggesting the potential role of the precuneus in maintaining visual continuity. This role of the precuneus in voluntary saccades was recently confirmed in fMRI studies by Berman et al (1998) and Tobler et al (2001). Apparently, voluntary blinks and saccades (see below) are associated with similar loci of activation patterns; however, the quantitative distribution of activation suggests that the middle part of the frontal gyrus and the precuneus cortex are of special significance for voluntary blinks. The results argue for the importance of considering quantitative distribution of parallel cortical activities associated with saccades and blinks. Precuneus activity has been shown in a task engaging visual recall (Le Bihan et al, 1993). It is unknown at present whether or not precuneus activity is needed in association with saccades and blinks in order to provide a memorized target location for voluntary saccades in the dark as well. Visual memory is often tested using oculomotor delayed response paradigms (ODR) or memory-guided saccadic tasks. For these tasks, single or multiunit recordings revealed memory-related activity in neuronal groups of the dorsolateral prefrontal cortex (Joseph and Barone, 1987; Boch and Goldberg, 1989; Funahashi et al, 1989), but the effects of blinks on precuneus activity were not studied. There are differences in blinks and saccades associated interruption of vision. DLPF studies most often use visuo-spatial tasks, whereas during blinking, details of the visual scene (i.e., stimulus features) need to be held in memory. This difference in tasks for blinks and for saccades suggests that precuneus activity may be representing short-term detailed visual information of visual field location and visual detail. Irrespective of the quantitative differential activity of the precuneus, the fMRI results suggest that the striate cortex is part and parcel of the distributed network associated with both saccades and blinks, (Bódis-Wollner et al, 1999).

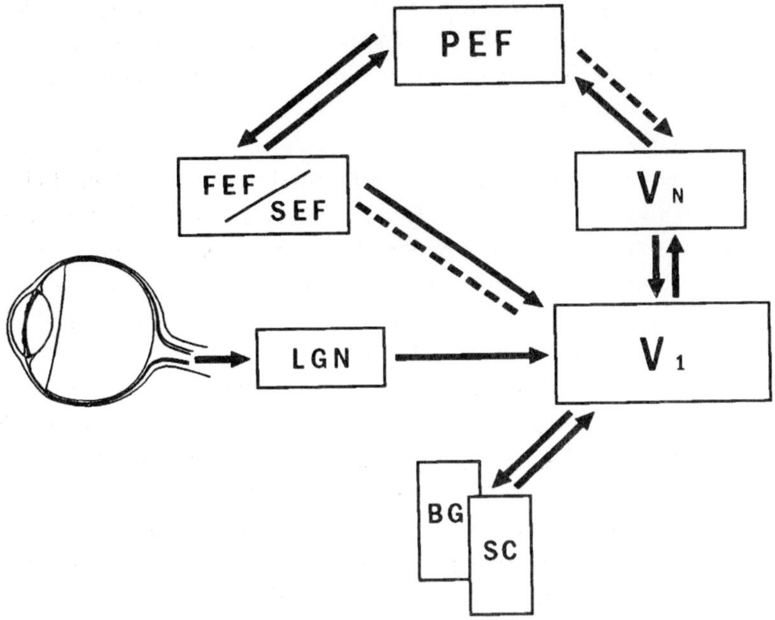

Figure 7.3. A diagram showing established (heavy lines) and inferred (interrupted) lines connections between principal cerebral structures involved in voluntary and visual cue directed saccades and fixational release.

7.3.4 The Effects of Saccades and Blinks on Visual Frame Stability and Visual Processing

7.3.4.1 Saccades

Perisaccadic V1 participation may be meaningful if it represents activity occurring as the result of FEF signals preparing the occipital cortex, which is consistent with the concept of corollary discharge (Sperry 1950, von Holst and Mittelstaedt 1950) or spatial rescaling (Ross et al 1997). Furthermore, it is likely that saccadic and blink suppression are associated with a lack of awareness of interrupted vision, and need to engage short-term memory and require the participation of the striate and deep medial intra-parietal cortex is required for the maintenance of visual stability during saccades and blinks.

Vision, prior to during and following saccades (Latour, 1962; Volkman, 1962) and blinks (Volkman et al, 1980; Ridder and Tomlinson, 1993; Wibbenmeyer et al, 1983) is "suppressed". The psychophysical evidence is consistent with the perceptual observation that when saccades bring the fovea to a new target, the objects in the overall scene do not appear to move. Following Helmholtz (1866), the notion of "corollary discharge" postulates that cortical signals prior to the saccades prepare occipital structures to cancel the shift of the visual frame and maintain a stable percept. A suggestion of a differential effect of saccades on different types on visual stimuli emanates from psychophysical masking studies (Breitmeyer and Ganz, 1976; Breitmeyer, 1978) concerning inhibition of "sustained channel" activity by "transient channels" during saccades. Cortical neuroanatomy, however, does not, reveal simple subtrates for "transient" and "sustained" channels. Current evidence suggests that the visibility of stimuli preferentially exciting "magnocellular" pathways are suppressed during saccades (Burr et al, 1994; Uchikawa and Sato, 1995). Morrone et al (1997) have shown that visual target position is uncertain in the perisaccadic window. They proposed a stimulus-specific spatial scale change during saccades. Where and how this non-linear dynamic transformation of the metric of the visual space takes place is unknown at present, but it is likely that it occurs in distributed areas of the cortex and subcortex. Their model assumes a shift in the external reference point for the fovea and non-linear compression of the visual space (Ross et al, 1997).

Cortical input to the superior colliculus (Gross, 1991) may be one part of the circuit recalibrating the reference points of the visual space (Schiller, 1999). Removing the superior colliculus silences eye movements evoked by stimulation of parietal and occipital eye fields (Keating et al, 1983). It is thought that disengagement from focused attention and saccade generation to a new target requires the participation of cells in the rostral pole of the superior colliculus (Wurtz and Goldberg, 1972). The superior colliculus has efferent connections to motor and premotor neurons, thereby controlling eye movements and orientation to multisensory inputs (Meredith and Stern, 1985). It was shown that spatial attention influences saccadic latency (Clark, 1999), and thus it appears that cortical input may be required to make or break fixation mediated by neuronal groups in the superior colliculus. Collicular neurons may be inhibited by cortical activity to prevent interactions between fixational reference mediated by V1 neurons and other cortical areas. Modulatory cortical connections of the superior colliculus (Wurtz and Mohler, 1976b) may allow an orderly and efficient map transformation by attention and saccades (Robinson et al, 1991).

7.3.4.2 Blinks

Compared to the wealth of studies linking saccades and visual spatial attention, much less research has been devoted to the effects of blinks on attention, and the relation of blinks to subcortical structures. Moreover, the relationship between blinks and cortical attentional activity has not been explored. Although it is known that mental load influences blink rate (Holland and Tarlow, 1972 and 1975) and blinks have been related to higher cognitive process (Fogarty and Stern, 1989) it is not known whether focused visual attention suppresses blinks. In our studies, fixation was maintained without a foveal visual cue. Our data, using infrared eye movement recordings showed, fixation is better than +/-1 degrees that when using head-centered coordinates. Although saccades and blinks may occur conjointly during non-laboratory conditions (von Cramon et al, 1969; Collewijn et al, 1985; Zee et al, 1983; Watanabe et al, 1980; Evinger et al, 1994), saccades associated with blinks are of small amplitude. Their conjoint occurrence however does not simply produce an additive metric; i.e. the interaction is non-linear and appears to be centrally determined (Rottach et al 1998).

In electrophysiological studies, particularly for studying event-related potentials (ERP), blinks are considered a nuisance, as their large signals occur in approximately in the same distribution as frontal components of the P300 signal. Results using source derivation techniques suggest that their distributions differ (Berg and Scherg, 1991; Lins et al, 1993). However, it is conceivable that blinks form an integral part of the process of stimulus contingent focused attention (Fogarty and Stern, 1989; Orchard and Stern, 1991) and V1 activity is a necessary component for visual attentional stability during blinks.

As mentioned earlier, blinks require a different logic for maintaining visual stability than saccades. During a blink, the visual frame may be occluded (Hung et al 1977). Moreover, a blink interrupts vision for a considerable length of time and yet one is unaware of any blanking out of vision. It is conceivable, however, that lack of awareness of blanking out of vision during blinks is a necessary prerequisite of maintaining visual continuity in time, somewhat analogous to maintaining visual continuity in space during saccades. Under normal visual conditions, blinks and saccades may occur conjointly. Current evidence suggests that anatomically the same cortical structures are active during blinks and saccades (Bódis-Wollner et al, 1999), but the weight within the same discrete cortical areas is different for the two tasks. FEF and V1 are more active for saccades while SEF and precuneus are more active for blinks (Bódis-Wollner et al, 1999). Human fMRI (Müri et al, 1996; Bódis-Wollner et al, 1999) data argue for the weighted importance of the posterior parietal cortex and precuneus for blinks. As discussed elsewhere, the precuneus may serve as a short-term memory "hold" device. However its limits for short-term spatial and temporal memory need to be established.

7.4 DISCUSSION

With the advent of new methodology of PET, fMRI, TMS and high resolution EEG, it has become possible to study functional cortical neuroanatomy in humans. In this chapter, the neuro-physiological and neuro-anatomical evidence linking the striate cortex to processes associated with visual attention and voluntary saccadic eye movements was highlighted. Current data strongly argue for the presence of contextual modulation in V1 via laminar connections, while spatially focal attention may occur via feedback circuits to V1 unto the laminar organization mediating contextual modulation.

7.4.1 Interactions Between Saccades and Attention

There is a fair amount of interest (Stevens et al, 1976; Shepherd et al, 1986) and controversy in the psychophysical literature concerning the precise relationship of visuo-spatial attention and saccades (Sereno, 1992). It is generally agreed that an eye movement preparation involves a shift of attention to the intended locus. However the focus of attention may be wider than the saccadic target. He and Kowler (1989 and 1991) argued that "saccades were programmed while attention was distributed across wide regions of the visual field in an attempt to locate the designated target accurately" (Kowler et al 1995). They suggested that spatially selective attention determines the effective input to saccades and they share the same neuronal organization. In the following, the suggested shared anatomy of saccades and attention in V1 will be discussed.

The evidence from neurophysiological studies summarized in the first part of this chapter suggests that spatially selective attention may affect contextually modulated V1neurons via feedback to V1 from higher cortices. Sharing of the gross anatomical substrate of V1 for saccades and attention is supported by isolated imaging data. Some studies show V1 activity for spotlight of attention, while other studies reveal saccade related V1 activity. At first, it seems curious that the only study directly comparing the effect of saccades and spotlight of attention found no fMRI activity for either in V1 (Corbetta et al 1998). This conclusion, however, may be the result of subtracting fMRI signals across conditions resulting in cancellation for each. Are the same V1 neural circuits active for attention without eye movements and vice versa? Few studies have put into evidence individual striate neurons showing attentional and saccadic modulation. It is, however, not only possible but likely that if a shared attentional/saccadic neuronal mechanism exists in V1, it is not occurring in identical single cells but in their interactive processes. Such a process may by dynamic and may depend on intrinsic laminate circuits in V1 and on loops of V1 to other cortices.

Several lines of evidence suggest that spatial metric is changed differently depending on whether the observer is performing a spotlight of attention paradigm or when actual saccades is executed. It has been suggested that the distance between the location of fixation and attended locus does change during saccades. The suggestion of Ross and colleagues (1997) is that in saccadic traverse the spatial attributes collapse. A "collapse" may mean that individual receptive fields (RF's) become smaller. Alternatively, the overlap between neighboring fields may become more extensive. This higher overlap may be mediated by excitatory lateral interconnections between neurons, which "normally" do not connect. Morrone et al (1997) point out that spatial localization of targets in the scanpath become uncertain. It has been also shown by Honda (1999) that visual localization in a frame of visual reference is contingent on saccades. Is there a change of spatial metric during attention similar to the change observed when saccades are executed? Henriques et al (1998) contrasted the accuracy of pointing to a brief visual foveal target in an otherwise dark room under the conditions of: straight-forward gaze was maintained; observers fixated peripherally; and following a saccade from center to periphery. The results showed that retinotopical space is shifted not as a function of eye position per se. It would thus appear that the spatial metric may change from absolute to weighted object distances in attentional and saccadic paradigms.

A plethora of microelectrode studies show attentional processes in V1. V1 may provide the neuronal organization to compute not only fixation and retinotopic location, but also a dynamic and object weighted version of visual space for attention and saccades. Parietal neurons, especially in LIP, may use V1 dynamic information for saliency maps (Kusunoki et al 2000). It remains to be seen, however, whether V1 computes a transformation of retinotopically based dynamic vectors representing shifts in attention with or without eye movement. A shared vectorial representation may direct attention and/or eye movements to a location in a transformed space. This system must be able to dynamically modulate the computed vectors by attention and saccades. It has been also suggested (Reynolds, Pasternak and Desimone, 1996) that facilitation between visuo-spatial attention (location) and attention to stimulus features such as color, orientation contrast etc., and "perceptual" grouping, utilize common mechanisms. Where these interactions take place in the cortex remains to be investigated. Recent studies by Ito and Gilbert (1999) raise the possibility that V1 is a target for ascending and descending pathways to mediate interactions between stimulus feature-dependent "contextual" modulation and spotlight of attention. Based on data summarized in this chapter, dynamic properties of V1 networks modulated during by saccades and by attention via extrastriate feedback need to be considered.

The importance of timing for the effects of attention on saccades, based on temporal model of attention by Reeves and Sperling (1986) was considered by Kowler et al (1995). They concluded that:

> (1) The endpoint of the saccade is determined by the locus of attention during a "critical segment of the saccadic latency period" when a saccadic GO signal is issued.
>
> (2) Directing attention away from the saccadic goal during non-critical portions of the latency period (perhaps early in the latency period) would have no ill effects.
>
> (3) Saccade errors will result when a saccade is initiated while attention is still at a non-goal location while saccades are accurate whenever the saccade is initiated after the shift of attention is computed.

Unfortunately, most studies have not examined in detail the timing of interactions of saccades and attention. At this point, there is not sufficient quantitative evidence for quantifying temporal interactions in models of attention and saccades, although some timing constraints can be suggested. (see Fig. 7.4).

Timing constraints and forward recurrent information flow is shown in Fig. 7.4 for major cortical areas involved in saccadic eye movements. Saccades and anti-saccades were elicited by a parafoveal visual cue (not target) in the TMS and masking experiments (Lalli et al 2000). When voluntary saccades are performed without visual cues, for example in the dark, V1 is presumably activated from all or any of the cortical areas: posterior parietal cortex (PPC), frontal eye fields (FEF) or even higher visual areas (HVA). Cortico-cortical connections to and from V1 have been shown in anatomical (Rockland and Pandiya, 1970; Perkel et al, 1986; Clarke and Miklossy, 1990) multiunit electrophysiological and temporary deactivation (Kennedy and Bullier 1985) experiments. The earliest V1 responses to visual stimulation in the monkey are recordable around 40-50 milliseconds while the earliest contextual attentional responses in V1 occur around 80 milliseconds. In humans, occipital VEP responses to suprathreshold patterned visual stimuli start around 50 milliseconds, peaking around 70-80 milliseconds over the occiput (Bódis-Wollner et al, 1981). MEG source analysis reveals overlapping activity between V1 and V4, with a peak shift of 20-40 milliseconds. Mari et al, (2000) showed that the onset of gamma range EEG activity following a saccade starts more than 80 milliseconds either in dark or light. Eighty milliseconds may be sufficient for completing

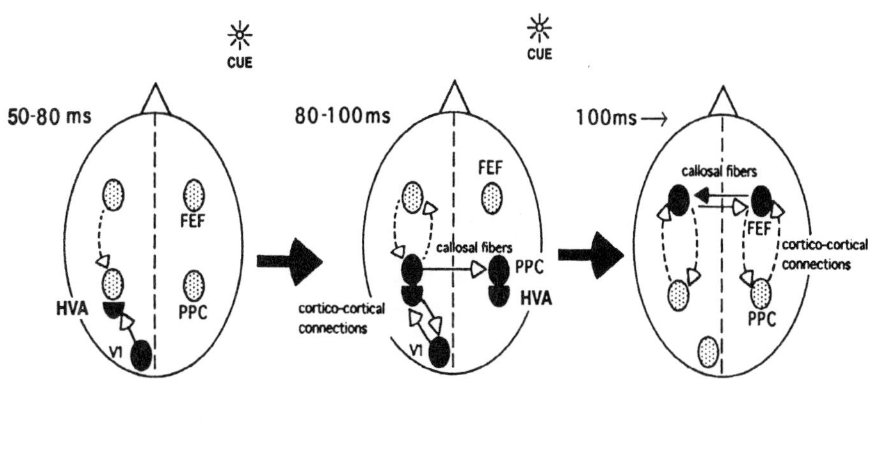

Figure 7.4 Show-down time in V1? An alternative to a simple forward information flow is presented here based on our studies using fMRI, EEG and TMS. See text for details.

a double loop of V1 to and from V1 to and from higher visual areas when no visual cue is presented and voluntary saccades are initiated. It is possible that gamma onset reflects V1 activity combining intrinsic V1 responses and extrastriate downstream feedback onto V1 (Ito and Gilbert, 1999), regardless of whether a visual cue is present. Such a loop may help to reconfirm the presence or absence of visual stimulation. This diagram thus may inferentially relate attentional spotlight and saccades. A recent fMRI study comparing spotlight of attention with saccades (Corbetta et al, 1998) showed an overlap of attention and saccade activity in extrastriate, but not the striate, cortex. Further experiments are needed to answer whether attentional mechanisms are reflected in postsaccadic V1 activity.

7.4.2 Striate Cortex and Conscious Perception

"Attention" has been a concern of early experimental psychology since its beginnings. Wundt (1862) proposed that attention turned perception into apperception (i.e., introspective self-consciousness). From the viewpoint of current studies summarized throughout this chapter, it is worth considering the role of V1 as a cortical area mediating this crucial step toward consciousness.

A number of imaging studies support a role of the occipital cortex in spatial attention (Somers et al, 1999; Ghandi et al, 1999; Brefczynski and DeYoe, 1999; Mangun et al, 1997; Hillyard et al, 1997; Corbetta, 1998). Some recent fMRI studies (Ghandi et al, 1999; Somers et al, 1999) even suggest that the primary visual cortex is strongly and specifically influenced by spatially specific "window of attention". During recall of a visual stimulus, fMRI activity was shown in V1 (and V2) that corresponded to the same area which was activated by the veridical stimulus (LeBihan et al 1993). A definite role of V1 in conscious perception of visual cue-directed saccades have been shown by TMS studies (Lalli et al 2000).

The effects of TMS on saccade metrics in humans has been studied in various laboratories. Surprisingly, when TMS was applied as a single pulse that was aimed at the FEF and SEF, there was little influence on saccades. Frontal and parietal TMS influences periodic saccades (Beckers et al, 1992) and anti-saccades (Brandt et al, 1998), while frontal TMS is reported to have an effect on anti-saccades only in women (Müri et al, 1991). Anti-saccades are saccades executed in the direction opposite of the locus of visual stimulus. In the parietal cortex (Kapoula et al, 1999), TMS affects the latency of saccades and vergence (Hung et al, 1990).

Electrical stimulation of the visual cortex in man (Brindley, 1973) elicits visual phosphenes and eye movements. Microelectrode electrical stimulation applied directly to neurons of area 17 in cats and monkeys (Ilg and Thier, 1999) evokes saccades with a latency of about 80 msec (McIlwain, 1988), thereby leaving ample time for V1 to be re-activated by feedback circuits. The saccades generally occur in retinotopic space with correct direction but often short of the calculated retinotopic distance. McIlwain inferred that saccades occurred to visual targets evoked by micro-electrode stimulation. However latencies of TMS masking at V1 are longer than visual response latencies of V1 in the monkey (Maunsell and Gibson, 1992; Zipser et al, 1996). These latency effects suggest that feedback circuits could be the targets of TMS.

The visual effects of single pulse occipital TMS has been more extensively studied. TMS can, but does not necessarily, induce phosphenes (Meyer et al, 1991). A letter stimulus, when masked by another visual stimulus, can be unmasked by TMS over the occipital cortex (Amassian et al, 1989). It has been suggested (Amassian et al, 1998) that occipital TMS, when using a figure of eight coil, primarily affects the calcarine cortex, or more precisely, incoming thalamocortical fibers rather than intrisic neurons. It has been variously proposed that saccadic suppression may be akin to visual masking. Lateralized occipital TMS induces a scotoma, which is perceptually ovoid, perpendicularly to the orientation of the background grating stimulus in the visual field quadrant opposite the TMS (Kamitami and Shimojo, 1999). The authors proposed that the filling-in effect of long range connections made the temporary scotoma smaller, parallel to the grating orientation and hence the ovoid shape. The observation of Morrone, Ross and Burr (1997) was that a grating consisting of four cycles, and perceptually "collapses" into a single bar in the pre-saccadic 50 milliseconds. This result may be related to the TMS results of Kamitami and Shimojo (1999) because both may affect spatial connections via lateral interconnecting circuits of V1. Lalli et al (2000) studied the effects of single pulse TMS applied at the occipital cortex and visual masking on visual cue directed saccades and anti-saccades. A visual cue was presented either to the left or right of fixation at 5 degrees eccentricity, and a larger amplitude saccade had to be executed either towards or against (anti-saccade) of the cue. Voluntary saccades were abolished along with the visibility of the cue when TMS was applied to the contralateral striate cortex. The effect was maximal about 80 msec following the presentation of the cue. When the cue is visually masked (not seen) about 120 msec following its onset, the observer failed to make a saccade or makes a very delayed one, randomly to left or right. The results of Lalli et al (2000) could be interpreted that TMS interferes with the registration of the visual cue and not with saccades per se.

In most observers the conscious perception of the visual cue was a necessary prerequisite for executing a cued saccade, suggesting that V1 is an essential component of visual orienting circuits. However the authors also reported in two observers a dissociation between perception and saccades at intermediate TMS intervals between peak effect and full recovery. The demonstration of conscious visual memory-related V1 activity and the necessity of V1 for conscious perception of visual cues to elicit a saccade nevertheless does not resolve the controversy concerning the role of V1 in visual consciousness (Crick and Koch 1995a, b; He et al, 1996;, Pollen, 1996, 1999; Lee et al; 1998, Lamme et al; 2000). Our results showed that TMS may disrupt visual perceptions about 80 milliseconds following the onset of the brief visual cue. This TMS effect may occur by disrupting intrinsic V1 circuits or by affecting cortical feedback to V1. From studying behavioral and visual evoked responses in a blind boy with partially preserved V1 (Bódis-Wollner et al; 1977) we inferred that V1 connections are necessary for conscious visual perception. The role of V1 connections with higher visual cortices, in particular recurrent connections (Pollen, 1999) and subcortical loops (Benevento and Port, 1995) need to be considered for modelling conscious visual perception.

7.5 INFERENCES: A SUMMARY

The revised literature leaves little doubt that the occipital cortex, including V1 is part and parcel of the posterior stream (Schiller 1999) of eye movement control. Its functional role goes beyond being a simple way-station to give signals related to retinotopical location of visual stimuli.

The anatomy of striate cortical participation in intention, attention and looking consist of lateral interconnections and feedback circuits modulating receptive field (RF) properties. Laminate circuits may mediate signals necessary for collating stimuli outside the traditional RF of single cells. Spotlight of attention and voluntary saccades may be subserved by both lateral and feedback circuits of V1. The candidate electrophysiological mechanisms modulating attention in V1 may rely on oscillatory rhythms, reflected in high frequency EEG components. Lower frequency EEG signals may be important markers of feedback circuits. Signals which "bind" together neurons at a high rate allow changes to take place at tens of milliseconds range.

High frequency EEG oscillatory waves may just have the required frequency range (30-100 Hz) (Eckhorn et al, 1993; Kruse and Eckhorn, 1996) to reflect short term changes in neuronal connectivity. Studies of coherence (Classen et al, 1998) of these signals may be rewarding, and may also elucidating differences in neurotransmitter kinetics at these frequencies. Most microelectrode but not all imaging studies suggest that paradigms eliciting "spotlight of attention" are accompanied by striate cortical activity. Spotlight of attention is likely to require feedback signals from higher visual cortices to V1 in order to activate laminate circuits.

Blinks interrupt vision for a considerable time yet one is unaware of the blanking of vision. For maintaining temporal visual continuity during blinks and spatial continuity during saccades the distributed activity of the same frontal parietal and occipital areas, and the precuneus are required; however, the contribution of each individual area is differently weighted for saccades and for blinks. The relationship of neuronal mechanisms of V1 involved in saccades, blinks, and attention to consciousness is an intriguing possibility.

For understanding a role of V1 in consciousness, neuronal ensemble connections to other cortices and feedback loops must be considered. Feedback circuits to and from V1 operate to enhance and possibly reconfirm attended signals. Primary dichoptic visual cortical responses may start in humans (Bódis-Wollner et al 1989) as early as 50 milliseconds following visual stimulation. The extensive effects of feedback on V1 are shown already around 80 milliseconds in the monkey. In humans TMS and masking effects peak from 80 to 120 milliseconds following visual stimulus presentation. These timing effects suggest that recurrent V1 processing, rather than input to V1 alone, may provide a necessary condition for conscious visual perception. However, several timing questions, in particular, the context dependent temporal properties of V1 neurons and their modulation by feedback and intrinsic circuits of V1 remain open.

The response of humans by either intentional eye movements or/and by intentional reaching to visual targets may be mediated by distributed neurons receiving patterned visual cortical input and having frontal, pulvinar and collicular connections. Neurons in the posterior parietal cortex (Mountcastle et al, 1975; Colby et al, 1996 and 1999, Snyder, Batista and Andersen, 1997 and 2000) with reciprocal connections to the frontal cortices and the primary visual cortex, may have some of the required inputs to participate in the volitional act. V1, with its intrinsic and extrinsic connections to efferent and other afferent visual areas, appears to be a necessary part of the circuit for intention to act in the visual world.

7.6 ACKNOWLEDGMENTS

I thank Drs. Ken Ciuffreda, Charles Gilbert, George Hung, John Z. Leigh and William Lytton for their review and critique of the manuscript, and for Ms. M. Maxwell for her devoted work on this manuscript and making sense of my nearly undecipherable notes.

7.7 REFERENCES

Amassian, V. E., Cracco, R. Q., Maccabee, P. J., Cracco, J. B., Rudell, A. P., and Eberle, L., 1989, Suppression of visual perception by magnetic coil stimulation of human occipital cortex, *Electroencephalogr. Clin.. Neurosphysiol.* **74**: 458-62.

Amassian, V. E., Cracco, R. Q., Maccabee, P. J., Cracco, J. B., Rudell, A. P., and Eberle, L., 1998, Transcranial magnetic stimulation in study of the visual pathway, *J. Clinical Neurophysiol.* **15**: 288-304.

Andersen, R. A., 1989, Visual and eye movement functions of the posterior parietal cortex, *Annu. Rev. Neurosci.* **12:** 377-403.

Andersen, R. A.., 1995, Encoding of intention and spatial location in posterior parietal cortex, *Cereb. Cortex.* **5**: 457-469.

Beckers, G., Canavan, A. G. M., Zangemeister, W. H., and Hömberg, V., 1992, Transcranial magnetic stimulation of human frontal and parietal cortex impairs programming of periodic saccades, *Neuro-ophthalmol.* **12**: 289-295.

Benevento, L. A., and Port, J. D., 1995, Single neurons with both form/color differential responses and saccade-related responses in the nonretinotopic pulvinar of the behaving monkey, *Vis Neurosci.* **12**: 523-44.

Berg, P., and Scherg, M., 1991, Dipole models of eye movements and blinks, *Electroenceph. Clin. Neurophysiol.* **79**: 36-44.

Berman, R. A., Colby, C. L., Genovese, C. R., Voyvodic, J. T., Luna, B., Thulborn, K. R., and Sweeney, J. A., 1999, Cortical networks subserving pursuit and saccadic eye movements in humans: an FMRI study, *Human Brain Mapping.* **8**: 209-225.

Boch, R. A., and Goldberg, M. E.., 1989, Participation of prefrontal neuron's in the preparation of visually guided eye movements in the rhesus monkey, *J. Neurophysiol.* **61**: 1064-1084.

Bódis-Wollner, I.., 1973, A distractive effect of peripheral attention on foveal trigram recognition, *Perception.* **2**: 407-413.

Bódis-Wollner, I., Atkin, A., Wolkstein, M., and Raab, E., 1977, Visual association cortex and vision in man: pattern evoked occipital potentials in a blind boy, *Science.* **198:** 629-631.

Bódis-Wollner, I., Barris,M. C., Mylin, L. H., Julesz, B., and Kropfl. W., 1981, Binocular stimulation reveals cortical components of the human visual evoked potential, *Electroencephalography and Clin. Neurophysiol.* **52**: 298-305.

Bódis-Wollner, I., Bucher, S. F., Seelos, K. C., Paulus, W., Reiser, M., and Oertel, W. H.., 1997, Functional MRI mapping of occipital and frontal cortical activity during voluntary and imagined saccades, *Neurology.* **49**: 416-420.

Bódis-Wollner, I., Bucher, S. F., and Seelos, K. C., 1999, Cortical activation patterns during voluntary blinks and voluntary saccades, *Neurology.* **53**: 1800-1805.

Bódis-Wollner, I., Mylin, L., and Frkovic, S., 1989, The topography of the N70 component of the visual evoked potential in humans, in: *Topographic Brain Mapping of EEG and Evoked Potentials. (ed),* Springer-Verlag, Berlin and Heidelberg, pp. 396-406.

Bódis-Wollner, I., Pollen, D. A., and Ronner, S., 1976, Responses of complex cells in the visual cortex of the cat as a function of the length of moving slits, *Brain Res.* **116**: 205-216.

Brandt, S. A., Ploner C. J., Meyer B.-U, Leistner S., and Villringer. A., 1998, Effects of repetitive transcranial magnetic stimulation over dorsolateral prefrontal and posterior parietal cortex on memory-guided saccades, ***Expt. Brain Research.*** **118**: 197-204.

Brefczynski, J. A., and DeYoe, E. A., 1999, A physiological correlate of the 'spotlight' of Visual attention, *Nature Neuroscience.* **2**: 370-374.

Breitmeyer, B. G., and Ganz, L., 1976, Implications of sustained and transient channels for theories of visual pattern masking, saccadic suppression, and information processing, *Psychological Review.* **83:** 1-36.

Breitmeyer, B. G., 1978, Disinhibition in metacontrast masking of vernier acuity targets: sustained channels inhibit transient channels, *Vis. Res.* **18**: 1401-1405.

Brindley, G. S., 1973, Sensory effects of electrical stimulation of the visual and paravisual cortex in man, in: *Visual Centers in the Brain.* Springer-Verlag, Chap. 8, pp. 583-594.

Broadbent, D. A., 1958, *Perception and Communication,* Pergamon Press, New York.

Büchel, C., Josephs, O., Rees, G., Turner, R., Frith, C. D., and Friston, K. J., 1998, The functional anatomy of attention to visual motion. A functional MRI study, *Brain.* **121**: 1281-1294.

Burr, D. C., Morrone, M. C., and Ross, J., 1994, Selective suppression of the magnocellular visual pathway during saccadic eye movements, *Nature.* **371**: 511-513.

Carter, N., and Zee, D. S., 1997, The anatomical localization of saccades using functional imaging studies and transcranial magnetic stimulation, *Current Opinion in Neurology.* **10**: 10-17.

Chelazzi, L., Miller, E. K., Duncan, J., and Desimone, R., 1993, A neural basis for visual search in inferior temporal cortex, *Nature.* **363**: 345-347.

Clark,J. J., 1999, Spatial attention and latencies of saccadic eye movements, *Vis. Res.* **39**: 585-602.

Clarke, S., and Miklossy, J., 1990, Occipital cortex in man: organization of callosal onnections, related myelo- and cytoarchitecture, and putative boundaries of functional visual areas, *J. Comp. Neurol.* **298**: 188-214.

Colby, C. L., Duhamel, J.-R., and Goldberg, M. E., 1993, The analysis of visual space by the lateral intraparietal area of the monkey: the role of extraretinal signals, *Prog. Brain. Res.* **95**: 307-316.

Colby, C. L., Duhamel, J.-R., and Goldberg, M. E., 1996, Visual, persaccadic and cognitive activation of single neurons in monkey lateral intraparietal area, *J. Neurophysiol.* **76**: 2841-52.

Colby, C. L., and Goldberg, M. E.., 1999, Space and attention in parietal cortex, *Ann. Rev. Neurosci.* **22**: 319-49.

Collewijn, H., Van Deer Steen, J., and Steinman, R. M., 1985, Human eye movements associated with blinks and prolonged eyelid closure, *J. Neurophysiol.* **54**: 11-27.

Corbetta, M., 1998, Frontoparietal cortical networks for directing attention and the eye to visual locations: Identical, independent, or overlapping neural systems? *Proc. Natl. Acad. Sci.* **95**: 831-838.

Corbetta, M., Akbudak, E., Conturo, T. E., Snyder, A. Z., Ollinger, J. M., Drury, H. A., Linenweber, M. R., Petersen, S. E., Raichle, M. E., Van Essen, D. C., and Shulman, G. L.., 1998, A common network of functional areas for attention and eye movements, *Neuron.* **21**: 761-773.

Corbetta, M., Kincade, M. J., Ollinger, J. M., McAvoy, M. P., Akbudak, E., Conturo, T.E., Snyder, A. Z., Petersen, S.E., and Shulman, G. L., 1999, Event-related fMRI of visuospatial attention: Cue, delay, and validity effects, 6.4 *Soc. for Neurosci.* **25**:1.

Crick, F., 1984, Function of the thalamic reticular complex. The searchlight hypothesis, *Proc. Natl. Acad. Sci.* **81**: 4586-4590.

Crick, F., and Koch, C., 1995a, Are we aware of neural activity in primary visual cortex? *Nature.* **375**: 121-123.

Crick, F., and Koch, C., 1995b, Cortical areas in visual awareness, *Nature.* **377**: 293-295.

Daubechies, I., 1990, The wavelet transform, time-frequency localization and signal analysis, *IEEE Trans. Inform. Theory.* **36**: 961-1004.

Diamond, M. E., Armstrong-Jones, M., Budway, M. J., and Ebner, F. F., 1992, Somatic sensory responses in the rostral sector of the posterior group (Pom) and in the ventral posterior medial nucleus (VPM) of the rat thalamus: Dependence on the barriel field cortex, *J. Comp. Neurol.* **319**: 66-84.

Dubner, R., and Zeki, S. M., 1971, Response properties and receptive fields of cells in an anatomically defined region of the superior temporal sulcus in the monkey, *Brain Res.* **35**: 528-532.

Duffy, F. H., and Burchfiel, J. L., 1975, Eye movement-related inhibition of primate neurons, *Brain Res.* **89**: 121-132.

Duncan, J., 1984, Selective attention and the organization of visual information. *J. Exp. Psychol. Gen.* **113**: 501-517.

Eckhorn, R., Bauer, R., Jordan, W., Brosch, M., Kruse, W., Munk, M., and Reitboeck, H. J.., 1988, Coherent oscilliations: a mechanism of feature linking in the visual cortex? Multiple electrode and correlation analysis in the cat, *Biol. Cybern.* **60**: 121-130.

Eckhorn, R., Frien, A., Bauer, R., Woelbern, T., and Kehr, H.., 1993, High frequency (60-90 Hz) oscillations in primary visual cortex of awake monkey, *Neuro. Report.* **4**: 243-246.

Engel, S. A., Glover, G. H., and Wandell, B. A., 1997, Retinotopic organization in human visual cortex and the spatial precision of functional MRI, *Cerebral Cortex.* **7**: 181-192.

Everling, S., and Fischer, B., 1998, The antisaccade: a review of basic research and clinical studies, *Neuropsychologia.* **36**: 885-899.

Evinger, C., Manning, K. A., and Pellegrini, J. J., et al., 1994, Not looking while leaping: the linkage of blinking and saccadic gaze shifts, *Exp. Brain Res.* **100**: 337-344.

Fischer, B., Boch, R., and Bach, M., 1981, Stimulus versus eye movements comparisons of neural activity in the striate and prelunate visual cortex (A17 and A19) of the trained rhesus monkey, *Exp. Brain Res.* **43**: 69-77.

Fischer, W. H., Schmidt, M., Stuphorn, V., and Hoffman, K. P., 1996, Response properties of relay cells in the A-laminae of the cat's dorsal lateral geniculate nucleus after saccades, *Exp. Brain Res.* **110**: 435-445.

Fletcher, P. C., Frith, C. D., Baker, S. C., Shallice, T., Frackowiak, R. S. J., and Dolan, R. J., 1995, The mind's eye- precuneus activation in memory-related imagery, *Neuroimage.* **2**: 195-200

Fogarty, C., and Stern, J., 1989, Eye movements and blinks: their relationship to higher cognitive processes, *Int. J. Psychophysiol.* **8**: 35-42.

Funahashi, S., Bruce, C. J., and Goldman-Rakic, P. S., 1989, Mnemonic coding of visual space in the monkey's dorslateral prefrontal cortex, *J. Neurophysiol.* **61**: 331-349.

Galambos, R., 1955, Suppression of the auditory wave activity by stimulation of efferent fibers to the cochlea, *Fed. Proc.* **14**: 53.

Gancarz, G., and Grossberg, S., 1999, A neural model of saccadic eye movement control explains task-specific adaptation, *Vis. Res.* **39**: 3123-3143.

Gentle, A., and Rushell, G., 1997, Pathway of the primary afferent nerve fibers serving proprioception in monkey extraocular muscles, *Ophthal. Physiol. Opt.* **17**: 225-231.

Ghandi, S. P., Heeger, D. J., and Boynton, G. M.., 1999, Spatial attention affects brain activity in human primary visual cortex, *Biological Sciences Proc. Natl. Acad. Sci.* **96**: 3314-3319

Ghosh, S., Murray, G. M., Turman, A. B., and Rowe, M. J., 1994, Corticothalamic influences on transmission of tactile information in the ventroposterolateral thalamus of the cat: effect of reversible inactivation of somatosensory cortical areas I and II, *Exp Brain Res.* **100**: 276-286.

Gaymard, B., Ploner, C. J., Rivaud, S., Vermersch, A. I., and Pierrot-Deseilligny, C., 1998, Cortical control of saccades, *Exp Brain Res.* **123**: 159-163.

Gilbert, C. D., and Wiesel, T. N., 1983, Clustered intrinsic connections in cat visual cortex, *J. Neuroscience.* **3**: 1116-1133.

Gilbert, C. D., and Wiesel, T. N., 1990, The influence of contextual stimuli on the orientation selectivity of cells in primary visual cortex, *Vis. Res.* **30**: 1689-1701.

Gray, C. M., Konig, P., Engel, A. K., and Singer, W.., 1989, Oscillatory responses in cat visual cortex exhibit inter-columnar synchronization which reflects global stimulus properties, *Nature.* **338**: 334-336.

Greenlee, M. W., 2000, Human cortical areas underlying the perception of optic flow: Brain Imaging Studies, *International Review of Neurobiology.* **44**: 269-292.

Grindley, G. C., and Townsend, V., 1968, Voluntary attention in peripheral vision and its effect on acuity and differential thresholds, *Q. J. Exp Psychol.* **20**: 11-19.

Grinvald, A., Lieke, E. E., Frostig, R. D., and Hildesheim, R., 1994, Cortical point-spread function and long-range lateral interactions revealed by real-time optical imaging of macaque monkey primary visual cortex, *Neurosci.* **14**: 2545-2568.

Gross, C. G., 1991, Contribution of striate cortex and superior colliculus to visual function in area MT, the superior temporal polysensory area and inferior temporal cortex, *Neuropsychologia.* **29**: 497-515.

Grossberg, S., and Raizada, R. D. S., 2000, Contrast-sensitive perceptual grouping and object-based attention in the laminar circuits of primary visual cortex, *Vis. Res.* **40**: 1413-1432.

Guthrie, B. L., Porter, J. D., and Sparks, D. L., 1983, Corollary discharge provides accurate eye position information to the oculomotor system, *Science.* **221**: 1193-1195.

Hadjikhani, N., Liu, A. K., Dale, A. M., Cavanagh, P., and Tootell, R. B., 1998, Retinotopy and color sensitivity in human visual cortical area V8, *Nature Neurosci.* **1**: 235-241.

Haenny, P. E., and Schiller, P. H., 1988, State dependent activity in monkey visual cortex. I single cell activity in V1 and V4 on visual tasks, *Exp. Brain Res.* **69**: 245-259.

Haupé, J. M., James, A. C., Payne, B. R., Lomber, S. G., Girard, P., and Bullier, J., 1998, Cortical feedback improves discrimination between figure and background by V1, V2 and V3 neurons, *Nature.* **394**: 784-787.

He, P., and Kowler, E., 1989, The role location probability in the programming of saccades: Implications for 'center-of-gravity' tendencies, *Vis. Res.* **29**: 1165-1181.

He, P., and Kowler, E., 1991, Saccadic localization of eccentric forms, *J. Opt. Soc. Am.* **8**: 440-449.

He, S., Cavanagh, P., and Intriligator, J., 1996, Attentional resolution and the locus of visual awareness, *Nature,* **383**: 334-337.

Helmholtz, H. v., 1866, *Handbuch der Physiologischen Optik*, in *A Treatise on Physiological Optics,* J. P. C. Southall, ed. with translation, 1963, Dover, New York.

Henriques, D. Y., Klier, E. M., Smith, M. A., Lowy, D., and Crawford, J. D., 1998, Gaze-centered remapping of remembered visual space in an open-loop pointing task, *J. Neuroscience.* **18**: 1583-94.

Hillyard, S. A., Hinrichs, H., Tempelmann, C., Morgan, S. T., Hansen, J. C., Scheich, H., and Heinze, H.-J., 1997, Combining steady-state visual evoked potentials and fMRI to localize brain activity during selective attention, *Hum. Brain Mapping.* **5**: 287-292.

Holland, M. K., and Tarlow, G., 1972, Blinking and neural load, *Psych Reports.* **3l**: 119-127.

Holland, M. K., and Tarlow, G., 1975, Blinking and thinking, *Perceptual and Motor Skills,* **41**: 303-306.

Honda, H., 1999, Modification of saccade-contingent visual mislocalization by the presence of visual frame of reference, *Vis. Res.* **39**: 51-57.

Hung, G. K., Sun, L., Semmlow, J. L., and Ciuffreda, K.J., 1990, Suppression of sensitivity to change in target disparity during vergence eye movements, *Exp. Neurol.* **110**: 291-297.

Hung, G., Hsu, F., and Stark, L., 1977, Dynamics of the human eye blink, *Am. J. Opt. Physiol. Opt..* **54**: 678-690.

Ilg, U. J., and Their, P., 1999, Eye movements of rhesus monkeys directed towards imaginary targets, *Vis. Res.* **39**: 2143-2150.

Ito, M., and Gilbert, C. D., 1999, Attention modulates contextual influences in the primary visual cortex of alert monkeys, *Neuron.* **22**: 593-604.

Ito, M., Westheimer, G., and Gilbert, C. D., 1998, Attention and perceptual learning modulate contextual influences on visual perception, *Neuron.* **20**: 1191-1197.

Joseph, J. P., and Barone, P., 1987, Prefrontal unit activity during a delayed oculomotor task in the monkey, *Exp. Brain Res.* **67**: 460-468.

Kamitani, Y., and Shimojo, S., 1999, Manifestation of scotomas created by transcranial magnetic stimulation of human visual cortex, *Nature Neurosci.* **2**: 767-771.

Kapadia, M. K., Ito, M., Gilbert, C. D., and Westheimer, G., 1995, Improvement in visual sensitivity by changes in local context: Parallel studies in human observers and in V1 of alert monkeys, *Neuron.* **15**: 843-856.

Kapadia, M. K., Westheimer, G., and Gilbert, C. D., 1999, Dynamics of spatial summation in primary visual cortex of alert monkeys, *Proc. Nat. Acad.Sci.* **96**: 12073-12078.

Kapoula, Z., Isotalo, E., Bocci, M. P., Rivaud-Péchoux, Müri R., Gaymard, B., Leboucher, P., and Pierrot-Desseilligny., 1999, Transcranial magentics stimulation (TMS) of the parietal cortex: effect on the latency of saccades vergence and combined eye movements, *221.3 Soc.for Neurosci.* **25**: 545.

Kastner, S., Nothdurft, H.-C., and Pigarev, I. N.., 1997, Neural correlates of pop-out in cat striate cortex, *Vis. Res.* **37**: 371-376.

Kastner, S., Pinsk, M., Desimone, R., and Ungerleider, L.G., 1999, Directed attention increases activity in human visual cortex in the absence of visual stimulation, *6.2 Soc. for Neuruosci.*. **25**: 1.

Keating, E. G., Gooley, S. G., Pratt, S., and Kelsey, J., 1983, Removing the superior colliculus silences eye movements normally evoked from stimulation of the parietal and occipital eye fields, *Brain and Research*. **269**: 145-148.

Kennedy, H., and Bullier, J., 1985, A double-labelling investigation of the afferent connectivity to cortical areas VI and V2 of the macaque monkey, *J. Neurosci.* **5**: 2815-2830.

Kowler, E., Anderson, E., Dosher, B., and Blaser, E., 1995, The role of attention in the programming of saccades, *Vis. Res.* **35**: 1897-1916.

Kruse, W., and Eckhorn, R., 1996, Inhibition of sustained gamma oscillations (35-80 Hz) by fast transient responses in cat visual cortex, *Proc Natl Acad Sci USA*. **93**: 6112-6117.

Kusunoki, M., Gottlieb, J., and Goldberg, M., 2000, The lateral intraparietal area as a saliencemap: the representation of abrupt onset, stimulus motion, and task relevance, *Vis. Res.* **40**: 1459-1468.

LaBerge, D., 1983, Spatial extent of attention to letters and words, *J. Exp. Psychol. Hum. Precept. Perform.* **9**: 371-379.

LaBerge D, and Buchsbaum MS., 1990, Position emission tomographic measurements of pulvinar activity during an attention task, *J. Neurosci.* **10**: 613-619.

Lalli, S., Ayub, A., Ahmad, A., Hussain, Z., Cracco, R. Q., Bódis-Wollner, I., and Amassian, V. E., 2000, Cued saccades and antisaccades depend on calcarine cortex, *J. Physiol.* **526**: 153-154.

Lamme, V. A. F., Supèr, H., Landman, R., Roelfsema, P. R., and Spekreijse, H., 2000, The role of primary visual cortex (V1) in visual awareness, *Vis. Res.* **40**: 1507-1521.

Lamme, V. A. F., 1995, The neurophysiology of figure-ground segregation in primary visual cortex, *J. Neurosci.* **15**: 1605-1615.

Lamme, V. A. F., and Roelfsema, P. R., 2000, The distinct modes of vision offered by feedforward and recurrent processing, *Trends. Neurosci.* **23**: 571-579.

Latour, P. L., 1962, Vision during voluntary saccadic eye movements, *Vis. Res.* **2**: 261-262.

Law, I., Svarer, C., Holm, S., and Paulson, O. B., 1997, The activation pattern in normal humans during suppression, imagination, and performance of saccadic eye movements, *Acta Physiol. Scand.* **161**: 419-434.

Law, I., Svarer, C., Rostrup, E., and Paulson, O. B., 1998, Parieto-occipital cortex activation during self-generated eye movements in the dark, *Brain*. **121:** 2189-2200.

Le Bihan, D., Turner, R., Zeffiro, T. A., Cuénod, Jezzard, P., and Bonnerot, V., 1993, Activation of human primary visual cortex during visual recall: A magnetic resonance imaging study, *Neurobiology*. **90**: 11802-11805.

Le, T. H., Pardo, J. V., and Hu, X., 1998, 4 T-fMRI Study of nonspatial shifting of selective attention: cerebellar and parietal contributions, *J. Neurophysiol.* **79**: 1535-1548.

Lee, T. S., Mumford, D., Romero, R., and Lamme, V. A. F., 1998, The role of the primary visual cortex in higher level vision, *Vis. Res.* **38**: 2429-2454.

Leigh, R. J., and Zee, D. S., 1999, *The Neurology of Eye Movements.* Third edition, Oxford University Press.

Li, C.-Y., and Li, W., 1994, Extensive integration field beyond the classical receptive field of cat's striate cortical neurons-classification and tuning properties, *Vis. Res.* **34**: 2337-2355.

Lins, O. G., Picton, T. W., Berg, P., and Scherg, M., 1993, Ocular artifacts in recording EEGs and event-related potentials II: source dipoles and source components, *Brain Topogr.* **6**: 65-78.

Luck, S. J., Chelazzi, L., Hillyard, S. A., and Desimone, R., 1997, Neural Mechanisms of spatial selective attention in areas V1, V2 and V4 of Macaque visual cortex, *J. Neurophysiol.* **77**: 24-42.

Maffei, L., and Fiorentini, A., 1976, The unresponsive regions of visual cortical receptive fields,,*Vis. Res.* **16**: 1131-1139.

Mallat, S., 1989, A theory for multi-resolution signal decomposition: the wavelet representation, *IEEE Trans. Pattern Anal. Machine Intel.* **11**: 674-693.

Mangun, G. R., Hopfinger, J. B., Kussmaul, C. L., Fletcher, E. M., and Heinze, H.-J., 1997, Covariations in ERP and PET measures of spatial selective attention in human extrastriate visual cortex, *Human Brain Mapping.* **5**: 273-279.

Mari, Z., Mima, T., Gerloff, C., Hallett, M., and Bódis-Wollner, I., 2000, Perisaccadic high frequency EEG changes in frontal and occipital regions are similar in light and dark, *J. Physiology (London).* **526P**: 25S.

Martinez-Conde, S., Macknik, S. L., and Hubel, D. H.., 2000, Microsaccadic eye movements and firing of single cells in the striate cortex of macaque monkeys, *Nature Neuroscience.* **3**: 251-258.

Maunsell, J. H. R., and Gibson, J. R.., 1992, Visual response latencies in striate cortex of the macaque monkey, *J. Neurophysiol.* **68**: 1332-1344.

McClurkin, J. W., Optican, L. M., and Richmond, B. J., 1994, Cortical feedback increases visual information *transmitted by monkey parvocellular lateral geniculate nucleus neurons, Visual* Neurosci. **11**: 601-617.

McGuire, B. A., Gilbert, C. D., Rivlin, P. K., and Wiesel, T. N., 1991, Targets of horizontal connections in macaque primary visual cortex, *J. Comp. Neurol.* **305**: 370-392.

McIlwain, J.T., 1964, Receptive fields of optic tract axons and lateral geniculate cells:Peripheral, extent and barbiturate sensitivity, *J. Neurophysiol.* **27**: 1154-1173.

McIlwain, J. T., 1966, Some evidence concerning the physiological basis of the periphery effect in the cat's retina, *Exp. Brain Research.* **1**: 265-275.

Mendola, J. D., Dale, A. M., Fischl, B., Liu, A. K., and Tootell, R. B., 1999, The representation of illusory and real contours in human cortical visual areas by functional magnetic resonace imaging, *J. Neurosci.* **19**: 8560-8572.

Meredith, M. A., and Stein, B. E., 1985, Descending efferents from the superior colliculus relay integrated multisensory information, *Science.* **8**: 657-659.

Merton, P. A., 1964, Absence of conscious position sense in the human eyes, in *The Oculomotor System*, M. Bender, ed., Hoeber Medical Division, Harper and Row Publishers, New York, Evanston and London, pp. 314-320.

Meyer,B.-U., Diehl, R., Steinmetz, H., Britton, T. C., and Benecke, R.., 1991, Magnetic stimuli applied over motor and visual cortex: influence of coil position and field polarity on motor responses, phosphenes, and eye movements, *Electroencephalogr. Clin. Neurophysiol. Suppl.* **43**: 121-134.

Mignard, M., and Malpeli, J. G., 1991, Paths of information flow through visual cortex, *Science.* **251**: 1249-1251.

Mohler, C. W., and Wurtz, R. H., 1977, Role of striate cortex and superior colliculus in visual guidance of saccadic eye movements in monkeys, *J. Neurophysiol.* **40**: 74-94.

Moran, J., and Desimone, R., 1985, Selective attention gates visual processing in the extrastriate cortex, *Science.* **229**: 782-784.

Moray, N., 1969, *Attention Selective Processes in vision and hearing*, Hutchinson Educational Ltd. London, Melbourne, Sydney, Auckland, Bombay, Toronto, Johannesburg and New York, pp. 1-194.

Morrone, M. C., Ross, J., and Burr, D. C.., 1997, Apparent position of visual targets during real and simulated saccadic eye movements, *J. Neurosci.* **17**: 7941-7953.

Motter, B. C., 1993, Focal attention produces spatially selective processing in visual cortical areas V1, V2, and V4 in the presence of competing stimuli, *J. Neurophysiol.* **70**: 909-919.

Mountcastle, V. B., Lynch, J. C., Georgopoulos, A., Sakata, H., and Acuna, C., 1975, Posterior parietal association cortex of the monkey: command function for operations within extrapersonal space, *J. Neurophysiol.* **38**: 871-908.

Müri, R. M., Iba-Zizen, M. T., Derosier, C., Cabanis, E. A., and Pierrot-Deseilligny, C., 1996, Location of the human posterior eye field with functional magnetic resonance imaging, *J. Neurology, Neurosurgery and Psychiatry.* **60**: 445-448.

Müri, R. M., Hess, C. W., and Meienberg, O., 1991, Transcranial stimulation of the human frontal eye field by magnetic pulses, *Exp. Brain Res.* **86**: 219-223.

Naidich,T. P., and Brightbill, T. C., 1995, The intraparietal sulcus: a landmark for localization of pathology on axial CT scans, *Int. J. Neuroradiol.* **1**: 3-16.

Nelson, J. I., and Frost, B. J., 1978, Orientation-selective inhibition from beyond the classic visual receptive field, *Brain Res.* **139**: 359-365.

Niebur, E., Koch. C., and Rosin, C., 1993, An oscillation-based model for the neuronal basis of attention, *Vis. Res.* **18**: 2789-2802.

Nobre, A. C., Sebestyen, G. N., Gitelman, D. R., Mesulam, M. M., Frackowiak, R. S. J., and Frith, C. D., 1997, Functional localization of the system for visuospatial attention using positron emission tomography, *Brain.* **120**: 515-533.

Noda, H., 1975, Depression of the excitability of relay cells of lateral geniculate nucleus following saccadic eye movements in the cat, *J. Physiol.* **249**: 87-102.

Ogren, M. P., and Hendrickson, A. E., 1977, The distribution of pulvinar terminals in visual areas 17 and 18 of the monkey, *Brain Res.* **137**: 343-350.

Olshausen, B. A., Anderson, C. H., and Van Essen, D. C., 1993, A neurobiological model of visual attention and invariant pattern recognition based on dynamic routing of information, *J. Neuroscience.* **13**: 4700-4719.

O'Craven, K. M., Rosen, B. R., Kwong, K. K., Triesman, A., and Savoy, R. L., 1997, Voluntary attention modulates fMRI activity in human MT-MST, *Neuron.* **18**: 591-598.

Orban, G. A., Kato, H., and Bishop, P. O., 1979, End-zone region in receptive fields of hypercomplex and other striate neurons in the cat, *J. Neurophysiol.* **42**: 818:832.

Orchard, L., and Stern, J. A., 1991, Blinks as an index of cognitive activity during reading, *Integr. Phys. Behav. Sci.* **26**: 108-116.

Paus, T., 1995, Location and function of the human frontal eye-field: A selective review, *Neuropsychologia.* **39**: 475-483.

Paus, T., Marrett, S., Worsley, K. J., and Evans, A. C., 1995, Extraretinal modulation of cerebral blood flow in the human visual cortex: Implications for saccadic suppression, *J. Neurophysiol.* **74**: 2179-2183.

Payne, B. R., Lomber, S. G., Villa, A.E., and Bullier, J., 1996, Reversible deactivation of cerebral network components, *Trends in Neurosciences*. **12**: 535-542.

Payne, B. R.., 1993, Evidence for visual cortical area homologs in cat and macaque monkey. *Cereb. Cortex*. **3**: 1-25.

Perkel, D. J., Bullier, J., and Kennedy, H., 1986, Topography of the afferent connectivity of area 17 of the macquae monkey: a double labelling study, *J. Comp. Neurol.* **253**: 374-402.

Perry, R. J., and Zeki, S., 1999, An event-related functional MRI study of saccades and covert shifts in spatial attention, *221.1 Soc. for Neurosci.* **25**: 545.

Petersen, S. E., Robinson, D. L., and Morris, J. D., 1987, Contributions of the pulvinar to visual spatial attention, *Neuropsychologia*. **25**: 97-105.

Polat, U., Mizobi, K., Pettet, M. W., Kasamatsu, T., and Norcia, T., 1998, Collinear stimuli regulate visual responses depending on cell's contrast threshold, *Nature*. **391**: 580-584.

Pollen, D. A., 1995, Cortical areas in visual awareness, *Nature*. **377**: 293-294.

Pollen, D. A., 1999, Feature article on the neural correlates of visual perception, *Cereb. Cortex*. **9**: 4-19.

Posner, M. I., and Petersen, S. E., 1990, The attention system of the human brain, *Annu. Rev Neurosci.* **13**: 25-42.

Reeves, A., and Sperling, G., 1986, Attention gating in short-term visual memory, *Psych. Review*. **93**: 180-206.

Ridder, W. H.., and Tomlinson, A., 1993, Suppression of contrast sensitivity during eyelid blinks, *Vis. Res.* **33**: 1795-1802.

Robinson, D. L., McClurkin, J. W., Kertzman, C., and Petersen, S. E. , 1991, Visual responses of pulvinar and collicular neurons during eye movements of awake, trained macaques, *J. Neurophysiol.* **66**: 485-496.

Rockland, K. S., and Pandya, D. N., 1970, Laminar origins and terminations of cortical connections of the occipital lobe in the rhesus monkey, *Brain Res*. **179**: 3-20.

Rockland, K. S., and Van Hoesen, G. W., 1994, Direct temporal-occipital feedback connections to striate cortex (V1) in the macaque monkey, *Cereb. Cortex*. **4**: 300-313.

Roelfsema, P. R., Lamme, V. A. F., and Spekreijse, H., 1998, Object-based attention in the primary visual cortex of the macaque monkey, *Nature*. **395**: 376-381.

Rose, D., 1977, Responses of single units in cat visual cortex to moving bars of light as a function of bar length, *J. Physiol., London*. **271**: 1-23.

Ross, J., Morrone, M. C., and Burr, D. C., 1997, Compression of visual space before saccades, *Nature*. **386**: 598-601.

Rottach, K. G., Das, V. E., Wohlgemuth, W., Zivotofsky, A. Z., and Leigh, R. J., 1998, Properties of horizontal saccades accompanied by blinks, *J. Neurophysiol.* **79**:2895-2502.

Salin, P..-A., and Bullier, J., 1995, Corticocortical connections in the visual system: structure and function, *Physiol. Rev.* **75**: 107-154.

Sandell, J. H., and Schiller, P. H., 1982, Effect of cooling area 18 on striate cortex cells in the squirrel monkey, *J. Neurophysiol.* **48**: 38-48.

Sasaki, Y , Hadjihjani, N., Hibino, H., Dale, A. M., Rosen, B. R., and Tootell, R. B. H., 1999, Local and global attention are represented retinotopically, not lateralized, in human occipital cortex, *6.1 Soc. for Neurosci.* **25**: 1.

Schall, J. D., and Thompson, K. G., 1999, Neural selection and control of visually guided eye movements, *Annu. Rev. Neurosci.* **22**: 241-259.

Schiller, P. H., 1999, The neural control of visually guided eye movements, in J. Richards, ed., *Cognitive Neuroscience of Attention*. Erlbaum.

Selemon, L. D., and Goldman-Rakic, P. S., 1988, Common cortical and subcortical targets of the dorsolateral prefrontal and posterior parietal cortices in the rehesus monkey: Evidencefor a distributed neural network subserving spatially guided behavior, *J. Neurosceince.* **8**: 4049-4068.

Sereno, A. B., 1992, Programming saccades: the role of attention, in K. Rayner, ed, *Eye Movements and Visual Cognition: Scene Perception and Reading*. Springer-Verlag, New York, pp. 89-107.

Sharpe, J. A., 1998, Cortical control of eye movements, *Current Opinion in Neurology*. **11**: 31-39.

Shepherd, M., Findlay, J. M., and Hockey, R. J., 1986, The relationship between eye movements and spatial attention, *Q. .J. Exp. Psychol.* **38**: 475-491.

Shulman, G. L., Fiez, J. A., Corbetta, M., Buckner, R. L., Miezin, F.M., Raichle, M. E., and Petersen, S. E., 1997, Common blood flow changes across visual tasks. II. Decreases in cerebral cortex *J. Cogn. Neurosci.* **9**: 648-663.

Sillito, A. M., Jones, H. E., Gerstein, G. L., and West. D., 1994, Feature-linked synchronization of thalamic relay cell firing induced by feedback from the visual cortex, *Nature.* **369**: 479-482.

Singer, W., and Gray, C. M., 1995, Visual feature integration and the temporal correlation hypothesis, *Annu. Rev. Neurosci.* **18**: 555-586.

Skrandies, W., and Luschke, K., 1997, Topography of visually evoked brain activity during eye movements: lambda waves, saccadic suppression, and discrimination performance, *International J. Psychophysiology*. **27**: 15-27.

Snyder, L. H., Batista, A. P., and Andersen, R. A., 2000, Intention-related activity in the posterior parietal cortex: a review, *Vis. Res.* **40**: 1433-1441.

Snyder, L. H., Batista, A. P., and Andersen, R. A., 1997, Coding of intention in the posterior parietal cortex, *Nature*. **386**: 167-170.

Sokolov, A., Lutzenberg, W., Pavlova, M., Preissl, H., Braun, C., and Birbaumer. N., 1999, Gamma-band MEG activity to coherent motion depends on task-driven attention, *Neuro. Report.* **10**: 1997-2000.

Somers, D. C., Dale, A. M., Seiffert, A. E., and Tootell, R. B. H., 1999, Functional MRI reveals spatially specific attentional modulation in human primary visual cortex, *Proc. Natl. Acad. Sci.* **96**: 1663-1668.

Sperry, R. W., 1950, Neural basis of the optokinetic response produced by visual inversion, *J. Comp. Psych. Physiol.* **43**: 482-489.

Stevens, J. K., Emerson, R. C., Gerstein, G. L., Kallos, T., Neufeld, G. R., Nichols, C. W., and Rosenquist, A. C., 1976, Paralysis of the awake human: visual perception, *Vis. Res.* **16**: 93-98.

Sweeney, J. A., Mintun, M. A., Kwee, S., Wiseman, M. B., Brown, D. L., Rosenberg, D. R., and Carl, J. R., 1996, Positron emission tomograpy study of voluntary saccadic eye movements and spatial working memory, *J. Neurophysiol.* **75**: 454-468.

Tallon-Baudry, C., and Bertrand, O., 1999, Oscillatory gamma activity in humans and its role in object representation, *Trends in Cognitive Sciences.* **3**: 151-162.

Their, P., and Andersen, R. A., 1998, Electrical microstimulation distinguishes distinct saccade-related areas in the posterior parietal cortex, *J. Neurophysiol.* **80**: 1713-1735.

Tobler, P. N., Felblinger, J., Bürki, M., Nirkko, A. C., Ozdoba, C., and Müri, R. M., Functional organisation of saccadic reference system processing extraretinal signals in humans, *Vis. Res.*, in press, 2001.

Tootell,R. B. H., Hadjikhani, N., Hall, E. K., Marrett, S., Vanduffel, W., Vaughan, J.T., and Dale, A. M., 1998, The retinotopy of spatial attention, *Neuron.* **21**: 1409-1422.

Treisman,A., 1966, Our limited attention, *Adv. Sci.* 600-611.

Treue, S., and Maunsell, J. H. R., 1996, Attentional modulation of visual motion processing in cortical areas MT and MST, *Nature.* **382**: 539-541.

Tzelepi, A., Bezerianos, T., and Bódis-Wollner, I., 2000, Functional properties of sub-bands of oscillatory brain waves to pattern visual stimulation in man, *Clin. Neurophysiol.* **111**: 259-269.

Uchikawa, K., and Sato, M., 1995, Saccadic suppression of achromatic and dichromatic responses measured by increment-threshold spectral sensitivity, *J. Opt. Soc. Am.* **12**: 66l-666.

Valente, M., Naidich, T. P., Abrams, K. J., and Blum, J. T., 1998. Differentiating the pars marginalis from the parieto-occipital sulcus in axial computed tomography sections, *Int. J. Neuroradiol.* **4**: 105-111.

Vanduffel, W., Payne, B. R., Lomber, S. G., and Orban, G. A., 1997, Functional impact of cerebral connections, *Proc. Natl. Acad, Sci. USA.* **94**: 7617-7620.

Velay, J. L., Roll, R., Demaria, J. L., Bouquerel, A., and Roll, J. P., 1995, Human eye muscle proprioceptive feedback is involved in target velocity perception during smooth pursuit, *Vis. Res.* **35**: 79-85.

Velay, J. L., Roll, R, Lennerstrand, G., and Roll, J. P., 1994, Eye proprioception and visual localization in humans: influence of ocular dominance and visual context, *Vis. Res.* **34**: 2169-2176.

Vidyasagar, T. R., 1998, Gating of *neuronal responses in macaque primary visual cortex by an attentional spotlight, Neuro*-Report. **9**: 1947-1952.

Villa, A. E. P., Rouiller, E. M., Simm, G. M., Zurita, P., de Ribaupierre, Y., and de Ribaupierre, F., 1991, Corticofugal modulation of the information processing in the auditory thalamus of the cat, *Exp. Brain Res.* **86**: 506-517.

Volkman, F., 1962, Vision during voluntary saccadic eye movements, *J. Opt. Soc. Am.* **52**: 571-578.

Volkmann, F. C., Riggs, L. A., and Moore, R. K., 1980, Eyeblinks and visual suppression, *Science.* **207**: 900-902.

Von Cramon,M., Schmid, R., and Vogel, M. W., 1969, Über einige Bedingungen des Zusammenhanges von Lidschlag and Blickwendung. *Psychologische Forschung*, **33**: 68-78.

Von Holst, E., and Mittelstaedt, H., 1950, Das Reafferenzprinzip, *Naturwissenschaften*. **37**: 464-476.

Watanabe, Y., Fujita, T., and Gyoba, J.., 1980, Investigation of blinking contingent upon saccadic eye movements, *Tohoku Psychologica Folia,*. **39**: 121-129.

Wenzel, R., Wobst, P., Heekeren, H. H., Kwong, K. K., Brandt, S. A., Kohl, M., Obrig, H., Dirnagl, U., Villringer, A., 2000, Saccadic suppression induces focal hypo-oxgenation in the occipital cortex, *J. Cerebral Blood Flow and Metabolism*. **20**: 1103-1110.

Weyand, T. G., and Malpeli, J. G., 1993, Responses of neurons in primary visual cortex are modulated by eye position, *J. Neurophysiol.* **69**: 2258-2260.

Wibbenmeyer, R., Stern, J. A., and Chen, S. C., 1983, Elevation of visual threshold associated with eyeblink onset, *Int. J. Neurosci.* **18**: 279-286.

Wundt, W., *Beiträge zur Theorie der Sinneswahrnehmung*. Leipzig: C.F. Winter 1862.

Wurtz, R. H., and Goldberg, M. E.., 1972, Activity of superior colliculus in behaving monkey: IV. Effects of lesions on eye movements, J. Neurophysiol. **35**: 575-586.

Wurtz, R. H., and Mohler, C. W., 1976a, Enhancement of visual responses in monkey striate cortex and frontal eye fields, *J. Neurophysiol.* **39**:766-772.

Wurtz, R. H., and Mohler, C. W., 1976b, Organization of monkey superior colliculus: enhanced visual response of superficial layer cells, *J. Neurophysiol.* **39**: 745-765.

Zee, D. S., Chur, F. C., Leigh, R. J., Savino, P. J., Schatz, N. J., Reingold, D. B., and Cogan, D. G., 1983, Blink saccade synkinesis, *Neurology*. **3**: 1233-1236.

Zipser, K., Lamme, V. A. F., and Schiller, P. H., 1996, Contextual modulation in primary visual cortex, *J. Neurosci.* **5**: 7376-7389.

III

OCULOMOTOR SYSTEM MODELS

Chapter 8

Models of Accommodation

George K. Hung[1], Kenneth J. Ciuffreda[2], Madjid Khosroyani[3], Bai-Chuan Jiang[4]

[1] *Dept. of Biomedical Engineering, Rutgers University, 617 Bowser Rd., Piscataway, NJ 08854-8014, PH: (732) 445-4137, FX: (732) 445-3753, EM: shoane@rci.rutgers.edu*

[2] *Dept. of Vision Sciences, State University of New York, State College of Optometry, 33 West 42nd St. New York, NY 10036; PH: (212) 780-5132, FX: (212) 780-5124; EM: kciuffreda@sunyopt.edu*

[3] *Tarbiat Modarres University, P. O. Box 14155-4838, Tehran, Iran; EM: khosro_m@net1cs.modarres.ac.ir*

[4] *College of Optometry, Health Professions Division, Nova Southeastern University, 3200 South University Dr., Ft. Lauderdale, FL 33328, PH: (954) 262-1444, FX: (954) 262-1818, EM: bjiang@nova.edu*

8.1 INTRODUCTION

The ability to see clearly at different distances is one of the most important functions of the human visual system. This is performed routinely and effortlessly in daily life by the process called accommodation. During this process, the accommodation system must sense that a new target is defocused beyond a blur threshold, develop the appropriate neurological control signal based on blur magnitude, and then adjust relatively rapidly the optics of the eye via the ciliary muscle until the target is once again in focus. Thus, it involves feedback regulation of visual optics via the sensing of retinal image blur. In addition, since blur per se does not provide the light vergence direction (Stark, 1968), the accommodation system must use perceptual cues and other sources of information to determine the appropriate direction of focus (Ciuffreda, 1991, 1998). It does this remarkably well, so that rarely does accommodation occur in the wrong direction under natural viewing conditions. Moreover, accommodation takes place repeated in daily life, so that the system must be continually available to provide clear vision in the performance of a variety of tasks at a range of different distances.

Helmholtz (1855) was the first to elucidate the ocular changes during accommodation. The basic structures involved are shown in Fig. 8.1a. The only active element in this process is the ciliary muscle, whereas all other elements act in a passive manner. During increased accommodation driven by parasympathetic innervation (Kaufman, 1992; Fig. 8.1b), the ciliary muscle contracts, which causes it to shift forward and inward, as well as stretch the choroid and posterior zonules (Ciuffreda, 1991, 1998). In this process, the tension in the anterior zonules is reduced, thus relaxing the inherent forces of the elastic lens capsule, and returning the lens to its more natural rounded shape (Fig. 8.1b). This increases the optical power of the crystalline lens to focus the near target's image on the retina. On the other hand, during a decrease in accommodation, the ciliary muscle reduces its state of contraction, and the passive restoring forces of the spring-like choroid and posterior zonules return each element towards its former position. The tension on the anterior zonules is now increased, thus pulling on the capsule and lens centrifugally, thereby causing the front surface of the lens to flatten (Fig. 8.1b). This decreases the optical power of the crystalline lens to focus a distant target's image on the retina.

Higher cortical centers are involved in accommodation. Briefly, the summated blur signals are transmitted through the magnocellular layer of the lateral geniculate nucleus to arrive at area 17 of the visual cortex (Moses, 1981; Ciuffreda, 1991, 1998). The summated cortical cell responses form a sensory blur signal. The signal is then transmitted to the parietal-temporal areas for processing and dissemination (Jampel, 1959; Harrison, 1987; Ohsuka et al, 1988). This supranuclear signal then goes on to the midbrain Edinger-Westphal nucleus, where the motor command is formulated (Gamlin, 1999). The motor command is transmitted to the ciliary muscle via the oculomotor (III) nerve, the ciliary ganglion, and then the short ciliary nerve (Warwick, 1954). This causes a change in the state of contraction of the ciliary muscle, which in turn deforms the crystalline lens to attain a clear and in-focus retinal image of the target.

However, nearly 150 years after Helmholtz (1855), accommodation was still being discussed in qualitative or descriptive terms only. For example, Heath (1956a) divided and categorized accommodation into four components: reflex (blur), convergence, proximity, and tonic. These are important aspects of accommodation, but they remained as isolated components that had not been linked together in a systematic manner. Such

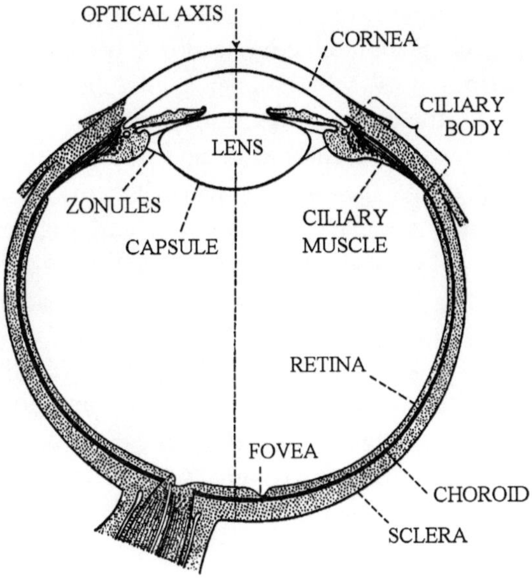

Fig. 8.1a.

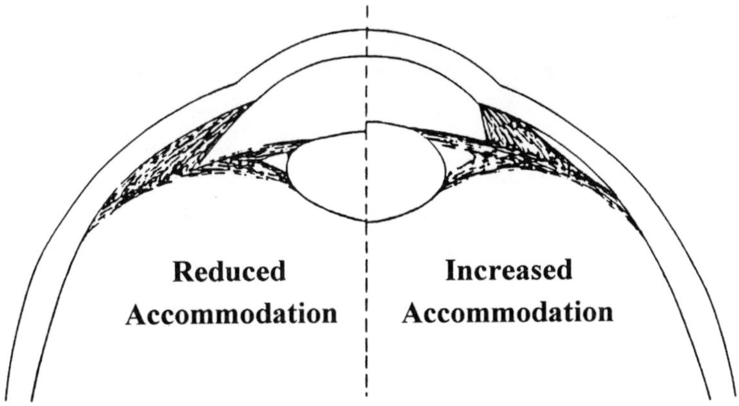

Fig. 8.1b.

Figure 8.1. (a) Horizontal section of the eye showing the major ocular components for accommodation. Adapted from Last (1968), pg. 30, Fig. 21, with permission of Kluwer Academic/Plenum Publishers. (b) Split view of the eye showing (left) relatively flat lens front surface curvature associated with reduced accommodation, and (right) relatively steep front surface curvature associated with increased accommodation. Adapted from Ciuffreda (1998), pg. 78, Fig. 4-1A, with permission of Harcourt Health Sciences.

conceptualization of these components is essential because, as noted above, accommodation involves an interactive feedback process. Yet, it is unclear, based on the descriptions of the components, how they would interact within a feedback loop. Moreover, a descriptive analysis of such component interactions would be very cumbersome and imprecise. Hence, a method was needed to quantify these individual elements and their feedback interactions in a more compact, precise, and unifying manner. Such a technique, called control systems theory, was first used qualitatively by Westheimer (1963) for the accommodation and vergence systems, and was also more formally introduced mathematically for the accommodation system by Stark and others (Stark and Takahashi, 1962; Stark et al, 1962). The use of control systems theory provided a new and powerful tool in the investigation of the accommodation system by allowing the development of homeomorphic and physiologically-realistic representations of the various components of the accommodation system, as well as the quantification of their feedback interactions. A number of models have been proposed since the 1960's, with each model advancing our understanding of the intricate and elegant process of accommodation.

These models have attempted to address two important issues: (1) How does the accommodation system use retina-image defocus, which provides an even-error signal that contains blur magnitude but not direction information, to accommodate in the appropriate direction?, and (2) Is the accommodation system governed by continuous or discrete feedback control?

In this chapter, we will first present some basic static and dynamic accommodative response information. Then, many of the earlier models of accommodation will be presented, and each model's configuration and applicability will be critically discussed. Finally, we will present our recent static and dynamic models which capture the most important characteristics of the accommodation system, and furthermore provide answers to the two major questions posed above.

8.2 BASIC PROPERTIES OF ACCOMMODATION

8.2.1 Static Behavior

The basic static property of the accommodation system can be seen in a plot of accommodative response (AR) versus accommodative stimulus (AS) (Ciuffreda and Kenyon, 1983; Fig. 8.2). This can be obtained by varying target distance from far to near in increments of diopters (D, a unit of optical power equal to the reciprocal of the distance of the target from the observer in meters) and measuring the steady-state accommodative response at each AS level. Inspection of Fig. 8.2 shows that the accommodative stimulus-response (AS/R) function does not fall on the 1:1 line, and moreover is not a simple linear function. Instead, the response can be divided into at least four zones (Ciuffreda and Kenyon, 1983; Ciuffreda, 1991, 1998). (1) The initial non-linear portion of the curve from 0 to 1.5 D of AS over which the AR is approximately constant; (2) the linear manifest zone over which a change in AS produces a proportional change in AR; this response is typically less than the stimulus, producing the so-called "lazy lag of accommodation" (Morgan, 1968); (3) the nonlinear transition zone (region of soft saturation)

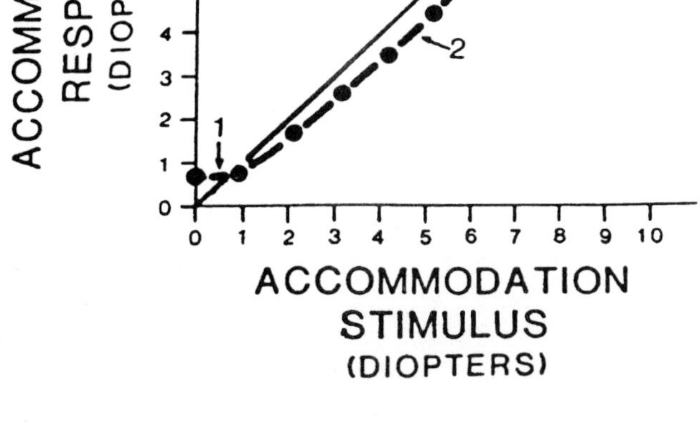

Figure 8.2. Schematic plot of the accommodative stimulus-response function. Equivalence of accommodative stimulus and response is represented by the diagonal (1:1) line. Reprinted from Ciuffreda and Kenyon (1983), pg. 102, Fig. 5.1, with permission of K. J. Ciuffreda, the copyright holder.

over which further increases in the stimulus produce a change in response, but progressively smaller than would be found for the same stimulus change over the manifest zone; and (4) the nonlinear latent zone (region of hard saturation), which defines the amplitude of accommodation, over which still further increases in the accommodative stimulus fails to produce additional increases in the lens response; however, ciliary body force can still increase in this zone, but the biomechanics of the overall accommodation system impedes further rounding of the lens to increase lenticular dioptric power; thus, functional presbyopia is attained (Saladin and Stark, 1975).

8.2.2 Dynamic behavior

Accommodative responses to pulse, step, and ramp stimuli are presented in Fig. 8.3a-c, respectively (Campbell and Westheimer, 1960). The pulse response (Fig. 8.3a) follows the pulse stimulus after a delay, and it has a duration approximately equal to that of the stimulus. This has been cited as evidence for continuous processing in the accommodation system (Campbell and Westheimer, 1960). However, as discussed in Section 8.5, a non-continuous process that responds to rapid changes in the stimulus could also account for the accommodative behavior (Hung et al, 1986; Khosroyani, 2000). The step response (Fig. 8.3b) typically exhibits a latency of 350-400 msec and an exponential rise with a time constant of about 250 msec (Campbell and Westheimer, 1960; Stark et al, 1965; Tucker and Charman, 1979; Krishnan and Stark, 1975). The ramp response (Fig. 8.3c) shows small wavering movements both for its rising and falling portions. A more systematic study using ramp stimuli of various velocities (Fig. 8.4; Hung and Ciuffreda, 1988) found dynamic characteristics that were contingent upon the stimulus ramp velocity. For relatively slow ramps (0.5 D/sec), the responses followed the target reasonably well. On the other, for intermediate velocities (1 to 2 D/sec), the responses consisted of mutiple-step movements in which the end of the step approximately coincided with the instantaneous position of the target. Further, for higher velocities ($\geq$ 3 D/sec), the responses consisted primarily of steps and step-ramps. These results indicated that the accommodation system behaved differently based on the target velocity, and furthermore suggested a two-part or dual-mode control process consisting of open- and closed-loop components (see Sub-Section 8.4.2 Fig. 8.23, for details). Sinusoidal responses to various temporal frequencies exhibit a decrease in response amplitude and increase in phase lag with increasing stimulus frequency (Fig. 8.5) (Kasai et al, 1971; Fujii et al, 1970).

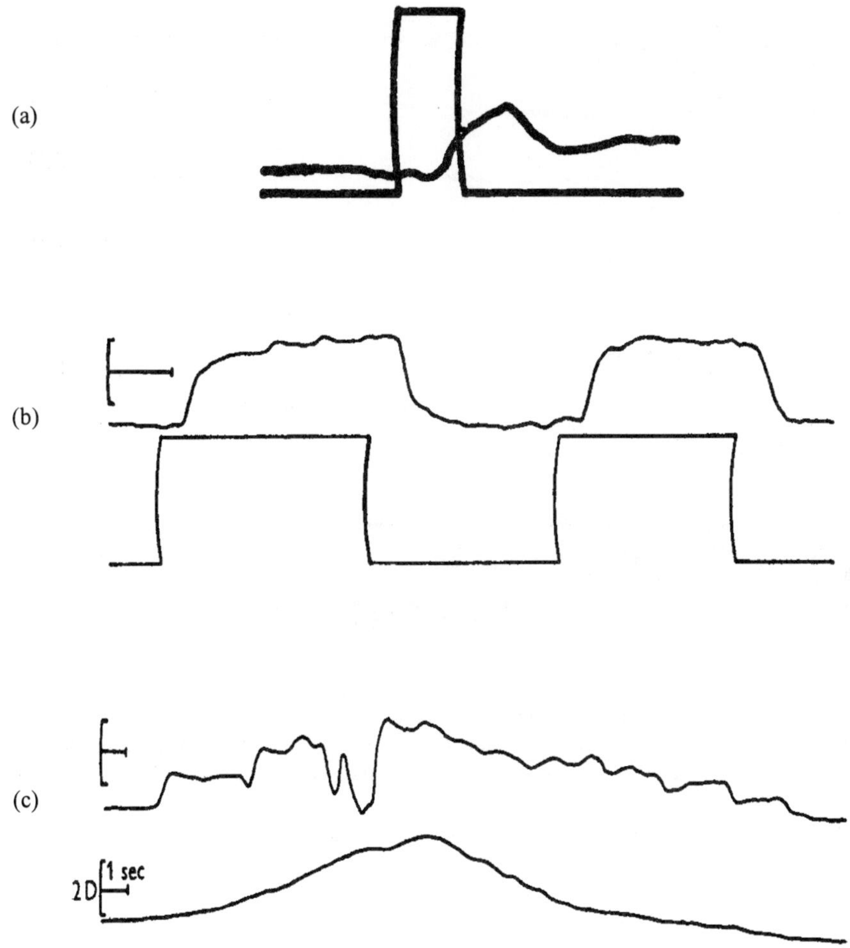

Fig. 8.3. Accommodative responses to: (a) 2 D pulse stimulus of 0.32 sec duration; (b) 2 D step and return to zero level of accommodation stimulus. Marker length: horizontal - 1sec; vertical - 1D; (c) gradually-changing stimulus, similar to an up- and then down-ramp stimulus. Note the irregular bumpy responses to the smoothly changing ramp stimulus. For a-c, response on top and stimulus on bottom. Reprinted with from Campbell and Westheimer (1959), pp. 288, 291, 294, Figs. 3, 7, 10, with permission of The Physiological Society.

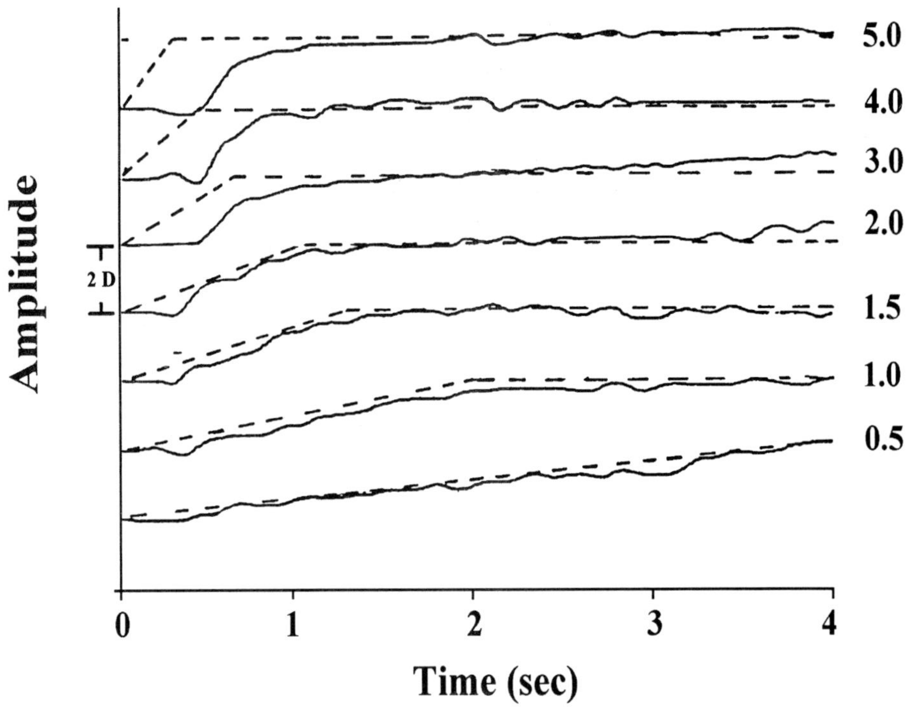

Figure 8.4. Accommodative responses (solid) to different ramp stimuli (dashed) with velocities (shown at right of curve) ranging from 0.5 to 5.0 D/sec. Maximum stimulus amplitude is 2D. Subj. GH. Note the multiple-steps in the responses to 1.0, 1.5 and 2.0 D/sec ramps, and progressive shift from step-ramp to step responses for the 3.0, 4.0 and 5.0 D/sec ramp stimuli. Reprinted from Hung and Ciuffreda (1988), pg. 330, Fig. 5, with permission of Elsevier Science.

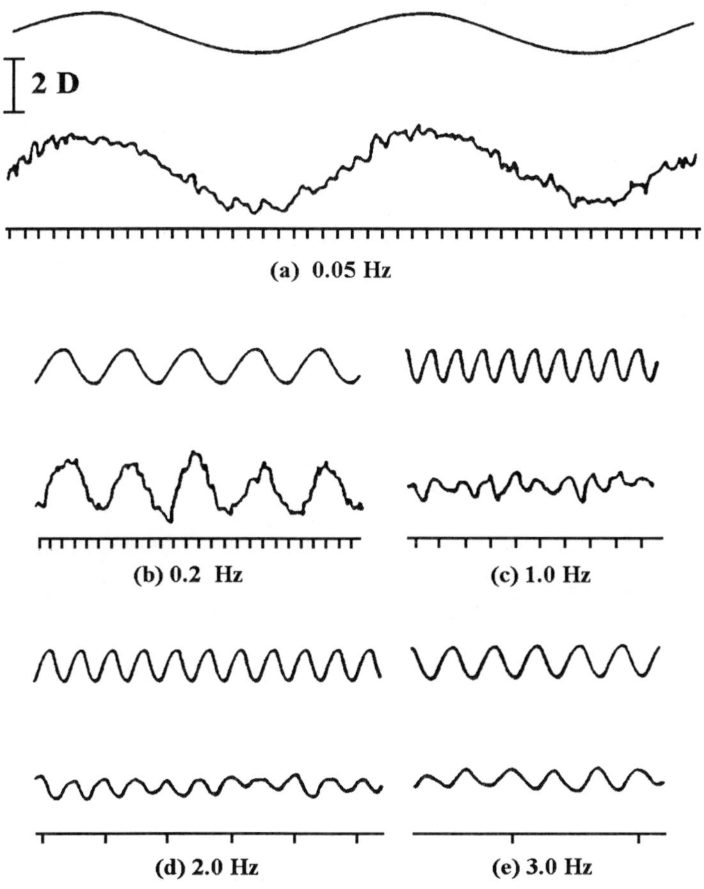

Figure 8.5. Accommodative responses to sinusoidal stimuli at various temporal frequencies. In each case, the upper record shows the stimulus change, the middle trace the corresponding response, and the bottom line is marked in seconds. Reprinted from Kasai et al (1971), pg. 569, with permission of Osaka University.

8.3 PREVIOUS MODELS OF ACCOMMODATION

Early models had some success in describing the behavior of the accommodation system. The following models are presented based on a conceptually progressive rather than historical sequence.

8.3.1 Westheimer model

Westheimer (1963) proposed a descriptive feedback model for accommodation showing the focus error (i.e., the difference between dioptric target distance and the accommodative response) driving the accommodation center (Fig. 8.6). This in turn is input to the peripheral mechanism, or plant, to generate the accommodative response. A similar feedback loop for vergence is also shown. Based on optical and pharmacological

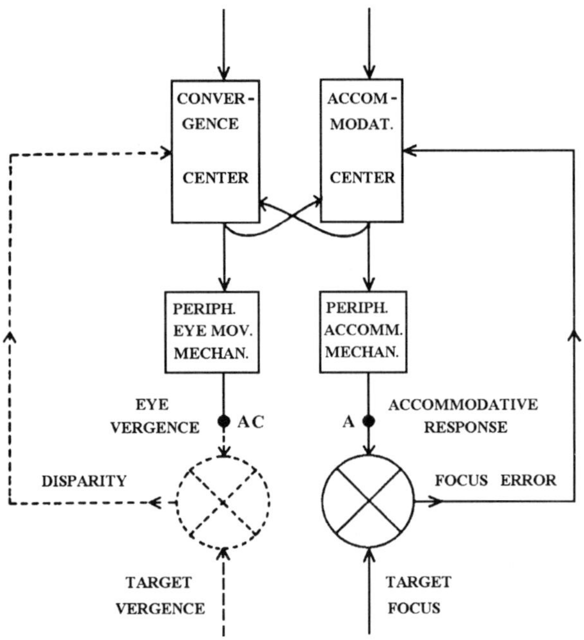

Figure 8.6. Westheimer's descriptive model of accommodation (right) and vergence (left) and their interactive connections. The peripheral accommodation mechanism represents the ciliary muscle, zonule, and lens, whereas the peripheral eye movement mechanism represents the extraocular muscles. Symbols: A = accommodative response; AC = accommodative convergence response when the vergence feedback loop is disabled (see dashed portion). Labels A and AC added for clarity. Adapted from Westheimer (1963), pg. 834, Fig. 4, with permission of American Medical Association.

experiments, he proposed central interactions between accommodative and vergence centers. Thus, if the convergence feedback loop is disabled, for example with monocular viewing, accommodation can still drive convergence via its central interactive connection (i.e., accommodative convergence). This provides a measure of the accommodative convergence (AC) to accommodation (A), or the AC/A ratio. Similarly, if the accommodative loop is disabled, for example with pinhole viewing, convergence can still drive accommodation via its central interactive connection (i.e., convergence accommodation). The details of these interactive processes, however, are the topics for Chapter 9 in this volume. In the present chapter, the models will concentrate primarily on the accommodation system.

8.3.2 Toates model

Toates (1972a) proposed a similar descriptive block diagram model of the accommodation system, but using a more conventional configuration (Fig. 8.7a). He incorporated elements for the depth-of-focus (DOF), neural controller, and ciliary muscle for driving the lens response. In addition, he proposed a more detailed descriptive model of the accommodation system (Fig. 8.7b; Toates 1972a,b).

Although he did not simulate the model, he reasoned that based on a general negative-feedback control system (Fig. 8.8a), there are two possibilities for control of accommodation. First, the system can operate with proportional control in the forward loop (Fig. 8.8b), so that the output is proportional to the error. That is,

$$\text{output} = \text{K} \bullet \text{error} \tag{8.1}$$

where K is the forward-loop gain. Toates derived equations in the feedback model for both the output and error as a function of input:

$$\text{output} = \frac{\text{K}}{1+\text{K}} \bullet \text{input} \tag{8.2}$$

$$\text{error} = \frac{1}{1+\text{K}} \bullet \text{input} \tag{8.3}$$

Substituting Eq. 8.3 into 8.2 gives Eq. 8.1, which provides a consistency check for these equations.

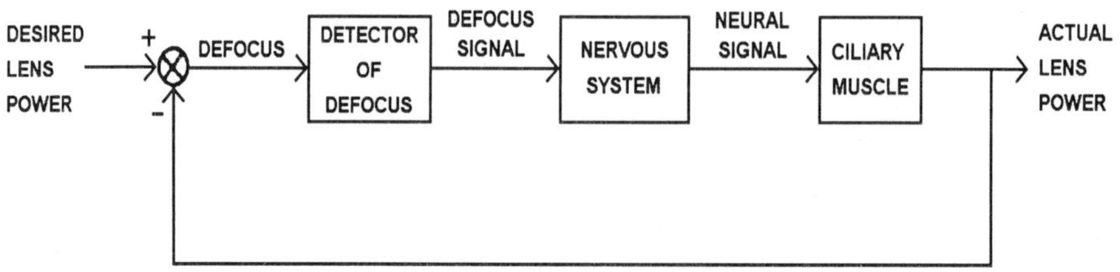

Fig. 8.7a. Descriptive model of the accommodation system. Reprinted from Toates (1972a), pg. 843, Fig. 2, with permission of Physiological Reviews.

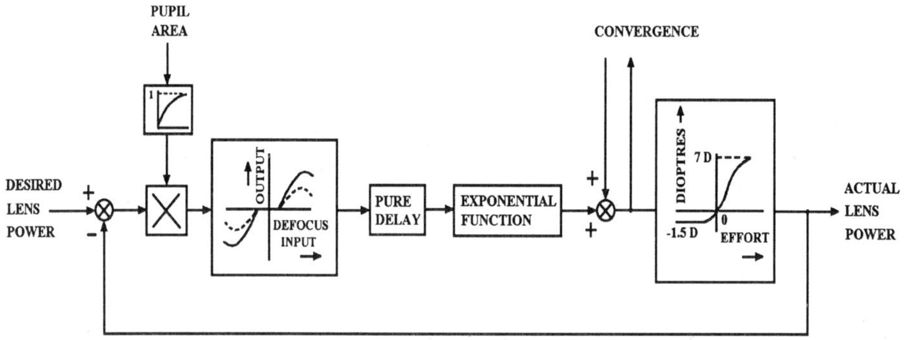

Fig. 8.7b. Toates' (1972a) detailed descriptive model of the accommodation system. The actual lens power is subtracted from the desired lens power to provide the accommodative error. The "detector of defocus" element shown in Fig. 8.7a is replaced here by a multiplicative element and a defocus input-output element. Thus, the accommodative error is multiplied by a value associated with pupil area to result in retinal defocus. The defocus is input to a functional relationship whose neural signal output is dependent on whether the illumination and/or visual acuity is high (solid) or low (dashed). Note that the output declines for large defocus values. The "nervous system" element in Fig. 8.7a is represented here by the delay and exponential function elements. The pure delay represents the amount of time the system takes to process the stimulus information. The exponential function is meant to produce an exponentially-shaped response to a step input. The arrows to and from "convergence" represent the interactive signals to and from the vergence system. Finally, the "ciliary muscle" output in Fig. 8.7a is replaced by a plant element with a negative limit at -1.5 D and a saturation limit at 7D. Adapted from Toates (1972a), pg. 847, Fig. 5, with permission of Physiological Reviews.

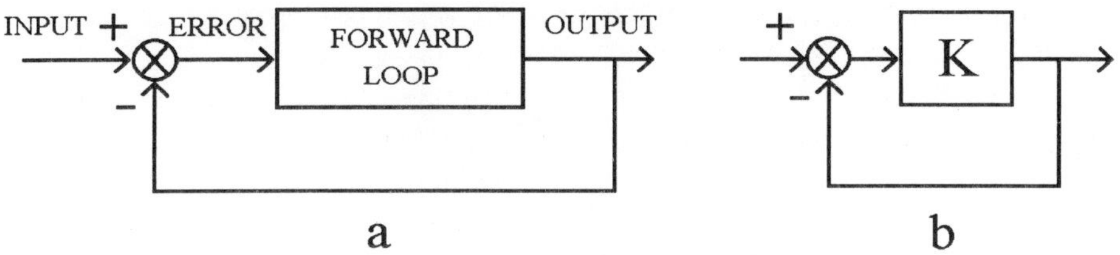

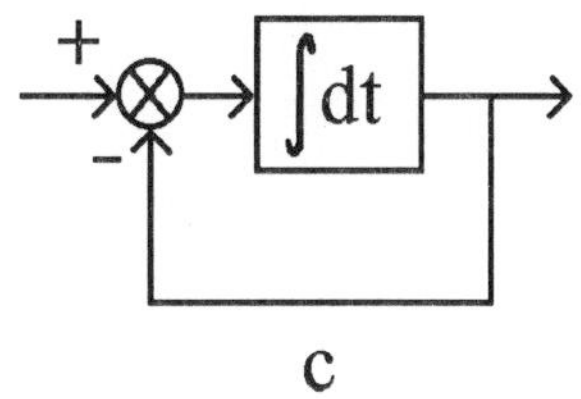

Fig. 8.8. Toates (a) general; (b) proportional; and (c) integral negative-feedback control systems. Reprinted from Toates (1972a), pg. 845, Fig. 3, with permission of Physiological Reviews.

The evidence for proportional control in the accommodation system can be seen in the lag of accommodative response in the linear region of the AS/R curve (Fig. 8.2), where the steady-state AR is approximately a fixed fraction of AS (see Eq. 8.2). If one were then to plot AR vs. AE, using the data from Fig. 8.2, it can be shown that AR is proportional to AE. Thus, the steady-state accommodative data are consistent with Eq. 8.1, which provides support for a proportional control system.

Second, the system can operate with integral control in the forward loop, so that the output is the integral of the error (see Fig. 8.8c, where the integrator in Laplace notation would be equal to 1/s):

$$\text{output} = \int_0^t \text{error} \bullet dt \tag{8.4}$$

It can be seen in Eq. 8.4 that if the error is a non-zero constant, the output would be equal to the integral of that constant and would continue to grow. Conversely, the only way the output can remain at a constant value is for no additional integration to take place. This can only occur if the error is zero (i.e., AR is on the 1:1 line). Toates (1972) reasoned that since the steady-state AS/R function exhibits a non-zero residual AE (i.e., AR is either above or below the 1:1 line), there could not have been an integrator in the model. Thus, he concluded that an appropriate model for the accommodation system used proportional rather than integral control.

However, Toates did not consider that a purely proportional control system would give instantaneous rather than relatively slowly-changing responses, as seen in the experimental time traces (see Fig. 8.3a-c). Indeed, Krishnan and Stark (1975) indicated that computer simulation of Toates' proportional control model exhibited unstable rather than smooth responses to step stimuli. It turns out that there is a compromise solution between proportional and integral controller. A controller that produces both a slowly-changing response and a non-zero steady-state error would have the form:

$$\frac{K}{\tau s+1} \tag{8.5}$$

which is a "leaky integrator" (see definition below) with controller gain K and time constant τ (Krishnan and Stark, 1975).

8.3.3 O'Neill model

O'Neill (1969) proposed an integral control model of accommodation (Fig. 8.9). Simulation step response shape was similar to experimental curves. However, as discussed above, a pure integral control model would not be able to simulate accurately the steady-state response.

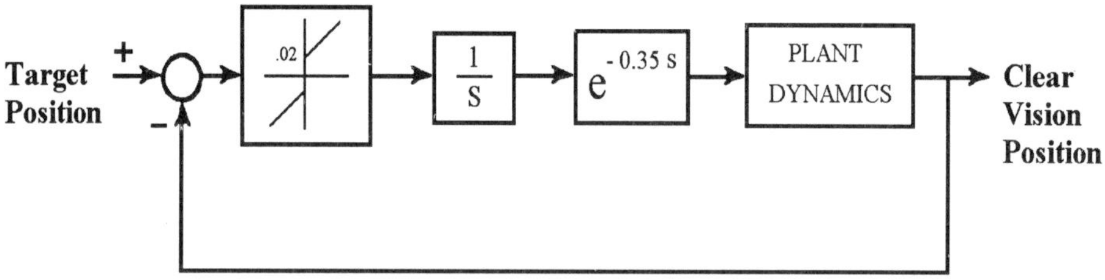

Fig. 8.9. Simplified O'Neill (1969) integral model of the accommodation system. The forward loop consisted of a threshold element, integrator, time delay, and plant dynamics. Target position and eye position are measured in diopters. Adapted from O'Neill (1969), pg. 654, Fig. 8, with permission of Elsevier Science.

8.3.4 Stark, Takahashi and Zames model

Stark, Takahashi and Zames (1965) derived an analytical model based on experimental frequency response data, which resulted in a rather complex transfer function (Fig. 8.10). However, to maintain stability in the simulation responses and to provide a good fit to the experimental data, they had to reduce the latency to 0.1 sec. This was much shorter than the 0.37 sec found experimentally (Campbell and Westheimer, 1960). Thus, while the model exhibited frequency response charâcteristics similar to experimental data, its latency was physiologically unrealistic.

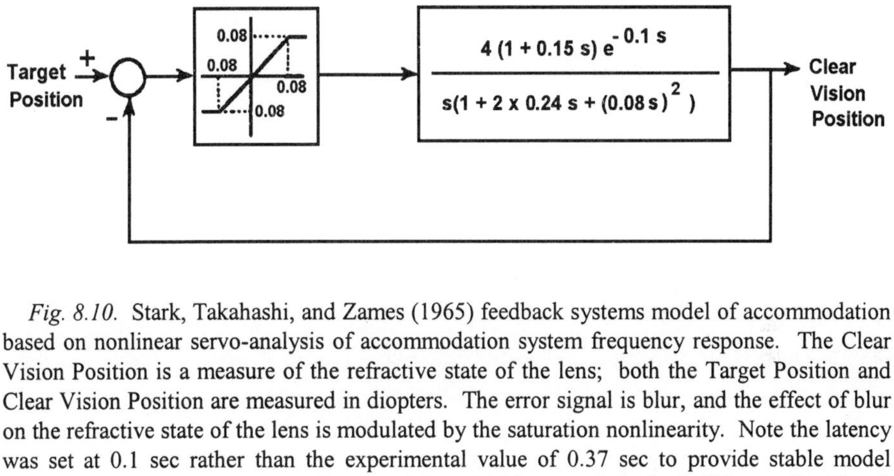

Fig. 8.10. Stark, Takahashi, and Zames (1965) feedback systems model of accommodation based on nonlinear servo-analysis of accommodation system frequency response. The Clear Vision Position is a measure of the refractive state of the lens; both the Target Position and Clear Vision Position are measured in diopters. The error signal is blur, and the effect of blur on the refractive state of the lens is modulated by the saturation nonlinearity. Note the latency was set at 0.1 sec rather than the experimental value of 0.37 sec to provide stable model simulation frequency responses. Reprinted from Krishnan and Stark (1975), pg. 78, Fig. 6, with permission of Elsevier Science.

8.3.5 Brodkey and Stark model

Brodkey and Stark (1967) also based their model on the above experimental frequency response data (see Sub-Section 8.3.4; Fig. 8.11) A derivative element, a static nonlinear element (i.e., a piecewise linear element), and a dynamic element with latency of 0.3 sec were introduced into the forward loop. Simulation results showed, however, that the phase predicted by this model lead the experimental phase-lag by 90 deg at low frequencies and lagged the experimental phase-lag by 90 deg at high frequencies. Hence, the dynamics of the model were problematic.

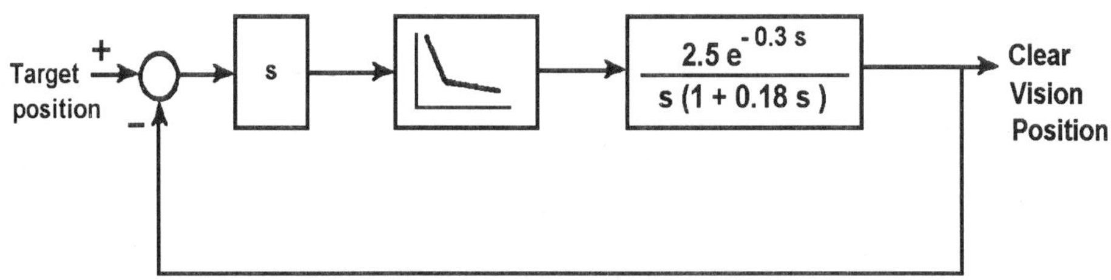

Fig. 8.11. Brodkey and Stark (1967) model of the accommodation system. This model is also based on frequency response data. Reprinted from Krishnan and Stark (1975), pg. 130, Fig. 24, with permission of Elsevier Science.

8.3.6 Krishnan and Stark model

Krishnan and Stark (1975) continued with the proportional and integral controller ideas and developed a model containing a "leaky" integrator along with other nonlinear elements (Fig. 8.12). Although the model exhibited simulation step responses with larger amplitude oscillations than the experimental data, it was able to account reasonably well for the decay of accommodation towards the tonic or empty-field position with a time constant of 6 sec.

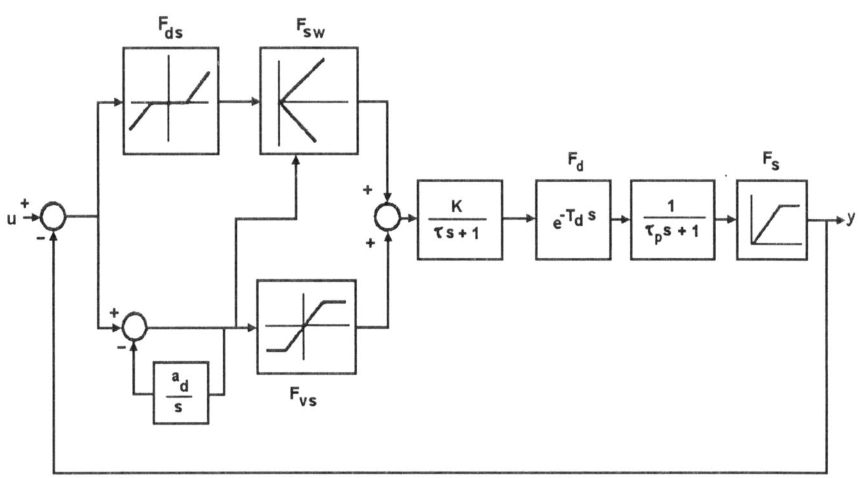

Fig. 8.12. Krishnan and Stark (1975) leaky-integrator model of the accommodation system. Symbols represent: u , target position; a_d = 10; τ = 10 sec; k = 4τ; τ_d = 0.38 sec; τ_p = 0.4 sec; F_{ds} , dead zone nonlinear element; F_{sw}, nonlinear switching element; F_{vs} , saturarion nonlinearity; y, clear vision position. Reprinted from Krishnan and Stark (1975), pg. 245, Fig. 8, with permission of Elsevier Science.

The numerous elements in the model are briefly describe as follows:

F_{ds} (deadspace operator). This represents the depth-of-focus, which allows for small neurosensory-based system errors to be tolerated without the perception of blur. If such neural tolerance were not permitted, we would be forced to have a perfect motor response at all times for clarity of vision, which is obviously an unrealistic expectation. Only if the input error exceeds this threshold level does it proceed to drive the system. The depth-of-focus is a function of pupil diameter, ranging from ± 0.15 D for an 8 mm diameter pupil to ± 0.85 D for a 1 mm diameter pupil (Campbell, 1957).

Nonlinear switching element (F_{SW}). Because blur is an even-error signal (i.e. , it lacks directional information), this element uses the sign information from the derivative operator to determine its direction. It generates a signal that is directionally correct and proportional to the magnitude of blur.

Derivative controller ($\frac{s}{s + a_d}$, derived from feedback loop containing $\frac{a_d}{s}$ in the feedback path). This parallel pseudo-derivative controller is a velocity operator. It generates the derivative of the error signal (i.e., the instantaneous velocity) for use by the control process. Such a controller improves the transient stability as well as the speed of the response.

Nonlinear saturation element (F_{VS}). This element is a velocity-sensitive component that prevents the response velocity from exceeding a certain limit. This, too, facilitates dynamic response stability and limits the amplitude of oscillations of the accommodative response.

"Leaky" integrator $\left(\frac{\tau}{\tau s + 1}\right)$. The "leaky" integrator is a "charge/discharge" element. It represents a central neurological integrating circuit that is rapidly activated ("charged" like an electronic capacitor) by the visual input and stores this information, thus providing for the initial steady-state maintenance of the response in the dark without visual information related to the target. Moreover, this circuit decays ("discharges") exponentially according to the value of its time constant, with the accommodative response shifting to the tonic accommodative bias level in 10 to 15 sec. In the model, the leaky integrator term is multiplied by 4, and thus the numerator in Fig. 8.12 is designated by K (= 4τ; where τ = 10 sec).

Time delay (F_d). This represents the combined neural and biomechanical transmission time delays, or latency.

Ciliary muscle/lens dynamics $\left(\frac{1}{\tau_p s + 1}\right)$. This represents the biomechanical response characteristics of the ciliary muscle/zonules/lens capsule complex, or "plant".

Saturation element (F_S). The saturation element limits the accommodative response imposed by the lens elasticity. In effect, it represents the maximum amplitude of accommodation (Hung, 1998; Ciuffreda, 1991, 1998).

8.3.7 Sun and Stark model

Sun and Stark (1990) followed up on Krishnan and Stark's (1975) nonlinear switching element concept and introduced a switching control model of accommodation (Fig. 8.13). The even-error nature of the blur signal is represented by a full-wave rectifying element. Thus, unlike previous models, the direction sense is assumed to be lost following the first element. Clearly, in a negative feedback control system, if the error signal is rectified, the system would rapidly go into instability oscillations. To prevent such a condition, they introduced a switching control technique. A switch-sensing element inserted after the rectifier block has three threshold levels, ON (0.7 D), OFF (slightly below 0.7 D), and "very large" (~4D, Neveu and Stark, 1995). If the absolute value of the error is either above ON or below "very large", thus representing the normal blur levels, the switch is closed, and a normal closed-loop response occurs. However, if the blur magnitude is either below OFF or above "very large", thereby representing either below blur threshold or very large blur conditions, respectively, the switch is opened, and the response decays towards the default tonic accommodative level.

The next element is a lag element with gain K=30 and time constant $\tau_1 = 2.5$ sec. This is followed by a lead-lag, or a leaky integrator, with a time constant $\tau_2 = 30$ sec. The output of this block is added with a white (i.e., broad-band) noise source (bandwidth = 0.3 sec^{-1}) having an average value of $\overline{n} = 0.5$ D. The last block is the plant, which represents the zonular-lens complex. It contains a pure time delay of $\tau_0 = 0.1$ sec. This latency is much less than that found experimentally, and presents the same problem in realistically simulating the accommodation system as the Stark et al (1965) model.

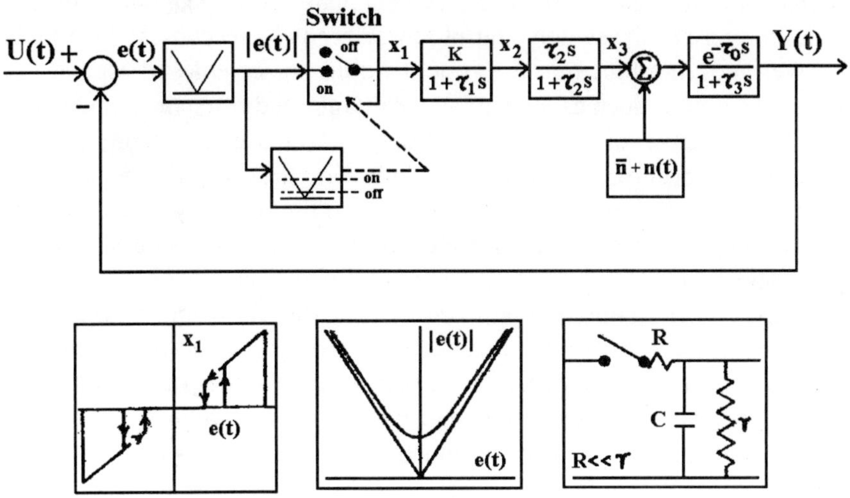

Fig. 8.13. Sun and Stark (1990) switching-control model of the accommodation system. Symbols represent: U= target position; Y = clear vision position, e = error between stimulus and response, or blur. Blur magnitude is input to a full-wave rectifier, resulting in even-error blur magnitude. The internal loop performs the switching function. Blur magnitude above ON threshold (0.7 D) closes the switch, resulting in normal closed-loop response; whereas blur magnitude that is either below OFF (representing the blur detection threshold; equal to "slightly less than 0.7 D") or is very large (with threshold set at ~ 4D) (Neveau and Stark, 1995), representing very large blur which also disables the accommodative response) opens the switch, resulting in an open-loop decay response. Noise n is added to the forward loop to mimic experimentally-observed high frequency accommodative oscillations. For blur magnitude outside this range, the switch is opened, and the response shifts towards the tonic level. Model parameter values: K=30; τ_0 = 0.1 sec; τ_1 = 2.5 sec; τ_2 = 30 sec; τn , the noise bandwidth = 0.3 (1/sec); $\bar{n}$ =0.5 D. Details of the switching mechanism in the inner loop are shown in the lower three panels. Lower left panel represents an imposed hysteresis relationship between X_1 and e(t) to reduce chattering (e.g., erratic rapid response changes) during switching. For positive e(t), switch ON follows the upward path, whereas switch OFF follows the downward path. Also, only |e(t)| is actually input into the switching operator. Lower middle panel represents the threshold based on geometrical and physical optics. Lower right panel represents equivalent electronic circuit diagram for the switch action showing an effective RC time constant during the switching operation. Reprinted from Sun and Stark (1990), pg. 76, Fig. 3, with permission of © IEEE.

Model simulation responses to increasing ramp stimuli were similar to experimental time traces. However, a closer analysis of the model indicated that the structure of the switch control sensor limits its responses. Due to the full-wave rectifying function, the system would not be able to respond to a decreasing stimulus such as a negative small amplitude step or ramp. This is because a negative error resulting from a decrease in accommodative stimulus would become a positive error following the rectifier function; and, if the amplitude is above ON, this would drive the response in the positive rather than negative direction. On the other hand, the model would appear to respond appropriately to a negative stimulus if its magnitude is above the "large" level. For example, a large downward step following a ramp, such as that seen in the simulations, causes the blur magnitude to be beyond this range, thus opening the switch, and the response will decay towards its tonic level. Hence, the large downward step movement seen following the ramp would not be a normal step response, but rather an open-loop decay movement.

8.3.8 Neveu and Stark model

The Neveu and Stark (Neveu and Stark, 1995; Stark et al, 2000) model is based on the Sun and Stark (1990) model, but with the addition of a schematic accommodation pathway (Fig. 8.14). Schematic accommodation represents the awareness of nearness of the target, or proximal accommodation, as well as other components such as voluntary accommodation and gaze/attention shifts. In addition, to overcome the inability to track negatively-directed accommodative stimuli (as discussed above for the Sun and Stark model), they used an odd- rather than even-error function for the "optics" component in the model (even though the symbol shown in Fig. 8.14 is for an even-error function). They simulated model responses to increasing and decreasing dioptric stimuli (in 1 D increments). For increasing stimuli (from 0 to 10 D), the simulated accommodative responses followed the target up to their pre-set presbyopic limit of 6 D. At that point, even as the stimulus steps continued to increase, the response decayed due to the presence of non-compensable retinal defocus and the resultant progressively reduced contrast gradient. However, when this was reversed, wherein the stimulus-decrease started from 10 D, the response remained at 1 D for stimuli above 5 D; but when the stimulus was $\leq$ 5 D, the response followed the stimulus (Fig. 8.15, left column figures). This directional difference in response based on its immediate past history is called hysteresis. Based on these simulation results, Neveu and Stark concluded that the behavior of the accommodation system depended both on the stimulus attributes and their immediate past history.

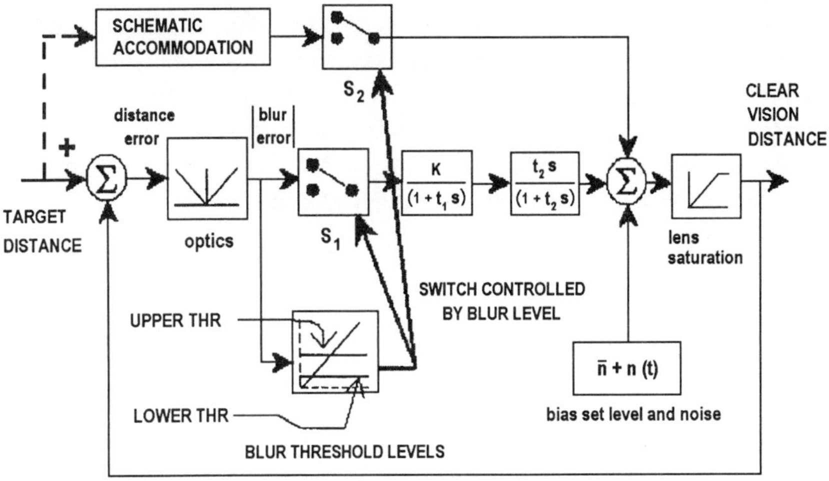

Fig. 8.14. Neveu and Stark (1995) blur switching control model of the accommodation system. Distance error, or blur, passes a rectifier to result in even-error blur magnitude. Mode 1: If blur magnitude is between the upper and lower thresholds (S1 closed) and schematic accommodation is absent (S2 open), normal feedback drives the accommodative response. Mode 2: If blur magnitude is outside of this region (S1 open) and schematic accommodation is absent (S2 open), the response drifts towards the tonic level. Mode 3: With voluntary effort for large accommodative error (i.e., blur error above upper threshold; S1 opened), schematic accommodation is activated (S2 closed) to drive the accommodative response. Reprinted from Neveau and Stark (1995), pg. 210, Fig. 6, with permission of Elsevier Science.

Analysis of this model indicates that there are two separate and distinct modes of action. For blur error magnitude between the upper and lower thresholds (S1 closed), the normal closed-loop response ensues, but without schematic accommodation (S2 open). On the other hand, for large blur error magnitude, voluntary or schematic accommodation (Neveu and Stark, 1995) is activated (S2 closed) to drive the response, but without normal closed-loop blur visual feedback (S1 open). This is consistent with the notion put forth by earlier investigators regarding the contribution from voluntary accommodation when blur error magnitude is very large (Fincham, 1951; Provine and Enoch, 1975; Ciuffreda and Kruger, 1988).

Simulation responses to decreasing step stimuli demonstrate the dichotomy between these two modes of action (Fig. 8.15). When only the closed-loop response is active (S1 closed; Fig. 8.15 left column figures), the range of responses is relatively small, but the error is also small. On the other hand, when only schematic accommodation is activated (S2 closed;

Fig. 8.15 right column figures), the range of responses is larger, but the error is also larger. These results are consistent with expected differences between normal closed-loop and voluntary accommodation (Fincham, 1951; Provine and Enoch, 1975; Ciuffreda and Kruger, 1988). However, examination of the simulated step responses reveal that they exhibit overshoots (Fig. 8.15), which are not seen in normal experimental step responses (see Fig. 8.3b; Campbell and Westheimer, 1959).

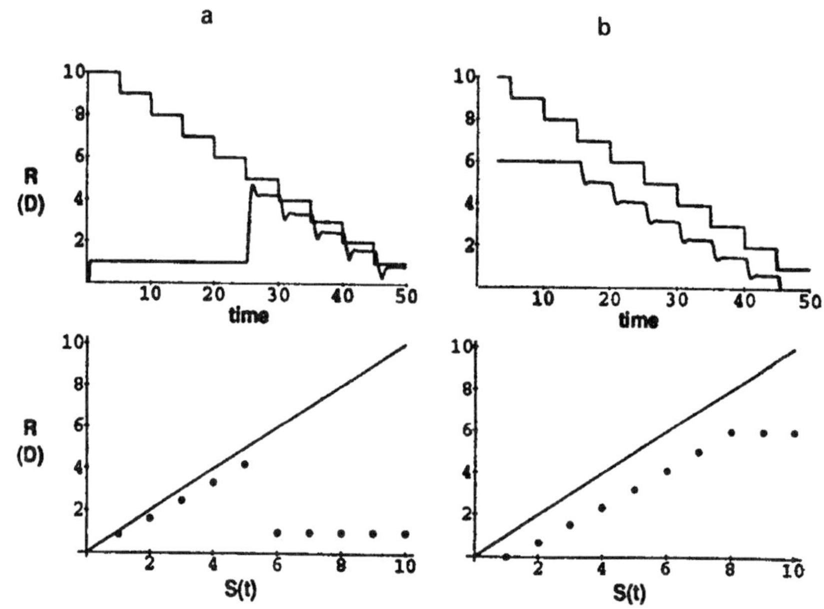

Fig. 8.15. Neveau and Stark (1995) model push-down simulation (10 to 0 D in 1 D increments). Top figure is simulation time course, and bottom figure is AS/R function. (a) Without schematic accommodation. Note an initial zero response until the stimulus is less than 5 D, at which point tracking begins and is fairly accurate. (b) With schematic accommodation. Note an initial 6 D until the stimulus is less than 6 D, at which point tracking begins, but the error is fairly large. In both (a) and (b), note the overshoot in the individual step responses, which is not seen experimentally. Reprinted from Neveu and Stark, (1995), 214, Fig. 10, with permission of Elsevier Science.

8.4 PRESENT MODELS OF ACCOMMODATION

8.4.1 Static models

There have been a few analysis and modeling attempts aimed at quantitatively generating an appropriate AS/R function (Fig. 8.2). They included a single index approach, incorporation of a sensory gain element in a feedback model, full feedback loop simulation, and finally, application of the model to amblyopia.

Chauhan and Charman (1995) proposed a single figure index for the steady-state accommodative stimulus/response profile (Fig. 8.2). It is based on the mean of the magnitude of the integrated accommodative error over the linear portion of the AR/S curve (i.e., the area bounded by the 1:1 line on top and AR below, over a range of AS values). They suggested that this index would provide a basis for comparison among different investigations. However, differences in the depth-of-focus (where one of its two boundaries is parallel to and just below the 1:1 line) among individuals would influence the effective area measures. Also, since this model can only be used over the linear range, it is somewhat limited as a general model of accommodation.

Hung (1997, 1998) proposed to simulate the static, or steady-state, AS/R function by driving a dynamic model with a series of step stimuli of various amplitudes and record the steady-state values. A descriptive block diagram of the model is shown in Fig. 8.16a. It is similar to Toates' (1972a) descriptive model (Fig. 8.7a), except that it also includes a latency and tonic element. A more detailed version of the model is shown in Fig. 8.16b. The deadspace element (with "breakpoints" at ±DSP) represents the depth-of-focus (DOF). The accommodative controller gain (ACG) represents the central neurological control of accommodation. The tonic term, ABIAS, represents the state of accommodation when the system is rendered open-loop, and it has been called "dark focus" and "night myopia" in the past (Westheimer, 1957; Liebowitz and Owens, 1978; Morgan, 1944; Hung and Semmlow, 1980), but is more appropriately called "tonic accommodation" (TA) since it is obtained under a wide range of conditions including darkness, empty field, and pinhole-viewing (Phillips, 1974). Gilmartin and Hogan (1985) found that parasympathetic innervation to the ciliary muscle plays a significant role in determining the TA position, and that the variation in TA among individuals is a consequence of parasympathetic rather than sympathetic ciliary muscle tone. The saturation element (Sat) of the plant limits the amplitude of the lens response. This decline in lens responsiveness with age represents the clinical condition of presbyopia (Mordi and Ciuffreda, 1998). Finally, a SIMILINK block diagram of the model was constructed (Fig. 8.16c). A unique feature of this program is the stimulus, which is composed of a staircase series of steps of increasing amplitude. The program performs the multi-step simulations automatically. It records one steady-state value at the end of each step response, and continues until the last step stimulus is completed. After the simulation is completed, the data are plotted as AS/R curves. This is repeated for different model parameter values.

Fig. 8.16. (See next page). Hung's (1998) accommodation system models. (a) Descriptive model. The difference between target distance and focus distance provides the retinal-image defocus whose sensory output, or blur, is processed by the accommodative controller following a time delay. The controller output is summed with the tonic signal to drive the accommodative plant, or lens. The feedback loop reduces the blur to a minimum to provide clear focus of the target image on the retina. (b) Parametric model. The difference between accommodative stimulus and response, AS – AR, provides the accommodative error (AE). The AE is input to the deadspace operator, which represents the depth-of-focus, with threshold limits or breakpoints at ± DSP. The output of the deadspace operator, AE1, drives the accommodative controller consisting of gain ACG and a unity-gain dynamic transfer function, following a time delay. The output of the controller is summed with the tonic level, ABIAS, to drive the accommodative plant. The plant is represented by a saturation element, Sat, whose saturation point or level decreases with age. (The dynamic elements are shown as dashed blocks). (c) MATLAB4.1/SIMULINK1.3 simulation model. This is the model used for the simulations. The difference between the stimulus (which has a gapped-staircase pattern with period = 10 sec and dioptric increment = 0.1 D) and the system response is input to the deadspace operator (breakpoints at ± DSP). The deadspace operator output is input to the controller with gain ACG and transfer function 1/(s+1) following a time delay. The controller output is summed with the tonic level, ABIAS, to drive the plant, which is represented by a soft saturation element. The steady-state levels are obtained by sampling once every 10 sec to give the static stimulus and response functions, x and y, respectively. The static data are then plotted for different parameter values. Reprinted from Hung (1998), pg. 336, Figs. 2A-C, with permission of © IEEE.

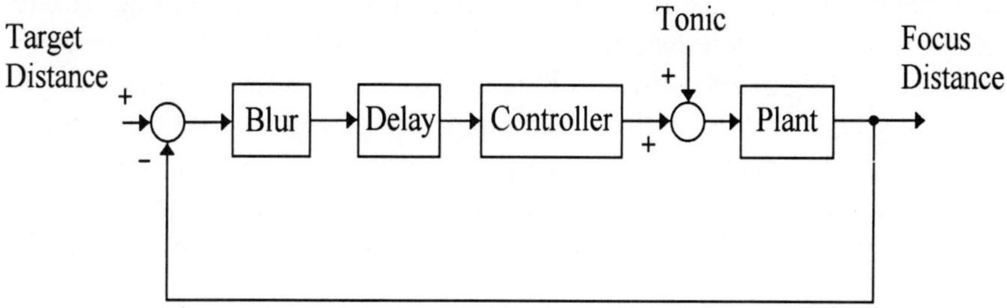
Target Distance
Tonic
Focus Distance
Blur
Delay
Controller
Plant

(a)

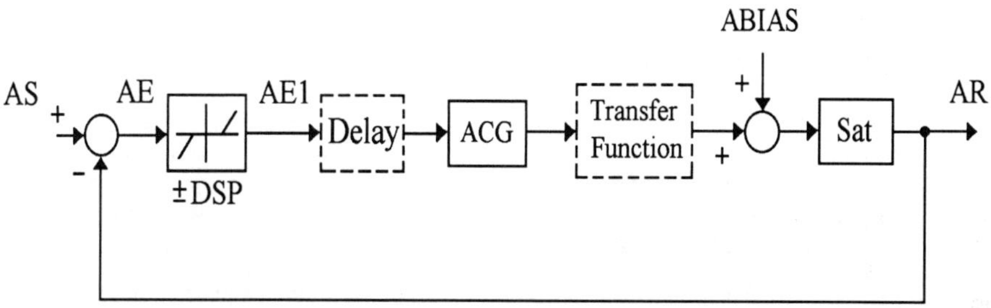
ABIAS
AS
AE
AE1
Delay
ACG
Transfer Function
Sat
AR
±DSP

(b)

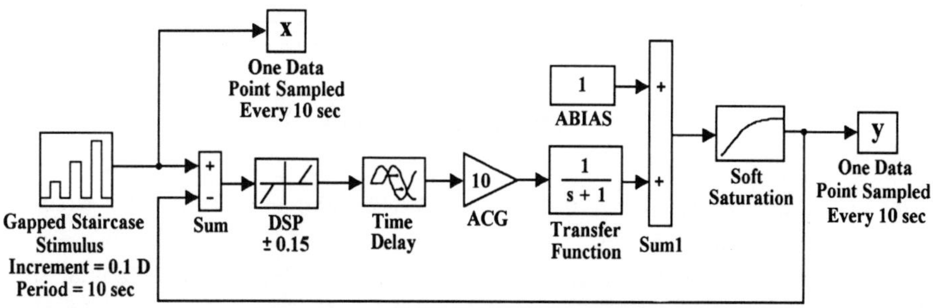
x
One Data Point Sampled Every 10 sec
1
ABIAS
1
s + 1
Soft Saturation
y
One Data Point Sampled Every 10 sec
Gapped Staircase Stimulus Increment = 0.1 D Period = 10 sec
Sum
DSP ± 0.15
Time Delay
10
ACG
Transfer Function
Sum1

(c)

Fig. 8.16

The simulated AS/R curve consisted of three regions demarcated by a deadspace region, which is the region bounded by two parallel lines (dashed) on either side and equidistant (= DSP) from the 1:1 line (e.g., see subplot with normal parameter values in Fig. 8.17c). For small AS levels, the AR is above the deadspace region, thus exhibiting the well-known "lead of accommodation". On the other hand, for large AS levels, the AR is below the deadspace region, thus exhibiting the well-known "lag of accommodation". Simulation results also show a transition region in which

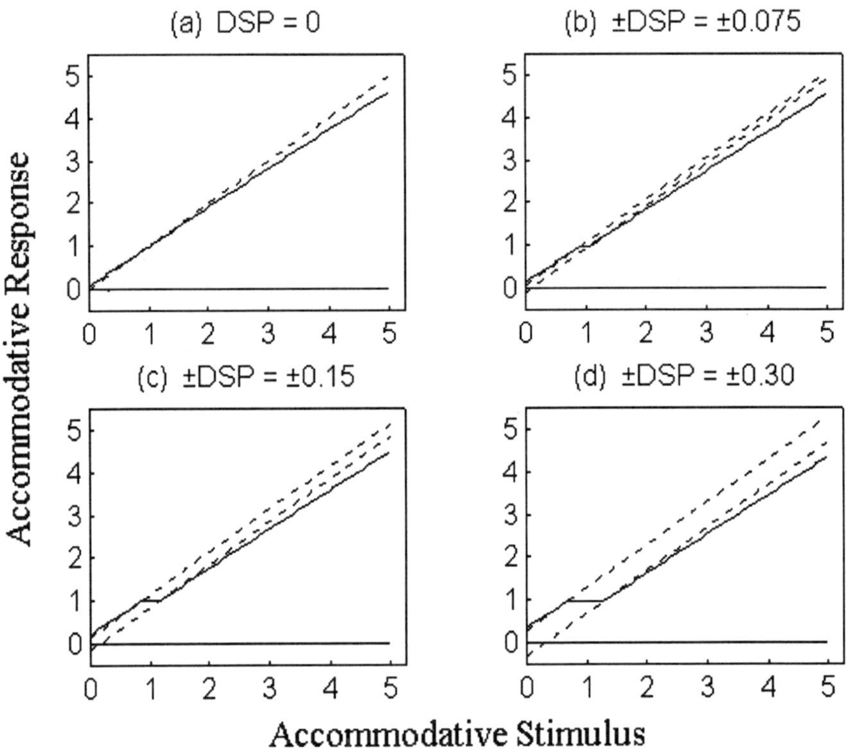

Fig. 8.17. Sensitivity analysis of the model AS/R relationship to variation in the parameter DSP (nominal values ±DSP = ±0.15 D, ACG = 10, ABIAS = 1.0 D, age ≤ 30 yrs.). For all four subplots, the dashed line respresents the limits of the deadspace (±DSP), and the solid lines represent the simulation results. The AS/R to the right of the deadspace line begins at the response level equal to ABIAS (Hung, 1998). The slope is equal to ACG/(1 + ACG). Similarly, the curve to the left of the deadspace lines begins at a response level equal to ABIAS and has the same slope. It can be seen in the four subplots that increasing DSP increases the horizontal width of the deadspaces. However, slope of the AS/R curve in the linear region remains the same. Reprinted from Hung (1998), pg. 338, Fig. 4A, with permission of © IEEE.

the response curve remains flat and equal to ABIAS, and this occurs entirely within the limits of the deadspace. The simulated AS/R curve (Fig. 8.17c) exhibits a shape similar to experimental results (see Fig. 8.2). At the crossover region, the experimental responses show more "rounded corners" than the model simulation curve. This can be explained, however, by the relatively wide spacing between stimulus values (e.g., 0, 1, and 2 D) that are typically taken for the AS/R curve. Smoothing of a curve through these relatively sparse experimental data points would obscure the underlying flat region, and result in the classical S-shaped AS/R curve. The relatively flat transition region can be more easily examined when the DOF is larger. Simulation results show that increasing ±DSP increases the DOF region about the 1:1 line (Figs. 8.17a-d), and produces a wider and flat transition region. Experimentally, this is observed in the AS/R curves for decreasing pupil size (Ripps et al., 1962) and increased target blur (Heath, 1956b).

The simulation results for varying ACG are shown in Fig. 8.18a. Increasing accommodative controller gain increases the slope of the AS/R curve. Note that the inflection of the curve occurs at AR level equal to ABIAS for all the different ACG curves. This is predicted from analysis of the model (Hung, 1998). Data for normal (+) and amblyopic (O) (an anomaly of vision in which there is a reduction in monocular visual acuity that is not correctable by refractive means and is not attributed to obvious structural or pathological ocular defects; Ciuffreda and Kenyon, 1983; Hung and Ciuffreda, 1983; Ciuffreda et al, 1991; Hung, 1998) eyes are presented. Since normal ACG (~ 10) is maintained by central neural control, the reduction in slope seen in the amblyopic eye indicates a central neural deficit of sensory origin. Indeed, values of ACG have been found to be a useful indicator of the amblyopic deficit (Ciuffreda et al, 1984, 1992).

The simulation results for varying ages are shown in Fig. 8.18b. With increasing age, the saturation level decreases, while maintaining the same normal AR and slope for values below the saturation level. This is consistent with the Hess-Gullstrand (Hess, 1904; Stark, 1987) theory of presbyopia, in which the lens and its surrounding capsule become less responsive with age, thus reducing the maximum level of accommodation. Representative experimental data for different age groups are also shown (Hung and Semmlow, 1980; Ripps et al, 1962; and Ciuffreda et al, 1997, 2000; Mordi and Ciuffreda, 1998). Age-related changes in model parameter values have been experimentally determined (Tables 8.1 and 8.2).

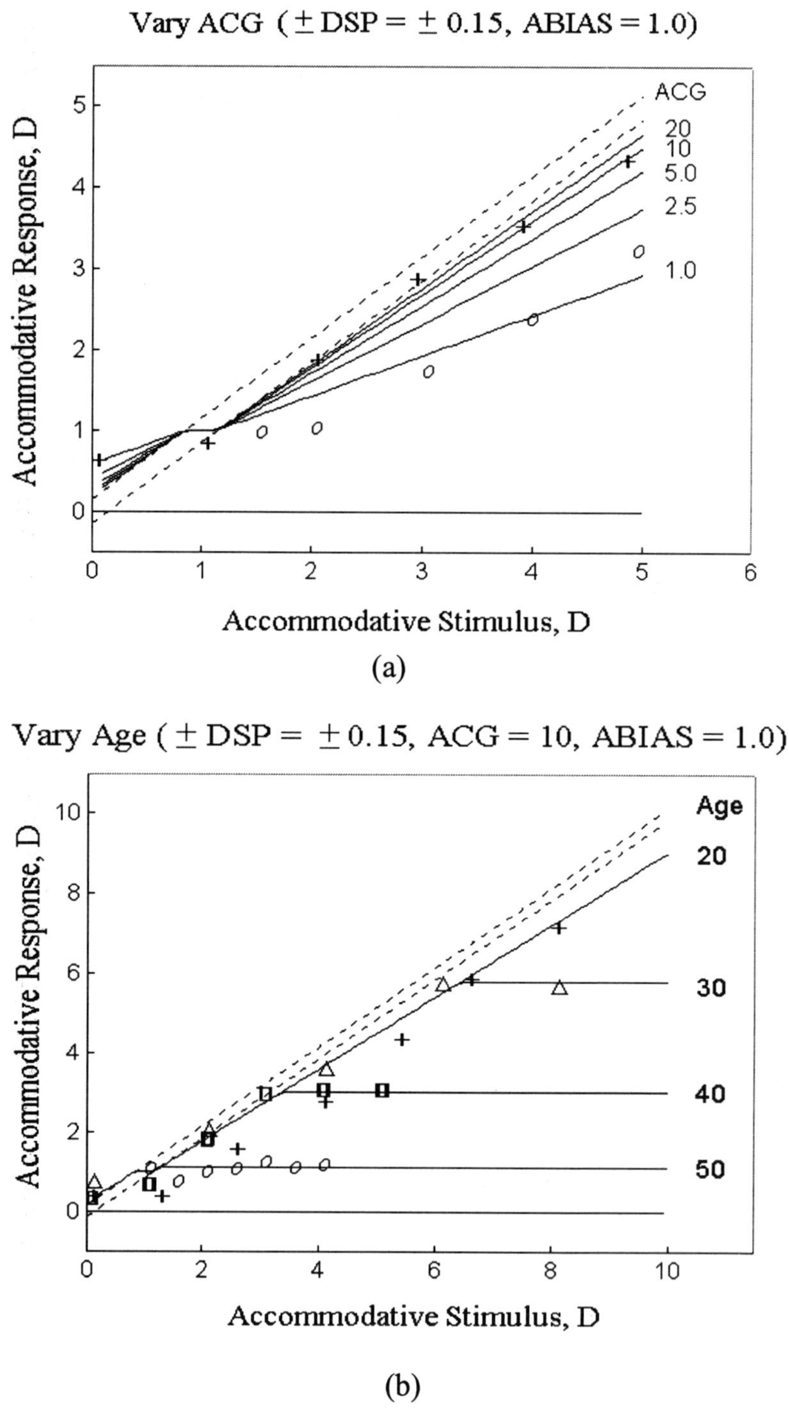
Vary ACG (± DSP = ± 0.15, ABIAS = 1.0)
ACG
20
10
5.0
2.5
1.0
Accommodative Response, D
Accommodative Stimulus, D
(a)
Vary Age (± DSP = ± 0.15, ACG = 10, ABIAS = 1.0)
Age
20
30
40
50
Accommodative Response, D
Accommodative Stimulus, D
(b)

Fig. 8.18

Fig. 8.18. (See previous page). Sensitivity analysis of AS/R function to variations in (a) ACG and (b) age. (a) Increasing ACG from 1.0 to 20 results in a progressive increase in slope, where the simulation slope is equal to ACG/(1 + ACG). The inflection of the curve at the level equal to ABIAS remains the same. Also plotted are experimental data for normal (+ symbol; corresponding to ACG ~ 10) and amblyopic (O symbol; corresponding to ACG ~ 1) subjects. The reduction of gain in the amblyopic eye indicates a central neural deficit. (b) Increasing age from 20 to 50 years results in a progressive decline in the saturation level, without a change in slope in the linear region. Also plotted are experimental data for subjects aged 20 (+ symbol), 32 (Δ symbol), 37 (□ symbol), and 49 (O symbol) years. The constant slope of the AS/R curve up to the saturation level for each of the groups supports the Hess-Gullstrand theory of presbyopia (Ciuffreda et al, 1997). Reprinted from Hung (1998), pg. 338, Figs. 4C-D, with permission of © IEEE.

Table 8.1. Oculomotor parameters and age. Reprinted from Ciuffreda et al (2000) with permission of Birkhäuser Verlag.

Increase	Decrease	Constant
Subjective depth-of-focus	Accommodative amplitude	Objective depth-of-focus
Accommodative latency	Tonic accommodation	Open-loop gain
	Accomm. microfluctuations	Closed-loop gain
	Accommodative adaptation	Accommodative time const.
	CA/C ratio	Stimulus AC/A ratio
	Proximally-induced accomm.	Response AC/A ratio
		Peak vel./amp. relation
		Proximally-induced vergence
		Tonic vergence
		Vergence adaptation

Table 8.2. Rate of parameter change per year with increasing age. Reprinted from Ciuffreda et al (2000), pg. 197, Tables 1 and 2, with permission of Birkhäuser Verlag.

Parameter	Annual rate of change
Subjective depth-of-focus	0.027 D
Accommodative latency	2.5 msec
Accommodative amplitude	0.34 D
Tonic accommodation	0.04 D
Accommodative microfluctuations	-
Accommodative adaptation	0.034 D
CA/C ratio	0.006 D/Δ
Proximally-induced accommodation	0.008 D

Jiang (2000) used the accommodative loop portion of the Hung and Semmlow (1980) model (Fig. 8.19a) to simulate the AS/R function. In the Hung and Semmlow model (1980), the depth-of-focus (DOF) was represented by a deadspace element with breakpoints at ±DSP, and the value for DSP was assumed for simplicity to be a constant. Jiang noted, however, that target luminance (Johnson, 1976) affected the pupil size, which in turn modified the depth-of-focus. Moreover, variation in target blur (Heath, 1956b), spatial frequency (Ciuffreda and Hokoda, 1983) and contrast (Ciuffreda and Rumpf, 1985) influenced the effective DOF, without changing the pupil size. To account for these influences, he introduced an accommodative sensory gain (ASG) element in front of the deadspace element (Fig. 8.19b). Thus, the accommodative error, AE, is multiplied by ASG, and the product is input to the deadspace element. The model was simulated by Hung (2001, unpublished) using a modified form of Fig 8.16c by inserting an ASG gain element between the summing junction and the deadspace operator. Simulations were performed for various values of ASG (0.2 to 1.6 in 0.2 increments) while holding the other parameters constant (±DSP = ±0.15, ACG = 10, and ABIAS = 1.0) (Fig. 8.20). The ASG value was limited to 1.6 because higher values resulted in instability oscillation.

The simulation results revealed that the ASG affects both the effective deadspace limits and the slope of the linear portion of the AS/R function. The effect on the deadspace limits is due to the multiplication of DSP and 1/ASG, so that the effective deadspace limits are ±AE/ASG. That is, the "effective" deadspace limits determine the actual boundaries of the deadspace region. Thus, for example, if ASG=0.2, and the nominal deadspace limits are equal to ±0.15 D, the effective deadspace limits become ±0.75 D. This means that to drive the accommodative controller, the accommodative error (AE) needs to be greater than +0.75 D or less than -0.75 D.

The increase in slope of the linear portion of the AS/R curve with increasing ASG (see Fig. 8.20) is predicted by the following equation (Jiang, 2000):

$$\text{slope} = \frac{\text{ASG} * \text{ACG}}{1 + \text{ASG} * \text{ACG}} \tag{8.6}$$

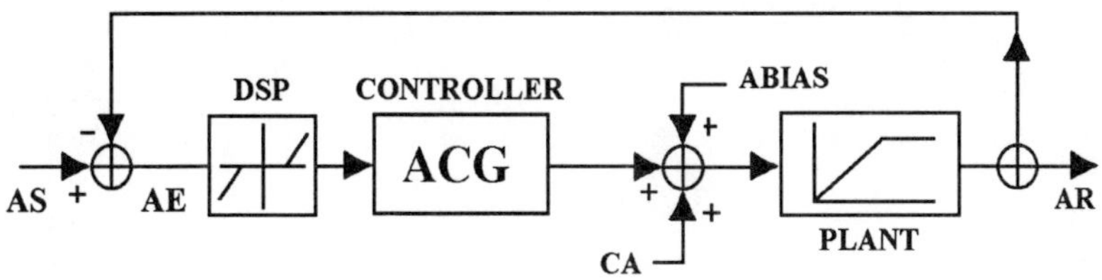

(a)

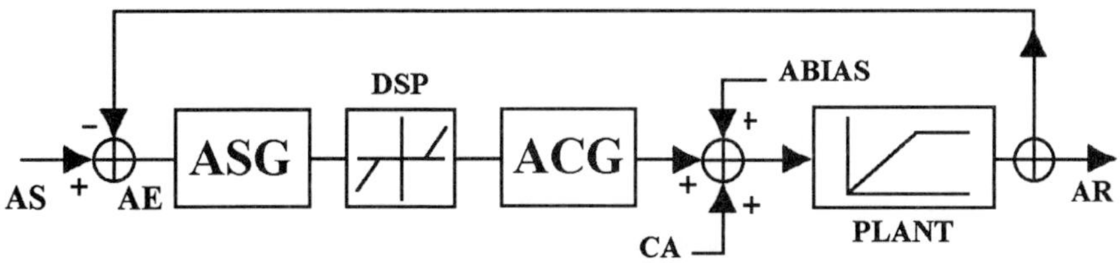

(b)

Fig. 8.19. (a) Accommodative loop of the Hung and Semmlow model (1980) showing deadspace operator DSP, accommodative controller gain ACG, tonic accommodation ABIAS, accommodative saturation element PLANT, and convergence accommodation crosslink input CA. (b) An accommodative sensory gain (ASG) element was placed in front of the deadspace element to account for the effect of reduced stimulus effectiveness (blurring; decreased spatial frequency, contrast, or luminance) on the effective DOF. Reprinted from Jiang (2000), pp. 236, 239, Figs. 1, 2, with permission of Birkhäuser Verlag .

Table 8.3 lists the slope as a function of ASG. It can be seen that for ASG from 0.2 to 0.4, there is a substantial increase the slope, whereas for larger ASG values from 0.4 to 1.6, there is a progressive but relatively smaller increase in slope from 0.80 to 0.94.

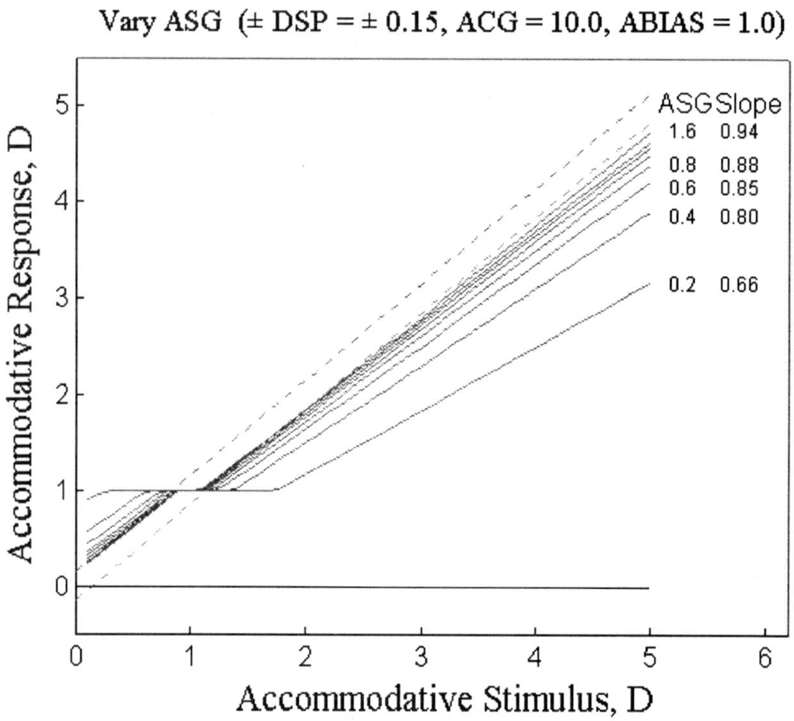

Fig. 8.20. Sensitivity of AS/R function to variations in ASG (0.2 to 1.6 in 0.2 increments) while other parameters of Jiang's (2000) model were held constant. Dashed lines represents the limits of the nominal deadspace region (DSP = ±0.15 D) about the 1:1 line. Note the increase in effective deadspace with decreasing ASG. Also, except for a substantial increase in slope for ASG between 0.2 to 0.4, the slope increases progressively but slightly as ASG is increased from 0.4 to 1.6 (see Table 8.3). From Hung, personal communication (2001).

Table 8.3 - Effect of ASG on slope of the linear portion of the AS/R curve

ASG	0.2	0.4	0.6	0.8	1.0	1.2	1.4	1.6
Slope	0.67	0.80	0.86	0.89	0.91	0.92	0.93	0.94

In addition to the above models, a conceptual model was developed by Ciuffreda et al (1991; Fig. 8.21) to examine the factors associated with the amblyopic accommodative deficit. It was based experimental findings in amblyopic eyes focusing on sine- and square-wave spatial stimuli (Ciuffreda et al., 1991). They showed an overall reduction in the accommodative response level in the amblyopic eye as compared with the fellow normal eye, but normal consensual responses in the amblyopic eye when driven by the fellow eye. This suggested that the site of the accommodative dysfunction was neither the motor controller nor the peripheral apparatus, but rather was in the sensory controller. Such a sensory deficit was presumed to be due to the early, prolonged abnormal visual experience. In terms of the model of steady-state accommodation (Hung, 1998), the accommodative defect was

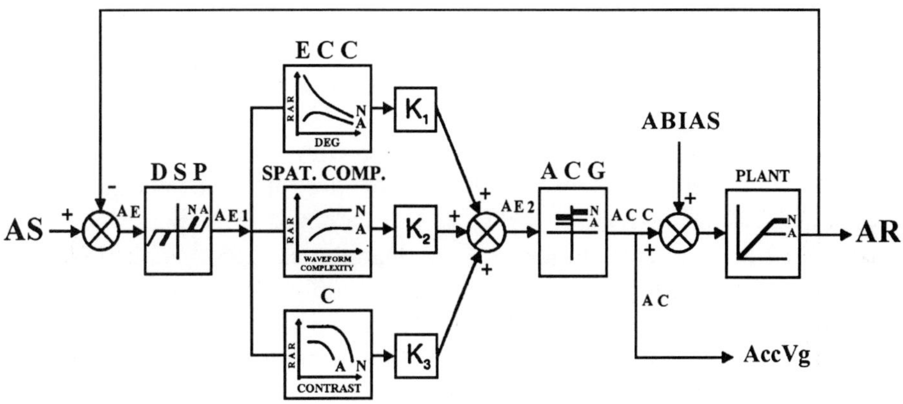

Fig. 8.21. Ciuffreda et al (1991) static model of the accommodation system, with inclusion of several of the sensory factors that account for much of the reduced accommodative responses found in amblyopic eyes. These include target retinal eccentricity (ECC), spatial composition (SPAT. COMP.), and contrast (C), with K_1, K_2, and K_3 being gain terms, and RAR being the relative accommodative response. Accommodative error (AE) is the difference between accommodative stimulus (AS) and accommodative response (AR). Deadspace (±DSP) reflects depth-of-focus of the eye. Output (AE1) from the deadspace operator goes into the accommodative controller, which exhibits nonlinear accommodative gain (ACG). Output (ACC) from the accommodative controller is summed at the summing junction (⊗) and also cross-linked to the vergence system (Acc Vg) by means of gain AC. Accommodative bias (ABIAS) or tonic accommodation under the "no stimulus" (i.e., open-loop accommodation and vergence) condition is also summed here. Output from the summing junction goes through a saturation element, which reflects plant saturation of the accommodation system. N and A are normal and amblyopic eyes, respectively. Reprinted from Ciuffreda et al (1991), pg. 299, Fig. 6.30, with permission of K. J. Ciuffreda, the copyright holder.

believed to be localized at the site of the accommodative controller gain (ACG in Fig. 8.16b), resulting in reduced response amplitude. In addition, such factors as abnormal fixational eye movements, defective contrast perception, and/or eccentric fixation may contribute to the reduced responsivity. Thus, in the conceptual model (Fig. 8.21), empirically-derived relationships for target-eccentricity, spatial composition, and contrast were placed in the forward loop to represent the various sensory factors that affect accommodative controller gain. This configuration is similar to the ASG element proposed more recently by Jiang (2000), except the sensory elements were placed after rather than before the deadspace operator. Such a model may be used clinically to simulate more accurately the various stimulus attributes that affect accommodative behavior in the amblyopic eye.

8.4.2 Dynamic models

8.4.2.1 Stability Analysis Using Root Locus

Campbell et al. (1959) analyzed the frequency spectrum of the steady-state accommodative response and found two spectral peaks, a larger peak at 2 Hz and a smaller peak at 0.5 Hz. The cause of these peaks had been unresolved for over two decades. Then, Hung et al. (1982) applied the root locus analysis technique (D'Azzo and Houpis, 1988) to a dynamic accommodation system (Fig. 8.22a) to determine whether these peaks were caused by system instability oscillations. This technique is a general method for quantifying the stability characteristics of a feedback system, with the forward loop as the controlling parameter. This method is well suited for evaluating the stability of the accommodation system and its dependence on ACG. However, the root locus method is only applicable to linear systems; hence, for this analysis a linear approximation is used in place of the nonlinear deadspace operator, or the DOF. As the accommodation system generally operates on one side of the DOF (the low-output side for normal near viewing, leading to the descriptive term "lag of accommodation") (Morgan, 1968), the deadspace operator can be appropriately replaced by a linear bias or "offset" term (Hung and Semmlow, 1980).

A root locus computer program developed by Krall and Fornaro (1967) was used to analyze the dynamic stability characteristics of the accommodation model (Hung et al, 1982). The unique feature of this root locus program is that it allows for a delay element (to simulate latency) in the feedback loop. The program plots the location of the closed-loop poles (equal to the roots of the denominator of the transfer function, so that when the system operates near the root value, the transfer function output rapidly grows to an infinite value, hence the name pole) as a function of gain K for forward-loop transfer function of the form, $K e^{-\tau s} G(s)$, or overall closed-loop transfer function of the form

$$\frac{K e^{-\tau s} \, G(s)}{1 + K e^{-\tau s} \, G(s)} \tag{8.7}$$

For the analysis of the accommodative system model, the following parameter values used were: controller time constant = 6 sec, plant time constant = 300 msec, and a total time delay = 350 msec. Thus,

$$K e^{-\tau s} \, G(s) = \frac{K e^{-0.35 s}}{(1+6s)\,(1+0.3s)} \tag{8.8}$$

The root locus plot of the linearized accommodation system with forward-loop transfer function given by Eq. 8.8 is shown in Fig. 8.22b, where the horizontal and vertical axes are the σ and $j\omega$ axes, respectively. From this plot we note that the gain corresponding to the root locus at the $j\omega$ axis is equal to 21, which gives the maximum gain before instability oscillation occurs (i.e., poles in the right half of the s-plane, or to the right of the vertical jw axis, corresponding to a time domain function which grows rapidly, and would mean the system is unstable). The predicted frequency for instability oscillation of 0.45 Hz (or 2.8 rad/sec) is near the lower accommodative spectral peak at 0.5 Hz, but far below the higher peak at 2 Hz (Campbell et al, 1959). This indicates that loop instability is not a source of higher frequency accommodative oscillations. However, the smaller peak at 0.5 Hz may in fact correspond to the closed-loop oscillation frequency of 0.45 Hz indicated in the root locus plot. Moreover, Winn et al (1990) found a significant correlation between the arterial pulse frequency and the accommodative higher frequency peak. In addition, Gray et al (1993) analyzed the accommodative spectrum as a function of pupil diameter and found that whereas the low frequency peak varied with pupil diameter, the high frequency peak did not. Thus, they concluded that the lower accommodative frequency peak which was predicted by the root locus analysis is most likely associated with neurologically-controlled feedback instability oscillations, whereas the high frequency peak is an epiphenomenon due to the effect of arterial pulse on lens motion that is detected by the recording optometer.

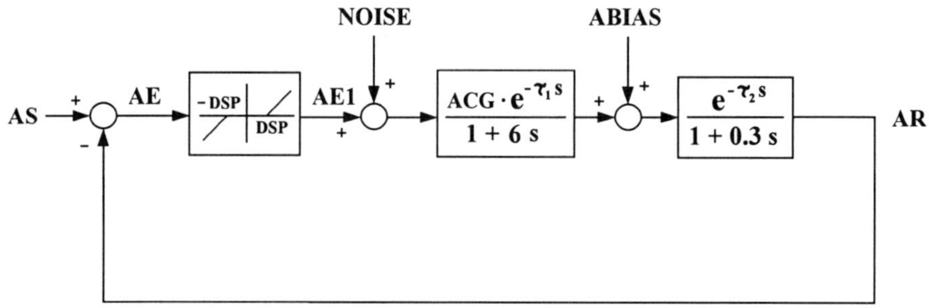

Fig. 8.22a. Accommodation model block diagram showing the difference between accommodative stimulus (AS) and accommodative response (AR) produces the accommodative error (AE). The output from the deadspace operator (± DSP = ± 0.30, which represents the depth of focus), AE1, is added to a 2 Hz noise signal and processed through the accommodative controller with gain ACG = 15, experimentally determined time constant of 6 sec, and delay τ_1 = 180 msec. The controller output is summed with tonic accommodation ABIAS = 0.5 D and processed through the plant representing the neuro-muscular apparatus with time constant 0.3 sec and time delay τ_2 = 170 msec to give the accommodative response. Reprinted from Hung et al. (1982), pg. 595, Fig. 2, with permission from © IEEE.

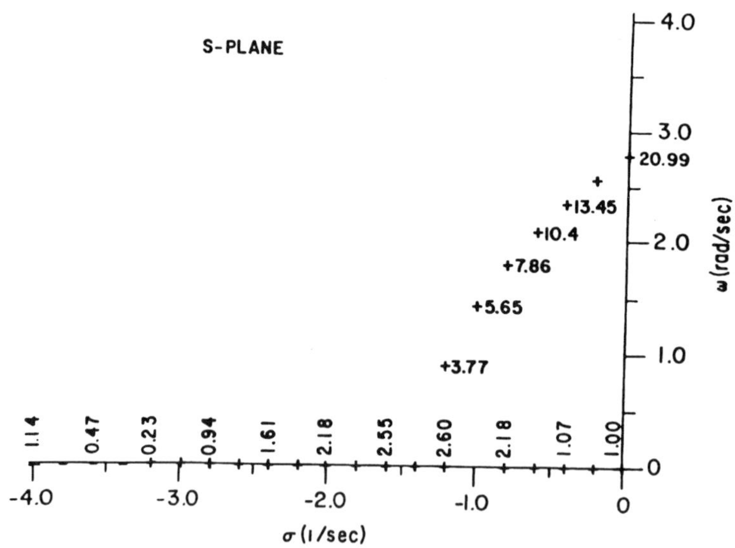

Fig. 8.22b. Root locus plot showing stability characteristics of the accommodation system. The accommodative controller gain, ACG, is used as a parameter in the root locus analysis. It can be shown that a closed-loop pole in the right half plane results in instability oscillations. This correspond to the ACG value of 21 with an instability frequency of ω = 2.8 rad/sec, or 0.45 Hz. Reprinted from Hung et al. (1982), pg. 597, Fig. 6, with permission from © IEEE.

8.4.2.1 Dual-Mode Dynamic Model of Accommodation

The difficulties encountered in previous continuous models of accommodation are not surprising (see Section 8.3). This is due to the inherent problem of having relatively slow dynamics (time constant = 250 msec) and a long time delay (350-400 msec) in a feedback loop. That is, the observed instantaneous accommodative output is actually a response to a controller signal (AE = AS - AR) that had occurred 370 msec earlier. If AE had changed sign (e.g., from positive AE, or lag of accommodation, to negative AE, or lead of accommodation) during the intervening delay interval, the accommodative output would be in the inappropriate direction. For dynamically changing accommodative stimuli, this could lead to repeatedly inappropriate responses, and in turn instability oscillations.

Switching control models were introduced to alleviate this problem. The idea was to switch the control mode based on a criterion level of a parameter, such as the blur magnitude. Although these models provided some reasonable simulations, they were not able to simulate a wide range of realistic and commonly encountered stimulus conditions. There were two main problems. First, they used incompletely defined criteria regimes, so that for example the model cannot respond to a negative ramp stimulus. Second, the model configuration did not provide appropriate dynamics, thereby resulting in unrealistic overshoots in the multiple-step responses for both closed-loop and voluntary or schematic accommodation.

To overcome all these difficulties, a dual-mode model of accommodation was developed (Fig. 8.23a-c; Hung and Ciuffreda, 1988; Khosroyani, 2000). It was based on the dual-mode model of vergence developed previously by Hung et al (1986) (see Chapter 9 in this volume). The underlying principle for the dual-mode model was evident in the experimental vergence responses to slow and fast ramp stimuli. For slow ramps, the responses followed the stimulus. However, for fast ramps, the responses consisted of a staircase series of step-like movements in which the end value of each step matched the ongoing ramp stimulus position. This indicated that there were two modes of operation. Thus, the key basis of this vergence model was that there were two mutually-exclusive modes of response: a fast, open-loop movement that corrected most of the error, followed by a slow, closed-loop movement that reduced the residual error to a minimum. Simulation of the dual-mode model showed accurate fit to a variety of experimental pulse, step, and ramp responses

Hung and Ciuffreda (1988) designed an experiment to determine whether dual-mode behavior for vergence as described above could also be present in the human accommodation system. Responses were recorded for accommodative ramp stimuli ranging from 0.5 to 5.0 D/sec. The results showed that for ramp stimuli below 1.5 D/sec, the responses consisted of smooth tracking movements (see Fig. 8.4). On the other hand, for ramp stimuli above this value, the responses exhibited a staircase-like step behavior similar to that for the vergence system. Thus, the accommodative system was also shown to have dual-mode characteristics. However, Hung and Ciuffreda (1988) did not construct a dynamic model for accommodation.

Khosroyani (2000) constructed the first dual-mode dynamic model of the accommodation system using MATLAB/SIMULINK. Since both accommodation and vergence have been shown to exhibit dual-mode behavior (Hung et al, 1986; Hung and Ciuffreda, 1988), the same program used in the vergence dual-mode model (Hung et al, 1986; Hung, 1998) was used for these accommodative simulations, with appropriate parameter value changes. The overall block diagram of the model is shown in Fig. 8.23a. The first block is a deadspace operator which represents depth-of-focus, with limits equal to ± 0.12 D. The controller consists of both fast and slow components. The fast component is derived from the sum of the visual feedback error signal and the neurological efference copy signal from the fast component output. The efference copy signal takes into account the effect of plant dynamics. This results in an open-loop signal that is nearly equal to the original stimulus amplitude. This open-loop drive is important for two reasons. First, it maintains stability in the presence of a relatively long latency (370ms); and second, it meets the requirement of an accurate initial step response. Such accuracy corresponds to very high gain in a feedback control system, which would have otherwise resulted in instability oscillations. The open-loop fast component movement accounts for most of the step response amplitude, with the reminder being taken up by the slow closed-loop component.

Fig. 8.23. (See next page). (a) Block diagram of the accommodation system used in MATLAB simulations. The difference between AS and AR, or AE, is input to a deadspace element, whose output is summed with the efference copy signal, resulting in a signal equal to the actual stimulus. This signal is used to drive the fast component. The output of the deadspace element also drives the slow component. The outputs of the slow and fast components are summed with the output of the sine-wave generator, which represents the microfluctuations, to drive the plant. The output of the plant provides the accommodative response. It is fed back and is subtracted from the accommodation stimulus to provide the error signal to the deadspace element. (b) The fast component operates in an open-loop manner, and it uses a sampler and has predictive capability for periodic stimuli. (c) The slow component operates under the closed-loop condition over a smaller range of accommodative error amplitudes and velocities. Overall, the fast and slow components operate over different stimulus regimes, so that when one is active, the other is disabled. This provides robustness in the model response (Hung, 1998; Khosroyani, 2000). Reprinted from Hung et al (1986), pg. 1023, Fig. 1, with permission of © IEEE.

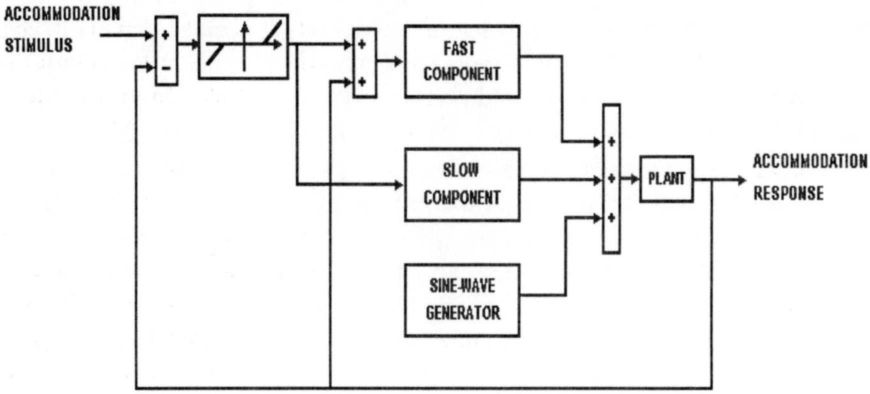

Fig. 8.23a. Overall block diagram of dual-mode accommodation model. See legend for details.

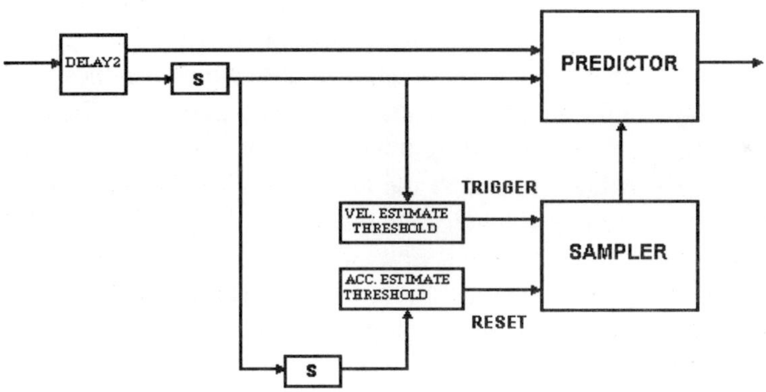

Fig. 8.23 b. Fast component.

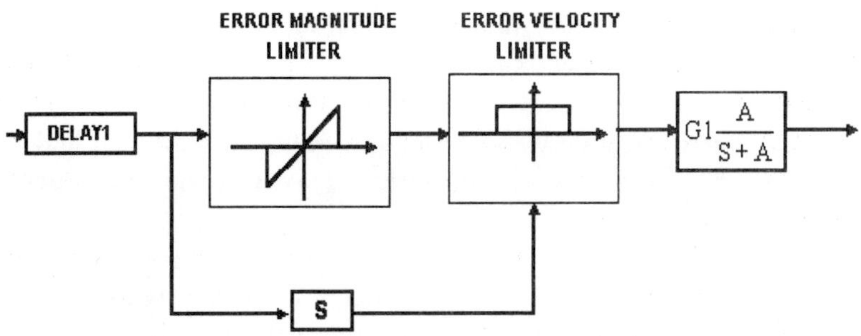

Fig. 8.23c. Slow component.

The block diagram for the fast component is shown in Fig. 8.23b. It receives a signal that represents the perceived target change. This is delayed by element 'DELAY2', which represents the effective delay throughout the fast component. The sampler and predictor act in conjunction to provide the sampling and predictive capabilities seen in the experimental ramp and sinusoidal responses. The sampler has a sensory threshold to account for the range of stimuli that elicit experimental sampling behavior. The sampler is triggered by a change in velocity of the perceived target above a given threshold. However, the sampler can be reset by a sudden change in target velocity, such as in a pulse stimulus. If the target velocity drops below a certain value, as in a step stimulus, the sampler is stopped. The sampler provides the timing control for the predictor. The predictor is a calculating unit that uses the target position and velocity information to estimate the future position of the target. The predictor estimates within the response latency where the target will be after a sampling interval, and then generates a step signal to correspond to the predicted new target position. For example, for a ramp stimulus, the predictor uses the target position and velocity information to estimate where the target position will be after a sampling interval and generates a characteristic step response which matches the target at the end of the sampling interval. If the ramp target continues, the predictor must recalculate after each sampling interval, so that the resulting staircase-like step response will again match the ramp stimulus. For step and ramp-step stimuli, which quickly reach and remain at the amplitude limit, the predictor determines the final value, and then generates a step signal to drive the response directly to the final position. For signals that regularly alternate, as in sinusoidal stimuli, the predictor serves another function. It reduces the time required for estimating the target position by reducing DELAY2, and hence decreases the phase lag between the accommodation stimulus and response. Thus, it represents the effect of anticipation of target motion.

The block diagram for the slow component is shown in Fig 8.23c. The slow component is driven by accommodation error, delayed by 370ms, and it has magnitude and velocity limiters to reflect the range of operation of the slow process. Its dynamics are modeled by a first-order lag element. The slow component acts over small amplitude and velocity ranges, and it uses negative feedback to provide the error signal for the controller. The fast and slow components operate under separate stimulus regimes, so that when one is active, the other one is disabled. This provide robustness in the accommodative response.

The other block in the overall model (Fig. 8.23a) is a sine-wave generator with 0.2D amplitude and 2 Hz frequency, which represents the high frequency microfluctuations as a plant noise (Campbell, 1959; Gray et al, 1993). This has been shown, however, to be derived from the cardiac pulse signal that is picked as an artifact of the accommodation recording process (Winn et al, 1990). The outputs of the fast and slow components, and the sine wave generator, are summed and fed to the plant, which represents the dynamics of zonule, ciliary muscle, lens capsule, and lens The plant has a time constant of 0.3 sec based on ciliary muscle stimulation in the monkey (Thompson, 1975).

The simulation results are shown in Fig. 8.24a-c for: (a) pulse and square-wave; (b) ramp; and (c) sinusoidal stimuli. They are in reasonably good agreement with experimental results. The pulse response follows the on- and off-portions of the stimulus. (Fig. 8.24a). The ramp responses exhibit a transition from smooth tracking for slow ramp stimuli to steps and multiple-steps for faster ramp stimuli (Fig. 8.24b). Sinusoidal responses show a transition from smooth tracking by the slow component (stimulus frequencies at 0.05 and 0.1 Hz) to combined fast and slow component movements for higher frequency stimuli. (Fig. 8.24c). For the periodic stimuli (Figs. 8.24a and c), a reduction in latency (to represent prediction) is evident after a few cycles in many of the responses. Thus, the dual-mode model of accommodation is able to simulate reasonably accurately a wide variety of stimuli.

Fig. 8.24. (See next two pages). Dual-model accommodation model responses to: (a) pulse (top trace, 0.32 sec stimulus duration) and square-wave stimulation (frequency, in Hz, is shown at right of traces) of 2D amplitude; (b) ramp stimulation (velocity, in D/sec, is shown at right of traces) with maximum amplitude of 2D; and (c) sine-wave stimulation (frequency, in Hz, is shown at right of traces) for +/-2D peak-to-peak amplitude (Khosroyani, 2000). Dashed lines represent the stimulus, and continuous lines represent the response. Horizontal lines represent zero level of accommodation. For graphs (a)-(c), the ordinate is designated in diopters (D).

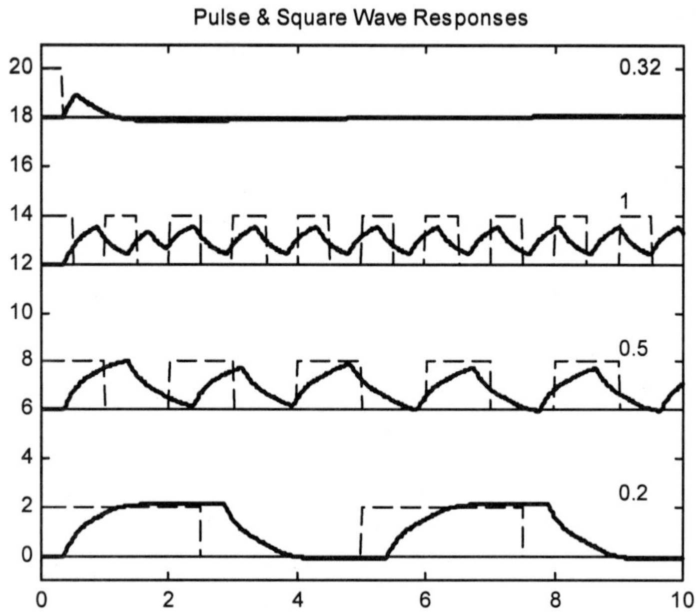

Fig. 8.24a.

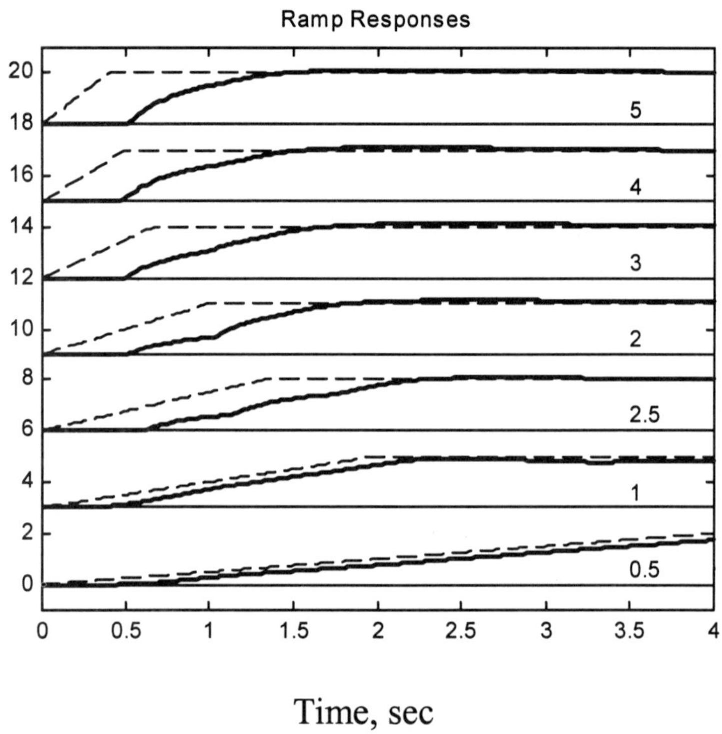

Fig. 8.24b.

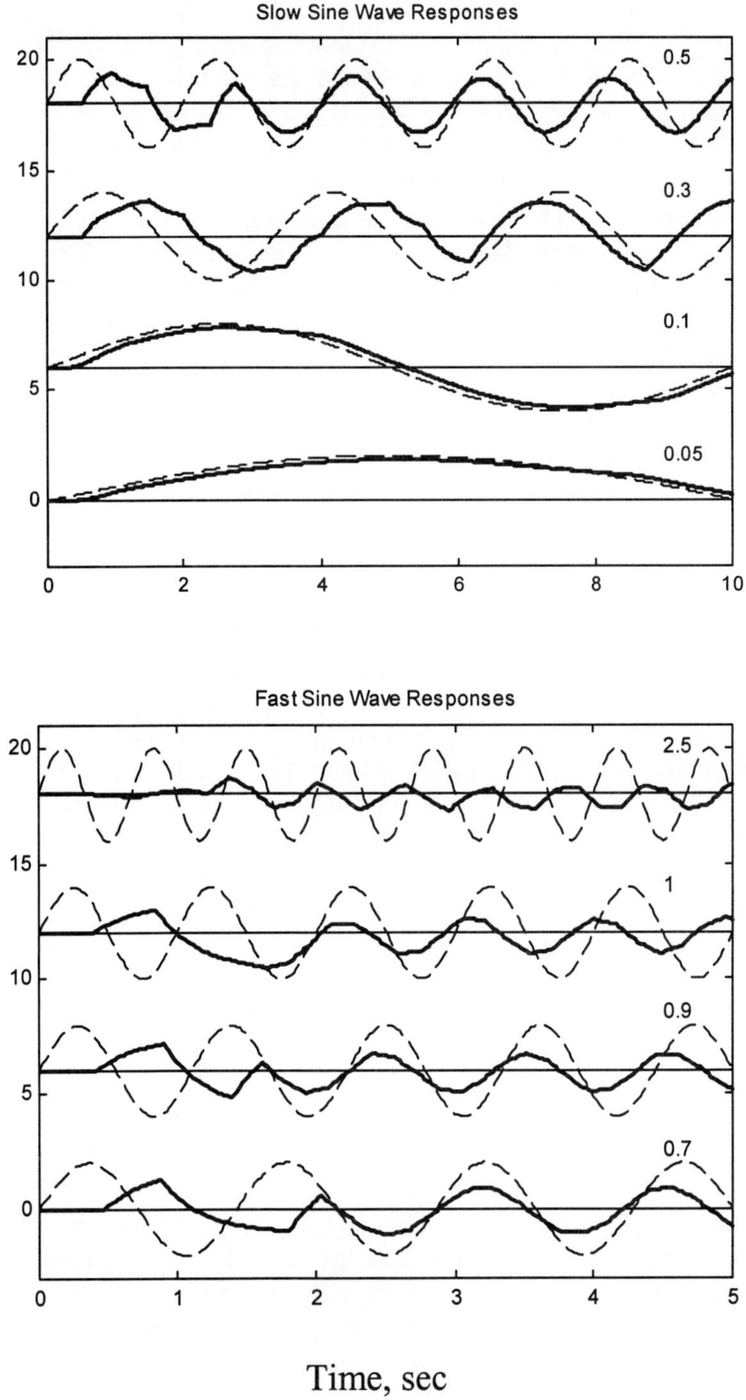

Slow Sine Wave Responses
0.5
0.3
0.1
0.05
20
15
10
5
0
0
2
4
6
8
10
Fast Sine Wave Responses
2.5
1
0.9
0.7
20
15
10
5
0
0
1
2
3
4
5
Time, sec

Fig. 8.24c.

In addition to the above dynamic models, Hung and Ciuffreda (Hung et al, 1996; Ciuffreda et al, 2000) developed a descriptive comprehensive model of accommodation and vergence that summarized the dual interactive nature of these systems (Fig. 8.25a; Hung and Semmlow, 1980; also see Chapter 9 in this volume). Accommodative (AS) and vergence (VS) stimuli are input to the accommodative (upper loop) and vergence (lower loop) system, respectively. The thresholds for initiating a response are represented by the depth-of-focus and Panum's fusional area, respectively. The proximal input (PS), representing the awareness of nearness, goes through gain elements and is input to the accommodative and vergence controllers, respectively (Hung et al, 1996). The proximal component has been shown to play a relatively greater role under the accommodative open-loop condition, with only a negligible role under the accommodative closed-loop condition which reflects everyday viewing conditions. The accommodative adaptive element "charges up" during near work, resulting in a slow decay, or "discharge", when the accommodative loop is subsequently opened (Hung and Ciuffreda, 1992). It is this "charge/discharge" property of the adaptive element that is responsible for producing nearwork-induced transient myopia (Hung and Ciuffreda, 1999; Ong and Ciuffreda, 1995, 1997). The adaptive element appears to assist the accommodative system during nearwork, such as reading, by maintaining a more sustained near accommodative response level, so that less change in accommodation is needed to return to the near level following very brief viewing at far. This occurs due to the slowed decay of accommodation following prolonged drive of the adaptive component during nearwork (see Chapter 18 of this volume). Therefore, by adapting to a higher accommodative level, less accommodative drive/effort is needed to return to the nearwork response level. However, excessively slowed decay, as seen in symptomatic individuals, may hinder normal accommodative responsivity. The interactive accommodative convergence component, represented by the AC/A ratio, serves to drive the vergence output. Conversely, convergence accommodation, represented by the CA/C ratio, serves to drive the accommodative output. These interactive effects are also discussed in detail in Chapter 9 of this volume. The tonic components represent the bias or "default" response levels in the absence of blur, disparity, or proximal stimulation (Rosenfield et al, 1993, 1994; Hung and Ciuffreda, 1991). Thus, when both systems are rendered open-loop, the systems return to their respective default tonic values. The accommodative plant represents the ciliary muscle/lens complex and provides the output of the accommodation system. The vergence plant represents the extraocular muscles and the eyeball, and it provides the output of the vergence system.

The descriptive model above provides an overview of the detailed comprehensive model of accommodation and vergence developed by Hung and Ciuffreda (1999) (see Fig. 8.25b). The detailed model consists of two feedback control loops driven by target defocus and binocular disparity, respectively. The two loops are connected via the accommodative convergence (AC) and convergence accommodation (CA) crosslinks. In the accommodative loop, the difference between the accommodative stimulus (AS) and response (AR), or accommodative error (AE) (i.e., retinal defocus), is input to the nonlinear deadspace element (±DSP) representing the DOF. If this input exceeds the DOF, then the output, which is now retinal-image blur, is input to the accommodative controller having gain ACG. The accommodative controller output is summed with tonic accommodation (ABIAS) and the crosslink signal via convergence accommodation to provide the aggregate accommodative response. Also, the accommodative controller output is multiplied by the crosslink gain, AC, to provide the accommodative convergence signal. Similarly, in the vergence loop, the difference between the vergence stimulus (VS) and response (VR), or vergence error (VE) (i.e., retinal disparity), is input to the nonlinear deadspace element (±DSP), representing Panum's fusional area (Panum, 1858). If this input exceeds Panum's fusional area, then the output from the deadspace element is input to the vergence controller having gain VCG. The vergence controller output is summed with tonic vergence (VBIAS) and the crosslink signal via accommodative convergence to provide the aggregate vergence response. Also, the vergence controller output is multiplied by the crosslink gain, CA, to provide the convergence accommodation signal.

In addition to the basic dual-interactive model, the unique feature of this model is the incorporation of both proximal (Hung et al, 1996) and adaptive (Hung, 1992) elements. The input to the proximal component is represented by a distance stimulus (DS), which drives the perceived distance gain (PDG). The output of PDG is input to both the accommodative proximal gain (APG) and vergence proximal gain (VPG) elements, which are summed with the respective controller outputs. It has been shown that while the proximal component constitutes a considerable percentage (up to about 80%) of the accommodative response under open-loop conditions, it provides a negligible contribution (<4%) under normal closed-loop conditions when visual feedback is present (Hung et al, 1996).

The adaptive element in each loop receives its input signal from the controller output, which in turn modifies the time constant of the controller itself. Although this configuration is unique among near-response oculomotor models, the modification of a component's time constant is seen in other systems. For example, in the saccadic system, Optican and Miles (1985) simulated adaptation using modification of time constants. In our model, the accommodative controller output is input to a multiplier, m_A, and compression element, CE, to drive the adaptive element having gain, K_A, and time constant, T_{A1}. The multiplier and compression elements are necessary to provide a saturation effect for large inputs that is seen in the adaptation experiments (Fisher et al, 1990; Rosenfield and Gilmartin, 1989). The adaptive element output, a, modifies the time constant of the accommodative controller via the term, $T_{A2}+ |a^3|$, where T_{A2} is the fixed portion of the time constant. The cubic relationship was obtained empirically to provide negligible increase in time constant for smaller amounts of adaptation, but a large increase in time constant for larger amounts of adaptation. Similar to the accommodative adaptive element, the vergence adaptive component consists of multiplier, m_V, compression element, CE, adaptive gain, K_V, adaptive time constant, T_{V1}, adaptive element output, b, and controller time constant $T_{V2}+ |b^3|$.

The nominal parameter values are: PDG=0.212, APG=2.1, ACG = 10, AC = 0.80 MA/D, ABIAS = 0.61 D, $m_A = 3$, $T_{A1} = 25$ sec, $T_{A2} = 4$ sec, and VPG=0.067, VCG = 150.0, CA = 0.37 D/MA, VBIAS = 0.29 MA, $m_V = 05$, $T_{V1} = 50$ sec, $T_{V2} = 8$ sec, $K_V = 9$ (Hung, 1992; Hung et al, 1996).

This comprehensive dynamic model has been used to simulate nearwork-induced transient myopia (NITM; Ong and Ciuffreda, 1997) in different refractive groups (Hung and Ciuffreda, 1999a) as well in a model of refractive error development (Hung and Ciuffreda, 1999b; also see Chapter 18 of this volume).

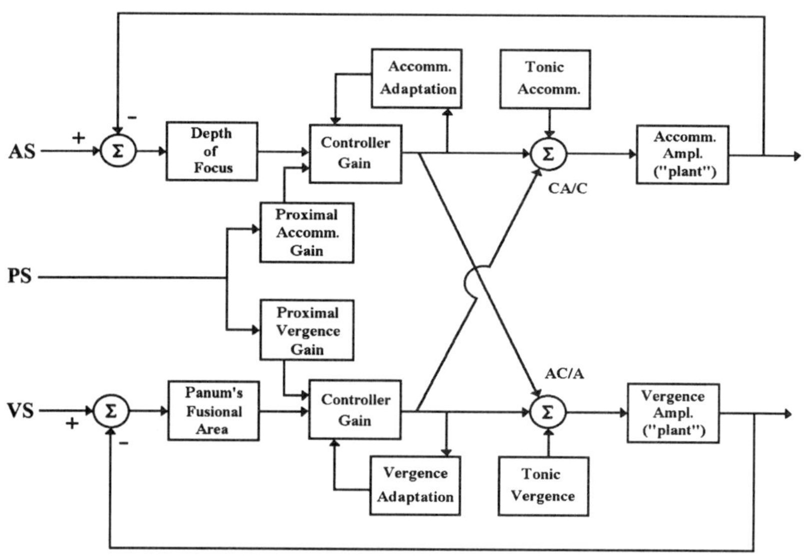

Fig. 8.25a. Hung and Ciuffreda descriptive comprehensive model of the accommodation and vergence system (Hung et al, 1996; Ciuffreda et al, 2000). Reprinted from Ciuffreda et al (2000), pg. 196, Fig. 2, with permission of Birkhäuser Verlag.

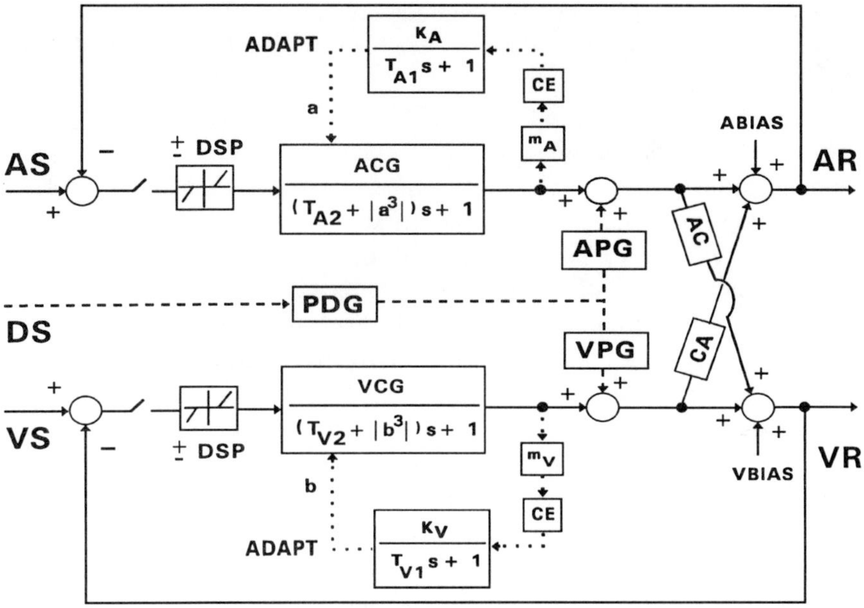

Fig. 8.25b. Hung and Ciuffreda detailed comprehensive model of the accommodation and vergence system (Hung and Ciuffreda, 1999a). Reprinted from Hung and Ciuffreda (1999a), pg. 151, Fig. 1, with permission of Elsevier Science.

8.5 SUMMARY

The basic biomechanical aspects of accommodation were understood nearly 150 years ago (Helmholtz, 1855). Yet, it was not until the introduction of feedback control theory nearly 40 years ago that true quantitative understanding of the accommodative control process began to develop. Proportional control was used to describe the steady-state behavior, but it exhibited nearly instantaneous, or extremely rapid dynamics, since no lag terms were associated with a purely proportional control element. Then, integral control was used to provide smoother dynamic behavior, but it could not simulate steady-state behavior. Later, proportional-integral control was used to attain both dynamic and steady-state behavior, but it had to contend with feedback instability due to the long latency and slow dynamics. More recently, switching control was introduced which restricted the response regions so that accurate simulation was possible. But these models could only be applied under limited conditions, since simulated responses outside of the restricted regions resulted in unrealistic responses. Finally, dual-mode control overcame these difficulties by providing only two possible modes. The fast open-loop mode operated for fast, large amplitude stimuli. It used rate of change, as well as magnitude, of retinal-image defocus to determine the appropriate direction for the accommodative response. In support of this, Mays and Gamlin (2000) have found velocity signals in the Edinger-Westphal nucleus that could be used by the accommodation system. The fast component thus attained most of the response amplitude and overcame the instability problem associated with a relatively long latency. The slow, closed-loop mode operated for slow and small amplitude stimuli. It further reduced the residual error via proportional control to provide an accurate steady-state response.

Simulations of the dual-mode model have provided answers to the two major problems posed in the beginning of this chapter. First, the accommodation system overcomes the even-error problem by using in part the rate of change of retinal defocus to determine the target direction. Of course, in daily life, other cues such as binocular disparity, vergence innervation, chromatic aberration, spherical aberration, size, perspective, and overlap all assist to provide a relatively robust system for determining target direction. Second, experimental and modeling results support a dual-mode, or discontinous, process for accommodation consisting of a fast open-loop movement followed by slow closed-loop control to reduce the error to a minimum. This explains rather well the pure step responses, and the staircase-like step responses to both ramp and sinusoidal stimuli. However, the pulse responses, which shows durations similar to pulse stimuli and therefore are suggestive of a continuous system, appear to present a problem for a discontinuous model. However, the dual-mode model can account for the pulse response data based on re-triggering during the downward transition of the pulse. Therefore, the overall response has the shape of a pulse with a duration approximately equal to that of the pulse stimulus, and thus resembles that of a simple continuous system, even though the underlying process is discontinuous.

Thus, these models have not only clarified our thought processes regarding the normal control of the accommodation system, but have also provided quantitative solutions to some of the fundamental problems that have perplexed investigators for over a century. Some questions still remain regarding both normal and abnormal control of accommodation. First, what are the key features of a dynamic, combined interactive model of accommodation and vergence that will provide all the normal response characteristics? Second, what static and dynamic model parameters can describe and be responsible for abnormal accommodative conditions, as found in patients with accommodative infacility (Ciuffreda, in press)? Lastly, can normal and abnormal accommodation systems be trained, can model parameters associated with such training be identified, and can optimal training paradigms be developed based on current and future models? It is clear from what has been achieved up to this point that modeling and simulation of these clinical conditions will continue to play a significant role in resolving clinical problems by providing insight into the underlying mechanisms of accommodative deficits. This will benefit clinicians and vision scientists, as well as bioengineers, by providing a common, clear, and concise language of models in gaining a deeper understanding of the intricate and elegant control process of accommodation.

8.6 REFERENCES

Bahill, A. T. and Stark, L., 1979, The trajectory of saccadic eye movements, *Sci. Am.* **240**: 108-117.

Benjamin, W. J., ed., 1998, *Borish's Clinical Refraction: Principles and Practice*, W. B. Saunders and Co., Philadel., PA.

Brodkey, J. D., and Stark, L., 1967, Accommodative convergence - an adaptive nonlinear system, *IEEE Trans. Sys. Sci. Cyber.* **3**: 121-133.

Campbell, F. W., 1957, The depth of field of the human eye, *Optica Acta.* **4**: 157-164.

Campbell, F. W., 1959, Fluctuations of accommodation under steady viewing conditions, *J. Physiol.* **145**: 579-594.

Campbell, F. W., and Westheimer, G., 1960, Dynamics of the accommodation responses of the human eye, *J. Physiol.* **151**: 285-295.

Chauhan, K., and Charman, W. N., 1995, Single figure indices for the steady-state accommodative response, *Ophthal. Physiol. Opt.* **15**: 217-221.

Ciuffreda, K. J., 1991, Accommodation and its anomalies, in: *Vision and Visual Dysfunction: Visual Optics and Instrumentation, vol. 1*, W. N. Charman, ed., Macmillan, London, pp. 231-279.

Ciuffreda, K. J., 1998, Accommodation, pupil, and presbyopia, in: *Borish's Clinical Refraction: Principles and Practice*, W. J. Benjamin, ed., W. B Saunders, Philadel., PA, pp. 77-120.

Ciuffreda, K.J., Scientific basis for and efficacy of optometric vision therapy in non-strabismic vergence and accommodative dysfunction, *Optometry*, in press.

Ciuffreda, K. J., Hokoda, S. C., Hung, G. K., and Semmlow, J. L., 1984, Accommodative stimulus/response function in human amblyopia, *Doc. Ophthalmol.* **56**: 303-326.

Ciuffreda, K. J., and Kenyon, R. V., 1983, Accommodative vergence and accommodation in normals, amblyopes, and strabismics, in: *Vergence Eye Movements: Basic and Clinical Aspects*, C. M. Schor, and K. J. Ciuffreda, eds., Butterworths, Boston, pp. 101-173.

Ciuffreda, K. J., and Kruger, P. B., 1988, Dynamics of human voluntary accommodation, *Am. J. Optom. Physiol. Opt.* **65**: 365-370.

Ciuffreda, K. J., Levi, D. M., and Selenow, A., 1991, *Amblyopia: Basic and Clinical Aspects.* Butterworths, Boston, MA.

Ciuffreda, K. J., Rosenfield, M., and Chen, H.-W., 1997, The AC/A ratio, age, and presbyopia, *Ophthal. Physiol. Opt.* **17:** 307-315.

Ciuffreda, K. J., Rosenfield, M., and Chen, H. W., 2000, Accommodation, age, and presbyopia, in: *Accommodation and Vergence Mechanisms in the Visual System*, O. Franzén, H. Richter, and L. Stark, eds., Birkhäuser Verlag, Basel, Switzerland, pp. 193-200.

Ciuffreda, K. J., and Rumpf, D., 1985, Contrast and accommodation in amblyopia, *Vis. Res.* **25**: 1445-1457.

D'Azzo, J. D., and Houpis, C. H., 1988, *Linear Control System Analysis and Design, Conventional and Modern*, McGraw Hill, New York, pp. 209-251.

Fincham, E. F., 1951, The accommodation reflex and its stimulus, *Br. J. Ophthalmol.* **35**: 381-393.

Fisher, S. K., and Ciuffreda, K. J., and Bird, J. E., 1990, The effect of stimulus duration on tonic accommodation and tonic vergence, *Optom. Vis. Sci.* **67**: 441-449.

Fujii, K., Kondo, K., and Kasai, T., 1970, An analysis of the human eye accommodation system, *Osaka Univ. Tech. Report No. 925, Vol. 20*, pp. 221-236.

Gamlin, P. D. R., 1999, Subcortical neural circuits for ocular accommodation and vergence in primates, *Ophthal. Physiol. Opt.* **19**: 81-89.

Gilmartin, B., and Hogan, R. E., 1985, The relationship between tonic accommodation and ciliary muscle innervation, *Invest. Ophthal. Vis. Sci.* **26**: 1024-1028.

Gray, L. S., Winn, B., and Gilmartin, B., 1993, Accommodative fluctuations and pupil diameter, *Vis. Res.* **33**: 2083-2090.

Harrison, R. J., 1987, Loss of fusional vergence with partial loss of accommodative convergence and accommodation following head injury, *Bino. Vis.* **2**: 93-100.

Heath, G. G., 1956a, Components of accommodation, *Am. J. Optom. Arch. Am. Acad. Optom.* **33**: 569-579.

Heath, G. G., 1956b, The influence of visual acuity on accommodative responses of the eye, *Am. J. Optom. Arch. Am. Acad. Optom.* **33**: 513-524.

Helmholtz, H., 1855, Uber die Accommodation des Auges, *Albrecht v. Graefes Arch. Ophthal.* **1**: 1-74.

Hess, C., 1904, Observations concerning accommodation organs, *Klin. Mbl. Augenheilk.*, **42**: 309-315.

Hokoda, S. C., and Ciuffreda, K. J., 1983, Theoretical and clinical importance of proximal vergence and accommodation, in: *Vergence Eye Movements: Basic and Clinical Aspects*, C. M. Schor, and K. J. Ciuffreda, eds., Butterworths, Boston, pp. 75-97.

Hung, G. K., 1998, Sensitivitiy analysis of the stimulus/response function of a static nonlinear accommodation model, *IEEE Trans. Biomed. Engin.* **45**: 335-341.

Hung, G. K., 1992, Adaptation model of accommodation and vergence, *Ophthal. Physiol. Opt.*, **12**: 319-326.

Hung, G. K., 1997, Quantitative analysis of the accommodative convergence to accommodation ratio: linear and nonlinear static models, *IEEE Trans. Biomed. Engin.*, **44**: 306-316.

Hung, G. K., 1998, Dynamic model of the vergence eye movement system: simulation using MATLAB/SIMULINK, *Comp. Meth. Prog. Biomed.* **55**: 59-68.

Hung, G. K., Semmlow, J. L., and Ciuffreda, K. J., 1986, A dual-mode dynamic model of the vergence eye movement system, *IEEE Trans. Biomed. Engin.* **33**: 1021-1028.

Hung , G. K., and Ciuffreda, K. J., 1988, Dual-mode behaviour in the human accommodation system, *Ophthal. Physiol. Opt.* **8**: 327-332.

Hung, G. K., and Ciuffreda, K. J., 1991, Model of accommodation after sustained near focus, *Optom. Vis. Sci.*, **68**: 617-623.

Hung, G. K., and Ciuffreda, K. J., 1999a, Adaptation model of nearwork-induced transient myopia, *Ophthal. Physiol. Opt.* **19**: 151-158.

Hung, G. K., and Ciuffreda, K. J., 1999b, Model of human refractive error development, *Curr. Eye Res.* **19**: 41-52.

Hung, G., K., Ciuffreda, K. J., and Rosenfield, M., 1996, Proximal contribution to a linear static model of accommodation and vergence, *Ophthal. Physiol. Opt.* **16**: 31-41.

Hung, G. K., Ciuffreda, K. J., Semmlow, J. L., and Hokoda, S. C., 1983, Model of static accommodative behavior in human amblyopia, *IEEE Trans. Biomed. Engin.* **30:** 665-672.

Hung , G. K. and Semmlow , J. L. , 1980, Static behavior of accommodation and vergence : computer simulation of an interactive dual-feedback system, *IEEE. Trans. Biomed. Eng.* **27**: 439-447.

Hung, G. K., Semmlow, J. L., and Ciuffreda, K. J., 1982, Accommodative oscillation can enhance average accommodative response: a simulation study, *IEEE Trans. Sys. Man Cyber.* **12**: 594-598.

Hung, G. K., Semmlow, J. L., and Ciuffreda, K. J., 1986, A dual-mode dynamic model of the vergence eye movement system, *IEEE Trans. Biomed. Engin.* **33**: 1021-1028.

Jampel, R. S., 1959, Representation of the near-response on the cerebral cortex of the macaque, *Am. J. Ophthal.* **48**: 573-582.

Jiang, B.-C., 2000, A modified control model for steady-state accommodation, in: *Accommodation and Vergence Mechanisms in the Visual System*, O. Franzén, H. Richter, and L. Stark, eds., Birkhäuser Verlag, Basel, Switzerland, pp. 235-243.

Johnson, C. A., 1976, Effects of luminance and stimulus distance on accommodation and visual resolution, *J. Opt. Soc. Am.* **66**: 138-142.

Kasai T., Unno, M., Fujii. K., Sekiguchi, M., Shinohara, K., 1971, Dynamic characteristics of human eye accommodation system, *Osaka Univ. Tech. Report, Vol. 21*, pg. 569.

Kaufman, P. L., 1992, Accommodation and presbyopia. Neuromuscular and biophysical aspects, in: Adleer's Physiology of the Eye, 9th Ed., W. M. Hart, ed., Mosby-Year Book, St. Louis, MO., pg. 397.

Krall, A. M., and Fornaro, R., 1967, An algorithm for generating root locus diagrams, *Commun. of the Assoc. on Computing Machinery.* **10**: 186-188.

Khosroyani, M., 2000, *Computer Simulation of Ocular Accommodation and Vergence Models.* M. S. Thesis. Tarbiat Modarres University, Tehran, Iran.

Krishnan , V. V. , and Stark , L., 1975, Integral control in accommodation, *Comp. Prog. Biomed.* **4**: 237-255.

Liebowitz, H. W., and Owens, D. S., 1978, New evidence for the intermediate position of relaxed accommodation, *Doc. Ophthalmol.* **46**: 133-147.

Mays, L. E., and Gamlin, P. D. R., 2000, Neuronal circuits for accommodation and vergence in primates, in: *Accommodation and Vergence Mechanisms in the Visual System*, O. Franzén, H. Richter, and L. Stark, eds., Birkhäuser Verlag, Basel, Switzerland, pp. 1-9.

Mordi, J. A., and Ciuffreda, K. J., 1998, Static aspects of accommodation: age and presbyopia, *Vis. Res.* **38**: 1643-1653.

Morgan, M. W., 1957, The resting state of accommodation, *Am. J. Optom. Arch. Am. Acad. Optom.* **34**: 347-353.

Morgan, M. W., 1968, Accommodation and vergence, *Am. J. Optom. Arch. Am. Acad. Optom.* **45**: 417-454.

Moses, R. A., ed., 1981, *Adler's Physiology of the Eye, Clinical Applications*, C. V. Mosby Co., St. Louis, pp 440-455.

Neveu, C., and Stark, L., 1995, Hysteresis in accommodation, *Ophthal. Physiol. Opt.* **15**: 207-216.

Ohtsuka, K., Maekawa, H., Takeda, M., Uede, N., and Chiba, S., 1988, Accommodation and convergence insufficiency with left middle cerebral artery occlusion, *Am. J. Ophthalmol.* **105**: 60-64.

O'Neill, W. D., 1969, An interactive control systems analysis of the human lens accommodative controller, *Automatica.* **5**: 645-654.

Ong, E., and Ciuffreda, K. J., 1995, Nearwork-induced transient myopia - a critical review, *Doc. Ophthalmol.* **91**: 57-85.

Ong, E., and Ciuffreda, K. J., 1997, *Accommodation, Nearwork, and Myopia*, Optometric Extension Program Foundation, Inc., Santa Ana, CA.

Optican, L M., and Miles, F. A., 1985, Visually induced adaptive changes in primate saccadic oculomotor control signals, *J. Neurophysiol.* **54**: 940-958.

Panum, P. L, 1858, *Physiologische Untersuchungen uber das Sehen mit zwei Augen, Schwersche Buchhandlung*, Kiel, Germany.

Phillips, S. R., 1974, *Ocular Neurological Control Systems: Accommodation and the Near Response Triad*, Ph.D. Dissertation, Dept. of Mechanical Engin., Univ. of Calif., Berkeley, CA, U.S.A.

Provine, R. R., and Enoch, J. M., 1975, On voluntary ocular accommodation, *Percept. Psychophys.* **17**: 209-212.

Ripps, H., Chin, N. B., Siegel, I. M., and Breinin, G. M., 1962, The effect of pupil size on accommodation, convergence, and the AC/A ratio, *Invest. Ophthal. Vis. Sci.* **1**: 127-135.

Rosenfield, M., Ciuffreda, K. J., Hung, G. K., 1991, The linearity of proximally induced accommodation and vergence, *Invest. Ophthal. Vis. Sci.* **32**: 2985-2991.

Rosenfield, M., Ciuffreda, K. J., Hung, G. K., and Gilmartin, B., 1993, Tonic accommodation: a review. I. Basic aspects, *Ophthal. Physiol. Opt.* **13**: 266-284.

Rosenfield, M., Ciuffreda, K. J., Hung, G. K., and Gilmartin, B., 1994, Tonic accommodation: a review. II. Accommodative adaptation and clinical aspects, *Ophthal. Physiol. Opt.* **14**: 265-277.

Rosenfield, M., and Gilmartin, B., 1989, Temporal aspects of accommodative adaptation, *Optom. Vis. Sci.* **66**: 229-234.

Saladin, J. J. and Stark, L., 1975, Presbyopia: a new evidence from impedance cyclography supporting the Hess-Gullstrand theory, *Vis. Res.* **15**: 537-541.

Stark, L., 1987, Presbyopia in light of accommodation, in *Presbyopia, Recent Research and Reviews from the Third International Symposium*, L. Stark, and G. Obrecht, eds, Professional Press, New York, pp. 264-274.

Stark, L., Kong., R., Schwartz, S., and Hendry, D., 1976, Saccadic suppression of image displacement, *Vis. Res.* **16**: 1185-1187.

Stark, L. W., Neveu, C. and Krishnan, V. V., 2000, Mode switching in control of accommodation, in: *Accommodation and Vergence Mechanisms in the Visual System*, O. Franzén, H. Richter, and L. Stark, eds., Birkhäuser Verlag, Basel, Switzerland, pp. 225-234.

Stark, L., and Takahashi, Y., 1962, Accommodative tracking, *Quart. Prog. Rept., Res. Lab. of Electronics, M.I.T.* **67**: 220.

Stark, L., Takahashi, Y., and Zames, G. 1962, The dynamics of the human lens system, *Quart. Prog. Rept., Res. Lab. of Electronics, MIT.* **66**: 337.

Stark, L., Takahashi, Y., and Zames, G. 1965. Nonlinear servo-analysis of human lens accommodation. *IEEE Trans. Sys. Sci. Cyber.* **1**: 75-83.

Sun, F., and Stark, L., 1990, Switching control of accommodation: experimental and simulation responses to ramp inputs, *IEEE Trans. Biomed. Engin.* **37**: 73-79.

Thompson, H. E., 1975, *The Dynamics of Accommodation in Primates*, Ph.D. dissertation, Dept. of Biomed. Engin., Univ. of Illinois Medical Center, Chicago, IL.

Toates, F. M., 1972a, Accommodation function of the human eye, *Physiol. Reviews.* **52**: 828-863.

Toates, F. M., 1972b, Studies on the control of accommodation and convergence, *Measurement and Control.* **5**: 58:61.

Tucker, J., and Charman, W. N., Reaction and response times for accommodation, *Am. J. Optom. Physiol. Opt.* **56**: 490-503.

Warwick, R., 1954, The ocular parasympathetic nerve supply and its mesencenphalic sources, *J. Anat., Lond.* **88**: 71-93.

Westheimer, G., 1963, Amphetamines, barbiturates and accommodative convergence, *Arch. Ophthalmol.* **70**: 830-836.

Winn, B., Pugh, J. R., Gilmartin, B., and Owens, H., 1990, Arterial pulse modulates steady-state ocular accommodation, *Curr. Eye Res.* **9**: 971-974.

Chapter 9

Models of Vergence and Accommodation-Vergence Interactions

Bai-chuan Jiang[1], George K. Hung[2], Kenneth J. Ciuffreda[3]

[1]College of Optometry, Health Professions Division, Nova Southeastern University, 3200 South University Dr., Ft. Lauderdale, FL 33328, PH: (954) 262-1444, FX: (954) 262-1818, EM: bjiang@nova.edu

[2]Department of Biomedical Engineering, Rutgers University, 617 Bowser Rd. Piscataway, NJ 08854, PH: (732) 445-4137, FX: (732) 445-3753, EM: shoane@rci.rutgers.edu

[3] Dept. of Vision Sciences, State University of New York, State College of Optometry, 33 West 42nd St. New York, NY 10036; PH: (212) 780-5132, FX: (212) 780-5124; EM: kciuffreda@sunyopt.edu

9.1 INTRODUCTION

Vergence (or disjunctive) eye movements provide single vision by bringing the images of a bifixation target onto corresponding retinal points in the two eyes. When a target moves in depth, the brain recognizes the change in position of the retinal images and drives the extraocular muscles to bring these images into proper register on the retinas. Since vergence eye movements reflect the function of the brain, quantitative assessment of these movements can reveal fundamental information regarding the brain's underlying neural control strategy. For this reason, an understanding of how vergence is controlled in both normal and symptomatic individuals has been one of the most important goals of vision scientists, clinicians, and bioengineers. This chapter provides a summary of some of the most significant research on modeling of the vergence system, as well as the interactions between accommodation and vergence.

9.1.1 Components of Vergence

According to Maddox (1893), the overall or aggregate vergence eye movement is composed of four linearly additive and sequential components: tonic, accommodative, reflex, and voluntary. Tonic vergence is the initial component of vergence which shifts the eyes from an unknown anatomic resting position to a more convergent physiological position of rest; it probably reflects baseline midbrain neural activity. Accommodative vergence is the blur-driven component, which is then added to the tonic vergence component. Maddox recognized that accommodative vergence was due to the amount of accommodation in force during near vision. He was not aware, however, that vergence affected accommodation as well, and hence did not consider interactive feedback effects. Reflex vergence responds to the presence of retinal disparity, or the angular difference between target angle and bifixation angle, and is considered "supplemental" to the other two components. It is now commonly referred to as fusional vergence by clinicians who emphasize its function (Hofstetter, 1945), and disparity vergence by bioengineers who emphasize its stimulus control properties (Stark et al, 1980), which reduces the residual amount of vergence error to a minimum (see Fig. 9.1). The fourth component, voluntary convergence, is attributed to "knowledge of nearness" of the target. It is also referred to as psychic or proximal vergence. Maddox "apparently had as

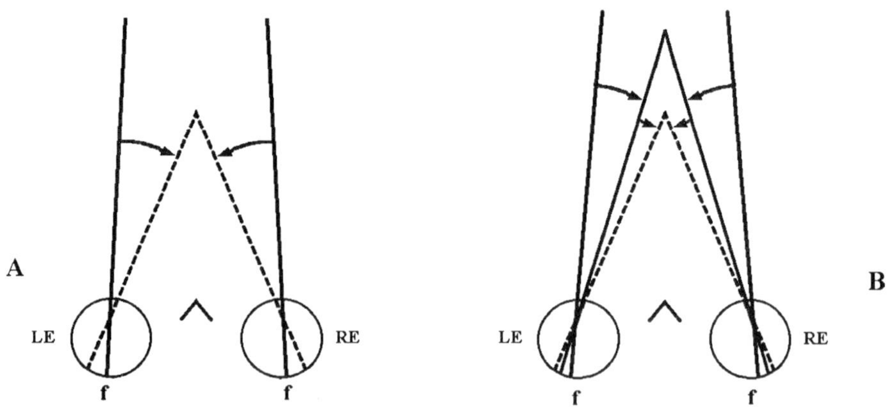

Figure 9.1. (a) Pure symmetrical vergence response showing a complete movement from a far fixation position to a near fixation position. (b) The same pure symmetric response showing sequential components based on the Maddox hierarchy: starting from the tonic vergence position (outer pair of solid lines), it proceeds with an initial accommodative vergence component movement (inner pair of solid lines), which is then followed by the reflex component (now commonly referred to as fusional or disparity convergence) movement (pair of dashed lines) to finally bifixate on the near target. The voluntary vergence component is not shown. The symbol f = fovea.

much difficulty placing psychic or proximal vergence as modern investigators" (Morgan, 1983). Maddox emphasized accommodative convergence and underestimated the role of disparity vergence in the overall vergence movement. However, Maddox's classification remains a useful technique for the analysis of binocular vision problems often encountered in optometric clinics (Morgan, 1983), as well as for the understanding of the interaction between accommodative vergence and the compensatory fusional vergence movement. It has also provided an elementary step in the modeling of the vergence system.

9.1.2 Vergence Dynamics

Vergence dynamics exhibit specific characteristics for each type of stimulus. The pulse response reflects the stimulus duration and amplitude (Fig. 9.2; Rashbass and Westheimer, 1961). That is, a longer duration pulse stimulus elicits a longer response; and a higher amplitude pulse results in a larger response. Step responses have a reaction time ranging from 160 to 200 msec for convergence and from 180 to 210 msec for divergence (Rashbass and Westheimer, 1961; Zuber and Stark, 1968) (Fig. 9.3 and Fig. 9.4 top trace). They exhibit relatively slow and smooth dynamics, having a time constant of about 200 msec for convergence and 240 msec for divergence (Semmlow and Wetzel, 1979). Moreover, for ramp stimulus velocities greater than about 2.7 deg/sec (Semmlow et al, 1986), vergence responses show multiple step movements even though the stimulus is a smooth ramp (Fig. 9.4; Rashbass and Westheimer, 1961).

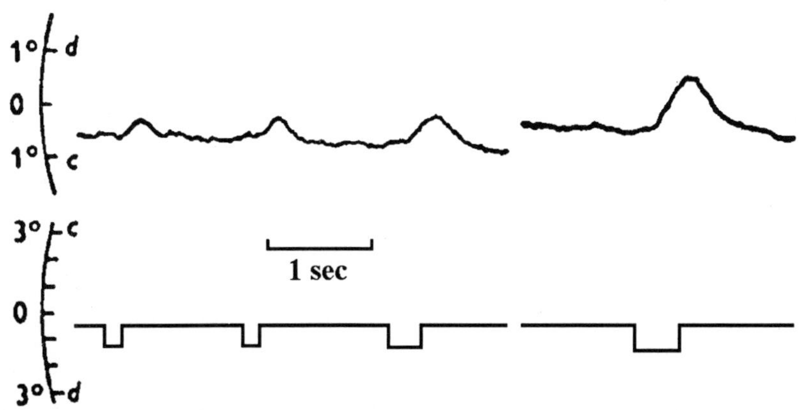

Figure 9.2. Eye movement responses (top trace) to disparity pulse stimuli (bottom trace) of various durations. Reprinted from Rashbass and Westheimer (1961), pg. 345, Fig. 7, with permission of The Physiological Society.

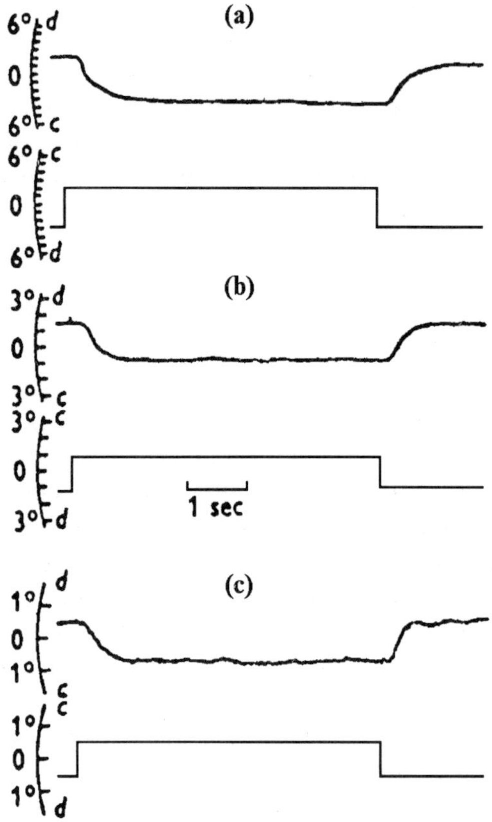

Figure 9.3. Eye movement responses (top traces) to disparity step stimuli (bottom traces) of amplitudes: (a) 4 deg., (b) 2 deg., and (c) 1 deg. Symbol "c" denotes convergence and "d" denotes divergence. Note changes in vertical scale. Reprinted from Rashbass and Westheimer (1961), pg. 344, Fig. 5a, with permission of The Physiological Society.

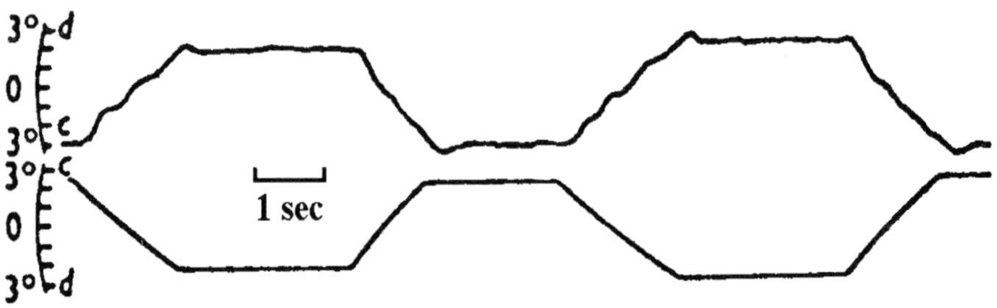

Figure 9.4. Eye movement responses (top trace) to divergent (d) and convergent (c) ramp disparity stimuli (bottom trace). Note the staircase-like multiple-step responses to the ramp stimulus. Reprinted from Rashbass and Westheimer (1961), pg. 346, Fig. 9, with permission of The Physiological Society.

Finally, sinusoidal responses show smooth tracking for the lower frequency sinusoids, but multiple step-like responses to the higher frequency sinusoids (Fig. 9.5).

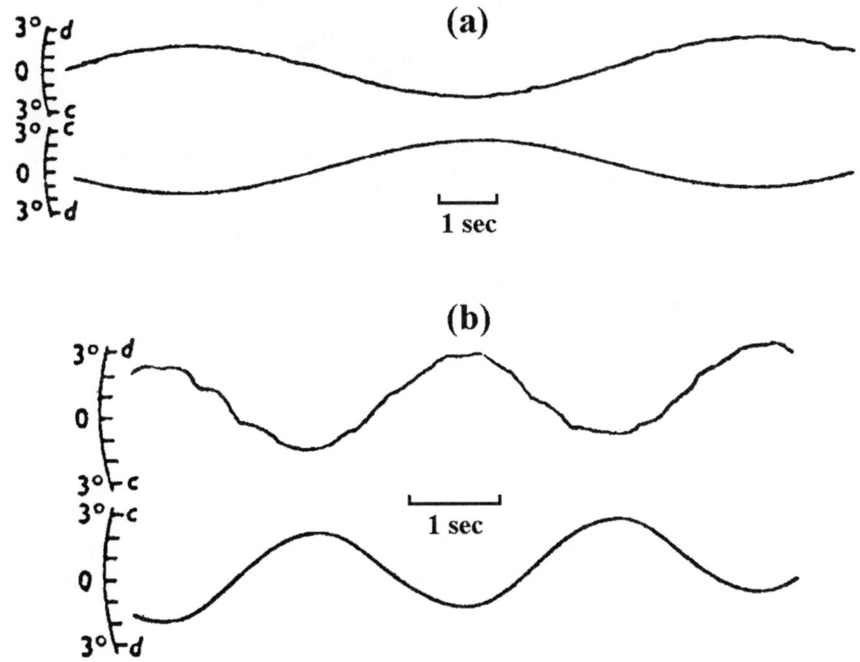

Figure 9.5. Eye movement responses (top traces) to disparity sinusoidal stimuli (bottom traces) at two different amplitudes and frequencies (a and b). Symbol "c" denotes convergence and "d" denotes divergence. Reprinted from Rashbass and Westheimer (1961), pg.. 346, Fig. 10, with permission of The Physiological Society.

The above records were obtained under normal closed-loop conditions. To investigate the control mechanism further, Rashbass and Westheimer (1961) used instrument feedback to maintain a constant vergence error while recording the response. Their results showed a proportionality between the amplitude of the clamped disparity and the velocity of the induced vergence response (Fig. 9.6). To account for these open-loop results, they proposed in their model an integrator in the forward loop of the vergence system. This was because integration of a step of clamped disparity would result in a ramp, or constant velocity, output. Moreover, increasing the clamped disparity step amplitude would increase proportionally the velocity of the response, as seen in Fig. 9.6.

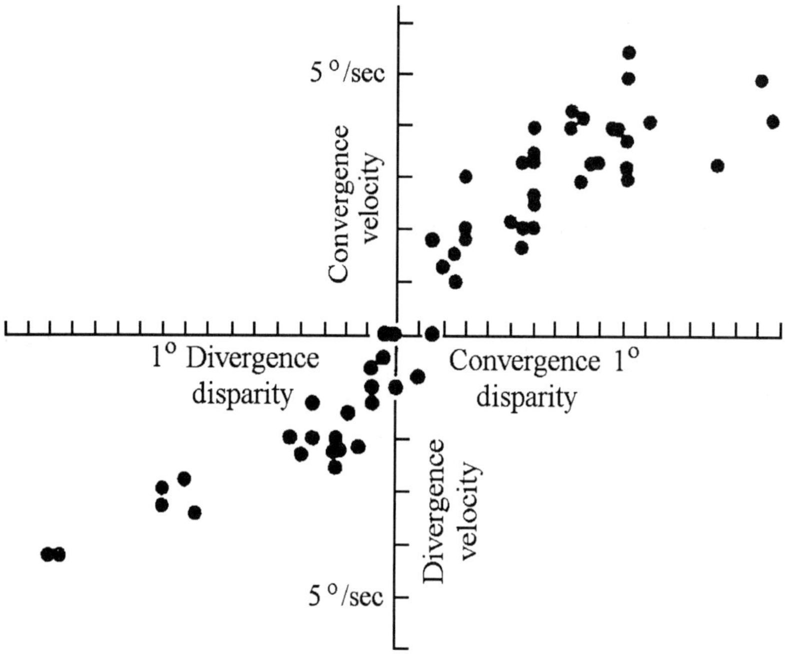

Fiigure 9.6. Ralationship between the magnitude of constantly maintained disparity and velocity of vergence response induced by it. Reprinted from Rashbass and Westheimer (1961), pg. 349, Fig. 13, with permission of The Physiological Society.

9.1.3 Basic Vergence Control Model

The basic configuration of the model of the vergence system is that of a feedback control system (described in greater detail below). It consists of a combination of dynamic elements that act in concert to accomplish an objective, i.e., bringing the images to corresponding points on the retinas to obtain sensory fusion. The model elements are designed to represent physiological components in the vergence system. The accuracy of the each model element can be confirmed by modifying its input and monitoring the output, and then comparing the resultant input-output relationship with that obtained experimentally. The model parameters can then be fine-tuned to provide a more accurate representation of vergence behavior. An important aspect of the vergence model is that of feedback control. The role of feedback control is to ensure that the output of the system attains and then maintains a desired value in the presence of variation produced by the external environment. Mathematically, this kind of control system can be described by differential equations, in which time is the independent variable. Generally, these differential equations are transformed into the Laplace domain, where they become simple algebraic equations. The resultant solution is then inverse-Laplace transformed back into the time domain for plotting and analysis. Thus, these models provide relatively easy manipulation of the elements, and more significantly, a deeper quantitative understanding of the vergence system.

9.1.4 Overview of Chapter

This chapter is intended for the bioengineer, vision researcher, and advanced clinician working in this area. The modeling and analysis of dynamic control systems have interested engineers for many decades. In recent years, faster computers and sophisticated simulation software have enabled us to simulate complex biological systems such as the vergence system. Models of vergence and accommodation have become popular among vision scientists and clinicians because they provide useful tools for explaining the data obtained in laboratories and clinics. An example is the explanation of the discrepancy identified in the early theories of near triad control. Maddox (1886) proposed that both accommodative and vergence responses were driven primarily by blur, with the contribution from disparity being relatively small. On the other hand, Fincham and Walton (1957) suggested that disparity vergence dominated the near triad responses. Both theories were eventually subsumed after they were replaced by an interactive and comprehensive dual-feedback model (see Hung and Semmlow, 1980). We hope that continued study of the modeling of the vergence and accommodation systems will provide a better understanding of the nature of these systems.

9.2 DISPARITY VERGENCE SYSTEM MODELS

Models of vergence have been developed by various investigators to provide insight into its mechanism of control. These models are discussed below in a logical-developmental rather than chronological order.

9.2.1 Continuous Feedback Models

9.2.1.1 Rashbass and Westheimer Model

In the absence of other cues, the dominant input to the vergence control system is retinal disparity (Westheimer and Mitchell, 1956; Fincham and Walton, 1957; Stark *et al.* 1980). Binocular disparity discussed in this chapter is defined as the difference between the desired vergence angle and the actual vergence angle in the horizontal meridian of the eyes. The goal of the vergence system is to reduce the amount of retinal disparity. Thus, the system can be considered a negative feedback control system (Westheimer, 1963). Under negative feedback control with its high controller gain, the vergence response matches the vergence demand very well, with a small residual error of a few minutes of arc called "fixation disparity" (Ogle et al, 1967). Experimental results showed that vergence responses could be modified during the reaction time as well as during the vergence movement itself. This suggested that the vergence system was under continuous feedback control (Rashbass and Westheimer, 1961). However, later modeling research indicated that a "re-triggerable" discontinuous system could also produce similar pulse responses (Hung , 1998a; see below).

The first control model approximated the vergence system as a linear control system (Fig. 9.7). Rashbass and Westheimer (1961) used a ramp stimulus while holding the disparity input constant (i.e., in the open-loop

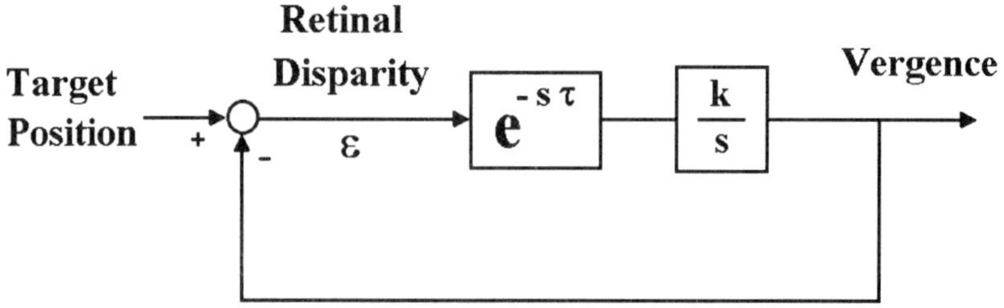

Figure 9.7. Block diagram illustrating the Rashbass and Westheimer model (1961). In this model, an integrator (*k/s*) follows a delay element ($e^{-s\tau}$). The vergence plant was assumed to have a zero-order unity gain. Disparity (ε) is the angular difference between target position and vergence position. From Patel et al (1995) with permission of the author.

condition) to study the vergence system. The advantage of operating under the open-loop condition is that it gives the experimenter complete control over the system input. They found that for small disparities up to 0.2 deg, the vergence velocity was proportional to the retinal disparity that existed one reaction time (about 160 ms) earlier (see Fig. 9.6). The following equation, then, was suggested:

$$\frac{dr}{dt} = k\varepsilon(t-\tau) \tag{9.1}$$

where r is the vergence response, ε the retinal disparity, τ the reaction time, and k the constant of proportionality. Another way to look at this equation is to convert it to the Laplace domain[1]. In this domain, Eq. 9.1 can be written as

$$sR(s) = kE(s)e^{-s\tau} \tag{9.2}$$

Under the open-loop condition, the error *E(s)* equals the input. Therefore, the forward-loop transfer function of the system, *H(s)*, is given by

$$H(s) = \frac{R(s)}{E(s)} = \frac{k}{s}e^{-\tau s} \tag{9.3}$$

Eq. 9.1 was used to predict the vergence response to an open-loop sinusoidal disparity stimulus ($\varepsilon = a\cos\omega t$). The result was that

$$r = \frac{ka}{\omega}\sin\omega(t-\tau) = \frac{ka}{\omega}\cos[\omega(t-\tau) - \frac{1}{2}\pi], \tag{9.4}$$

where a constant of integration was omitted. This result implies that if the system is linear, then a cosinusoidal disparity stimulus will evoke a sinusoidal vergence response having the same frequency as the stimulus. The vergence response will have an amplitude proportional to the amplitude of the stimulus. In addition, the gain of the system, which is defined as the ratio of the amplitude of the vergence response to the amplitude of the disparity stimulus (ka/ω in Eq. 9.4) will be inversely proportional to the frequency of the disparity stimulus. These predictions were verified by the experiment of Rashbass and Westheimer (1961). However, the prediction of the phase delay for the vergence responses to sinusoidal disparity stimuli did not agree with the experimental results. From the model, it appears that the phase lag is equal to 90^0 plus 160 ms delay. But, the measured latencies of

[1] The Laplace transform converts a differential equation for function *f(t)* to an algebraic equation for function *F(s),* in which *s* is a complex variable.

the vergence response were substantially shorter than expected. Using regular alternation of convergent and divergent step disparities, they found no reduction in phase. They therefore concluded that anticipation (i.e., prediction) did not play a role in phase reduction in the vergence system, but the system might use other information such as velocity and perhaps acceleration of the sinusoidal stimulus to reduce the phase.

9.2.1.2 Krishnan and Stark Model

Krishnan and Stark (1977) suggested a modified model that consisted of an integral-derivative controller, a time-delay element, and a third-order plant (Figure 9.8). They analyzed experimental convergence step responses and found that a parallel operator was needed; the derivative-integral element provided the fast initial response, and the "leaky" integral element (i.e., element of the form $\tau/(\tau s + 1)$, where τ is the time constant; Krishnan and Stark, 1975) provided the sustained response. The derivative element can also be thought of as a neural network that is sensitive to changes in retinal disparity rather than to the magnitude of the disparity. This parallel arrangement significantly improved the phase characteristics as compared to those of Rashbass and Westheimer.

The Krishnan and Stark model was the first to account for the difference in dynamic responses between convergence and divergence. The convergent response is generally faster than the divergent response. This represents a major non-linearity of the vergence system. Following their work, other

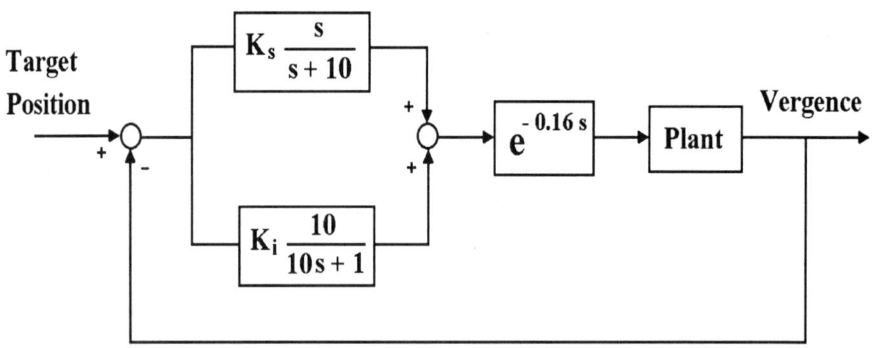

Figure 9.8. Krishnan and Stark model (1977) of the vergence system consisting of an integral-derivative controller parallel to a leaky integrator. The element $e^{-0.16s}$ represents the 160 msec delay in this system. The plant used in this model is the Cook-Stark model developed for the versional system (Cook and Stark, 1967). Reprinted from Krishnan and Stark, (1977), pg. 46, Fig. 6, with permission of © IEEE.

important non-linearity features in the vergence system, such as the saturation limits in different components and the threshold (or dead-zone) of disparity, were for the first time included in different models (Schor, 1979; Hung and Semmlow, 1980).

Under static viewing conditions (obtained by setting s = 0 in the model), the Krishnan and Stark model has a forward-loop gain of $10K_iK_p$, in which K_i represents the gain of the leaky integral controller and K_p (not shown in Fig. 9.3) represents the gain of the plant. The static gain determines the linear relationship between vergence error and the vergence response under open-loop conditions.

9.2.1.3 Schor Model

The concept of two primary control components, i.e., transient and sustained, in the vergence response was also suggested by Schor's early work (1979). However, unlike the Krishnan and Stark model, Schor used two parallel leaky integral elements for modeling the vergence controller (Figure 9.9). He simulated the vergence closed-loop response to a step stimulus. The step response of the system is the sum of the responses of the two integrators (although the output of the fast integrator is first filtered through the slow integrator; see Fig. 9.9). Schor stated that the fast integrator provided the initial step response, and then the slow integrator

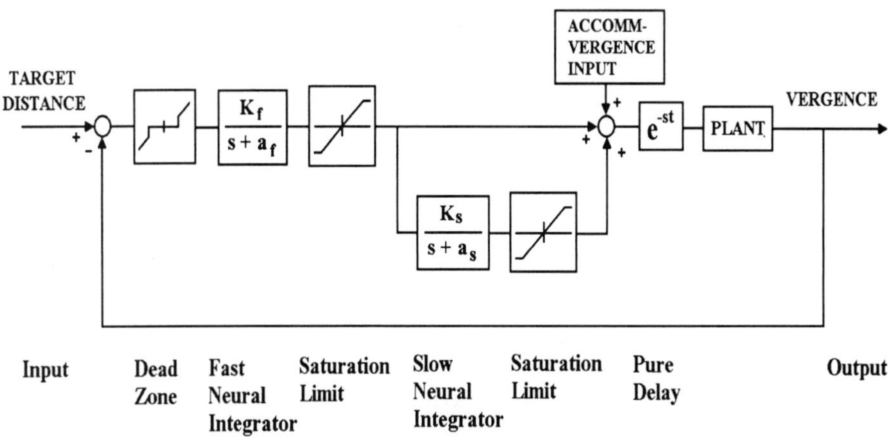

Figure 9.9. Schor model (1979) of the vergence system consisting of parallel leaky integrators. The vergence plant is assumed as a second order system. $K_f = 2.5$, $a_f = 0.1$, $K_s = 3$, $a_s = .03$, delay t is assumed to be 160 msec (not given), and dead zone limits are assumed to be ±0.25 deg (not given; although it can be shown that this has relatively little effect on the simulation responses). Reprinted from Schor (1979), pg. 834, Fig. 4, with permission of Elsevier Science.

gradually dominated the response and replaced the fast integrator drive, thus allowing it to be available for a rapid response to subsequent stimuli. In the Schor model, non-linear elements were added to represent the dead-zone and saturation limits of the fast and slow integrators.

A recent MATLAB/SIMULINK simulation of the Schor model indicated that the response to a 10 deg step stimulus consisted of large overshoot followed by oscillations over approximately the next 4 to 5 sec (Fig. 9.10). However, experimental vergence step responses did not exhibit multiple oscillations, and the step responses were completed in about 1 sec (Rashbass and Westheimer, 1961; Semmlow et al, 1986). Moreover, the duration of the oscillations in the Fast and Slow Neural Integrators were similar (Fig. 9.10), thus, calling into question the justification for distinguishing between these two components based on their dynamics.

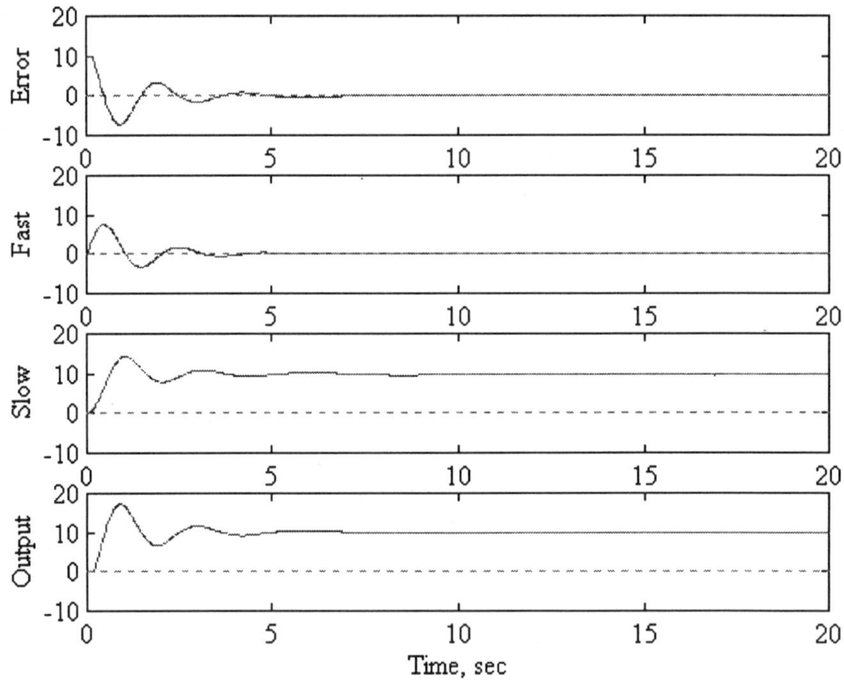

Figure 9.10. MATLAB/SIMULINK simulation of Schor's (1979) model (see Fig. 9.9) showing time courses following a 10 deg step stimulus for the vergence error, the Fast Neural Integrator output, the Slow Neural Integrator output, and the vergence response. In this simulation, the plant dynamic element (not given in the Schor model) was approximated by an all pass filter (or 1 in the Laplace domain), since it has much faster dynamics than the other forward loop elements (i.e., $1/a_f = 10$ sec, and $1/a_s = 33.3$ sec). From Hung (personal communication, 2001).

Ludvigh *et al.*(1964) presented experimental evidence to support the idea that the vergence controller has a leaky nature. In their experiment, the subject was asked to converge to a target. Then, one of his eyes was occluded (i.e., an open-loop condition for disparity vergence). Immediately after this, the vergence response started to decay to its phoria position. This provided the time constant of decay of the leaky integrator. In Schor's (1979) model, leaky integrators of the form $\tau/(\tau s + 1)$ (Krishnan and Stark, 1975) were used for the fast and slow vergence controllers. The response of a leaky integrator to a constant steady-state error input is a stable and constant vergence output. On the other hand, the response of a non-leaky (e.g., pure) integrator would be a continued integration of the constant error, resulting in an ever-growing response (Toates, 1975). Therefore, a leaky integrator is the appropriate form for the vergence controller. Moreover, non-zero fixation disparity is a necessary input signal to the controller to maintain a static vergence response under closed-loop conditions.

Schor (1979) derived an equation for the steady-state error, or fixation disparity:

$$f.\,d. = \frac{x}{kT+1} \tag{9.5}$$

where x = vergence stimulus, k = gain of the integral controller, and T = open-loop decay time constant. However, the predicted fixation disparities based on this equation are over twice as large as those found experimentally (Schor, 1979; Hung and Semmlow, 1980).

9.2.1.4 Pobuda and Erkelens Model

Instead of using derivative-integral or leaky integral elements, Pobuda and Erkelens (1993) proposed parallel channels, each of which was sensitive to a specific range of retinal disparities with its own low-pass filter characteristics (Fig. 9.11). In each channel, the disparity signal was first selected by a range detector, and then was filtered by a correspondingly tuned low-pass filter. The output from parallel filters feeds into a slow integrator with a pure delay element (100 ms), similar to the Schor model (1979), to simulate sustained binocular fixation. A second-order low-pass filter was used to simulate the vergence plant. In their model, the open-loop phase improvement is a result of these parallel low-pass filters. However,

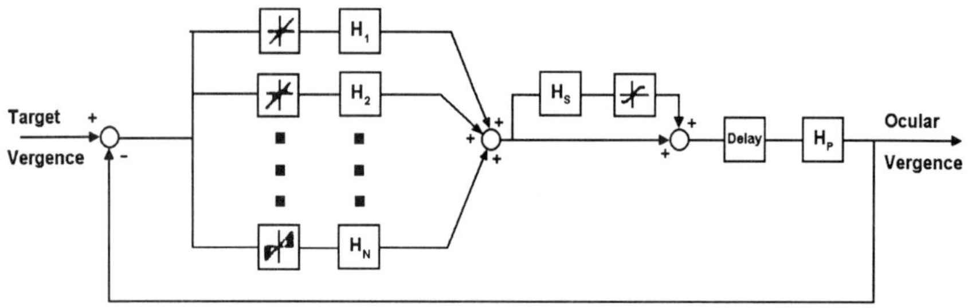

Figure 9.11. Pobuda and Erkelens model (1993) of disparity vergence. Its essential feature is the processing of disparity in several (5 were used) parallel channels, each of which is sensitive to a specific range of disparities and has its own lowpass filter characteristics ($H_1, H_2, \ldots, H_N$). The model also contains a slow integrator (H_S) and a pure delay of 100 msec. The plant (H_P) is modeled by a second-order lowpass filter (i.e., a series of 2 first-order filters) with time constants of 8 and 150 msec. Reprinted from Pobuda and Erkelens (1993), pg. 227, Fig. 6, with permission of Biol. Cybernetics.

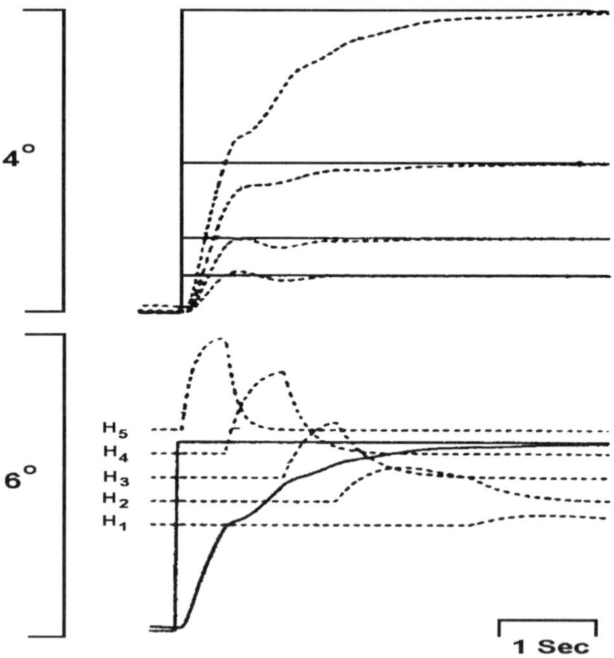

Figure 9.12. Simulation of Pobuda and Erkelens' (1993) model responses to step changes in target vergence. The upper panel shows four ocular vergence responses (dashed lines) to steps in target vergence (continuous lines) of 0.5, 1, 2, and 4 deg. The lower panel shows an ocular vergence response to a step of 4 deg in target vergence (continuous lines). The outputs of the individual lowpass filters are indicated by the dashed lines. Reprinted from Pobuda and Erkelens (1993), pg. 227, Fig. 8, with permission of Biol. Cybernetics.

model simulation step responses (Fig. 9.12) showed multiple step-like movements, which were not seen in experimental step responses such as those by Rashbass and Westheimer (1961) (Fig. 9.3). Moreover, their responses took several seconds for completion rather than one sec (Rashbass and Westheimer, 1961; Semmlow et al, 1986). This was especially evident for the larger vergence amplitudes, which were processed successively through the range of disparity filters.

9.2.1.5 Patel et al's Neural Network Model

Control theory modeling uses a "black box" approach, in which the system characteristics are derived from the relationship between the input and output of the system. For example, in the continuous vergence models discussed above (excluding Pobuda and Erkelens, 1993) and the non-continuous vergence model below (Hung et al, 1986), disparity and vergence eye-position are thought to be the input and output, respectively, without specifying how disparities are computed from retinal processing and how motoneurons are driven to generate the desired vergence response. Models built within a neural network aim to capture both the architecture and the function of the system under study, and hence overcome limitations of other models.

Patel *et al.* (1997) developed a neural network model for simulating short-term disparity vergence dynamics. The general structure of the model consists of seven functional stages as shown in Fig. 9.13. The model assumes the existence of retinotopic maps where localized and normalized activities are generated corresponding to the retinal locations of the vergence target in the two eyes. In the first stage, the localized activities in the retinotopic maps are used to detect instantaneous disparity by a pool of neurons called retinal disparity detectors. In the second stage, the detected disparities are used to generate a one-dimensional spatial map of disparity. In the third stage, the activity in the disparity map is converted to a velocity signal. The disparity encoders in the one-dimensional map activate the velocity elements with weights proportional to the encoded disparity (Rashbass and Westheimer, 1961; Krishnan and Stark, 1977). In the fourth stage, the velocity signals are converted to position signals by a push-pull architecture in which the convergence (divergence) velocity element excites the convergence (divergence) position element, and inhibits the fellow divergence (convergence) position element. The position elements are modeled as non-leaky integrators with non-linear dynamics. In the fifth stage, the position signals from the convergence and divergence position elements are converted to moto-neural activities. These moto-neural elements innervate the medial and lateral muscles of both eyes. It is important to note that in this model, the moto-neural elements also receive inputs from the corresponding velocity elements. This velocity input plays an important role in improving the phase characteristics of the sinusoidal vergence responses. In the sixth stage, a velocity overdrive circuit with discrete parallel channels, each gated by a pre-set velocity threshold, is used to provide a velocity-dependent signal to the corresponding moto-neural element. This stage is similar to that by Pobuda and Erkelens (1993), except here velocity rather than disparity range determines the channel input that passes this stage. In the final stage, an active turn-off circuit is used to provide a discharge path for the otherwise non-leaky position elements. The discharge circuit is used to move the eyes to their tonic levels in the absence

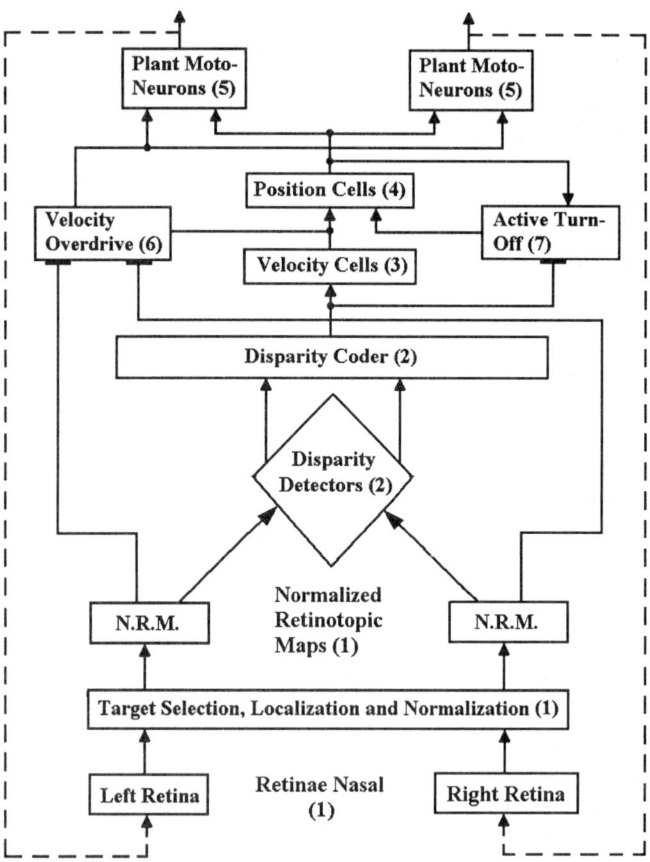

Figure 9.13 Patel *et al.* model (1997) of the vergence system. The solid lines with arrows represent the primary signal path. The solid lines with rectangular connections show modulatory signals. The dotted lines represent the external visual feedback. The numbers in the boxes correspond to those of the stages described in the text. Reprinted from Patel et al. (1997), pg. 1385, Fig. 1, with permission of Elsevier Science.

of activity in the disparity encoders. This discharge mechanism results in disassociation of the tonic vergence dynamics from the stimulus-driven dynamics. Although there is experimental evidence that velocity-driven signals contribute to the control of vergence eye movements (Mays, 1984), there is as yet no physiological evidence for the proposed velocity overdrive and velocity gate control circuits (Patel et al, 1997). Using this model, the authors simulated closed-loop (normal binocular viewing) and open loop (disparity clamped viewing) symmetric step, sinusoidal, pulse, staircase, square, and ramp wave responses. The simulation results closely resembled experimental data obtained in their laboratory and those reported in the literature (Rashbass and Westheimer, 1961; Semmlow et al, 1986, 1993; Zuber and Stark, 1968). The notable difference in their faster ramp responses (> 2 deg/sec) is that they do not exhibit the multi-step movements (which are more easily observed in the velocity traces) seen in the experimental responses (Hung et al., 1986).

9.2.2 Non-Continous Feedback Model

In contrast to continuous models, the development of the two-component (or non-continuous) theory was based on experiments that used specially-designed stimuli. For example, Westheimer and Mitchell (1956) noticed that the initial transient response amplitude could differ from that required for precise binocular fixation by as much as a degree or more. A subsequent slower movement reduced this error to attain accurate binocular fixation. Jones and Kerr (1971, 1992; Jones, 1980) used non-fusable targets, that is, a horizontal line segment presented to one eye and a vertical line segment presented to the other eye, and observed that the system initiated a transient vergence movement, with subsequent shift to the phoria or open-loop vergence position due to the lack of a fusable target. This was also shown by Semmlow et al (1986). Thus, non-fusable, dissimilar targets could initiate but not sustain the vergence response. They showed that the transient vergence component had a special property. They studied vergence responses to ramp disparity stimuli ranging from 0.7 degree/sec to 36 degree/sec with amplitudes of up to 4 degree. For lower ramp velocities (< 2 degree/sec), the vergence response consisted of a smooth following movement. Above 9 degree/sec, the vergence response was similar to a step response. When the ramp velocity was between 2 and 9 degree/sec, the vergence response consisted of a series of step responses. These small step responses had the same main sequence (peak velocity/amplitude) characteristics as the actual disparity vergence responses to step stimuli (Bahill and Stark., 1979). This result suggested a sampling and prediction mechanism, with the system predicting the future position of the target based on stimulus velocity.

9.2.2.1 Hung, Semmlow, and Ciuffreda Model

Hung et al. (1986) developed a dual-mode model of the vergence system, which consisted of a fast open-loop component and a slow closed-loop component. The fast component had both sampling and prediction mechanisms, which provided the staircase-like step responses to fast ramp stimuli. The slow component under negative feedback control accounted for the smooth following of slowly-moving stimuli with small residual error (Figure 9.14). Simulation responses to pulse, step-pulse, ramp (Fig. 9.15), and sinusoidal (Fig. 9.16) stimuli demonstrated good fit between the simulations and experimental results. For ramp stimuli, note the multiple-step movments to faster ramp stimuli in both experimental and model simulation responses (Fig. 9.15). Also, for both step and ramp stimuli, model simulation responses exhibited faster dynamics for convergence than divergence, similar to those in experimental responses (not shown, see Hung et al, 1986, 1997). Moreover, for the sinusoidal stimuli, note the step-like responses in the higher frequencies for both experimental and model simulation responses (Fig. 9.16).

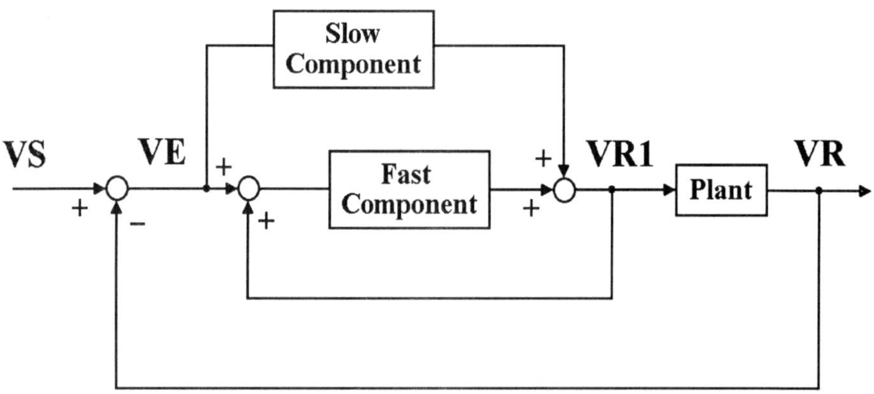

Figure 9.14. Hung *et al.* model (1986) of the vergence system. Slow and fast components are included in the forward loop of the model along with delay elements and plant. The sum of the slow and fast components, VR_1, provides a positive feedback to the fast component, as well as drives the plant. The vergence response from the plant is then compared with the vergence stimulus through the negative feedback loop to create an error signal, *VE*. Based on the VR_1, and vergence error *VE*, the fast component is able to estimate the position of target with its predictor and sampler. Reprinted from Hung et al (1986), pg. 1023, Fig. 1a, with permission of © IEEE.

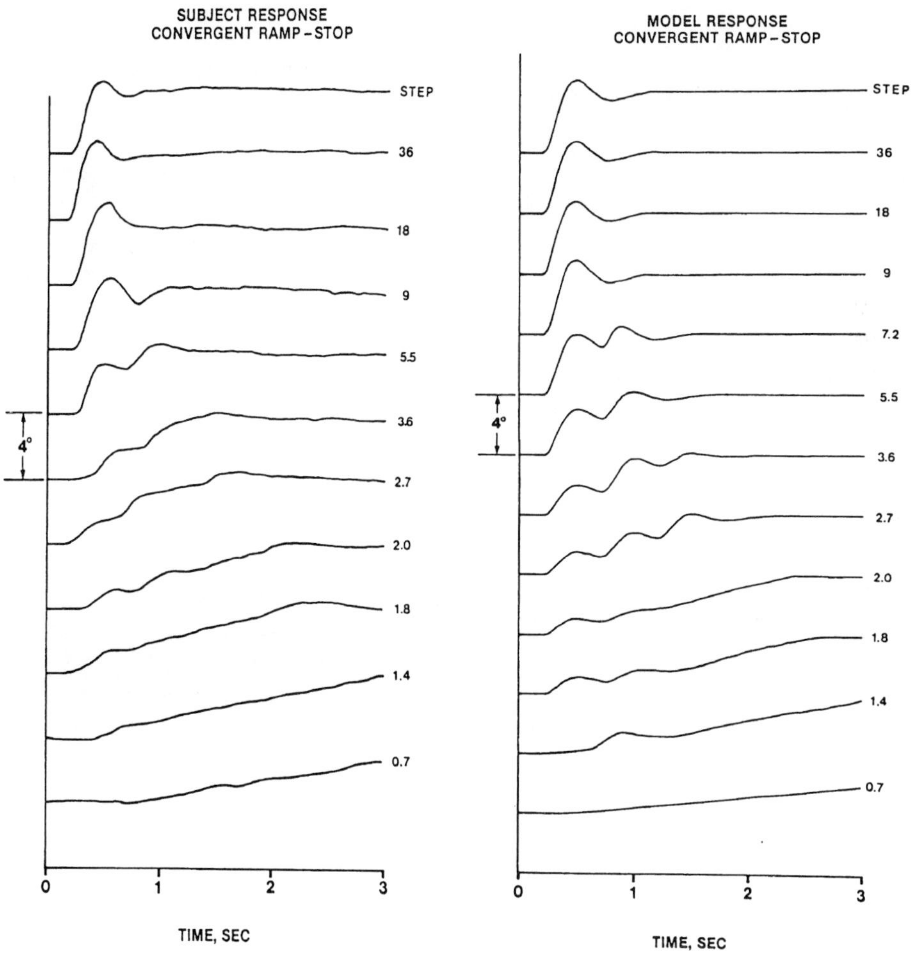

(a) (b)

Figure 9.15. Responses to convergent ramp stimuli (up to 4 deg amplitude) are shown for (a) experimental and (b) model simulation conditions. Experimental curves are individual responses. Ramp velocity, in deg/sec, is shown next to each curve. Top curves are for convergent step responses. Reprinted from Hung et al (1986), pg. 1025, Fig. 3, with permission of © IEEE.

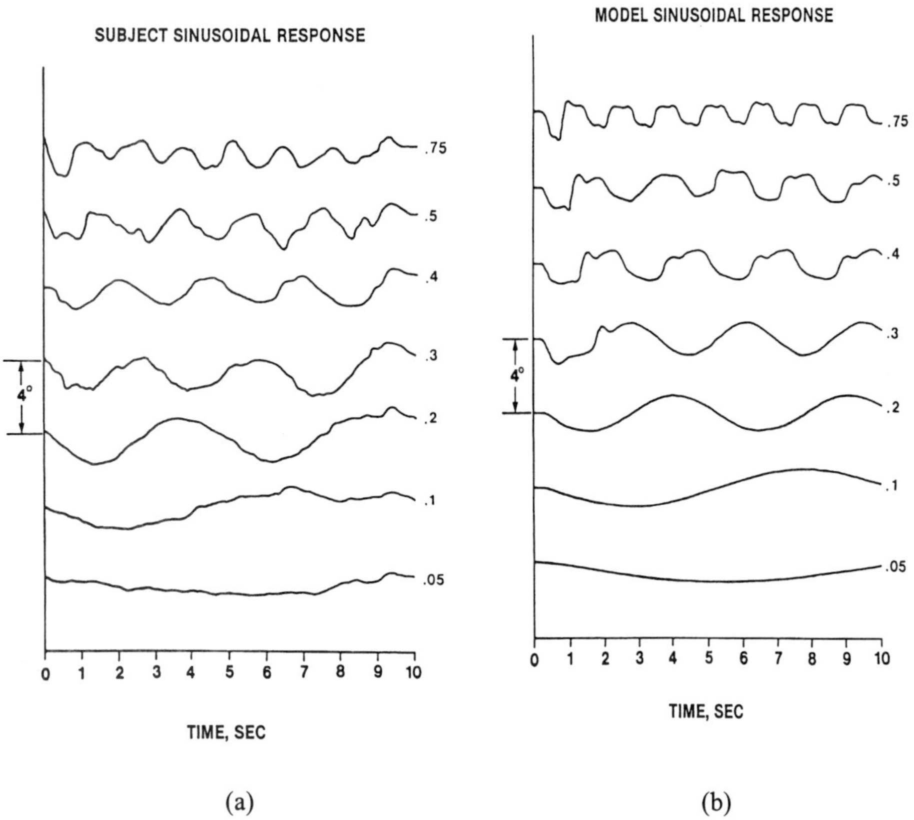

Figure 9.16. Sinusoidal responses of 2 deg peak-to-peak amplitude are shown for experimental (a) and model simulation (b) conditions. Experimental curves are individual subject responses. Sinusoidal frequency of the stimulus, in Hz, is shown next to each curve. Note that for higher frequency stimulation, the model responses exhibit step-like movements similar to those in the experimental responses, which also resemble multi-step movements to faster ramp stimuli (see Fig. 9.15). Reprinted from Hung et al (1986), pg. 1026, Fig. 5, with permission of © IEEE.

9.3 ACCOMMODATION-VERGENCE INTERACTIONS

9.3.1 Cross-Links and Tonic Components

The existence of interactions between vergence and accommodation has been known at least since the work of Porterfield in 1759 (Hofstetter, 1945). Similar to the Maddox (1893) classification of vergence components, Heath (1956) suggested that there were four components in the accommodative response: blur, vergence, tonic, and proximal accommodation. After feedback control theory was used to develop models of accommodation and vergence systems (Westheimer. 1963), a basic feature of all candidate models was that blur-driven accommodation and disparity-driven vergence were controlled by two negative feedback loops. And, interactions between the two systems were represented by two feed-forward cross-links from the controller outputs, so that the accommodative controller could initiate a vergence response (accommodative vergence or AC), and conversely, the vergence controller could initiate an accommodative response (vergence accommodation or CA). The accommodative convergence/accommodation (AC/A) ratio (Fry, 1939) measures the vergence magnitude produced by a unit change in accommodation and is expressed in terms of prism diopters (or meter angles) per diopter. The convergence accommodation/convergence (CA/C) ratio (Fincham and Walton, 1957) measures the accommodation magnitude produced by a unit change in vergence and is expressed in terms of diopters per prism diopter (or meter angle).

With progress in research on the tonic positions of accommodation and vergence (Owens, 1984), these tonic components have been added to the accommodation and vergence loops to represent the stimulus-free states of each system (Hung and Semmlow, 1980). Accommodative and vergence changes and final positions in darkness are thought to represent tonic accommodation and tonic vergence, respectively (Owens and Leibowitz, 1980). However, conflicting models had been suggested as to whether the tonic elements were located before (Ebenholtz and Fisher, 1982) or after (Schor and Kotulak, 1986) inputs to the feed-forward cross-links. If both reflex and tonic component signals arrived before the cross-links and drove the complementary system, then the accommodative (or vergence) response measured under the dual open-loop condition would represent not only a part of its own system's tonic level, but also a portion of the tonic vergence (tonic accommodation) as relayed by the cross-link. However, this was not seen experimentally. For example, Jiang (1996) found experimentally that higher dark focus (i.e., tonic values obtained in darkness) values were associated with lower, rather than higher, near dissociated phorias. This result suggested that the tonic elements were located after the cross-links.

Therefore, the values of accommodation and vergence with both feedback loops opened represent the tonic positions of the two systems.

9.3.2 Static Interactions Between Accommodation and Vergence

9.3.2.1 Hung and Semmlow Model

Hung and Semmlow (1980) suggested a model to simulate quantitatively the static behavior of accommodation and vergence. A block diagram of this model is shown in Fig. 9.17. The difference between the accommodative stimulus (AS) and the accommodative response (AR) forms the accommodative error (AE), or defocus, signal. To drive the accommodative controller, this error signal first has to pass a deadspace (DSP), a non-linear threshold element. The accommodative controller has a gain ACG. This static gain is based on the assumption that the accommodative system has an open-loop transfer function $ACG/(\tau s+1)$ as a first-order system. The output from the accommodative controller is summed with tonic accommodation (ABIAS) and vergence accommodation at a summing junction, and also provides an accommodative vergence signal to the vergence system through the cross-link. The output from the summing junction goes through a saturation element, which simulates the accommodative plant, to provide the

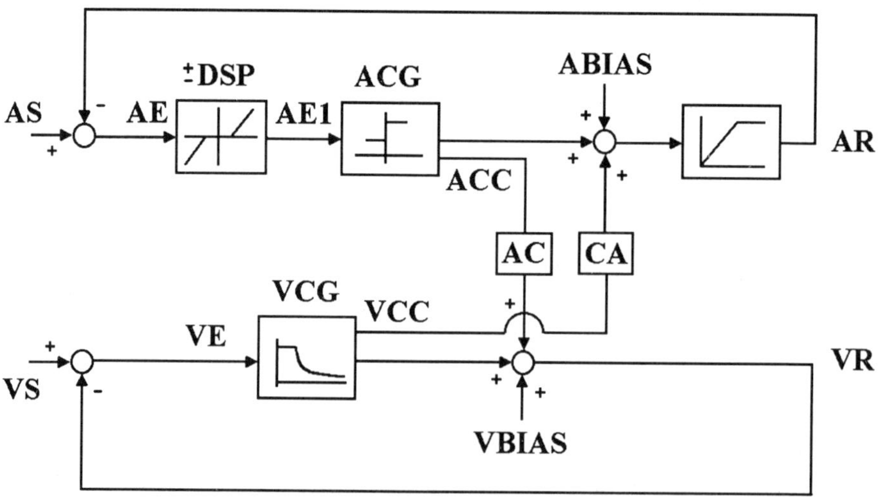

Figure 9.17 Hung and Semmlow (1980) model of the static interactive dual feedback accommodation and vergence systems. Reprinted from Hung and Semmlow (1980), pg. 441, Fig. 1, with permission of © IEEE.

accommodative response (AR). The difference between the vergence stimulus (VS) and the vergence response (VR) is the vergence error (VE). Static vergence error is referred to as fixation disparity under binocular viewing conditions. Vergence error goes through a vergence controller with gain VCG. The output from the vergence controller is summed with tonic vergence (VBIAS) and accommodative vergence to drive the vergence plant (omitted in the model) and provide the vergence response (VR). The cross-links receive a signal from one system's controller and link to another system at the point between the controller and tonic component with a gain AC or CA. Here, the gains AC and CA are proportional to AC/A and CA/C ratios, respectively, (see Hung and Semmlow, 1982).

The Hung and Semmlow model is concerned with short-term static responses. They demonstrated the ability of the model to predict human static near response behavior under various circumstances (Hung and Semmlow, 1980). They also investigated the controller sensitivities of accommodation and vergence to changes in stimulus (Hung and Semmlow, 1982). The results indicated that the contributions of the outputs of the accommodation and vergence controllers to accommodative and vergence responses in normal binocular vision were dependent on the individual's cross-link parameters, i.e. the AC and CA values. This suggested that normal AC and CA values lay in a continuum between the two extremes of accommodative-dominated (Maddox, 1893) and vergence-dominated (Fincham-Walton, 1957) control. This study (Hung and Semmlow, 1982) also indicated that the stability of the model (Hung and Semmlow, 1980) required the term $(1+ACG)(1+VCG)-AC*CA*ACG*VCG$ to not equal zero. When ACG and VCG are high enough to have $ACG \cong 1+ACG$ and $VCG \cong 1+VCG$, this stability requirement suggests that AC and CA should not be reciprocals of each other in subjects with normal binocular vision. More recent research has demonstrated that they have an inverse but not reciprocal relationship (Rosenfield et al, 1995).

It can be shown that under the open-loop vergence condition, the accommodative response AR, accommodative controller output ACC, and vergence response VR are given by (Hung and Semmlow, 1980):

$$AR = (AS \mp DSP) * \frac{ACG}{1+ACG} + ABIAS * \frac{1}{1+ACG} \tag{9.6}$$

$$ACC = (AS \mp DSP - ABIAS) * \frac{ACG}{1+ACG} \tag{9.7}$$

and

$$VR = AC * ACC + VBIAS \tag{9.8a}$$

or

$$VR = AC * (AS \mp DSP - ABIAS) * \frac{ACG}{1 + ACG} + VBIAS \tag{9.8b}$$

9.3.2.2 Jiang and Woessner Analysis of the Hung and Semmlow Model

Jiang and Woessner (1996) defined AR_0 and VR_o as the accommodative and vergence responses, respectively, under the vergence open-loop condition and for AS = 0. Thus, from Eq. 9.6,

$$AR_0 = (\mp DSP) * \frac{ACG}{1 + ACG} + ABIAS * \frac{1}{1 + ACG} \tag{9.9}$$

and from Eqs. 9.8b,

$$VR_0 = AC * (\mp DSP - ABIAS) * \frac{ACG}{1 + ACG} + VBIAS \tag{9.10}$$

It can be shown that by substituting Eqs. 9.6, 9.9 and 9.10 into Eq. 9.8b, and re-arranging, we obtain

$$VR = AC * (AR - AR_0) + VR_0 \tag{9.11}$$

Jiang and Woessner (1996) also sought to find the open-loop vergence response for the specific condition on the AS/R curve when

$$AR = ABIAS \tag{9.12}$$

Substituting Eq. 9.12 into Eq. 9.6 and rearranging gives

$$ABIAS = AS \mp DSP \tag{9.13}$$

Substituting Eq. 9.13 into Eq. 9.7 results in an accommodative controller output (ACC) value of zero, which is normally associated with open-loop of accommodation. However, the above equations are based on closed-loop of accommodation. A possible solution to this dilemma is that the accommodation system, under the conditions of AR=ABIAS and open-loop vergence, may operate just at the boundary between being open- and closed-loop (and eventually settling on the nearest closed-loop value). Thus, greater variability would be expected under this condition. Supportive evidence for this can be seen in the data of Kotulak and Schor (1968b; pg. 226, Fig. 4), which showed in 2 of 3 monocularly viewing subjects a slightly larger accommodative variability for a stimulus near the subject's tonic level.

The zero ACC value results in a zero crosslink drive, and thus VBIAS is the only remaining drive of vergence (see model, Fig. 9.17). Indeed, substituting Eq. 9.13 into Eq. 9.8b gives

$$VR = VBIAS \tag{9.14}$$

Substituting Eqs. 9.12 and 9.13 into Eq. 9.11 gives

$$VBIAS = AC * (ABIAS - AR_0) + VR_0 \tag{9.15}$$

Using the more descriptive terms tonic vergence (TV) for VBIAS, and tonic accommodation (TA) for ABIAS, and also calling VR_0 the distance heterophoria (Jiang and Woessner, 1996), the equation becomes

$$TV = AC/A * (TA - AR_0) + distance\ heterophoria. \tag{9.16}$$

The authors experimentally verified this equation. Before this study, O'Shea *et al.* (1988) and Wolf *et al.* (1990) predicted the TV from measurements of the TA, the distance heterophoria, and the AC/A ratio; Owens and Tyrrell (1992) predicted the distance heterophoria from measurements of the TV, the AC/A ratio, and the accommodative relaxation from the TA to the distant target. The Jiang and Woessner (1996) study extended the results of these studies to a general situation and gave these results a uniform explanation.

9.3.2.3 Sensitivity Analysis of Relative Accommodation and Vergence

To investigate the effect of parameter variation on accommodative and vergence responses, a sensitivity analysis was performed on the static dual-interactive model (Hung and Ciuffreda, 1994). Model simulation responses were computed under two conditions using parameter values obtained in a previous study (Hung and Semmlow, 1980). The first condition was to calculate the accommodative error when the accommodative demand was systematically varied between ± 2.5 D, while the vergence demand was held at 2.5 MA. The second condition was to calculate the vergence error when the vergence demand was systematically varied between 25 BI and 25 BO, while the accommodative demand was held at 2.5 D. Moreover, each parameter was repeated at 50% and 150% of the nominal value, while all other parameters were held constant at their nominal values.

The results indicated that the model was most sensitive to variations in the cross-link gains AC and CA. These gain elements interconnect the two feedback systems, and thus determine the extent of mutual interaction between accommodation and vergence. Since the output from each controller is multiplied with its cross-link gain element to serve as an input to the fellow system, it effectively increases these interactive influences. That is, the multiplicative effects of each parameter, both within and across each motor system, contribute to and therefore influence the overall response to achieve steady-state system stability. It can be shown (Hung and Ciuffreda, 1994) that the main effect of increased AC or CA is to increase the accommodative and vergence errors, with this occurring to a much greater extent than from variation in the other parameters. Moreover, high AC and CA values may lead to instability in the interactive system and result in abnormalities such as strabismus (Hung and Semmlow, 1982)

On the other hand, it was found that the model was only moderately sensitive to variations in the controller gains ACG and VCG. This can be explained by the fact that for an isolated feedback control system, the overall closed-loop gain attributed to the controller is of the form G/(1+G). Such a closed-loop term is inherently only moderately sensitive to changes in G. Indeed, it is a basic feature of a negative feedback system to maintain stability despite larger fluctuations in the controller gain. In addition, since each controller primarily governs its own feedback loop and has only an indirect influence on the fellow loop, changes in gain would have a relatively small influence on the fellow system's response. However, low controller gain values, such as a low ACG, may be associated with central deficits such as amblyopia and congenital nystagmus (Hung and Semmlow, 1982; Hung and Ciuffreda, 1994).

Also, the model was only moderately sensitive to changes in the tonic terms ABIAS and VBIAS, and was quite insensitive to variations in the deadspace elements. This is in agreement with earlier experimental results (Ogle, 1972; Ogle, *et al.*, 1967; Ripps *et al.*, 1962).

Blackie and Howland (2000) used a state-space technique (i.e., a derivative-based simultaneous-equation mathematical approach) to performed a stability analysis of the Hung and Semmlow (1980) model. Consistent with Hung and Semmlow's (1982) earlier results, they found that the dual-interactive feedback system became unstable when the product AC*CA was greater than one.

9.3.2.4 Deadspace, a Nonlinear Element in the Accommodation System Model

The deadspace (DSP) in the control theory model of accommodation represents the depth-of-focus of the eye (Hung and Semmlow, 1980). Kotulak and Schor (1986a) found that the defocus threshold for eliciting a motor response from accommodation was about 0.12 – 0.14 D. Based on this result, Schor (1992) suggested that the value of DSP (or threshold) in his model should be much smaller than the depth-of-focus. Hung and Ciuffreda (1994) called the threshold measured with the Kotulak and Schor method (or a similar method) the "objective threshold", and they called the threshold based on the probability of detection the "subjective threshold". In reviewing previous studies, they found that the subjective-threshold was consistently larger than the objective-threshold. Mordi and Ciuffreda (1998; Mordi, 1991; Ciuffreda et al, 2000) confirmed this finding, and moreover extended it to show that the subjective threshold increased with age whereas the objective threshold did not.

In the case where vision is not normal (i.e., worse than 20/20 visual acuity), the effective threshold for the depth-of-focus may increase. To investigate this quantitatively, Jiang (2000a,b) modified the deadspace element in the Hung and Semmlow (1980) model. He first clarified the definitions of different defocus thresholds and discussed the differences between these thresholds. The depth-of-focus of the human eye is not only related to the optical system of the eye, but also the sensory system of the eye. Quantitatively, the depth-of-focus can be estimated by using both geometric and physical optics methods (Green, *et al.*, 1980). One equation is:

$$\Delta D = 17.45\phi / p, \tag{9.17}$$

where ΔD is the depth-of-focus in diopters, p the pupil diameter in millimeters, and ϕ the minimal resolvable angle in degrees. This equation shows that the depth-of-focus is related to both pupil size and visual resolution of the eye. Jiang (2000a,b) suggested naming the depth-of-focus measured with a psychophysical procedure the "perceptual defocus threshold" (i.e., the subjective-threshold in the previous example) to distinguish it from the depth-of-focus estimated from the above equation. The depth-of-focus or perceptual-defocus-threshold was thought to be the non-linear operator (DSP) in control theory models that simulate the static behavior of accommodation (e.g., Hung and Semmlow, 1980). Any defocus signal has to be larger than DSP to drive the controller and change the accommodative response. Based on these models, DSP does not affect the slope of the accommodative response function, which is only determined by the gain of the accommodative controller (ACG). However, not all changes in the slope of the AS/R function are related to changes in the ACG alone. For example, when the pupil size becomes very small (≤ 1 mm), the accommodative feedback loop is opened, and the slope of the AS/R function becomes flat. In this case, the only change is an increase in the optical depth-of-focus. Jiang (1997) suggested adding a linear operator with gain ASG (accommodative sensory gain) in front of DSP to account for the sensory performance of the accommodative system (see accommodation model in Chap. 8 of this volume). From this model, he derived the following equation:

$$AR = K * (AS - \frac{DSP}{ASG}) + (1 - K) * ABIAS \tag{9.18}$$

where AS is the accommodative stimulus, AR the accommodative response, ABIAS the tonic accommodation, and K (= (ASG*ACG)/(1+ASG*ACG)) the slope of the AS/R function. These parameters can be measured experimentally. The ratio (DSP/ASG) serves the same role as DSP in the Hung and Semmlow (1980) model, and represents the amount of retinal defocus signal (error) needed by the accommodative system to maintain the static response. Jiang and his co-workers found that this threshold for emmetropic subjects was about 0.40–0.45 D and for myopic subjects, whose refractive errors were progressing, was about 0.78-0.88 D (Jiang, 1997; Jiang and Morse, 1999). Jiang (2000a,b) named this the sensory-motor threshold.

He defined the "accommodative stimulus/response (AS/R)", threshold as the smallest change in accommodative stimulus that creates a detectable change in accommodative response (i.e., the objective-threshold; Mordi, 1991; Mordi and Ciuffreda, 1998). This objective threshold is different from the subjective perceptual-defocus threshold because it depends on the accommodative response instead of the subjectively perceived blur (Hung and Ciuffreda, 1994). From the modified model (see Eq. 9.17; Jiang, 1997), it can be seen that the change in response depends only on the change in stimulus and the slope of the accommodative AS/R function, i.e. that

$$\Delta AR = K \Delta AS, \tag{9.19}$$

where Δ represents a change. In support of this, Winn et al. (1989) had shown that the threshold for the detection of defocus was close to the root-mean-square (r.m.s.) value of the accommodative microfluctuations. Their result suggested that the change of accommodative response must be larger than the intrinsic noise level of the system for it to be detected. Also, Wong and Jiang (2000a,b) found a high correlation between the AS/R threshold and the average standard deviation of the accommodative response measurements among 36 subjects ($r = 0.70$, $p < 0.005$).

Jiang's distinction between these thresholds is supported by the finding that the AS/R threshold (i.e., the objective threshold) is smaller than the perceptual-defocus threshold (i.e., the subjective threshold) (Mordi and Ciuffreda, 1998). Moreover, the objective threshold was found to be unchanged with age, whereas the subjective threshold increased at a rate of 0.027 D/year (Mordi and Ciuffreda, 1998). These results indicate that the oculomotor mechanism associated with the objective threshold is different from the perceptual-defocus process associated with the subjective threshold.

9.3.2.5 Hung, Ciuffreda, and Rosenfield Proximal Model of Accommodation and Vergence

The Hung et al (1996) proximal model of accommodation and vergence is shown in Fig. 9.18. For the first time, it takes into account the contribution of proximity, or awareness of nearness (Toates, 1974; Maddox, 1893), on the total system steady-state response. The positions of the components in the model were based on open- and closed-loop accommodative and vergence experimental results (Hung and Semmlow, 1980; Fisher, 1988; Rosenfield and Gilmartin, 1990; Rosenfield et al, 1991).

The interactive model equations were solved in the form of a ratio between proximal contribution and either the overall accommodative or vergence response. It was found that the relative contribution of proximal accommodation to the overall accommodative response under accommodation CL and vergence OL was given by:

$$\left[\frac{\text{Proximal Accommodation Term}}{\text{Overall Accommodative Response}}\right]_{\text{Acl, Vol}} = \frac{0.0409 \ * \text{DS}}{0.950 \ * \text{DS} + 0.0555} \tag{9.20}$$

where DS is the distance stimulus. Also, the relative contribution of proximal accommodation to the overall accommodative response under accommodation CL and vergence CL was given by:

$$\left[\frac{\text{Proximal Accommodation Term}}{\text{Overall Accommodative Response}}\right]_{\text{Acl, Vcl}} \tag{9.21}$$

$$= \frac{[\text{APG}*(1+\text{VCG}) - \text{APG}*\text{VCG}*\text{AC}*\text{CA} + \text{VPG}*\text{CA}]*\text{PDG}*\text{DS}}{\begin{gathered}[\text{ACG}*(1+\text{VCG}) - \text{ACG}*\text{VCG}*\text{AC}*\text{CA}]*\text{AS} + \text{VCG}*\text{CA}*\text{VS} \\ + [\text{APG}*(1+\text{VCG}) - \text{APG}*\text{VCG}*\text{AC}*\text{CA} + \text{VPG}*\text{CA}]*\text{PDG}*\text{DS} \\ + (1+\text{VCG})*\text{ABIAS} - \text{VCG}*\text{CA}*\text{VBIAS}\end{gathered}}$$

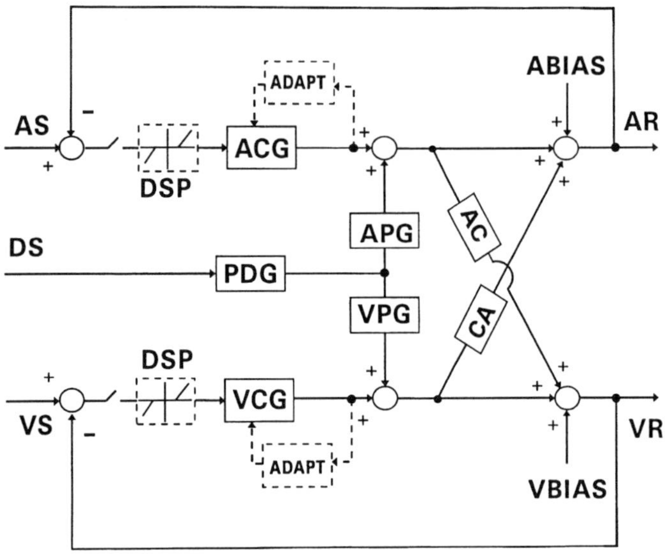

Figure 9.18. Hung, Ciuffreda, and Rosenfield (1996) proximal model of accommodation and vergence. Most of the model parameters are the same as those in Fig. 9.17. For the proximal components, the distance stimulus (DS) is input to the perceived distance gain (PDG) element, which represents the subjective apparent distance estimate. It then goes through the accommodative proximal gain (APG) and vergence proximal gain (VPG) elements, which represent the contribution from target proximity to the two systems, respectively. Moreover, the outputs of these elements are summed with ACG and VCG, respectively, and the summed signals proceed in the forward-loop paths. The adaptation component (ADAPT) is shown in this static model for completeness only. Reprinted from Hung et al (1996), pg. 32, Fig. 1, with permission of Elsevier Science.

Similar expressions were obtained for the vergence contribution under the different conditions. The definition of the parameters and their nominal values are given in Table 9.1. Upon substitution of the parameter values from Table 9.1 into the model equations, it was found that under the dual open-loop (OL) condition, the contribution of proximal accommodation (PA) to the overall accommodative output ranged from 42.5 to 81.6%, whereas the contribution of proximal vergence (PV) to the overall vergence output ranged from 56.1 to 88.5%. In contrast, under nearly all other stimulus conditions, the relative contributions were much smaller, ranging from 0.04 to 7.0%. These results are consistent with experimental and clinical findings (Hokoda and Ciuffreda, 1983; Wick, 1985; and North et al, 1993). Thus, although the relative contributions of PA and PV were large under the dual-OL condition, they were generally very small under the various closed-loop conditions that simulated more naturalistic viewing situations. Nevertheless, Hung et al (1996) suggested that proximity still plays an important role, with these cues reinforcing the dominant blur and disparity motor responses.

Table 9.1 - Proximal Model Parameter Values

Reprinted from Hung et al (1996), pg. 34, Table 1, with permission of Elsevier Science.

Perceived Distance Gain	PDG	0.212
Accommodative Proximal Gain	APG	2.100
Vergence Proximal Gain	VPG	0.067
Accommodative Controller Gain	ACG	10.0
Vergence Controller Gain	VCG	150.0
Accommodative Convergence	AC	0.80 MA/D
Convergence Accommodation	CA	0.37 D/MA
Tonic Accommodation	ABIAS	0.61 D
Tonic Vergence	VBIAS	0.29 MA

9.3.2.6 Eadie, Carlin, and Gray Fuzzy Set Model of Accommodation and Vergence

Eadie et al. (1999) applied a fuzzy set approach to represent the accommodative and vergence proximal gain elements such as those in Fig. 9.18 (see Fig. 9.19). Fuzzy logic control is a mapping procedure which relates the input to the output using a fuzzy set (Zadeh, 1965, 1994). A list of "If... then..." statements define the fuzzy set. For example: "If distance is far, then proximal accommodation is low; If distance is near, then proximal accommodation is high." The relative impreciseness, or fuzziness, of the input-output relationship provides a response characteristic, over the range of normal response variability, that is meant to be a more realistic representation of the actual physiological control process. Their simulation responses showed that the proximal contribution was high under the dual open-loop condition, but was negligible under the dual closed-loop condition, similar to that found previously by Hung et al. (1996).

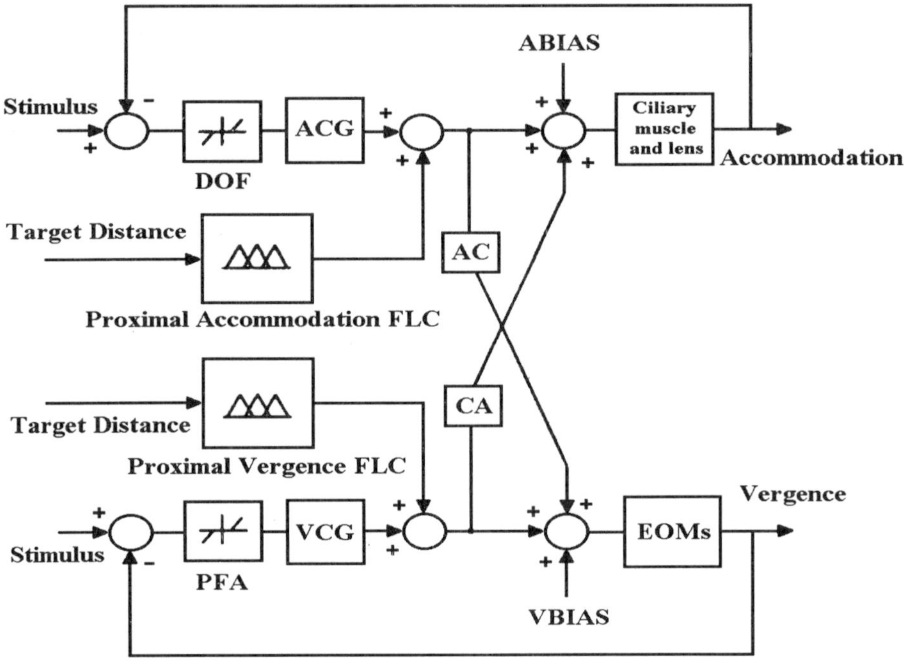

Figure 9.19. Eadie, Carlin, and Gray (1999) fuzzy set model of accommodation and vergence. Most of the model parameters are the same as those in Fig. 9.18. Proximal accommodation and vergence fuzzy logic control (FLC) replace PDG*AGP and PDG*VPG in Fig. 9.18. Reprinted from Eadie et al. (1999), pg. 179, Fig. 1, with permission of Plenum Press.

9.3.3 Dynamic Interactions Between Accommodation and Vergence

9.3.3.1 Schor Model

In 1992, Schor developed his dynamic model of the interaction between accommodation and vergence (Fig. 9.20). This model differed from his previous model (Schor, 1979) in that, in the fast component part of this model, an additional proportional gain path (K_b) was added in parallel with the leaky integrator to provide a fast-slow combination for the phasic response. But as the simulation of his earlier model showed (see above), this may not actually provide a fast-slow combination of movements in the response. The output clipper element for the fast integrator in the previous model was removed. In the slow component part, the saturation limit non-linear element, located after the slow integrator, remained. A non-linear filter that served as a stimulus funnel was added to limit the input to the slow integrator. Both the fast and slow components used leaky integrators. These integrators were characterized by their gains (*K*) and decay time constants (*1/a*). In the plant part, the new model used first-order control elements for both accommodation and vergence plant mechanisms. The configuration and parameters of this model were justified based upon his observations of the dynamic behavior of accommodation, vergence, and their cross-link interactions (Schor, 1992). In contrast to the static models (Hung and Semmlow, 1980; Schor, 1985), this model has only unity-gain elements in the cross-links. However, Schor (1999) recently has added non unity-gain AC/A and CA/C elements into the cross-links of his model.

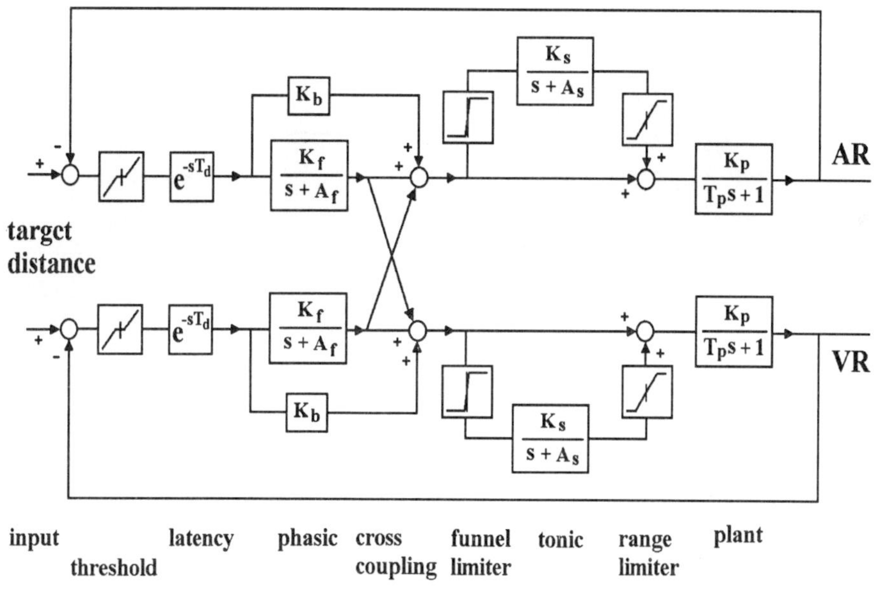

Figure 9.20. Schor (1992) dynamic interactive model of accommodation and vergence. This model includes phasic and tonic control components. In the tonic component, signal passes a funnel limiter first before it reaches the tonic integrator, then the signal is limited by a saturation limiter. Both limiters are non-linear elements in the model. Reprinted from Schor (1992), pg. 261, Fig. 1, with permission of Optom. Vis. Sci.

Schor's model simulation closed-loop step responses are shown for accommodation and vergence in Fig. 9.21. The vergence responses exhibit high frequency oscillations. Although Schor claims that these high frequency oscillations are a characteristic of the closed-loop step response (Schor, 1992), they are not observed experimentally (Rashbass and Westheimer, 1961; Hung et al, 1986; see Figs. 9.3 and the step response in Fig. 9.15). Moreover, no pulse, ramp or sinusoidal simulations were presented.

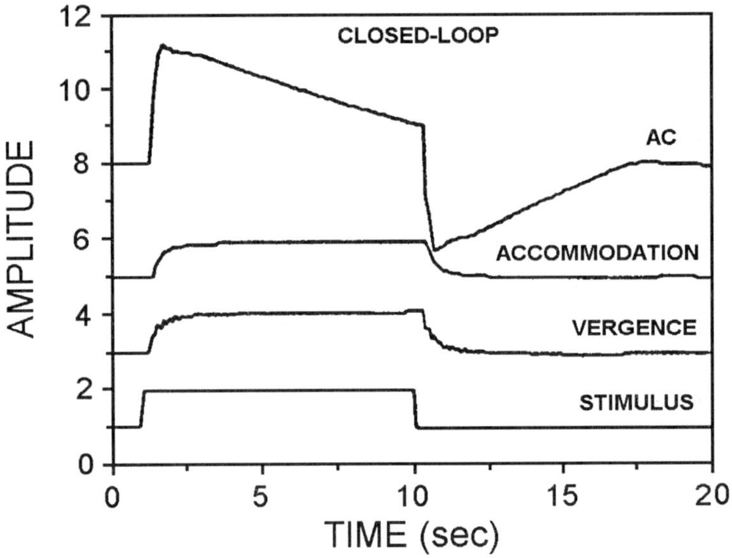

Figure 9.21. Schor (1992) model simulation of closed-loop step responses for accommodation and vergence. Also shown is the accommodative convergence drive (AC). Note the rapid oscillations in the vergence response, which is not seen experimentally (Rashbass and Westheimer, 1961; see Fig 9.3). Reprinted from Schor (1992), pg. 263, Fig. 2, with permission of Optom. Vis. Sci.

9.4 ADAPTATION MODELS OF ACCOMMODATION AND VERGENCE

Both accommodation and vergence have been shown to exhibit adaptation after extended near viewing. Normally, when the stimulus to accommodation is removed, the accommodation system returns rapidly towards its tonic position. However, if the stimulus is removed after sustained focusing effort, the decay is much slower, and in some cases the initial steady-state level is directionally biased for a substantial period of time. A similar effect can be observed in the vergence system. For example, after prolonged wearing of horizontal prisms, occlusion of one eye results in a much slower decay of the vergence output towards its tonic value.

9.4.1 Schor Model

In the slow component of his model, Schor (1992) used a leaky integrator with two non-linear elements, a funnel limiter and a range limiter, to simulate the adaptive process. Patel (1995) used the vergence part of this model (Schor, 1992) to simulate the open-loop response and found that if a constant disparity of 0.1 MA were held for 20 seconds, the vergence output would saturate at 1.25 MA in conjunction with the vergence velocity signal decaying to zero. If the disparity stimulus were 1 MA, then the open-loop response would saturate at about 4.5 MA. However, this saturating phenomenon in the simulations is not consistent with experimental observation. Furthermore, Hung (1992b) investigated the overall transfer function of Schor's model and found that the form of the transfer function did not exhibit adaptive properties.

9.4.2 Hung Model

Hung (1992b) developed a model to simulate accommodative and vergence adaptive responses. The model was based on the static dual-interactive model of the accommodation and vergence systems (Hung and Semmlow, 1980). The components of the basic accommodation and vergence systems are similar. Both contain a controller in the forward-loop. The controller consists of a fast and a slow component. The fast component drives the initial dynamic portion of the response, whereas the slow component maintains closed-loop feedback of the steady-state level (Westheimer and Mitchell, 1956; Jones, 1980; Hung and Ciuffreda, 1988). The tonic component represents the value of accommodation or vergence when the feedback loop is opened (Toates, 1972; Hung and Semmlow, 1980). To account for the change in decay characteristics following sustained fixation, an adaptive component is incorporated which receives its input from the output of the controller. This is consistent with the concept that adaptation is related to the effort of accommodation or vergence (Hung, 1998b). Neurologically, adaptation may take place in the cerebellum (Windhorst, 1999) in response to the sustained effort, which in turn modifies the neural integrator circuitry controlling the oculomotor system. The unique feature of the model is that the sustained element time constant is modified by the adaptive component output. Thus, in the accommodation and vergence sub-systems, as the controller output increases with increased effort, the adaptive component increases the time constant of the sustained element. If at that point the feedback loop is opened, the decay of the sustained element will take longer, thereby simulating the longer decay time following prolonged adaptation found experimentally.

The combined accommodation and vergence adaptation model is shown in Fig. 9.22 (Hung, 1992b). Consider first the accommodative loop. The deadspace element (DE) represents the neuro-optical depth-of-focus. The controller output is multiplied by factor m_A and input to a tanh function, which serves as a compression element (CE). The factor m_A is used to provide an appropriate range on the abscissa of the tanh function. The compression element reduces the controller output for large magnitude inputs, so that the adaptation effect is not drastically different at various adapting stimulus levels. The adaptive component is represented by the first-order dynamic element

$$\frac{1}{T_{A1}s+1} \tag{9.22}$$

where T_{A1} is the time constant of accommodative adaptation. The accommodative adaptation gain, K_A, controls the magnitude of the adaptive

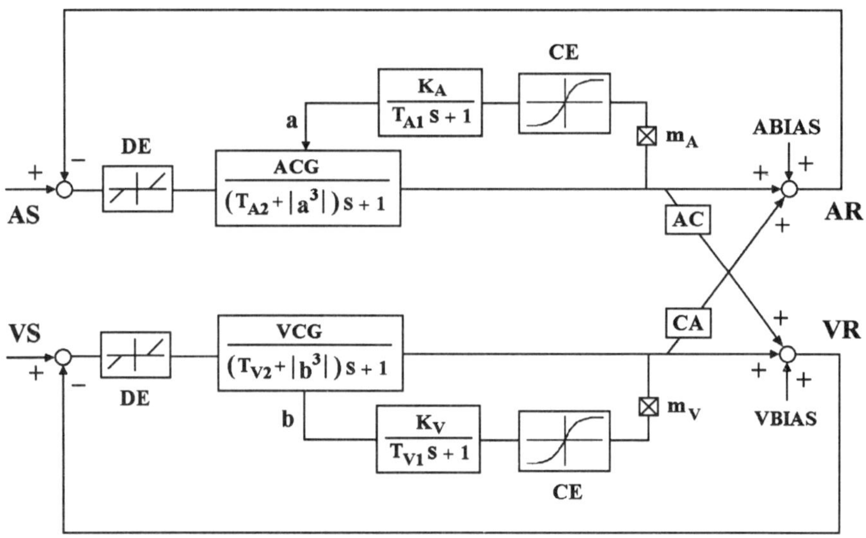

Figure 9.22. Hung (1992b) adaptation model of combined accommodation and vergence system. The time constants T_{A2} and T_{V2} of the accommodative and vergence controllers, respectively, are modified by their adaptive components. In each feedback loop, the adaptive component consists of a constant gain (m_A or m_V) element, a compression element (CE), and a first-order dynamic controller. Reprinted from Hung (1992b), pg. 322, Fig. 3, with permission of Elsevier Science.

component output level. The adaptive element output, *a*, modifies the time constant of the accommodative controller via the term

$$T_{A2} + \left| a^3 \right| \tag{9.23}$$

where T_{A2} is the fixed portion of the time constant. The cubic relationship was obtained empirically to provide a negligible increase in the time constant for a small amount of adaptation, but a very long time constant for a large amount of adaptation. A similar configuration applies to the vergence system, where the deadspace element (DE) represents Panum's fusional area. The vergence adaptive component consists of multiplier *m*, compression element CE, adaptive gain K_V, adaptive time constant T_{V1}, adaptive element output *b*, and controller time constant $T_{V2}+ \left| b^3 \right|$.

A number of adaptation experimental results (Henson and North, 1980; Sethi and North, 1987) were accurately simulated using parameter values of the subjects in a previous study (Hung, 1992b). The adaptation simulation results can also explain how accommodative hysteresis occurs. Following sustained near viewing, the adaptive element output remains high as it begins its decay towards the tonic level, or ABIAS. Hence, immediate post-task accommodative response measured in the dark is higher than the tonic value. On the other hand, following sustained far viewing, the adaptive element output remains low as it begins to rise towards the ABIAS value, and the immediate post-task response is lower than the tonic value. This dependence of the post-task accommodative value on the path, or initial condition (either near or far sustained viewing), gives rise to the phenomenon of accommodative hysteresis (Fisher et al, 1987).

9.4.3 Nearwork-Induced Transient Myopia (NITM) Model

The development of myopia has both genetic and environmental components (Goss and Wickham, 1995). Although genetic factors appear to play a larger role in early-onset myopia (myopia with onset before age 15 years), modern-day work styles clearly demonstrate that environmental factors may play a significant role in the development of late-onset myopia (myopia with onset after age 15 years). A particularly important environmental factor is that of prolonged nearwork, which has been especially implicated in the development of late-onset myopia (Ong and Ciuffreda, 1995, 1997).

Quantitative measures of oculomotor parameters and dynamic responses would be helpful to differentiate between, and perhaps even predict, those who will develop myopia versus those who will not. Such a measure can be found by stimulating the accommodation system during near viewing, which produces lenticular-based pseudo-myopia, and then measuring the closed-loop temporal course of decay of the lens response back to the original far point of accommodation. This is referred to as the nearwork-induced transient myopia (NITM) paradigm (Ehrlich, 1987; Rosenfield, *et al.*, 1992a; Ong and Ciuffreda, 1995, Jiang, 1999), with the difference between post- and pre-task values representing the NITM.

The Hung adaptation model (Hung, 1992b) discussed above quantified the effect of prolonged nearwork on the accommodative response. This model served as the basis for the simulation of NITM dynamics (Hung and Ciuffreda, 1999). An important parameter in the model, the adaptive gain, K_A, was used previously to modify the time constant of the accommodative controller, and thus controlled the rate of decay in the dark following an adaptation period. Simulations were therefore performed to determine whether variation in K_A could also account for the differences in the dynamic decay timecourse in the light following near work in the different refractive groups. In addition, the computer-simulated effect of higher K_A values on retinal defocus was examined over a 160 hr period, thus representing one work-month with 40 hours of nearwork per week, to assess its influence on the long-term development of myopia. The accommodative adaptation gain played a crucial role in inducing NITM due to its effect on the accommodative controller time constant. That is, after sustained nearwork, the increased accommodative adaptation element output would result in an increase in controller time constant. This in turn would result in a slower than normal return, or decay, of the NITM toward the pre-task baseline. Larger adaptive element gains produce slower decay rates, so that different gain values could be used to simulate NITM in the various refractive groups. It was found that the K_A values of 2.0, 2.5, 4.0 and 5.5 simulated reasonably accurately the experimental NITM time courses (Ciuffreda and Wallis, 1998) for subjects with hyperopia (HYP), emmetropia (EMM), early-onset myopia (EOM), and late-onset myopia (LOM), respectively. For simulation of the HYP group, an additional constraint was imposed wherein the accommodative response to distant stimuli was biased on the under-accommodated side of the deadspace operator. This was done for consistency with the experimental results (Ciuffreda and Wallis, 1997). In contrast, for the other three refractive groups, no such constraint was imposed so that the accommodative responses exhibited the normal slight (~ 0.25 D) over-accommodation for the far target (Rosenfield, *et al.*, 1992b).

In addition, the model was used to investigate long-term effects of nearwork. There are typically alternating periods of prolonged nearwork and brief distance viewing, which is representative of our everyday activities. Under this condition, both under-accommodation at near (lag of accommodation) and over-accommodation at far (lead of accommodation) typically occur (Ciuffreda, 1991, 1998). A useful measure of the long-term effect of the resultant retinal defocus on an individual, regardless of how it is generated, is that of the root-mean-square (rms) of the accommodative error. The rms error is essentially equal to the standard deviation about the mean value, so that a larger value is associated with greater variability in the retinal defocus. This measure was used in the prolonged nearwork simulation. This was simulated by alternating 1 hr of nearwork (3 D, 3 MA) and 5 min of far viewing (0.25 D, 0.25 MA) over a 160 hr period, which represents one work-month with 40 hours of nearwork per week. The final steady-state rms value of the overall (i.e., combined for distance and near conditions) accommodative error was measured and plotted as a function of K_A. The results show a small but progressive increase in the rms of the accommodative error with increased K_A, hence implicating this parameter in the development of myopia.

The emphasis of this model-based study was the role of the adaptive component on the NITM time course. Indeed, the NITM simulation results demonstrated that the accommodative adaptation gain K_A was clearly different for the four refractive groups. The lower adaptive gain values (K_A = 2.0 and 2.5) corresponded to the HYP and EMM groups, respectively, whereas the higher adaptive gain values (K_A = 4.0 and 5.5) corresponded to the EOM and LOM groups, respectively. The adaptive element represented a neural-oculomotor feedback process that controlled the effect of sustained stimulation of the accommodative system during nearwork. The slowed decay of the distance accommodative response following prolonged nearwork was due to an increase in output from the adaptive element, thus resulting in an increase in the accommodative controller time constant. This decay was slower for larger K_A values. The long-term increase in exposure to accommodative error, and the resultant retinal defocus, may induce increased axial length and in turn produce permanent myopia (Ong and Ciuffreda, 1997; Jiang, 1997, 1999).

9.5 SUMMARY

This chapter has provided an extensive overview of the application of bioengineering techniques to the study of the vergence system and the interactions between vergence and accommodation. In the accommodation and vergence systems, static linear and nonlinear elements serve important roles in shaping the steady-state responses and providing insight into clinical abnormalities. For example, decreased lens saturation level while maintaining normal ACG accurately simulates the AS/R curve as a function of age (Ciuffreda, 1998). Also, decreased accommodative controller gain and increased sensory-motor threshold of accommodation are associated with the visual deficits of amblyopia and congenital nystagmus, respectively.

Moreover, high AC or CA crosslink gains can lead to frank eye misalignment called strabismus, or more subtle binocular dysfunction of fusion (Hung and Ciuffreda, 1994).

Dynamic characteristics of these models have provided important insights into how these systems attain both stability and rapid motor responsivity. When each system is studied in isolation, its response characteristics provide fundamental clues regarding the system's neural control strategy. For both the accommodation and vergence systems, whose sensory latencies are long relative to their motor dynamics, a continuous feedback control process would lead to instability oscillations. It turns out that the strategy used is to respond with an initial fast open-loop movement that provides a large portion of the response amplitude, followed by a slow closed-loop movement that reduces the residual error to a minimum. In this way, both dynamic responsivity and accuracy are attained without introducing instability oscillations. However, when these systems operate together, as is generally the case in daily life, their responses are not just simple summations of their isolated motor responses, but are rather nonlinear in nature. For example, the neural linkage between the accommodation and vergence control processes results in a combined dual-interactive feedback control system that is quite complex. Additional complexity is introduced when the model is used to investigate the adaptive control of accommodation and vergence, as well as the related processes that may be involved in refractive error development. It was shown that these problems could be both conceptualized and solved using engineering feedback control systems techniques.

In the past three decades, much has been learned from these models regarding normal oculomotor control processes. Future extensions of the models include the detailed quantitative investigations of the development of myopia and ocular abnormalities such as strabismus and amblyopia, the adaptive control in the cross-links of the vergence and accommodation systems, and the interactions between visual, auditory, and perceptual cues in complex multi-media displays as well as in real-life situations. Finally, computer simulations of neurons and neural networks have the potential of improving the traditional techniques in modeling the accommodation and vergence systems.

9.6 REFERENCES

Bahill, T., and Stark, L., 1979, The trajectories of saccadic eye movements, *Sci. Am.* **240**: 108-117.

Blackie, C. A., and Howland, H., 2000, Stability analysis of two linear accommodation and convergence models, *Opt. Vis. Sci.* **77**: 608-615.

Cook, G., and Stark, L., 1967, Derivation of a model for the human eye-positioning mechanism, *Bull. Math. Biophys.* **29**: 153-175.

Ciuffreda, K. J., 1991, Accommodation and its anomalies, in: *Vision and Vision Dysfunction: Visual Optics and Instrumentation, Vol. 1,* W. N. Charman ed., Macmillan, London, pp. 231-279.

Ciuffreda, K. J., 1998, Accommodation, the pupil, and presbyopia, in: *Borish's Clinical Refraction*, W. J. Benjamin ed., W. B. Saunders Company, Philadel., PA., pp. 77-120.

Ciuffreda, K. J., and Wallis, D., 1997, Myopes show increased susceptibility to nearwork aftereffects, *Invest. Ophthal. Vis. Sci.* **39**: 1797-1803.

Eadie, A. S., Carlin, P., and Gray, L. S., 1999, Modelling vergence eye movements using fuzzy logic, in: *Current Oculomotor Research, Physiological and Psychological Aspects*, eds., W. Becker, H. Deubel, and T. Mergner, eds., Kluwer Academic/Plenum Publishers, New York, pps. 179-181.

Ebenholtz, S. M., and Fisher, S. K., 1982, Distance adaptation depends upon plasticity in the oculomotor control system, *Percept. Psychophys.* **31**: 551-560.

Ehrlich, D. L., 1987, Near vision stress: vergence adaptation and accommodative fatigue, *Ophthal. Physiol. Opt.* **7**: 353-357.

Fisher, S. K., Ciuffreda, K. J., 1988, Accommodation and apparent distance, *Perception.* **17**: 609-621.

Fisher, S. K., Ciuffreda, K. J., and Levine, S., 1987, Tonic accommodation, accommodative hysteresis, and refractive error, *Am. J. Optom. Physiol. Opt.* **64**: 799-809.

Fincham, E. F., and Walton, J., 1957, The reciprocal actions of accommodation and convergence, *J. Physiol.* **137**: 488-508.

Fry, G. A., 1939, Further experiments on the accommodation convergence relationship, *Am. J. Optom.* **16**: 325-336.

Goss, D. A., and Wickham, M. G., 1995, Retinal-image mediated growth as a mechanism for juvenile onset myopia and for emmetropization, *Doc. Ophthalmol.* **90**: 341-375.

Green, D. G., Powers, M. K., and Banks, M. S., 1980, Depth of focus, eye size and visual acuity, *Vis. Res.* **20**: 827-835.

Heath, G. G., 1956, Components of accommodation, *Am. J. Optom. Arch. Am. Acad. Optom.* **33**: 569-579.

Henson, D. B., and North, R., 1980, Adaptation to prism-induced heterophoria, *Am. J. Optom. Physiol. Opt.* **57**: 129-137.

Hofstetter, H. W., 1945, The zone of clear single binocular vision: Part 1, *Am. J. Optom. Arch. Am. Acad. Optom.* **22**: 301-333.

Hokoda, S. C., and Ciuffreda, K. J., 1983, Theoretical and clinical importance of proximal vergence and accommodation, in *Vergence Eye Movements: Basic and Clinical Aspects,* C. M Schor and K. J. Ciuffreda, eds, Butterworths, Boston, pp. 75-97.

Hung, G. K., 1998a, Dynamic model of the vergence eye movement system: simulations using MATLAB/SIMULINK, *Comp. Meth. Prog. Biomed.* **55**: 59-68.

Hung, G. K., 1998b, Sensitivity analysis of the stimulus-response function of a static nonlinear accommodation model, *IEEE. Trans. Biomed. Eng.* **45**: 335-341.

Hung, G. K., 1992a, A simple equation for relating AC/A ratio to accommodative controller gain, *Ophthal. Physiol. Opt.* **12**: 106-108.

Hung, G. K., 1992b, Adaptation model of accommodation and vergence, *Ophthal. Physiol. Opt.* **12**: 319-326.

Hung, G. K., and Ciuffreda, K. J., 1988, Dual-mode behaviour in the human accommodation system, *Ophthal. Physiol. Opt.* **8**: 327-332.

Hung, G. K., and Ciuffreda, K. J., 1994, Sensitivity analysis of relative accommodation and vergence, *IEEE. Trans. Biomed. Eng.* **41**: 241-248.

Hung, G. K., Ciuffreda, K. J., and Rosenfield, 1996, Proximal contribution to a linear static model of accommodation and vergence, *Ophthal. Physiol. Opt.* **16**: 31-41.

Hung, G. K., and Semmlow, J. L., 1980, Static behavior of accommodation and vergence: computer simulation of an interactive dual-feedback system, *IEEE. Trans. Biomed. Eng.* **27**: 439-447.

Hung, G. K., and Semmlow, J. L., 1982, A quantitative theory of control sharing between accommodative and vergence controllers, *IEEE. Trans. Biomed. Eng.* **29**: 364-370.

Hung, G. K., Semmlow, J. L., and Ciuffreda, K. J., 1986, A dual-mode dynamic model of the vergence eye movement system, *IEEE. Trans. Biomed. Eng.* **33**: 1021-1028.

Hung, G. K., Zhu, H.-M., and Ciuffreda, K. J., 1997, Convergence and divergence exhibit different response characteristics to symmetric stimuli, *Vis. Res.* **37**: 1197-1205.

Jiang, B.-C., 1996, Accommodative vergence is driven by the phasic component of the accommodative controller, *Vis. Res.* **36**: 97-102.

Jiang, B.-C., 1997, Integration of a sensory component into the accommodation model reveals differences between emmetropia and late-onset myopia, *Invest. Ophthal. Vis. Sci.* **38**: 1511-1516.

Jiang, B.-C., 1999, Oculomotor function in nearwork – induced transient and permanent myopia, *Chin. J. Optom. Ophthalmol.* **1**: 58-61 and 125-128.

Jiang, B.-C., 2000a, A modified control model for steady-state accommodation, in: *Accommodation and Vergence Mechanisms in the Visual System*, O. Franzén, H. Richter, and L. Stark, eds., Birkhäuser Verlag, Basel, Switzerland, pp. 235-243.

Jiang, B.-C., 2000b, Defocus threshold. *The VIII International Conference on Myopia, Boston, MA, 291-295.*

Jiang, B.-C., and Morse, S. E., 1999, Oculomotor functions and late-onset myopia, *Ophthal. Physiol. Opt.* **19**: 165-172.

Jiang, B.-C., and White, J. M., 1999, Effect of accommodative adaptation on static and dynamic accommodation in emmetropia and late-onset myopia, *Optom. Vis. Sci.* **76**: 295-302.

Jiang, B.-C., and Woessner, W. M., 1996, Dark focus and dark vergence: an experimental verification of the configuration of the dual-interactive feedback model, *Ophthal. Physiol. Opt.* **16**: 342-347.

Jones, R. 1980, Fusional vergence: sustained and transient components, *Am. J. Optom. Physiol. Opt.* **57**: 640-644.

Jones, R., and Kerr, K., 1971, Motor responses to conflicting asymmetrical vergence stimulus information, *Am. J. Optom.* **48**: 989-1000.

Jones, R., and Kerr, K., 1972, Vergence eye movements to pairs of disparity stimuli with shape selection cues, *Vis. Res.* **12**: 1425-1430.

Kotulak, J. C., and Schor, C. M., 1986a, The accommodative response to subthreshold blur and to perceptual fading during the Troxler phenomenon, *Perception.* **15**: 7-15.

Kotulak, J. C., and Schor, C. M., 1986b, Temporal variations in accommodation during steady-state conditions, *J. Opt. Soc. Am. A.* **3**: 223-227.

Krishnan , V. V. , and Stark , L., 1975, Integral control in accommodation, *Comp. Prog. Biomed.* **4**: 237-255.

Krishnan, V. V., and Stark, L., 1977, A heuristic model for the human vergence eye movement system, *IEEE. Trans. Biomed. Eng.* **24**: 44-49.

Ludvigh, E., McKinnon, P., and Zartzeff, L., 1964, Temporal course of the relaxation of binocular duction (fusion) movements, *Arch. Ophthalmol.* **71**: 389-399.

Maddox, E. E., 1893, *The Clinical Use of Prisms; and the Decentering of Lenses*. 2nd ed. Bristol, England: John Wright & Sons.

Mordi, J. A., 1991, *Accommodation, Age and Presbyopia*, Ph.D. Dissertation. State Univ. of New York, State College of Optometry, New York, U.S.A.

Mordi, J. A., and Ciuffreda, K. J., 1998, Static aspects of accommodation: age and presbyopia, *Vis. Res.* **38**: 1643-1653.

Morgan, M. W., 1983, The Maddox analysis of vergence, in: *Vergence Eye Movements: Basic and Clinical Aspects*, C.M. Schor and K. J. Ciuffreda eds., Butterworths, Boston, pp. 15-21.

North, R. V., Henson, D. B., and Smith, T. J., 1993, Influence of proximal, accommodative and disparity stimuli upon the vergence system, *Ophthal. Physiol. Opt.* **13**: 239-243.

Ogle, K. N., 1972, *Researches in Binocular Vision*. Hafner, New York, pp. 59-93.

Ogle, K. N., Martens, T. G., and Dyer, J. A., 1967, *Binocular Vision and Fixation Disparity*. Lea and Febiger, Philadel., PA, pp. 9-119.

Ong, E., and Ciuffreda, K. J., 1995, Nearwork-induced transient myopia – a critical review, *Doc Ophthalmol.* **91**: 57-85.

Ong, E., and Ciuffreda, K. J., 1997, *Accommodation, Nearwork, and Myopia,* Optometric Extension Program Foundation, Inc., Santa Ana, CA.

O'Shea, W. F., Ciuffreda, K. J., Fisher, S. K., Tannen, B., and Super, P., 1988, Relation between distance heterophoria and tonic vergence, *Am. J. Optom. Physiol. Opt.* **65**: 787-793.

Owens, D. A., 1984, The resting state of the eyes, *Am. Sci.* **72**: 378-387.

Owens, D. A., and Leibowitz, H. W., 1980, Accommodation, convergence, and distance perception in low illumination, *Am. J. Optom. Physiol. Opt.* **57**: 540-550.

Owens, D. A., and Tyrrell, R. A., 1992, Lateral phoria at distance: contribution of accommodation, *Invest. Ophthal. Vis. Sci.* **33**: 2733-2743.

Patel, S. S., 1995, A neural network model of short-term dynamics of human disparity vergence system, Ph.D. Dissertation, Univ. of Houston, Houston, TX.

Patel, S. S., Ogmen, H., White, J. M., and Jiang, B., 1997, Neural network model of short-term horizontal disparity vergence dynamics, *Vis. Res.* **37**: 1383-1399.

Pobuda, M., and Erkelens, C. J., 1993, The relationship between absolute disparity and ocular vergence, *Biol. Cyber.* **68**: 221-228.

Rashbass, C., and Westheimer, G., 1961, Disjunctive eye movements, *J. Physiol.* **159**: 339-360.

Ripps, H., Chin, N. B., Siegel, I. M., and Breinin, G. M., 1962, The effect of pupil size on accommodation, convergence, and the AC/A ratio, *Invest. Ophthal. Vis. Sci.* **1**: 127-135.

Rosenfield, M., Ciuffreda, K. J., and Chan, H.-W., 1995, Effects of age on the interaction between AC/A and CA/C ratios, *Ophthal. Physiol. Opt.* **15**: 451-455.

Rosenfield, M., Ciuffreda, K. J., and Hung, G. K., 1991, The linearity of proximally induced accommodation and vergence, *Invest. Ophthal. Vis. Sci.* **32**: 2985-2991.

Rosenfield, M., Ciuffreda, K. J., and Novogrodsky, L., 1992a, Contribution of accommodation and disparity-vergence to transient nearwork-induced myopic shifts, *Ophthal. Physiol. Opt.* **12**: 433-436.

Rosenfield, M., Ciuffreda, K. J., and Rosen, J., 1992b, Accommodative response during distance optometric test procedures. *J. Am. Optom. Assoc.* **63**: 614-618.

Rosenfield, M., and Gilmartin, B., 1990, Effect of target proximity on the open-loop accommodative response, *Optom. Vis. Sci.* **67**: 74-79.

Schor, C. M., 1979, The relationship between fusional vergence and fixation disparity, *Vis. Res.* **19**: 1359-1367.

Schor, C. M., 1985, Models of mutual interactions between accommodation and convergence, *Am. J. Optom. Physiol. Opt.* **62**: 369-374.

Schor, C. M., 1992, A dynamic model of cross-coupling between accommodation and convergence: simulation of step and frequency responses. *Optom. Vis. Sci.* **69**: 258-269.

Schor, C. M., 1999, The influence of interactions between accommodation and convergence on the lag of accommodation, *Ophthal. Physiol. Opt.* **19**: 134-150.

Schor, C. M., and Horner, D., 1989, Adaptive disorders of accommodation and vergence in binocular dysfunction, *Ophthal. Physiol. Opt.* **9**: 264-268.

Schor, C. M., and Kotulak, J. C., 1986, Dynamic interactions between accommodation and convergence are velocity sensitive, *Vis. Res.* **26**: 927-942.

Semmlow, J. L., and Hung, G. K., 1983, The near response: theories of control, in: *Vergence Eye Movements: Basic and Clinical Aspects*, C.M. Schor and K. J. Ciuffreda eds., Butterworths, Boston, pp. 175-195.

Semmlow, J. L., Hung, G. K., and Ciuffreda, K. J., 1986, Quantitative assessment of disparity vergence components, *Invest. Ophthal. Vis. Sci.* **27**: 558-564.

Semmlow, J. L., Hung, G. K., Horng, J.-L., and Ciuffreda, K. J., 1993, Initial control component in disparity vergence eye movements, *Ophthal. Physiol. Opt.* **13**: 48-55.

Sethi, B., and North, R. V., 1987, Vergence adaptive changes with varying magnitudes of prism-induced disparities and fusional amplitudes, *Am. J. Optom. Physiol. Opt.* **64**: 263-268.

Stark, L., Kenyon, R. V., Krishnan, V. V., and Ciuffreda, K. J., 1980, Disparity vergence: A proposed name for a dominant component of binocular vergence eye movements, *Am. J. Optom. Physiol. Opt.* **57**: 606-609.

Toates, F. M., 1972, Further studies on the control of accommodation and convergence, *Measurement and Control.* **5**: 58-61.

Toates, F. M. 1974, Vergence eye movements, *Doc. Ophthal.* **37**: 153-214.

Toates, F. M., 1975, *Control Theory in Biology and Experimental Physiology,* Hutchinson, London.

Westheimer, G., 1963, Amphetamine, barbiturates and accommodation-convergence, *Arch. Ophthal.* **70**: 830-836.

Westheimer, G., Mitchell, A. M., 1956, Eye movement responses to convergence stimuli, *Arch. Ophthalmol.* **55**: 848-856.

Wick, , B., 1985, Clinical factors in proximal vergence, *Am. J. Optom. Physiol. Opt.* **62**: 1-18.

Windhorst, U., 1996, Spinal cord and brainstem: Pattern generators and reflexes, in Greger R and U. Windhorst, eds., *Comprehensive Human Physiology: From Cellular Mechanisms to Integration, Vol. 1*, Springer-Verlag, Berlin, pp. 1007-1032.

Wolf, K. S., Bedell, H. E., and Pedersen, S. B., 1990, Relations between accommodation and vergence in darkness, *Optom. Vis. Sci.* **67**: 89-93.

Wong, L., and Jiang, B.-C., 2000, Stimulus/response threshold of accommodation in emmetropes and myopes, *Invest. Ophthal. Vis. Sci.* **41**: S816.

Zadeh, L., 1965, Fuzzy sets, *Inf. Control.* **8:** 338-353.

Zadeh, L., 1994, The Role of Fuzzy Logic in Modeling, Identification and Control, *Modeling Identification and Control.* **15**: 191-203.

Zuber, B., and Stark, L., 1968, Dynamical characteristics of fusional vergence eye movement system, *IEEE Trans. Sys. Sci. Cybern.* **4**: 72-79.

Chapter 10

Models of the Saccadic and Smooth Pursuit Systems

Jordan Pola

Department of Vision Sciences, State University of New York, State College of Optometry, 33 West 42nd St., NewYork, NY 10036; PH: (212) 780-5118; FX: (212) 780-5124; EM: jpola@sunyopt.edu

10.1 PRELIMINARY CONSIDERATIONS

Saccadic and smooth pursuit movements are among several eye movements that we make as we visually attend to objects in our environment. An important function of these two movements is to shift the direction of gaze (i.e., an imaginary line directed outward from the central fovea) to clearly view an object of interest. Both movements are concerned with the horizontal-vertical coordinates of objects. This is in contrast to vergence movements that deal with the proximal location of objects (see Chapter 11: Models of Saccadic-Vergence Interactions). A *saccade* is a rapid movement, perhaps the fastest of skeletal muscle movements, that quickly takes our direction of gaze from an initial point in space to some other point or target. A s*mooth pursuit* movement is a slow to medium velocity movement that allows us to visually follow a moving target, and thus maintain our gaze on or near the target. It is generally thought that the stimulus for a saccade is target position with respect to the fovea, whereas the stimulus for smooth pursuit is target velocity relative to the retina. Saccades can occur without pursuit, and vice versa, but in many circumstances, the two types of movement act conjointly. For example, during visual following of a target moving at moderate to high velocity, smooth pursuit is typically supplemented by saccades. These saccades quickly reduce target offset from the fovea that develops when pursuit velocity is less than target velocity.

Normal saccadic eye movement has a somewhat stereotypic step-like trajectory, as shown in Figure 10.1A. Two important features of saccades are that their peak velocity increases exponentially and their duration increases linearly as a function of amplitude (Fig. 10.1B). Saccades sometimes end with "dynamic" overshoot (a 5-25 msec, small-amplitude overshoot), although this occurs more often with small to medium than with larger saccades. In addition, saccades show "static" undershoot or overshoot of their intended target, followed, within about 50 to 200 msec, by small corrective saccades to the target (some examples of static undershoot appear in Fig. 10.1A).

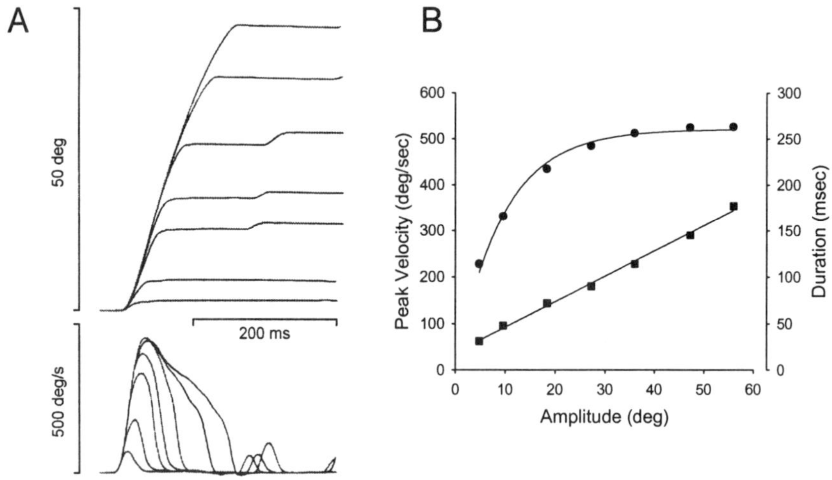

Figure 10.1. (A) Saccadic eye movements with amplitudes ranging from small to large, and the corresponding saccade velocity. (B) Duration (filled squares) and peak velocity (filled circles) of saccadic movement plotted against amplitude. Adapted from Becker (1991).

Unlike saccadic movement, smooth pursuit does not have a fixed trajectory, but instead is able to conform itself approximately to the time-varying position of a moving target. Pursuit movement often begins with overshoot (relative to target velocity) and oscillation (Fig. 10.2), although there are individuals who consistently show no oscillation (Wyatt, Pola, Fortune and Posner, 1994). Ongoing pursuit typically follows target motion with a velocity gain (the ratio of pursuit to target velocity) ranging from 0.6 to 0.95 (Robinson, Gordon and Gordon, 1986, Wyatt and Pola, 1987; Pola and Wyatt, 1997). Maximum pursuit velocity has not been determined under normal everyday circumstances, but in controlled experiments with highly experienced subjects, smooth pursuit can follow target motion (with gain of about 0.95) up to about 100 deg/sec (Meyer, Lasker and Robinson, 1985).

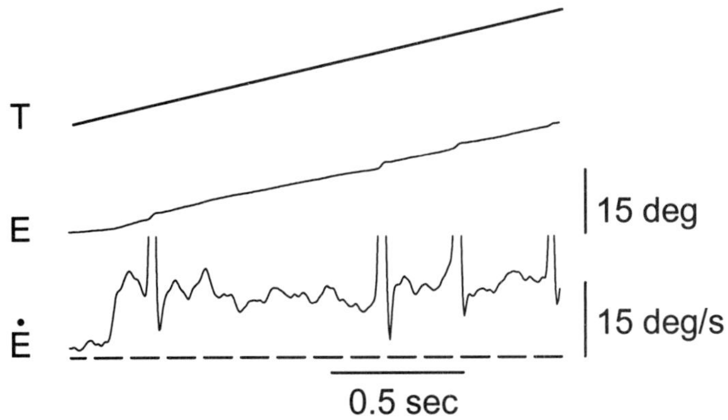

Figure 10.2. Smooth pursuit eye movement E and velocity $\dot{E}$ in response to target T moving at constant velocity. Notice the occasional saccades.

An important feature of research on the saccadic and smooth pursuit systems has been the vigorous development and utilization of models, perhaps more so than in virtually any other area of neuroscience. There are at least two reasons for this: First, the output generated by each system is relatively simple. This suggests that the mechanisms underlying the eye movements involve a relatively small number of functional components, and thus offer the possibility of being understandable within the framework of a "simple" deterministic model. Second, the movements are the result of an interaction of a variety of factors, e.g., input stimuli, perceptual information processing, motor mechanisms, the mechanics of the oculomotor plant, and visuo-motor feedback. Corresponding to these, a number of different disciplines have been central to the study of these movements, e.g., psychology, neurophysiology, neuroanatomy, muscle physiology and bioengineering. A convenient way of bringing together various findings from the different disciplines is with a common language provided by the models. (It should be noted that models of the saccadic and pursuit systems have not only served to integrate current data, but have also been important in the process of creating new data, i.e., making predictions and providing a basis for the formulation of new experiments.)

Models of oculomotor function typically have been concerned with system dynamics, information processing, or stimulus prediction. The development of models in each of these three areas is essential if we are to obtain a good appreciation of how oculomotor systems function in the face of the complex stimuli that occur in daily life. However, reviewing all of these areas is beyond the scope of this chapter, and thus the chapter concentrates on system dynamics, the most well developed area in the study of saccades and pursuit movements. The chapter reviews the characteristics of a variety of models in a quasi-historical context, and the ways in which the models generate oculomotor behavior.

10.2 MODELS OF THE SACCADIC SYSTEM

10.2.1 Early "Ballistic" Models

Raymond Dodge (Dodge and Cline, 1901; Dodge, 1903) was the first to make an objective record of saccadic eye movements. However, it was not until about 50 years later that Westheimer (1954) recorded in detail some of the response features of saccades and proposed the first model of the saccadic system. In essence, he suggested that a saccade is generated by a step-change in muscle force applied to an oculomotor plant with slightly underdamped second-order dynamics. The high velocity of the saccade comes from the step of muscle force, and "dynamic overshoot" at the end of the saccade is a consequence of the underdamped plant dynamics.

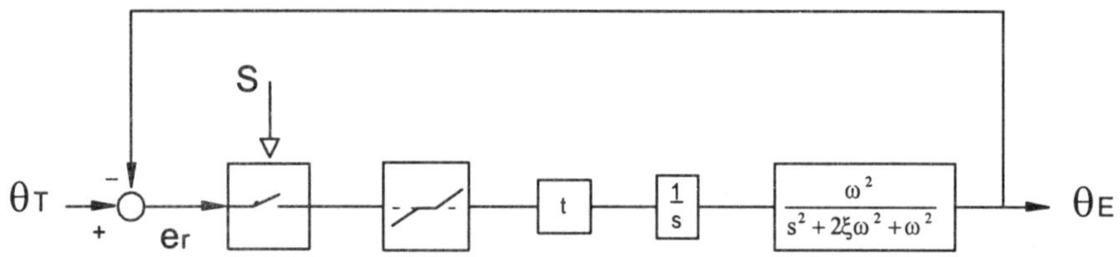

Figure 10.3. Model of the saccadic system according to Young and Stark. In conformance with some of the previous literature, for this and all following models, *t* signifies a time delay, although it should be understood that in a model with Laplace variables, e^{-ts} is the correct notation for a time delay. Adapted from Young and Stark (1963) with permission of the IEEE.

Young and Stark (1963) formalized Westheimer's ideas in a closed-loop negative-feedback model as illustrated in Fig. 10.3. The input drive is the retinal-error signal *er*, which is equal to the offset between the eye position θ_E and the target position θ_T. A central feature of the model is a sampled data mechanism (S) that briefly closes an input switch every 200 msec. During this brief sampling instant, *er* gives rise to a short pulse whose height equals the amount of target difference. If the pulse height exceeds a critical amount (given by the dead-zone non-linearity), it passes through the integrator 1/*s*, yielding a step. The step, after a time delay *t* of 200 msec, drives the second-order plant dynamics to produce a saccade. This model may be regarded as ballistic in the sense that the brief closing of the switch compels the generation of a saccade, without further internal regulation or feedback control. (Although the internal mechanism of the model may be regarded as ballistic, the saccadic movement itself is not ballistic since the drive signal continues throughout the duration of the saccade.)

Although this model can make credible simulations of saccades, some experimental findings show that it is incompatible with at least two features of the actual saccadic system. First, a number of experiments indicate that the saccadic system, instead of simply a sampled data system, may involve some form of continuous information processing (Wheeless, Boynton and Cohen, 1966; Becker and Jurgens, 1979). Second (and most important in the context of this chapter), studies of how the eyes move, with either mechanical displacement or neural activation, show that the oculomotor plant is not underdamped. In an early study on the mechanics of saccadic eye movements, Robinson (1964) found that a step of force applied to the oculomotor plant does not result in saccade-like eye movement, but instead generates a slow exponential-like movement that lasts several hundred milliseconds. The simplest account for this response is that it comes from an oculomotor plant involving a combination of elastic restoring force and viscous drag. Consistent with these findings, experiments using awake, behaving monkeys (Robinson, 1970) showed that the relation between the firing rate R of individual eye movement motoneurons and the eye rotation θ can be characterized by

$$R = k\theta + r\frac{d\theta}{dt}. \tag{10.1}$$

The coefficients k and r are slopes relating firing rate to eye position and velocity, respectively. The eye position signal $k\theta$ may be seen as moving the eye against the plant elastic restoring force, and the velocity signal $r(d\theta/dt)$, as opposing the plant viscous drag. A more detailed analysis (Keller and Robinson, 1972) showed that

$$R = k\theta + r\frac{d\theta}{dt} + m\frac{d^2\theta}{dt^2}, \tag{10.2}$$

and assuming that $r/k >> m/r$, and $k \cong 1$, this yields the second-order transfer function

$$\frac{\theta}{R} = \frac{1}{(sT_1 + 1)(sT_2 + 1)}, \tag{10.3}$$

where the time constant $T_1 = r/k$ and the time constant $T_2 = m/r$. Averaging over many neurons, the value of T_1 is 0.2, and the value of T2 is 0.007. (Robinson, 1973). In short, the oculomotor plant, instead of being slightly underdamped, is overdamped, and can be approximated by a transfer function with two real-axis poles.

What sort of neural signal is required to produce a saccade, given this transfer function? Recordings from individual eye movement motoneurons in awake behaving monkeys shows a pulse-step neural discharge for each saccade (Fuchs and Luschei, 1970; Robinson, 1970). The pulse (corresponding to $r(d\theta/dt)$) generates the high velocity change in eye position, and the step (corresponding to $k\theta$) holds the eye in its new position. Based on these and related findings, Robinson (1973) proposed a ballistic-type model (in the sense defined above), with forward path saccadic circuitry similar to what is shown in Fig. 10.4. Important features of this model include the pulse generator PG, whose output signal goes to both the neural integrator $1/s$, and the parallel (feed-forward) pathway with the gain T_1. According to the model, when a "decision" is made to make a saccade, the trigger-signal *Trig* briefly closes a switch, allowing the target retinal error signal e_r to activate the pulse generator. The resulting pulse P travels both around (via the parallel path) and into the integrator. The pulse from the parallel path and the resulting step output from the integrator come together at the summing point as a pulse-step. The effect of the pulse-step acting on the oculomotor plant is to generate a rapid change in eye position, i.e., a saccade.

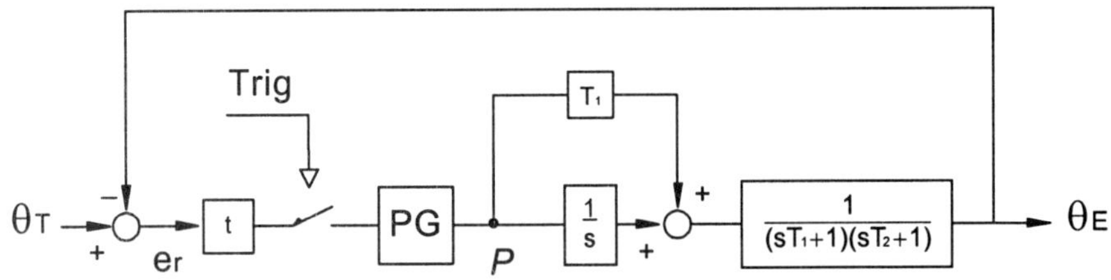

Figure 10.4. Robinson's early model of the saccadic system.

This transformation from pulse to saccade can be analytically derived from the model, where:

$$\theta_E = P \cdot \left(T_1 + \frac{1}{s}\right)\left(\frac{1}{(sT_1+1)(sT_2+1)}\right) \tag{10.4}$$

$$= P \cdot \left(\frac{sT_1+1}{s}\right)\left(\frac{1}{(sT_1+1)(sT_2+1)}\right)$$

$$= P \cdot \left(\frac{1}{s}\right)\left(\frac{1}{sT_2+1}\right).$$

In effect, T_1 in parallel with $1/s$ cancels the first-order term $1/(sT_1 + 1)$ in the oculomotor plant. Moreover, since T_2 is very small

$$\theta_E \approx P \cdot \frac{1}{s}. \tag{10.5}$$

The model shows that the resulting eye movement is the time integral of the pulse, which is roughly a step, and thus the desired saccade.

10.2.2 Early "Internal Feedback" Models

In the above models, the pulse generator is simply a switch or a black box, neither of which provides insight into how the spatial signal for target location is transformed into the temporal pulse for the saccade. An important step in rectifying this situation was provided by Robinson's "internal feedback" model (Fig. 10.5A), which, besides suggesting how some features of the pulse are generated, substantially modified our conception of how the

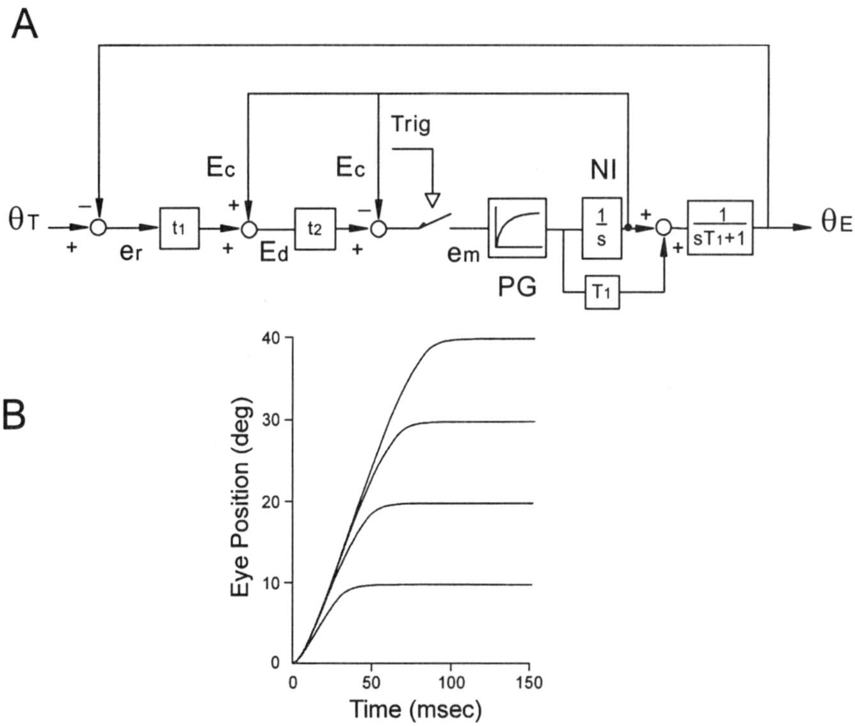

Figure 10.5. (A) Internal feedback model of the saccadic system as originally proposed by Robinson. (B) Simulation of medium to large amplitude saccades. Adapted from Zee, Optican, Cook, Robinson and Engels (1976) with permission of the American Medical Association.

oculomotor system creates eye movements (Robinson, 1975; Zee, Optican, Cook, Robinson and Engel, 1976). In this model, the pulse is not pre-programmed before a saccade, but instead is generated and controlled throughout the course of the saccade. In essence, a motor error signal drives a pulse generator, with this signal arising from the difference between a reference signal of desired eye position and an efference copy signal of current eye position.

An internal feedback loop, enclosing much of the model's forward path, carries the efference copy of the current eye position signal Ec coming from the neural integrator $1/s$. (An important feature of the model is that this integrator also provides the step signal for eye position.) At the front end of the forward path, Ec is *added* to the retinal-error signal er, to give the desired eye position signal Ed. Following this, Ec appears once more, but is *subtracted* from Ed, the result being the motor error signal em. (In this and the following model, the plant is shown as a first-order lag, a reasonable simplification given that $T2$ is very small – see Eq. 10.5.)

To make a saccade, the trigger signal *Trig* closes a switch, and the switch remains closed for the saccade's duration (as opposed to the "ballistic" models above where the switch only closes briefly). Consequently, em activates the pulse generator PG whose output P can be characterized by

$$P = P_{\max}\left[1 - \exp\left(\frac{-em}{k}\right)\right]. \tag{10.6}$$

PG's output passes both into and around the neural integrator $1/s$, initiating a saccade. However, as the saccade proceeds, the magnitude of Ec rapidly increases, subtracting from Ed. The result is that em decreases, and thus the output P of PG decreases (Eq.10.6). The decrease in em continues until it goes to zero, and at that moment, the saccade is complete. (It should be emphasized that PG is simply a soft saturation element, and that PG depends upon the action of the trigger, the switch, and the internal feedback loops to generate the pulse for a saccade.) Saccadic eye movements produced by this type of model (Robinson, 1975) are shown in Fig. 10.5B. The main features of these simulated saccades compare favorably to those of actual saccades (see Fig. 10.1).

This model is able to account for a number of neurophysiological and behavioral findings, but it has at least two problematic features. First, the neural integrator provides both the efference copy signal Ec and the eye position signal that goes to the oculomotor plant. However, integrator dysfunction does not interfere with the generation of saccades, although such dysfunction is responsible for a wide range of oculomotor difficulties (Cannon and Robinson, 1987). This strongly suggests that the integrator is not the source of *Ec*. Second, the reference signal *Ed* represents the desired eye position. However, there is no evidence from neurophysiology that such a signal exists, at least in areas of the brain that appear to be concerned with the generation of saccadic eye movements.

A way to avoid these problems was initially suggested by Abel and colleagues (Abel, Dell'Osso and Daroff, 1978; Abel, Dell'Osso, Schmitt and Daroff, 1980), and shortly thereafter by Jurgens, Becker and Kornhuber (1981). First, instead of an internal feedback signal coming from the neural integrator that activates the oculomotor plant, feedback arises from a second integrator located (in most versions of the model) within the internal feedback loop. Second, the reference signal does not represent the desired eye position, but instead the desired *change* in eye position. Actually, in the model as originally presented by Jurgens et al. (1981), the reference signal represents the target retinal error, which is quantitatively equivalent to the desired change in eye position.

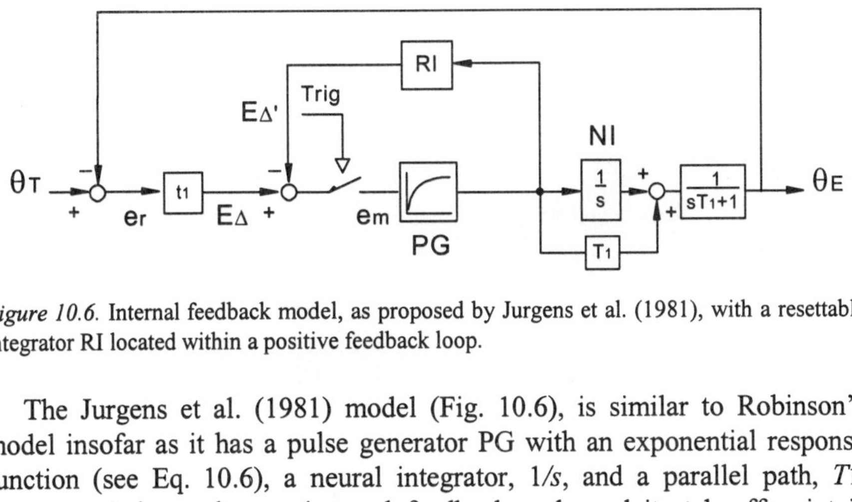

Figure 10.6. Internal feedback model, as proposed by Jurgens et al. (1981), with a resettable integrator RI located within a positive feedback loop.

The Jurgens et al. (1981) model (Fig. 10.6), is similar to Robinson's model insofar as it has a pulse generator PG with an exponential response function (see Eq. 10.6), a neural integrator, $1/s$, and a parallel path, T_1. However, it has only one internal feedback path, and its takeoff point is located before the neural integrator. Thus, PG sends an efference copy signal directly to the internal feedback loop. A second integrator located within the loop, typically called a resettable integrator (RI), integrates from zero at the onset of a saccade, and is reset to zero at the end of each saccade. Thus, the feedback signal E_Δ' reflects change in eye position.

The overall action of this model is similar to that of the previous model. Saccade amplitude is specified by the desired change-in-eye-position signal E_Δ. When the trigger signal *Trig* closes its switch, *em* (initially the same as E_Δ) activates PG, whose output both travels forward to the oculomotor plant and feeds back to RI. As the resulting saccade proceeds, the signal E_Δ' increases, subtracting from E_Δ and thus decreasing *em*. When the difference between $E\Delta$ and $E\Delta'$ goes to zero, the saccade has arrived at its goal location and the eye movement is completed.

10.2.3 "Internal Feedback" Models and Neurophysiology

10.2.3.1 The Oculomotor Brainstem

Recent models of the saccadic system are generally concerned with how the dynamics of saccadic eye movement arise from the detailed connectivity of the underlying neural mechanisms. Thus, these models are a product of both functional systems concepts and current knowledge in neuroanatomy and neurophysiology. Before discussing the models, its is useful to briefly review some relevant features and components of the "oculomotor brain."

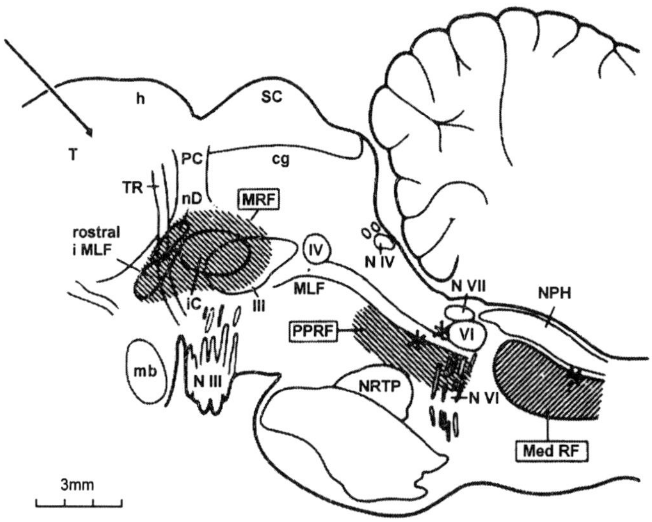

Figure 10.7. Sagittal section of monkey brainstem showing some of the structures involved in the generation of saccades and smooth pursuit. Abbreviations relevant to text: PPRF, paramedian pontine reticular formation; MRF, mesencephalic reticular formation; III, oculomotor nucleus; IV, trochlear nucleus; VI, abducens nucleus; N III, oculomotor nerve; N IV, trochlear nerve; N VI, abducens nerve; MLF, medial longitudinal fasciculus; rostral i MLF, rostral interstitial nucleus of the MLF; iC, interstitial nucleus of Cajal; NPH, nucleus prepositus hyperglossi; NRTP, nucleus reticularis tegmenti pontis; SC, superior colliculus; T, thalamus. The arrow indicates the Horsley-Clark plane. Adapted from Buttner-Ennever and Horn (1996) with permission of The New York Academy of Sciences.

The mechanism for generating saccades involves most of the major areas of the brain. Frontal and parietal cortical areas (e.g., frontal and parietal eye fields), concerned with planning and initiating saccades, project directly and indirectly to the superior colliculus. The indirect projection goes to the superior colliculus via the basal ganglia, i.e., the caudate nucleus and the substantia nigra pars reticulata. The frontal areas also project to the nucleus reticularis tegmenti pontis (NRTP), which, in turn, projects to the cerebellum. The cerebellum and the superior colliculus play a significant role in shaping and generating motor signals for saccades, and are connected to mesencephalic and pontine brainstem structures.

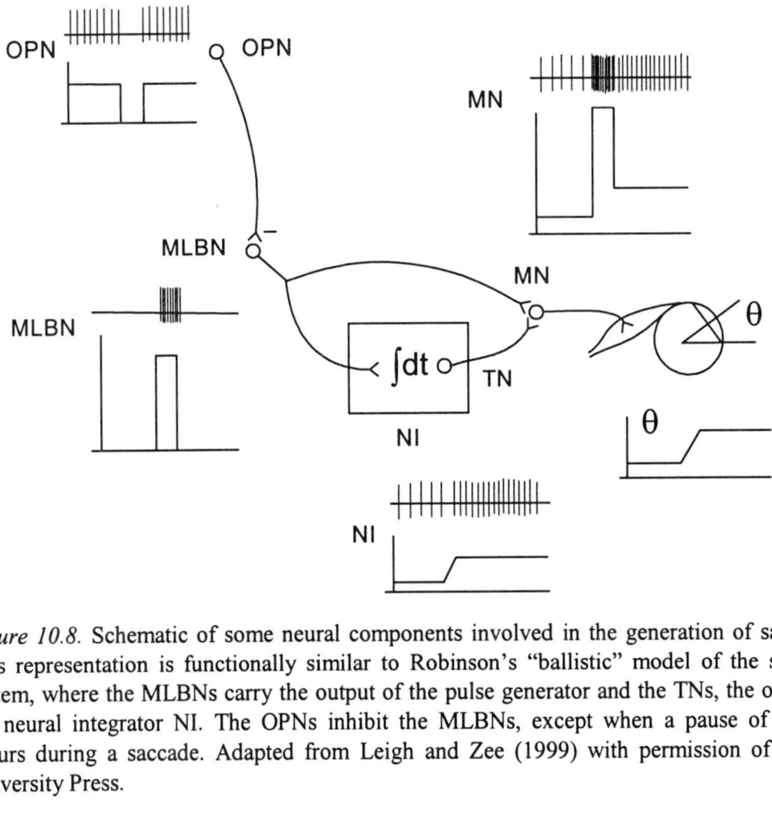

Figure 10.8. Schematic of some neural components involved in the generation of saccades. This representation is functionally similar to Robinson's "ballistic" model of the saccadic system, where the MLBNs carry the output of the pulse generator and the TNs, the output of the neural integrator NI. The OPNs inhibit the MLBNs, except when a pause of activity occurs during a saccade. Adapted from Leigh and Zee (1999) with permission of Oxford University Press.

Some of the important brainstem areas for saccadic eye movement are shown in Fig. 10.7. Motoneurons in the oculomotor, trochlear and abducens nuclei send pulse-step neural activity for a saccade to the extraocular muscles. (The oculomotor nucleus innervates the ipsilateral medial rectus, superior rectus, inferior rectus, and inferior oblique; the trochlear nucleus innervates the contralateral superior oblique; and the abducens nucleus innervates the ipsilateral lateral rectus.) The pulse component of the pulse-step in the motor nuclei comes from medium-lead burst neurons (MLBNs). These neurons are located in two premotor areas – the paramedian pontine reticular formation (PPRF) and the rostral nucleus of the medial longitudinal fasciculus (riMLF). The first is responsible for horizontal saccades and the second, for vertical saccades. The step comes from tonic neurons (TNs) found in areas concerned with neural integration – the nucleus prepositus hypoglossi (NPH) and the medial vestibular nucleus for horizontal movements, and the intersitial nucleus of Cajal (iC) for vertical movements. The functional relation of these neurons, shown in Fig. 10.8, bears a correspondence to Robinson's early "ballistic" model (see Fig. 10.4). The MLBNs and the TNs can be seen as output neurons of the pulse-generator and the neural integrator, respectively, with the signals from these two coming together at the motoneurons (MNs). Also shown are omnidirectional pause neurons (OPNs) whose tonic discharge inhibits the MLBNs, except during a saccade, when this discharge pauses. These neurons can be seen as being a component of the switch that gates saccade onset.

10.2.3.2 Reticulo-centric Models

Neural internal feedback models of the saccadic system fall into three groups: reticulo-centric, colliculo-centric and cerebello-centric (see Quaia, Lefevre and Optican, 1999). The earliest reticulo-centric model (Fig. 10.9A) was presented by Robinson (1973), based on his original internal feedback

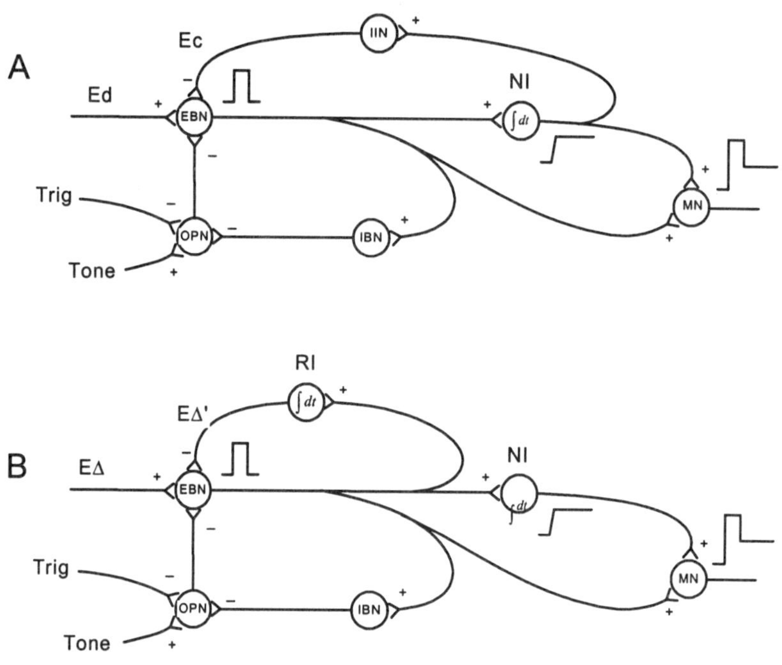

Figure 10.9. Reticulo-centric models of the saccadic system (see text). Adapted from Leigh and Zee (1999) with permission of Oxford University Press.

model (see Fig. 10.5A). The model, as represented here, includes only the second internal feedback loop. Thus, the feedback (efference copy) signal *Ec*, coming from the neural integrator NI via the inhibitory interneurons (IINs), subtracts from the desired eye position signal *Ed*. The model includes two types of MLBNs: excitatory and inhibitory burst-neurons (EBNs and IBNs). OPN discharge, providing inhibition to the EBNs between saccades, is due to a tonic excitatory signal (*Tone*). To make a saccade, the trigger signal *Trig* inhibits the OPNs, allowing *Ed* to activate the EBNs. The consequent output of the EBNs does three things: it goes to the MNs, the IBNs, and the NI. Activation of the IBNs inhibits the OPNs for the duration of the saccade, and the NI presents a position signal to both the MNs and the internal feedback loop. The EBNs continue to discharge until the feedback signal cancels *Ed*, at which point the saccade is complete.

Another reticulo-centric model is shown in Fig. 10.9B, based on Jurgens et al.'s model (see Fig. 10.6). This has the same basic structure and operation as the model in Fig. 10.9A, except that the reference signal is the desired change-in-eye-position signal $E\Delta$, and the resettable integrator RI generates the internal feedback signal $E\Delta'$.

10.2.3.3 Colliculo-centric models

The monkey superior colliculus (SC), located at the roof of the mesenchephalic brainstem (see Fig. 10.7), is a laminated structure consisting of superficial, intermediate and deep layers. The superficial layers contain visual neurons arranged as a retinotopic map of visual space. Visual neurons respond when a visual stimulus is presented to a localized retinal region (a cell's receptive field), and may show enhanced activity if the stimulus is of motivational significance (Wurtz and Mohler, 1976). The intermediate and deep layers contain saccade-related neurons that form a motor map. Roughly speaking, three different types of saccade-related neurons have been found in the superior colliculus: burst, buildup and fixation cells (Munoz and Wurtz, 1995). Burst cells discharge around the time of a saccade, the activity of each burst cell being maximal for a saccade of a given size and direction. Discharge of buildup cells typically begins well before a saccade and may continue for a short period after the saccade. The overall pattern of buildup cell activity during a saccade has the appearance of a spread of excitation from caudal to rostral colliculus (Munoz and Wurtz, 1995). Fixation cells (located in the rostral colliculus) discharge vigorously between saccades, but are less responsive or silent during a saccade.

Scudder (1988) was among the first to offer a model of the saccadic system that explicitly included the superior colliculus. His model, however, might be regarded as reticulo-centric insofar as the superior colliculus simply provides an input reference signal, but otherwise all of the neural machinery and internal feedback circuitry is post-collicular in the brainstem. Waitzman and colleagues (Waitzman, Optican, Ma and Wurtz, 1988; Waitzman, Ma, Optican and Wurtz, 1991) proposed a colliculo-centric model in which collicular burst cells are enclosed by an internal feedback loop (Fig. 10.10A). They found that the decrease in collicular burst cell activity during a saccade appears to change according to the error between the eye and the target as the saccade proceeds. Based on this result, they suggested that the burst cell activity is modulated by an internal feedback signal. In their model, formally similar to that of Jurgens et al. (1981) (see Fig. 10.6), collicular burst cells are located within the region where the feedback signal $E\Delta'$ subtracts from the reference signal $E\Delta$, and the output of the burst cells provides the motor error *em*. The model's burst cell response compares well with the actual response (Fig. 10.10B).

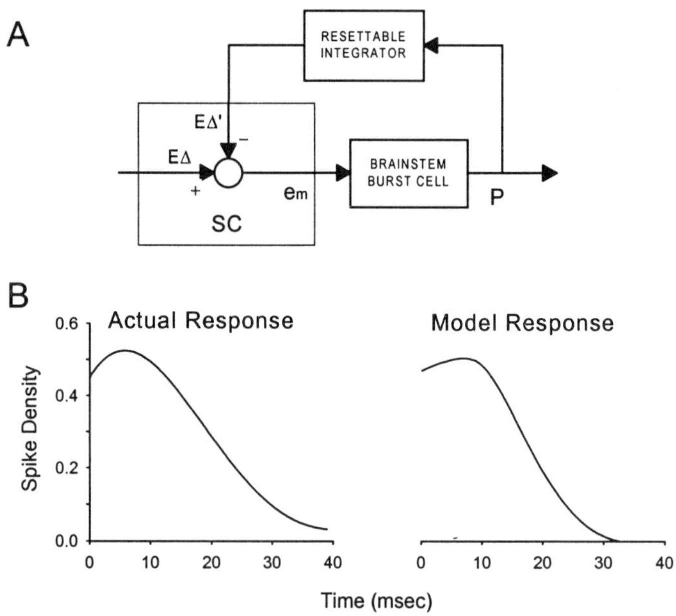

Figure 10.10. (A) Colliculo-centric model of the saccadic system as proposed by Waitzman et al. (1991). (B) Actual collicular burst cell response and model response. Adapted from Waitzman, Ma, Optican and Wurtz (1991) with permission of The American Physiological Society.

The laminar structure of the superior colliculus has been incorporated as a central feature of a variety of "neural net" models (Droulez and Berthoz, 1988; Lefevre and Galiana, 1992; Van Opsal and Kappan, 1993; Arai, Keller and Edelman, 1994; Grossberg, Roberts, Aquilar and Bullock, 1997). As one example, Arai et al. (1994) developed a model that in some respects is similar to the Waitzman et al. (1988; 1991) model. A simple version of this model (Fig. 10.11) represents the superior colliculus as two layers of cells, each layer consisting of 20 cells (with λ indicating the lateral coordinate in each layer). The top layer is comprised of the visual cells *Vi*, and the bottom layer, the motor (burst) cells *Mi*. The visual cells are not interconnected, and each visual cell sends its signal to a corresponding motor cell below. On the other hand, the motor cells signal each other via lateral interconnections (both inhibitory and excitatory), and their output signals come together at the summing junction Σ. This junction generates the motor error *em*, which goes to a saccadic burst generator. The burst generator provides a feedback signal to the motor cell layer, where it impinges on all of the motor cells.

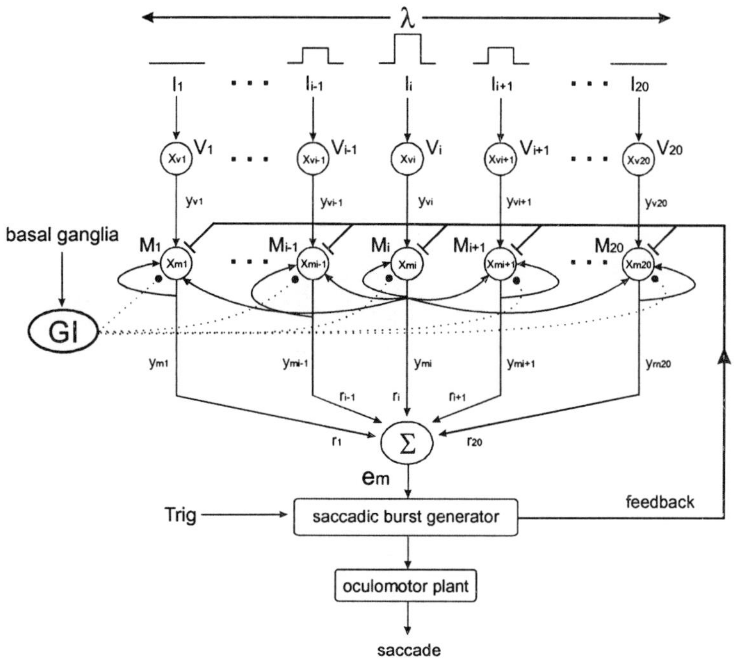

Figure 10.11. Two layered colliculo-centric neural-net model. The top layer represents visual cells and the bottom layer, motor cells. In the bottom layer, solid-line arrows represent both excitatory and inhibitory interconnections, and dotted lines show inhibitory inputs. Adapted from Arai, Keller and Edelman (1994) with permission of Elsevier Science.

The dynamics of the visual layer are given by a set of 20 coupled first-order equations:

$$T_v \frac{dx_{vi}}{dt} = -x_{vi} + g\frac{dI_i(t - \Delta t)}{dt} \tag{10.7}$$

where x_{vi} is visual neuron activity; T_v is a time constant; I_i is visual input with a time delay Δt; and g is gain. Similarly, in the motor layer:

$$T_m \frac{dx_{mi}}{dt} = -x_{mi} + w_{ii} y_{vi} - w_{fi} \cdot FB + \sum_j w_{ij}(\lambda) y_{mj} \tag{10.8}$$

where x_{mi} is motoneuron activity; T_m is a time constant; w_{ii} is the weight of connections between the visual and motor layers; w_{fi} is the weight of the feedback signal FB from the burst generator to the motor layer; and $w_{ij}(\lambda)$ is a weighting function for lateral interconnections from neurons j to i in the motor layer. Output from the visual layer neurons is y_{vi}, and output from the motor layer is y_{mi} (see Fig. 10.11). Both of these signals are non-linear functions of neuron activity, where $y_{vi} = f(x_{vi})$, and $y_{mi} = f(x_{mi})$.

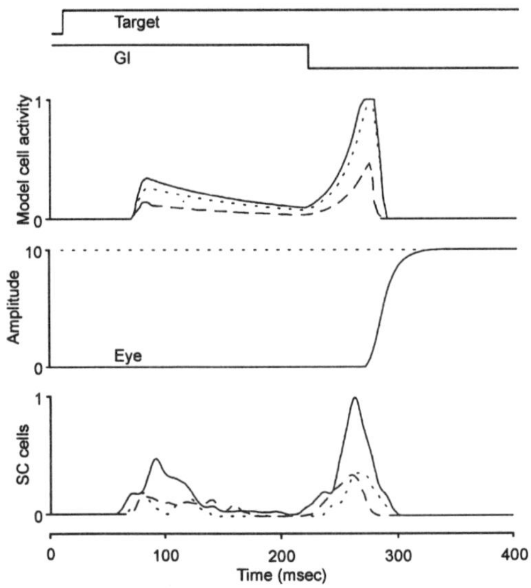

Figure 10.12. Model cell activity (second set of traces) and superior colliculus (SC) cell discharge (lower set of traces) for a 10 deg saccade. For both the model and SC, the solid trace shows activity of a motor cell that responds most strongly for a 10 deg saccade; and the dotted and dashed traces show activity of cells that respond best for saccades larger and smaller than 10 deg, respectively. The eye trace shows a simulated saccade. Adapted from Arai, Keller and Edelman (1994) with permission of Elsevier Science.

According to the model, the initial stage in generating a saccade is a visual stimulus, I_i, activating a local cluster of visual cells. This activation travels to corresponding motor cells below, but the global shunting inhibition GI (corresponding to activity known to come from the basal ganglia) allows only minimal activation of the motor cells (Fig. 10.12). The global inhibition is removed from the motor cells at about the same time that the trigger allows *em* to act on the burst generator. As the motor cell activity increases (Fig. 10.12), the saccadic burst generator sends its output to both the oculomotor plant and the dynamic feedback path. The effect of the feedback is to inhibit and thus reduce the motor cell activity. Fig. 10.12 shows that the model cell response is very similar to the actual collicular response.

10.2.3.4 Cerebello-centric model

Although there can be no doubt that the superior colliculus is involved in the generation of saccades, it seems unlikely that the saccadic pulse generator is colliculo-centric in the manner suggested above. A primary reason for this is that monkeys are able to make reasonably accurate saccades after lesions of their superior colliculus (Schiller, True and Conway, 1980). On the other hand, it has been known for some time that the cerebellum has several areas (e.g., the dorsal vermis and the fastigial nucleus) that exhibit neural discharge in association with saccades. Furthermore, after lesions of these areas, saccades become dysmetric and highly variable in amplitude and direction over successive trials (Ritchie, 1976; Optican and Robinson, 1980; Robinson, Straube and Fuchs, 1993; Takagi, Zee and Tamargo, 1998). Such findings suggest that the cerebellum

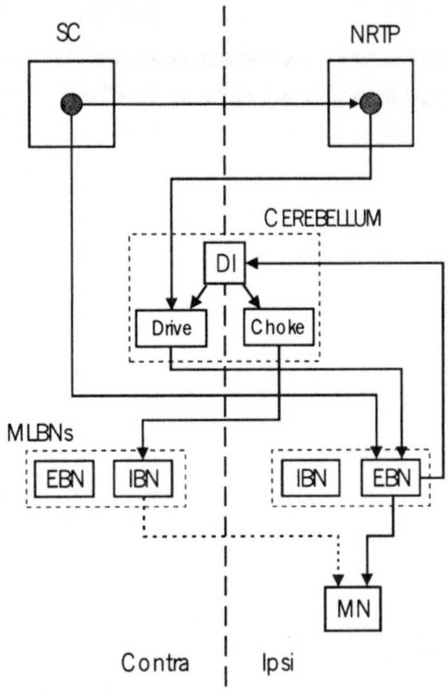

Figure 10.13. Some components and connections of a cerebello-centric model. Adapted from Quaia, Lefevre and Optican (1999) with permission of The American Physiological Society.

has a central role in the generation of saccades. Based on this, a model has recently been proposed (a combination control system and neural network model) in which an internal feedback loop closes on the cerebellum (Quaia, Lefevre and Optican, 1999).

Some of the main features of the model are shown in Fig. 10.13. Ipsilateral (ipsi) designates structures on the side of saccade direction, and contralateral (contra), on the opposite side. Visual input signals (not shown) come from the cortex and go to both the superior colliculus (SC) and the cerebellum. To initiate a saccade, the contralateral SC and the cerebellum send excitatory signals to the ipsilateral EBNs. In response to this, EBN discharge activates the agonist motoneurons (MNs), and feeds back to the cerebellum to activate the displacement integrator (DI). DI's output signal helps in driving the saccade to its goal target, and also goes to the contralateral IBNs, whose discharge inhibits the MNs. Thus, as the eye approaches the target, DI's output "chokes off" EBN activation of the MNs. In contrast to all previous internal feedback models, this scheme suggests that active shutting off of motoneuron discharge stops the saccade.

10.2.4 Oblique Saccades

The models reviewed thus far are concerned with the generation of horizontal saccades that have a straight-line trajectory. However, under everyday circumstances, saccadic direction is often oblique, and consists of a combination of horizontal and vertical components. Oblique saccades have two interesting properties: (1) when the two components are not the same size, their different durations (the smaller component having a shorter duration) causes the resulting oblique trajectory to curve; (2) the duration of the smaller component is often "stretched" toward the duration of the larger component.

In attempting to account for the generation of oblique saccades, several different types of models have been devised (Fig. 10.14A-C). In the common source model (A), the input signal, vectorial motor error e_{vec}, goes to a common pulse generator CPG. The consequent vectorial pulse, P_{vec}, is decomposed into horizontal and vertical pulses, P_h and P_v. These pulses go to horizontal and vertical burst neurons, HBN and VBN, and the output of the burst neurons activates horizontal and vertical motoneurons, HMN and VMN. In the independent model (B), the vectorial motor error e_{vec} is decomposed into horizontal and vertical errors, e_h and e_v. These signals go to horizontal and vertical pulse generators, HPG and VPG, which activate horizontal and vertical motoneurons. The cross-coupled model (C) is similar to the independent model, except for a linkage between the horizontal and vertical pulse generators, where the action of one generator can modulate that of the other. It should be noted that, although not explicit, the pulse generation in all of these models involves internal feedback.

The simplest version of each model, as presented in Fig. 10.14, makes unique predictions. The common source model produces perfect stretching (the duration of the smaller component matches that of the larger), and thus the model's oblique saccades are perfectly straight. In contrast, the independent model does not produce "stretching," with the consequence that its oblique saccades are highly curved, more so than is actually observed. In the cross-coupled model, it is possible to adjust the amount by which the horizontal pulse generator modulates the vertical, or vice-versa, so that resultant "stretching" causes a modest amount of oblique saccadic curvature, similar to what occurs.

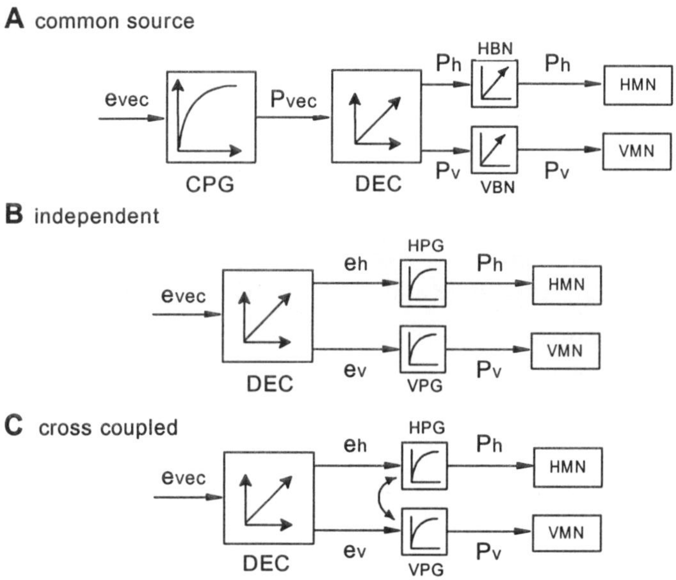

Figure 10.14. Three models for the generation of oblique saccades (see text). Adapted from Quaia and Optican (1997) with permission of The American Physiological Society.

During actual experiments (Becker and Jurgens, 1990), saccadic curvature varies between subjects and also between successive trials for a given subject. Becker and Jurgens (1990) devised a version of the common-source model to show how changes of horizontal and/or vertical motor parameters might underlie such intra-subject variability. The model, presented in Fig. 10.15, consists of a motor-map MM (perhaps in the superior colliculus) whose output feeds into an array of vector pulse generators (VPG). The resulting pulse goes to motor nuclei with gains goh and gov, and to neural integrators with gains gih and giv. If the neural time constants $T'h = goh/gih$ and $T'v = gov/giv$ are the same as the corresponding

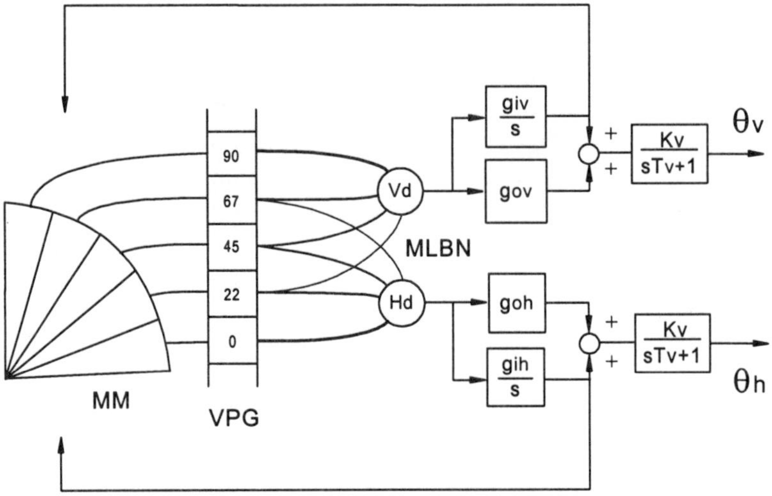

Figure 10.15. Common source model that can generate oblique saccades with variable curvature. Adapted from Becker and Jurgens (1990) with permission of Elsevier Science.

plant time constants, then $(sT'h + 1)/(sTh + 1) = (sT'v + 1)/(sTv + 1)$, and saccadic trajectory is straight. However, if *goh* is low on a given trial, the saccadic trajectory is convex; and if *gih* is low, the trajectory is concave.

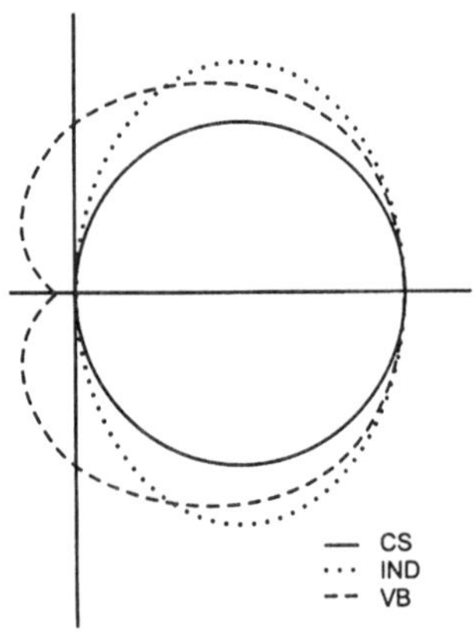

Figure 10.16. Tuning curves, plotted in polar coordinates, for horizontal burst neurons in the common source (CS), independent (IND) and vectorial burster (VB) model. The horizontal axis shows discharge for horizontal eye movement, and the vertical axis shows discharge for vertical movement. Horizontal burst neurons fire during vertical saccades only in the VB model. Reprinted from Quaia and Optican (1997) with permission of The American Physiological Society.

Recently, Quaia and Optican (1997) suggested that virtually all of the models describing oblique saccades are inconsistent with some aspects of basic brainstem physiology. The issue is that the modeled horizontal burst neurons, for example, do not fire during pure vertical saccades, whereas actual horizontal burst neurons do fire during vertical saccades. Using the models in Fig. 10.14, it is easy to appreciate the reason for this lack of neural activity during vertical movement. In the common source model (see Fig. 10.14A), the discharge produced by the vectorial pulse generator is given by Eq. 10.6. The input to this pulse generator, vectorial motor error evec, consists of a magnitude e and angle θ. To find the model's horizontal MLBN discharge, the vectorial pulse is multiplied by the cosine of θ (where $\theta = 0$ signifies rightward horizontal direction). Thus, the horizontal MLBN discharge for the vectorial motor error (e, θ), normalized with the discharge for error $(e, 0)$, is given by:

$$R = \frac{P\left[1-\exp\left(\frac{-e}{k}\right)\right]\cdot\cos(\theta)}{P\left[1-\exp\left(\frac{-e}{k}\right)\right]} = \cos(\theta). \tag{10.9}$$

This expression yields a circular tuning curve, with the result that the common source horizontal MLBNs are silent during vertical saccades (Fig. 10.16).

In the independent model (see Fig. 10.14B), the normalized horizontal MLBN discharge for the motor error signal (e, θ) is given by:

$$R = \frac{P\left[1-\exp\left(\frac{-e\cdot\cos(\theta)}{k}\right)\right]}{P\left[1-\exp\left(\frac{-e}{k}\right)\right]}. \tag{10.10}$$

For a value of $k = 8$ (Quaia and Optican, 1997), the tuning curve is an ellipse, and, as in the above, the independent horizontal MLBNs do not fire during vertical saccades (Fig. 10.16). In the cross-coupled model (see Fig. 10.14C), the horizontal MLBNs have a tuning curve that can vary from that of the common source model to that of the independent model. Thus, they also do not fire with vertical movement (Quaia and Optican, 1997).

Quaia and Optican (1997) suggest that, in contrast to Eqs. 10.9 and 10.10, a Gaussian function provides the best approximation of actual burst neuron discharge. The expression for Gaussian MLBN discharge is

$$R = \exp\left[-\frac{(\theta-\psi)^2}{2\sigma^2}\right] \tag{10.11}$$

where ψ is the neuron's on-direction. A plot of this function for $\psi = 0$ (horizontal on-direction) in Fig. 10.16 yields a curve resembling a cartoid. The important point is that a Gaussian horizontal MLBN discharges during vertical saccades, which is in line with experimental data (Van Gisbergen, Robinson and Gielen, 1981).

Using the Gaussian function and variation in MLBN on-directions, Quaia and Optican (1997) proposed a vectorial burster (VB) model as shown in Fig. 10.17. In essence, it is a common source model with a vectorial drive signal distributed to both horizontal and vertical MLBNs. The summed outputs of the MLBNs yields an almost circular tuning curve for horizontal and vertical motoneurons, HMN and VMN. An important feature of the model is that the cartoid discharge characteristics of the MLBNs play a role in the component stretching of saccadic eye movement (Fig. 10.18, left). Furthermore, when the distribution of MLBN on-directions is asymmetric, the model is able to produce curved trajectories for oblique saccades (Fig. 10.18, right).

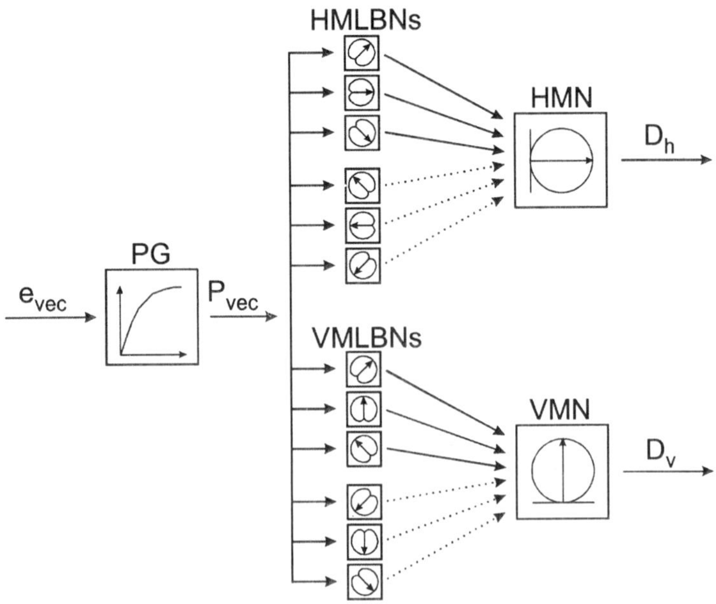

Figure 10.17. VB model of the saccadic system. This representation of the model shows only the feed-forward path. Discharge of the horizontal and vertical MLBNs is given by cartoid tuning curves, and discharge of the horizontal and vertical MNs, by circular tuning curves. The solid-line arrows represent excitatory connections and the dotted-line arrows represent inhibitory connections. Adapted from Quaia and Optican (1997) with permission of The American Physiological Society.

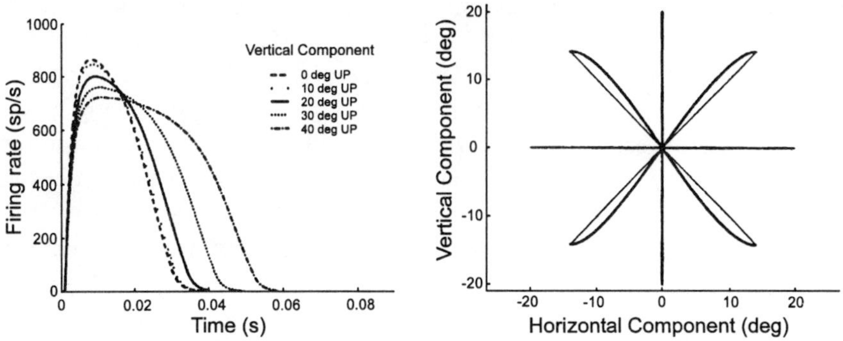

Figure 10.18. (Left) Discharge of horizontal MLBN's in the VB model for saccades with a 20 deg horizontal component and different vertical components. The discharge reflects stretching of the horizontal component as the vertical component angle (and size) increases. (Right) VB model saccades have curved trajectories when there are asymmetries in the model's MLBN on-directions. Adapted from Quaia and Optican (1997) with permission of The American Physiological Society.

10.3 MODELS OF THE SMOOTH PURSUIT SYSTEM

10.3.1 Early Models

Young and Stark (1963) were the first to simulate smooth pursuit eye movements using a negative feedback, sampled-data model. Fig. 10.19A gives essential features of the model. The retinal error signal e_r, coming from the difference between eye and target, is sampled every T seconds. Following this, an "error rate sample" is determined from the difference between each two successive samples divided by T. The resulting error rate pulse is integrated yielding a desired eye velocity signal, and a second integration gives a desired eye position signal. This eye position signal drives the (underdamped) oculomotor plant to generate smooth pursuit movement. The first limiter in the model prevents the system from responding to abrupt changes in target position, and the second prevents the system output velocity from exceeding 30 deg/sec. Although this model can simulate a variety of experimental results, it is inconsistent with some aspects of the actual smooth pursuit system, i.e., the pursuit system exhibits continuous rather than discrete behavior (Robinson, 1965), and the plant is overdamped rather than slightly underdamped.

A simple schematic model of the pursuit system with continuous data processing and an overdamped plant is shown in Fig. 10.19B. The forward path of the model is composed of three main subdivisions: sensory input mechanisms, central processing mechanisms, and motor output mechanisms. An important stimulus for smooth pursuit seems to be target velocity (Rashbass, 1962; Lisberger and Westbrook, 1985), and thus the sensory mechanism at the front end of the model is represented by the differentiator *s*. (A velocity limiter is omitted from this and most of the following models to simplify their presentation.) To be able to simulate pursuit response time and some aspects of pursuit dynamics, the model's central processing involves time delay *t*, followed by high gain *G* and first order lag $1/(sT+1)$. At the motor end of the system is the neural integrator $1/s$ along with the parallel path gain T_1.

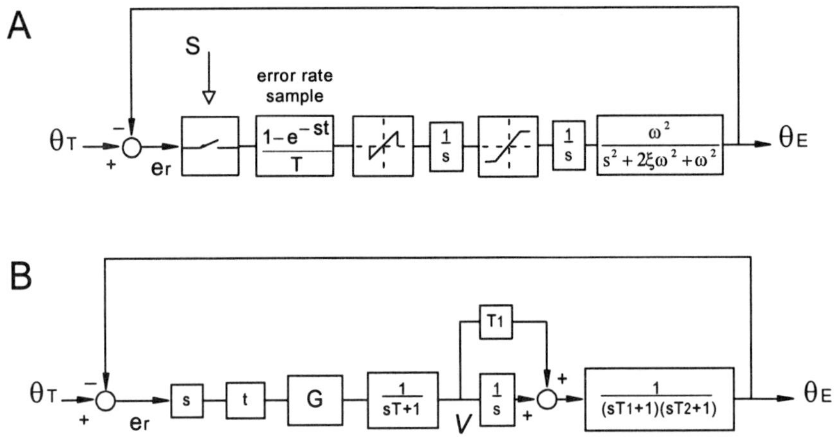

Figure 10.19. (A) Young-Stark sampled-data model of the smooth pursuit system. Adapted from Young and Stark (1963) with permission of the IEEE. (B) Simple schematic model of the pursuit system with continuous processing of input.

When target motion occurs, the resulting change in the retinal error signal e_r passes through *s*, producing a velocity error signal. After the time delay *t*, the velocity signal is amplified by *G*, filtered by the lag, and sent to the integrator. The integrator output (an eye position signal) is added to the parallel path signal (an eye velocity signal), and the sum goes to the plant, resulting in movement of the eye. The overall effect of the forward path mechanisms together with the presence of negative feedback is that the eye is able to follow target motion with reasonable accuracy.

It should be pointed out that just as the motor signal for a saccadic eye movement consists of a pulse and a step in order to effectively drive the viscoelastic mechanics of the plant, the motor signal for smooth pursuit involves an eye velocity signal and an eye position signal. Thus, using the velocity signal *V* from Fig. 10.19B and substituting it for *P* in Eq. 10.5 we obtain

$$\theta_E \approx V \cdot \frac{1}{s}. \tag{10.12}$$

This shows that eye position is simply the integral of V.

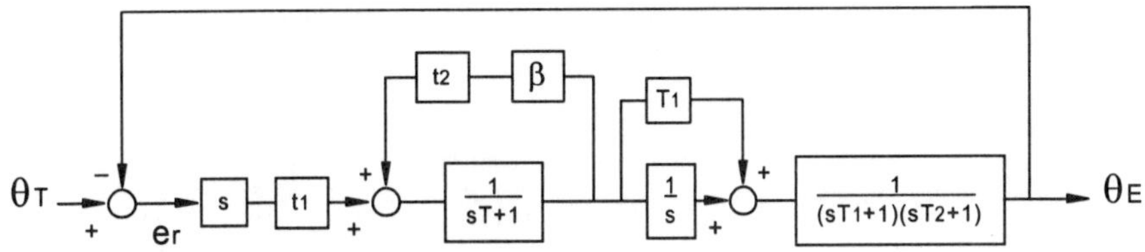

Figure 10.20. Model of the smooth pursuit system with an internal positive feedback path carrying an efference copy signal.

A major change in pursuit system modeling occurred when Yasui and Young (1975) proposed that, besides the pursuit efferent signal to the extraocular muscles, an efference copy signal is fed back as input to the pursuit motor mechanisms. This efference copy is carried by an internal positive feedback loop as shown in Fig. 10.20. The forward path components are similar to those in the schematic model (Fig. 10.19B), although the parameters are different. The internal loop does at least two things: it provides an intuitively reasonable way of obtaining a high internal gain, and to the extent that it compensates for the effects of external negative feedback, it contributes to system stability.

One of the first attempts to account for the details of smooth pursuit behavior using internal feedback was by Robinson, Gordon and Gordon (1986). A striking feature of pursuit onset velocity is that it rapidly increases with damped oscillations of about 2-4 Hz (see Fig. 10.2). To attempt to simulate this response, Robinson et al. (1986) developed a pursuit system model with two internal feedback loops (Fig. 10.21A), the outer one positive and the inner one negative. The basic structure of the model is similar to that of the models we have already considered, consisting of sensory, central, and motor mechanisms. (To simplify the presentation of this model and all of the following models, the motor mechanism is reduced to $1/s$. As Eq. 10.12 shows, this is functionally equivalent to the motor mechanism with the parallel path T_1 and the two-pole oculomotor plant.) Of note, the model has essentially the same functional architecture as Robinson's internal feedback model of the saccadic system (Fig. 10.5). Thus, the eye velocity signal V_c is added to the error velocity signal V_e. The sum of these two signals is the target velocity signal V_t and, after a time delay t_1 (15 msec), the desired eye velocity signal V_d. Following this, V_c subtracts from V_d, yielding the motor error e_m. An interesting feature of this model is that e_m is proportional to the desired change in eye velocity, and thus, is a form of acceleration signal.

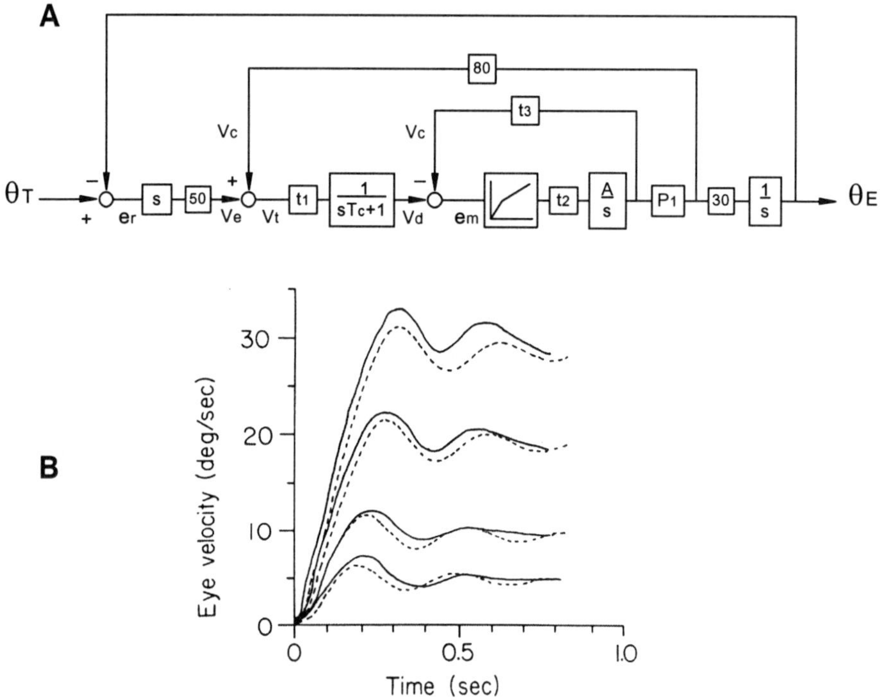

Figure 10.21. (A) Robinson's model of the pursuit system with positive and negative internal feedback. (B) Model onset oscillation (dashed traces) compared with actual onset oscillation (solid traces). Adapted from Robinson, Cannon and Cannon (1986) with permission of Springer-Verlag.

The time delays (t_1, t_2 and 80 msec) and dynamics that are responsible for the signal in the internal positive feedback loop are the same as the time delays (t_1, t_2, 50 and 30 msec) and dynamics for the signal in the external negative loop. Thus, these two loops cancel each other, and by themselves are not responsible for any significant time varying behavior. On the other hand, the delay t_2 (35 msec) together with the internal loop delay t_3 (30 msec) generate system oscillations at 3.8 Hz. Fig. 10.21B shows simulated onset oscillations coming from the model along with actual pursuit oscillations.

10.3.2 The Open-Loop Pursuit System

Under normal circumstances, smooth pursuit occurs with a closed-loop negative feedback of -1.0. Negative feedback tends to mask pursuit forward-path dynamics. This is a virtue insofar as it "protects" normal pursuit from the effects of forward-path parameter variability (normal and abnormal). However, the masking limits the use of normal pursuit as a research tool to investigate pursuit internal mechanisms. Fortunately, it is possible to study pursuit without negative feedback, i.e., in the open-loop condition.

The relation between the closed-loop and the open-loop response of a linear system is

$$G_{cl} = \frac{G_{ol}}{1 + G_{ol}}, \tag{10.13}$$

where G_{cl} and G_{ol} represent complex valued functions of frequency. Solving this formula in terms of real-valued gain and phase lag,

$$|G_{cl}| = \frac{|G_{ol}|}{\sqrt{1 + |G_{ol}|^2 + 2|G_{ol}|\cos\phi_{ol}}} \tag{10.14}$$

and

$$\tan\phi_{cl} = \frac{\sin\phi_{ol}}{|G_{ol}| + \cos\phi_{ol}}, \tag{10.15}$$

where $|Gcl|$, $|Gol|$, cl, and ol are closed-loop gain, open-loop gain, closed-loop phase lag, and open-loop phase lag, respectively.

Wyatt and Pola (1983) made the first systematic investigation of normal closed-loop smooth pursuit movement and its relation to open-loop pursuit behavior. In this study, they asked subjects to follow 3 deg peak-to-peak sinewave target motion. In the closed-loop condition, subjects followed the target motion as usual. However, to obtain the open-loop condition, the 3 deg peak-to-peak motion was centered on the fovea and stabilized relative to the retina, i.e., the target motion moved from side to side of the fovea no matter how or where the eyes moved (and thus the normal relation between eye and target motion was decoupled).

Pursuit responses of two subjects (open circles and squares) in both the open-loop and closed-loop conditions are plotted in Fig. 10.22 as a function of frequency of sinewave target motion. These data show high open-loop gain and modest phase lag at 0.5 Hz, with gain decreasing and phase lag increasing over frequency. These values of gain and phase at each frequency were inserted into Eqs. 10.14 and 10.15 to predict closed-loop gain and phase. The predictions (indicated by the solid lines) are shown along with actual closed-loop responses, and clearly are very similar to the actual data. This has at least two implications: (1) Forward path dynamics are the same in the open-loop condition (using a stabilized target) as in the closed-loop condition. As a consequence, (2) the open-loop condition would seem to be a valuable way to reveal the system's internal dynamic properties.

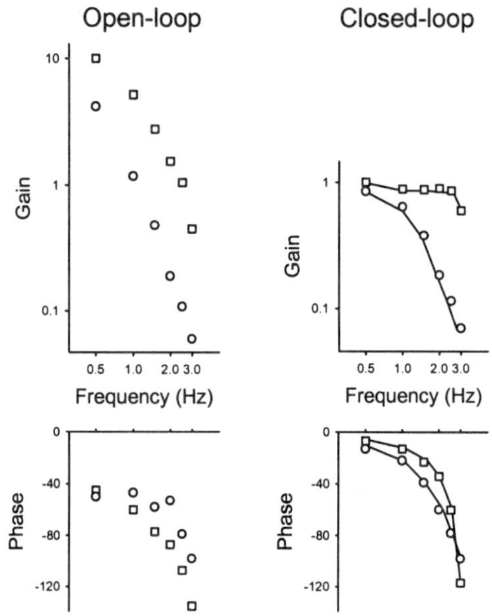

Figure 10.22. Smooth pursuit gain and phase lag for two subjects in the open-loop and closed-loop feedback conditions. Predictions of closed-loop gain and phase are shown by solid traces. Adapted from Wyatt and Pola (1983) with permission of Elsevier Science.

10.3.3 The Stimulus for Pursuit

In all of the pursuit system models presented thus far the stimulus for smooth pursuit is target velocity (see Preliminary Considerations). However, a variety of experiments (Kommerell and Taumer, 1972; Pola and Wyatt, 1980; Wyatt and Pola, 1981; Neary, 1986; Neary, Pola and Wyatt, 1987; Morris and Lisberger, 1987; Segraves and Goldberg, 1994) with both humans and monkeys have produced compelling data suggesting that, along with target velocity, target position (target offset from the fovea) may also be an important stimulus for pursuit. A simple schematic model of the smooth pursuit system with target velocity and target position input channels, and an internal positive feedback loop, is shown in Fig. 10.23A. It is of interest to derive the model's open-loop and closed-loop response features. First, if we eliminate the external negative feedback loop and ignore the time delays $t1$ and $t2$, the open-loop response θ_E is given by

$$\theta_E = \theta_T \cdot (as + b)\left(\frac{\frac{1}{sT+1}}{1 - \frac{1}{sT+1} \cdot \beta} \right)\left(\frac{1}{s}\right). \tag{10.16}$$

To simplify the forward path dynamics on the right side of 10.16, we factor out b in the first term, and multiply the top and bottom of the second term by $(sT + 1)$. This yields the expression

$$b \cdot (sT' + 1)\left(\frac{1}{sT + 1 - \beta} \right)\left(\frac{1}{s}\right), \tag{10.17}$$

where $T' = a/b$. We can also divide the top and bottom of the second term by $(1 - \beta)$ to obtain

$$b \cdot (sT' + 1)\left(\frac{A}{sT'' + 1} \right)\left(\frac{1}{s}\right) \tag{10.18}$$

in which $A = 1/(1 - \beta)$ and $T'' = T/(1 - \beta)$. If the value of T' is close to T'' and $bA = G$, then finally,

$$\theta_E = \theta_T \cdot \frac{G}{s}. \tag{10.19}$$

Thus, the open-loop transfer function of this model reduces to a simple integrator. Integrator gain, shown with both high and low G in Fig. 10.23B, has no low-frequency asymptote and decreases as frequency increases. Of importance here is that the open-loop gain of this model bears a good resemblance to experimental open-loop pursuit gain (Fig. 10.22).

The presence of a target position channel is the primary reason for the model's integrator-like response. At low stimulus frequencies ($\omega < 1/T'$), target position is the "dominant" stimulus ($b > as$), and becomes more so as the frequency gets lower. At high frequencies ($\omega > 1/T'$), target velocity is dominant ($as > b$). If there were no position channel, the model's open-loop transfer function would be $G/(sT' + 1)$. In that case, the model's response would have a low frequency asymptote G, in contrast to the actual data (Fig. 10.22).

In obtaining the model's transfer function, the time delays $t1$ and $t2$ were ignored. Although this might seem a bit presumptuous, it may approximate what actually occurs when the pursuit system is tracking periodic stimulus motion. Under many everyday circumstances, the pursuit system has a relatively fixed response time (the forward path time delay) between a change in stimulus motion and a change in pursuit behavior. However, a well known feature of the pursuit system is that it is able to respond in a predictive manner when the stimulus motion is periodic (Fender and Nye, 1961; Stark, Vossius and Young, 1962; Dallos and Jones, 1963; Michael and Mellville-Jones, 1966; Bahill and McDonald, 1983). Simply speaking, this involves an *apparent* decrease in the response time, and for sinewave target motion, a decrease in the phase lag of the eye movement with regard to the target motion. The amount of the decrease may be as much as, or more than, the value of the forward path delay. In effect, prediction may involve a process that approximates a cancellation of the forward path delay.

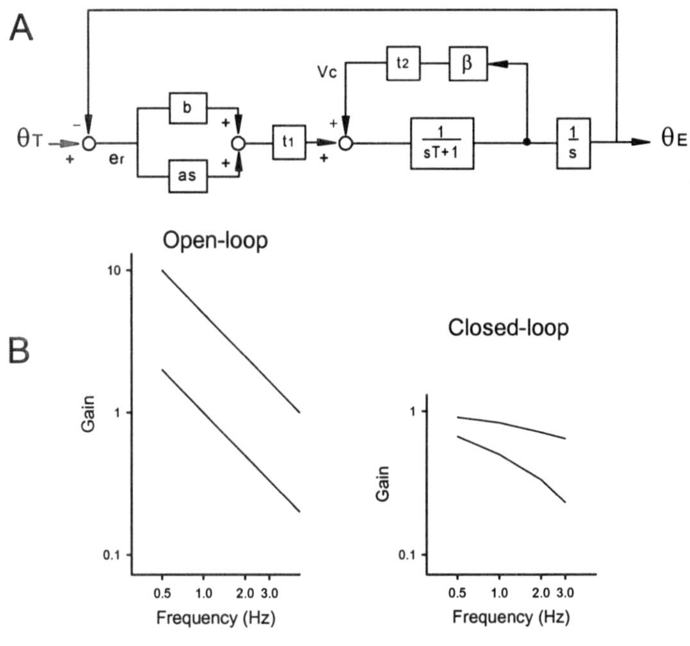

Figure 10.23. (A) Model of the pursuit system with input channels for both target position and velocity. (B) The model's open-loop and closed-loop response for both high and low forward path gain (see text).

To obtain the closed-loop transfer function of the model, we insert the open-loop transfer function from Eq. 10.19 into 10.13. This gives

$$\frac{\frac{G}{s}}{1+\frac{G}{s}}, \qquad (10.20)$$

and multiplying the top and bottom by s/G gives

$$\frac{1}{sT_c + 1} \tag{10.21}$$

where $Tc = 1/G$. Thus, the closed-loop transfer function of the model in Fig. 10.23A is a first-order lag. Gain for this transfer function, with both high and low G, is shown in Fig. 10.23B. As with the open-loop gain, the closed-loop gain of the model bears a good resemblance to experimental findings (Fig. 10.22).

This representation of the pursuit system with target position as a stimulus might appear to put pursuit movements into competition with saccades, which also respond to target position. However, such competition seems unlikely for at least two reasons: (1) the pursuit response to target position develops at a relatively slow rate compared to the rapidity with which a saccade eliminates target offset (Pola and Wyatt, 1980; de Bie and van den Brink, 1986); and (2) the pursuit response may mostly be concerned with minimizing small position errors, whereas a saccade reduces larger errors (see below).

10.3.4 Current Models and Pursuit Termination

Much of our understanding of smooth pursuit mechanisms comes from investigations of pursuit initiation and maintenance. Recently, however, our appreciation of how the pursuit system works has been supplemented by studies of the features of pursuit termination. Robinson, Gordon and Gordon (1986) were the first to observe that the pursuit termination response in the normal closed-loop condition does not necessarily involve oscillation. The importance of this is that virtually all models of the pursuit system predict oscillation whenever a change in pursuit velocity occurs, including pursuit termination. To account for the difference between the model prediction and actual pursuit, Robinson et al. (1986) suggested that while the pursuit system is responsible for pursuit initiation, it has little or nothing to do with pursuit termination. Instead, a separate fixation system is responsible for pursuit termination and its lack of oscillation.

Huebner, Leigh, Seidman, Thomas, Billian, DiScenna and Dell'osso (1992) developed a model based on this idea. Essentially, the model consists of a pursuit system with oscillatory dynamics for pursuit initiation, and a fixation system with relatively sluggish dynamics for pursuit termination. In their scheme, the pursuit system is located in parallel with the fixation system. During pursuit termination, a "neural switch" (activated by an eye velocity signal for velocity $\leq$ 5 deg/sec) disables the pursuit system and engages the fixation system, resulting in an exponential decrease in smooth velocity.

Krauzlis and colleagues (Krauzlis and Lisberger, 1994; Krauzlis and Miles, 1996) proposed another approach to account for the difference between pursuit initiation and termination. The main features of their model include velocity, acceleration and motion transient input channels, and a positive feedback loop encircling first-order dynamics. To eliminate oscillation during pursuit termination, the value of a central gain component either decreases slightly (Krauzlis and Miles, 1996) or goes to zero (Krauzlis and Lisberger, 1994). (The signal responsible for the gain decrease is not specified.) This model is able to simulate pursuit termination when the pursuit target jumps ahead of the fovea, both predictably and non-predictably, with normal visual feedback of -1.0.

Although the above two models have substantially different implications for the type of neurophysiology underlying the initiation and termination of smooth pursuit, they are the same in one important respect: both are governed by target motion relative to the retina; that is, both are target "retinal-slip" servomechanisms.

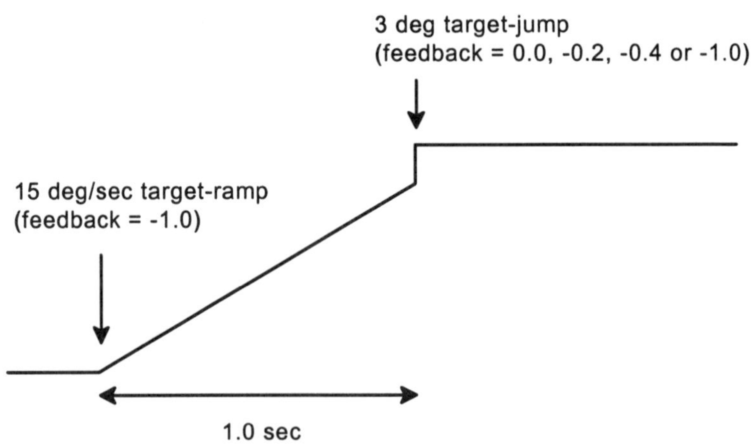

Figure 10.24. Stimulus motion for pursuit deceleration and termination. Each trial began with target motion of 15 deg/sec for 1 sec, followed by the target jumping 3 deg ahead of the fovea. At the time of the target jump, target velocity became zero and target visual feedback became either 0, -0.2, -0.4 or -1.0.

In contrast to this, a number of studies (see Section 10.3.3) indicate that besides target retinal-slip, the smooth pursuit system uses target position as a stimulus. If this is the case, then one would expect that target position is involved in the process of pursuit termination. Recently, Pola and Wyatt (2001) investigated smooth pursuit deceleration and termination under a variety of visual feedback conditions as a means of evaluating pursuit system models with target position as a stimulus. In these studies, subjects initially made pursuit movements with a target moving at 15 deg/sec in the normal closed-loop condition (Fig. 10.24). After the subjects pursued the target for about 1 sec, the target jumped 3 deg ahead of the fovea with a target velocity of 0, and visual feedback of either 0 (open-loop condition with target stabilized ahead of the fovea) or one of three negative values, -0.2, -0.4, or -1.0 (closed-loop conditions). Smooth pursuit and occasional saccades occurred both during the initial ramp target motion, and after the target jump with feedback of 0, -0.2, and -0.4, but not with feedback of -1.0. With 0 feedback, the target moved in space at the same velocity as the eyes, and thus, neither pursuit nor saccadic movement could reduce the 3 deg "error" between the fovea and the target. In this case, the occurrence of a saccade amounted to a futile attempt to foveate the target. However, as the negative feedback increased from -0.2 to -1.0, the ability of the eye movement to reduce the error increased proportionately.

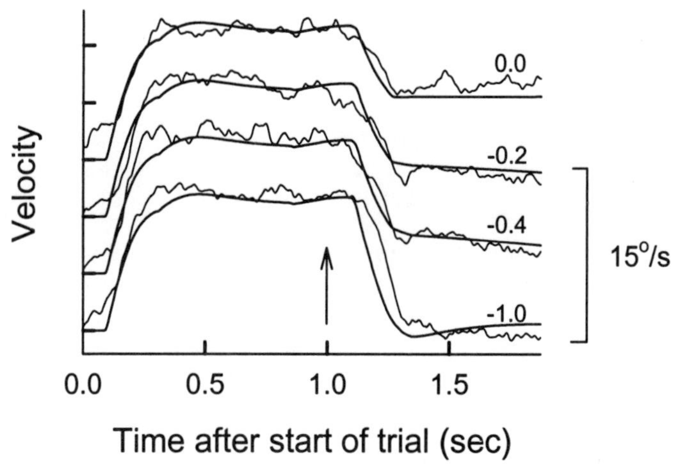

Figure 10.25. Mean pursuit velocity (light traces) and model pursuit velocity (dark traces) during trials in which the target jumped 3 deg ahead of the fovea after 1 sec of closed-loop ramp target motion (indicated by arrow). At the time of the target jump, visual feedback became either 0, -0.2, -0.4, or -1.0. Only smooth velocity is shown, with saccadic responses removed. Reprinted from Pola and Wyatt (2001) with permission of Elsevier Science.

Fig. 10.25 presents smooth pursuit velocity in each feedback condition for one subject, with saccadic responses *removed.* The light traces show actual eye velocity, and the arrow indicates the time of the target jump. Dark traces are simulated eye velocity (to be discussed below). In all conditions, pursuit for this subject began with little or no oscillation. When the target jumped 3 deg ahead of the fovea, the initial response was a fast decrease in eye velocity. With 0 visual feedback, the fast decrease was followed by a relatively constant velocity movement, an apparent response to target position. (Since there was no target retinal-slip, the only "error-signal" generating pursuit was target offset from the fovea.) With -0.2 and -0.4 feedback, the fast decrease was followed by a slow decrease, and with -1.0 feedback, by the eye quickly stopping.

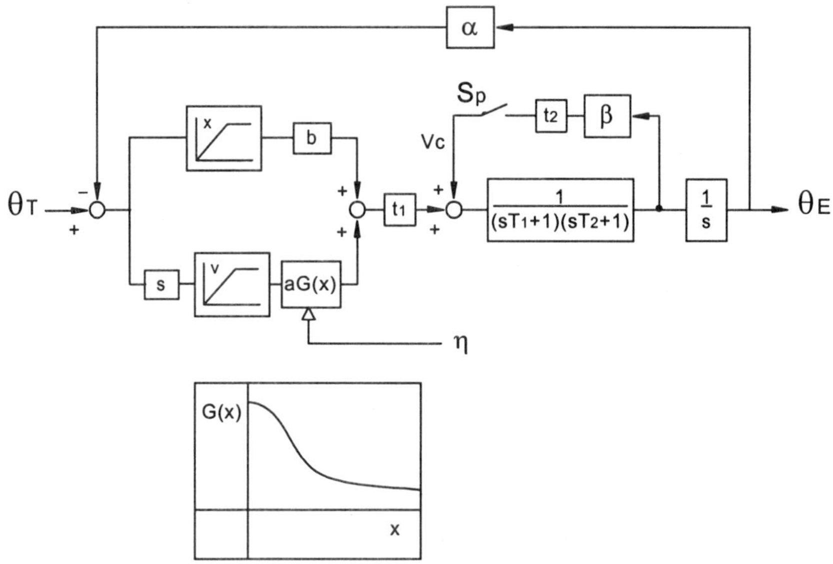

Figure 10.26. Pursuit system model with non-linear target position and velocity channels. See text for details. Adapted from Pola and Wyatt (2001) with permission of Elsevier Science.

To account for these results, Pola and Wyatt (2001) proposed a new model of the smooth pursuit system (Fig. 10.26). The model's main features include: (a) a target position channel whose signal saturates when the target is 1 to 3 deg from the fovea, depending on the subject; (b) a target velocity channel whose signal decreases with target retinal eccentricity; (c) a signal to the velocity channel that switches the channel gain from high to low when target velocity is 1 to 5 deg/sec, depending on the subject; and (d) an internal positive feedback loop that is enabled by target presence and disabled by its absence.

The position channel has two components: a simple saturation non-linearity followed by a gain constant *b*. The velocity channel is more complex and has four components: a differentiator *s*; a saturation non-linearity; a gain constant *a*; and a non-linear decrease in the target velocity signal as a function of retinal eccentricity, represented by *G*(*x*) (Pola and Wyatt, 2001). The expression for this decrease in signal is

$$G(x) = G_{low} + \frac{[1 - G_{low}]}{[1 + (x / x_h)^{\mu}]}, \tag{10.22}$$

where G_{low} is the mininum eccentric gain, x is retinal position, x_h is the retinal eccentricity at which the gain is halfway from 1 to G_{low}, and μ gives the rate of gain change from the fovea to the eccentric retina. A schematic of $G(x)$ is shown in the lower portion of Fig. 10.26. The input signal η reduces the velocity gain a during pursuit termination.

Central dynamics of the model include a second-order transfer function with the time constants T_1 and T_2. This function is enclosed by a positive feedback loop with the gain β. When the switch S_p is opened, the positive feedback loop is disabled. At the output end of the model, the integrator $1/s$ is responsible for eye position. Time delays in the forward and feedback paths are t_1 and t_2, respectively. In the external negative feedback path α can have values of 0, -0.2, -0.4 or -1.0, consistent with the feedback values in the above experiment.

An optimization procedure was used to simulate responses in the different feedback conditions. Fig. 10.25 shows that in all of the conditions the model pursuit responses (dark traces) are remarkably similar to the actual pursuit responses. The model initiation velocity, showing only slight overshoot, and subsequent maintenance velocity are both similar to the actual initiation and maintenance velocity. For both initiation and maintenance, the model suggests that the gain of the velocity channel ($a \cong 1.0$) is larger than the gain of the position channel ($b \cong 0.6$). The model's response after the target jumps ahead of the fovea – corresponding to the actual data – can be seen as consisting of an initial fast decrease in velocity followed by a slow decrease in velocity. The way in which the model accounts for the initial fast decrease, although speculative, is that when the target unpredictably jumps ahead of the fovea, it briefly disappears as far as the pursuit system is concerned. This brief disappearance causes the switch in the positive feedback loop to open for 25 -100 msec (depending on the subject), and for that short period, the duration of the system transient response decreases. The result is a brief but rapid drop in eye velocity.

The characteristics of the slow decrease are determined by several factors, depending on the amount of visual feedback. In the 0 feedback condition, the slow steady velocity comes exclusively from a response to target position (since there is no target velocity signal, and the only stimulus is target position). But with -0.2 and -0.4 feedback, the slow response reflects the combined influence of the velocity and position channels and their non-linear components. After the target jumps ahead of the fovea, the presence of negative feedback causes target retinal-slip back towards the fovea, retarding ongoing eye movement. However, with the target offset from the fovea, the velocity signal, due to $G(x)$, is weak (see insert in Fig. 10.26), and the target position signal is able to maintain slow eye movement at an almost steady velocity. In addition, with feedback of -0.4 and -1.0, the gain of the velocity channel decreases (due to the action of η), eliminating any possibility of oscillation. In effect, the importance of target position as a stimulus increases during pursuit deceleration and termination.

Pursuit system models without a target position channel, such as the Huebner et al. (1992) and the Krauzlis and Miles (1996) models, discussed above, are not able to simulate most of the features of the pursuit response that occur in the "target-jump" conditions. That is, such models cannot produce the constant velocity response that occurs with 0 visual feedback or the near constant velocity response with -0.2 and -0.4 feedback.

In summary, the Pola and Wyatt (2001) model suggests that both target velocity and target position are involved in driving smooth pursuit eye movement. In particular, the model suggests that during pursuit initiation and maintenance, target velocity gain is higher than target position gain, but during pursuit termination, the velocity gain decreases and the relative importance of target position as a stimulus increases.

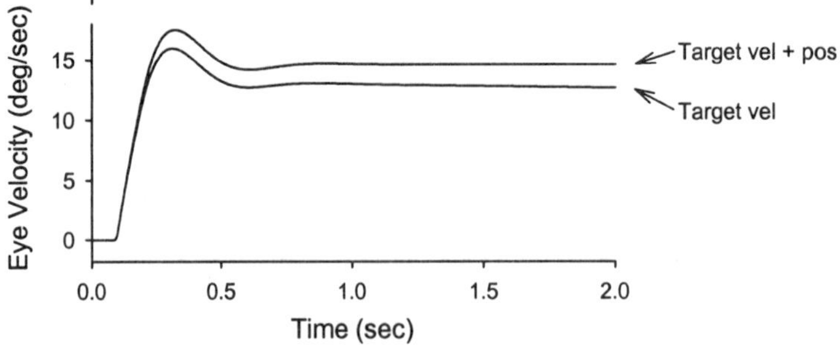

Figure 10.27. Response of the model presented above (in Fig. 10.24) with target velocity and position as stimuli versus only target velocity as stimulus. Target motion is shown by the dashed trace.

Although the above results provide support for target position as a stimulus for smooth pursuit, they do not indicate how target position serves pursuit under normal closed-loop tracking. To provide an answer to this, the above model (Fig. 10.26) was run with ramp target motion under two circumstances: (1) with target velocity as the sole stimulus (i.e., the target position channel gain was set to 0), and (2) with both target velocity and position as stimuli. Both simulations consisted of pursuit behavior without saccades, so that the features of the pursuit would be clearly visible. The model's responses (Fig. 10.27) show that a smaller error accumulates between the eye and the target when both target velocity and position are stimuli than when only target velocity is the stimulus. This suggests that the pursuit system uses target position to *minimize* the error between the eye and the target. Presumably, as the offset increases, the probability of a saccade occurring also increases.

Perhaps a deeper appreciation for the existence of both target velocity and position as pursuit stimuli may be gained by considering the open-loop schematic model response (Fig. 10.23B) and the corresponding actual subject response (Fig. 10.22). As suggested above, the absence of a low frequency asymptote in both the model response and the actual response comes from target position as a stimulus. On the other hand, the middle to higher frequency portions of the response are largely the result of target velocity. In general, then, target position may be responsible for mediating the pursuit response to low frequency components of target motion, whereas target velocity may be responsible for the pursuit response to medium and higher frequency components. The implication is that target velocity would be especially important as a stimulus during (a) onset of target motion, and during (b) changes in target velocity. However, target position would play a significant role as a stimulus when (c) target velocity is relatively constant and (d) target velocity goes to zero (as during pursuit termination and visual fixation). Support for (a) comes from the classic Rashbass (1961) experiment in which pursuit onset with step-ramp target motion (the step opposite to the ramp) ignored the step (target position) and simply followed the ramp (target velocity). Support for (b) is suggested by Luebke and Robinson's (1988) study in which pursuit oscillations (arising from target velocity as stimulus) occurred as a consequence of a change in the velocity of ramp target motion. Support for (c) can be found in Fig. 10.27 in which the simulation of pursuit matches actual pursuit best when both target velocity and position serve as stimuli. Support for (d) appears in Fig. 10.25 where the simulation of pursuit deceleration and termination calls for both target position and velocity as stimuli.

10.3.5 Pursuit System Models and Neurophysiology

In contrast to saccadic system models, pursuit system models are based largely on behavioral and psychophysical data. Nonetheless, there is some correspondence between the main features of pursuit system models and neural components known to underlie pursuit. In the above discussion, it was suggested that pursuit system models, in general, could be regarded as consisting of three main subdivisions: sensory, central, and motor. At the sensory (or front) end of the models, the input mechanisms respond to target velocity or both target velocity and position. In the actual visual system, visual signals travel from the retina to the lateral geniculate nucleus (located in the thalamus – see Fig. 10.7) arriving at the striate cortex (located in the occipital cortex, the most posterior portion of the brain). From there, visual information is distributed to a variety of cortical regions, including the middle temporal visual area (MT) whose neurons are especially responsive to target velocity. Both neural recording and lesion studies suggest that MT provides a velocity signal for both smooth pursuit and for perception (for a review see Leigh and Zee, 1999). Contiguous with MT is the medial superior temporal visual area (MST) which also responds to target velocity. Besides these cortical areas, several subcortical regions may also provide pursuit velocity signals (Leigh and Zee, 1999). These include the accessory optic system (AOS) and the nucleus of the optic tract (NOT). Retinal signals travel directly to both of these areas. The AOS consists of several subnuclei: the dorsal terminal nucleus (DTN); the lateral terminal nucleus (LTN); the medial terminal nucleus (MTN); and the interstitial terminal nucleus (ITN). The DTN responds to horizontal stimulus motion whereas the LTN and the MTN respond to vertical motion.

Compared with target velocity signals, little is known about the areas or pathways that might mediate target position signals. Recently, Krauzlis, Basso and Wurtz (1997) found that fixation cells located in the rostral superior colliculus of monkey respond to target offset from the fovea. The firing rate of these neurons increases in association with an offset that leads to either saccadic or smooth pursuit movement. Thus, their signal may be involved in the generation of both types of movement.

Central components of pursuit system models include positive feedback and the dynamic elements it encloses. Area MST in the cortex may be located within a positive feedback circuit insofar as it shows activity related to the occurrence of eye movement as well as to target motion. Velocity signals from MST and MT descend monosynaptically to the dorsolateral pontine nuclei and the nucleus reticularis tegmenti pontis (NRTP) and from there to the cerebellum (see Fig. 10.7). The pursuit neural integrator appears to be (similar to saccades) in the nucleus prepositus hypoglossi (NPH) and the medial vestibular nucleus for horizontal movement, and in the interstitial nucleus of Cajal (iC) for vertical movement (Fig. 10.7).

10.4 SOME IMPLICATIONS OF THE MODELS

A primary emphasis of this chapter has been on the structure of saccadic and pursuit system models and how these models generate normal oculomotor behavior. However, besides commonly observed behavior, the models provide a view of some of the factors that may underlie variability in eye movements, atypical movements, and in some instances, abnormal movements. The models also include components that can explain certain features of space perception. Some ways in which the models account for diverse aspects of oculomotor behavior are presented below.

In all but the earliest models of the saccadic system, a pulse-step generates a saccade, the pulse coming from a pulse-generator and the step, from a neural integrator. These models represent the two mechanisms as being functionally distinct, and in line with this, it turns out that the pulse and step can vary independently of each other. One example is a pulse-step "mismatch" that produces an eye movement called a "glissade" (Bahill and Stark, 1979). When the pulse is too small relative to the step, an "undershoot" saccade occurs followed by a rising exponential movement ($T \cong 200$ msec) towards the desired target position. When the pulse is too large, an "overshoot" saccade occurs, with a decaying exponential movement back towards the desired target position. Pulse and step independence has also been shown in experiments where saccadic eye movements adapt to the effects of weakening the horizontal recti muscles by tenectomy (Optican and Robinson, 1980). Abnormalities of the pulse generator appear in eye movement dysfunction, ranging from chronic undershoot and/or overshoot saccades (Ritchie, 1976; Optican and Robinson, 1980) to abnormally slow saccades (Zee, Optican, Cook, Robinson and Engels, 1976). Problems with the integrator may underlie a common form of jerk nystagmus known as gaze-evoked nystagmus (Abel, Dell'Osso and Daroff, 1978; see Chap. 20).

Just as the pulse is responsible for generating a saccade, a velocity signal is of primary importance in driving smooth pursuit. In models of the pursuit system, velocity channel gain has an influence on both transient and steady state features of pursuit. Low gain means that a reduced and perhaps insufficient signal about target motion goes to the central mechanisms and thus pursuit movement, in general, suffers. Consistent with this, lesions of area MT, concerned with visual target motion (see above), result in deficits in pursuit velocity as well as in problems with motion perception (Newsome, Wurtz, Dursteler and Mikami, 1985; Dursteler, Wurtz and Newsome, 1987).

An increase in velocity channel gain increases both the amplitude and the duration of oscillation. Wyatt and Pola (1987) found that a subject's mode of attending to pursuit target motion may influence velocity gain. When subjects "passively" observe the target motion, pursuit onset shows no oscillation and low velocity. However, when subjects "actively" attend to the target, pursuit onset oscillates and pursuit velocity is very close to target velocity.

Both saccadic and pursuit system models involve internal feedback as an important functional component. In pursuit system models, positive feedback makes a substantial contribution to the magnitude of forward path gain. This gain, although normally hidden, is revealed when the pursuit target is stabilized in relation to the retina (the open-loop condition – see Fig. 10.22). The amount of positive feedback may notably affect pursuit behavior. Some subjects, for example, never seem to show pursuit onset oscillation (see, for example, Fig. 10.25), whereas others always show strong oscillation (see Fig. 10.21). The former may have low positive feedback gain whereas the latter, relatively high gain. An interesting possibility is that excessively high positive feedback may be a factor contributing to chronic oculomotor oscillation, as occurs in some forms of congenital nystagmus (Dell' Osso, 1982).

Internal feedback in the saccadic and smooth pursuit systems, besides being involved in generating eye movements, may serve perception. When a person makes either a saccadic or pursuit movement, he or she perceives the location of stationary objects in space to be stable, even though the location of the retinal image of the objects changes, rapidly in the case of a saccade, and slowly for pursuit. The classic explanation of this "stability of visual space" is that during an eye movement, change in the location of the retinal image is nullified in perception by the presence of an extraretinal signal. In other words, the retinal signal reporting change is cancelled by an "equal and opposite" extraretinal signal. In line with this account, psychophysical experiments on perception of direction during saccades (Matin, 1972; Pola, 1976; Honda, 1991) and smooth pursuit (Pola and Wyatt, unpublished observations) suggest that an extraretinal signal does exist, and that the manner in which it changes before, during and following eye movement is compatible with perceived stability of visual space.

What is the source of this signal? A traditional view is that the extraretinal signal arises as a copy of the efferent signal to the extraocular muscles. One possibility is that an efference copy is provided either entirely or in part by internal feedback as represented in both the saccadic and pursuit system models. Thus, during an eye movement, the feedback signal is sent both to a "summing point" concerned with motor behavior, and also to some location involved in perception of space and/or motion. In line with this, recent experiments (Pola and Wyatt, unpublished observations) suggest that the temporal features of the positive feedback signal in pursuit system models bear a striking resemblance to a psychophysical determination of subjects' extraretinal signal before, during, and following pursuit.

10.5 SUMMARY

The saccadic and pursuit system models presented in this chapter are able to account for many of the important features of normal oculomotor behavior and some forms of atypical behavior. The earliest models consisted of serial forward-path mechanisms enclosed by a single external negative feedback loop, with the input drive signal arising from a simple retinal stimulus. In subsequent models, both the forward path and drive signal increased in complexity. In the saccadic system models, internal feedback was added to the forward path, involving a combination of positive and negative feedback loops, or a single negative feedback loop. Along with this internal feedback, the drive signal evolved from a simple retinal stimulus to include an eye position signal, either desired eye position or desired change in eye position. In the pursuit system models, internal feedback was also added to the forward path, and the drive signal was modified to include two or more attributes of the retinal stimulus, e.g., some combination of target velocity, target position, and target acceleration. Recent models (particularly the saccadic system models) involve neural networks or a composite of neural network and control systems. Overall, models of the saccadic and pursuit systems have made a significant contribution to our understanding of the mechanisms controlling eye movements. Evolution in the complexity of the models mirrors our growing appreciation of the complexity of brain structure and function, and promises that models will continue to play an important role in revealing the mechanisms underlying oculomotor behavior.

10.6 ACKNOWLEDGMENTS

I would like to thank the editors of this volume, George Hung and Kenneth Ciuffreda, for their insightful discussions on several aspects of this manuscript. I am also grateful to Harry Wyatt and Nancy Oley, both of whom provided many valuable comments on an early draft of the paper.

10.7 REFERENCES

Abel, L. A., Dell'Osso L. F., and Daroff, R. B., 1978, Analog model for gaze-evoked nystagmus, *IEEE Trans.**Biomed. Engin.*** **25**: 71-75.

Abel, L. A., Dell'Osso, L. F., Schmitt D., and Daroff, R. B., 1980, Myaesthenia gravis: analog computer model, *Exp. Neurol.* **68**: 378-389.

Arai, K., Keller E. L., and Edelman, J. A., 1994, Two dimensional neural network model of the primate saccadic system, *Neural Networks.* **7**: 1115-1135.

Bahill, A. T., and McDonald, J. D., 1983, Smooth pursuit eye movements in response to predictable target motions, *Vision Res.* **23**: 1573-1583.

Bahill, A. T., and Stark, L., 1979, The trajectories of saccadic eye movements, *Sci. Am.* **240**: 108-117.

Becker, W., 1991, Saccades, in: Vision and Visual Dysfunction, Vol. 8, Eye Movements, R. H. S. Carpenter, ed., The Macmillan Press Ltd., London, pp. 95-137.

Becker, W., and Jurgens, R., 1979, An analysis of the human saccadic system by means of double step stimuli, *Vision Res.* **19**: 967-983.

Becker, W., and Jurgens, R., 1990, Human oblique saccades: quantitative analysis of the relation between horizontal and vertical components, *Vision Res.* **30**: 893-920.

Buttner-Ennerver, J. A., and Horn, A. K. E., 1996, Pathways from the cell groups of the paramedian tracts to the floccular region, *Ann. NY Acad. Sci.* **781**: 532-540.

Cannon, S. C., and Robinson, D. A., 1987, Loss of the neural integrator of the oculomotor system from brain stem lesions in monkey, *J. Neurophysiol.* **57**: 1383-1409.

de Bie, J., and van den Brink, G., 1986, A model for the slow control system during monocular fixation, *Vision Res.* **26**: 1129-1142.

Dallos, P. J., and Jones, R. W., 1963, Learning behavior of the eye fixation control system, *IEEE Trans. Automatic Control.* **8**: 218-227.

Dell'Osso, L. F., 1982, Congenital nystagmus: basic aspects, in: *Functional Basis of Ocular Motility Disorders*, G. Lennerstrand, D. S. Zee and E. L. Keller, eds., Pergamon Press, Oxford, pp. 129-138.

Dodge, R., 1903, Five types of eye movements in the horizontal meridian plane of the field of regard, *Am. J. Physiol.* **8**: 307-329.

Dodge, R., and Cline, T. S., 1901, The angle velocity of eye movements, *Psychol. Rev.* **8**: 145-157.

Droulez, J., and Berthoz, A., 1988, Spatial and temporal transformation in visuo-motor coordination, in: *Neural Computers*, R. Eckmiller and C. von der Malsburg, eds., Springer-Verlag, Berlin, pp. 345-357.

Dursteler, M. R., Wurtz, R. H., and Newsome, W. T., 1987, Directional pursuit deficits following lesions of the foveal representation within the superior temporal sulcus of the macaque monkey, *J. Neurophysiol.* **57**: 1262-1287.

Fender, D. H., and Nye, P. W., 1961, An investigation of the mechanism of eye movement control, *Kybernetik.* **1**: 81-88.

Fuchs, A. F., and Luschei, E. S., 1970, Firing patterns of abducens neurons of alert monkeys in relationship to horizontal eye movements, *J. Neurophysiol.* **33**: 382-392.

Grossberg, S., Roberts, K., Aquilar M., and Bullock, D., 1997, A neural model of multimodal adaptive saccadic eye movement control by superior colliculus, *J. Neurosci.* **17**: 9706-9725.

Honda, H., 1991, The time courses of visual mislocalization and of extraretinal eye position signals at the time of vertical saccades, *Vision Res.* **31**: 1915-1921.

Huebner, W. P., Leigh, R. J., Seidman, S. H., Thomas, C. W., Billian, C., DiScenna, A. O., and Dell'Osso, L. F., 1992, Experimental tests of a superposition hypothesis to explain the relationship between the vestibuloocular reflex and smooth pursuit during horizontal combined eye-head tracking in humans, *J. Neurophysiol.* **68**: 1775-1792.

Jurgens, R., Becker, W., and Kornhuber, H. H., 1981, Natural and drug-induced variations of velocity and duration of human saccadic eye movements: evidence for a control of the neural pulse generator by local feedback, *Biol. Cybern.* **39**: 87-96.

Keller, E. L., and Robinson, D. A., 1972, Abducens unit behavior in the monkey during vergence movements. *Vision Res.* **12**: 369-382.

Lefevre, P., and Galiana, H. L., 1992, Dynamic feedback to the superior colliculus in a neural network model of the gaze control system, *Neural Networks.* **5**: 871-890.

Leigh, R. J., and Zee, D. S., 1999, *The Neurology of Eye Movements, Third Edition*, Oxford University Press, New York.

Lisberger, S. G., and Westbrook, L. E., 1985, Properties of visual inputs that initiate horizontal smooth pursuit eye movements in monkeys, *J. Neurosci.* **5**: 1662-1673.

Luebke, A. E., and Robinson, D. A., 1988, Transition dynamics between pursuit and fixation suggest different systems, *Vision Res.* **28**: 941-946.

Kommerell, G., and Taumer, R., 1972, Investigation of the eye tracking system through stabilized retinal images, in: *Cerebral Control of Eye Movements and Motion Perception*, J. Dichgans and E. Bizzi, eds., S. Karger, Basel, pp. 288-297.

Krauzlis, R. J., and Lisberger, S. G., 1994, A model of visually-guided smooth pursuit eye movements based on behavioral observations, *J. Comp. Neurosci.* **1**: 265-283.

Krauzlis, R. J., and Miles, F. A., 1996, Transitions between pursuit eye movements and fixation in the monkey: dependence on context, *J. Neurophysiol.* **76**: 1622-1638.

Krauzlis, R. J., Basso, M., A., and Wurtz, R., H., 1997, Shared motor error for multiple eye movements, *Science.* **276**: 1693-1695.

Meyer, C. H., Lasker, A. G., and Robinson, D. A., 1985, The upper limit of human smooth pursuit velocity, *Vision Res.* **25**: 561-563.

Michael, J., and Melville Jones, G., 1966, Dependence of visual tracking capability upon stimulus predictability, *Vision Res.* **6**: 707-716.

Morris, E. G., and Lisberger, S. G., 1987, Different responses to small visual errors during initiation and maintenance of smooth-pursuit eye movements in monkeys, *J. Neurophysiol.* **58**: 1351-1369.

Munoz, D. P., and Wurtz, R. H., 1995, Saccade-related activity in monkey superior colliculus. II. Spread of activity during saccades, *J. Neurophysiol.* **73**: 2334-2348.

Newsome, W. T., Wurtz, R. H., Dursteler, M. R., and Mikami, A., 1985, Deficits in visual motion processing following ibotenic acid lesions of the middle temporal visual area of the macaque monkey, *J. Neurosci.* **3**: 825-840.

Neary, C., 1986, *Control of Monkey Smooth Pursuit Eye Movements in Open-Loop and Closed-Loop Conditions*. Unpublished doctoral dissertation, State University of New York State College of Optometry.

Neary, C., Pola J., and Wyatt, H. J., 1987, Target position: a stimulus for smooth pursuit eye movement in the monkey, in: *From Physiology to Cognition*, J. K. O"Regan and A. Levy-Schoen, eds., Elsevier, Amsterdam, pp. 257-363.

Optican, L. M., and Robinson, D. A., 1980, Cerebellar-dependent adaptive control of primate saccadic system, *J. Neurophysiol.* **44**: 1058-1076.

Pola, J., 1976, Voluntary saccades, eye position, and perceived direction, in: *Eye Movements and Psychological Processes*, R. A. Monty and J. W. Senders, eds., Lawrence Erlbaum Associates, Hillsdale, NJ, pp. 245-254.

Pola, J., and Wyatt, H. J., 1980, Target position and velocity: the stimuli for smooth pursuit eye movement, *Vision Res.* **20**: 523-534.

Pola, J., and Wyatt, H. J., 1997, Offset dynamics of human smooth pursuit eye movements: effects of target presence and subject attention, *Vision Res.* **37**: 2579-2595.

Pola, J., and Wyatt, H. J., 2001, The role of target position in smooth pursuit deceleration and termination, *Vision Res.* **41**: 655-669.

Quaia, C., Lefevre, P., and Optican, L. M., 1999, Model of the control of the saccades by superior colliculus and cerebellum, *J. Neurophysiol.* **82**: 999-1018.

Quaia, C., and Optican, L. M., 1997, A model with distributed vectorial premotor bursters accounts for the component stretching of oblique saccades, *J. Neurophysiol.* **78**: 1120-1134.

Rashbass, C., 1961, The relation between saccadic and smooth tracking eye movements, *J. Physio., Lond.* **159**: 326-338.

Ritchie, L., 1976, Effects of cerebellar lesions on saccadic eye movements, *J. Neurophysiol.* **39**: 1246-1256.

Robinson, D. A., 1964, The mechanics of human saccadic eye movements, *J. Physiol., Lond.* **174**: 245-264.

Robinson, D. A., 1970, Oculomotor unit behavior in the monkey, *J. Neurophysiol.* **33**: 393-404.

Robinson, D. A., 1973, Models of the saccadic eye movement control system, *Kybernetik.* **14**: 71-83.

Robinson, D. A., 1975, Oculomotor control signals, in: *Basic Mechanisms of Ocular Motility and Their Clinical Implications*, G. Lennerstrand and P. Bach-y-Rita, eds., Pergamon Press, Oxford, pp. 337-374.

Robinson, D. A., Gordon, J. L., and Gordon, S. E., 1986, A model of the smooth pursuit eye movements system, *Bio. Cybernet.* **55**: 43-57.

Robinson, F. R., Straube, A., and Fuchs, A. F., 1993, Role of the caudal fastigial nucleus in saccade generation. II. Effects of muscimol inactivation, *J. Neurophysiol.* **70**: 1741-1758.

Segraves, M. A., and Goldberg, M. E., 1994, Effect of stimulus position and velocity upon the maintenance of smooth pursuit eye velocity, *Vision Res.* **34**: 2477-2482.

Schiller, P. H., True, S. D., and Conway, J. L., 1980, Deficits in eye movements following frontal eye field and superior colliculus ablations, *J. Neurophysiol.* **44**: 1175-1189.

Scudder, C. A., 1988, A new local feedback model of the saccadic burst generator, *J. Neurophysiol.* **59**: 1455-1475.

Stark, L., Vossius, G., and Young, L. R., 1962, Predictive control of eye tracking movements, *IRE (IEEE) Trans. Human Factors in Electronics.* **3**: 52-57.

Takagi, M., Zee, D. S., and Tamargo, R. J., 1998, Effects of lesions of the oculomotor vermis on eye movements in primate: saccades, *J. Neurophysiol.* **80**: 1911-1931.

Van Gisbergen, J. A. M., Robinson, D. A. and Gielen, S., 1981, A quantitative analysis of generation of saccadic eye movements for burst neurons, *J. Neurophysiol.* **45**: 417-442.

Van Opstal, A. J., and Kappan, H., 1993, A two-dimensional ensemble coding model for spatial-temporal transformation of saccades in monkey superior colliculus. *Network* **4**: 19-38.

Waitzman, D. M., Ma, T. P., Optican L. M., and Wurtz, R. H., 1991, Superior colliculus neurons mediate the dynamic characteristics of saccades, *J. Neurophysiol.* **66**: 1716-1737.

Waitzman, D. M., Optican, L. M., Ma, T. P., and Wurtz, R. H., 1988, Superior colliculus neurons provide the saccadic error signal, *Exp. Brain Res.* **72**: 649-652.

Westheimer, G., 1954, Mechanism of saccadic eye movements, *A. M. A. Arch. Ophthal.* **52**: 710-724.

Wheeless, L. L., Boynton, R. M., and Cohen, G. H., 1966, Eye-movement responses to step and pulse-step stimuli, *J. Opt. Soc. Am.* **56**: 956-960.

Wurtz, R. H., and Mohler, C. W., 1976, Organization of monkey superior colliculus: enhanced visual response of superficial layer cells. *J. Neurophysiol.* **39**: 745-765.

Wyatt, H. J., and Pola, J., 1981, Slow eye movements to eccentric targets, *Invest. Ophthal. Visual Sci.* **21**: 477-483.

Wyatt, H. J., and Pola, J., 1983. Smooth pursuit eye movements under open-loop and closed-loop conditions, *Vision Res.* **23**: 1121-1131.

Wyatt, H. J., and Pola, J., 1987, Smooth pursuit eye movements with step-ramp stimuli: the influence of attention and stimulus extent, *Vision Res.* **27**: 1565-1580.

Wyatt, H. J., Pola, J., Fortune, B., and Posner, M., 1994, Smooth pursuit eye movements with imaginary target defined by extrafoveal cues, *Vision Res.* **34**: 803-820.

Yasui, S., and Young, L. R., 1975, Perceived visual motion as effective stimulus to pursuit eye movement system, *Science.* **190**: 906-908.

Young, L. R., and Stark, L., 1963, Variable feedback experiments testing a sampled data model for eye tracking movements, *IRE (IEEE) Trans. Human Factors in Electronics.* **4**: 38-51.

Zee, D. S., Optican, L. M., Cook, J. D., Robinson, D. A., and Engel, W. K., 1976, Slow saccades in spino-cerebellar degeneration, *Arch. Neurol.* **33**: 243-251.

Chapter 11

Models of Saccade-Vergence Interactions

George K. Hung[1] and Kenneth J. Ciuffreda[2]

[1] *Dept. of Biomedical Engineering, Rutgers University, 617 Bowser Rd., Piscataway, NJ 08854-8014, PH: (732) 445-4137, FX: (732) 445-3753, EM: shoane@rci.rutgers.edu*

[2] *Dept. of Vision Sciences, State University of New York, State College of Optometry, 33 West 42nd St. New York, NY 10036; PH: (212) 780-5132, FX: (212) 780-5124; EM: kciuffreda@sunyopt.edu*

11.1 INTRODUCTION

Eye fixations in daily life are controlled by two types of eye movements. Saccades rotate the two eyes in the same direction (i.e., conjugately; Fig. 11.1A), such as during reading and scanning of a scene, whereas vergence movements rotate the two eyes in opposite-directions (i.e., disjunctively; Fig. 11.1B), such as during tracking of objects moving in depth. Together, they provide control of binocular fixation in three-dimensional space

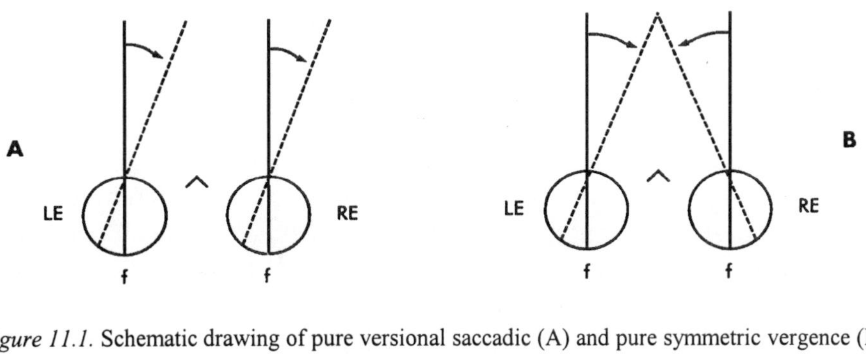

Figure 11.1. Schematic drawing of pure versional saccadic (A) and pure symmetric vergence (B) eye movements. LE, left eye; RE; right eye , f, fovea. Reprinted from Ciuffreda and Tannen (1995), pg. 2, Fig. 1-1, with permission of Harcourt Health Sciences.

(Ciuffreda and Tannen, 1995). Since nearly all of our naturally-occurring target shifts are asymmetrrical in nature, responses generally involve both saccade and vergence and their interactions (see Fig. 11.2), rather than either purely saccadic or vergence eye movements (Fig. 11.1).

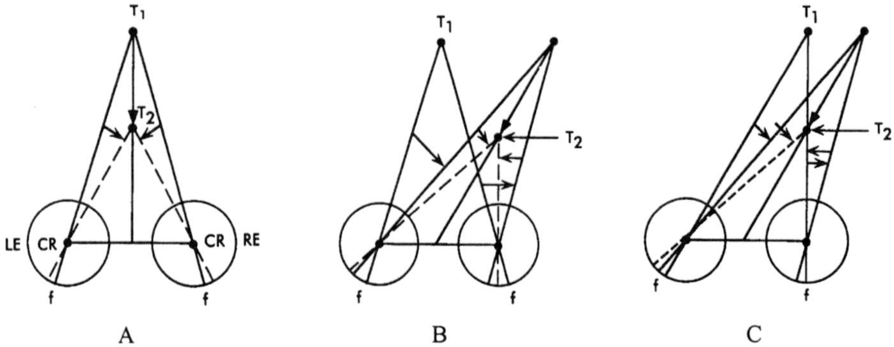

Figure 11.2. Symmetrical versus asymmetrical disparity stimuli and overall response patterns. A. Symmetric vergence. B. Asymmetric vergence. C. Line-of-sight asymmetric vergence. T_1, Initially fixated target; T_2, subsequently fixated target; f, fovea; *CR*, center of rotation of the eye; *LE*, left eye; *RE*, right eye. Reprinted from Ciuffreda and Tannen (1995), pg. 132, Fig. 6-6, with permission of Harcourt Health Sciences.

Eye movements have been monitored non-invasively using a variety of techniques, including infrared reflection, video, and magnetic search coil (Ciuffreda and Tannen, 1995). The measured responses accurately reflect the brain's control strategy for directing the eye movements, thereby providing insight into the underlying neuronal control of the oculomotor responses. It would appear that the saccadic and vergence neuronal control programs can be inferred directly from the oculomotor responses. Indeed, when the saccadic and vergence movements occur at different times, it is easy to distinguish between the rapid (typically 20 to 60 msec) conjugate eye rotations of the saccade (Yarbus, 1967; Bahill 1981; Fig. 11.3A) and the relatively slow (typically 800-1200 msec) disjunctive eye rotations of vergence (Hung et al, 1997; Fig. 11.3B). However, since the saccadic and vergence systems share a common plant (i.e., the extraocular muscles and eyeball), when these movements occur simultaneously, their individual control programs may not be so readily discerned. Moreover, some simultaneously-occurring eye movements contain a component that cannot be accounted for by linear superposition of saccade and vergence contributions (Kenyon et al, 1980; Miller et al, 1980).

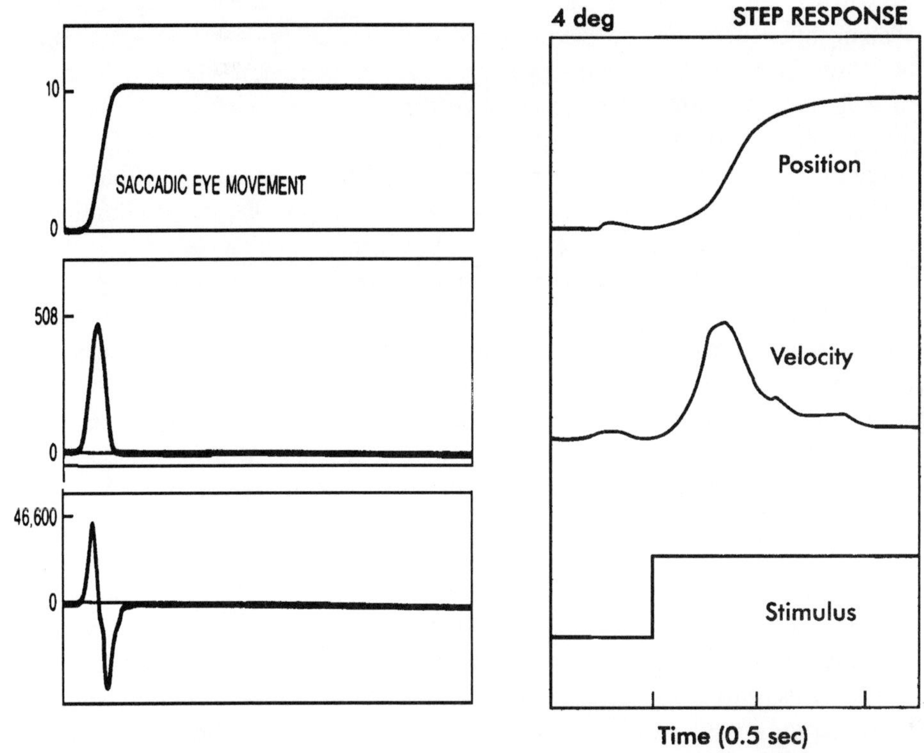

Figure 11.3. (Left) Model simulation showing dynamics of saccadic eye movement. Eye position (top, deg), velocity (middle, deg/sec), and acceleration (bottom, deg/sec^2) traces are shown over a time range of 500 msec. Reprinted from Bahill and Stark (1979), pg. 112, with permission of Sci. Am. (*Right*) A 4-degree convergence step stimulus (*bottom*) resulted in disparity vergence response (*top*) with latency of 180 msec and peak velocity of 12 deg/sec (*middle*) occurring 150 msec after the vergence response begins. The total time range for the plot is 2 sec. Reprinted from Schor et al (1986), pg. 615, Fig. 4, with permission of Optom. Vis. Sci.

Some investigators (Enright, 1984, 1986, 1992; Erkelens, 1989; Zee et al, 1992) believed that this was due to vergence being facilitated by saccades, as evidenced by the transient, higher-velocity, disjunctive time course seen during combined saccade-vergence responses to asymmetrical targets (Kenyon and Ciuffreda, 1978). Yet, it is difficult to reconcile this facilitation notion with the approximately 200 msec latency of saccades, since it would take another 200 msec to respond to the transient dynamics when saccade and vergence movements coincided. If facilitation based on feedback were to occur, the ongoing vergence movement would be augmented by a signal meant to assist its position 200 msec earlier. Thus, instead of facilitating the vergence movement, it could actually lead to system instability. Others have argued that facilitation is pre-programmed based on the target positions. However, this cannot explain why in a number of situations in which there is an asymmetrical target change, a transient divergence occurs when convergence would be required to facilitate the vergence movement. Moreover, saccadic (Zuber and Stark, 1966) and vergence (Hung et al, 1989, 1990) suppression during the actual combined movements preclude accurate guidance of vergence movements by saccades. Hence, there is considerable evidence against the facilitation notion.

To clarify these apparently contradictory results, a more systematic approach is needed. Models of saccade and vergence movements and their interactions can provide the necessary quantitative analysis for the following reasons: First, a saccade-vergence model requires a formal structuring of the important components in the system. This demands a deep conceptual understanding of visual optics, anatomy, muscle biomechanics, and neurophysiology of the oculomotor components. Also, the interconnections within a model require understanding and insight into the unique anatomical and neurological connectivity between the saccadic and vergence systems. Second, a model is needed because of the complexity of the system. Since eye movement responses involve temporal dynamic interactions between the two systems, only a model can keep track of the multitude of rapid changes in parameter values within the model. And, third, model simulations may provide new insight and understanding of these systems that would not have been possible using traditional descriptive or graphical techniques.

Some of the earlier saccade-vergence models are discussed below. In addition, a recent robust quantitative model of saccade-vergence interactions, which provides significant insight into the underlying mechanisms in helping to resolve the previous controversies, is discussed in detail (Hung, 1998a,b). The control processes developed in this model were based on known neurophysiological signal pathways and extraocular muscle innervations. In essence, it was found that the small difference in latency between ipsilateral and contralateral signal pathways was responsible for the transient divergence mentioned earlier. Model responses under different stimulus conditions were found to be consistent with known experimental results. Perhaps most significantly, the model did not require complicated higher-center processes to modify the eye movement dynamic timecourse, as others (Enright, 1984, 1986, 1992; Erkelens, 1989; Zee et al, 1992) have hypothesized. Thus, this saccade-vergence model, which is discussed later in the chapter, provides a realistic, relatively simple, clearly-defined, and robust framework for both the understanding and quantitative assessment of saccade-vergence interactions.

11.2 BACKGROUND

11.2.1 Stimulus Configuration

Target displacements equivalent to those occurring in free-space can be re-created optically in the laboratory (Fig. 11.4). Linear displacements of the targets on the oscilloscopes presented to each eye result in the percept of a target shift in 3-D space. In this way, various types of target movements can be presented to the subject.

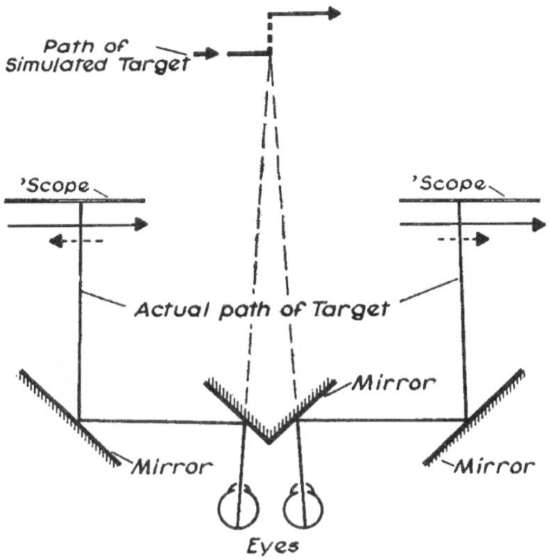

Figure 11.4. Mirror stereoscope through which stimuli are viewed. Stimuli moving at the same velocity but in different directions on the scope (dotted arrows) appear as a single stimulus moving toward or away from the viewer. Stimuli moving at the same velocity in the same direction (solid arrows) appear as a single stimulus moving to the left or right in front of the viewer. Reprinted from Ono (1983), pg. 382, Fig. 11.3, with permission of K. J. Ciuffreda, the copyright holder.

11.2.2 Saccadic Eye Movements

Saccadic eye movements refer to the rapid conjugate rotations of the eyes in response to target changes in the same depth, or isovergence plane (see Figs. 11.1A and 11.4). They have been found to exhibit a latency of about 180 to 220 msec (Stark, 1968), and durations ranging from 25 msec for 0.1 deg amplitude to 125 msec for 50 deg amplitude (Bahill & Stark, 1979). Furthermore, saccade dynamics exhibit general characteristics described by its peak velocity-amplitude relationship, or main sequence (Bahill & Stark, 1979), having a slope of about 50 (deg/sec)/deg.

Two types of saccadic models have been proposed. In the sampled-data model, saccades consist of open-loop movements, i.e., once they are initiated, their dynamics cannot be modified over the intervening 200 msec interval (Young and Stark, 1962; Semmlow & Venkiteswaren, 1976). This is applicable under normal viewing conditions. On the other hand, in the continuous feedback model, target changes can alter the saccadic responses during the 200 msec interval. This occurs under special conditions in which target pulses are introduced (Aslin & Shea, 1987; Becker & Jurgens, 1979; Hou & Fender, 1979; Wheeles et al, 1966).

To test whether the open- (i.e., sampled-data model) or the closed-loop (continuous feedback) model of the controller is more applicable, a technique in which the variability of amplitude and duration of saccadic step responses has been used (Jurgens et al, 1981). The source of the variation has been attributed to changes in firing frequency or recruitment of the neurons that constitute the pulse generator (Keller, 1977; Keller & Robinson, 1972; King & Fuchs, 1977; van Gisbergen et al, 1981). A preprogrammed, open-loop model would be expected to show variations in both amplitude and duration, whereas a local feedback closed-loop model would be expected to vary in duration only. The results for both normal and Diazepam (a tranquilizer known to reduce saccadic velocity) experiments favored the local feedback model.

In reality, experimental results suggest that probably <u>both</u> open- and closed-loop models are correct. This is because the refractory period of the saccadic system appears to have an early relative portion, in which rapid stimulus changes can re-trigger and modify the saccade, and a later absolute portion (during the last 75 msec or so) in which the saccade is not re-triggerable (Ciuffreda and Tannen, 1995).

11.2.3 Vergence Eye Movements

Vergence eye movements are relatively slow, oppositely-directed, or disjunctive, rotations of the eyes that bring about a single, binocular 3-dimensional percept of objects moving in depth (Riggs and Niehls, 1960; Rashbass & Westheimer, 1961b; Westheimer & Mitchell, 1956) (see Figs. 11.1B & 11.3B). These disjunctive movements have been thought previously as being controlled by a simple continuous feedback system (Krishnan & Stark, 1977; Rashbass & Westheimer, 1961b; Schor, 1979; Zuber & Stark, 1968). However, such a continuous model could not account for a number of experimental results. For example, the model response to sinusoidal input exhibited a phase lag that was much larger than that found in the experimental data (Rashbass & Westheimer, 1961b). Also, the response to a step followed by a ramp (i.e., smooth, continuous target change of disparity) in the opposite direction exhibited a reversal in direction well before the target moved past the zero disparity level, which indicated that velocity information was used to anticipate target position (Rashbass & Westheimer, 1961b). Furthermore, the response to a ramp of disparity exhibited multiple movements that could not be simulated simply as visual feedback oscillations (Hung et al, 1986). The primary problem with a continuous feedback model was that the time constant of a step response (about 200 msec) was equal or nearly equal to the response latency (160 to 200 msec) (Hung et al, 1983). Thus, the response to a rapidly changing periodic stimulus, such as a sinusoid, would not be fast enough to account for the measured short experimental latencies if it simply consisted of a passive continuous-feedback slow movement following a 200 msec time delay. Furthermore, due to its relatively long latency and time constant, a conflict could arise between instantaneous error and actual target position, so that the model responses to either steps or ramps would show marked instability oscillations.

Over the past decade, we have accumulated a body of evidence clearly demonstrating that the vergence system is composed of two components or subsystems: a fast, open-loop movement which brings the eyes near the target position (Hung et al, 1986; Semmlow et al, 1986, 1993, 1994), followed by a much slower movement under visual feedback control that reduces the residual error (or fixation disparity) to a few minutes of arc, i.e., within foveal Panum's fusional area (Panum, 1858). This was clearly shown using ramp disparity stimuli of various velocities (Hung et al, 1986; Semmlow et al, 1986) (see Fig. 11.5a). For velocities less than about 1.4 degrees of disparity per second (deg/sec), the response consisted mainly of smooth ramp tracking of the target, whereas for a velocity greater than 2.7 deg/sec, the response consisted of multiple-steps in which the termination of each step approximately matched the ramp target position. The latter indicated that the vergence system used an estimate of target velocity to generate the appropriate vergence step movement (Hung et al, 1986). Intermediate velocities resulted in responses containing a mixture of step and ramp movements. Also, the steps in the multiple-step responses were found to lie on the normal main sequence for vergence eye movements (Bahill & Stark, 1979), having a slope of about 4 (deg/sec)/deg. Thus, the open-loop component estimated the future position of the ramp stimulus, and then generated a normal step movement to match the position of the stimulus. As the ramp stimulus progressed, successive steps were generated to follow the ramp accurately. Moreover, when instrument feedback was used to open the vergence feedback loop (Semmlow et al, 1993, 1994), the response to a step disparity also consisted of a series of steps that were on the main sequence, which lent further support to the concept of an open-loop fast component.

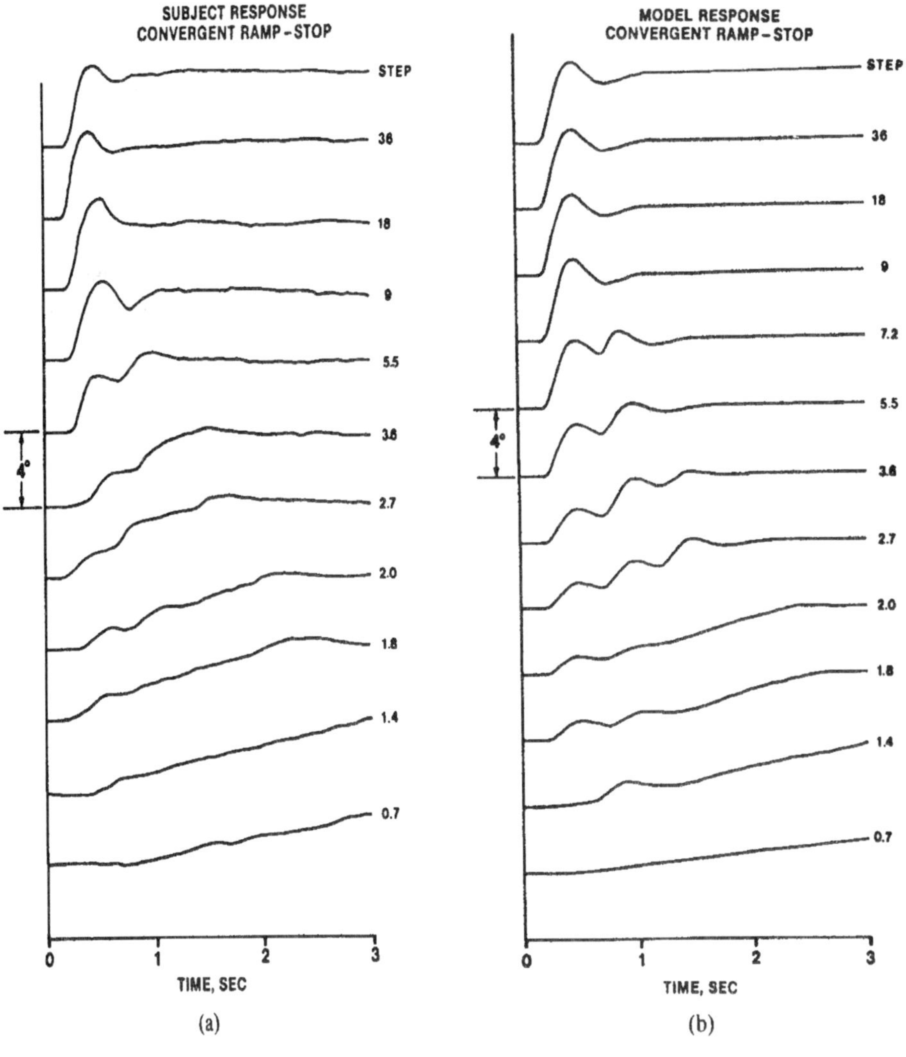

Figure 11.5. Convergent ramp responses up to 4 deg amplitude are shown for (a) experimental and (b) model simulation conditions. Experimental curves are individual responses of a subject. Ramp velocity, in deg/sec, is shown next to each curve. Top curve for convergent step responses. Reprinted from Hung et al (1986), pg. 1025, Fig. 3, with permission of © IEEE.

To simulate the experimental responses to various stimuli accurately, a dual-mode vergence model was developed by Hung et al (1984) (see Fig. 11.6a-c). In this model, both the fast and slow components have disparity (or vergence error) and velocity thresholds to provide operation within their respective ranges. The fast open-loop component used target position and velocity information to produce the appropriate step movement, and the slow closed-loop component used visual feedback to reduce the disparity and fuse the binocular retinal images. Simulation results showed accurate fit to the experimental responses for pulse, step, and ramp stimuli, and in addition showed reduced phase lag to the sinusoidal stimulation mentioned earlier (Hung et al, 1984) (see Fig. 11.5b). Thus, this dual-mode model provided an accurate representation of the vergence system that simulated vergence responses to a variety of stimuli. Moreover, recent findings in our laboratory showed that vergence responses to symmetric step disparities under a variety of naturalistic conditions (e.g., in the light and dark, under voluntary control, etc.) also fell on the main sequence for simple step disparities, thus indicating a common motoneuronal controller signal for the generation of symmetric vergence step responses regardless of the stimulus spatial pattern or mode of initiation (Hung et al, 1994).

Figure 11.6. (See next page). (a) The overall model of the vergence system showing slow and fast components in the forward loop. The slow and fast component responses are summed to give VR1. Internal feedback from VR1 is summed with vergence error VE to give an estimate of target position. The plant represents the mechanical properties of the eyeball and musculature and is assumed to have unity gain for the vergence stimulation. The vergence response VR is subtracted from the vergence stimulus VS to give the vergence error VE. *(b)* Slow component in the forward loop: delayed vergence error (VE1) is VE delayed by 200 msec (DELAY1). Error magnitude limiter (up to 1 deg) and error velocity limiter gain element (up to 2 deg/sec) simulate range of slow component dynamics. The time constant 1/A is 10 sec. Gain G1 was determined via simulation to be 30. (c) Fast component in the forward loop: the vergence error VE is summed with VR1 to give an estimate of the target position. The delay element (DELAY2) represents the effective delay throughout the fast component. The estimated target velocity above a threshold of 1.7 deg/sec is used to trigger the sampler. The sampler enables the predictor to use estimated target position and velocity to predict the future position of, for example, the ramp stimulus. After triggering, the threshold increases slightly to 2.1 deg/sec. This accounts for the initial step but subsequent smooth following seen in response to 1.8 deg/sec ramp stimulus. If estimated stimulus velocity remains constant, the sampler repeats every 0.5 sec. This accounts for the staircase step-like responses to ramp stimuli. Sudden large changes in velocity will reset the sampler. This accounts for the ramp-pulse data. The predictor also reduces its calculation time, thus reducing DELAY2, for repetitive stimuli such as sinusoids. This accounts for the small phase lag found in sinusoidal responses. Reprinted from Hung et al (1986), pg. 1023, Fig. 1, with permission of © IEEE.

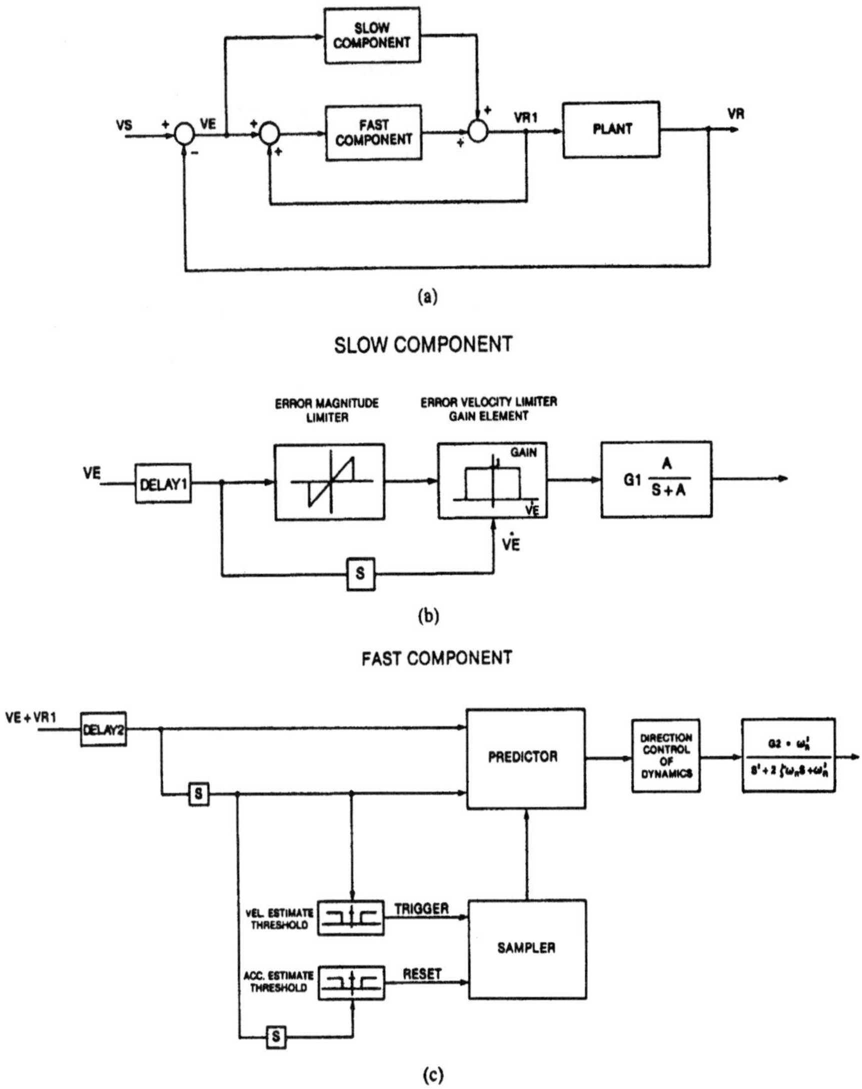

Figure 11. 6 - See figure legend on previous page.

11.2.4 Saccade-Vergence Interactions

If we had only one eye, the innervational pattern for directing one's gaze, for example to the right, would be relatively simple. It can be describe by the Descartes-Sherrington principle of reciprocal innervation (Ciuffreda and Stark, 1975) (see Fig. 11.7A), whereby the right extraocular muscle (EOM) receives a positive increment, and the left EOM simultaneously receives a negative increment, of innervation. However, since we have two eyes, both eyes need to be innervated to direct one's gaze in space appropriately (see Fig. 11.7B), and this could result in relatively complicated innervational patterns, especially for asymmetrical target changes. Two primary innervational scenarios are possible. In the first, the two eyes are innervated separately as two independent units. The brain pre-calculates the exact amount of innervation needed in the extraocular muscles of each eye, so that the two eyes are driven independently and directly to their final bifixation position. In the second, the innervations retain their separate drives to the saccadic and vergence systems. The brain in essence calculates the "cyclopean" (i.e., single common eye) amount of version (i.e., lateral shift) and vergence (i.e., disparity-based depth change) needed, and distributes half of the these version and vergence amounts to each eye, and then drives the two eyes using the separate saccade and vergence signals that are combined at each eye to arrive at the final bifixation position (see Figs. 11.1A,B. Therefore, although two sets of signals (saccade and vergence) are used, they innervate the two eyes together as if they were one unit.

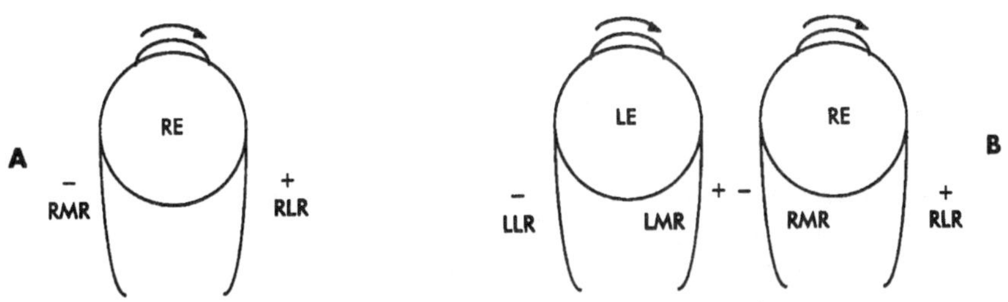

Figure 11.7. A. Descartes-Sherrington law of reciprocal innervation. B. Hering's Law of Equal Innervation. *RE*, right eye; *LE*, left eye; +, increased innervation or excitation; -, decreased innervation or inhibition; *RLR*, right lateral rectus; *RMR*, right medial rectus; *LLR*, left lateral rectus; and *LMR*, left medial rectus. Reprinted from Ciuffreda and Tannen (1995), pg. 6, Fig. 1-3, with permission of Harcourt Health Sciences.

Hering's theory of binocular vision is based on the second scenario (Hering, 1977), whereby the two eyes are innervated as though they were one. This concept of equal innervations for saccade and vergence movements in the two eyes was codified as Hering's Law of Equal Innervation (Hering, 1977; Ono, 1983; Findlay and Harris, 1993). It should be noted that Hering did not consider unequal movement of the two eyes to be inconsistent with his hypothesis (Ono, 1983). This is because summation of saccade and vergence signals could result in unequal interocular movement amplitudes, even though the underlying control signals were simply the required cyclopean-determined amounts of version and vergence innervations.

Hering's law, however, requires an intact vergence system. For example, Alpern and Ellen (1956) found that under normal binocular viewing conditions, the eye movements were consistent with Hering's law. However, for accommodative vergence movements (with the non-fixating eye occluded, and therefore with disparity vergence feedback rendered ineffective), the viewing eye remained relatively fixed during such a movement, thus suggesting a gross violation of Hering's law. The main difference is that the disparity vergence feedback loop is open during accommodative vergence, but is closed under binocular viewing.

Some investigators (Clark and Crane, 1978; Steinman and Collewijn, 1980; Ono, 1983; and Erkelens et al, 1989) have considered unequal movements of the two eyes under binocular viewing to be a violation of Hering's law. For example, Ono (1983) set out to investigate the implications of Hering's law on a combination of version and vergence movements. An example of one of his recorded asymmetrical eye movement traces along with a geometrical view of the movement is shown in Fig. 11.8. Ono (1983) concluded from his analyses that "... when the disjunctive movement combines with the slow conjunctive movement, the difference in velocities for the two eyes is slightly less than that predicted from the additivity proposition. When the disjunctive movement combines with the saccadic movement, the difference in magnitude and velocity are larger than those predicted." Thus, he observed several different patterns that did not fit the additivity hypothesis, which suggested a violation of Hering's law. Erkelens et al (1989) measured oculomotor responses to large target shifts of up to 40 and 50 deg and found that unequal saccades could account for 95% of the vergence required for divergence and 75% for convergence. Again, this suggested that the movements violated Hering's law. However, it can be shown that these apparent violations of Hering's law can be explained by the difference in latencies between the two eyes due to differences in their individual neuronal pathways (see Differential Latency

Theory, detailed in Section 11.3; Hung, 1998a,b). In fact, these movements can actually be shown to conform with Hering's law.

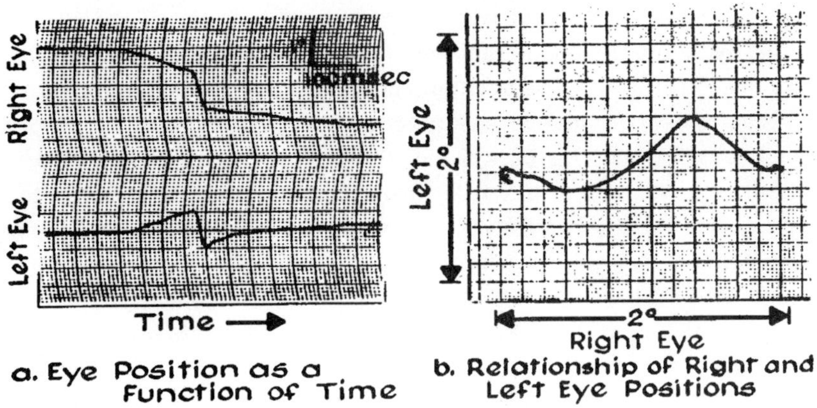

Figure 11.8. a) Sample record of eye movement when the target is abruptly moved to a different distance and different direction, equivalent to 0 deg in the LE and 2 deg leftward in the RE b) Representation of the same movement in Cartesian (LE - rightward is up, vs.RE - rightwards is to the right) coordinates. Movement begins at right with an initial convergence (RE leftward and LE rightward), followed by saccade (both RE and LE leftward), and then followed by additional convergence to the final position. Reprinted from Ono (1983), pg. 383, Fig. 11.4, with permission of K. J. Ciuffreda, the copyright holder.

Other investigators (Enright, 1984, 1986, 1992; Collewijn et al, 1988; Maxwell and King, 1992; Zee et al, 1992; Erkelens et al, 1989) proposed that not only some movements violated Hering's law, but that the vergence movement was in fact facilitated by the ongoing saccade. For example, Maxwell and King (1992) found in macaque monkeys that for symmetric stimuli, vergence peak velocities fell on the normal main sequence (Hung et al, 1997). However, for asymmetrical stimuli, the vergence peak velocities, which were taken simply as the difference between the right and left eyes, were much higher than main sequence values. They believed that the oculomotor system was able to use these disjunctive transients to shorten the length of time required to complete the vergence movements. Zee et al (1992) investigated oculomotor responses to step stimuli in space. They used a graphical method, which projected a combined vergence trace onto a pure vergence trace of the expected amplitude, and measured the difference. Any difference was presumed to reflect a facilitation process. They concluded that horizontal vergence was facilitated by both horizontal and vertical saccades, although more so with horizontal saccades. However, the transient vergence movements found in these studies often occurred in the opposite direction to that required for the intended vergence movement. Thus, they did not assist, and in many cases actually opposed, the ongoing vergence movement. As discussed in detail in Section 11.3 (Hung, 1998a,b), transient divergence during combined saccade-vergence movements is an inherent consequence of the latency difference between the neural pathways in the two eyes, and not due to facilitation of the vergence movement.

Enright (1984; 1986; 1992) examined changes in vergence movements during responses to asymmetrical targets, with normal disparity, blur, and proximal cues present. He used a video recording technique at a relatively slow frame interval of 33 msec to record eye movements. He found that a large portion of the total change in vergence occurred during saccades, with values ranging from 40% for certain tasks to 100% when large version was combined with a small vergence stimulus. Enright also noted that additivity of vergence and saccadic movements was violated, since the rate of vergence change during saccades was much greater than that found either before or after the saccade. He therefore concluded that vergence was facilitated by the saccade during the combined movement. However, since dynamic changes, such as small to moderate saccadic movements, could occur within the 33 msec. sampling interval he used, the presumed facilitation may have occurred, but it could also have been an artefact of under-sampling.

11.2.5 Saccadic Suppression/Omission and Vergence Suppression

The numerous oculomotor movements one makes in daily life produce rapid changes in the retinal image of the visual scene. For example, a saccade is accompanied by a complete lateral shift in the visual frame of reference, a blink briefly darkens the field of view, and a vergence movement changes the binocular reference to depth. Nevertheless, an observer perceives the world as being both continuous and stable. Much of this sensation of perceptual continuity can be attributed to the correlated transient reduction in visual sensitivity both immediately before, during, and after the motor response. This has been observed for saccades (Alpern, 1953; Stark, 1968, Campbell and Wurtz, 1978), blinks (Volkmann et al, 1980), and vergence (Manning and Riggs, 1986; Hung et al, 1989, 1990). Thus, despite the apparent continuity of perception, during a combined saccade-vergence movement, there is a true reduction in visual sensitivity for up to 100 msec or so. Such a disruption would preclude accurate visual feedback guidance and control of vergence by saccades. Thus, this too does not support the notion of dynamic facilitation.

11.2.6 The Fast Learning Mechanism for Asymmetrical Targets

van der Steen and Bruno (1995) noted that "When the objects of interest are at optical infinity, saccades are essentially conjugate, with the exception of a small transient divergence during the saccade (Collewijn et al, 1988; Zee et al, 1992). However, when we change fixation between objects at different directions and distances, as we do under natural circumstances, saccades become disconjugate which allows to accomplish most of the required vergence during the saccade (Erkelens et al, 1989) ". Recently, investigators proposed a fast learning mechanism to provide appropriate disconjugacy in response to asymmetrical targets following brief periods of training (Eggert and Kapoula 1995; Kapoula et al, 1995; van der Steen and Bruno, 1995; Averbauch-Heller et al, 1999; Bucci et al, 1999). The training consisted of, for example, viewing through different parts of prisms in the two eyes which displaced the images, or viewing different-size images in the two eyes. These results indicate that some type of motor learning may have occurred. Yet, these peripheral modifications, which occur perhaps at or beyond the final common pathway for each eye, are still consistent with Hering's law of equal central innervation to the two eyes.

11.2.7 The Neurophysiology of Saccade-Vergence Interactions

Tamler et al (1958) used electromyography to examine whether the lack of movement in the viewing eye during asymmetrical vergence violated Hering's law. This involved inserting a small microelectrode into selected extraocular muscles and observing changes in firing frequency of muscle motor units. They found that there was co-contraction of the opposing horizontal recti muscles of the viewing eye during asymmetrical vergence. Hence, they concluded that this confirmed Hering's view that there was a peripheral adjustment (i.e., cancellation) of opposing version and vergence central innvervations, and thus Hering's law was not violated during asymmetrical vergence.

Keller and Robinson (1972) found that in the monkey abducens units there was "no type of eye movements, version or vergence", that was "the exclusive product of a particular subset of oculomotor neurons". In an additional experiment, they tested accommodative vergence, in which the viewing eye remained relatively stationary while the fellow occluded eye varied in position due to blur-driven accommodative vergence. From these experiments, they concluded that the "version and vergence commands are summed centrally with the net result appearing as more global neural activity in a shared final common path. That is, during asymmetric vergence, the vergence and version neural commands obey reciprocal innervation and simply cancel in each motor nucleus of the horizontal recti of the non-moving eye" (see Fig. 11.7). Thus, their conclusion supports Hering's law.

Mays (1984), and Judge and Cummings (1986), found that premotor neurons in the mesencephalic reticular formation, 1-2 mm dorsal and dorsolateral to the oculomotor nucleus, were involved specifically in the control of vergence and discharged in relation to convergence angle. Also, Mays et al (Mays & Porter, 1984; Mays et al, 1986) found motoneurons that displayed signals related to both position and velocity of the movement. In addition, there were a small number of divergence burst neurons, and burst-tonic cells that combined vergence velocity and position, respectively, in their output (Zee & Levi, 1989). Anatomical studies of the oculomotor nucleus revealed three aggregates of cells: subgroup A was located in the ventral and rostral region, subgroup B was located in the dorsal and caudal region, and subgroup C located in the dorso-medial and rostral region. Zee and Levi (1989) speculated that subgroup C, comprising of the smallest cell bodies that were traced using radioactive label to the small outer layer fibers of the medial rectus muscle, was perhaps selective for vergence. Moreover, in comparing the vergence and pursuit subsystems, the position and velocity signals from these different subsystems were found to be generated independently and were distributed to motoneurons in a way that was not highly correlated between the subsystems (Gamlin & Mays, 1992). Finally, Gamlin and Clarke (1995) found cells in the medial nucleus reticularis tegmenti pontis (NRTP) that increased their activity for far responses and other cells that increased their activity for near responses. Although these cells were often encountered close to neurons that displayed saccade-related activity, their activity did not increase with conjugate eye movements. Therefore, there was anatomical and neurophysiolgical evidence for <u>separate control of vergence and version</u>, although it may be speculated that the physical proximity of neurons for these two systems allowed for interactive influences. Nevertheless, as further support for separation of control, Keller and Robinson (1972) showed that although the control signals for saccade and vergence travelled along the same motoneuron in the final common pathway, there was no evidence of (nonlinear) interaction between saccade and vergence signals. Therefore, these neurophysiological findings also support Herings law of equal <u>central</u> innervation to the two eyes.

11.2.8 Models of Saccade-Vergence Interactions

Zee et al (1992) proposed three models for saccade-vergence interactions. The first model hypothesized the existence of a separate class of saccade-related vergence burst neurons (SVBN) (see Fig. 11.9), which were gated by the omindirectional pause neurons (OPN) and generated premotor horizontal vergence commands but only during saccades. The second model hypothesized separate right and left eye saccade burst neurons that received not only conjugate, but also equal and oppositely-directed vergence error signals. The difference between the right and left eye burst neurons was used to produce a saccade-related vergence command. The third model proposed that the facilitation of vergence during saccades was a result of an increase in gain of premotor vergence velocity neurons caused by a lifting of the inhibition by the OPN during saccades. Their model simulation results favored the first and third models. They also proposed that disconjugate adaptation to, for example, wearing anisometropic spectacle correction, which creates unequal and also continually different amounts of prism vergence demand through different parts of the spectacle lens in each eye (thus resulting in asymmetrical saccades), could be accounted for in their model by modification of the SVBN to link a change

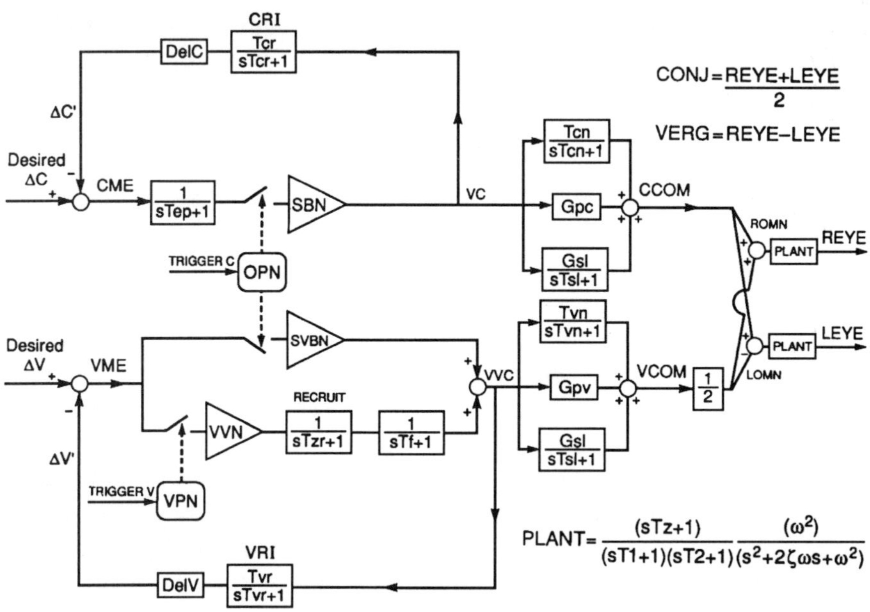

Figure 11.9. Saccade-Related Vergence Burst Neuron (SVBN) model of Zee et al (1992). Reprinted from Zee et al (1992), pg 1638, Fig. 19, with permission of The American Physiological Society.

in ocular alignment automatically with saccades, even in the absence of disparity cues. However, it should be noted that such linkage would require the SVBN to be continually changing and responding nearly instantaneously to each change of gaze. Such a mechanism seems to be physiologically unrealistic.

Mays and Gamlin (1995) presented a similar model in which the inhibition by the OPN is shared by the saccadic and vergence systems. They proposed that there is normally a "weak" inhibition of the vergence burst cells by the OPN to allow them to fire in the absence of saccades. The inhibition is released during a saccade in a mixed vergence-saccade movement to result in unequal saccades.

Both of the above models involve modification of neural signal processing of vergence during the saccade. However, since saccadic latency is approximately 200 msec, it would require 200 msec before the saccade can appropriately modify the vergence movement. That is, the modification would be for the stimulus disconjugacy that occurred 200 msec earlier. Thus, such a rapid-modification mechanism also appears to be physiologically unrealistic

Moreover, it may be argued that the saccade and vergence signals are pre-planned based on the target positions. Such a saccadic modification of vergence would appear to be purposeful, and therefore should assist vergence in arriving at it final position. However, it can be shown (see Hung, 1998a,b; Collewijn et al, 1997) that in a number of experimental conditions when a convergence movement is required, the transient movement is that of divergence, which is the opposite of that required for a purposeful assisted movement. Thus, neither timing nor directionality are consistent with these models.

This leads to a dilemma as to why the visual system would process and modify responses that do not lead to a benefit in responsivity. Indeed, to modify the response, processing would be required to determine whether a saccade and vergence are about to occur simultaneously, and then energy expended to provide the inhibitory signals, but without any apparent benefit. This would not be consistent with parsimony of neural signal processing exhibited in the nervous system (Glassman, 1999). The answer may lie in the fact the transient vergence during saccades is simply a result of neuronal latency difference between the two eyes, and thus there was no intended purposeful, higher-level, modification of vergence during saccades. Such a mechanism has been proposed by Hung (1998a,b) in the Differential Latency Theory, and is described in Section 11.3.

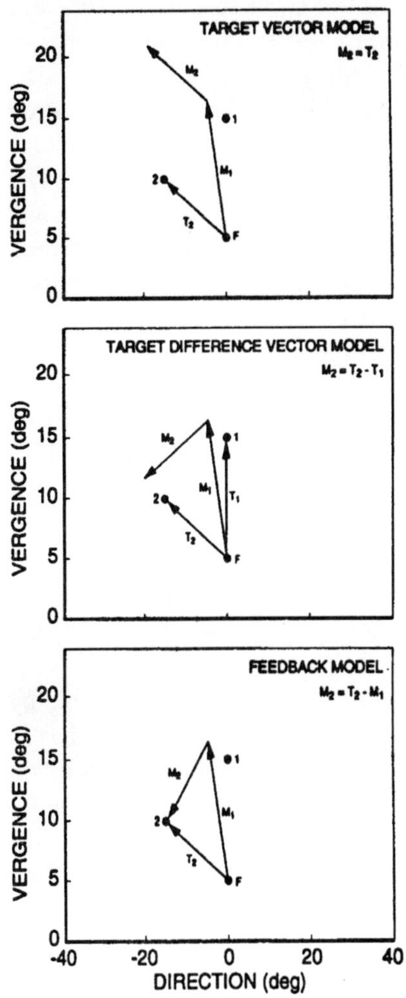

Figure 11.10. Models tested by Krommenhoek and van Gisbergen (1994). The three models have different predictions for the second movement, as illustrated by an example where the first movement was not precisely directed at the first target (1). Note that only the feedback model can correct the second movement to the final target (2) for an error in the first movement (T_1 first target position, T_2 second target position, M_1 first movement, M_2 second movement, F fixation position during target presentation). Reprinted from Krommenhoek and van Gisbergen (1994), pg. 96, Fig. 1, with permission of Springer-Verlag.

Krommenhoek and van Gisbergen (1994) examined eye movement responses in 5 subjects to test three hypothesized models regarding the control of version and vergence (see Fig. 11.10). They used a paradigm in which two target positions in direction and depth were presented in sequence; after the second target was extinguished, the subject was instructed to direct his fixation in the dark to the imagined position of the two targets in the same sequence. In all three models, the eyes would initially make a movement to the first target, T1. For Model 1 (Target Vector), it is predicted that the eyes would then make a movement to the second target, T2, based on the vector direction from the subject to T2 prior to the initiation of the movement sequence. For Model 2 (Target Difference), it is predicted that the eyes would move based on the vector difference between T1 and T2 estimated prior to initiation of the movement sequence. Finally, for Model 3 (Feedback), it is predicted that the eyes would move based on non-retinal local feedback loops (Scudder, 1988) derived from the efference copy signals. They found that the Feedback Model provided the best fit to the experimental data. Moreover, they proposed that a higher center, such as the lateral intraparietal (LIP) area in the parietal cortex, was involved in the control of saccade and vergence eye movements, where the commands for version and vergence components of binocular gaze shifts may be represented by a single control center.

Recently, Chaturvedi and van Gisbergen (1999) produced brief-duration electrical microstimulation of the deep layers of the right superior colliculus in the monkey to elicit a nearly pure leftward conjugate movement. The microstimulation was then applied during eye movements to asymmetrical targets in space to determine whether such stimulation influenced the timing and metric (amplitude and direction) of the vergence portion of the resultant response. Experimental results showed that whereas early electrical stimulation resulted in a clear vergence component, perturbation during midflight "markedly curtailed the ongoing vergence component." Thus, instead of facilitating the vergence response, electrical stimulation that should have mimicked a saccadic assist signal actually reduced the vergence response. They proposed a model in which stimulation of the superior colliculus elicited not only a conjugate signal, but also a "zero vergence-change" signal, which competed with the visually-driven convergence signal to result in a compound (step-slide-step shaped) time course. This simulated the diminution of the vergence response, but was not able to model any proposed facilitation.

The source of saccadic disconjugacy has also been described by Averbuch-Heller et al (1998). They ascribed it as being primarily due to faster "abducting saccades ... than adducting saccades", and is thus based on difference in saccadic speeds rather than latencies, as in Hung's theory (Hung, 1998a,b). However, in the reference that they cited, Abel et al (1979) actually found "no overall trend" regarding abducting and adducting saccades; moreover, they found that for both eyes, centering saccades were faster than eccentrically-directed saccades. Thus, this could not serve as a basis for disconjugacy between the two eyes during a saccade.

11.3 SACCADE-VERGENCE DYNAMIC MODEL: HUNG'S DIFFERNTIAL LATENCY THEORY

The Differential Latency Theory states that the transient divergence seen during saccade-vergence responses can be accounted for by a small difference in the latencies between the contralateral and ipsilateral neural pathways driving the conjugate eye movement (Hung, 1998a,b) (Fig. 11.11). The efferent pathway from the deep layers of the contralateral superior colliculus crosses the predorsal bundle (or tectospinal tract) to arrive at the ipsilateral abducens nucleus (Sparks, 1986), which in turn sends axons to the ipsilateral lateral rectus muscles as well as the contralateral medial rectus muscle via the medial longitudinal fasciculus and the oculomotor nucleus (Leigh & Zee, 1991). Due to the brief latency difference (~ 6 msec) between these two neuronal pathways (Smith et al, 1970), there is a transient difference in the movement in the two eyes, which is exhibited as transient divergence during conjugate eye movement in the ipsilateral direction (see Fig. 11.11).

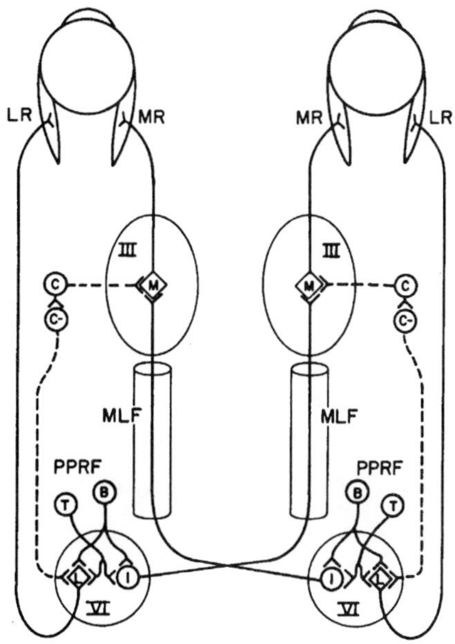

Figure 11.11. Neural pathways for saccade and vergence. For saccades, horizontal burst neurons (B) and tonic eye position neurons (T) in the paramedian pontine reticular formation (PPRF) provides similar input signals to both lateral rectus motoneurons (LR) and internuclear neurons (I) in the abducens nucleus (VI). The abducens internuclear neurons cross the midline and ascend in the medial longitudinal fasciculus (MLF) to drive the medial rectus motoneuron. Also, for vergence., the presumed complementary vergence signals, c and c-, innervate the ipsilateral MR and LR to drive the response (Reprinted from Mays (1983), pg. 656, Fig. 20.4, with permission of K. J. Ciuffreda, the copyright holder). Hung's Differential Latency Theory is based on a difference in latency between ipsilateral and contralateral neural pathways. For example, a signal from the right (contralateral) superior colliculus (not shown) crosses over to innervate the left (ipsilateral) VI nucleus. From here, neuronal signal are sent to the left LR as well as to the right MR via the MLF and the oculomotor nucleus (III). For a leftward saccade, there is a slightly longer latency (by about 6 msec; Smith et al, 1970) through the MLF to the right MR than the left LR, resulting in a brief transient divergence (Hung, 1998a,b).

In daily life, target changes are almost always asymmetrical in nature, and therefore must involve both saccade and vergence movements occurring in an interactive manner. Accommodation has a much longer latency (~350 msec) than those for saccades (~200 msec) and vergence (~180 msec), and therefore plays a relatively minor role in these early-occurring dynamic interactions. Saccadic and vergence eye movements direct gaze laterally and in depth, respectively. Indeed, the interactions between saccade and vergence have been of considerable interest in recent years, because of the erroneous hypothesis put forth by some investigators that saccades facilitated vergence dynamics, and that the extent of the facilitation was based on the configuration and richness of the visual scene (Erkelens et al, 1989; Collewijn et al, 1995; Enright, 1986; Zee et al, 1992). The following is a discussion of experimental and modeling development based on the Differential Latency Theory (Hung, 1998a,b):

Experiments were conducted to examine the trajectories of the bifixation point in space (called top-view trajectories) during combined saccade-vergence movements under both instrument-space (IS) and free-space (FS) conditions (Hung, 1998a,b) (Fig. 11.12). Then, a saccade-vergence model was constructed and simulations performed to test the Differential Latency Theory to determine whether the latency difference between neuronal pathways can result in the same top-view trajectories seen in the experiments (Hung, 1998a,b) (Fig. 11.13).

The experimental apparatus used under the instrument-space (IS) (Zhu, 1995; Hung et al, 1994, 1997) and free-space (FS) (Collewijn et al, 1997) environments are shown in Figs. 11.12A, B, respectively. The FS environment corresponds to the natural viewing of objects in a scene that consists of all the usual cues such as blur, disparity, size, perspective and overlap, etc., whereas the IS environment corresponds to a more restricted viewing of targets in an optical assembly that consists only of retinal disparity cues to the two eyes (Hung et al, 1994). It was found that under both IS and FS viewing environments, oculomotor responses to asymmetrical target displacements exhibited transient divergence, mainly for far-to-near target displacement, with the occurrence of large divergence transients being greater under the FS (Fig. 11.14A) (Enright, 1984; Erkelens et al, 1989; Zee et al, 1992) than the IS (Fig. 11.14B) environment (Hung and Ciuffreda, 1996; Hung, 1998a,b). These transient divergent movements were seen as large loops in the dynamic binocular fixation two-dimensional top-view plot (see Fig. 11.14A, bottom subplot, loop portion of the trajectory), which deviated from the classical iso-vergence curves (Ono and Nakamizo, 1977). The movements were generally followed by a pure vergence response, which is seen as radially-directed (i.e., towards the midpoint between the eyes) traces (Figs. 11.14 A,B). A variety of targets at different distances and directions were presented, the subjects' responses were recorded, and the trajectories plotted as top-view plots (Hung, 1998a,b).

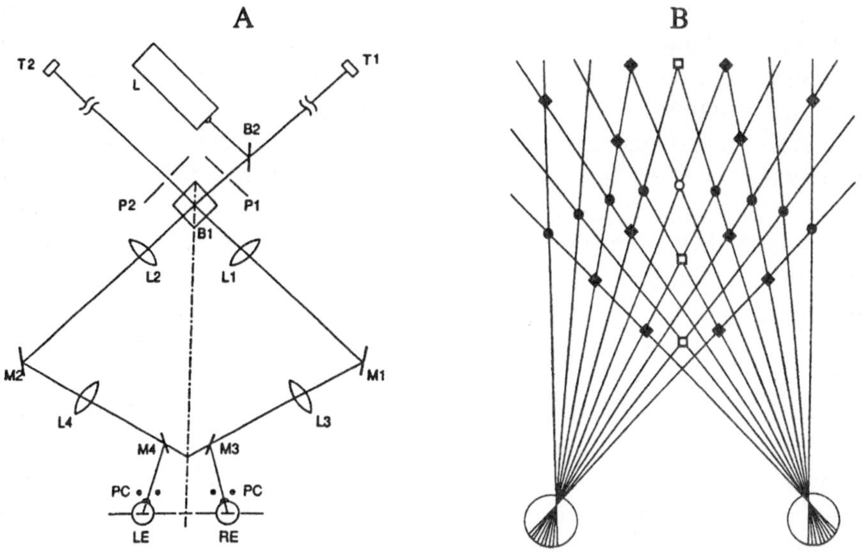

Figure 11.12. Experimental apparatus for the (A) Instrument-space environment: Dynamic Binocular Stimulator (DBS) (Semmlow and Venkiteswaren, 1976); Pinhole apertures P1 and P2 are optically conjugate to the pupils of the left and right eye, respectively. When the aperture of P1 and P2 are reduced to < 1 mm, blur stimulation is eliminated and cannot directly or indirectly (through accommodative feedback) influence the eye movement response. Since T1 and T2 are located 28.5 cm from their respective pinhole apertures, a horizontal target movement on an oscilloscope of 1 cm equals 2 deg of angular change in the eye. The target in each oscilloscope consists of a thin vertical line 0.25 deg wide and 8 deg high, presented as bright bars against a totally dark background. The relatively long bar targets assist in horizontal binocular fusion of the targets. The beam splitter B1 is used for alignment only and is removed during the experiments. Adapted from Hung et al (1994), pg. 3488, Fig. 2, with permission of Invest. Ophthal. Vis. Sci. (B) Free-space environment: Schematic diagram of the target positions used in the free-space experiments. The initial fixation point is indicated by the open circle (o) at 28.5 cm from the subject's eyes. Iso-vergence points are represented by the filled circles (•). Pure vergence points are represented by open squares (□), while asymmetrical target positions are represented by filled diamonds (♦). Not shown is a symmetric target at optical infinity (20 ft away). The separation between the lines is 4 deg (drawing not to scale).

To simulate the various experimental top-view trajectories, a saccade-vergence model was constructed based on Hung's Differential Latency Theory (Fig. 11.13). The two main characteristics of the model were: there was no central computation for assisting the vergence movement; and there was included in the pathway from the ipsilateral abducens nucleus to the contralateral medial rectus muscle an increased latency of 6 msec. It turned out that this latency was crucial in producing the transient divergence (which was reflected in the loops seen in the top-view trajectories) during asymmetrical eye movements. The dynamic saccade-vergence model consists of both conjugate pulse-step and disjunctive step controllers. Conjugate and disjunctive transfer functions are used to provide appropriate overall dynamics for saccade and vergence, respectively. Each conjugate (C1 to C4) and disjunctive (D1 to D4) gain has a nominal value of 0.5. The extraocular muscle plant is based on that by Robinson (1973) and Zuber and Stark (1968) and is given by 1/(0.064s + 1) (0.007s +1). The conjugate signal to the MR of the contralateral adducting eye is set to be 6 msec longer than that for the LR of the ipsilateral abducting eye. For example, for a rightward saccade, the delay at C2 is 6 msec longer than that at C4 (see Fig. 11.13). A 1st-order filter with a time constant of 20 msec is placed following the delay stage to provide smooth transients. The model simulation results for conditions similar to the experimental trials seen in Figs. 11.14A and B are shown in Figs. 11.15A and B, respectively. The correspondence between model and experimental results was shown for other target conditions as well, thus demonstrating the accuracy of the model in simulating experimental results under a variety of aymmetrical stimulus conditions (Hung, 1998a,b). Therefore, this relatively simple model was able to account for the previously unexplained vergence behavior during combined saccade-vergence movements.

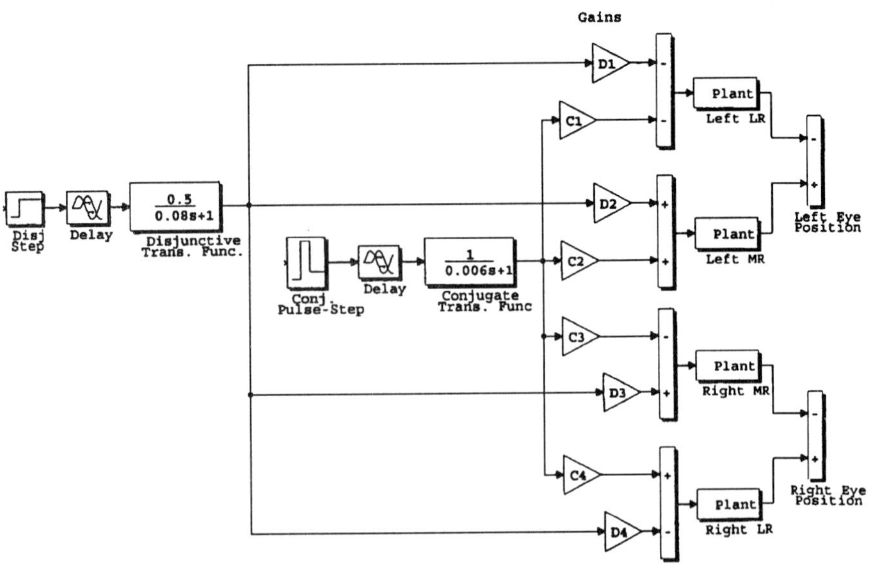

Figure 11.13. Dynamic saccade-vergence model. See text for details. Reprinted from Hung (1998a), pg. 10, Fig. 1, with permission of Med. Sci. Res.

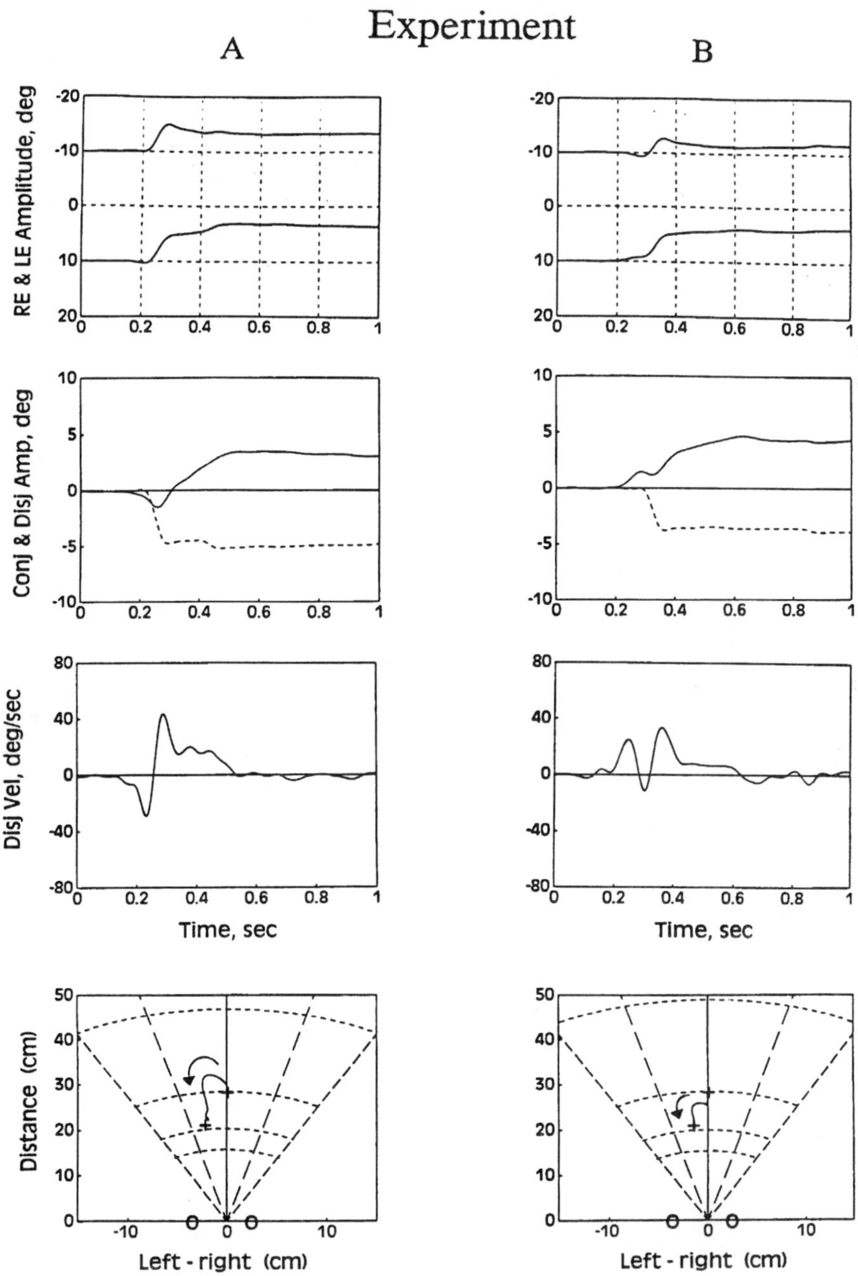

Figure 11.14 - See figure legend two pages down.

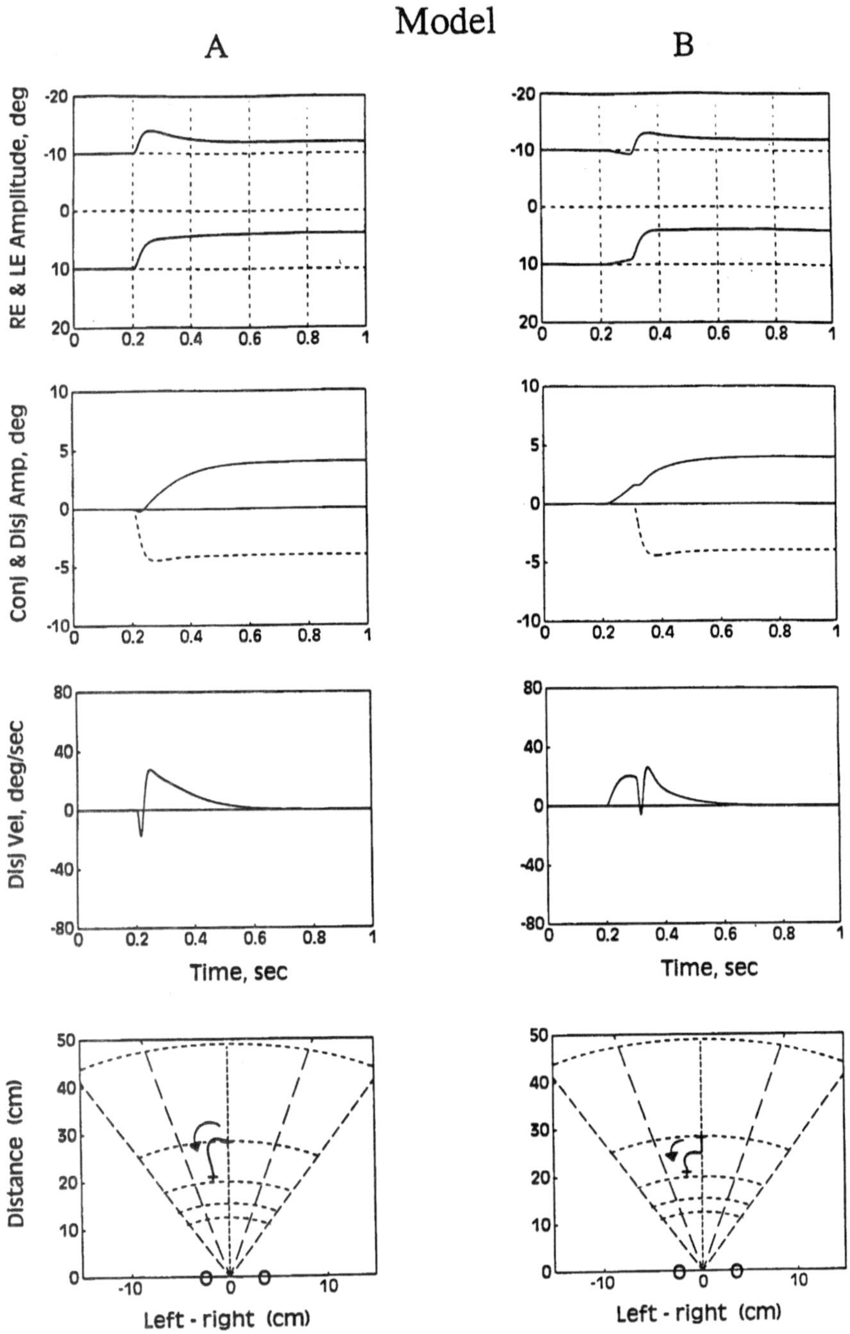

Figure 11.15 - See figure legend next page.

11.4 SUMMARY

In summary, the higher neural centers are concerned primarily with the generation of the appropriate vergence and version controller signals. The overall goal is to direct the eyes from the initial fixation point to the new target position. The various trajectory shapes can be accounted for by the difference in peripheral neural delays between the two eyes, as well as the timing between saccade and vergence controller signals. Nevertheless, all the different trajectories arrive accurately at the target. Thus, the parsimony and simplicity offered by the present Differential Latency Theory model (Hung, 1998a,b) reflects the essence of Hering's law (Hering, 1977), which states that the two eyes act as one, so that the separate conjugate and disjunctive controllers work together to drive the eyes toward the target in space.

Figure 11.14. (See 2 pages back). (A) Representative experimental time traces under the free-space (FS) environment for stimulus requiring a response of -4 deg in the LE and -8 deg in the RE (Positive and negative numbers represents rightward and leftward target displacement, respectively), corresponding to 4 deg of convergence and 6 deg of leftward versional movement (Subj. SW). (B) Representative experimental time traces under the instrument-space (IS) environment for stimulus requiring a response of -2 deg in the LE and -6 deg in the RE, corresponding to 4 deg of convergence and 4 deg of leftward versional movement (Subj. GH). Top Graph - Left eye (LE, upper) and right eye (RE, lower) time traces. Second Graph - Conjugate (dotted) and disjunctive (solid) amplitude time courses. Third Graph - Disjunctive velocity time course. Bottom graph - Top-view binocular fixation trajectories corresponding to the movements shown in top graph. The initial central fixation point and the target are shown as "+" symbols. The circular-shaped iso-vergence arcs (dotted) are separated at 5 deg intervals, whereas the radial lines (dashed) are separated at 10 deg intervals. Note that for the bottom graph under the FS environment (A), the trajectory, starting from a position indicated by the central fixation cross, consists of an overshoot loop followed by a radially-directed vergence movement towards the target. On the other hand, under the IS environment (B), the trajectory consists of an initial convergence (along the central radial line), followed by a saccadic trajectory, which is then followed by a final convergence movement (along another radial line). Reprinted from Hung (1998b), pg. 12, Fig. 2, with permission of Swets and Zeitlinger.

Figure 11.15. (See previous page). Model simulation responses for a target displacement requiring -2 in the left eye and -6 deg in the left eye, corresponding to 4 deg of convergence and 4 deg of leftward saccadic response for the conditions of (A) simultaneous (latency=200 msec) and (B) sequential (latency: disjunctive=200 msec; and conjugate=300 msec) onset of controller signals. The description of the traces are the same as those for Fig. 11.14. Reprinted from Hung (1998b), pg. 13, Fig. 3, with permission of Swets and Zeitlinger.

11.5 REFERENCES

Abel, L., Dell'Osso, L, Daroff, R.B. and Parker, L., 1979, Saccades in extremes of lateral gaze, *Invest. Ophthal. Vis. Sci.* **18**: 324-327.

Alpern, M., 1953, Metacontrast, *J. Opt. Soc. Am.* **29**: 631-646.

Alpern, M. and Ellen, P., 1956, A quantitative analysis of the horizontal movements of the eyes in the experiment of Johannes Mueller. I. Methods and results, *Am. J. Ophthalmol.* **42**: 289-303.

Aslin, R. N. and Shea, S. L., 1987, The amplitude and angle of saccades to double-step target displacements, *Vis. Res.* **27**: 1925-1942.

Averbauch-Heller, L., Lewis, R. F., and Zee, D. S., 1999, Disconjugate adaptation of saccades: contribution of binocular and monocular mechanisms. *Vis. Res.* **39**: 341-352.

Bahill, A. T. and Stark, L., 1979, The trajectory of saccadic eye movements, *Sci. Am.* **240**: 108-117.

Bahill, A. T., 1981, *Bioengineering: Biomedical, Medical, and Clinical Engineering*, Prentice-Hall, Englewood Cliffs, NJ, pg. 146.

Becker, W. and Jurgens, R., 1979, An analysis of the saccadic system by means of double step stimuli, *Vis. Res.* **19**: 967-982.

Bucci, M. P., Kapoula, Z., and Eggert, T, 1999, Saccade amplitude disconjugacy induced by aniseikonia: role of monocular depth cues, *Vis. Res.* **39**: 3109-3122.

Campbell F. W., and Wurtz, R. H., 1978, Saccadic omission: Why we do not see a grey-out during a saccadic eye movement, Vis. Res. **18**: 1297-1303.

Chaturvedi, V. and van Gisbergen, J. A. M., 1999, Perturbation of combined saccade-vergence movements by microstimulation in monkey superior colliculus, *J. Neurophysiol.* **81**: 2279-2296.

Ciuffreda K.J. and Stark, L., 1975, Descartes' law of reciprocal innervation, *Amer. J. Optom. Physiolo. Opt.* **52**: 662-673.

Ciuffreda, K.J. and Tannen, B., 1995, *Eye Movement Basics for the Clinician*, Mosby, St. Louis, MO, pp. 36-71, 127-160

Clark, M. R. and Crane, H. D., 1978, Dynamic interactions in binocular vision, in: *Eye Movements and the Higher Psychological Functions*, J. W. Senders, D. F. Fisher, and R. A. Monty, eds., Erlbaum, Hillsdale, NJ, pp.77-88.

Collewijn, H., Erkelens, C. J., and Steinman, R. M., 1988, Binocular co-ordination of human saccadic eye movements, *J. Physiol.* **404**: 157-182.

Collewijn, H., Erkelens, C. J., and Steinman, R. M., 1995, Voluntary binocular gaze-shifts in the plane of regard: dynamics of version and vergence, *Vis. Res.* **35**: 3335-3358.

Collewijn, H., Erkelens, C. J., and Steinman, R. M., 1997, Trajectories of the human binocular fixation point during conjugate and non-conjugate gaze-shifts, *Vis. Res.* **37**: 1049-1069.

Eggert, T. and Kapoula, Z., 1995, Position dependency of rapidly induced saccade disconjugacy, *Vis. Res.* **35**: 3493-3503.

Enright, J. T., 1984, Changes in vergence mediated by saccades, *J. Physiol.* **350**: 9-31.

Enright, J. T., 1986, Facilitation of vergence changes by saccades: influences of misfocused images and of disparity stimuli in man, *J. Physiol.* **371**: 69-87.

Enright, J. T., 1992, The remarkable saccades of asymmetrical vergence, *Vis. Res.* **32**: 2261-2276.

Erkelens, C. J., Steinman, R. M., and Collewijn, H., 1989, Ocular vergence under natural conditions: II: Gaze shifts between real targets differing in distance and direction, *Proc. R. Soc. London.(B).* **236**: 444-465.

Findlay, J. M., and Harris, L. R., 1993, Horizontal saccades to dichoptically presented targets of differing disparities, *Vis. Res.*, **33**: 1001-1010.

Gamlin, P. D. R. and Clarke, R. J., 1995, Single-unit activity in the primate nucleus reticularis tegmenti pontis related to vergence and ocular accommodation, *J. Neurophysiol.* **73**: 2135-2119.

Gamlin, P. D. R. and Mays, L. E., 1992, Dynamic properties of medial rectus motoneurons during vergence eye movements, *J. Neurophysiol.* **67**: 64-74.

Glassman, R. B., 1999, A working memory "theory of relativity": Elasticity in temporal, spatial, and modality dimensions conserves item capacity in radial maze, verbal tasks, and other cognition, *Brain Res. Bull.* **48**: 475-489.

Hering, E.. 1977 (originally 1868), *The Theory of Binocular Vision*, B. Bridgeman and L. Stark L, eds. trans., Plenum Press, New York.

Hou, R. L. and Fender, D. H., 1979, Processing of direction and magnitude by the saccadic eye movement system, *Vis. Res.* **19**: 1421-1426.

Hung, G. K., 1998a, Dynamic model of saccade-vergence interactions, *Med. Sci. Res.* **26**: 9-14.

Hung, G. K., 1998b, Saccade-vergence trajectories under free- and instrument-space environments, *Curr. Eye Res.* **17**: 159-164.

Hung, G. K. and Ciuffreda, K. J., 1996, Schematic model of saccade-vergence interactions, *Med. Sci. Res.* **24**: 813-816.

Hung, G. K., Ciuffreda, K. J., Semmlow, J. L., and Horng, J.- L., 1994, Vergence eye movements under natural viewing conditions, *Invest. Ophthal. Vis. Sci.* **35**: 3486-3492.

Hung, G. K., Li, S., Semmlow, J. L., and Ciuffreda K. J., 1990, Suppression of sensitivity to change in target disparity during vergence eye movements, *Exp. Neurol.* **110**: 291-297.

Hung, G. K., Semmlow, J. L., and Ciuffreda, K. J., 1983, Identification of accommodative vergence contribution to the near response using response variance,. *Invest. Ophthal. Vis. Sci.*, **24**: 772-777.

Hung, G. K., Semmlow, J. L., and Ciuffreda, K. J., 1984, The near response: modeling, instrumentation,and clinical applications, *IEEE Trans. Biomed. Engin.* **31**: 910-919.

Hung, G. K., Semmlow, J. L., and Ciuffreda, K. J., 1986, A dual-mode dynamic model of the vergence eye movement system, *IEEE Trans. Biomed. Engin.* **33**: 1021-1028.

Hung, G. K., Wang, T. J., Ciuffreda, K. J., and Semmlow, J. L., 1989, Suppression of sensitivity to surround displacement during vergence eye movements. *Exp. Neurol.* **105**: 300-305.

Hung, G. K., Zhu, H. M. and Ciuffreda, K. J., 1997, Convergence and divergence exhibit different response characteristics to symmetric stimuli, *Vis. Res.* **37**: 1197-1205.

Judge, S. J. and Cumming, B. G., 1986, Neurons in the monkey midbrain with activity related to vergence eye movement and accommodation, *J. Neurophysiol.* **55**: 915-930.

Jurgens, R., Becker, W., and Kornhuber, H. H., 1981, Natural and drug-induced variations of velocity and duration of human saccadic eye movements: evidence for a control of the neural pulse generator by local feedback, *Biol. Cyber.* **39**: 87-96.

Kapoula, Z., Eggert, T., and Bucci, M. P., 1995, Immediate saccade amplitude disconjugacy induced by unequal images, *Vis. Res.* **35**: 3505-3516.

Keller, E. L., 1977, Control of saccadic eye movements by midline brain stem neurons, in: *Control of Gaze by Brain Stem Neurons,* Baker R and Berthoz A, eds., Elsevier-North Holland, Amsterdam, pp. 327-336.

Keller, E. L. and Robinson, D. A., 1972, Abducens unit behavior in the monkey during vergence movements, *Vis. Res.* **12**: 369-382.

Kenyon, R. V. and Ciuffreda, K. J., 1978, Unequal saccades during vergence, *Am. J. Optom. Physiol. Opt.* **57**: 586-594.

Kenyon, R. V., Ciuffreda, K. J., and Stark, L. 1980, An unexpected role for normal accommodative vergence in strabismus and amblyopia, *Am. J. Optom. Physiol. Opt.* **57**: 655-577.

King, W. M. and Fuchs, A. F., 1977, Neuronal activity in the mesencephalon related to vertical eye movements, in: *Control of Gaze by Brain Stem Neurons*, R. Baker and A. Berthoz, eds., Elsevier-North Holland, Amsterdam, pp. 319-326.

Krishnan, V. V. and Stark, L., 1977. A heuristic model for the human vergence eye movement system. *IEEE Trans. Biomed. Engin.* **24**: 44-48.

Krommenhoek, K. P. and van Gisbergen, J. A. M., 1994, Evidence for nonretinal feedback in combined version-vergence eye movements, *Exp. Brain Res.* **102**: 95-109.

Leigh, R. J. and Zee, D. S., 1991, *The Neurology of Eye Movements*, 2nd Ed., F.A. Davis Company, Philadelphia, PA.

Manning, K. A. and Riggs, L. A., 1984, Vergence eye movements and visual suppression, *Vis. Res.* **24**: 521-526.

Maxwell, J. S. and King, W. M., 1992, Dynamics and efficacy of saccade-facilitated vergence eye movements in monkeys, *J. Neurophysiol.* **68**: 1248-1260.

Mays, L. E., 1983, Neurophysiological correlates of vergence eye movements, in: C. M. Schor and K. J. Ciuffreda, eds. *Vergence Eye Movements: Basic and Clinical Aspects.* Butterworths, Boston, pp. 649-670.

Mays, L. E., 1984, Neural conrol of vergence eye movements: convergence and divergence neurons in the midbrain, *J. Neurophysiol.* **51**: 1091-1108.

Mays, L. E. and Gamlin, P. D. R., 1995,. A neural mechanism subserving saccade-vergence interactions, in: J. M. Findlay et al, eds., *Eye Movement Research,* Elsevier Science, Oxford, pp. 215-223.

Mays, L. E. and Porter, J. D., 1984, Neural control of vergence eye movements: activity of abducens and oculomotor neurons, *J. Neurophysiol.* **52**: 743-761.

Mays, L. E., Porter, J. D., Gamlin, P. D. R., and Tello, C. A., 1986, Neural control of vergence eye movements: neurons encoding vergence velocity, *J. Neurophysiol.* **56**: 1007-1021.

Miller, J. M., Ono, H., and Steinbach, M. J., 1980, Additivity of fusional vergence and pursuit eye movements, *Vis. Res.* **20**: 43-47.

Ono, H. 1983, The combination of version and vergence, in: *Vergence Eye Movements: Basic and Clinical Aspects*, C. Schor and K. J. Ciuffreda, eds., Butterworths, Boston, pp. 373-400.

Ono, H. and Nakamizo, S., 1977, Saccadic eye movements during changes in fixation to stimuli at different distances, *Vis. Res.* **17**: 233-238.

Panum, P. L. 1858. Physiologische Untersuchungen ŭber das Sehen mit zwei Augen. Schwersche Buchandlung, Kiel, Germany.

Rashbass, C. and Westheimer, G., 1961a, The relationship between saccadic and smooth tracking eye movements, *J. Physiol.* **159**: 326-338.

Rashbass, C. and Westheimer, G., 1961b, Disjunctive eye movements, *J. Physiol.* **159**: 339-360.

Riggs, L. A. and Niehls, E. W., 1960, Eye movements recorded during convergence and divergence, *J. Opt. Soc Am.* **50**: 913-920.

Robinson, D. A., 1973, Models of the saccadic eye movement control system, *Kybernetic.* **14**: 71-83.

Schor, C. M., 1979, The influence of rapid prism adaptation upon fixation disparity, *Vis. Res.* **19**: 757-765.

Schor C. M., Robertson K. M., and Wesson, M., 1986, Disparity vergence dynamics and fixation disparity, *Am. J. Optom. Physiol. Opt.* **63**: 611- 616.

Scudder, C. A., 1988, A new local feedback model of the saccadic burst generator, *J. Neurophysiol.* **59**: 1455-1475.

Semmlow, J.L., Hung, G. K., and Ciuffreda, K. J., 1986, Quantitative assessment of disparity vergence components, *Invest. Ophthal. Vis. Sci.* **27**: 558-564.

Semmlow, J. L., Hung, G. K., Horng, J. L., and Ciuffreda, K. J. 1993, The initial control component in disparity vergence eye movements, *Ophthal. Physiol. Optics*, **13**: 48-55.

Semmlow, J. L., Hung, G. K., Horng, J. L., and Ciuffreda, K. J. 1994, Disparity vergence eye movements exhibit preprogrammed motor control, *Vis. Res.* **34**: 1335-1343.

Semmlow, J. L. and Venkiteswaren, N., 1976, Dynamic accommodative vergence in binocular vision, *Vis. Res.* **16**: 403-411.

Smith, K. U., Schmidt, J., and Putz V., 1970, Binocular coordination: feedback synchronization of eye movements for space perception, *Am. J. Optom. Arch. Am. Acad. Optom.* **47**: 679-689.

Sparks, D. L., 1986, Translation of sensory signals into commands for control of saccadic eye movements: role of primate superior colliculus, *Physiol. Rev.* **66**: 118-171.

Stark, L., 1968, *Neurological Control Systems: Studies in Bioengineering*. Plenum Press, New York.

Steinman, R. M. and Collewijn, H., 1981, Binocular retinal image motion during active head rotation, *Vis. Res.* **20**: 415-429.

Tamler, E., Jampolsky, A., and Marg, E., 1958, An electromyographic study of asymmetric convergence, *Am. J. Ophthalmol.* **46**: 174-182.

van der Steen, J. and Bruno, P., 1995, Unequal amplitude saccades produced by aniseikonic patterns: effect of viewing distance, *Vis. Res.* **35**: 3459-3471.

van Gisbergen, J. A., Robinson, D. A., and Gielen, S., 1981, A quantitative analysis of generation of saccadic eye movements by burst neurons, *J. Neurophysiol.* **45**: 417-442.

Volkmann, F. C., Riggs, L. A., Ellicott, A. G., and Moore, R. K. , 1982, Measurements of visual suppression during opening and closing, and blinking of the eyes, *Vis. Res.* **22**: 991-996.

Westheimer, G. and Mitchell, A. M. , 1956, Eye movement responses to convergence stimuli., *Arch Ophthalmol.* **55**: 848-856.

Wheeles, L., Boynton. R., and Cohen, G., 1966, Eye movement responses to step and pulse-step stimuli, *J. Opt. Soc. Am.* **56**: 956-960.

Yarbus, A. L., 1967 (originally 1956), *Eye Movements and Vision*, B. Haigh , trans. and L. A. Riggs, ed., Plenum Press, New York.

Young, L. R. and Stark, L., 1962. A sampled-data model for eye tracking movements, *Quart. Prog. Rept., Res. Lab. Electronics*, M.I.T. **66**: 370-383.

Zee, D. S., Fitzgibbon, E. J., and Optican, L. M., 1992, Saccade-vergence interactions in humans, *J. Neurophysiol.* **68**: 1624-1641.

Zee, D. S. and Levi, L., 1989, Neurological aspects of vergence eye movements,. *Rev. Neurol. (Paris).* **145**: 613-620.

Zhu, H. M., 1995,. *Investigation of interaction between vergence and saccadic eye movements.* M.S. Thesis, Biomedical Engineering, Rutgers University, pp. 35-38.

Zuber, B. L. and Stark, L. 1966. Saccadic suppression: elevation of visual threshold associated with saccadic eye movements, *Expt. Neurol.* **16**: 65-79.

Zuber, B. L. and Stark, L. 1968, Dynamical characteristics of fusional vergence eye movement system,. *IEEE Trans. Syst. Sci. Cybern.* **4**: 72-79.

IV

PERCEPTUAL SYSTEM MODELS

Chapter 12

Psychophysics and Modeling of Texture Segregation

Ramanujan Kashi[1], Thomas V. Papathomas[2], and Bela Julesz[2]

[1] *Avaya Labs Research, 233 Mt. Airy Road, Basking Ridge, NJ 07920, PH: (908) 696-5121, FX: (908) 696-5402, EM: ramanuja@research.avayalabs.com*

[2] *Dept. of Biomedical Engineering and Laboratory of Vision Research, Rutgers University, 152 Frelinghuysen Rd., Piscataway, NJ 08854-8020, PH: (732) 445-6533, FX: (732) 445-6715, EM: papathom@zeus.rutgers.edu*

[3] *Dept. of Psychology and Laboratory of Vision Research, Rutgers University, 152 Frelinghuysen Rd., Piscataway, NJ 08854-8020, PH: (732) 445-6520, FX: (732) 445-6715, EM: julesz@cyclops.rutgers.edu*

12.1 INTRODUCTION

Modern research in visual texture perception can be traced to the pioneering work of Julesz (1962), and Beck (1966). What exactly is visual texture? Even though visual texture is not easy to define, a good "stratified" definition was proposed by Gorea: "Visual texture is a 2D [two-dimensional] visual stimulus characterized by a visible *grain*. Visual grain consists of local modulations along dimensions such as luminance, color, and shape, which may or may not be discriminable. Two textures are visually different if they do not share the same grain and/or if they do not share, in the statistical sense, the same grain distribution across space." (Gorea 1995, pp. 55-56). Figure 12.1 shows an example of texture-based segregation. In Fig. 12.1a, the central region consisting of X's shows a marked segregation from the peripheral region composed of T's. This is an example of pre-attentive or "pop-out" texture segmentation. However, in Fig. 12.1b, the central region comprising of L's does not segregate preattentively from the peripheral regions composed of T's.

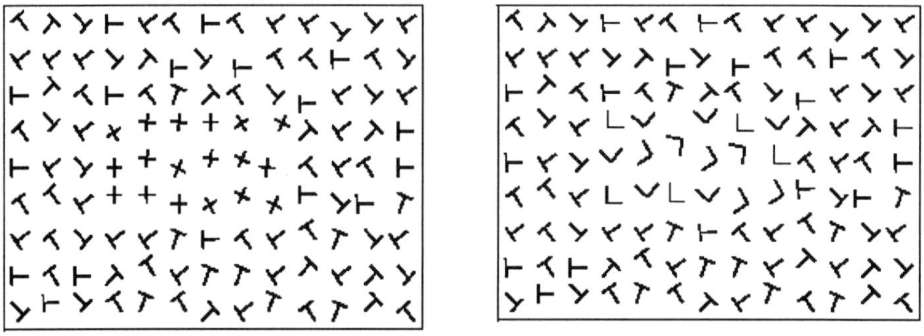

(a) (b)

Figure 12.1. Example of texture-based segregation. In (a) the central region composed of X's is easily segregated from the background containing T's. In (b) the central region composed of L's is not immediately discriminable from the background containing T's. Adapted from Julesz and Bergen (1983).

As with any definition, one must be careful in clarifying various terms. In this particular case, it must be pointed out that texture, as dealt with in this chapter, as well as in the vision research literature in general, is indeed a 2D stimulus (Kashi et al., 1996). The reader must not confuse this with references to 3D texture (e.g., Gallant, van Essen and Nothdurft, 1995). What is really meant by the term "3D texture" in such references is the mapping of a 2D texture pattern onto a planar surface that is located in 3D space, such as when viewing the outside corner of a building. This gives rise to the so-called "shape-from-texture" problem, i.e., determining the 3D spatial orientation of the surface from the foreshortening of the 2D texture pattern (Gibson, 1950; Rosenholtz and Malik, 1997; Gallant, van Essen and Nothdurft, 1995). Of course, this can be extended to the mapping of a 2D texture pattern onto a non-planar 3D surface, which results in a more demanding "shape-from-texture" problem. Ultimately, one can also speak of a genuinely 3D texture, namely the texture of a natural surface with protrusions, bumps, and dips, such as the surface of an orange or a sandy beach. In this chapter we are going to deal mostly with how humans process 2D visual textures, and how this processing can be simulated with biologically plausible models.

Why would researchers study visual texture perception and the associated biological neural processing mechanisms? It turns out that texture perception is a very important low-level task for visual information processing. In his excellent review on theories of visual texture perception, Bergen (1991) makes the point that, although we are not generally very aware of it, we use texture information to help unify surfaces and to distinguish objects from background." Watt (1995) has documented some interesting speculations on the important role that texture plays in everyday visual tasks by drawing some analogies with color perception. Texture perception has been receiving increasing attention in psychophysical, neurophysiological, and computational vision studies (Nothdurft and Parlitz, 1993; Wilson and Mast, 1993; Buckley and Frisby, 1993; Gorea and Papathomas, 1993; Gorea, 1995; Sagi, 1995; Gallant, van Essen and Nothdurft, 1995; Victor, Conte, Purpura and Katz, 1995; Victor and Conte, 1996; Rao and Loshe, 1996; Pessoa, Beck and Mingolla, 1996; Troje and Bulthoff, 1996; Caputo, 1996; Rosenholtz and Malik, 1997; Bach and Meigen, 1997; Li and Lennie, 1997; Durgin and Huk, 1997; Papathomas, Kashi and Gorea, 1997; Kashi, Papathomas and Gorea, 1994; Caputo, 1998; Knill, 1998a, 1998b; Kwan and Regan, 1998; Papathomas, Gorea, Feher and Conway, 1999; Gorea and Papathomas, 1997).

Historically, when compared to other low-level visual processes, such as color, motion and stereo, progress in neurophysiology, psychophysics and modeling has arrived late in the area of texture perception. *Color* mechanisms have long been identified and studied (Zeki, 1973; Gouras and Kruger, 1979; Michael, 1973). Also, there is almost universal agreement about the role of the MT area in *motion* perception (Mikami, et al., 1986; Livingstone and Hubel, 1988; DeYoe and Van Essen, 1988). Further, the site of the cyclopean retina for *stereopsis* has also been identified (G. Poggio, et al., 1985; see also Ohzawa, et al., 1990). On the other hand, we know little about the location of *texture* mechanisms in the visual system (Nothdurft, 1990; Bonds, 1989; Knierim and van Essen, 1992; Bach and Meigen, 1992; Lamme, van Dijk and Spekreijse, 1992). Finally, systematic work in the field of psychophysics in texture began rather late (Julesz, 1962, 1965; Beck, 1966, 1967), as compared to work on color, motion and stereo.

Also, with regard to neurophysiologically plausible computational models and algorithms, models for texture segregation were even later (Bergen and Adelson, 1988; Voorhees and T. Poggio, 1988; Sutter, Beck and Graham, 1989; Fogel and Sagi, 1989; Malik and Perona, 1989, 1990), although one must mention the early efforts of Caelli (1982, 1985), Grossberg and Mingolla (1985, 1986) and Turner (1986). By contrast, models for the other modalities had been developed much earlier. Thus, for motion we have the time-honored Reichardt model (Reichardt, 1961), its variants (Adelson and Bergen, 1985; van Santen and Sperling, 1985), the frequency-demodulation model of Watson and Ahumada (1985), and even more complex recent models for second-order motion (Chubb and Sperling,

1988, 1989). For stereo, we have Marr and Poggio's cooperative (1976), and non-cooperative (Marr and Poggio, 1979), algorithms and, more recently, computational models for disparity-tuned cells (Ohzawa, et al., 1990).

Some interrelated reasons for this lag in the progress of texture perception are: (1) That texture is more difficult to characterize than stereo, motion and color and, hence, more complex and difficult to study. (2) Studies on texture began much later than studies in the other modalities. (3) Cortical cells that respond selectively for texture-defined features were not encountered until very recently (Van Essen, et al., 1990; Knierim and VanEssen, 1992). However, one thing is certain, that texture perception offers a fertile area for research in low-level visual information processing.

12.2 PSYCHOPHYSICS OF TEXTURE PERCEPTION

Limited by the difficulties in studying the perception of natural textures in the laboratory, most researchers in this area have used artificial texture-like stimuli, composed of discrete elements, commonly called textels or texels. These are arranged in two-dimensional (2-D) space, usually at pseudo-random locations and orientations. Some representative papers on experiments of this type are by Julesz (1962, 1981a, 1981b); Beck (1966, 1967); Garner and Feldoldy (1970); Caelli and Julesz (1979); Beck, Prazdny and Rosenfeld (1983); Bergen and Julesz (1983); Nothdurft (1985); Enns (1986); Callaghan, Lasaga and Garner (1986); Gurnsey and Browse (1987); Sagi and Julesz (1987); Sutter, Beck and Graham (1989); Callahan (1989) and Gorea and Papathomas (1981). Examples of textels are: straight line segments, arcs, geometrical shapes, combinations of lines, letters such as "X", "T", "E", etc. (Julesz 1965, 1981a, 1981b; Beck 1966, 1982; Nothdurft 1985). Most often, the background is displayed at a uniform luminance (black or white), whereas the textels' luminance is fixed at a different level, thus resulting in bi-level images; gray-level images are far less frequent. One common method for studying pattern segregation is to form a target texture patch using one type of textel and to imbed the patch in a surround texture formed by a textel of a different type as shown in Fig. 12.1a and b. The performance of observers in discriminating the target is then used to assess the role of the textel properties in forming textural groupings.

Another common method is to display a single target micropattern among several distractor textels and study the ease of detectability (Treisman and Gelade, 1980; Bergen and Julesz, 1983; Nakayama and Silverman, 1986). However, it has been shown by Wolfe (1992) that the two tasks of texture segmentation and of visual search are not necessarily correlated; namely, given two textels A and B, the ease of performing the task of segmenting a texture patch of textels A set amidst a texture background composed of textels B cannot be predicted from the performance in detecting a single textel A against several distractors B.

Julesz was among the first psychophysicists to realize the importance of texture perception in early vision. His main contribution is that he introduced quantitative approaches to the study of texture discrimination, and he set the stage for the remarkable progress in this area of research over the last fifteen years or so. As early as 1962, Julesz published his studies on pre-attentive, or scrutiny-free, texture discrimination with precisely controlled texture statistics (Julesz, 1962).

Julesz had the insight to characterize texture in terms of Nth-order statistics, which he defined as the set of probabilities that the vertices of an N-gon, thrown randomly on the texture, end up on certain combinations of black or white. These statistics are defined for bi-valued textures, namely textures composed of two luminance levels, say black and white. Typically, such textures are built by using a black (or white) texture element (textel) against a white (or black) uniform background, as shown in Fig. 12.1. As an example, 2nd-order statistics characterize the probabilities that the two ends of a needle of a certain arbitrary length would fall both of white, or both on black, or one on black, the other on white. Similarly, 3rd- or 4th-order statistics would deal with the vertices of a certain arbitrary triangle or quadrilateral, respectively, and so on. Obviously, first-order statistics would give the probability of a single point landing on white or black, and it would provide the proportion of white and black areas in the texture.

Julesz noted that observers could not effortlessly discriminate two textures that had identical 2nd-order statistics. He tried several examples of such "iso-second-order" texture pairs, and it seemed that observers needed scrutiny for discriminating such pairs. This led him to pose the conjecture that *all* iso-second-order texture pairs are not pre-attentively discriminable. Subsequently, he motivated his mathematician colleagues to embark together on a long-term research project to test the conjecture. Section 12.3.1.1 deals with the results of this research.

Most of the early methods concentrated on differences in the figural, or form, features between the surround and target textels, while all other attributes of the textels (luminance, color, stereo disparity, etc.) remain fixed. A set of stimuli as shown in Fig. 12.2, developed by Gorea and Papathomas (1989, 1991) allowed studies on the *interaction* of several attributes, in particular color, luminance, and orientation, that vary simultaneously in the textural grouping process. These stimuli were also composed of discrete textels and, therefore, shared the disadvantage of other researchers' stimuli, i.e., that they were an oversimplification in the attempt to study texture perception mechanisms. They did, however, allow a step in the right direction, since systematic studies of multi-attribute texture grouping were conducted by judicious choices of the stimuli and the experimental parameters (Gorea and Papathomas, 1991, 1995, 1999; Papathomas, Kashi and Gorea, 1997; Papathomas, Gorea, Feher and Conway, 1999). This research also enabled the development of one of the first computational models of texture segregation that included the role of color (Kashi, Papathomas and Gorea, 1994). The inclusion of color in the texture formation process has since been studied to some extent (Pessoa, Beck and Mingolla, 1996; Li and Lennie, 1997).

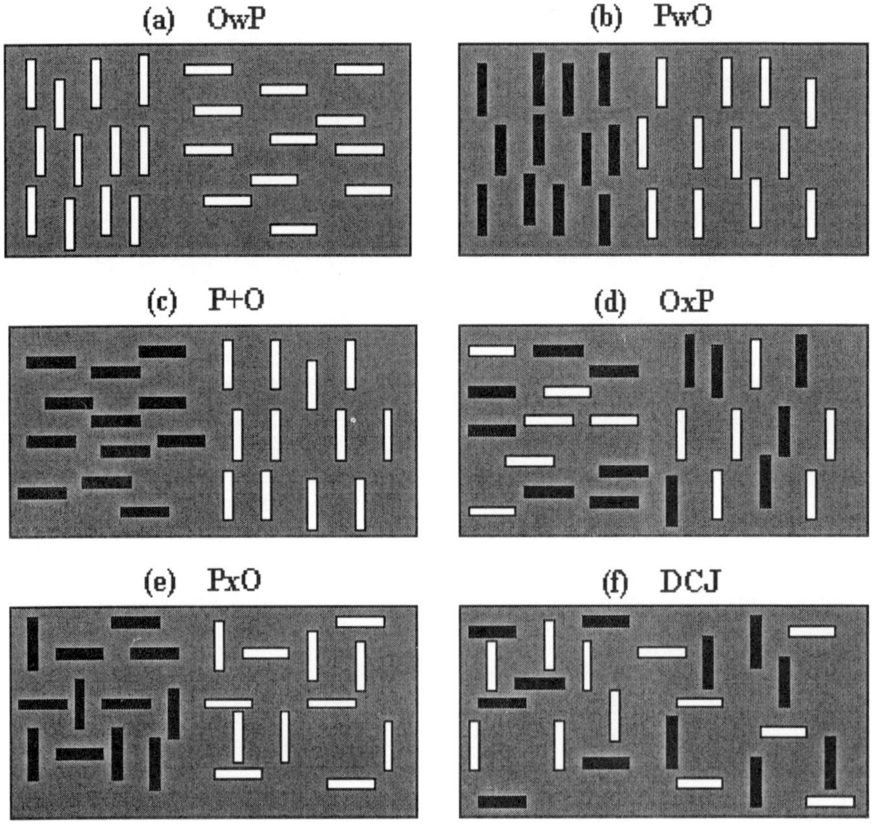

Figure 12.2. (See previous page). Examples of multi-attribute stimuli involving luminance-polarity (P) and orientation (O). In (a), the condition known as O within P (OwP), segregation is based on O (vertical versus horizontal), while the luminance-polarity (P) is the same (bright) for all the textels. In (b) which is the dual of (a), segregation is based on luminance-polarity (dark versus bright), while the orientation is kept constant (vertical). In (c), termed the plus condition (P+O), segregation is based on both P and on O. In (d), which is termed orientation across polarity (OxP) condition, segregation is based on O while the P attribute is randomly distributed across the stimulus. In (e), which is the dual of (d), segregation is due to the P attribute, the O attribute is now randomly distributed across the stimulus. Finally in (f), known as the DCJ, there is double conjunction of O and P. Segregation is difficult because both orientations and both luminance-polarities are present throughout the image. However, if one attends only to say, the white textels, then one can segregate the subpopulation of white textels on the basis of orientation differences; see Fig 12.3 for more details.

Recently, natural images have been studied to identify the relevant dimensions of texture. An interesting study using a limited set of images (Rao and Lohse, 1996) was performed to characterize texture through multi-dimensional scaling by a variety of three-dimensional representations. These authors identified three orthogonal dimensions. One axis is repetitive versus non-repetitive. The authors found that this dimension is the most important feature used by humans in distinguishing textures. The feature of repetitiveness also significantly correlated with that of regularity, uniformity and non-randomness. The other dimension revealed by their study was low-contrast and directional at one end vs. high-contrast and non-directional at the other end. The third dimension ranged from granular, coarse and low-complexity to non-granular, fine and high complexity grain textures.

Julesz (1981), along with others (Gurnsey and Browse, 1987; Treisman and Gormican, 1988; Beck, 1982), also recognized the so called "asymmetry problem" in texture discrimination. The asymmetry problem arises with some, not all, pairs of textels. If a pair which exhibits the problem involves textels A and B, then a texture patch composed of A is much easier to detect in the middle of a background texture composed of B than the other way around. Julesz's hypothesis that such a problem arises because of the blank spaces between textels was confirmed by a quantitative analysis (Rubenstein and Sagi 1990).

One of the most recent developments in texture perception deals with the issue of perceptual learning (Sagi, 1995; Ahissar and Hochstein, 1995). Some of the characteristics of the mechanisms involved in learning are that they are retinotopic, a slow time course of learning (four to five days) and a long-term retention (around 3 years). Also, the issue of attention in texture segregation has been studied only recently (Bravo and Nakayama, 1994; Papathomas et al., 1999). There are types of textures that are comprised of double conjunction of attributes such as luminance-polarity and orientation as shown in Fig. 12.3a (also shown in Fig. 12.2f), which can be segregated if the observer attends to one of the attributes. For instance, if one attends to only the luminance-polarity attribute (say dark) in Fig. 12.3a, the texture can be segregated based on orientation as seen in Fig. 12.3b. Also, if one attends to the orientation attribute (say vertical), the original original double-conjunction texture can now be easily segregated based on luminance polarity as seen in Fig. 12.3c.

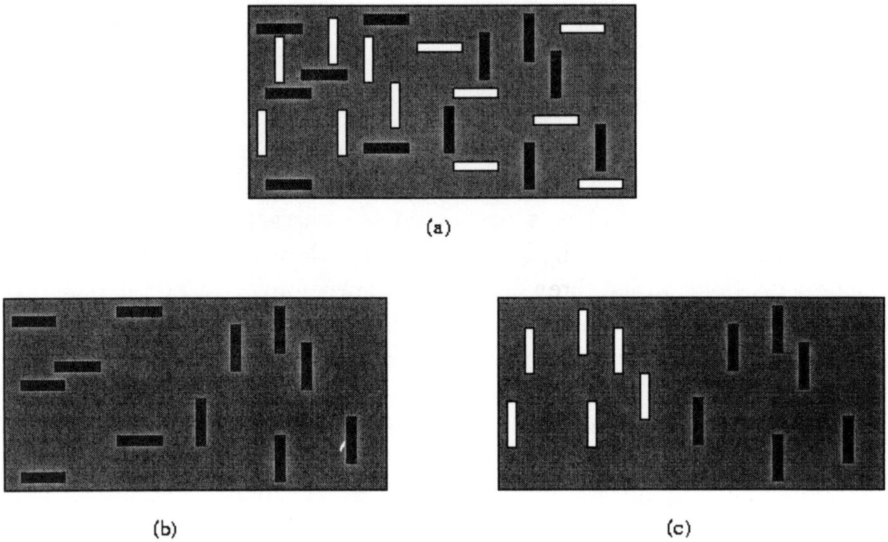

Figure 12.3. Double-conjunction stimulus comprised of orientation and luminance-polarity textels. Segregation is difficult, because both orientations and both luminance-polarities are present throughout the image. However, if one attends to luminance-polarity (say dark), segregation is possible as shown in (b). Similarly, if one attends to orientation (say vertical), segregation is shown in (c).

12.3 MODELS FOR TEXTURE PERCEPTION

As mentioned earlier in this chapter, the development of models for texture perception followed a slower pace than those for stereo and motion perception. Here is a brief representative list of papers which traces the early history of this endeavor: Shatz (1977); Caelli (1982, 1985) ; Julesz and Bergen (1983); Beck, Prazdny and Rosenfeld (1983); Grossberg and Mingolla (1985, 1986); Turner (1986); Beck, Sutter and Ivry (1987); Bovik, Clark and Geisler (1987); Daugman (1987, 1988); Bergen and Adelson (1988); Voorhees and T. Poggio (1988); Fogel and Sagi (1989); Malik and Perona (1989, 1990); Sutter, Beck and Graham (1989). Some of these models have been simulated on digital computers. The input to the program is the texture image and the output usually displays texture boundaries, i.e., the contours that separate two textures of different characteristics.

12.3.1 Stimulus Based Models of Texture Perception

12.3.1.1 The Work of B. Julesz and Texton Theory

Section 12.2 presented Julesz's quantitative techniques for generating textures with well-defined statistical characteristics. Motivated by his conjecture that iso-second-order texture pairs cannot be segregated preattentively, Julesz started a quest with his mathematician colleagues in 1962, which continued through 1975 (Julesz, 1995, p. 131). The long-term research was successful in disproving the conjecture. Caelli and Julesz (1978), were able to construct iso-second-order texture pairs that were effortlessly discriminable. With the conjecture proven wrong, the obvious new key question was: What is the highest N, Nmax, for which people can effortlessly discriminate iso-Nth-order texture pairs, but cannot discriminate iso-(N+1)th-order pairs? In fact, Julesz, Gilbert and Victor (1978), found even iso-*third*-order discriminable texture pairs, indicating that Nmax was at least 3. Julesz's quest produced excellent techniques for generating textures that had identical Nth-order statistics but different (N+1)th-order statistics. An excellent review of this research is provided in Julesz's scientific autobiography (Julesz, 1995).

The research described above provided Julesz with a key realization, which was the basis of this texton theory (Julesz, 1981). Namely, since texture segregation did not depend on global, or statistical, criteria, then it must be based on local features, which Julesz called "textons". He envisioned such textons to be local shape, orientation, flicker rate, collinearity, line terminators, motion characteristics, etc. The main insight from the texton theory was that, of the infinite variety of 2D textures, only a limited number of textons have perceptual significance and are evaluated locally in effortless texture discrimination. This was analogous to the trichromacy theory, the fact that the gamut of colors can be matched by just three colors.

The texton theory, appropriately modified through the years, provided the impetus for a large body of research in psychophysics, neurophysiology, and computational modeling. It persuaded vision scientists that the problem was a strategic one, and it attracted researchers in all of these areas to study texture perception systematically. Julesz's computational approach produced one of the first filter based models (Julesz and Bergen, 1983) that was able to discriminate automatically texture boundaries between "even" and "odd" textures (Julesz, Gilbert and Victor, 1978). It set the stage for more elaborate models that took the same approach in essence, i.e., to filter the image through a bank of linear filters, tuned to various orientations and spatial frequencies, and then apply some non-linear processing to the outputs of the linear filters (see Section 12.3.2).

12.3.1.2 The Work of J. Beck

Beck, along with Julesz, were one of the first few researchers to investigate systematically texture segregation. Contrary to Julesz's approach to texture segregation, Beck's initial approach relied on feature detectors and on "similarity grouping" (Beck 1967, 1982). His early work on texture segregation was based on identifying a small set of properties or primitives which help grouping or pooling of regions based on feature similarity. Using synthetic textures composed of line segments, Beck showed that the orientation attribute is a more salient characteristic which leads to discriminating textures than positional relationships. On the basis of numerous pscyhophysical experiments, Beck proposed that texture segregation occurred strongly on the basis of local differences in first-order statistics of simple primitives such as size, orientation, contrast and color that are processed in parallel (Beck 1982).

Beck's later approach shifted from identifying feature primitives to characterize the mechanisms involved in texture segregation. Using "element-arrangement" texture patterns, Beck explained many psychophysical experiments either using differential stimulation of spatial frequency channels or pre-attentive grouping processes. Along with Graham and Sutter, a filter based model on the lines of the generic three stage model explained in Section 12.3.2 was proposed to explain the texture segregation in element-arrangement patterns (Graham et al., 1992). Using psychophysical experiments, Beck concluded that the relevant variable for spatial frequency channels is stimulus contrast (Sutter et al., 1989) and the relevant variable for similarity grouping is perceived lightness (Beck et al., 1991).

12.3.2 Filter based models for texture perception

In the last few years, several models have been developed, some of which address directly the issue of neurophysiological relevance, and most attempt to achieve close agreement with results from psychophysical experiments (Bergen, 1991 offers a critical review).

Caelli (1982), developed a filter whose orientation tuning was a function of spatial frequency and used it to predict texture discrimination performances. The idea here was to mimic the response properties of cells in the mammalian striate cortex. Caelli's (1982), model, uses three stages as shown in Fig. 12.4: (a) convolution by frequency- and orientation-selective linear filters, followed by a nonlinearity; (b)"impletion", to spread of activity of similar proximal filters (the term "spatial pooling" is used currently); and, (c) grouping, to locate texture boundaries by attempting to maximize the correlation between activity levels within regions and minimize it between regions.

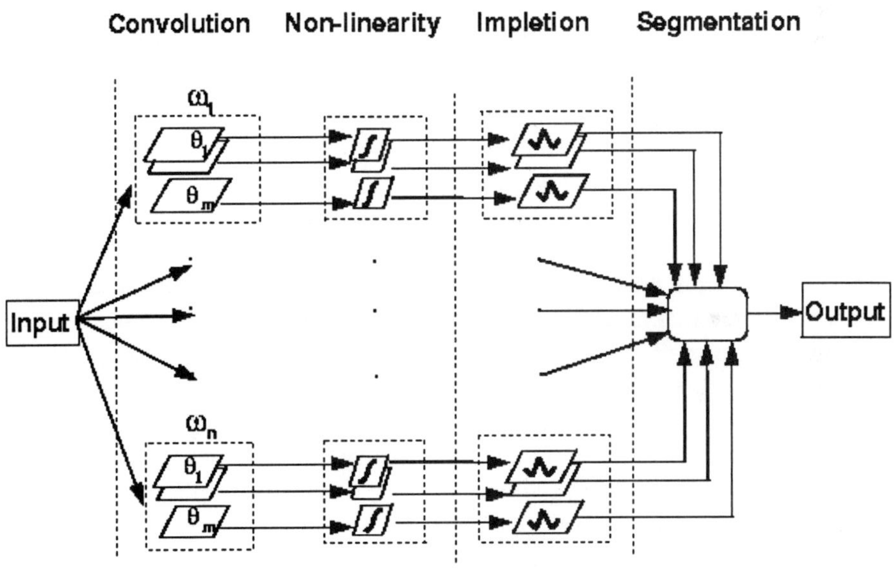

Figure 12.4. A generic three-stage texture segmentation model. The first stage consists of linear filters that are tuned to a set of spatial frequencies, $\omega_1, ..., \omega_n$. Within each frequency, there are filters that are tuned to orientations $\theta_1, ..., \theta_m$. The outputs are fed to non-linear units. The second stage performs spatial pooling by impletion. Segmentation is achieved in the third stage.

Bergen and Adelson (1986, 1988) built their model based on absolute scale invariance of texture perception. Their premise was that the tendency of a texture pair to segregate depends weakly on the distance from which it is viewed. They modeled local interactions within a pyramidal image representation at multiple scales. The model consisted of representing an input image by filtering it with a range of oriented band-pass filters. Each filter output was then squared and summed over a region to compute local energy. The hypothesis was that differences in spatial structure which yield segregating textures produce significant differences in the amplitudes of the energy responses for at least one type of filter within the pyramid structure.

Turner (1986) demonstrated the usefulness of Gabor filters in texture discrimination. The procedure involved convolving a sampled version of an image with a bank of Gabor filters tuned to a variety of orientations and frequencies. Applying the model on a variety of textures that were experimentally found to be preattentively discriminable produced good fits as predicted by the model. Good fits to psychophysical data were also achieved by Sagi and his colleagues (1989, 1990, 1995) who used Gabor functions. The Gabor power spectrum was computed by taking the sum of squares of quadrature filters. A distance function was defined between the power spectrum to obtain good correlation with human performance. Further, a model was presented that performs linear filtering by Gabor filters which enabled discrimination between features by means of intensity differences. The second stage involved finding the borders by using a Laplacian of Gaussian operator which was a form of an edge detector (Fogel and Sagi, 1989). Rubenstein and Sagi (1990) went a step further by explaining the assymetry in texture discrimination tasks using the Gabor filter model. Performance asymmetry was observed with respect to which texture represents the foreground and which represents the background. They attributed this asymmetry to noise in the texture gradient. They pointed out that the noise characteristics were caused primarily by spatial variability properties of the stimulus.

Sperling (1989) differentiated between first-order linear (Fourier) and second-order nonlinear (non-Fourier) rectifying regimes. He made an analogy between an object's texture and the carrier in amplitude modulated (AM) communication. The same analogy was made by Bovik, Clark and Geisler (1987), who used the spatially pooled amplitude and phase responses of Gabor filters. By comparing the channel amplitude envelopes, the boundaries between textured regions differing markedly in their spatial frequency content were accurately located. Discontinuities in texture phase, arising from surface discontinuities, were located by detecting large variations in the phase envelopes. They recognized that although their framework was inspired by physiological evidence of the receptive fields of striate visual cortex, their implementation offered primarily a machine vision method, as does that of Voorhees and T. Poggio (1988).

Sutter, Beck and Graham (1989), examined the discrepancies between data and their linear-filter model predictions to arrive at two possible nonlinearities (Graham, 1991; Graham, Beck and Sutter, 1992), which improve the fit substantially. One of the non-linearities suggested was a local (pointwise) non-linearity occurring early (before the linear spatial filters), and the second is a normalization process, which might have resulted from intracortical interaction, operating at the level of the spatial filters.

Malik and Perona (1990) modeled human pre-attentive texture perception in three stages which have close parallels with Caelli's model. i) Convolution of the image with a bank of orientation- and frequency-tuned filters, which are all even-symmetric. A persuasive argument was made against the use of odd-symmetric mechanisms at this stage. ii) Two non-linear mechanisms that included half-wave rectification followed by a non-linear inhibition. The half-wave rectification along with the linear filters modeled the ON and OFF V1 simple cells. The second non-linearity was the intra-cortical inhibition localized in space within and among the responses, to suppress weaker (noise) responses in the presence of stronger ones followed by half-wave rectification, to "model outputs of V1 simple cells." iii) Spatial pooling and texture edge extraction by odd-symmetric mechanisms. They obtained close fits with experimental data from Gurney and Browse (1987), and Krose (1987).

Landy and Bergen (1991) (also Bergen and Landy 1991) used multiple spatial resolutions to represent input images. These scales were further divided into four orientation-tuned components by convolving through a band of four orientation-selective filters. Spatial pooling was then performed by squaring the output of the linear units and taking a weighted average over a small region to compute the local energy. The next stage involved a novel idea of orientation opponency. Orientation opponency constructs opponent signals by subtracting the vertical energy response from the horizontal, and the right diagonal from the left diagonal. This opponency is similar to a co-extensive single-opponency found in the ganglion and geniculate neurons for the chromatic attribute. Finally, the opponent responses were normalized to achieve contrast gain control in order to separate the contrast information from the structure information. They also analyzed their model with filtered noise textures and obtained close matches with psychophysical experiments conducted with the same textures on human observers.

Finally, an elegant approach in texture perception was the "orthogonal distribution analysis" of Chubb and Landy (1991), which has the potential of estimating the parameters of texture segregation models and further establish their validity. They used textures composed by modulating pixel distributions by classes of orthogonal functions. Using such IID (identically independent distributed) textures, their analysis pointed towards the need of an early non-linearity in texture processing models (Nam and Chubb, 1998).

All of the above models accept a grey-level image as input, i.e. they work exclusively in the luminance domain. As a result, they concentrate on differences in the figural, or form, features of the texture elements (textels), while all other attributes of the textels (luminance, color, stereo disparity, etc.) remain fixed. A detailed description of a texture model which worked well with stimuli in which textels were defined by combinations of three attributes, i.e., color, luminance and orientation, was the double-opponency model of Papathomas et al (1997). This computational model shown in Fig. 12.5 was implemented in five stages: 1) Image decomposition into chromatic (C), and achromatic (luminance, L) signal components. 2) Processing of the chromatic and luminance signals by separate parallel pathways. Within both the color and luminance pathways there exist two domains. One comprising isotropic (CnO and LnO) and another comprising oriented linear filters (CO and LO) tuned to specific spatial frequencies. 3) A set of nonlinear operators (half-wave rectification followed by squaring) to eliminate spurious weak responses. 4) Double opponent units to extract texture edges. 5) Mechanisms that combined the outputs of the various domains. The novel aspect of this model was that it illustrated that the concept behind double-opponency could be implemented as a possible mechanism in texture edge extraction. As noticed from the color domain, the double-opponent cells in the cortex fire for red-green edges or blue-yellow edges. An analogous mechanism is detailed in the above model to extract orientation edges and luminance polarity (bright/dark) edges. The extension of the idea of double-opponency to other attributes (orientation, luminance-polarity), in addition to color, introduced a new research in modeling. This approach was supported by psychophysical (Cannon and Fullenkamp, 1991; Polat and Sagi, 1992) and neurophsyiological evidence (van Essen et al., 1990; Bonds, 1989; Knierim and van Essen, 1992) supporting spatial opponency (not necessarily double opponency) in various attributes. A good correlation between the model's simulation results and data from psychophysical experiments with a systematically selected set of visual stimuli with texture patterns defined by spatial variations in color, luminance and orientation, was achieved (Pearson correlation r^2 expressed as a percentage was 81.5, Papathomas et al., 1997).

Figure 12.5. (See next page). Chromatic texture segmentation model based on double-opponent mechanisms. L(x,y) and C(x,y) are the luminance and chrominance components of the input image respectively. A and B stand for luminance values above and below the background luminance, respectively. R and G denote red and green. Excitatory and inhibitory regions within each filter's support are denoted by + and -, respectively. The generalized double opponency (GDO) units combine the signals of two mechanisms in a center-surround opponency fashion (see inset). The outputs of the GDO's, multiplied by appropriate weights, are combined to produce the output signal. Reprinted from Papathomas et al. (1997) with permission of © IEEE.

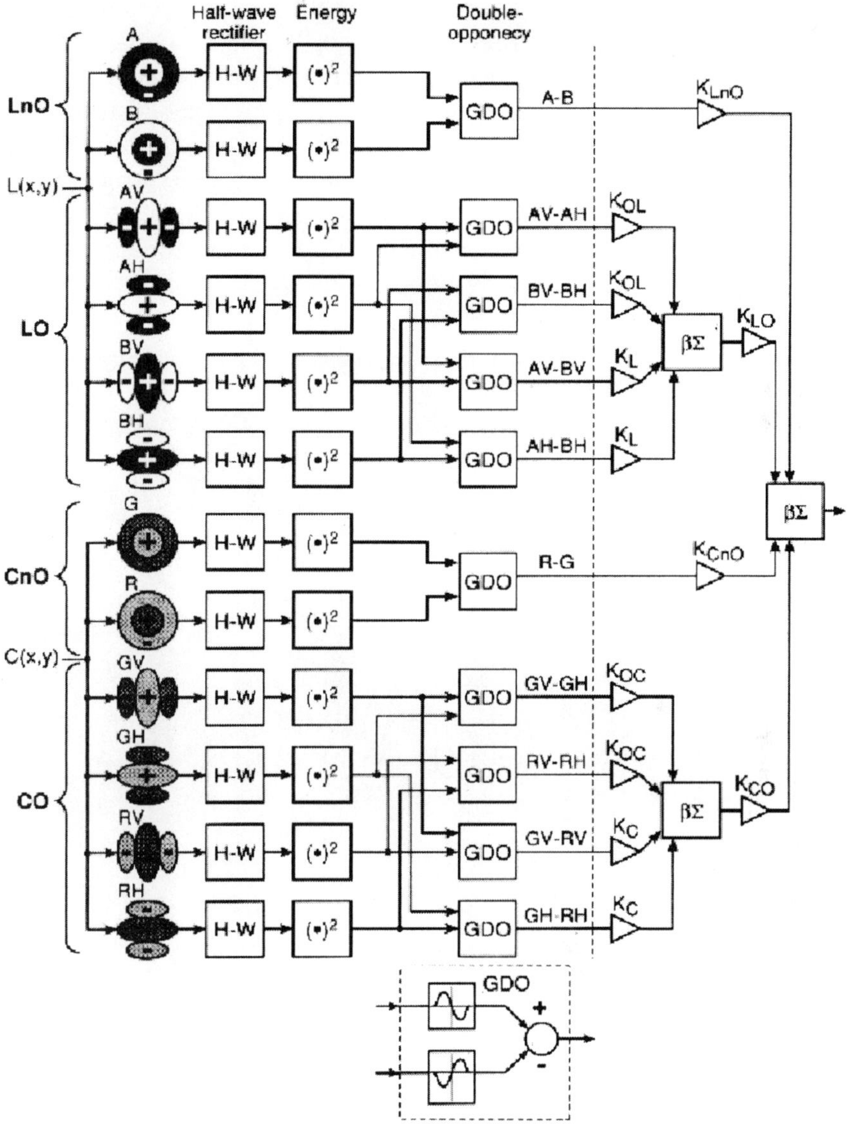

The double opponency model explained above predicts the strong segregation for a texture composed of textels that have a bright center and a dark surround from a texture made up of textels with a dark center and a bright surround. This example is shown in the top left panel in Fig. 12.6, and the success of the model can be attributed to the half-squaring operation. It also predicts the weak segregation of an odd antisymmetric texture pair composed of vertical Gabor patches with a single sinusoidal cycle, where one texture contains patches that are bright on the left and the other texture has patches that are dark on the left, shown in the right top panel of Fig. 12.6. It is to be noted that the mean luminance of all four types of textels is the same as the uniform luminance of the background. These two pairs are key tests for any texture model. The top two images in Fig. 12.6 are the even and odd stimuli, respectively, and below each are the responses of the double-opponency texture model. Note that there is a much stronger response at the center for the even stimuli (brighter edge corresponds to a stronger response) and a medium response for the odd stimuli.

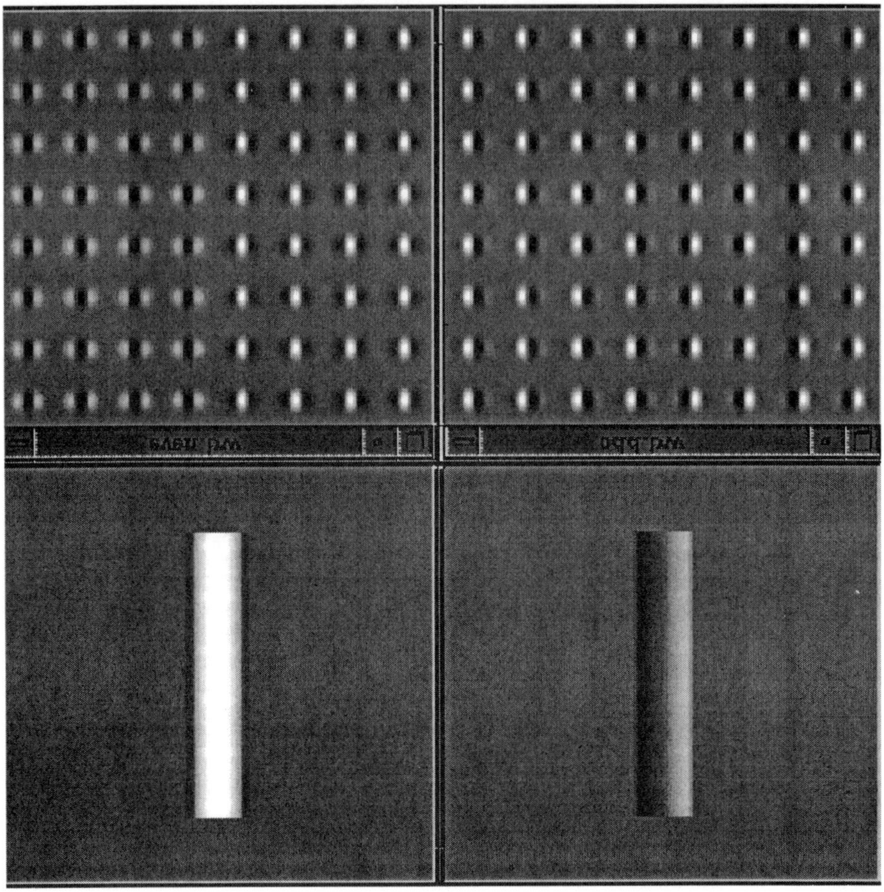

Figure 12.6. Even and odd texture stimuli containing texels whose cross-sections are even and odd functions, respectively. Below each stimulus is the response of the texture model, with bright areas signifying strong texture edges. The even texture is easily segregated into two halves by a global vertical edge. However, the top right image (odd case) cannot be easily segregated. These observations are also predicted by the double-opponency model. Reprinted from Papathomas et al. (1997) with permission of © IEEE.

12.4 SUMMARY

Texture perception has experienced considerable progress in the past four decades of research. Progress has been made in all fronts of neurophysiology, psychophysics, and in computational modeling of texture perception. As the chapter describes, modern day texture models have converged on the generic three-stage model to explain a variety of psychophysical stimuli and natural images. However, understanding of the various mechanisms involved in texture perception is far from being fully elucidated. We are now beginning to understand the effects of learning and attention in texture perception. We are also beginning to explain chromatic mechanisms involved in texture. The parallel channels explained in various models are combined in an ad-hoc basis, which needs considerably more refinement based on neurophysiological inputs. The quest for these solutions continues to keep texture segregation a fertile area of research.

12.5 REFERENCES

Adelson, E. H., and Bergen, J., 1985, Spatiotemporal energy models for the perception of motion, *Journal of the Optical Society of America*, **2A**: 284-299.

Ahissar, M., and Hochstein, S., 1991, Learning is both stimulus and task specific, *Investigative Ophthalmology and Visual Science Supplement*, **32**: 715

Bach, M., and Meigen, T., 1992, Electrophysiological correlates of texture segregation in the human visual evoked potential, *Vision Research*, **32**: 417-424.

Bach, M., and Meigen, T., 1997, Similar electrophysiological correlates of texture segregation induced by luminance, orientation, motion, and stereo, *Vision Research*, **37**: 1409-1414.

Beck, J., 1966, Perceptual grouping produced by changes in orientation and shape, *Science*, **154**: 538-540.

Beck, J., 1967, Perceptual grouping produced by line figures, *Perception and Psychophysics*, **2**: 491-495.

Beck, J., 1982, Textural segmentation, in: *Organization and Representation in Perception*, Hillsdale, NJ.

Beck, J., Graham, N., and Sutter, A., 1991, Lightness differences and the perceived segregation of regions and populations, *Perception and Psychophysics*, **49**: 257-269.

Beck, J., Prazdny, K., and Rosenfeld, A., 1983, A theory of textural segmentation, in: Human and Machine Vision, J. Beck, B. Hope, and A. Rosenfeld, ed., Academic Press, New York.

Beck, J., Sutter, A., and Ivry, R., 1987, Spatial frequency channels and perceptual grouping in texture segregation, *Computer Vision, Graphics and Image Processing*, **37**: 299-325.

Bergen, J. R., 1991, Theories of visual texture perception, in: Vision and Visual Dysfunction, D. Regan, ed., Vol 10, Macmillan, New York.

Bergen, J. R., and Adelson, E.H., 1986, Visual texture segmentation based on energy measures, *Journal of the Optical Society of America*, **3A**: 98.
Bergen, J. R., and Adelson, E.H., 1988, Early vision and texture perception, *Nature*, **333**: 363-364.
Bergen, J.R., and Julesz, B., 1983, Parallel versus serial processing in rapid pattern discrimination, *Nature*, **303**: 696-698.
Bergen, J.R., and Landy, M. S., 1991, Computational modeling of visual texture segregation, in: *Computational Models of Visual Perception*, M.S. Landy and J.A Movshon, editors, MIT Press, Cambridge, MA.
Bonds, A. B., 1989, Role of inhibition in the specification of orientation selectivity of cells in the cat striate cortex, Visual Neuroscience, **2**: 41-55.
Bovik, A. C., Clark, M., and Geisler, W. S., 1987, Computational texture analysis using localized spatial filtering, in: *Proceedings of the Workshop on Computer Vision*, IEEE Computer Society Press, Miami Beach, pp 201-206.
Bravo, M., and Nakayama, K., 1994, The role of attention in different visual-search tasks, *Perception and Psychophysics*, **51**: 465-472.
Buckley, D., and Fisby, J. P., 1993, Interaction of stereo, texture and outline cues in the shape perception of three-dimensional ridges, Vision Research, **33**: 919-933.
Caelli, T., 1982, On discriminating visual textures and images, *Perception and Psychophysics*, **31**: 149-159.
Caelli, T., 1985, Three processing characteristics of visual texture segmentation, *Spatial Vision*, **1**: 19-30.
Caelli, T., and Julesz, B., 1978, On perceptual analyzers of visual texture discrimination, *Biological Cybernetics*, **28**: 167-175.
Caelli, T., and Julesz, B., 1979, Psychophysical evidence for global feature processing in visual texture discrimination, *Journal of the Optical Society of America*, **69**: 675-677.
Callahan, T. C., 1989, Interference and dominance in texture segregation: Hue, geometric form, and line orientation, Perception and Psychophysics, **46**: 299-311.
Callahan, T. C., Lasaga, M. I., and Garner, W. R., 1986, Visual texture segregation based on orientation and hue, *Perception and Psychophysics*, **39**: 32-38.
Cannon, M. W., and Fullenkamp, S. C., 1991, Spatial interactions in apparent contrast: Inhibitory effects among grating patterns of different spatial frequencies, spatial positions and orientations. Vision Research, **31**: 1985-1998.
Caputo, G., 1996, The role of the background: Texture segregation and figure-ground Segmentation, *Vision Research*, **36**: 2815-2826.
Caputo, G., 1998, Texture brightness filling-in, *Vision Research*, **38**: 841-851.
Chubb, C., and Sperling, G., 1988, Drift-balanced random stimuli: A general basis for studying non-Fourier motion perception, *Journal of the Optical Society of America*, **5A**: 1986-2006.
Chubb, C., and Sperling, G., 1989, Second-order motion perception: Space/time separable mechanisms, in: *Proceedings of the Workshop on Visual Motion*, pp. 126-138.
Daugman, J. G., 1987, Image analysis and compact coding by oriented 2-D Gabor primitives, *S.P.I.E. Proceedings*, **758**: 19-30.
Daugman, J. G., 1988, Complete discrete 2-D Gabor transforms by neural networks for image analysis and compression, *IEEE Transactions on Acoustics, Speech and Signal Processing*, **36**: 1169-1179.
DeYoe, E. A., and VanEssen, D. C., 1988, Concurrent processing streams in monkey visual cortex, *Trends in Neuroscience*, **11**: 219-226.

Durgin, F. H., and Huk, A. C., 1997, Texture density after-effects in the perception of natural and artificial textures, *Vision Research*, **37**: 3273-3282.

Enn, J., 1986, Seeing textons in context, *Perception and Psychophysics*, **39**: 143-147.

Fogel, I., and Sagi, D., 1989, Gabor filters as texture discriminator, *Biological Cybernetics*, **61**: 103-113.

Gallant, J. L., van Essen, D. C., and Nothdurft, H. C., 1995, Two-dimensional and three-dimensional texture processing in visual cortex of the macaque monkey, in: *Early Vision and Beyond*, T.V. Papathomas, C. Chubb, A. Gorea, and E. Kowler, editors, pp 89-98, MIT Press, Cambridge, MA.

Garner, W. R., and Feldoldy, G. L., 1970, Integrality of stimulus dimensions in various types of information processing, *Cognitive Psychology*, **1**: 225-241.

Gibson, J., 1950, *The perception of the visual world*, Houghton Mifflin, Boston, MA.

Gorea, A., 1995, Visual texture, in: *Early Vision and Beyond*, T.V. Papathomas, C. Chubb, A. Gorea, and E. Kowler, editors, pp. 55-57, MIT Press, Cambridge, MA.

Gorea, A., and Papathomas, T. V., 1999, Luminance, color, and orientation: Local versus global contrasts in texture segregation, *Journal of the Optical Society of America*, **16A**(3): 728-741.

Gorea, A., and Papathomas, T. V., 1989, The role of color and orientation matching in texture discrimination, *Presented at Annual OSA Meeting, Technical Digest Series*, **18**: 161.

Gorea, A., and Papathomas, T. V., 1991, Texture segregation by chromatic and achromatic visual pathways: An analogy with motion processing, *Journal of the Optical Society of America*, **8A**(2): 386-393.

Gorea, A., and Papathomas, T. V., 1991b, Extending a class of motion stimuli to study multi-attribute texture perception, *Behavioral Research Methods, Instruments and Computers*, **23**(1): 5-8.

Gorea, A., and Papathomas, T. V., 1993, Double-opponency as a generalized concept in texture segregation illustrated with color, luminance and orientation defined stimuli, *Journal of the Optical Society of America A*, **10**: 1450-1462.

Gouras, P., and Kruger, J., 1979, Responses of cells in foveal visual cortex of the monkey to pure color contrast, *Journal of Neurophysiology*, **42**: 850-860.

Graham, N., 1991, Complex channels, early local nonlinearities, and normalization in texture segregation, in: *Computational Models of Visual Perception*, M.S. Landy and J.A Movshon, editors, MIT Press, Cambridge, MA.

Graham, N., Beck, J., and Sutter, A., 1992, Nonlinear processes in spatial-frequency channel models of perceived texture segregation: Effects of sign and amount of contrast, *Vision Research*, **32**: 719-743.

Grossberg, S., and Mingolla, E., 1985, Neural dynamics of perceptual grouping: Textures, boundaries and emergent segmentations, *Perception and Psychophysics*, **38**: 141-171.

Grossberg, S., and Mingolla, E., 1986, Computer simulation of neural networks for perceptual psychology. *Behavior Research Methods, Instruments and Computers*, **18**: 601-607.

Gurney, R., and Browse, R., 1987, Micropattern properties and presentation conditions influencing visual texture discrimination, *Perception and Psychophysics*, **41**: 239-252.

Julesz, B., 1962, Visual texture discrimination, *IRE Transactions on Information Theory*, **8**:84-92.

Julesz, B., 1965, Texture and visual perception, *Scientific American*, **212**: 38-48.

Julesz, B., 1981, Textons, the elements of texture perception and their interactions, *Nature*, 290: 91—97.

Julesz, B., 1981, A theory of preattentive texture discrimination based on first-order

statistics of textons, *Biological Cybernetics*, **41**: 131-138.
Julesz, B., 1995, *Dialogues on Perception*, MIT Press, Cambridge, MA.
Julesz, B., and Bergen, J. R., 1983, Textons, the fundamental elements in preattentive vision and perception of textures, *The Bell System Technical Journal*, **62**: 1619-1645.
Julesz, B., Gilbert, E.N., and Victor, J. D., 1978, Visual discrimination of textures with identical third-order statistics, *Biological Cybernetics*, **31**: 137-140.
Kashi, R. S., Papathomas, T. V., and Gorea, A. G., 1994, A perceptually based computational model for texture segregation of color images, *Proceedings of the IEEE Workshop on Visual Signal Processing and Communications*. IEEE Press.
Kashi, R. S., Papathomas, T. V., and Gorea, A. G., 1997, Grouping in sparse random-dot patterns: Linear and non-linear mechanisms, *Proceedings of the SPIE conference on human vision and electronic imaging*, **3016**: 420-429.
Kashi, R. S., Papathomas, T. V., Gorea, A. G., and Julesz, B., 1996, Similarities between texture grouping and motion perception: The role of color, luminance, and orientation. *International Journal of Imaging Systems and Technology*, **7**: 85-91.
Knierim, J. J., and van Essen, D. C., 1992, Neuronal responses to static texture patterns in area V1 of the alert macaque monkey, *Journal of Neurophysiology*, **67**: 4:961-980.
Knill, D. C., 1998, Discrimination of planar surface from texture: human and ideal observers compared, *Vision Research*, **38**: 1683-1711.
Knill, D. C., 1998b, Surface orientation from texture: ideal observers, generic observers, and the information content of texture cues, *Vision Research*, **38**: 1655-1682.
Krose, B. J. A., 1987, Local structure analyzers as determinants of preattentive pattern discrimination, *Biological Cybernetics*, **55**: 289-298.
Kwan, L., and Regan, D., 1998, Orientation-tuned spatial filters for texture-defined form, *Vision Research*, **38**: 3849-3855.
Lamme, V. A., van Dijk, B. W., and Spekreijse, H., 1992, Texture segregation is processed by primary visual cortex in man and monkey: Evidence from VEP experiments, *Vision Research*, **32**: 797-807.
Landy, M. S., and Bergen, J. R., 1991, Texture segregation and orientation gradient, *Vision Research*, **31**: 679-691.
Li, A., and Lennie, P., 1997, Mechanisms underlying segmentation of colored textures, *Vision Research*, **37**: 83-97.
Livingstone, M. S., and Hubel, D. H., 1988, Segregation of form, color, movement and depth: Anatomy, physiology and perception, *Science*, **240**: 740-749.
Malik, J., and Perona, P., 1989, A computational model of human texture perception, *Investigative Ophthalmology and Visual Science Supplement*, **30**: 161
Malik, J., and Perona, P., 1990, , Preattentive texture discrimination with early vision mechanisms, *Journal of the Optical Society of America*, **7A**: 923-932.
Marr, D., and Poggio, T., 1976, Cooperative computation of stereo disparity. *Science*, **194**: 283-287.
Marr, D., and Poggio, T., 1979, A computational theory of human stereo vision, *Proceedings of the Royal Society of London B*, **204**: 301-328.
Michael, C.R., 1979, Color-sensitive hypercomplex cells in monkey striate cortex, *Journal of Neurophysiology*, **42**: 726-744.
Mikami, A., Newsome, W.T., and Wurtz, R.H., 1986, Motion selectivity in macaque visual cortex: Spatio-temporal range of directional interactions in MT and V1, *Journal of Physiology*, **55**: 1328-1339.
Nakayama, K., and Silverman, G. H., 1986, Serial and parallel processing of visual feature conjunctions, *Nature*, **320**: 264-265.

Nam, J. H., and Chubb, C., 1988, Perceived texture contrast is determined by a negative half-wave rectifying mechanism, *Investigative Ophthalmology and Visual Science Supplement*, **39**: 649.

Northdurft, H. C., and Parlitz, D., 1993, Absence of express saccades to texture or motion defined targets, *Vision Research*, **33**: 1367-1383.

Northdurft, H. C., 1985, Sensitivity for structure gradient in texture discrimination tasks, *Vision Research*, **25**: 1957-1988.

Northdurft, H. C., 1990, Texture discrimination by cells in the cat lateral geniculate nucleus, *Experimental Brain Research*, **82**: 48-56.

Ohzawa, I., DeAngelis, G. C., and Freeman, R. D., 1990, Stereoscopic depth discrimination in the visual cortex: Neurons ideally suited as disparity detectors, *Science*, **249**: 1037-1041.

Papathomas, T. V., Gorea, A., Feher, A., and Conway, T. E., 1999, Attention-based texture segregation, *Perception and Psychophysics.*

Papathomas, T. V., Kashi, R. S., and Gorea, A., 1997, A human vision based computational model for chromatic texture segregation, *IEEE Transactions on Systems, Man and Cybernetics--Part B: Cybernetics*, **27**(3): 428-440.

Pessoa, L., Beck, J., and Mingolla, E., 1996, Perceived texture segregation in chromatic element-arrangement pattern: High intensity interference, *Vision Research*, **36**: 1745-1760.

Poggio, G. F., Motter, B. C., Squatrito, S., and Trotter, Y., 1985, Responses of neurons in visual cortex V1 and V2 of the alert macaque to dynamic random-dot stereograms, *Vision Research*, **25**: 397-406.

Polat, U., and Sagi, D., 1992, Lateral interactions between spatial channels: Suppression and facilitation revealed by lateral masking, Investigative Ophthalmology and Visual Science Supplement, **33**: 1345.

Rao, R. A., and Lohse, G. L., 1966, Towards a texture naming system: Identifying relevant dimensions of texture, *Vision Research*, **36**: 1649-1669.

Reichardt, W., 1961, Autocorrelation, a principle for the evaluation of sensory information by the central nervous system, in: *Sensory Communication*, W.A. Rosenblith, editor Wiley, New York.

Rosenholtz, R., and Malik, J., 1997, Surface orientation from texture: isotropy or homogeneity (or both)? *Vision Research*, **37**: 2283-2293.

Rubenstein, B. S., and Sagi, D., 1990, Spatial variability as a limiting factor in texture-discrimination tasks: Implications and performance, *Journal of the Optical Society of America*, **7A**: 1632-1643.

Sagi, D., 1995, The psychophysics of texture segmentation, in : *Early Vision and Beyond*, T.V. Papathomas, C. Chubb, A. Gorea, and E. Kowler, editors, pp. 89-98, MIT Press, Cambridge, MA.

Sagi, D., and Julesz, B., 1987, Short-range limitation on detection of feature differences, *Spatial Vision*, **2**: 29-49.

Shatz, B. R., 1977, The computation of immediate texture discrimination, *MIT AI Memo*, page 426.

Sperling, G., 1989, Three stages and two systems of visual processing, *Spatial Vision*, **4**: 183-207.

Sutter, A., Beck, J., and Graham, N., 1989, Contrast and spatial variables in texture segregation: Testing a simple spatial-frequency channels model, *Perception and Psychophysics*, **46**: 312-332.

Treisman, A., and Gelade, G., 1980, A feature integration theory of attention, *Cognitive Psychology*, **12**: 97-136.

Treisman, A., and Gormican, S., 1988, Feature analysis in early vision: Evidence from search asymmetries, *Psychological Reviews,* **95**: 15-48.

Troje, N. F., and Bulthoff, H. H., 1996, Face recognition under varying poses: The role of texture and shape, *Vision Research*, **36**: 1761-1771.

Turner, M. R., 1986, Texture discrimination by Gabor functions, *Biological Cybernetics*, **55**: 71-82.

van Essen, E. A., DeYoe, E. A., Olavarria, J. F., Krierim, J. J., Fox, J. M., Sagi, D., and Julesz, B., 1990, Neural responses to static and moving texture patterns in visual cortex of the macaque monkey, in : *Neural Mechanisms of Visual Perception*, D.M.K. Lam and C.D. Gilbert, editors, pp. 137-154. Porfolio Publishing, Woodlands TX.

Van Santen, J. P. H., and Sperling, G., 1985, Elaborated Reichardt detectors, *Journal of the Optical Society of America*, **2A**: 300-321.

Victor, J. D., and Conte, M. M., 1996, The role of high-order phase correlations in texture processing, *Vision Research*, **36**: 1615-1632.

Victor, J. D., Conte, M. M., Purpura, K., and Katz, E., 1995, Isodipole textures: a window on cortical mechanisms of form processing, in : *Early Vision and Beyond*, T.V. Papathomas, C. Chubb, A. Gorea, and E. Kowler, editors, pp. 89-98, MIT Press, Cambridge, MA.

Voorhees, H., and Poggio, T., 1988, Computing texture boundaries from images, *Nature*, **333**: 364-367.

Watson, A. B., and Ahumada, A. J., 1985, Model of human visual-motion sensing, *Journal of the Optical Society of America*, **2A**: 322-342.

Watt, R. J., 1995, Some speculations on the role of texture processing in visual Perception, in : *Early Vision and Beyond*, T.V. Papathomas, C. Chubb, A. Gorea, and E. Kowler, editors, pp. 89-98, MIT Press, Cambridge, MA.

Wilson, H., and Mast, R., 1993, Illusory motion of texture boundaries, *Vision Research*, **33**: 1437-1446.

Wolfe, J. M., 1992, Effortless texture segmentation and 'parallel' visual search are not the same thing, *Vision Research*, **32**: 757-763.

Zeki, S. M., 1973, Colour coding in rhesus monkey prestriate cortex, *Brain Research*, **53**: 422-427.

Chapter 13

Neural Models of Motion Perception

Thomas V. Papathomas[1], Amy S. Rosenthal[2], Bela Julesz[3]

[1] *Dept. of Biomedical Engineering and Laboratory of Vision Research, Rutgers University, 152 Frelinghuysen Rd., Piscataway, NJ 08854-8020, PH: (732) 445-6533, FX: (732) 445-6715, EM: papathom@zeus.rutgers.edu*

[2] *Alpha Technologies, 88 Centenial Ave., Piscataway, NJ 08854, PH: (732) 980-1800, EM: arosenthal@alpha88.com*

[3] *Dept. of Psychology and Laboratory of Vision Research, Rutgers University, 152 Frelinghuysen Rd., Piscataway, NJ 08854-8020, PH: (732) 445-6520, FX: (732) 445-6715, EM: julesz@cyclops.rutgers.edu*

13.1 INTRODUCTION

Apparent motion (AM) involves two or more "frames" of images, where elements are displaced from frame to frame to elicit the percept of motion. An important question is how a certain element, or feature, is matched across frames to yield the veridical motion percept without false target localization. This is known as the "correspondence problem" (Julesz, 1968; Ullman, 1979). One possibility is that bottom-up, hard-wired neural mechanisms compute motion automatically, without solving explicitly the correspondence problem. At the other extreme top-down motion-tuned mechanisms can first extract specific features of moving objects (such as edges, corners, etc.) and then track these features from one frame to the next to achieve a solution to the correspondence problem. It is quite possible that a wide variety of motion sensing units exist in the visual system, covering the entire spectrum that is defined between the two extremes outlined above. Neural mechanisms that are tuned to a specific direction of motion have been found in the striate cortex, as well as in the extrastriate areas, such as area V5, otherwise known as MT (middle temporal), and area MST (medial superior temporal) (Tovee, 1996).

This review chapter deals with computational models that can extract the motion signal from a temporal sequence of images, which would produce apparent motion if they were presented as stimuli to the visual system. Of course, we are going to deal only with models that are biologically plausible. Moreover, we are interested in models that are motivated from results obtained in psychophysical and neurophysiological studies.

Many models have been developed to explain how the visual system is able to perceive motion. Experiments in motion perception began over a century ago (Exner, 1875; Wertheimer, 1912) and investigated how humans can solve the correspondence problem to accurately determine the direction of motion (Julesz and Payne, 1968; Anstis, 1970; Braddick, 1974; Barlow, 1979; Marr and Ullman, 1981; Cavanagh and Mather, 1989; etc.). Many powerful techniques have been developed to study motion perception systematically. Among them, four approaches can be distinguished which played a key role in the development of computational models that can accurately predict biological visual motion perception. The first was Reichardt's seminal work with flies that led him to postulate a Fourier motion model based on correlation (Reichardt, 1961). Variants of this model are still in use (van Santen and Sperling, 1985; Lu and Sperling, 1995). Thus, Reichardt's modeling approach is one of the few biologically plausible approaches that has withstood the test of time, and is still in use today in modified forms. In contrast, a multitude of other models has been proposed for various sensory and perceptual processes, only to be eventually replaced by drastically different models. The second approach employed random-dot cinematograms (Julesz and Payne, 1968) which ultimately led to the proposal for the putative "short-range" and "long-range" motion systems (Braddick, 1974). The third approach used a set of dots that move coherently in one direction (signal dots), in the presence of motion noise, i.e., dots that move randomly in all possible directions (noise dots). This approach enabled researchers to develop quantitative methods for assessing the sensitivity of motion mechanisms in both psychophysical and neurophysiological experiments. The fourth approach employed the well-known motion after-effect (MAE) phenomenon, commonly known as the "waterfall illusion", to study the properties of the underlying mechanisms of motion perception (Hiris and Blake 1992; Verstraten et al., 1994).

This chapter reviews classical and recent models of motion perception, and is organized as follows: Section 13.2 presents some key stimuli that challenge us to find good models to predict human response to these stimuli. Subsequent sections deal with models that were inspired by the key stimuli of Section 13.2. Section 13.3 covers first-order (Fourier) luminance-responding motion mechanisms; Section 13.4 deals with second-order contrast-responding mechanisms; and Section 13.5 examines and how these can be combined into a model. Section 13.6 is devoted to third-order motion that is based on other attributes (binocular disparity, relative motion, inter-attribute AM, etc.). Color-driven motion is covered in Section 13.7, and global motion issues are presented in Section 13.8. Section 13.9 starts with a summary and concludes with some comments on alternate views of visual motion modeling.

13.2 SOME KEY STIMULI

In this section, we briefly review five classes of stimuli that offer insight into the inner workings of motion extracting mechanisms. These stimuli have proved to be valuable tools in helping to decide the overall structure of putative motion mechanisms. The first class (subsection 13.2.1) illustrates powerfully that, as far as motion is concerned, when an element in one frame can be matched to two possible targets in the next frame, it is the total energy that matters, rather than the physical similarity of the element to the targets (the issue of energy is examined in subsection 13.2.1). The second class (subsection 13.2.2) offers evidence that the spatial filters that pre-process the spatiotemporal distribution of luminance, which provide their output as input to bottom-up Fourier motion mechanisms, behave as linear filters, akin to the so called "simple cells" in the striate cortex. It also offers evidence for the possible existence of non-linear spatial filters that provide input to bottom-up non-Fourier motion mechanisms, akin to the so-called "complex cells" in striate cortex. The third class (subsection 13.2.3) distinguishes whether a certain type of AM is extracted by bottom-up neural units or is based on feature tracking. The fourth class (subsection 13.2.4) involves the important stimuli of moving plaids and their extensions. Finally, the fifth (subsection 13.2.5) class deals with key experiments on motion after-effect phenomena.

13.2.1 Competing Motion Paths - Similarity or Covariance?

The issue of what metric is employed by the visual system in matching targets across frames in a motion sequence can be posed by the following analysis: Assume that element Q in frame 1 has an equal probability of being matched to two different targets A and B in frame 2, where A and B are equidistant from Q. Suppose that match QA produces motion to the right, whereas QB produces motion to the left. The question is: which motion path, QA or QB, will be perceived? To simplify the problem, we set all the properties of A, B, and Q to be the same (size, color, shape, etc.), and we distinguish them only by virtue of their luminance L_A, L_B, and L_Q, respectively. Also, let DL be the difference between the target's luminance and that of the uniform background, L_0, so that $DL_A=L_A-L_0$, $DL_B=L_B-L_0$, and $DL_Q=L_Q-L_0$.

To test whether a similarity metric is used in deciding which motion path prevails, one simply makes element Q identical to A, namely, $DL_Q=DL_A$. In this case, one path is produced by the similarity match AA, whereas the competing path is produced by match AB. By keeping L_A fixed and varying L_B, the similarity metric would predict that motion would always be perceived along path AA, independently of L_B. Surprisingly, such experiments showed that the AB match dominates when its covariance is larger than that of AA. In other words, the visual system does not prefer path AA, even though such a match is perfect with respect to a similarity metric. Instead, it prefers the motion path that involves the maximum covariance. The covariance of path AB for any elements A, B with luminances L_A and L_B is some function $f(DL_A \cdot DL_B)$ of the product of DL_A and DL_B. The covariance, sometimes referred to as energy, turns out to be an excellent predictor of motion path strength Reichardt, 1961; Werkhoven et al., 1990, 1993; Papathomas et al., 1995; Nishida and Takeuchi, 1990). Namely, paths of greater covariance are perceived more dominantly than paths of lower covariance.

13.2.2 Non-competing Paths - Reverse Phi, Contrast-Defined Targets, and Inter-attribute Motion

The competing-path stimuli of the previous subsection have been used successfully to investigate the issue of similarity versus covariance. However, such stimuli can be very complicated, and simpler stimuli may be more suitable to illustrate some fundamental principles of motion processing. Simpler non-competing-path stimuli can be used to obtain useful data for building motion models. Such single-path stimuli are shown in Fig. 13.1, where a target Q_1 at location x_1 in the first frame is displaced to x_2 in the next frame, possibly altering its appearance to Q_2. The notation used in Fig. 13.1 for uni-dimensional motion will be used throughout this chapter. Motion is constrained along the horizontal axis, which is shown as the x-axis, and can be leftward or rightward. The temporal variable t is mapped along the vertical axis. It must be emphasized that the vertical axis is discretized. Each frame is shown in one row, with the next frame displayed just below it in the x-t plane. When actually displaying the frames in sequence as motion stimuli, each subsequent frame is displayed at the same location that the previous frame occupied, thus erasing the previous frame. Over time, the target will appear to be moving; in the case of these figures, the direction of physical displacement is to the right. In describing the stimuli for psychophysical experiments, each frame is displayed for duration Δt and then replaced by the next frame.

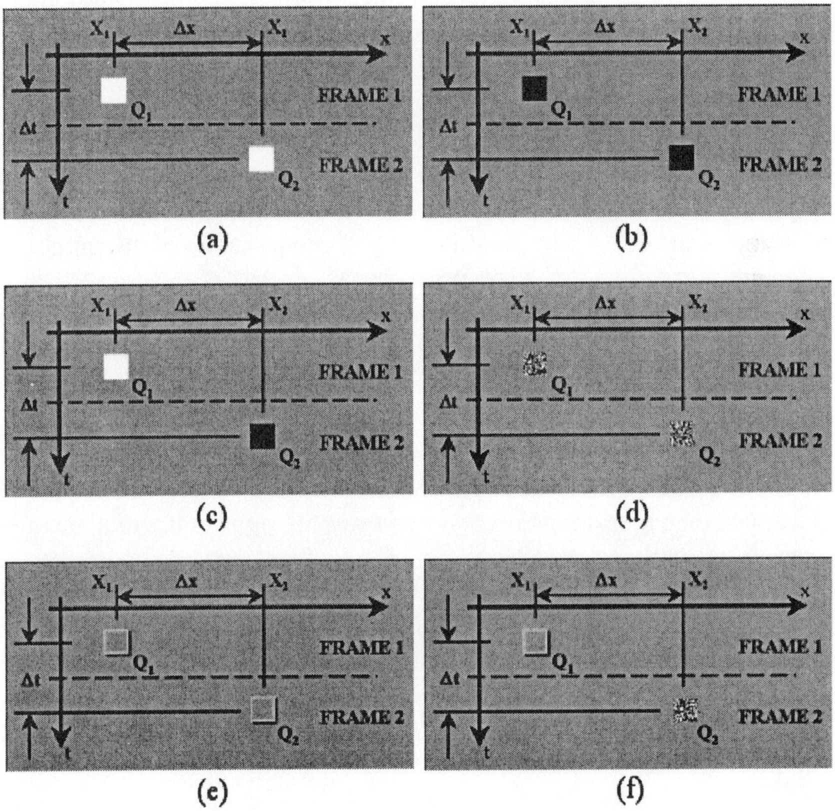

Figure 13.1. Schematic x-t (space-time) diagrams of simple apparent-motion (AM) stimuli. (a) Forward AM with luminance-defined bright targets. (b) Forward AM with luminance-defined dark targets. (c) Luminance-defined alternating bright-dark targets, resulting in reverse-phi motion. (d) Forward AM with purely contrast-defined targets. (e) Forward AM with purely disparity-defined targets. (f) Interattribute forward AM with targets that are defined by different attributes in different frames (here by disparity and by contrast). The stimuli of parts a-c contain Fourier energy with a preferred motion direction (see also Figs. 13.2 and 13.3); the stimuli of parts d-f do not.

The slope (dt/dx) is the ratio of the difference in vertical and horizontal co-ordinates. In general, the inverse of the slope of the line that connects the centers of targets in x-t space defines the speed v of the apparent motion. A line with a negative or positive slope produces motion to the right or to the left, respectively. Since $v = dx/dt$, the smaller the absolute value of the slope the greater the speed. As a special case, when the slope is infinite (vertical line), the speed is zero; indeed, in this case there is no change in position over time. A computational model can use a two-dimensional image representation of the x-t plane as a convenient representation of the spatiotemporal stimulus.

If both targets are of the same luminance polarity with respect to the background, as in Fig. 13.1a (both brighter than the background) and Fig. 13.1b (both dimmer), the perceived motion is called *forward phi*. The motion percept is in the same direction as the physical displacement in the stimulus. Furthermore, experiments have revealed that the perceived speeds of the ensuing motion with the stimuli of Figs. 13.1a and b are equal if the physical displacements, and hence the slopes, are equal. In other words, the speed is independent of the polarity (bright versus dark) of the target, as common sense may predict. (The perceived speed has been observed to be lower for targets of low contrast, as compared to that for moderate and high contrast, even for equal physical displacements, but the effect is markedly noticeable only for very low contrast values.)

However, a strange phenomenon is observed with the stimulus of Fig. 13.1c. The opposite-polarity stimulus of Fig. 13.1c appears on paper to be moving to the right. After all, this is the veridical displacement. One would expect to perceive a bright target that moves to the right as it simultaneously changes to a dark target. However, the actual percept under certain conditions is that of leftward motion. This phenomenon is called *reverse phi* or *reverse AM* (Anstis, 1970), and the conditions that favor it are: greater eccentricity, high spatial frequencies (i.e., small target sizes), high temporal frequencies (i.e., small frame durations), and low contrast. At first, reverse phi motion appears to be paradoxical, but it can be explained by Fourier motion mechanisms, as we shall see below.

Let us assume, for a moment, that we have been able to design a satisfactory Fourier motion mechanism that can reliably extract motion signals, as long as there is Fourier motion energy in the mechanism's preferred direction in the spatiotemporal plane. By Fourier energy we mean the magnitude of the Fourier transform of the spatiotemporal distribution of luminance. Such a Fourier motion extractor would be able, in principle, to predict human performance with the stimuli of Figs. 13.1a, b, and c, because they all possess a net energy with a preferred motion direction that coincides with that perceived by human observers (see Figs. 13.2 and 13.3). The problem is that it would not be able to predict performances with the stimulus of Fig. 13.1d, in which a purely contrast-defined target is displaced to the right in two frames. By a purely contrast-defined target, we mean that the average luminance of the target is the same as that of the background. In this case, the stimulus of Fig. 13.1d contains no net directional Fourier energy, and a Fourier model would predict ambiguous motion. Yet, humans perceive motion along the veridical displacement. This points to the inadequacy of Fourier motion units to model, by themselves, the perceived motion, and the need to extend the model with non-Fourier components.

Another challenge for developing models to analyze motion comes from cases where the targets are defined by attributes other than luminance or contrast. The attribute of color comes first in mind, and we will devote an entire section (Section 13.7) to it later. Other attributes that may be able to support motion are binocular (stereo) disparity, texture contrast, and relative motion. Fig. 13.1e shows schematically motion of targets defined purely by binocular disparity, moving to the right; this can be achieved by using random-dot stereograms, or RDS (Julesz, 1960). Fig. 13.1f illustrates the case where the attribute that distinguishes the target from the background changes from one frame to the next: binocular disparity in frame 1, contrast in frame 2, using RDS in both frames. We consider such models in Section 13.6.

13.2.2.1 Fourier Analysis

Figs.13.2 and 13.3 illustrate the above issues more clearly, with parts a, b, c, and d of Fig. 13.2 being more general versions of the stimuli shown in parts a, b, c, and d of Fig. 13.1, respectively. Instead of a single target in each frame, there is a row of targets, each of which is displaced physically to the right (remember, the vertical axis represents time, t, as in Fig. 13.1).

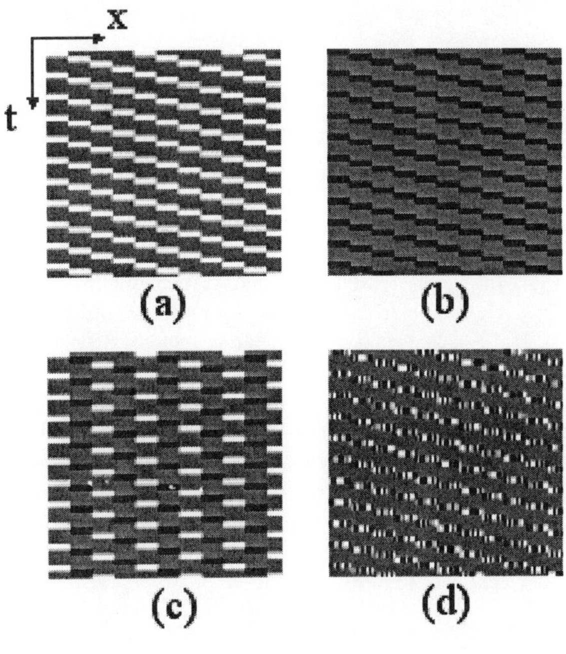

Figure 13.2. Schematic x-t diagrams of extended versions of the stimuli of Figs. 13.1a-d. The vertical axis represents time t. (a), (b) Luminance-defined bright and dark targets, respectively. (c) Alternating bright and dark targets. (d) Purely contrast-defined targets.

Figures 13.2a and b show bright and dark targets, respectively. Figure 13.2c shows targets that switch their polarity from bright to dark and back to bright as they are displaced to the right. Finally, Fig. 13.2d shows purely contrast-defined targets, with each target composed of random distributions of an equal number of dark and light dots to maintain a mean luminance equal to that of the background.

The magnitude of the two-dimensional Fourier transform of the stimuli in Fig. 13.2 is shown in the corresponding parts of Fig. 13.3 to demonstrate the Fourier components of first-order motion. The Fourier forward AM of Figs. 13.2a and b is reflected in the two dominant bright dots of Figs. 13.3a and b which are oriented perpendicular to the direction of stimulus motion. Reverse phi motion, shown in Fig. 13.2c, has a resulting Fourier transform (Fig. 13.3c) with two dominant bright dots which indicate motion in the direction opposite to that of the forward phi. The second-order motion shown in Fig. 13.2d contains no coherent Fourier energy, and the Fourier transform illustrates this in Fig. 13.3d, which contains no dominant components in any direction.

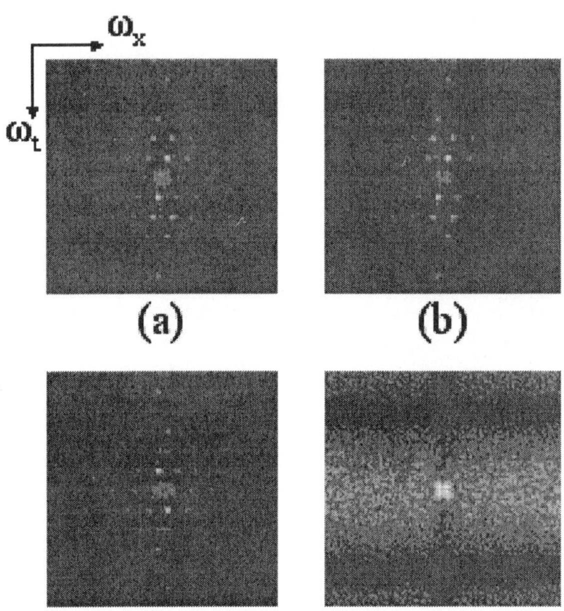

Figure 13.3. Fourier energy for the corresponding stimuli of Fig. 13.2. The common origin of the two axes is at the center of each panel. The horizontal axis represents spatial frequency ω_x, and the vertical axis represents temporal frequency ω_t. Positive values for ω_x and ω_t are to the right and below the origin, respectively. The pair of dominant dots located symmetrically around the origin indicates the dominant direction of motion. Parts a, b, and c possess Fourier components that can produce a motion percept for a specific direction; part d does not, as it contains motion energy in both directions.

The energy in Figs. 13.3a-c is in the same direction as that obtained by human observers when they are exposed to the corresponding stimuli of Fig. 13.2. As expected, the energy in parts Figs. 13.3a and b is identical despite the polarity difference, because the spatiotemporal parameters are the same, and the contrasts of the bright and dark elements are equivalent. Part 3c illustrates that the motion energy in the alternating-polarity stimulus of Fig. 2c is, in fact, in the opposite direction than the actual displacement (reverse phi). Thus, the reverse-phi phenomenon, unexpected as it may seem, is easily explained by Fourier motion analyzers. Thus we see that reverse-AM is the signature of Fourier motion analysis. The analysis of non-Fourier motion (Figs. 13.1d and 13.2d) seems to be more complex (Section 13.4).

13.2.3 Tests for Bottom-Up Versus Feature Tracking

It is very important to know whether motion based on a particular attribute, such as luminance, color, contrast, flicker rate, binocular disparity, etc., is extracted automatically by motion-energy tuned neural mechanisms, or requires the observer to track the features of the stimulus over time. Two powerful stimuli have been developed over the years for selecting experimentally between the two possibilities.

13.2.3.1 Fluted Square Wave

The fluted square wave is a pattern derived from a square-wave modulation in the horizontal direction, from which is subtracted the fundamental-frequency sinusoid. The pattern has sharp edges, and varies sinusoidally between edges. In the next frame of the animation sequence, the pattern is displaced by 90 degrees, namely a quarter-period, say to the left. A careful examination reveals that, if the resulting AM is based on tracking the sharp edges, then it will be perceived to move to the left; if, however, motion is extracted automatically by Fourier mechanisms, then rightward motion will be perceived (Adelson and Bergen, 1985).

13.2.3.2 Pedestal-Plus-Test

This is a compound stimulus with two spatially-modulated sinusoidal components: one that does not move (the pedestal); and another that is displaced by 45 degrees (an eighth of a period) from one frame to the next, say to the left. A careful analysis will show that tracking of a feature will result in an ambiguous direction of AM, whereas Fourier motion-energy-tuned mechanisms will extract leftward motion (Lu and Sperling, 1995).

13.2.4 Moving Plaids and Targeted Motion Displays

The moving plaid stimulus of Adelson and Movshon (1982) proved to be an invaluable stimulus for studying how motion components are integrated to obtain a global percept. A moving plaid is composed of two moving sinusoidal gratings of different orientations that form an angle between them. This results either in transparent motion (as if they are two surfaces that slide past each other), or in coherent motion (as if they form a plaid that moves as a single surface), depending on their spatial frequencies, speeds, angle, and other attributes. Recently, Schrater et al. (2000) created a stochastic moving stimulus that removes the undesirable nodes of moving plaids (Braun, 2000). Whereas the plaid nodes provide features that move coherently, the new stimuli contain spatiotemporally random features so that the nodes move over much shorter distances. The advantage is that motion mechanisms with specific spatiotemporal tuning can be isolated.

13.2.5 Key Motion After-Effect (MAE) Stimuli

Motion after-effect (MAE) is the phenomenon of perceiving movement of a static pattern after adapting to a moving pattern for several seconds. The after-effect produces motion in the opposite direction to that during adaptation. There is a rich literature on the motion after-effect (e.g., Mather et al., 1998). We have selected one key class of stimuli that challenges conventional motion models. This class consists of two families of randomly placed dots presented during the adaptation phase. Dots within each family move at the same speed in a coherent direction, and the two directions form an angle w between them. On the other hand, during the test phase, the dots are displayed in random locations, either statically, or with each dot moving in a random direction.

One key finding is that during the adaptation phase when w is small, all of the dots appear to move in one direction, which is the average of the two component directions. However, when w exceeds a threshold, then transparent motion is perceived, i.e., two families of dots moving in two different directions. Another key finding is that during the test phase a single direction of MAE is perceived, moving in the direction opposite to the average of the two adapting directions, independently of whether coherent or transparent motion was perceived during adaptation.

13.3 MODELS FOR FIRST-ORDER MOTION

Before we go into the details of first-order motion, let us briefly describe the three types of motion as formulated by Lu and Sperling (1995). 1) *First-order* motion is based on the spatiotemporal (x-t) distribution of luminance. 2) *Second-order* motion (see Section 13.4) is based on the x-t distribution of luminance contrast. 3) *Third-order* motion (see Section 13.6) is based on all other visual attributes (binocular disparity, spatial frequency, etc.), as well as inter-attribute motion.

We have demonstrated that if a model can extract the Fourier energy in the spatiotemporal domain, one will be able to predict the phenomena of forward AM and reverse AM, at least theoretically. As an example, Braddick's short-range system extracts motion from luminance-defined stimuli that possess clearly distinguishable components in the spatiotemporal frequency domain. This type of motion is called “first-order” or “Fourier” motion (Chubb and Sperling, 1988; Zanker, 1995; Papathomas et al., 1996). First-order motion (motion of luminance-defined elements) can be predicted using a wide variety of models that can extract motion in the spatiotemporal domain (Reichardt, 1961; Adelson and Bergen, 1985; van Santen and Sperling, 1985; Watson and Ahumada, 1985).

The objective of this section is to present a brief overview of such Fourier motion mechanisms. One must recognize that such models will not be adequate to duplicate human performance with "non-Fourier" contrast-defined stimuli, such as those of Figs. 13.1d and 2d. Non-Fourier motion models are covered in Section 13.4. The minimal requirements for such Fourier models are the following: 1) biological plausibility; 2) that they be consistent with the principle of covariance (see subsection 13.2.1) that they predict human performance with the forward-AM and reverse-AM stimuli.

13.3.1 Reichardt's Model

One of the earliest forms of this type of model is the standard correlation-based opponent Reichardt Standard Motion Analyzer (SMA), a simplified schematic of which is shown in Fig. 13.4 (Reichardt, 1961). The Reichardt detector is an example of a bilocal operator because it receives inputs from two locations in the retina. Its operation is based on autocorrelation. If a target moves, then the product of a frame's image with an appropriately delayed, i.e., time-translated, copy of the next frame's image will produce a strong signal. This multiplicative operation also satisfies the covariance metric (subsection 13.2.1). Accordingly, each of the "h" components represents a filter that will process the stimulus in the spatial domain. The “Δt” elements are temporal filters that essentially impose delays on the spatial input. Motion to the right (M_R) is computed as follows: multiply the delayed response at position x_1 with the undelayed response at x_2, as shown by the solid-line path. Leftward motion (M_S) is computed similarly via the dashed-line path. The strength of the opponent motion "M" is then evaluated by subtracting the leftward motion M_S from the rightward motion M_R. A positive result indicates rightward motion energy, while leftward motion is signified by a negative result. Because Δt is fixed in this model, the optimal response of this analyzer will occur if the stimulus travels a retinal distance Δx over time Δt, i.e., has a speed of $\Delta x/\Delta t$, where Δx is the difference between retinal location x_2 and x_1. A variety of such units in the visual system, with each tuned to different combinations of Δx and Δt, can be pooled to get a good estimate of the stimulus speed (see subsection 13.8.1)

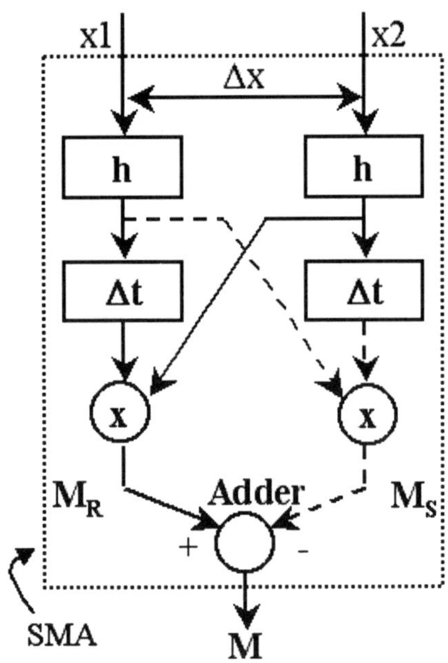

Figure 13.4. Simplified Reichardt model for motion extraction.

Figure 13.4 represents a simplified Reichardt model just to introduce the rationale of the main principle. The model has been studied and enhanced extensively by Sperling and his colleagues (van Santen and Sperling, 1985; Lu and Sperling, 1995) to produce the "elaborated Reichardt detector" (ERD), which achieves a remarkable degree of accuracy in predicting human performance. In the simple formulation of Fig. 13.4 there are two identical spatial filters h occupying two different locations. In one formulation of the ERD, there are two types of spatial filters h_1 and h_2 occupying the same location; also, the delay Δt is implemented roughly as the difference between the responses of two temporal filters g_1 and g_2. The rightward and leftward path outputs of the ERD are

$$M_R = [h_1(x) * I(x,t)] * g_1(t)] \times [h_2(x) * I(x,t)] * g_2(t)] \qquad (13.1)$$

$$M_S = [h_1(x) * I(x,t)] * g_2(t)] \times [h_2(x) * I(x,t)] * g_1(t)] \qquad (13.2)$$

where "*" and "×" denote convolution and multiplication, respectively, and $I(x,t)$ is the luminance distribution in the stimulus. The overall output is $M = M_R - M_S$, as before. NOTATION: We will use the symbols R and S consistently for rightward and leftward motion, respectively, where the S stands as a mnemonic code for "sinister" ("left" in Latin). We reserve the symbol L for "luminance".

It can be easily shown that the simplified Reichardt model of Fig, 13.4 will predict the response to the basic stimuli in Fig. 13.1. The notation h(Q) is assumed to describe the maximum response of the filter "h" to a well-aligned target "Q". Therefore, $h(Q_i)$ describes the maximum response at x_i. For the case of Fig. 13.1a, the model output h(Q) is positive for Q_1 (at time t_1) and also for Q_2 (at time t_2) because the luminance of both targets is positive with respect to the background. The solid path of the model in Fig. 13.4 will predict positive rightward motion for Fig. 13.1a because

$$h(Q_1) > 0 \text{ and } h(Q_2) > 0 \qquad \therefore h(Q_1) \times h(Q_2) > 0 \qquad (13.3)$$

where a positive product indicates rightward motion and a negative product indicates leftward motion. Similarly, for the stimuli in Fig. 13.1b, the Reichardt model will predict

$$h(Q_1) < 0 \text{ and } h(Q_2) < 0 \qquad \therefore h(Q_1) \times h(Q_2) > 0 \qquad (13.4)$$

Notice that the dashed-line path in the Reichardt model SMA of Fig. 13.4, which is tuned to leftward motion, will respond weakly for any of the stimuli of Fig. 13.1, all of which contain only rightward motion. The reason is that the delay imposed on the right-hand signal at x_2 is in such a direction as to produce a weak response for the correlator-multiplier that computes M_S. In general, if there is rightward or leftward Fourier motion energy in the spatiotemporal stimulus, the correlator that computes M_R or M_S is active, respectively, while the one responsible for M_S or M_R responds weakly. Notice also that the magnitude of the response M_R or M_S depends on how well the stimulus speed matches the preferred speed of the SMA, which is the optimal speed v* that the SMA is tuned to. Of course, $v^* = \Delta x/\Delta t$, where $\Delta x = (x_2 - x_1)$, and Δt is the SMA's characteristic delay.

The reason why the spatial filter's response to a dark target is the negative of that to a bright target of equal contrast is that these filters are assumed to be linear operators. Of course, the multiplying operation makes the entire SMA a nonlinear unit; nevertheless, this Reichardt analyzer has come to be termed as a Fourier motion analyzer because it extracts directional Fourier energy in the stimulus. This property of linearity makes the overall output M, which is a measure of the perceived speed, independent of the polarity (bright versus dark) of a target that maintains its contrast polarity as it moves; this is a desirable property, because this is consistent with psychophysical and physiological experiments.

This linearity of the spatial filter makes the model appropriate for predicting reverse AM as well. This is because $h(Q_1) > 0$ and $h(Q_2) < 0$ for the stimulus of Fig. 13.1c; hence the output M_R of the rightward-motion-tuned subunit (solid lines in Fig. 13.4) will be

$$M_R = h(Q_1) \times h(Q_2) < 0 \tag{13.5}$$

and the overall output M of the adder will be negative as well. Since M is negative, the unit signals leftward motion, namely reverse phi. It is interesting to note that a negative value of the adder's output M, signifying leftward motion, can result from two types of AM stimulation: 1) Forward AM: a target that maintains its polarity as it moves to the left; 2) Reverse AM: a target that switches its polarity in every frame as it moves to the right. The first results from a positive M_S and near-zero M_R; the second comes from a near-zero M_S and negative M_R.

We conclude this subsection with a note that is also valid for the elementary motion extractor of the next subsection. This concerns the limitation that the SMA's response, M, varies both as a function of speed v and target contrast C (univariance principle, see Section 13.8). Thus, given the response M, it is impossible to recover the desired v from the output of a single SMA without knowing the contrast C. Somehow, in a subsequent processing stage, we have to pool the outputs of many such units tuned to different speeds, and normalize the responses to eliminate the effect of contrast, so that we can get a good estimate of v. This is discussed in some detail in Section 13.8.

13.3.2 Spatiotemporal Energy Model

Another model, proposed by Adelson and Bergen (1985), is based on extracting the spatiotemporal energy of the motion. Just as for the Reichardt detector, feature tracking would not be necessary when using this kind of model. This is because the model treats the problem of extracting motion, tuned for a particular speed, as that of extracting orientation energy in the x-t plane, since slope in the x-t plane translates into speed. Accordingly, the model starts with spatiotemporal linear filters that have a specific orientation (slope) in the x-t plane. These are a quadrature pair of even-and odd-symmetric filters $G_{Re}(x,t)$ and $G_{Ro}(x,t)$ for rightward motion, and $G_{Se}(x,t)$ and $G_{So}(x,t)$ for leftward motion, where "e" represents even and "o" represents odd. Each one is either a sum or a difference of products of separable spatial filters (h_1 and h_2) and temporal filters (g_1 and g_2). h_1 and h_2 are even and odd spatial filters, respectively; g_1 and g_2 are temporal filters, with g_2 essentially being a delayed version of g_1. The oriented filters, which are shown in Fig. 13.5, are implemented as follows:

$$\begin{aligned} G_{Re}(x,t) &= h_1(x)\, g_1(t) - h_2(x)\, g_2(t) \\ G_{Ro}(x,t) &= h_1(x)\, g_2(t) + h_2(x)\, g_1(t) \\ G_{So}(x,t) &= h_1(x)\, g_2(t) - h_2(x)\, g_1(t) \\ G_{Se}(x,t) &= h_1(x)\, g_1(t) + h_2(x)\, g_2(t) \end{aligned} \quad (13.6)$$

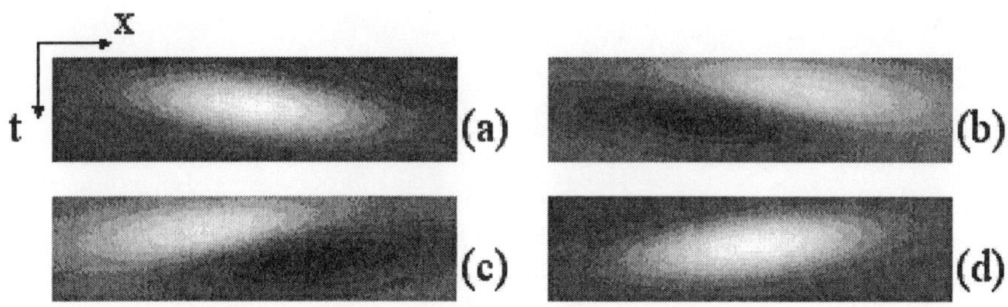

Figure 13.5. Directional receptive-field profiles of linear filters for energy models. (a) $G_{Re}(x,t)$ even-symmetric and (b) $G_{Ro}(x,t)$ odd-symmetric filters for rightward-tuned mechanisms; (c) $G_{So}(x,t)$ odd- and (d) $G_{Se}(x,t)$ even-symmetric filters for leftward-tuned mechanisms.

The outputs of the two leftward and two rightward units are squared and added together in quadrature fashion (M' does *not* denote differentiation).

$$M'_R(x,t) = \{G_{Re}(x,t) * I(x,t)\}^2 + \{G_{Ro}(x,t) * I(x,t))\}^2; \quad (13.7)$$

$$M'_S(x,t) = \{G_{Se}(x,t) * I(x,t)\}^2 + \{G_{So}(x,t) * I(x,t))\}^2 \quad (13.8)$$

This operation extracts the motion energy independently of the phase of the stimulus: if we did not square the outputs of the linear filters, then the response to bright targets would be the opposite of that to dark targets. The squaring operation accomplishes a similar task to that of the multiplier of the Reichardt detector. The final overall output is formed as in the Reichardt detector by the opponency operation of subtracting the two components, i.e., $M'(x,t) = M'_R(x,t) - M'_S(x,t)$. The note on univariance at the end of the previous subsection is also valid for this model.

13.3.3 Luminance Gradient Models

Suppose one looks at the problem of computing the speed v from an image-processing point of view. The question is how the speed v can be computed from a stimulus that has a spatiotemporal luminance distribution I(x,t). A luminance pattern moving at speed v has the form I(x,t) = I(x-vt). Taking the partial derivatives with respect to x and t, we get:

$$\partial I/\partial x = I'(x-vt), \qquad \text{and} \qquad \partial I/\partial t = -vI'(x-vt), \tag{13.9}$$

where I' denotes the derivative of I. The speed v can be computed from the above two equations:

$$v = -(\partial I/\partial t)/(\partial I/\partial x) \tag{13.10}$$

This gradient model computes the speed in one stage following two differentiations. As such, it has an advantage over the models of sections 13.3.1 and 13.3.2, which require an additional integrative stage to pool the outputs of the first-stage motion mechanisms. There is, however, the problem of dividing by very small, near-zero values for luminance profiles that vary slowly over space, which would lead to an unstable v value. This problem is solved by more complex formulations of the model (Johnston et al., 1992).

13.3.4 Comparison of the Three Models

Adelson and Bergen (1985) and van Santen and Sperling (1985) have shown that the ERD and energy models are mathematically equivalent. We have formulated the two models in the present chapter so that the reader can easily verify the equivalence of the two. The reader can show that M'(x,t) of the energy model is a multiple of M(x,t) of the Reichardt model. The real question is which one, if any, is actually implemented in the visual system. Emerson et al. (1992) studied directionally selective complex cells in cat and obtained evidence for the energy model, but without the opponent operation, i.e., without the adder for the very last stage; thus, their data are not consistent with any of the stages of a Reichardt model.

Even though the gradient model involves fewer stages for arriving at an estimate of the speed v, it appears to have less biological relevance than either of the other two models. This is because the operation of division may be more difficult to implement with neural units than are addition/subtraction or multiplication. Bruce et al. (1996) have an interesting formulation of the three models that indicates the mathematical equivalence, under certain assumptions, of the energy and the gradient models. Thus, at first glance, it appears that the three models are virtually the same. However, as they argue, when one gets to the details of each model, there are significant differences at the implementation level and other types of differences. Nevertheless, it is interesting to note that, even though each model has quite a different motivation and starting point, at the end the three of them share properties at a very fundamental level, most notably that they all extract Fourier motion energy.

13.4 MODELS FOR SECOND-ORDER MOTION

The three models of subsections 13.3.1-3 were meant to provide brief introduction to Fourier motion analyzers. None of them, however, can extract a motion signal from the purely contrast-defined stimuli of Figs. 13.1d and 13.2d. Additional evidence for some motion percepts that cannot be accounted for by Fourier mechanisms is what happens to the alternating-polarity stimulus of Figs. 13.1c and 13.2c. When observers view the stimulus under certain conditions, such as foveal viewing, large frame duration, non-zero inter-stimulus interval (ISI), large-sized targets, and high contrast, then the percept of reverse-AM gives way to forward-AM (Sperling 1989; Papathomas et al., 1996). In fact, Chubb and Sperling (1988) designed an entire class of stimuli that do not contain any net directional Fourier energy. They named these "microbalanced" stimuli; targets may have the same mean luminance as the background but differ from it only in contrast, in flicker rate, or in texture. A Fourier-based model, such as the Reichardt model of Fig. 13.3.1 with its linear spatial filters, h, cannot detect motion for "microbalanced" stimuli, but humans can. Obviously, the Fourier models of the previous section need to be extended to account for these psychophysical observations.

When targets are composed of texture patches with local variations in luminance (Figs. 13.1d and 13.2d), each target can be described by both its mean luminance, $\overline{L}$, and its contrast, C. If there are equal numbers of dots with only two local luminance values within such a target (L_{W1} and L_{W2}), the mean luminance and contrast are:

$$\overline{L} = \frac{L_{W1} + L_{W2}}{2}$$
$$C = \frac{L_{W1} - L_{W2}}{L_{W1} + L_{W2}} \qquad (13.11)$$

Fourier models are able to predict accurately the responses to purely luminance-based stimuli as previously shown. The targets in the stimuli of Figs. 13.1a-c and 13.2a-c are described only by their luminance values. There is no local contrast within any of the targets, and models which extract motion of stimuli that can be described in the frequency domain will accurately predict the motion. However, the stimuli shown in Figs. 13.1d and 13.2d represent pure second-order motion. To a Fourier (first-order) computational model, the targets are indistinguishable from the background because their mean luminance is equal to that of the background (L_0). Only the global luminances are "visible" to such large Fourier units. If, on the other hand, the size of the SMA is small compared to the dot size, then there is no consistent directional motion energy, and such units will signal a net zero motion, when seen as an ensemble.

Because the motion seen by humans using the stimulus of Figs. 13.1d and 2d is not carried by luminance, the models described in the previous section do not apply, due to their lack of ability to detect the global motion. The energy is averaged out of the system in cases where there is no net Fourier energy in the stimulus, thus causing no motion to be detected. Some of the stimulus types that are visible to humans but not to Fourier models include those defined by contrast, flicker rate, and texture gradients.

To force the contrast-based targets to be "visible" to a computational model, the local polarity differences must be removed to create Fourier energy in the resulting output. This led Chubb and Sperling (1988, 1989) to postulate a nonlinear spatial filter by preceding the linear filters by full-wave rectifiers (FWR). Such mechanisms would indeed extract motion for equiluminant contrast-induced targets, and have come to be called "second-order", or "non-Fourier" (NF) motion mechanisms. This type of nonlinearity will bring the average activity of the contrast-defined stimuli to above-background levels in the processed map, so they will have energy in the frequency domain. Chubb and Sperling (1988) proposed that the full-wave rectification be preceded by spatiotemporal linear preprocessing to filter the input. After imposing the nonlinearity, a standard motion analyzer will be able to predict the second-order motion (see Fig. 13.6). After FWR, the responses to the negative elements are forced to be positive, creating targets that are now "visible" to a global motion analyzer.

One issue is whether second-order motion is sensed by low-level units, or by feature-tracking, attention-driven mechanisms. The psychophysical (Chubb and Sperling 1991; Stoner and Albright 1992; Papathomas et al, 1996) and physiological (O'Keefe et al., 1993) evidence favors the low-level detectors.

13.5 COMBINING 1ST- AND 2ND- ORDER MOTION

There is a need for a model that predicts both reverse-AM and forward AM for the alternating-polarity stimuli of Figs. 13.1c and 13.2c depending on conditions. Such a model should be able to mimic human performance both with reverse-AM of Figs. 13.1c and 13.2c, as well as with the contrast-defined stimuli of Figs. 13.1d and 13.2d. It is this type of need that suggests a system with multiple visual pathways: i.e., a first-order (Fourier) pathway and a second-order (non-Fourier) pathway, with the two of them operating in parallel (Chubb and Sperling, 1988; Zanker, 1995; Papathomas and Rosenthal, 1996). When considering these parallel mechanisms, the variations in global luminance can be detected by the standard (Fourier) Reichardt motion detector as previously described. The input to the non-Fourier pathway must be preprocessed to produce an activity map that extracts the second-order features. Once these features are transformed so that they generate a first-order-like spatiotemporal neural activity at the output of the pre-processors, the same standard motion detector can be used on this new map to determine the second-order motion. It is yet unknown how these two separate motion signals combine, assuming that they do exist in this form.

To validate the dual-pathway hypothesis, Chubb and Sperling (1989) have developed additional psychophysical stimuli that do not contain any energy in the preferred direction. Many theories have been proposed on the nature of the multiple-pathway network. As an example, Wilson and Kim (1994) have developed a computational model that accounts for some of the experimental evidence.

We will present an implementation of a system that combines the effects of first- and second-order subsystems (Papathomas and Rosenthal, 1996). The computational model was developed using two parallel pathways shown schematically in Fig. 13.6. In the first-order path, the input images are fed directly into a motion analyzer as described in Fig. 13.4 to produce the overall response M_1. The second-order path receives the same inputs, but they are first preprocessed using a Gaussian smoothing kernel followed by a full-wave rectification mechanism before being used in the motion analyzer, which results in overall response M_2. Consequently, two separate motion signals are obtained from this system. In general, the two responses can be combined in some fashion to produce the overall response $M = u(M_1, M_2)$.

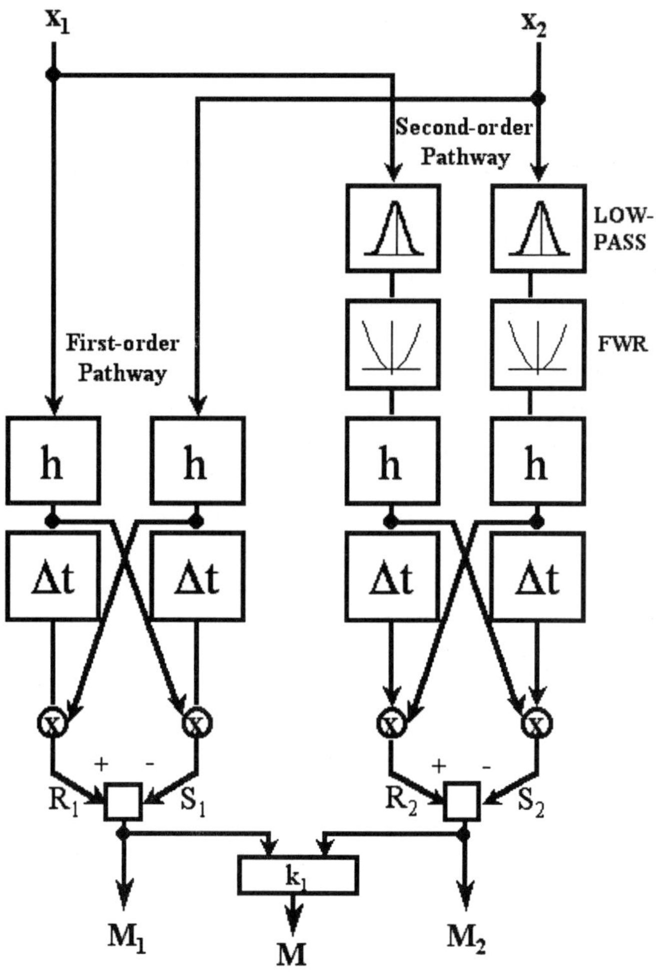

Figure 13.6. Simplified schematic for the combination of 1st- and 2nd-order systems.

The original stimulus image I(x,t) was preprocessed in the second-order pathway by two sequential one-dimensional convolutions, first in the spatial direction (filter P_σ), followed by the temporal direction (filter P_τ). The low-pass spatial pre-processing provides smoothing of the random contrast within the image. The smoothed image is determined by $I_S = (I{*}P_\sigma){*}P_\tau$, where "*" denotes convolution. After smoothing, the outputs are squared, thus producing an effect similar to a full-wave rectification (FWR) mechanism. Elements of the targets that have luminance values below the background now cause responses that are higher than the response resulting from the background. This non-linearity will allow the contrast-based targets to be "visible" to a motion correlator by producing a response that increases monotonically with contrast. The final preprocessed image I_2 is therefore $I_2 = I_S^2 = ((I*P_\sigma)*P_\tau)^2$. This image is then fed into the standard motion analyzer (SMA), as shown in Fig. 13.6.

The way in which the outputs M_1 and M_2 of the first- and second-order paths, respectively, are combined, determines how well the overall model will predict experimental data. Our approach is to change the relative sensitivities of the first- and second-order systems depending on spatial frequency, temporal frequency, and eccentricity. The model's output provides an excellent fit to the psychophysical results obtained with the stimuli of Figs. 13.1 and 13.2, as well as results from a wide variety of other experiments (Papathomas and Rosenthal, 1996; Rosenthal, 1996). In particular, the model correctly predicts reverse-AM or forward-AM with the stimuli of Figs. 13.1c and 13.2c, depending on the conditions.

At this point, we must mention that some researchers have developed computational models that do not make the distinction between first- and second-order motion analyzers (Benton et al., 1997). Their models, encompassing a single pathway, are able to account for most phenomena in motion perception. The debate on whether there is a single pathway or separate pathways for motion processing in the visual system is still not resolved. The neurophysiological evidence favors separate systems (O'Keefe et al., 1993; Vaina and Cowey, 1996; Clifford and Vaina, 1999; Vaina, Cowey et al., 1999; Vaina, Makris et al., 1999; Vaina et al., 2000).

13.6 MODELS FOR THIRD-ORDER MOTION

There are types of motion that are perceived by humans, yet neither the first-order nor the second-order models can account for them. Consider, for example, the stimuli shown schematically in Figs. 13.1e and f. Figure 13.1e shows schematically a target defined purely by binocular disparity moving to the right, where all the other attributes of the targets are exactly the same as the background. Such purely disparity-defined stimuli are called cyclopean stimuli, because they are perceived beyond each eye's retina, in what Julesz called the "cyclopean retina". Finally, the stimulus of Fig. 13.1f is an example of inter-attribute apparent motion: the attribute by which the target is segregated from the background changes from one frame to the next. In the case of Fig. 13.1f, the target is segregated by binocular disparity in frame 1, and by contrast in frame 2. Humans do perceive motion with disparity-defined targets (Fig. 13.1e) or in inter-attribute AM (Fig. 13.1f). The question is what type of a model can account for data with such stimuli.

Parenthetically, within each frame of Fig. 13.1e, a random-dot stereogram (Julesz, 1960) portrays the target in depth in front of the background; the random texture of the target has a different binocular disparity than that of the background, but the target is otherwise indistinguishable from the background within each eye's individual image. In the next frame, the random-dot patterns that define the background and the target change completely from what they were in the previous frame, but the depth relationship is maintained, and the depth-defined target is displaced slightly to the right to produce AM. This change of pattern is necessary to avoid possible Fourier motion components. (If the patterns were the same, then there would be the possibility of matching the luminance patterns across frames. A sketch of such a stimulus in x-t space would show Fourier components in the rightward direction.)

Some form of pre-processing, quite similar to that in the second-order system, followed by standard motion analysis (SMA), can create models for predicting the apparent motion of a target that is segregated from the background by the same attribute across frames. For example, is the target is defined by a different flicker rate, then some preprocessing by different linear temporal filters, followed by full-wave rectification can generate a signal that can subsequently be fed as input to a SMA. In a similar way, the apparent motion in Fig. 13.1e can be extracted by preceding SMA by a stage of disparity-tuned binocular neurons that respond differentially to targets and background. After extracting the disparity signal, the spatiotemporal distribution of activities in these disparity-tuned neurons has the same form as a luminance distribution in Fig. 13.1a: namely, one level of activity for the background, and a different level for the targets. Thus, when the signals of the disparity-tuned neurons are fed into a SMA, motion may be extracted just as it is in the case of Fig. 13.1a, where the luminance distribution is fed directly into a SMA. One can similarly obtain an appropriate spatiotemporal distribution of activities in neurons that respond differentially to any attribute that can segregate targets from background: texture contrast (e.g. Gorea and Papathomas, 1993), relative motion (see Zanker, 1995, for a review) and, possibly, color (Gorea et al., 1993a, b; Cavanagh, 1995; Farell, 1995).

However, things become progressively more complicated when we try to apply this model of preprocessing followed by SMA to inter-attribute motion, illustrated in Fig. 13.1f. Consider the possibilities of combining a target segregated by attribute A in one frame and attribute B in the next frame. Is there a SMA fed by low-level preprocessors for all possible pairings of A and B, say one for texture and disparity, another for texture and relative motion, etc.? This would make it necessary to have a large number of parallel specialized motion analyzers indeed. Another possibility, at the other extreme, is to hypothesize the existence of pre-processors for all possible attributes, with each providing its input to a common SMA. In this scheme, we need a very complex way of combining the contributions of texture, disparity, relative motion, etc., and then feeding their combined signal into a SMA, resulting in a very complex model.

There is still a debate on how the visual system processes all these types of motion stimuli, but one of the most widely accepted theories (Lu and Sperling 1995) is that there are three separate motion systems: The first- and second-order systems extract motion signals for targets that are defined by luminance and contrast modulations, respectively, as covered in the previous sections. A putative third-order system is responsible for motion of targets defined by all other attributes, as well as for inter-attribute motion. The main tenets of Lu and Sperling's (1995) theory is that the first- and second- order systems are fast (cut-off frequency at 12 Hz), bottom-up (automatic), exclusively monocularly driven, little affected by attention and other cognitive influences. The third-order system, by contrast, is slow (cut-off frequency at 3 Hz), tracks features, combines both eyes' inputs, and is significantly influenced by attention (Blaser et al., 1999) and, possibly, other cognitive processes. The third-order system favors saliency for tracking features. This saliency is either present in the stimulus (high-contrast targets are preferred automatically over low-contrast ones), or it is affected by attention (attention to red makes red targets more salient than green, even though the two colors are equisalient prior to the engagement of attention). A block diagram of the functional architecture of the three systems, as studied by Sperling and his colleagues, is given in Lu and Sperling (1995). Much research is needed to investigate such putative systems in detail.

In fact, if we concentrate on the attribute of disparity, the issue of whether disparity-driven motion is extracted in a purely bottom-up, automatic, fashion (namely second-order-like), or it is based on tracking of disparity features (namely third-order-like), is not resolved yet. Lu and Sperlings' (1995) experiments with the pedestal-plus-test paradigm have indicated that disparity-driven motion tracks features, whereas other experimenters have observed motion after-effects after exposure to such motion (Patterson et al., 1994), indicating bottom-up energy-based mechanisms.

13.7 COLOR AND MOTION

Luminance is the principal token for motion, whereas the role of color has been highly debated. At equiluminance, performances fall to a minimum (i.e., Ramachandran and Gregory, 1978), and apparent motion slows down (Cavanagh et al., 1984; Teller and Lindsey, 1993). This may be taken as evidence that the motion pathway is nearly color-blind (e.g., Livingstone and Hubel 1988). However, recent psychophysical (i.e., Krauskopf and Farell, 1990; Stromeyer et al., 1990; Papathomas et al., 1991; Cavanagh and Anstis, 1991; Gorea et al., 1993a,b; Farell, 1995; Cropper and Derrington, 1996) and physiological (DeYoe and VanEssen, 1988; Dobkins and Albright, 1993) evidence suggests that the color contribution is significant. The apparent contradiction between these two sets of studies, one set supporting the role of color in motion and the other set refuting it, can be best resolved by the difference in the stimuli employed in the studies. There is agreement that motion is severely compromised with equiluminant chromatic stimuli, which were the stimuli used in one set of studies; thus, equiluminance seems to be a singular mode of stimulation. However, most of the experiments in the other set employed chromatic stimuli that contained luminance components; these studies provided evidence for the role of color in motion by comparing across conditions. Thus the question of whether color contributes to motion seems to have been answered affirmatively.

Research has entered a second phase of debate on how color contributes to motion. This is a new area of research, and opinions are diverse. For example, Lu et al. (1999) used the pedestal-plus-test paradigm to obtain evidence that color motion processing is feature-based, whereas Gorea et al. (1993a,b) and Cropper and Derrington (1996) present evidence for a fast first-order-like color motion system. As another example of this debate, Krauskopf and Farell (1990) presented evidence that motion is processed separately on luminance and color channels, whereas Cavanagh and Anstis (1991) argued for a common motion pathway. Of course, these are two extremes among several possibilities, and our own experiments (Gorea et al., 1993a,b) indicate that both architectures may be implemented, but each dominates in different spatiotemporal domains. More systematic experimental studies are needed to resolve the issue.

The model that emerges from our own and others researchers' studies is very similar to that of Fig. 13.6. In addition to the fast first-order luminance-based and second-order luminance-contrast-based systems, there are two types of color-related systems: 1) Separate fast Fourier systems for color, one for the red-green (long-medium wavelength, L-M) axis, the other for the blue-yellow (short/long+medium, (S-(L+M)) axis. These have the same architecture as the Fourier motion analyzers in the luminance domain, and are parallel to the other pathways that we have presented so far. Instead of being sensitive to luminance gradients, they are sensitive to spectral gradients along the respective chromatic axes. 2) A slower common color/luminance system, which combines the contributions of the achromatic and chromatic signals, each processed through full-wave rectification, and then feeding the combined signal into a motion analyzer. The influence of the slow system is significantly reduced beyond about 7.5Hz, in agreement with Lu and Sperling's (1995) data. Preliminary results indicate that such a model will provide a reasonable fit with psychophysical data. A very interesting question that needs to be investigated experimentally is whether the common color/luminance system is the same system as the second-order system of section 13.4.

13.8 GLOBAL MOTION AND TOP-DOWN INFLUENCES

13.8.1 Pooling the Outputs of Motion Analyzers - Speed Invariance with Contrast

In subsection 13.3.1, we alluded to a limitation of the Reichardt-type and spatiotemporal energy extractors when it comes to computing an estimate for the velocity v. The problem is that the response M(v) covaries with the target's contrast. For each value of contrast C there is a different tuning curve M(v, C) that is a multiple of M(v, C*), to a first-order approximation, where C* is some reference contrast. This is known as the "univariance principle", and it is quite general in sensory units in which the response is tuned to the value of an attribute, but it covaries with luminance variations in the unit's receptive field: orientation-tuned, spatial-frequency-tuned and wavelength-tuned units all obey the univariance principle.

The problem of computing the speed from univariance-obeying units is related to the problem of "color constancy". In the former case, the goal is to recover the speed v independently of the stimulus contrast C, even though their contribution is intricately combined to produce the response M(v, C). In the latter case, the goal is to perceive the fixed color of a surface, independently of the lighting conditions; one possible solution to achieve color constancy is to have a stage of processing that pools the responses of the three classes of cones (S, M and L), each tuned to a different part of the spectrum, and identify a color from ratios of responses of pairs of cone classes. Assuming that the responses scale with illumination intensity in the same fashion for each class of cones, such ratios will be constant as the level of illumination varies, achieving color constancy (this is an oversimplified scheme). In an analogous manner, the goal of "speed constancy" can be achieved by pooling together the outputs of speed-tuned units, with each being tuned to a different speed, and sensing the ratios of responses in pairs of such speed-tuned units. Even though the response of the first-stage SMA varies with the moving target's contrast, these ratios eliminate this variation, thereby achieving a stable estimate of speed for a wide range of stimulus contrast.

An excellent model to achieve the above goals is the one developed by Simoncelli and his colleagues (Simoncelli, 1993; Simoncelli and Heeger, 1998), containing two-stages with divisive normalization (Heeger, 1991) in each stage. The first stage models V1 units starting with linear filters that are selective for orientation and motion direction. Their outputs are half-wave rectified (HWR) and squared, and then summed to obtain a normalization factor for the HWR output of each unit. The second stage models MT units in exactly the same fashion, but it starts with filters that are selective for velocity (direction and speed). Their model is the best available to date, in that it is both physiologically plausible, and provides a good fit for a very wide range of experimental data. In turn, the model has inspired experimental studies (Schrater et al, 2000). It is a prime example of the desirable collaboration among physiologists, psychophysicists, and computational modelers toward understanding of the visual system.

13.8.2 Motion After-Effect Models

Most models attribute MAE to the "fatiguing" of units that are activated during adaptation. Early explanations theorized that MAE occurs within a Reichardt-type detector due to the fatiguing of the active path; the opponent path, which is not active, takes over upon motion termination, thus resulting in the MAE. Later explanations of the MAE are based on a two-stage motion analysis: motion detectors and integrators (see subsection 13.8.1): 1) The ratio model assumes that detectors provide excitatory and inhibitory inputs to integrators of the same and opposite directions, respectively (Barlow and Hill, 1963). 2) The distribution-shift model (Mather, 1980) does not involve inhibition, but only excitation of similar-direction-tuned integrators; it theorizes that, after adaptation, detectors of the adapted direction excite the integrators with below-baseline signal, which favors the integrators tuned for the opposite direction.

However, such models fail to predict the phenomena described in subsection 13.2.5, involving two families of dots that move in two different directions. A broad-inhibition model recently developed by Grunewald (1996) assumes that motion detectors send excitatory inputs to integrators in a narrow band that are centered on the adapted direction; they also send inhibitory input to integrators in a broad band, centered on the opposite direction. This model is in agreement with the results described in subsection 13.2.5, and was later extended to cover cases where the two directions are presented to separate eyes (Grunewald and Mingolla, 1998). What is more important is that this model predicted an interesting phenomenon in MAE, which was subsequently verified experimentally: when adapting to dots moving transparently in opposite directions, say to the left and right, observers perceive orthogonal MAE during the test phase, i.e., up and down (Grunewald and Lankheet, 1996).

13.9 SUMMARY - CONCLUSIONS

We have presented a review of biologically relevant models that have been developed to account for experimental findings in both psychophysics and neurophysiology. We began the chapter by noting observers' responses to a brief list of some key stimuli. The next step was to present the development of models that could predict the responses to such stimulation, starting with simple stimuli and progressing to more complex ones. One of the most interesting aspects of modeling is the realization that three approaches to Fourier energy motion extraction, starting from different formulations, have been shown to have deep essential similarities. It remains to be seen which of the three is actually implemented in the human visual system, or if some other scheme is adopted.

We presented the models under the general architecture of three putative motion systems (Lu and Sperling, 1995): a first-order luminance-based Fourier system, a second-order contrast-based system that pre-processes the input by the nonlinear operator of full-wave rectification, and a third-order system that covers motion that is driven by all other attributes (binocular disparity, spatial frequency, relative motion, etc., including, possibly, color). The first two are thought to be fast, monocularly driven, and little affected by attention, unlike the third-order system, which has the opposite properties: slow, binocular, and influenced by attention.

We must note that not all researchers in the field accept this formulation. There are several issues that have been raised. As a first example, there is evidence of MAE effects after exposure to purely cyclopean (disparity-driven) motion, when tested with cyclopean motion noise. This MAE transfers to luminance-based test stimuli (Patterson et al., 1994; Shorter et al., 2000), suggesting low-level motion extracting units. This is not in agreement with Lu and Sperling's assessment, using the pedestal-plus-test paradigm, that cyclopean motion is based on feature tracking.

Existing models attribute MAE to the fatiguing of low-level motion mechanisms; it follows that the presence of MAE signifies the existence of low-level motion mechanisms. Hence, it is widely believed that third-order AM does not produce MAE because, by definition, it is based on feature tracking, not on low-level energy detectors. Of course, the hypothesis that a low-level motion mechanism is necessary for MAE needs to be investigated. Although highly unlikely, it is possible that even a feature-tracking motion system exhibits MAE.

A second example of debates about the functional architecture of the motion system is the issue of the role of color in motion perception. Even if there is wide agreement that color does contribute to motion perception, the nature of the neural mechanisms for color-based motion is still debated (see Section 13.7). For color-based motion, there is conflicting evidence for feature tracking (Lu and Sperling, 1995) and for low-level mechanisms (Morikawa and Zaidi, 1999), even within the pedestal-plus-test paradigm. Different paradigms have revealed fast Fourier-like systems for pure color motion, observing reverse-AM for moving targets that alternate between red and green and are equiluminant to a yellow background (Gorea et al., 1993a). Since reverse-AM is the signature for low-level Fourier-energy motion extractors, this finding is at odds with Lu and Sperling's (1995) conclusion that color-based motion is feature based.

We conclude with the challenge of modeling the top-down influences on the perception of motion. In particular, there are three-dimensional (3-D) objects that are perceived to have the reverse (opposite) depth arrangement to the actual arrangement. Two examples are a hollow human facial mask (which is perceived as a convex normal face), and reverspectives, i.e., 3-D sculptured paintings by Patrick Hughes where the perspective cues conflict with the surface structure (Papathomas, 2000). Because of this depth reversal, these objects appear to turn vividly as viewers move in front of them. This percept can be explained by assuming that the visual system first constructs a 3-D representation of the object, and then uses the retinal optical flow and the knowledge of the viewer's own motion to update the 3-D representation. In the case of the hollow mask, Papathomas and DeCarlo (1999) showed that a model-based algorithm by DeCarlo and Metaxas (1998, 2000), that incorporates the schema of a normal 3D-face model, predicts the percepts obtained by humans. A similar model can be developed for the perception of reverspectives under self-motion.

13.10 REFERENCES

Adelson, E. H. and Bergen, J., 1985, Spatiotemporal energy models for the perception of motion, *J. Opt. Soc. Am.* **2A**: 284-299.

Adelson, E. H. and Movshon, J. A., 1982, Phenomenal coherence of moving visual patterns, *Nature.* **300**: 523-525.

Anstis, S. M., 1970, Phi movement as a subtraction process, *Vision Res.* **10**: 1411-1430.

Barlow, H. B., 1979, Reconstructing the retina image in space and time, *Nature.* **279**: 189-190.

Barlow, H. B. and Hill, R. M., 1973, Evidence for a physiological explanation of the waterfall phenomenon and figural after-effects, *Nature.* **200**: 1345-1347.

Benton, C. P., Johnston, A., and McOwan, P. W., 1997, Perception of motion direction in luminance- and contrast-defined reversed-phi motion sequences, *Vision Res.* **37**: 2381-2399.

Blaser, E., Sperling, G., Lu, Z. L., 1999, Measuring the amplification of attention, *Proc. Natl. Acad. Sci. U S A.* **96**: 11681-11686.

Braddick, O., 1974, A short-range process in apparent motion, *Vision Research,* **25**: 839-847.

Braun, J., 2000. Targeting visual motion, *Nature Neurosci.* **3**: 9-11.

Bruce, V., Green, P. R., and Georgeson, M. A., 1996, *Visual Perception: Physiology, Psychology, and Ecology*, 3rd edition, Psychology Press, East Sussex, UK.

Cavanagh, P., 1995, Is there low-level motion processing for non-luminance-based stimuli?, in *Early Vision and Beyond*, T.V. Papathomas, C. Chubb, A. Gorea, and E. Kowler, eds, MIT Press, Cambridge, Massachusetts, pp. 113-119.

Cavanagh, P. and Anstis, S. M., 1991, The contribution of color to motion in normal and color-deficient observers, *Vision Res.* **31**: 2109-2148.

Cavanagh, P. and Mather, G., 1989, Motion: The long and the short of it, *Spatial Vis.* **4**:103-129.

Cavanagh, P., Tyler, C.W., and Favreau, O. E., 1984, Perceived velocity of moving chromatic gratings, *J. Opt. Soc. Am. A.* **1**: 893-899.

Chubb, C. and Sperling, G., 1988, Drift-balanced random stimuli: A general basis for studying non-Fourier motion perception, *J. Opt. Soc. Am. A.* **5**: 1986-2006.

Chubb, C. and Sperling, G., 1989, Second-order motion perception: Space-time separable mechanisms, *Proc. Workshop on Visual Motion*, March 20-22, 1989, Irvine, CA, IEEE Computer Society Press, Washington, D.C., pp. 126-138.

Chubb, C. and Sperling, G., 1991, Texture quilts: Basic tools for studying motion-from-texture, *J. Math. Psych.* **35**: 411-442.

Clifford C. W. and Vaina L. M., 1999, A computational model of selective deficits in first and second-order motion processing. *Vision Res.* **39**: 113-130.

Cropper, S. J. and Derrington, A. M., 1996, Rapid colour-specific detection of motion in human vision, *Nature.* **379**: 72-74.

DeCarlo, D. and Metaxas, D., 1996, The integration of optical flow and deformable models with applications to human face shape and motion estimation, *Proc. CVPR '96*, pp. 231-238.

DeCarlo, D. and Metaxas, D., 2000, Optical flow constraints on deformable models with applications to face tracking, *Internat. J. Computer Vision.* **38**: 99-127.

DeYoe, E. A. and VanEssen, D. C., 1988, Concurrent processing streams in monkey visual cortex, *Trends in Neurosci.* **11**: 219-226.

Dobkins, K. R. and Albright, T. D., 1994, What happens if it changes color when it moves? The nature of chromatic input to Macaque visual area MT, *J. Neurosci.* **14**: 4854-4870.

Emerson, R. C., Bergen, J., and Adelson, E. H., 1992, Directionally selective complex cells and the computation of motion energy in the cat visual cortex, *Vision Res.* **32**: 203-218.

Exner, S., 1875, Experimentelle Untersuchung der einfachsten psychischen Processe, *Archiv für die Gesamte Physiologie des Menschen und der Tiere.* **11**: 403-432.

Farell, B., 1995, Spatial structure and the perceived motion of objects of different colors, in *Early Vision and Beyond*, T. V. Papathomas, C. Chubb, A. Gorea, and E. Kowler, eds, MIT Press, Cambridge, Massachusetts, pp. 121-131.

Gorea, A. and Papathomas, T. V., 1993, Double-opponency as a generalized concept in texture segregation illustrated with color, luminance and orientation defined stimuli, *J. Opt. Soc. Am. A.* **10**: 1451-1462.

Gorea, A., Papathomas, T. V., and Kovacs, I., 1993a, Motion perception with spatiotemporally matched chromatic and achromatic information reveals a 'slow' and a 'fast' motion system, *Vision Res.* **33**: 2515-2534.

Gorea, A., Papathomas, T. V., and Kovacs, I., 1993b, Two motion systems with common and separate pathways for color and luminance, *Proc. Natl. Acad. Sci. USA.* **90**: 11197-11201.

Grunewald, A., 1996, A model of transparent motion and non-transparent motion aftereffects, in: *Advances in Neural Information Processing Systems 8*, D. S. Touretzky, M. C. Mozer, and M. E. Hasselmo, eds, MIT Press, Cambridge, Massachusetts, pp. 837-843.

Grunewald, A. and Lankheet, M. .J. .M., 1996, Orthogonal motion after-effect illusion predicted by a model of cortical motion processing, *Nature.* **384**: 358-360.

Grunewald, A. and Mingolla, E, 1998, Motion after-effect due to binocular sum of adaptation to linear motion, *Vision Res.* **38**: 2963-2971.

Heeger, D. J., 1991, Nonlinear model of neural responses in cat visual cortex, in *Computational Models of Visual Processing*, M. Landy, and J. A. Movshon, eds., MIT Press, Cambridge, Massachusetts.

Hiris, E. and Blake, R., 1992, Another perspective in the visual motion after-effect, *Proc. Natl. Acad. Sci. USA.* **89**: 9025-9028.

Johnston, A., McOwan, P. W., and Buxton, H., 1992, A computational model of the analysis of some first-order and second-order motion patterns by simple and complex cells, *Proc. Royal Society of London, B.* **250**: 297-306.

Julesz, B., 1960, Binocular depth perception of computer-generated patterns, *Bell System Technical Journal.* **39**: 1125-1162.

Julesz, B., 1968, Experiments in perception, *Psychology Today.* **2**: 16-23.

Julesz, B. andPayne, R. A., 1968, Differences between monocular and binocular stroboscopic movement perception, *Vision Res.* **8**: 433-444.

Krauskopf, J. and Farell, B., 1990, Influence of colour on the perception of coherent motion, *Nature.* **348**: 328-331.

Livingstone, M. S. and Hubel, D. H., 1988, Segregation of form, color, movement and depth: Anatomy, Physiology and Perception, *Science.* **240**: 740-749.

Lu, Z. L., Lesmes, L. A., and Sperling, G., 1999, The mechanism of isoluminant chromatic motion perception, *Proc. Natl. Acad. Sci. U S A.* **96**: 8289-8294.

Lu, Z. L. and Sperling, G., 1995, The functional architecture of human visual motion perception, *Vision Res.* **35**: 2697-2722.

Marr, D. and Ullman, S., 1981, Directional selectivity and its use in early visual processing, *Proc. R. Soc. London B.* **211**: 151-180.

Mather, G., 1980, The movement after-effect and a distribution-shift model for coding the direction of visual movement, *Perception.* **9**: 379-392.

Mather, G., Verstraten, F., and Anstis, S. M., 1998, *The Motion Aftereffect: A Modern Perspective*, MIT Press, Cambridge, MA.

Merigan, W. H. and Maunsell, J. H., 1993, How parallel are the primate visual pathways?, *Annual Review of Neurosci.* **16**: 369-402.

Morikawa, K. and Zaidi, Q., 1999, Minimum velocity limits for motion energy detection, *Invest. Ophthal. Vis. Sci.* **40**: S191.

Nishida, S. and Takeuchi, T., 1990, The effects of luminance on affinity of apparent motion, *Vision Res.* **30**: 709-721.

Nishida, S. and Sato, T., 1993, Two kinds of motion aftereffects reveal different types of motion processing, *Invest. Ophthal. Vis. Sci.* **34**: 1363.

O'Keefe, L. P., Caradini, M., Beusmans, J. M. H., and Movshon, J. A., 1993, MT neuronal responses to 1st- and 2nd-order motion, *Soc. for Neuroscience Abstracts*, **19**: 1283.

Papathomas T. V., 2000, See how they turn: False depth and motion in Hughes's reverspectives, Human Vision and Electronic Imaging, Proc. IS&T/SPIE Symposium on Electronic Imaging: Science and Technology, Conference on Human Vision and Electronic Imaging, San Jose, CA. **3959**: 506-517.

Papathomas, T. V. and DeCarlo, D., 1999, Top-down and bottom-up processes in 3-D face perception: Psychophysics and computational model, *Perception.* **28**: 112-113.

Papathomas, T. V., Gorea, A., and Chubb, C., 1996, Precise assessment of the effective mean luminance of texture patches: An approach based on reverse-phi apparent motion, *Vision Res.* **36**: 3775-3784.

Papathomas, T. V., Gorea, A., and Julesz, B., 1991, Two carriers for motion perception: Color and luminance, *Vision Res.* **31**: 1883-1891.

Papathomas, T.V., Kovacs, I., Gorea, A., and Julesz, B., 1995, A unified approach to the perception of motion, stereo, and static flow patterns, *Behavior Res. Methods, Instruments and Computers.* **27**: 419-432.

Papathomas, T. V. and Rosenthal, A. S., 1996, Unified computational model for Fourier and non-Fourier motion, *Proc.IEEE 22ndAnnual Northeast Bioengineering Conference*, Piscataway, N.J., K.-J. Li and S. S. Reisman, eds., pp. 44-45.

Patterson, R., Bowd, C., Phinney, R., Pohndorf, R., Barton-Howard, W.J. and Anqilletta, M., 1994, Properties of the stereoscopic (cyclopean) motion after-effect, *Vision Res.* **34**: 1139-1147.

Ramachandran, V. .S. and Gregory, R. L., 1978, Does colour provide an input to human motion perception? *Nature.* **275**: 55-56.

Reichardt, W., 1961, Autocorrelation, a principle for the evaluation of sensory information by the central nervous system, in W.A. Rosenblith, ed., *Sensory Communication*, New York: Wiley.

Rosenthal, A. S., 1996, Computational models for Fourier and non-Fourier motion, Master's Thesis, Rutgers University.

Schiller, P. H., Logothetis, N. K., and Charles, E. R., 1990, Role of color-opponent and broad-band channels in vision, *Visual Neuroscience.* **5**: 321-346.

Schrater, P. R., Knill D. C., and Simoncelli, E. P., 2000, Mechanisms of visual motion detection, *Nature Neuroscience.* **3**: 64-68.

Shorter, S., Bassetti, T., Tamura, E., Donnelly, M., Bowd, C., and Patterson, R., 2000, Stereoscopic (cyclopean) and luminance aftereffects induced by adaptation to global motion displays, *Invest. Ophthal. Vis. Sci.* **41**: S737.

Simoncelli, E. P., 1993, Distributed analysis and representation of visual motion, Ph.D. thesis, MIT, Department of Electrical Engineering and Computer Science, Cambridge, Massachusetts.

Simoncelli, E. P. and Heeger, D. J., 1998, A model of neuronal responses in visual area MT, *Vision Res.* **38**: 743-761.

Sperling, G., 1989, Three stages and two systems of visual processing, *Spatial Vis.* **4**: 183-207.

Stoner, G. R. and Albright, T. D., 1992, Motion coherency rules are form-cue invariant, *Vision Res.* **32**: 465-475.

Stromeyer, C. F. III, Eskew, R. T., and Kronauer, R. E., 1990, The most sensitive motion detectors in humans are spectrally opponent, *Invest. Ophthal. Vis. Sci.* **31**: 241.

Teller, D. Y. and Lindsey, D. T., 1993, Motion at isoluminance: motion dead zones in three-dimensional color space, *J. Opt. Soc. Am.* **10A**: 1324-1331.

Tovee, M. J., 1996, *An Introduction to the Visual System*, Cambridge University Press, Cambridge, MA.

Ullman, S., 1979, *The Interpretation of Visual Motion*, MIT Press, Cambridge, MA.

Vaina, L. M. and Cowey, A., 1996, Impairment of the perception of second order motion but not first order motion in a patient with unilateral focal brain damage. Proc. Royal Society of London B Biol Sci. **263**: 1225-1232.

Vaina, L. M., Cowey, A., and Kennedy, D., 1999, Perception of first- and second-order motion: separable neurological mechanisms? *Hum Brain Mapp.* **7**: 67-77.

Vaina, L. M., Makris, N., Kennedy D., and Cowey, A., 1999, The selective impairment of the perception of first-order motion by unilateral cortical brain damage. *Visual Neurosci.* **15**: 333-348.

Vaina, L. M., Soloviev, S., Bienfang, D. C., and Cowey, A., 2000, A lesion of cortical area V2 selectively impairs the perception of the direction of first-order visual motion, *Neuroreport.* **11**: 1039-1044.

van Santen, J. P. H. and Sperling, G., 1985, Elaborated Reichardt detectors, *J. Opt. Soc. Am.* **2A**: 300-321.

Verstraten, F. A. J., Fredericksen, R. R., and van der Grind, W.A., 1994, Movement after-effect of bivectorial transparent motion, *Vision Res.* **34**: 349-358.

Watson, A. B. and Ahumada, A. J., Jr., 1985, Model of human visual-motion sensing, *J. Opt. Soc. Am.* **2A**: 322-342.

Werkhoven, P., Snippe, H. P., and Koenderink, J. J., 1990, Metrics for the strength of low-level motion perception, *J. Visual Communication and Image Representation.* **1**: 176-188.

Werkhoven, P., Sperling, G., and Chubb, C., 1993, The dimensionality of texture-defined motion: A single-channel theory, *Vision Res.* **33**: 463-485.

Wertheimer, M., 1912, Experimentelle Studien über das Sehen von Bewegung, *Zeitschrift für Psychologie.* **61**: 161-265.

Wilson, H. R. and Kim, J., 1994, A model for motion coherence and transparency, *Visual Neuroscience.* **11**: 1205-1220.

Zanker, J. M., 1995, Of models and men: mechanisms for human motion perception, in *Early Vision and Beyond*, T. V. Papathomas, C. Chubb, A. Gorea, and E. Kowler, eds., MIT Press, Cambridge, MA, pp. 155-165.

Chapter 14

Computational Models of Visual Attention

Scott B. Steinman[1] and Barbara A. Steinman[2]

[1] *Biomedical Sciences Dept., Southern College of Optometry, 1254 Madison Ave., Memphis, TN 38104-2222. PH: (901) 722-3380, FX: (901) 722-3325, EM: steinman@sco.edu;*

[2]*Software In Motion, 172 Island Place, Memphis, TN 38103-1716. PH: (901) 529-1562, FX: (901) 529-1562, EM: barb@midsouth.rr.com*

14.1 INTRODUCTION

14.1.1 What is Attention?

It seems intuitively obvious what visual attention is, so much so that the first person to study attention, William James, did not provide a definition for attention, but simply made the assumption that "we all know what attention is" (James, 1890). However, the research of the past few decades has shown that this assumption is far from true. Attention is actually a complex process with several interacting subprocesses. It is carried out by extensive neural networks that interconnect many regions of the brain. A fundamental question we must ask is: why is so much neural machinery devoted to attention? A simple answer would be that so much sensory information is presented simultaneously to the brain that it would be impossible for all the information to be processed quickly enough to ensure its immediate use. For example, there are practical limits on how many distinct orientations, colors, sizes, depths and velocities at any possible retinal locus that can be processed in parallel (for example, see Shiffrin and Gardner, 1972). Visual attention is therefore necessitated by the limited computational capacity of the human brain. Attention allows for the selective processing of one or more important visual stimuli while filtering out less critical information.

Visual attention is a broad term that encompasses several basic components. These include: (1) engaging or activating attention in preparation to detect a target of interest (Yantis and Jonides, 1984); (2) directing or orienting attention to a specific location in the field (Posner et al., 1987; Steinman et al., 1985); (3) locking attention on that locus (Rafal and Posner, 1987); and (4) suppressing irrelevant information from other locations (Downing and Pinker, 1985; Steinman et al., 1995). All of the above processes are crucial to shifting attention. In addition, we must at times maintain attention at a given location, either to wait for an expected new stimulus or to completely process all of a present target's visual information (Nakayama, 1990). Finally, we must be able to disengage attention in preparation for processing new information, possibly at another location (Morrow and Ratcliff, 1988; Mackeben and Nakayama, 1993).

Attention is directed to one particular location in the field, called the *attentional focus*, which can be moved independently of eye movements, allowing the observer to attend to a non-fixated location (Sagi and Julesz, 1985a; Rafal et al., 1988; Breitmeyer, 1991). Within the attentional focus, response times (as measured by reaction times) for detection and discrimination of visual stimuli in that region are faster (Ericksen and Collins, 1969; Posner et al., 1982; Ericksen and St. James, 1986). Event-related cortical potentials also reveal the effects of attention by exhibiting larger amplitudes when the focus of attention coincides with a visual stimulus than when it does not (Mangun and Hillyard, 1991); a similar response is noted in PET scans (Corbetta et al., 1991). Concurrently, response times are slower for stimuli located in the remaining portions of the visual field (Posner and Boies, 1971; Posner et al., 1982). Visual attention also increases the visibility of the attended target, while reducing the visibility of stimuli outside of the attentional focus (Ericksen and St. James, 1986). Visual attention therefore serves to prioritize our visual sensory input so that important information is enhanced while relatively irrelevant information is inhibited (Allport, 1980).

14.1.2 The Need for Models of Visual Attention

Attention has a ubiquitous role in vision, having been demonstrated to be involved in such diverse visual processes as motion perception (Cavanagh, 1992), surface perception (Nakayama, He and Shimojo, 1995), texture segregation (Treisman, 1985; Julesz, 1986), eye movements (Sereno, 1992; Fischer and Weber, 1993) and binocular rivalry (Ooi and He, 1999). It is therefore critical when studying vision to understand how attention shapes our perception. In addition to its effects on visual perception, attention plays a critical role in the planning and execution of eye movements. Before we can make a saccadic eye movement, visual attention must first be shifted to the intended new fixation point (Sereno, 1992). The role of attention in eye movements has only recently been explored.

Although its influence is widespread, attention operates "behind the scenes" in vision. It is therefore harder to test its limits via psychophysical measurement than is sensory processing. It is also difficult to tease apart using psychophysical or neurophysiological techniques alone the underlying mechanisms of attention, especially when interactions exist between these mechanisms.

The study of attention is a relatively young discipline. Several models of attention have been proposed, but the scientific community has not yet been able to narrow them down by proving or disproving the models with psychophysical and physiological data. Many models address either psychophysical or neurophysiological data alone. A comprehensive computational model would need to account for both sources of experimental data. In the 1980's, David Marr's computational models of visual processing helped to fuel a wealth of experiments that sought to find proof for his model. This same process of discovery is now needed for attention.

14.2 HISTORICAL OVERVIEW

14.2.1 Filtering ("Bottleneck") Theories

Despite introspective writings on attention at the turn of the century by Helmholtz (1867) and James (1890), it was not until Broadbent's work in the 1950's that a comprehensive qualitative model of attention was proposed that relied upon empirical behavioral data. Broadbent (1958) viewed attention as a means of completely filtering out unselected information. Attention could therefore be thought of as a "bottleneck" that restricts the amount of information reaching cognitive circuits of limited capacity (Nakayama, 1990). Broadbent's filter theory marked the emergence of the *early-selection* model of attention. In this model, the sensory system first processes all basic visual features in the visual field in parallel. The higher-level mechanisms that interpret, recognize and identify the stimulus are band-limited, such that they may only process one stimulus at a time in a serial fashion. Attention plays the role of a selective filter that preferentially allows visual information from within the attentional focus to reach the capacity-limited identification mechanisms and reach consciousness. The term "early-selection" connotes that the filtering takes place at an early stage of visual processing. The serial nature of the identification system suggests that more time would be needed to recognize two stimuli, since the observer would need to attend to one, then the other, stimulus of regard. As is common with suppressive mechanisms in the visual system, the attenuating effects of attention were later found to be relative rather than absolute (Treisman, 1960; Eriksen and Hoffman, 1973). Broadbent's early-selection model was therefore modified to reflect attentive selection whose filtering merely reduces the effectiveness of unselected targets, rather than removing them completely from being processed or consciously perceived (Eriksen and Hoffman, 1973). Nevertheless, this concept of limited processing capacity by the visual system has been upheld by numerous subsequent studies and remains a fundamental principle in major models of attention (for example, see Kahneman, 1973; Treisman and Gelade, 1980; Posner, 1980; Nakayama, 1990).

Others have instead provided evidence for a *late-selection* model in which the processing bottleneck occurs subsequent to categorization of the visual stimulus (Schneider and Schiffrin, 1977; Posner, 1982; Hung et al., 1995); that is, stimuli are both processed and identified prior to the influence of attention, this suggesting that identification is not capacity-limited. In this case, identification is involuntary, and it is the active cognitive processes of awareness, context, and memory that determine which stimuli reach consciousness. Multiple stimuli would be both processed and identified in parallel.

This extreme dichotomous argument over the location at which attentional exerts its influence no longer is relevant. Studies of subjects with lesions of visual cortex suggests that there must be more than one type of visual attention due to the varied roles that attention plays, that they are mediated by different regions of the brain, and that these different attentional processes interact with one another. Both early and late-selection models only explain selective attention, and they do not account for other roles of attention such as shifting attention or divided attention.

14.2.2 Resource Sharing Models

One set of models that attempts to bridge the gap between early-selection and late-selection theories assumes that higher-order processing requires the use of limited cognitive resources. Rather than attention selectively limiting access to these processes for a single stimulus, these models suggest that attention may be divided up and shared between multiple stimuli (for example, see Kahneman, 1973). Each stimulus therefore ties up a portion of the available cognitive resources. The consequence of the reduced degree of resources allocated to each stimulus is that the processing of each stimulus requires more time to complete. Resource sharing models therefore explain well the difficulties subjects have in *dividing* attention, while early- and late-selection models do not. Resource sharing models do not totally disagree with either early- or late-selective theories; they simply do not explicitly assume at which stage the limited cognitive resources are allocated or how they are allocated.

14.2.3 The Attentional Spotlight and Zoom Lens Models

Most models of attention assumed that attention could only be allocated to a single location in the visual field, the attentional focus. Posner (1980) introduced the concept of attention acting like a "spotlight" which illuminates only a small region of the visual world at any given moment. Processing is enhanced at the locus of the spotlight. For attention to be focused on a new location, the spotlight of attention must be shifted to that location. Shulman et al. (1979) found increased reaction times for cued locations at 18 deg versus 8 deg of eccentricity and concluded that a "window of attention" had to be moved across the visual field. Tsal (1983) estimated the velocity of the movement to be about 1 deg per 8 ms. Ericksen and Murphy (1987) later questioned this analog movement model, a view that was confirmed by Miyauchi et al. (1992), who showed that when two cues are presented asynchronously, attention does not "slide" from one cued location to the next as predicted by Tsal. Instead, the attentional response builds and ebbs at the first locus, then builds and declines at the second locus, in agreement with Ericksen and Murphy's model.

Ericksen and co-workers (Ericksen and Yeh, 1985; Ericksen and St. James, 1986) embellished the spotlight metaphor to include a "zoom lens". Once the attentional spotlight has been shifted to a particular location, the observer could then zoom in to focus on small details in that area. Within the focus of attention, there is an even distribution of cognitive resources. The size of the focus can be manipulated experimentally by the choice of attentional cues. As the size of the focus shrinks, the distribution of resources becomes denser, allowing finer discrimination and increased performance (Ericksen and Yeh, 1985), but longer reaction times. The zoom lens model has since been questioned by Kröse and Julesz (1989), who were unable to manipulate the size of the attentional focus by manipulating the cued area. They proposed a cooperative parallel and serial processing model, in which visual information is processed in parallel across the visual field. Fundamental stimulus properties direct the attentive process to the location most closely matching the feature to be searched, then the marked area of the field can be serially searched for finer detail. The transient/sustained model of attention (Section 14.2.6) concurs with this model.

14.2.4 Feature Integration Theory

Recently, many studies of visual attention have focused upon the role of attention in feature binding, as proposed by the Feature Integration Theory (FIT) of Treisman and co-workers (Treisman and Gelade, 1980). Evidence for this model comes from visual search tasks, in which a target must be detected among several distracting targets. Searching for a target defined by a single feature such as color (for example, a red target among several green distracting targets) yields little to no increase in reaction time as a function of the number of distractors, a phenomenon termed parallel search. Conversely, searching for a target defined by a conjunction of features, such as a red square target among distracting green squares, red circles and green circles requires an exhaustive serial search of one target after another until the red square is located, resulting in reaction times that increase significantly with the number of distractors. According to FIT, the visual system retains independent retinotopic maps of the locations of individual visual primitives (features, or textons [Julesz, 1984]) across the visual field. The features in each are accessed in parallel by early visual processes. Attention then serves to bind these different feature maps together to form a unified percept of objects containing conjunctions of features at any given location. FIT therefore explains the difference in slope of the reaction time function for simple and conjunctive searches as reflecting two underlying mechanisms: preattentive, or "early" vision (Neisser, 1967), and attentive. Parallel search reflects independent early visual processing of all targets in parallel within any given single feature map, thus a mechanism requiring a fixed processing time no matter how many distractors are present. Serial search reflects the need to deploy and shift attention to examine each target in turn. The relevant target is no longer unique within a single feature map; the search for a target defined by a conjunction therefore requires more time as the number of targets is increased.

The prospect of having an easy means to categorize visual features as either low-level primitives processed by feature detectors versus features which require higher-order processing sparked nearly four decades of research utilizing visual search. However, Feature Integration Theory, at least in its initial form (Treisman and Gelade, 1980), was limited in its capacity to explain visual attention. First, it could not explain phenomena that blur the distinction between parallel and serial search. For example, learning allows observers to switch from serial to parallel search even with conjunctive stimuli (Steinman, 1987). This suggests that the trigger for parallel search could be a learned ability to segregate targets perceptually into two different groups – one group containing the single unique target and a few distractors, and the other containing only distractors – allowing a rapid focusing of attention on the unique target. Parallel and serial search may therefore reflect two different attentional processes, as suggested by the transient/sustained attention model discussed below (Nakayama and Mackeben, 1989; Steinman et al., 1994, 1995, 1997b). The presence of attention in parallel search is supported by experiments demonstrating that dividing attention can cause degradation of performance in even a simple orientation search (Joseph et al., 1997). It should also be noted that Pashler (1987) proposes instead that all searches are in fact parallel, but that there is an upper limit on the number of elements that can be processed in parallel, yielding increased search times for additional items. Second, FIT assumed a single attentional bottleneck at an early stage in the visual system prior to the stage of visual recognition. The reliance of FIT on a strict dichotomous view of the location at which attention exerts its influence is not supported by empirical evidence. Finally, the underlying principle that the visual search data reflect a dichotomy between early feature detection and attentive processing is too simplistic. There are dozens of stimulus parameters that have been labeled preattentive, including some very complex features such as faces or shape from shading (Ramachandran, 1988), with each requiring its own feature map. Conversely, even basic features known to be processed by neurons early in the visual system can fail to be detected in parallel when stimulus context is manipulated (He and Nakayama, 1992). These exceptions to the rule have prompted a new viewpoint that visual attention is engaged not following early feature detection, but after the processing of visual surfaces and edges (Bravo and Blake, 1990; Enns and Rensick, 1990; He and Nakayama, 1992). Each of the arguments listed here has required FIT to be modified to take into account attentional influences on parallel search [including divided attention and inhibitory effects (see Steinman et al., 1995)] and the existence of both early selection and late selection (Treisman and Sato, 1990; Treisman, 1993). The resulting ad hoc model addresses many of these shortcomings, but at a price of being more complex and more difficult to test quantitatively.

Wolfe and co-workers (Wolfe et al., 1989; Wolfe, 1994) offer a modification of FIT called the Guided Search Theory to explain how parallel search can occur even with conjunction stimuli (Steinman, 1987). They propose that if the activity in the feature map for one conjunctive feature is added to that for the second conjunctive feature, only the stimulus with both features – the conjunction – will be doubly activated, and that location in the visual field, by virtue of its salience relative to those of the distractors that stimulate only one or the other feature map, will attract the attentional focus.

Revised FIT (Treisman and Sato, 1990; Treisman, 1993) proposed to the contrary, that inhibitory interactions between distractors allows for rapid conjunctive searches.

14.2.5 Center-Surround Model

The faster processing time and enhanced detectability within the focus of attention is balanced by slower processing and decreased detectability outside of this region (Posner and Cohen, 1984; Ericksen and Murphy, 1987; Kröse and Julesz, 1989). Using an illusory motion percept induced by attentional precueing called the *line motion illusion*, Steinman et al., (1995) were able to map out directly two major zones of differing attentional response: (1) a narrow excitatory region at and near the cued location, in which visual processing is enhanced or accelerated, with this being analogous to the attentional spotlight of earlier models (Ericksen and St. James, 1986), and (2) a large inhibitory surrounding zone in which visual processing is suppressed and slowed down. Steinman et al. modeled these attentional "perceptive fields" (Jung and Spillmann, 1970; Spillmann et al., 1987) using the difference of Gaussian (Enroth-Cugell and Robson, 1966; Jung and Spillmann, 1970) functions used in neurophysiological studies; such functions provided an excellent fit to their attentional enhancement/inhibition plots. The zoom lens model (Ericksen and St. James, 1986) would predict that when the focus of attention is constricted, the magnitude of the attentional enhancement within this central region should increase. However, an inverse relationship between the perceptive field center diameter and the height of the center was not supported by the data of Steinman et al., thereby contradicting the zoom lens model.

Such concentric center-surround organization is found in receptive fields throughout the visual system (Kuffler, 1953; Allman et al., 1985). However, in the case of the attentional perceptive field, the overall field size is very large, with it more closely matching receptive field diameters of neurons in higher cortical areas such as MT cortex (Allman et al., 1985) or inferotemporal (IT) cortex (Gross, 1973).

14.2.6 Transient-Sustained Model

The time course of attention remains unchanged regardless of the retinal locus stimulated (Mackeben and Nakayama, 1993; Steinman and Steinman, 1997b). After the onset of a cue, visual performance exhibits a rapid rise to a sharp peak, with a minimum latency of 50 ms (Steinman et al., 1995; Tsal, 1983; Ericksen and St. James, 1986; Downing, 1988). After reaching its peak, the improvement in performance gradually dissipates until the beneficial effects of attention disappear by 200 to 300 ms (Posner and Cohen, 1984; Steinman et al., 1995; Steinman and Steinman 1997b). A second activation of attention can be seen after 400 ms following the onset of the cue (Hikosaka et al., 1993b), reaching a maximum effectiveness at 600 ms and lasting at least 1800 ms (Steinman and Steinman, 1997b); it has not been measured at longer durations. This second activation of attention is only about half as effective as that of the first activation (Steinman and Steinman, 1997b).

The two activations of attention have been called *transient attention* and *sustained attention*, respectively, by Nakayama and Mackeben (1989); others have used the terms reflexive attention or the orienting response (Posner and Cohen, 1984; Kahneman, 1973) versus voluntary attention (Hikosaka et al., 1993a). The orienting response is an automatic and involuntary focusing of attention towards any sudden change in the visual field, with it being stronger for the sudden appearance of a stimulus than for its extinction (Yantis and Jonides, 1984; Jonides and Yantis, 1988). It is such a strong response that even the sudden appearance of very minor differences between targets can elicit it (Siddle, Stephenson and Spinks, 1983). The automaticity of orienting concurs with the ability to divide one's attention among several targets, thus allowing simultaneous parallel processing of each target. Transient attention is thought to be a bottom-up process as suggested by its retinotopic mapping; that is, if one moves the eye, the region of the visual field that is enhanced by transient attention moves with the eye and remains at the same locus relative to the retina (Posner and Cohen, 1984). In addition, the size of the attentional focus scales with retinal eccentricity (Sagi and Julesz, 1986). Transient visual attention is dominated by the magnocellular pathway (Steinman et al., 1997).

Conversely, sustained attention requires top-down voluntary cognitive effort at a single location, which makes it difficult to divide one's attention. Sustained attention does not require precuing at the intended location of the attentional focus. However, it does require a discrimination or interpretation, that is, a cognitive decision to determine where to locate the attentional focus (Sagi and Julesz, 1985b; Hikosaka et al., 1993a). Discrimination of fine spatial details requires long processing times, which may partly reflect the involvement of sustained visual attention (Steinman et al., 1995; Steinman and McKee, 1988). This shift from transient to sustained attention has also been demonstrated by means of the visual evoked potential. Steinman and Levi (1992) noted that VEPs to suprathreshold hyperacuity stimuli had latencies over twice as long as VEPs to grating or checkerboard stimuli. In addition, the scalp topography of the hyperacuity VEPs was consistent with a generator site well outside of striate cortex. These findings suggest that while gross temporally-changing targets such as gratings can be processed using short-latency transient visual attention alone, the ability to complete the processing of fine details such as hyperacuity stimuli requires the activation of long-latency sustained visual attention.

Much of the evidence for the transient/sustained model of attention has come from experiments with the line motion illusion (Steinman et al., 1995; Steinman and Steinman, 1997). It should be noted that the line-motion illusion has been the subject of heated debate, with some researchers arguing that the illusory motion is simply a side-effect of preattentive apparent motion (motion impletion) arising from the asynchronous presentation of the attentional cue and probe line target used for measurement (for example, see Tse and Cavanagh, 1995; Baloch and Grossberg, 1997). However, compelling evidence exists to disprove this argument. The visual illusion can be demonstrated in response either to auditory or somatosensory cues for which motion impletion would not be expected to occur (Shimojo et al., 1992). Additional evidence comes from Steinman and Steinman (1997b), who examined changes in the shape of the envelope of the attentional perceptive field as modeled by symmetrical Gabor functions (Marčelja, 1980):

$$S_s = \exp\left[\frac{-(x - x_m)^2}{4\sigma^2}\right]\cos\left[2\pi f_n (x - x_m)\right] \tag{14.1}$$

where x_m is the lateral position of the middle of the function, f is the frequency and σ^2 is the standard deviation of the function. They demonstrated that the line motion illusion in fact measures two coincident processes – the first varies the baseline level of the attentional perceptive field as a function of time, while the second varies the width of the center region (focus). That is, the illusory motion is produced by a combination of two factors. The first is a simple modulation of the degree of responsiveness of the motion detectors that produce the motion percept, being akin to how injecting current into MT cortical neurons biases their baseline activity (Salzman et al., 1990), while the second factor modifies the extent of the visual field over which attention exerts its excitatory influence. The temporal response of this second mechanism is in accordance with that of the transient and sustained mechanisms hypothesized by Nakayama and Mackeben (1989) from their experiments on target detectability. However, Nakayama and Mackeben's tentative transient mechanism has a longer peak time (150-1200 ms) and decay than the 33-50 ms found by Steinman et al. (1995, 1997), or the 17-50 ms times noted for attentional shifts (Sagi and Julesz, 1985b; Bergen and Julesz, 1983). The center-surround organization of the attentional perceptive field is in agreement with cost-benefit effects of attention (Posner and Cohen, 1984). In other words, this second mechanism underlying the line motion illusion produces the same attentional effects found using paradigms not involving motion perception. The line-motion illusion therefore does not simply reflect motion impletion, but also quantifies the robustness of the visual attentional response. However, without such an analysis it is not possible to separate these two influences.

Steinman and Steinman (1997a) demonstrated that a prior activation of attention can alter a subsequent transient attentional response at the same location. Instead of the precued location exhibiting a further increase in processing speed with the introduction of a second cue at the same location (with a concurrent pronounced decrease in processing speed outside of the attentional focus), the precued location is inhibited for a refractory period up to 1000 to 2000 ms, while the attentional surround is enhanced in its response. The differential response between the center and surround of the attentional perceptive field is therefore reduced, hence making it less likely that repeated stimulation at one location will continue to draw attention, and thus ensuring the visual system is primed to detect sudden changes in the environment at other locations.

While the parallel/serial dichotomy of visual search might reflect a preattentive-attentive difference, it is tempting to suggest that both forms of visual search are in fact attentive, with the seeming dichotomy reflecting the automatic or transient versus the cognitive or sustained attentional processes of Posner (1980), Nakayama and Mackeben (1989) and Steinman et al. (1995, 1997b). In accordance with this view, Treisman (1993) has modified Feature Integration Theory to include two forms of attentive processing, an early-selection process and a late-selection process.

14.3 CURRENT MODELS OF VISUAL ATTENTION

14.3.1 Early Connectionist Models

While the early theories of attention have explained some basic properties of attention and the role of attention in form perception, they fall short in that they are not quantitative models and therefore cannot easily be tested either experimentally or with computer simulations. Quantitative computational models of attention have only been available since the mid-1980's.

Many of the initial attempts to model the visual system using connectionist or parallel distributed processing (PDP) models (now called "neural networks") did little to include attention as an active process (for example, see Rumelhart et al., 1986). Cohen (1993) notes that artificial intelligence machines such as PDPs have difficulty in self-initiating actions such as directed attention based upon stimulus relevancy, although some models assume stochastic thermodynamic processes to "self-start" attention and avoid this problem. Therefore, in the early connectionist models, attention is not directly addressed by the model, but rather is simply one of the outcomes of the modeled neural processes.

Neural network models are particularly well-suited to incorporate learning processes (Grossberg, 1988). The competition between neighboring filters in a network results in the strengthening of connections that convey information about a stimulus of interest (i.e., the "signal"), while weakening those connections conveying "noise", that is, any stimulus that is not of interest. Such learning processes could be involved in facilitating the activation of the attentional focus (Hinton, 1981). Learning of a pattern template for a particular stimulus feature from prior exposures would alter the weights associated with that stimulus feature in the neural network's filters, thereby allowing the observer to direct feature-specific attention quickly and accurately. Sokolov (1963) modeled the orienting response, that is the early selective attentional process that responds to novel stimuli, as a mapping of the degree of mismatch between a novel stimulus and the existing learned pattern templates for stimuli of that type; the greater the mismatch, the larger the attentional response.

14.3.2 Selective Routing or Input-Gating Models

Koch and Ullman (1985) proposed a pyramid neural network consisting of a large number of center-surround input receptive fields that process all stimuli and all stimulus characteristics in parallel and a higher-level representation that "sees" only one stimulus at a given moment (see Fig. 14.1). Koch and Ullman's pyramidal organization must be differentiated from that of Nakayama's (1990) iconic bottleneck model, in which the pyramid consists solely of a multi-resolution retinotopically-organized feature map much like the Laplacian pyramids used in computer vision and image processing algorithms. The pyramid in the iconic bottleneck model is situated entirely prior to the focal selective attention process. Koch and Ullman's pyramidal neural network, on the other hand, begins with a set of parallel feature maps forming the processing units of the lowest layer alone, and it converges upon the upper output layer of neural elements that non-topographically represents a single attended location in the field. The shift from parallel processing of all features in the visual field to preferential processing of a single stimulus is possible due to a *winner-take-all* (Feldman and Ballard, 1982) algorithm that passes the output from the most active neuronal element signaling the most salient stimulus, via mutual inhibition of neighboring units, upward to higher-level centers responsible for stimulus recognition and visual memory. In this regard, the output of the pyramid neural network is analogous to the attentional spotlight proposed by Posner (1980) within which visual processing is enhanced. The Koch-Ullman model is therefore feed-forward or bottom-up only. Once the most active output neuron responds, it is then inhibited, and attention shifts then to the location represented by the next-strongest responding output unit (see also the psychophysical data of Steinman and Steinman, 1997a). In other words, parallel feature maps (Treisman and Gelade, 1980; Koch and Ullman, 1985) are scanned for their most salient feature relative to other features in the map (Nothdurft, 1993). The location of the elements in the input layer that have detected the stimulus in question therefore determines the routing of the signal to a particular "attentive" neuron in the output layer, hence the term selective routing. Tsotsos (1995) notes that the winner-take-all algorithm that is common to most neural network models may not be appropriate for models of selective attention. First, Koch and Ullman state that the response strength of the winning upper-layer unit is relayed to a central representation, yet the presence of lateral inhibition as the sole process would greatly attenuate the response of the winning neuron. Second, only saliency information is passed on to higher representations, with the representation of features from the saliency map being lost. Third, the winner-take-all algorithm would predict that attention requires time to shift from one location to another proportional to the distance between the two loci, but Kröse and Julesz (1989) demonstrated that such shifts occur in constant time no matter the distance between attended locations. Finally, such models may account for engaging spatial attention at a particular visual field location, but cannot account for division of attention which would require more than one winner, each of whose response would be severely attenuated, nor the sharpening of feature specificity at the attended locus. Even so, Koch and Ullman's model and winner-take-all selection scheme have been used as the basis for other computational models of attention (Olshausen et al., 1994; Niebur et al., 1993; Grossberg et al., 1994; however, see Tsotsos, 1993 for an alternative winner-take-all algorithm).

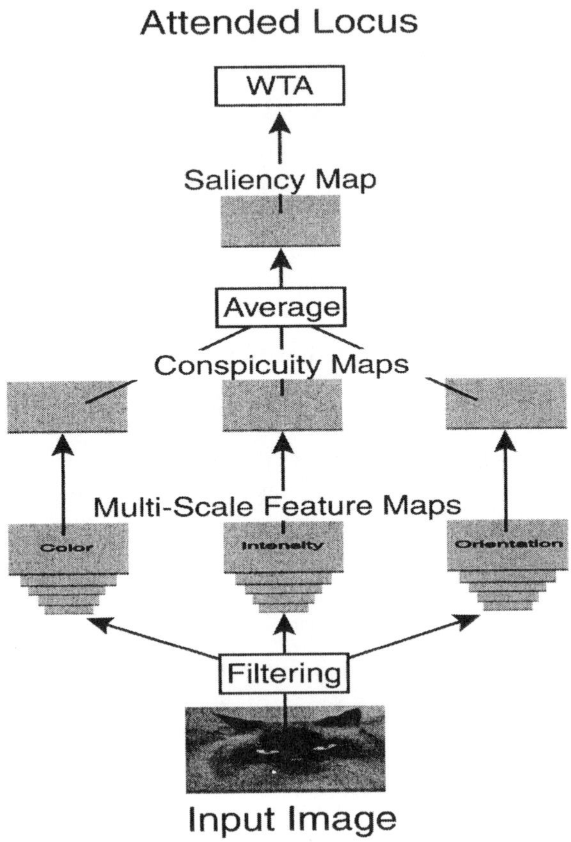

Figure 14.1. Koch and Ullman selective routing model. The input image is selectively filtered by feature detectors of different sizes to yield parallel multi-resolution feature maps. The conspicuousness of features in each map is averaged to yield a saliency map that marks the positions of prominent objects or parts of objects. A winner-takes-all algorithm is used to find the location of the most salient item, which is chosen to be the attended locus. Adapted from Itti and Koch (2000) with permission of Elsevier Science.

14.3.2.1 The Shifter Circuit Model

Shifter models (Anderson and van Essen, 1987; Olshausen et al., 1993) are another form of selective routing model, in which shifter circuits perform the routing. Starting with a Koch and Ullman-like (1985) set of receptive fields, signals from within the attentional focus are routed or switched by diverging excitatory inputs to larger higher-level units, presumably in visual area V4 and inferotemporal cortex (IT), while preserving their spatial mapping relative to the visual field. Inhibitory relay interneurons arranged in several layers are linked by ascending excitatory connections that diverge, with smaller degrees of divergence in the lower layers of the neural network and progressively larger increments in the upper layers. This ensures the greatest possible magnitude of shift at the output layer and prevents the need for a quadratic increase in the number of output cells as a function of the number of input cells. The inhibitory neurons redirect information by allowing only one diverging branch to be active, that is, the attentional focus, thereby selectively suppressing non-attended inputs to the higher-level cortical areas. The result is that the representation of the object is shifted and scaled as it progresses to IT cortex, but the local spatial relationships within the stimulus are maintained, thereby creating a higher resolution zoomed object-based mapping from the input retinotopic mapping. The degree of spatial shift of the activity from the input layer to the output layer can be any amount up to 2^m neural units, where m is the number of layers under control by a shift control "black box" presumably controlled by the pulvinar. The focus of attention can also be either shifted or zoomed. Anderson and van Essen (1998) argue that macaque V1 cortex contains the necessary excitatory and inhibitory interneurons to implement the shifter circuit, with GABA-ergic neurons providing the inhibitory inputs from the shift controller. To achieve the different shift scales in the lower and upper layers, the original shifter circuit model (Anderson and van Essen, 1987) required a multitude of connections; the subsequent reworking of the model by Olshausen et al. (1993) corrects this oversight by incrementing the spacing between neighboring units in sequential layers. The predictions of a computer simulation of this model are in accordance with the neurophysiological data of Moran and Desimone (1985). However, the model hinges on two assumptions. The first is an active outside control of the inhibitory signals by the pulvinar which modifies the network's synaptic weights. The second is a saliency map in posterior parietal cortex that in turn modulates the pulvinar control neurons.

There are several drawbacks to the shifter circuit model. First, an active independent attentional system, as proposed by Posner and modeled by Anderson and Van Essen, has not been proven. In addition, the neurophysiological evidence for dynamic receptive field positional shifts is mixed (Motter and Poggio, 1982; Parker and Hawken, 1985; Connor et al., 1994). Finally, the shifter circuit model has a strict assumption that spatial relationships of the image representation are preserved at several stages in the visual system, an assumption that may not be true in higher-level cortical areas.

14.3.2.2 The Pulvinar Neural Circuit Model

The pulvinar, a thalamic nucleus, is interconnected with several visual cortical areas and the superior colliculus (Ungerleider et al., 1983), and exhibits both retinotopic organization (Allman et al., 1972; Bender, 1981) and selectivity for visual features such as color, orientation and motion. Single unit recordings in the pulvinar demonstrate that its neurons fire more briskly when attending a target (Petersen et al., 1985). Both these single units and the attentional response are inhibited by injections of the GABA agonist muscimol (Petersen et al., 1987). Thalamic nuclei such as the pulvinar are organized in a columnar organization. In the pulvinar, principal cells and their reticular network inputs conduct information only within columns or between near neighboring columns (Crick, 1984), preserving its topographic organization (see Fig. 14.2). The reticular network cells allow for recurrent lateral inhibition between columns, sharpening the difference in firing rates between neighboring principal cells, much like lateral inhibition in the retina enhances small differentials in activity arising from lightness differences. In addition, interneurons create feed-forward lateral inhibition between principal cells.

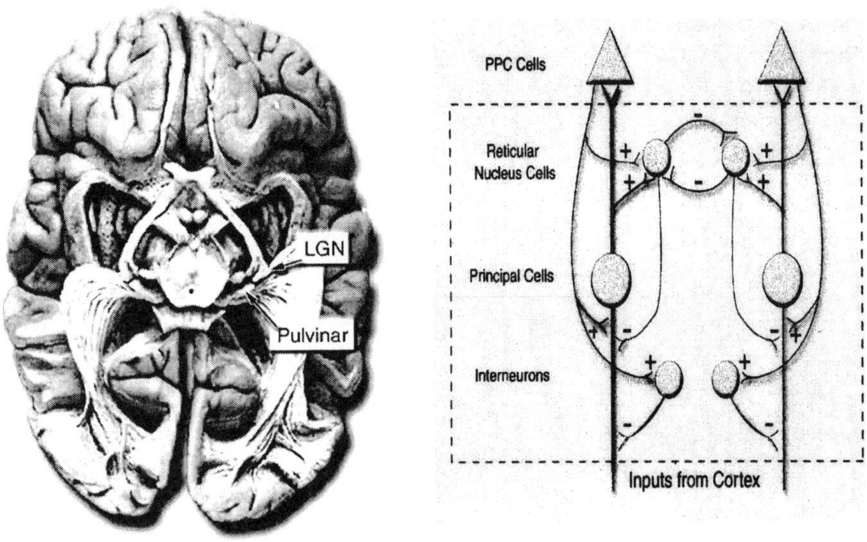

Figure 14.2. Pulvinar neural circuit model. The left side of the figure shows the inferior aspect of the brain, with the dorsal lateral geniculate nucleus (dLGN) and pulvinar labelled. The right side of the figure depicts the neurocircuitry within the pulvinar, shown as those neurons within the dotted lines. Principal cells form columns, with reticular nucleus cells and interneurons forming lateral interconnections between adjacent columns. Adapted from LaBerge (1995) with permission of Harvard University Press.

Laberge and co-workers (1992) performed simulations of this pulvinar neural circuit by presenting a stimulus of regard to one principal cell and a flanking or distractor stimulus to its neighbor. The sole difference between the stimuli was that the afferent input corresponding to the target was set to 38 units of activity, while that for the flanker was 37. The pulvinar circuit model produced a principal cell efferent output that differed by over 20 units between target and flanker. The main factor yielding this increased differential was an enhancement of the activity of the neural unit processing the target. These simulations suggest that the pulvinar serves to produce a pronounced concentric focus of attention and surround region similar to that reported by Steinman et al. (1995). It should be noted, however, that this model does not attempt to explain other aspects of attention such as shifts of attention or divided attention. In addition, while this study suggests that the attentional perceptive field is a function of solely an excitatory response to the target, other researchers have shown that inhibitory processes can also play a role in selective attention (Koch and Ullman, 1985; Treisman and Sato, 1990; Olshausen et al., 1993; Steinman et al., 1995; Mozer and Sitton, 1998).

14.3.2.3 Feature Binding Models

The FAÇADE model

Although originally intended to model boundary and surface representations in the visual system, Grossberg et al.'s extended FAÇADE (Form-And-Color-And-DEpth) model (1994) distinguishes itself from most other computational models of attention in that it directly addresses Treisman's Feature Integration Theory (Treisman and Gelade, 1980; Treisman, 1993) and Wolfe's Guided Search model (Wolfe et al., 1989; Wolfe, 1994). In this model, boundary and surface segregation occurs prior to an object recognition that relies upon visual attention (Neisser, 1967; Grossberg, 1987a,b; He and Nakayama, 1992), but the two processes mutually interact. Stated in terms of the model's implementation, the FAÇADE model of boundary and surface segmentation is used as the input front end to the attention-dependent recognition algorithms. Reciprocal interactions allow attention and object recognition processes to reorganize the input perceptual units following their preattentive detection of features. Visual search is therefore applied recursively on segmented surface groupings that may contain multiple items (Nakayama and Silverman, 1986; Steinman, 1987; Wolfe et al., 1989). The selective tuning model of attention (Tsotsos, 1993), discussed below, is another example of recursive attentional processing.

Grossberg's model therefore combines three classes of neural networks that (1) create a three-dimensional boundary and surface representation (i.e., FAÇADE), (2) control object recognition, memory-dependent learning and priming of feature-based attention (a "what" representation), and (3) create a spatial representation within the attention system (a "where" representation), respectively. The object recognition system is an adaptive resonance theory-based network (Carpenter and Grossberg, 1991). It is the interactions between these networks that allow modeling of visual search. For example, the boundary and contour segmentation systems must interact with the object recognition system to bind features to objects (Treisman and Gelade, 1980), and the contour segmentation system must interact with the spatial attentional map to determine the location in space of the bound features.

Despite the seeming complexity of this system, it can be summarized as a four-step algorithm. The first step involves preattentive processing of stimulus features in the visual field to form a retinotopic map of features. In step two, boundary and contour segregation mechanisms use these features to segment the visual scene into separate regions. In step three, a particular region is selected for further analysis by either bottom-up salience or top-down priming, a step suggestive of the transient-sustained model of attention (Nakayama and Mackeben, 1989; Steinman and Steinman, 1997b, 1998). Finally, in step four, the features in the selected region are compared to a memory store. If an appropriate target is not identified, steps three and four are repeated, or step two is repeated to further segment the visual scene.

Simulations of a visual search task were performed by determining the time required for each step and combining them to yield an aggregate reaction time. A key assumption is that each stage need not exhaustively process its input before its output can be used by the next layer of the model. In fact, in some scenes, the processes may occur together, thereby allowing their durations to be combined into a single constant duration. The mean search time for a target-present condition is given by:

$$RT = R + (N+1) \times (S+M) \tag{14.2}$$

where R is the duration needed to complete step one, S is the time needed to segment and select a candidate region of the scene (steps two and three), M is the duration to match the candidate region with the target representation in step four, and N is the mean number of candidate regions on an initial segmentation. On average, only half of the candidate regions would need to be recursively searched before the target is located. The reaction time for target-absent presentations is given by:

$$RT = R + N \times 2 \times (S+M) \tag{14.3}$$

Here, each region would need to be searched, thus making the total search time for target-absent longer by a factor of two. By manipulating the values of *R*, *N*, *S* and *M*, as well as saliency (a quantity *C*, since the simulations involved the saliency of a color target) and *P*, the percentage of groupings permitted for a given salience, Grossberg's model produces simulated reaction time data that agrees well with the psychophysical data of Treisman and Gelade (1980) and Wolfe et al. (1989). The lower the saliency of the target, the steeper the slope of the reaction time versus number of elements plot for both the target-present and target-absent conditions.

Grossberg argues that his model incorporates biologically plausible mechanisms. For example, boundary and contour segmentation are carried out by striate and extrastriate cortex, feature-based attention employs the inferotemporal cortex, and spatial attention is a function of the parietal cortex. Memory storage is presumed to involve frontal cortex. The neural networks that are combined to form this visual search model have been previously used in other studies to model the processing properties of those cortical areas; moreover, interactions between widely separated areas of cortex are crucial to visual attention (see Section 14.4.2).

A Hybrid Model

Mozer and co-workers (Mozer, 1991; revised by Mozer and Sitton, 1998) have formulated a neural network model to explain letter recognition in a visual search-like task that synthesizes several of the properties of other laboratories' models (Koch and Ullman, 1985; LaBerge and Brown, 1989; Sandon, 1990; Ahmad, 1991). The model employs a set of parallel retinotopic feature maps and a retinotopic saliency map whose state determines where to focus attention (see Fig. 14.3). Once attention is directed to a particular locus, all of the features at that locus are processed (Kahneman and Henik, 1981). The gating of attention is via a simple multiplicative combination of the activity in the saliency and feature maps:

$$\hat{r}_{qxy} = g(a_{xy})(r_{qxy} - \bar{r}) + \bar{r} \tag{14.4}$$

where the activity from a feature map's unit r_{qxy} is passed to a higher-order recognition unit. In the gating function, $\bar{r}$ is the resting potential of neurons in the feature maps, a_{xy} is the activity of a unit in the saliency map at a location (x,y) constrained to lie between 0 and 1, and r_{qxy} represents the activity of the single unit at that same retinotopic location (x,y) within one feature map. Therefore, the activity at a given position in a feature map is maximal if that given feature is present at that location and the attentional focus is coincident; otherwise, even strong activity in the feature map will be reduced to a range somewhere between the optimal response and the neuron's baseline firing rate. Since the response to unattended stimuli is attenuated rather than entirely suppressed, a monotonic gating function g is included:

$$g(a) = \lambda + (1 - \lambda)a^{\phi} \tag{14.5}$$

where λ is a weighting factor that determines the percentage of unattended information that will pass through and ϕ helps to remove the influence of weakly-responding units. The model assumes a rigid rather than a zoom-lens attentional spotlight. To address visual search for letters presented at one of four possible locations, Mozer's model simply employs only four output units, one for each quadrant of the retina. The output units of the neural network therefore serve as the "bottleneck" that limits how much information reaches the object recognition processors.

The model incorporates both exogenous (stimulus-evoked) and endogenous (cognitive learning, expectation or priming) inputs, much like that suggested by the transient-sustained model (Nakayama and Mackeben, 1989; Steinman and Steinman, 1998). The exogenous inputs are provided by retinotopically-organized connections from the feature maps to the saliency map such that the more features present at the location the greater the exogenous attentional effect, but the source of the endogenous attentional influences on the saliency map are not specified. A winner-take-all algorithm (Feldman and Ballard, 1982) selects the most active retinal quadrant. To help this process along, a looping connection is included from each output unit to itself. The net input net_{xy} to each output neuron at a given time t is given by:

$$net_{xy}(t) = ext_{xy}(t) + \alpha\, a_{xy}(t) - \beta \sum_{q,r \neq x,y} a_{qr}(t) \tag{14.6}$$

where ext_{xy} is an external attentional influence, either exogenous or endogenous, α is the weight of the loop connection, and β is the strength of lateral inhibition between neighboring output units. The update rule a_{xy} for the activity of each output unit is governed by the above factors and h, a linear threshold:

$$a_{xy}(t+1) = \tau h(net_{xy}(t)) + (1 - \tau)\, a_{xy}(t) \tag{14.7}$$

The output unit activity level (or response r) at any given time may be converted to a detection reaction time by assuming that the probability $P(r,t)$ of a response r at a given time t is:

$$P(r,t) = \frac{\exp(\xi e_r(t))}{\Sigma s \exp(\xi e_s(t))} \tag{14.8}$$

where $e_r(t)$ is the evidence for a response at that time and ξ is the constant factor that converts this evidence into a response strength. Once sufficient evidence for a response has been collected, a response is made at that time. A small constant is added to the reaction time to account for low-level visual processing and motor response times to bring the model's total reaction time closer to those seen in psychophysical studies.

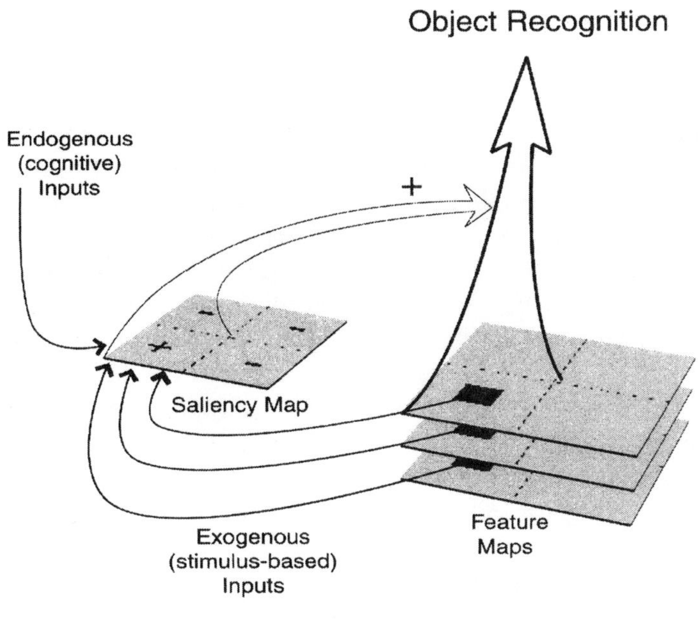

Figure 14.3. Mozer and Sitton's hybrid model is comprised of a set of feature maps and a saliency map that gates the output of the feature maps to the object recognition mechanism. The saliency map is controlled by either endogenous inputs from higher brain centers or stimulus-driven exogenous inputs from the feature maps themselves. Adapted from Mozer and Sitton (1998) with permission of Psychology Press Ltd.

Simulations of letter detection using this model demonstrate that precuing a stimulus location results in more rapid detection for a subsequently presented letter target, in agreement with experiments on transient attention

(Posner, 1980). In addition, precuing with a valid cue (one at the correct location) correctly produces faster reaction times than responses to invalid cues (at inappropriate locations). Unfortunately, the simulation only addresses visual reaction time rather than accuracy of detection, so the enhancing influence of the attentional focus on detection thresholds (Nakayama and Mackeben, 1989) cannot be directly determined. The focusing of attention on a particular location while attenuating activity elsewhere has been shown to improve detection thresholds within the attentional focus, with discrimination thresholds unaffected for the same set of stimuli (Steinman and McKee, 1988). Shiu and Pashler (1994) noted similar results in reaction time experiments. Mozer's model fails to reproduce the different effects on detection and discrimination. Therefore, we cannot extrapolate from the reaction time predictions of Mozer's model to detectability predictions or receptive field selectivity.

Even so, the model does predict reaction time quite well for a variety of tasks. Uncertainty about stimulus position results in prolonged reaction times for the model, thus mirroring the decreased detectability noted by Steinman and McKee (1988), presumably because the observer cannot predetermine where to place the attentional focus. The flanker effects of visual attention, in which conflicting information in neighboring targets has a detrimental effect on processing time, but compatible information in the flankers can enhance reaction time (Ericksen and Ericksen, 1974), are also accurately predicted by the model. Finally, by modifying the model slightly to allow for zoom-lens-like control of the size of the attentional focus, instantaneous shifts rather than analog "sliding" of attention from one locus to another can be correctly modeled (Ericksen and Murphy, 1987; Miyauchi et al., 1992).

Like Grossberg's FAÇADE model, Mozer and Sitton's model (1998) can be applied to simulations of a visual search task. These simulations, however, require a few additional assumptions beyond those discussed above. First, the targets and distractors must be known in advance. This assumption is relatively minor, as human observers learn the identity of the target feature and distractor features after just a few trials of a visual search experiment. Secondly, the control signals for attention are not fixed. Similarly, the width of the attentional spotlight and the criterion activity level corresponding to a response are adjusted. Such changes in weight may underlie learning effects in humans during visual search tasks (Steinman, 1987). Third, as in other computational models of attention, a feature saliency map is used to drive the attentional network exogenously. Fourth, the region of the display selected by attention is not fixed in size, as in the zoom-lens model, but instead depends upon display element density and arrangement. In some cases, several targets may be selected, each of which would then need to be examined by the object recognition network in series.

Finally, if the feature in question is not found within the attentional spotlight, attention is inhibited at that locus, and the network is reset to allow the attentional selection process to begin again.

The stimuli used in the simulation were of three possible configurations: a vertical bar, a horizontal bar, or a conjunction of the two forming a plus sign target. The target could appear in one of nine possible locations. The network was trained on 450 sample displays containing from one to nine targets. During the simulations, a single target was presented along with a variable number of distractors. A criterion threshold of 0.5 for a unit in the output layer was used to denote a detection response. One hundred cycles of the neural network model were allowed before triggering a "target absent" response. The low threshold and high number of cycles kept the false alarm rate near zero.

On a simple search task, for example, searching for a single vertical bar among horizontal distractor bars, the control signal weight for the discriminated feature increased from 0.8 to 1.0, thus driving attention more strongly and increasing the likelihood that attention would be drawn to the locus of the target. Simultaneously, the control signal weight for the distractors diminished from 0.8 to 0.0, thus making it less likely that the spotlight of attention would be placed on them. The modulation of attentional control signals in this model is critical to the successful achievement of a unique solution for the neural network, which suggests that attention is required even on a simple search task, as was noted psychophysically by Joseph et al. (1997). The network correctly predicts a constant reaction time as a function of the number of elements in the display.

During conjunction search, the model would predict that units responding to a conjunction would be activated twice as strongly as units responding to a single feature alone, as suggested by Wolfe's Guided Search Theory (Wolfe et al., 1989; Wolfe, 1994). However, an additive saliency response would lead to parallel search for conjunction stimuli. While this has been demonstrated for human subjects for some target features (Nakayama and Silverman, 1986) or following learning (Steinman, 1987), parallel search does not occur immediately for many other target features. In this regard, Mozer and Sitton's model exceeds human performance. To bring the model in line with human psychophysical data, an additional additive intrinsic noise was introduced to both the attention and recognition systems, which must be overcome to detect a conjunction target. A simple additive combination of activity from two feature maps may not be strong enough to overcome the noise level, and so serial inspection of each target would be required. With this additional proviso, the model's predictions of reaction time for conjunction search match human psychophysical data well. The additional time needed to locate the target when the number of distractors is increased could be explained by increasing interference from the unattended targets. A role for influence by unattended stimuli is not addressed in most of the other models of attention; even in Treisman's modified FIT model (Treisman, 1993), the role of distractors is simply that of being inhibited when attention is focussed on the target. When the influence of attention is removed from Mozer and Sitton's model, or when attention is allowed to be divided among several stimuli via a wide attentional focus, the error rates for the neural network increase for both target-present and target-absent conditions, such that the target cannot be located at all. In summary, this ambitious model does a commendable job of accounting for many of the known properties of simple and conjunctive visual search, spatial attention, and feature-based attention.

14.3.3 The Temporal Tagging or Cell-Gating Model

Rather than encode the attended stimulus on the basis of the spatial locus of a higher-level neuron, temporal tagging models assume that the awareness of a stimulus is encoded by a widespread network of neurons whose temporal response structure is modulated by a synchronous 30-50 Hz oscillation (Crick and Koch, 1990; Niebur et al., 1993). Such synchronous oscillations in firing rate have been noted in visual cortex of the cat (Gray and Singer, 1989) and monkey (Kreiter and Singer, 1992). Focal attention is therefore manifested not by which particular neuron fires, but instead is "tagged" by the presence of imposed synchronous oscillations in the firing rate of those neurons that do respond. The mean firing rates of these neurons, however, remains the same. The input of such models is the saliency map that encodes where in the visual field conspicuous objects lie using a winner-take-all mechanism to determine the most salient object. The mechanism by which the most salient stimulus is selected is not addressed; the model only addresses how this selection affects the responses of visual neurons.

Desimone (1992) assumed that mutual inhibition between cells in area V4 could bias processing towards neurons in an attended locus. The aim of temporal tagging models is to describe how the responses of single neurons in visual cortex are modified by attention (Moran and Desimone, 1985). Here, the mutual competition is between tagged and untagged neurons, leading to inhibition of the untagged neurons.

In Niebur et al.'s model, the input layer is an array of overlapping selective receptive fields in area V1, each represented by a 10 by 10 pixel array, and under the influence of a subcortical saliency map presumed to be located in the superior colliculus, pulvinar (LaBerge and Buchsbaum, 1990; Robinson and Petersen, 1992) or represented in multiple subcortical regions (see Fig. 14.4). Tagged neurons in V1 are subjected to synchronous temporal oscillations if they lie within the attentional focus. The action potentials of these units are created by a Poisson process with a 2-5 ms refractory period. Attention modulates the discharge rate of the V1 neurons via an inhomogeneous Poisson process whose mean rate varies over time:

$$\lambda(t) = \lambda_0 (1 + A \sin(\omega t + \phi)) + \lambda_{spont} \tag{14.9}$$

where $\lambda(t)$ is the mean firing rate, λ_0 is the stimulus-dependent response dependent upon the overlap between the stimulus and the receptive field, λ_{spont} is the spontaneous firing rate, ϕ represents a time-independent phase shift, and $\omega/2\pi$ is the modulating frequency. A is a factor proportional to the spatial overlap between the neuron in question with the focus of attention, such that the greater the separation, the less the amplitude of oscillation.

Each group of 100 V1 units projects to both an excitatory pyramidal cell and frequency-specific (band-pass filtering) inhibitory interneurons within a column in V4. The interneurons detect the temporal oscillations of the "attended neurons" and inhibit neighboring neurons at non-attended loci and overlapping neurons with non-attended stimulus features. The inhibitory interneurons are described as having a quasi-active cell membrane that acts as a three-branch RLC circuit, which creates a resonance frequency of 67 Hz. Spikes are produced in the interneurons if the intracellular voltage exceeds a time-varying threshold Θ_e = 10mV, determined in part by a 2-5 ms refractory period. The time-course of activity in the pyramidal cells of V4 resulting from excitatory inputs is represented by a decaying exponential:

$$\frac{d\rho_e(t)}{dt} = -\frac{\rho_e(t)}{\tau_{exc}} + W_e \sum_j \delta(t - t_j) + \frac{W_e}{4} \sum_k \delta(t - t_k) \tag{14.10}$$

where τ_{exc} is an excitatory synaptic time constant of 1.5 ms, We is a synaptic weight factor of 0.025, t_j are the spike arrival times from excitatory V1 input cells with the same stimulus specificity as the V4 cell in question, and t_k are the spike arrival times from excitatory V1 input cells with stimulus specificity differing from that of the V4 cell. The contribution from inhibitory inputs to the V4 cell is also modeled as a decaying exponential, but with a longer time constant (τ_{inh}=40 ms) and a synaptic weight (W_i) of 40 to account for the relatively fewer number of inhibitory V1 inputs:

$$\frac{d\rho_i(t)}{dt} = -\frac{\rho_i(t)}{\tau_{inh}} + W_i \sum_l \delta(t - t_l) \tag{14.11}$$

Plots of the predicted neuronal activity resulting from this model suggest that V1 neurons would exhibit no change in mean firing rate when inside or outside of the attentional focus, but a modulation of their frequency of firing that would not be evident from examining spike trains alone. Conversely, neurons in V4 would show vast differences in activity as a result of attention. Neurons within the attentional focus would respond strongly, while inhibitory connections to non-attended neurons would nearly eliminate activity in those units. The net effect is a constriction of the receptive fields of V4 units around the attended stimulus feature. The activity in the surrounding region is unaffected. Such findings, although only for single neurons, are in agreement with the center-surround model of attention, in which manipulation of stimulus features results in changes in perceptive field center strength and width, and not in the relative influence of the surround (Steinman et al., 1995, 1997b; Steinman and Steinman, 1997). The temporal tagging model can also be used to model shifts of attention simply by extinguishing the frequency modulation at one locus and initiating it elsewhere. The relocation of the attentional focus leads to a change in V4 neuronal firing rates within 30 ms, with the previously attended stimulus now being suppressed.

The output of Niebur et al.'s (1993) neural network is single-unit spike trains in areas V1 and V4 that match well the single-unit recordings of Moran and Desimone (1985). Attention is shown to have an obvious effect on the spike trains of V4, while producing changes in V1 that could only be revealed by frequency analysis of the spike train. In addition, the model provides a function for the 40 Hz modulation noted in the spike trains of visual neurons. It also provides for distributed processing of visual attention, as neuronal tagging may occur in several cortical areas, and for a role of the pulvinar in selecting the attentional focus, either in terms of locus in the visual field or saliency in a feature map. Such a distributed scheme would be needed to implement Treisman and Gelade's Feature Integration Theory.

One problem with both the shifter circuit and temporal tagging models is that they assume both early selection alone and bottom-up control of a cognitively-selected target arising from the pulvinar. However, it is sustained attention that reflects cognitive input to attention (Hikosaka et al., 1993a), via top-down control. Steinman and Steinman (1998) recognized the rich reciprocal connections between the pulvinar and several visual cortical areas (and noted previously by LaBerge, 1995) and postulated that the pulvinar could act as an intermediate controller for *both* transient and sustained attention, although this would not be to the exclusion of simultaneous inputs from other brain regions such as the superior colliculus and posterior parietal lobe.

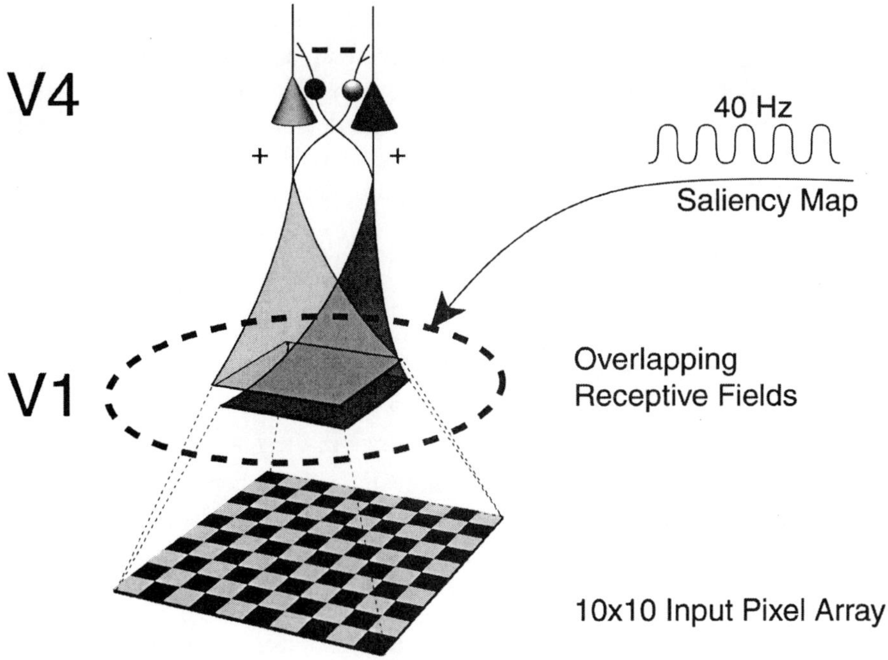

Figure 14.4. Temporal tagging model. Overlapping receptive fields in area V1, each tuned to different stimulus features, project to pyramidal cells in V4. A saliency map determines the placement of the focus of attention, indicated by the dotted circle. V1 neurons within the focus of attention are "tagged" with a synchronous oscillation in firing rate. Inhibitory interneurons in V4 are selectively tuned to this oscillation frequency. The net result is that a tagged V1 neuron responding to an attended feature will excite its corresponding V4 neuron, while inhibiting untagged V4 neurons that encode unattended locations and features. Adapted from Niebur et al. (1993) with permission of Elsevier Science.

14.3.4 The Selective Tuning Model

Tsotsos' (1993) selective tuning model attempts to account for both location-selective and feature-specific attention via the inhibition of connections from non-attended loci or stimulus features. This model, like other neural network models, starts with a parallel set of image processing receptive fields at the lowest layer and uses a winner-take-all algorithm to select the most salient location and/or feature in the visual field. However, the model uses a modified winner-take-all algorithm, and this spotlight-like selection process occurs in the three middle interpretive layers and the upper output layer of the neural network. At that point, a second *inhibitory* process proceeds in a recursive top-down fashion back via gating and bias units to the lower layers to suppress information from other receptive fields that do not lie upon the direct path from the winning lowermost unit to its upper layer counterpart (Tsotsos, 1993). Rather than a simple winner-take-all algorithm alone, this modified winner-take-all algorithm results in active selection of the most salient unit coupled with decreased weight of the other inputs, without attenuation of the response of the winner as occurs in the Koch and Ullman (1985) algorithm. The inhibitory interactions are given by:

$$s_i' = R\left[s_i - \frac{1}{\sum_{j\in M, j\neq i} w_{ij}}\left(\sum_{j\in M, j\neq i} w_{ij} R\left[s_j - s_i\right]\right)\right] \tag{14.12}$$

where R is a half-wave rectifying function, w_{ij} is a competitive weighting function between 0 and 1, s_i is the response prior to the winner-take-all process and M represents responsive locations in the feature map. The most active neuron inhibits its neighbors more than the reverse. This function, unlike Koch and Ullman's, is guaranteed to converge upon a winner without oscillation. The contribution of the top-down gating function of the winner-take-all process is given by $\Delta_{i,j} = G_{l,k,i}^{t-1} - G_{l,k,j}^{t-1}$ when $0 < \theta < G_{l,k,i}^{t-1} - G_{l,k,j}^{t-1}$ and $\Delta_{i,j} = 0$, otherwise, where $G_{l,k,i}^{t}$ is the response of the gating unit at time t and θ is a threshold equal to:

$$\theta = \frac{Z}{2^{\gamma} + 1} \tag{14.13}$$

with Z being its ceiling value. The top-down bias function is defined as:

$$b_{l,k} = \min_{\hat{a}\in B_{l+1,k}} \{a\} \tag{14.14}$$

where b is the bias for a given interpretive unit in assembly k within layer l, and $B_{l+1,k}$ is the set of bias units in layer l+1 that feed back to the interpretive unit. The net result when considering spatial attention is a center-surround attentional receptive field with both localized excitation and more widespread inhibition which increases the signal-to-noise ratio of the attended stimulus with respect to non-attended stimuli. Such a neural arrangement was postulated by Steinman et al. (1995), but the inhibitory surround of Tsotsos' model is more localized than that found psychophysically by Steinman and co-workers. Such size mismatches between neurophysiological receptive fields and psychophysical perceptive fields may be due to the integration of several overlapping receptive fields in a single perceptive field (Fiorentini et al., 1990). The sequential timing between the winner-take-all process and the recursive inhibition would predict a rapid rise of facilitation by attention with a subsequent slower decrease in effectiveness of attention as the inputs to the higher-level unit are pruned. The sustained-transient model of attention predicts this response pattern (Nakayama and Mackeben, 1989); although it has been demonstrated for transient attention (Nakayama and Mackeben, 1989; Steinman et al., 1995, 1997b), insufficient data exists at long cue-target asynchronies to confirm a similar pattern for sustained attention. It should also be noted that the sequential bottom-up then top-down processes of this model are in agreement with the transient-sustained model (Hikosaka et al., 1993b; Steinman and Steinman, 1998a).

The selective tuning model of Tsotso and co-workers is unique in that it is the only model that incorporates not just a pyramidal neural network, but also a variety of types of neurons within the network. The input layer consists of interpretive units that are equivalent to the typical center-surround receptive fields of early visual cortex. The winner-take-all process is implemented by gating neurons, while gating control neurons provide the top-down construction of the spatial inhibitory surround. Finally, bias units allow for feature-specific top-down multiplicative attentional inhibition. A grouping of each of these units forms the basic assembly of the model. Therefore, although the neural network only contains three major layers, the addition of the gating, gating control and bias neurons introduce intervening sublayers to the network. Even so, the model still requires relatively fewer connections than the shifter circuit model.

14.4 RELATION OF THE MODELS TO CURRENT STATE OF KNOWLEDGE IN NEUROPHYSIOLOGY

14.4.1 Single-Unit Activity

The effects of attention are demonstrated neurophysiologically by a change in the firing rate or selectivity of visual neurons. When required to attend to a target with a particular feature, such as orientation, neurons in area V4 of the monkey maintain a higher firing rate and exhibit narrower tuning for the attended stimulus feature (Spitzer et al., 1988). The size of the receptive field may be altered by attention, that is, neuronal receptive fields decrease in size when attention is directed to its corresponding locus in space (Wurtz et al., 1984; Desimone et al., 1990). In addition, Moran and Desimone (1985) demonstrated that neurons selective for an unattended stimulus feature are inhibited. Direct evidence for attentional influence on the visual system comes from single-unit studies of visual cortex which demonstrate that visual attention can directly alter the responses of visual neurons. The effects are of two types: feature-specific and location-specific. Neurons in V4 of the form-processing pathway respond more to a precued orientation than to an uncued one; that is, the activation of attention serves to enhance the detection of subsequent targets of the same orientation (Maunsell et al., 1991). Neurons in striate cortex (V1), on the other hand, respond best to a specific orientation whether or not it is attended (Hubel and Wiesel, 1974). Similarly, neurons in area 7A of the motion-processing pathway respond better to a cued direction of motion, as do neurons in area MT (Maunsell and Ferrara, 1995). In both cases, attention improves detection and discrimination. What is more surprising is that attention can make neurons that normally do not respond to a particular stimulus feature responsive. For example, cells in cortical area V4 do not typically respond to motion, yet they will exhibit increases in firing rate in response to an attended direction of motion (Maunsell and Ferrara, 1995); simultaneously, other neurons in V4 are inhibited in their processing of unattended stimulus features. Spatial attention also directly affects neuronal responsivity. Neurons whose receptive fields fall within an attended locus may be activated more rigorously and their tuning may be enhanced (Spitzer et al., 1988), while receptive fields at unattended loci may be inhibited (Moran and Desimone, 1985). In addition, spontaneous firing rates in areas V2 and V4 are 30 to 40% higher when attention is directed inside the receptive field than when it is directed outside of it, regardless of the presence or absence of a stimulus within the receptive field (Luck et al., 1997). All of these findings suggest that neural signals are biased in favor of an attended locus. Such findings are in agreement with the center-surround model of attention (Steinman et al., 1995) in which attention accentuates visual processing at the attentional focus and inhibits it elsewhere.

The temporal tagging model represents the single-unit recordings obtained by Moran and Desimone (1985) quite well, with observable differences in the spike trains of V4 units as a function of the location of the attentional focus, but not of V1 units. However, it must be noted that subsequent neurophysiological experiments have demonstrated that attentional influences can in fact be demonstrated as early as visual area V1 and the neighboring area V2 (Motter, 1993). There are two possible reasons for this discrepancy. The first is that Moran and Desimone's study investigated feature-specific attention, while Motter's examined spatially selective attention. In addition, Motter used much different interstimulus spatial separations than did Moran and Desimone, which may have resulted in differing degrees of inhibitory competition between neighboring neurons.

It is well established that the pulvinar plays a role in visual attention (Petersen et al., 1985, 1987). Pharmacological studies by Petersen et al. (1987) demonstrate that inhibition of lateral pulvinar neurons decreases the ability to focus attention onto a target, while exciting the same neurons enhanced the ability to orient attention. In addition, GABA agonist injections in the pulvinar decrease the ability to discriminate color or form in the presence of distracting stimuli. LaBerge's (1992) model of lateral inhibition between principal cells of the pulvinar provides a mechanism by which the pulvinar helps to direct the focus of attention onto a target and filter out distracting information. Such a filtering mechanism is critical to the modified Feature Integration Theory (Treisman and Sato, 1990; Treisman, 1993), the center-surround model (Steinman et al., 1995), Mozer's hybrid model (Mozer, 1991; Mozer and Sitton, 1998), and the temporal tagging model (Niebur et al., 1993).

14.4.2 Neuroanatomical Networks

While there is evidence for attentional modulation of neurophysiological activity from single-unit recordings, the search for the mechanisms of attention has been more elusive. It is thought that selective visual attention is mediated by large networks of neurons rather than individual neurons, but little is known about the activity of neural networks during selective visual attention. Studies of alert animals (Moran and Desimone, 1985; Colby, 1991), lesioned animals (Carli et al., 1983), brain-injured human patients (Rafal and Posner, 1987; Rafal et al., 1987) and the recent advent of PET scans (LaBerge and Buchsbaum, 1990; Corbetta et al., 1993) have added to our knowledge of the neuroanatomical loci and physiological responses of the neural networks affected by or involved in attention. However, a fundamental question that remains unanswered is whether attention is an active process requiring a separate neural system (Posner and Petersen, 1990) or is simply an emergent property arising from lateral inhibitory processes in sensory processing (Mesulam, 1990; Desimone and Duncan, 1995). Posner assumes that cognitive input is required to orient the attentional focus, since his largely anatomical model is characterized by rich interconnections between several cortical and subcortical brain regions, several of which are involved to some extent in cognition (see Fig. 14.5). For example, shifting attention, a process that involves disengaging, moving, then re-engaging the focus of attention, requires three neural areas to accomplish its task—the posterior parietal lobe, the superior colliculus, and the pulvinar, respectively, as demonstrated in patients with brain lesions (Posner et al., 1988). The parietal lobe is considered the main cortical area involved in visual attention because damage to the parietal region from strokes or injury severely limits the ability to attend to targets (Posner and Cohen, 1984; Husain, 1991). Patients with posterior parietal cortex (PPC) lesions are unable to accurately locate objects, even though they can identify them (Holmes, 1918). This difficulty in locating objects may be attributed in part to the overall role of the parietal cortex in spatial localization, with attention serving as the mediator of the transformation between sensory and motor maps (Stein, 1992); patients with parietal lobe lesions exhibit a shift in the reference point for egocentric localization (Karnath, 1997). The superior colliculus may form the link between shifting attention and oculomotor activity (Rafal et al., 1988). The pulvinar, a thalamic structure neighboring the dLGN, is important for restricting attention to a new locus and filtering out irrelevant locations (LaBerge and Buchsbaum, 1990). Finally, the frontal lobes determine the trade-off between how much of the visual field we attend versus how much we ignore. Damage to the frontal lobes affects this balance. Posner's model may be supported by this widespread involvement of several cortical regions. However, a similar neuroanatomical model, involving nearly identical brain regions as Posner's model, was proposed by Mesulam (1990) that does not require the assumption of top-down cognitive input. Both Posner's and Mesulam's models simply note the anatomical areas and pathways involved in attention; they fail to elucidate the mechanisms by which attention is allocated. Desimone and Duncan (1995) also argue against a distinct attentional system. Such a view fits better with parallel distributed processing and neural network models of visual attention that do not assume a distinct attentional system.

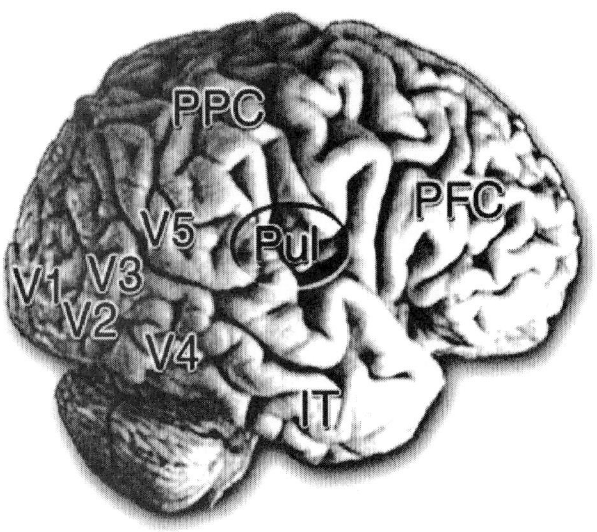

Figure 14.5. Brain areas involved in attention. V1 = striate cortex, V2-V5 (MT) = extrastriate visual cortical areas, IT = inferotemporal cortex, PPC = posterior parietal cortex, PFC = prefrontal cortex, Pul = pulvinar (in thalamus). The superior colliculus, within the brainstem, is not shown.

Most neural network models assume bottom-up processing only, although the human attention system exhibits both bottom-up and top-down processing. Only the transient-sustained (Steinman and Steinman, 1998) and selective tuning (Tsotsos, 1993) models incorporate top-down attentional selection. In addition, each model presumes a different location for the saliency map. The shifter circuit model places the saliency map in the posterior parietal cortex, the temporal tagging model assumes it lies in the pulvinar and superior colliculus, and the selective tuning model places the saliency system internal to the neural network itself; that is, attention is an emergent property of the model. To date, the anatomical location of the saliency map has not been determined. Although damage to any of these areas affects different aspects of attention, the underlying circuitry and representation of a saliency map is still a mystery. It is also possible that separate saliency maps exist for different subtypes of attention. For example, Braun and Sagi (1990, 1991) suggest that only bottom-up attention is governed by a retinotopic saliency map, whereas top-down attention is governed by a separate mechanism with an egocentric coordinate system. The widespread neural networks devoted to attention might suggest that multiple saliency maps might exist (Robinson and Petersen, 1992; Mountcastle et al., 1981; Goldberg and Wurtz, 1972). Alternatively, as proposed by Desimone and Duncan (1995), a separate saliency map might not even be necessary for attention, arguing that interactions between feature maps could by themselves result in selective attention. Finally, each of the modern neural network models of attention relies upon the presumption of specific types of interconnection circuits or interneurons. Shifter circuits require a specific pattern of spacing of neural units at various stages of processing. Temporal tagging must be implemented by frequency-specific inhibitory interneurons, while the selective tuning model requires several types of gating interneurons to restrict the number of inputs to the higher layers. It has been demonstrated that each of these schemes could, in theory, work. However, direct neurophysiological evidence for such circuits is lacking, and even if such circuits exist in the visual system, it does not necessarily follow that they are in fact part of an attentional mechanism rather than serving some other function.

14.5 BASIC AND CLINICAL IMPLICATIONS OF THE MODELS, AND FUTURE DIRECTIONS

To date, there is no inclusive, comprehensive model of visual attention. Although several models of attention exist, they have either simply described the brain areas or processes involved in attention, focused on local cell assemblies, or attempted to explain circumscribed sets of experimental data. For example, the spotlight, zoom lens, and center-surround models explain the benefits and costs of attention at various locations in the field, but do not explain the mechanisms that produce the center-surround perceptive field. The transient-sustained and Feature Integration Theory models also suffer from this same problem.

While neural network models of attention can explain subsets of psychophysical and neurophysiological data, they do not yet offer a physiologically-sound simulation of the widespread neural circuitry known from medical imaging and brain lesion patients to be involved in the attentional process. Lesions in the brain structures subserving attention, whether due to vascular accidents, injury or tumors, can lead to profound attentional abnormalities. The nature of the abnormality is dependent upon which brain area is affected. One of the most pronounced attentional deficits, which arises from posterior parietal lobe damage, more often to the right parietal lobe, is the visual neglect syndrome (Holmes, 1918; Posner and Cohen, 1984; Hussain, 1991; Rafal and Robertson, 1985). In visual neglect, the patient fails to notice a visual stimulus in a complex or natural environment; neglect is typically restricted to one hemifield. If stimuli are presented to both hemifields simultaneously, only the stimulus in the unaffected hemifield will be noticed, while the other stimulus undergoes extinction and fails to reach consciousness. These effects occur even though with the presentation of single stimuli during visual field testing, visual detection is not impaired in the neglected hemifield. Similarly, if subjects are presented with an attentional cue in the unaffected hemifield, then required to detect a stimulus in the affected hemifield, reaction times are abnormally prolonged, which suggests a role for the parietal lobe in disengagement of attention (Posner et al., 1984). Extinction is thought to involve mostly sustained, voluntary attention, since instructing the subject to disregard stimuli in the unaffected hemifield reduces the extinction effect (De Pellegrino and De Renzi, 1995). With bilateral parietal and occipital lobe damage Balint's syndrome occurs, in which divided attention to multiple objects is diminished; attention can only be directed to single objects even when the objects are in close proximity (Humphreys and Riddoch, 1992). Balint's syndrome therefore reflects a problem with object-based rather than location-based attention. Injuries to the pulvinar affect the ability to engage attention in response to a novel stimulus (Rafal and Posner, 1987). This is in keeping with LaBerge and Brown's (1992) model of pulvinar neural circuitry. If a target is not easily distinguished from a distractor in terms of its salience, difficulties in orienting to that target would be expected. Lesions to the pulvinar might affect the saliency difference-enhancing role of the pulvinar. It is not surprising that Posner's and Mesulam's models of attention are supported by studies of patients with brain damage, since these models arose partly in response to data obtained on such patients. While there have been some attempts to test the tenets of Feature Integration Theory with neglect symdrome (Arguin et al., 1993; Humphreys and Riddoch, 1992) and Balint's syndrome (Rafal and Robertson, 1995) patients, the results of these studies have been inconclusive.

Alzheimer's disease may also affect the posterior parietal cortex, thereby interfering with selective spatial attention even in the early stages of the disorder (Parasuraman et al., 1992; Parasuraman and Haxby, 1993). Sensory functions remain intact. The valid versus invalid cueing effects of Posner (1980) are preserved in Alzheimer's disease (Parasuraman et al., 1992), which suggests that transient visual attention is intact. Conversely, voluntary sustained attention was impaired. The transient-sustained model (Nakayama and Mackeben, 1989; Steinman and Steinman, 1998) would predict these results because Alzheimer's disease affects top-down cognitive function rather than bottom-up covert orienting. The role of the posterior parietal cortex in the disengagement of attention (Posner et al., 1984) is also supported by difficulties in disengagement in early Alzheimer's disease (Oken et al., 1994). Some Alzheimer's patients also demonstrate difficulties in shifting attention or searching for targets. The thalamic nuclei of these patients exhibit less radiodensity than those of normal subjects (Forstl and Sahakian, 1993), findings that are consistent with the pulvinar neural circuit model (Laberge et al., 1992). It should be noted that Parkinson's disease, which also may result in dementia, affects sustained but not transient attention (Rafal et al., 1984). Selective deterioration of cognitive, top-down sustained attention in Alzheimer's disease is also consistent with the transient-sustained model (Kahnemann, 1973; Posner and Cohen, 1984; Nakayama and Mackeben, 1989; Hikosaka et al., 1993a; Steinman et al., 1997).

Attentional processing may be diminished even in the normal aging population (Ball et al., 1993). It has been suggested that only cognitive-based effortful attention is affected (Hasher and Zacks, 1979; Greenwood and Parasuraman, 1997); however, transient visual attentional loss in normal aging was confirmed by Steinman et al. (1994), who examined aging patients which were prescreened for the absence of Alzheimer's disease. The possible selective impairment of the magnocellular pathway in aging (Spear, 1993) and the predominance of the magnocellular pathway in transient attention (Steinman et al., 1997) may provide a basis for such attentional losses. It is the magnocellular component of transient visual attention that is degraded in normal aging, with the parvocellular component left unaffected (Steinman et al., 1994).

Young individuals can be affected by alterations in visual attention. This is evident in reading disability (Tarver and Hallihan, 1974; Keogh and Margolis, 1976; Conte, 1991). In these patients, the attentional focus is excessively constricted, and the inhibitory surround of the attentional perceptive field is more pronounced (Steinman et al., 1998). The accentuated inhibition is more extreme to the right of the attentional cue, and it may be one factor in the difficulty that reading disabled subjects have in planning saccadic eye movements when reading (Sereno, 1992; Breitmeyer and Ganz, 1976). In addition, the activation of transient visual attention decays more rapidly than in normal readers, with a prolonged period of profound inhibition of sustained attention, thereby preventing the disabled reader from fully processing the fixated text.

What we now know about the nature of attention deficits arising from brain lesions, Alzheimer's disease, normal aging, and reading disabilities has primarily come mostly from the application of models of visual attention. Despite decades of research and clinical findings, however, we still know very little about the mechanisms of attention. What is needed now is to take the next step in modeling – to combine the existing models of attention to yield a more comprehensive neural network simulation of attention. Such a model would contain: (1) the global neural assemblies that provide each subtype of visual attention (Koch and Ullman, 1985; Olshausen et al., 1993; Niebur et al., 1993; Tsotsos, 1993; Grossberg et al., 1994; Mozer and Sitton, 1998), (2) selective inputs from the magnocellular and parvocellular visual streams (Steinman et al., 1997), plus (3) the local neural circuitry of individual cortical and subcortical areas (LaBerge et al., 1992). This comprehensive model should then be applied to the body of knowledge on object perception (Treisman and Gelade, 1980; Treisman, 1993; Wolfe, 1994) and existing psychophysical and physiological data on attention (for example, Tsal, 1983; Posner and Cohen, 1984; Ericksen and Yeh, 1985; Nakayama and Mackeben, 1989; Steinman et al., 1995; Steinman and Steinman, 1998). Only then may the parameters of subsystems of attention be manipulated and "lesioned" to determine if simulations utilizing a model of attention mimics human spatial vision or the attentional losses seen in brain disorders and normal aging.

14.6 REFERENCES

Ahmad, S., 1991, *VISIT: An efficient computational model of human visual attention, ICSI Technical Report* vol. 91-049. International Computer Science Institute, Berkeley, CA.

Allman, J., Kaas, J. H., Lane, R. H., and Miezin, F. M., 1972, A representation of the visual field in the inferior nucleus of the pulvinar in the owl monkey (aotus trivirgatus), *Brain Res.* **40**: 291-302.

Allman, J., Miezin, F. M., and McGuinness, E., 1985, Stimulus specific responses from beyond the classical receptive field: neurophysiological mechanisms for local-global comparisons in visual neurons, *Ann. Rev. Neurosci.* **8**: 407-430.

Allport, D. A., 1980, Attention and performance, In: *Cognitive Psychology: New Directions.*, G. Claxton, ed., Routledge and Kegan Paul, London.

Anderson, C., and Essen, D. V., 1987, Shifter circuits: a computational strategy for dynamic aspects of visual processing, *Proc. Natl. Acad. Sci. , U.S.A.* **84**: 6297-6301.

Arguin, M., Joanette, Y., and Cavanagh, P., 1993, Visual search for feature and conjunction targets with an attentional deficit, *J. Cognit. Neurosci.* **5**: 436-452.

Ball, K., Owsley, C., Soane, M. E., Roenker, D. L., and Bruni, J. R., 1993, Visual attention problems as a predictor of vehicle crashes in older drivers, *Inv. Ophthalmol. Vis. Sci.* **34**: 3110-3123.

Baloch, A. A., and Grossberg, S., 1997, A neural model of high-level motion processing: line motion and formotion dynamics, *Vis. Res.* **37**: 3037-3059.

Bender, D. B., 1981, Retinotopic organization of the macaque pulvinar, *J. Neurophysiol.* **46**:672-693.

Bergen, J., and Julesz, B., 1983, Parallel versus serial processing in rapid pattern discrimination, *Nature.* **303**: 696-698.

Braun, J., and Sagi, D., 1990, Vision outside of the focus of attention, *Percept. Psychophys.* **48**: 45-58.

Braun, J., and Sagi, D., 1991, Texture-based tasks are little affected by a second task which requires peripheral or central attentive fixation, *Perception.* **303**: 45-58.

Bravo, M., and Blake, R., 1990, Preattentive vision and perceptual groups, *Perception.* **19**: 515-522.

Breitmeyer, B. G., and Ganz, L., 1976, Implications of sustained and transient channels for theories of visual pattern masking, saccadic suppression and information processing, *Psychol. Rev.* **83**: 1-36.

Breitmeyer, B. G., 1991, Reality and relevance of sustained and transient channels in reading and reading disability, In: *Oculomotor Control and Cognitive Processes.*, R. Schmidt and D. Zambarbieri, eds., Elsevier/North-Holland, Amsterdam.

Carli, M., Robbins, T. W., Evenden, J. L., and Everitt, B. J., 1985, Effects of lesions to ascending noradregenergic neurones on performance of a 5-choice serial reaction time task; implications for theories of dorsal noradregenergic bundle function based on selective attention and arousal, *Behav. Brain Res.* **9**: 361-380.

Carpenter, G. A., and Grossberg, S., 1991, *Pattern recognition by self-organizing neural networks.* MIT Press, Cambridge, MA.

Cavanagh, P., 1992, Attention-based motion perception, *Science.* **257**: 1563-1565.

Cohen, R. A., 1993, *The neuropsychology of attention.* Plenum Press, New York.

Colby, C. L., 1991, The neuroanatomy and neurophysiology of attention, *J. Child Neurol., Suppl.* **6**: S90-S118.

Connor, C. E., Gallant, J. L., and Van Essen, D., 1994, Modulation of receptive field properties in area V4 by shifts in focal attention, *Inv. Ophthalmol. Vis. Sci. Suppl.* **35**: 2147.

Conte, R., 1991, Attention disorders: reviews, In: *Learning about Learning Disabilities.*, B. Y. L. Wong, ed., Academic Press, San Diego, CA, pp. 60-101.

Corbetta, M., Miezin, F., Dobmeyer, S., Shulman, G., and Peterson, S., 1991, Selective and divided attention during visual discrimination of shape, color, speed: functional anatomy by positron emission tomography, *J. Neurosci.* **11**: 2383-2402.
Crick, F., 1984, The function of the thalamic reticular complex: the searchlight hypothesis, *Proc. Natl. Acad. Sci. , U.S.A.* **81**: 4586-4590.
Crick, F., and Koch, C., 1990, Towards a neurobiological theory of consciousness, *Seminars in the Neurosciences*. **2**: 263-275.
Desimone, R., Wessinger, M., Thomas, L., and Schneider, W., 1990, Attentional control of visual perception: cortical and subcortical mechanisms, In: *Cold Spring Harbor Symposium on Quantitative Biology*, Vol. 60.,, Cold Spring Harbor Laboratory Press, New York, pp. 963-971.
Desimone, R., 1992, Neural circuits for visual attention in the primate brain, In: *Neural Networks for Vision and Image Processing*., G. A. Carpenter and S. Grossberg, eds., MIT Press, Cambridge, MA.
Desimone, R., and Duncan, J., 1995, Neural mechanisms of selective attention, *Ann. Rev. Neurosci.* **18**: 193-222.
DeValois, R. L., Albrecht, D. G., and Thorell, L. G., 1982, Spatial frequency selectivity of cells in macaque visual cortex, *Vis. Res.* **22**: 545-559.
Downing, C. J., and Pinker, S., 1985, The spatial structure of visual attention, In: *Attention and Performance*, Vol. XI., M. I. Posner and O. S. M. Marin, eds., Lawrence Erlbaum Associates Inc., Hillsdale, NJ, pp. 171-187.
Downing, C. J., 1988, Expectancy and visual-spatial attention: effects on perceptual quality, *Journal of Experimental Psychology and Human Perception*. **14**: 188-202.
Enroth-Cugell, C., and Robson, D. G., 1966, The contrast sensitivity of retinal ganglion cells of the cat, *J. Physiol.* **187**: 517-552.
Ericksen, C. W., and Collins, J. F., 1969, Temporal course of selective attention, *Journal of Experimental Psychology and Human Perception*. **80**: 254-261.
Ericksen, C. W., and Hoffman, J. E., 1973, Selective attention: noise suppression or signal enhancement?, *Bull. Psychonom. Soc.* **4**: 587-589.
Ericksen, B. A., and Ericksen, C. W., 1974, Effect of noise letters upon identification of a target letter in a nonsearch task, *Percept. Psychophys.* **16**: 143-149.
Ericksen, C. W., and Yeh, Y., 1985, Allocation of attention in the visual field, *Journal of Experimental Psychology and Human Perceptual Performance*. **11**: 583-597.
Ericksen, C. W., and James, J. D. S., 1986, Visual attention within and around the field of focal attention: a zoom lens model, *Percept. Psychophys.* **40**: 225-240.
Ericksen, C. W., and Murphy, T. D., 1987, Movement of the attentional focus across the visual field: a critical look at the evidence, *Percept. Psychophys.* **42**: 299-305.
Feldman, J., and Ballard, D., 1982, Connectionist models and their properties, *Cog. Sci.* **6**: 205-254.
Fiorentini, A., Baumgartner, G., Magnussen, S., Schiller, P. H., and Thomas, J. P., 1990, The perception of brightness and darkness: relations to neuronal receptive fields, In: *Visual Perception: The Neurophysiological Foundations*., L. Spillmann and J. S. Werner, eds., Academic Press, San Diego, CA, pp. 129-161.
Fischer, B., and Weber, H., 1993, Express saccades and visual attention, *Behav. Brain Sci.* **16**: 553-567.
Fortstl, H., and Sahakian, B. J., 1993, Thalamic radiodensity and cognitive performance in mild and moderate dementia of the Alzheimer type, *J. Psychiatr. Neurosci.* **18**: 33-37.
Goldberg, M. E., and Wurtz, R. H., 1972, Activity of superior colliculus in behaving monkey II. The effect of attention on neuronal responses, *J. Neurophysiol.* **35**: 560-574.
Gray, C. M., and Singer, W., 1989, Stimulus-specific neuronal oscillations in orientation columns of cat visual cortex, *Proc. Natl. Acad. Sci. , U.S.A.* **86**: 1698-1702.
Greenwood, P. M., and Parasuraman, R., 1997, Attention in aging and Alzheimers disease: behavior and neural systems, In: *Attention, Development and Psychopathology*., J. Enns

and J. Burack, eds., Guilford Press, New York.
Gross, C. G., 1973, Inferotemporal cortex and vision, *Physiol. Psychol.* **5**: 77-115.
Grossberg, S., 1987a, Cortical dynamics of three-dimensional form, color and brightness perception: I. monocular theory, *Percept. Psychophys.* **41**: 87-116.
Grossberg, S., 1987b, Cortical dynamics of three-dimensional form color and brightness perception: II. binocular theory, *Percept. Psychophys.* **41**: 117-158.
Grossberg, S., 1988, *Neural networks and natural intelligence*. MIT Press, Cambridge, MA.
Grossberg, S., Mingolla, E., and Ross, W. D., 1994, A neural theory of attentive visual search: interactions of boundary, surface, spatial and object representations, *Psychol. Rev.* **101**: 470-489.
Hasher, L., and Zacks, R. T., 1979, Automatic and effortful processes in memory, *Journal of Experimental Psychology General*. **108**: 356-388.
He, Z. J., and Nakayama, K., 1992, Surface vs. features in visual search, *Nature*. **359**: 231-233.
Helmholtz, H., 1867, *Handbuch der Physiologischen Optik*. L Voss, Hamburg.
Hikosaka, O., Miyauchi, S., and Shimojo, S., 1993a, Voluntary and stimulus-induced attention detected as motion sensation, *Perception*. **22**: 517-526.
Hikosaka, O., Miyauchi, S., and Shimojo, S., 1993b, Focal visual attention produces illusory temporal order and motion sensation, *Vis. Res.* **33**: 1219-1240.
Hinton, G. E., 1981, Implementing semantic networks in parallel hardware, In: *Parallel Models of Associative Memory*., G. E. Hinton and J. A. Anderson, eds., Lawrence Erlbaum Associates Inc., Hillsdale, NJ, pp. 161-188.
Holmes, G., 1918, Disorders of visual orientation, *Brit. J. Ophthalmol.* **2**:449-468, 506-516.
Hubel, D., and Wiesel, T. N., 1974, Sequence regularity and geometry of orientation columns in the monkey striate cortex, *J. Comp. Neurol.* **158**: 267-294.
Humphreys, G. W., and Riddoch, M. J., 1992, Interactions between objects and space-vision revealed through neuropsychology, In: *Attention and Performance*, Vol. XIV., D. E. Meyers and S. Kornblum, eds., Lawrence Erlbaum Associates Inc., Hillsdale, NJ, pp. 143-162.
Hung, G. K., Wilder, J., Curry, R., and Julesz, B., 1995, Simultaneous better than sequential for brief presentations, *J. Opt. Soc. Am.* **12**: 441-449.
Husain, M., 1991, Visuospatial and visuomotor functions of the posterior parietal lobe, In: *Vision and Visual Dyslexia*, Vol. 13., J. F. Stein, ed., MacMillan Reference Ltd., London.
Jackson, S. R., Marrocco, R., and Posner, M. I., 1994, Networks of anatomical areas controlling visuospatial attention, *Neural Networks*. **7**: 925-944.
James, W., 1890, *Principles of Psychology*. Holt, New York.
Jonides, J., and Yantis, S., 1988, Uniqueness of abrupt visual onset in capturing attention, *Percept. Psychophys.* **43**: 346-354.
Joseph, J. S., Chun, M. M., and Nakayama, K., 1997, Attentional requirements in a "preattentive" feature search task, *Nature*. **387**: 805-807.
Julesz, B., 1984, Toward an axiomatic theory of preattentive vision, In: *Dynamic Aspects of Neocortical Function*., G. M. Edelman W. E. Gall and W. M. Cowan, eds., Neurosciences Research Foundation, New York.
Julesz, B., 1986, Texton gradients: the texton theory revisited, *Biol. Cybern.* **54**: 245-251.
Jung, R., and Spillmann, L., 1970, Receptive field estimation and perceptual integration in human vision, In: *Early Experience and Visual Information Processing in Perceptual and Reading Disorders*., F. A. Young and D. B. Lindsley, eds., National Academy Press, Washington, D.C.
Kahneman, D., 1973, *Attention and Effort*. Prentice-Hall, Englewood Cliffs, NJ.
Kahneman, D., and Henik, A., 1981, Perceptual organization and attention, In: *Perceptual Organization*., M. Kubovy and J. R. Pomerantz, eds., Lawrence Erlbaum Associates Inc., Hillsdale, NJ.
Karnath, H. O., 1997, Neural encoding of space in egocentric coordinates? Evidence for and

limits of a hypothesis derived from patients with parietal lesions and neglect, In: *Parietal Lobe Contributions to Orientation in 3-D Space.*, P. Their and H. O. Karnath, eds., Springer, Heidelberg, pp. 497-520.
Keogh, B., and Margolis, J., 1976, Learn to labor and wait: attentional problems of children with learning disorders, *J. Learing. Disability.* **9**: 276-286.
Khurana, B., and Kowler, E., 1987, Shared attentional control of smooth eye movement and perception, *Vis. Res.* **27**: 1603-1618.
Kleffner, D. A., and Ramachandran, V. S., 1992, On the perception of shape from shading, *Percept. Psychophys.* **52**: 18-36.
Koch, C., and Ullman, S., 1985, Shifts in selective visual attention: towards the underlying neural circuitry, *Hum. Neurobiol.* **4**: 219-227.
Kreiter, A. K., and Singer, W., 1992, Oscillatory neuronal responses in the visual cortex of the awake macaque monkey, *European J. Neurosci.* **4**: 369-375.
Kröse, B., and Julesz, B., 1989, The control and speed of shifts of attention, *Vis. Res.* **29**: 1607-1619.
Kuffler, S. W., 1953, Discharge patterns and functional organization of mammalian retina, *J. Neurophysiol.* **16**: 37-68.
LaBerge, D., and Brown, V., 1989, Theory of attentional operations in shape identification, *Psychol. Rev.* **96**: 101-124.
LaBerge, D., and Buchsbaum, M. S., 1990, Positron emission tomographic measurements of pulvinar activity during an attentional task, *J. Neurosci.* **10**: 613-619.
LaBerge, D., Carter, M., and Brown, V., 1992, A network simulation of thalamic network operations in selective attention, *Neural Comp.* **4**: 318-331.
LaBerge, D., 1995, Computational and anatomical models of selective attention in object identification, In: *The Cognitive Neurosciences.*, M. S. Gazzaniga, ed., MIT Press, Cambridge, MA.
Luck, S. J., Hillyard, L., and Desimone, R., 1997, Neural mechanisms of spatial selective attention in areas V1, V2 and V4 of macaque visual cortex, *J. Neurophysiol.* **77**: 24-42.
Mackeben, M., and Nakayama, K., 1993, Express attentional shifts, *Vis. Res.* **33**: 85-90.
Mangun, G. R., and Hillyard, S. A., 1991, Modulations of sensory-evoked brain potentials during visual-spatial priming, *Journal of Experimental Psychology Human Perceptual Performance.* **17**: 1057-1074.
Marcelja, S., 1980, Mathematical description of the responses of simple cortical cells, *J. Opt. Soc. Am.* **70**: 1297-1300.
Maunsell, J. H., Sklar, G., Nealy, T. A., and De Priest, D. D., 1991, Extraretinal representations in area V4 in the macaque monkey, *Vis. Neurosci.* **7**: 561-573.
Maunsell, J. H., and Ferrara, V. P., 1995, Attentional mechanisms in visual cortex, In: *The Cognitive Neurosciences.*, M. S. Gazzaniga, ed., MIT Press, Cambridge, MA.
Meslam, M. M., 1990, Large-scale neurocognitive networks and distributed processing for attention, language and memory, *Ann. Neurol.* **28**: 597-613.
Miyauchi, S., Hikosaka, O., and Shimojo, S., 1992, Visual attention can be assessed by illusory line motion sensation, *Inv. Ophthalmol. Vis. Sci. Suppl.* **33**: 1262.
Moran, J., and Desimone, R., 1985, Selective attention gates visual processing in the extrastriate cortex, *Science.* **229**: 782-784.
Morrow, L. A., and Ratcliff, G. C., 1988, The disengagement of covert attention and the neglect syndrome, *Psychobiology.* **16**: 261-269.
Motter, B. C., and Poggio, G. F., 1982, Spatial invariance of receptive field location in the presence of eye movements of fixation for neurons in monkey's striate cortex, *Soc. Neurosci. Abstr.* **8**: 707.
Motter, B. C., 1993, Focal attention produces spatially selective processing in visual cortical areas V1, V2 and V4 in the presence of competing stimuli, *J. Neurophysiol.* **70**: 909-919.
Mountcastle, V., Andersen, R., and Motter, B. C., 1981, The influence of attentive fixation upon the excitability of the light-sensitive neurons of the posterior parietal cortex, *J.*

Neurosci. **1**: 2118-1232.

Mozer, M. C., 1991, *The Perception of Multiple Objects: A Connectionist Approach*. MIT Press, Cambridge, MA.

Mozer, M. C., and Sitton, M., 1998, Computational modelling of spatial attention, In: *Attention*., H. Pashler, ed., Psychology Press, East Sussex, England.

Nakayama, K., and Silverman, G. S., 1986, Serial and parallel processing of visual feature conjunctions, *Nature*. **320**: 264-265.

Nakayama, K., and Mackeben, M., 1989, Sustained and transient components of focal visual attention, *Vis. Res.* **29**: 1631-1647.

Nakayama, K., 1990, The iconic bottleneck and the tenuous link between early visual processing and perception, In: *Vision: Coding and Efficiency*., C. Blakemore, ed., Cambridge University Press, Cambridge, England.

Nakayama, K., He, Z. J., and Shimojo, S., 1995, Visual surface representation: A critical link between lower-level and higher-level vision, In: *Visual Cognition*, Vol. 2., S. M. Kosslyn and D. N. Osherson, eds., MIT Press, Cambridge, MA, pp. 1-70.

Neisser, U., 1967, *Cognitive Psychology*. Appleton-Century-Crofts, New York.

Niebur, E., Koch, C., and Rosin, C., 1993, An oscillation-based model for the neuronal basis of attention, *Vis. Res.* **33**: 2789-2802.

Nothdurft, H. C., 1993, Saliency effects across dimensions in visual search, *Vis. Res.* **33**:839-844.

Oken, B. S., Kishiyama, S., Kaye, J. A., and Howieson, D. B., 1994, Attention deficit in Alzheimer's disease is not simulated by an anticholinergic/antihistaminergic drug and is distinct from deficits in healthy aging, *Neurology*. **44**: 657-662.

Olshausen, B., Anderson, C., and Van Essen, D., 1994, A neurobiological model of visual attention and invariant pattern recognition based on dynamic routing of information, *J. Neurosci.* **13**: 4700-4719.

Ooi, T. L., and He, Z. J., 1999, Binocular rivalry and visual awareness: the role of attention, *Perception*. **28**: 551-574.

Parasuraman, R., Greenwood, P. M., Haxby, J. V., and Grady, C. L., 1992, Visuospatial attention in dementia of the Alzheimer type, *Brain*. **115**:711-733.

Parasuraman, R., and Haxby, J. V., 1993, Attention and brain function in Alzheimer's disease: a review, *Neuropsychologia*. **7**: 243-273.

Parker, A., and Hawken, M., 1985, Capabilities of monkey cortical cells in spatial-resolution tasks, *J. Opt. Soc. Am.* **A2**: 1101-1114.

Pashler, H., 1987, Detecting conjunctions of color and form: reassessing the serial search hypothesis, *Percept. Psychophys.* **41**: 191-201.

Pelligrino, G. D., and Renzi, E. D., 1995, An experimental investigation on the nature of extinction, *Neuropsychologia*. **33**: 153-170.

Peterson, S. E., Robinson, S. I., and Keys, W., 1985, Pulvinar nuclei of the behaving rhesus monkey visual responses and their modulation, *J. Neurophysiol.* **54**: 867-886.

Peterson, S. E., Robinson, D. L., and Morris, J. D., 1987, Contributions of the pulvinar to visual spatial attention, *Neuropsychologia*. **25**: 97-105.

Posner, M. I., and Boies, S. J., 1971, Components of attention, *Psychol. Rev.* **78**: 391-408.

Posner, M. I., 1980, Orienting of attention, *Quart. J. Exp. Psychol.* **32**: 3-25.

Posner, M. I., 1982, Cumulative development of attentional theory, *Am. Psychol.* **37**: 168-179.

Posner, M. I., Cohen, R. A., and Rafal, R. D., 1982, Neural systems control of spatial orienting, *Phil. Trans. Roy. Soc. Lond.* **B298**: 187-198.

Posner, M. I., and Cohen, Y., 1984, Components of visual orienting, In: *Attention and Performance*, Vol. X., H. Bouma and D. Bouhuis, eds., Lawrence Erlbaum Associates Inc., Hillsdale, NJ.

Posner, M. I., Walker, J. A., Friedrich, F. J., and Rafal, R. D., 1984, Effects of parietal injury on covert orienting, *J. Neurosci.* **4**: 1863-1874.

Posner, M. I., Walker, J. A., Friedrich, F. J., and Rafal, R. D., 1987, How do the parietal lobes

direct covert attention?, *Neuropsychologia*. **25**: 135-145.
Posner, M. I., and Peterson, S. E., 1990, The attention system of the human brain, *Ann. Rev. Neurosci.* **13**: 25-42.
Rafal, R. D., Posner, M. I., Walker, J. A., and Friedrich, F. J., 1984, Cognition and the basal ganglia, *Brain*. **107**: 1083-1094.
Rafal, R. D., and Posner, M. I., 1987, Deficits in human visual spatial attention following thalamic lesions, *Proc. Natl. Acad. Sci. , U.S.A.* **84**: 7349-7353.
Rafal, R. D., Friedman, J. H., Inhoff, A. W., and Bernstein, E., 1988, Orienting of visual attention in progressive supranuclear palsy, *Brain*. **111**: 267-280.
Rafal, R. D., and Robertson, L. C., 1995, The neurology of visual attention, In: *The Cognitive Neurosciences*., M. S. Gazzaniga, ed., MIT Press, Cambridge, MA, pp. 625-648.
Robinson, D. L., and Peterson, S. E., 1992, The pulvinar and visual salience, *Trends Neurosci.* **15**: 127-132.
Rummelhart, D. E., Hinton, G. E., and Mc Clelland, J. L., 1986, A general framework for parallel distributed processing, In: *Parallel Distributed Processing: Explorations in the Microstructure of Cognition*, Vol. I., J. L. Mc Clelland and D. E. Rummelhart, eds., MIT Press, Cambridge, MA, pp. 45-77.
Sagi, D., and Julesz, B., 1985a, Fast noninertial shifts of attention, *Spatial Vis.* **1**: 141-149.
Sagi, D., and Julesz, B., 1985b, "Where" and "what" in vision, *Science*. **228**: 1217-1219.
Sagi, D., and Julesz, B., 1986, Enhanced detection in the aperture of focal attention during simple discrimination tasks, *Nature*. **321**: 693-695.
Salzman, C. D., Britten, K. H., and Newsome, W. T., 1990, Cortical microstimulation influences perceptual judgements of motion direction, *Nature*. **346**: 174-177.
Sandon, P. A., 1990, Stimulating visual attention, *Cog. Neurosci.* **2**: 213-231.
Schneider, W., and Shiffrin, R. M., 1977, Controlled and automatic human information processing: I. Detection, search and attention, *Psychol. Rev.* **84**: 1-66.
Sereno, A. B., 1992, Programming saccades: the role of attention, In: *Eye Movements and Visual Cognition*., K. Rayner, ed., Springer-Verlag, New York.
Shiffrin, R. M., and Gardner, G. T., 1972, Visual processing capacity and attentional control, *Journal of Experimental Psychology*. **93**: 72-83.
Shimojo, S., Miyauchi, S., and Hikosaka, O., 1992, Visual motion sensation yielded by non-visually driven attention, *Inv. Ophthalmol. Vis. Sci. Suppl.* **32**: 1354.
Shiu, L. P., and Pashler, H., 1994, Negligible effect of spatial precuing on identification of single digits, *J. Exp. Psychol.: Human Percept. and Perf.* **20**: 1037-1054.
Shulman, G., Remington, R. W., and McLean, J. P., 1979, Moving attention through visual space, *Hum. Percept. Perform.* **5**: 522-526.
Siddle, D., Stephenson, D., and Spinks, J. A., 1983, Elicitation and habituation of the orienting response, In: *Orienting and Habituation: Perspectives in Human Research*., D. Siddle, ed., Wiley, New York, pp. 109-182.
Sokolov, 1963, *Perception and the Conditioned Reflex*. Pergamon Press, Oxford, England.
Spear, P. D., 1993, Neural bases of visual deficits during aging, *Vis. Res.* **33**: 2589-2609.
Spillmann, L., Ransom-Hogg, A., and Oehler, R., 1987, A comparison of perceptive and receptive fields in man and monkey, *Hum. Neurobiol.* **6**: 51-62.
Spitzer, H., Desimone, R., and Moran, J., 1988, Increased attention enhances both behavioural and neuronal performance, *Science*. **240**: 388-340.
Stein, J. F., 1992, The representation of egocentric space in the posterior parietal cortex, *Behav. Brain Sci.* **15**: 691-700.
Steinman, S. B., 1987, Serial and parallel search in pattern vision?, *Perception*. **16**: 389-398.
Steinman, S. B., and McKee, S. P., 1988, A low-level approach to the discrimination of natural stimuli, *Inv. Ophthalmol. Vis. Sci. Suppl.* **29**: 400.
Steinman, S. B., and Levi, D. M., 1992, Topographic mapping of evoked potentials to spatial localization stimuli, *Vis. Neurosci.* **8**: 281-294.
Steinman, S. B., Steinman, B. A., Trick, G. L., and Lehmkuhle, S., 1994, A visual explanation

for attentional deficits in the elderly, *Optom. Vis. Sci.* **71**: 743-749.
Steinman, B. A., Steinman, S. B., and Lehmkuhle, S., 1995, Line-motion illusion reveals that visual attention mechanisms have a center-surround organization, *Vis. Res.* **35**: 1859-1869.
Steinman, S. B., and Steinman, B. A., 1996, Long-term excitatory and inhibitory interactions in visual attention, *Optom. Vis. Sci.* **72**.
Steinman, B. A., Steinman, S. B., and Lehmkuhle, S., 1997, Transient visual attention is dominated by the magnocellular stream, *Vis. Res.* **36**: 17-23.
Steinman, B. A., and Steinman, S. B., 1997a, Inversion of the attentional perceptive field by a preceding cue, *Inv. Ophthalmol. Vis. Sci.* **38**: S371.
Steinman, S. B., and Steinman, B. A., 1997b, The line motion illusion: visual attention or apparent motion?, *Inv. Ophthalmol. Vis. Sci.* **38**: S372.
Steinman, S. B., and Steinman, B. A., 1998, Vision and attention I: Current models of visual attention, *Optom. Vis. Sci.* **75**: 146-155.
Steinman, B. A., Steinman, S. B., and Garzia, R. P., 1998, Vision and attention II: Is visual attention the mechanism through which a deficient magnocellular stream causes reading disability?, *Optom. Vis. Sci.* **75**: 674-681.
Tarver, S., and Hallihan, D., 1974, Attention deficits in children with learning disabilities: a review, *J. Learn. Disabil.* **7**: 560-569.
Treisman, A. M., 1960, Contextual cues in selective listening, *Quart. J. Exp. Psychol.* **12**: 242-248.
Treisman, A. M., and Gelade, G., 1980, A feature integration theory of attention, *Cognitive Psychology*. **12**: 97-136.
Treisman, A. M., 1985, Preattentive processing, *Comput. Vision Graphics Image Proc.* **31**: 156-177.
Treisman, A. M., and Sato, S., 1990, Conjunction search revisited, *J. Exp. Psychol.: Human Percept. and Perf.* **16**: 459-478.
Treisman, A. M., 1993, The perception of features and objects, In: *Attention: Selection, Awareness and Control.*, A. Baddeley and L. Weiskrantz, eds., Clarendon Press, Oxford, England.
Tsal, Y., 1983, Movement of attention across the visual field, *Journal of Experimental Psychology Human Perception*. **9**: 523-530.
Tse, P., and Cavanagh, P., 1995, Line motion occurs after surface parsing, *Inv. Ophthalmol. Vis. Sci.* **36**: 1919.
Tsotsos, J. K., 1993, An inhibitory beam for attentional selection, In: *Spatial Vision in Humans and Robots.*, L. Harris and M. Jenkins, eds., Cambridge University Press, Cambridge, MA, pp. 313-331.
Tsotsos, J. K., 1995, Toward a computational model of attention, In: *Early Vision and Beyond.*, T. V. Papathomas C. Chubb A. Gorea and E. Kowler, eds., MIT Press, Cambridge, MA, pp. 207-218.
Ungerleider, L. G., Galkin, T. W., and Mishkin, M., 1983, Visuotopic organization of projections from striate cortex to inferior and lateral pulvinar in rhesus monkey, *J. Comp. Neurol.* **217**: 137-157.
Wolfe, J. M., Cave, K. R., and Franzel, S. L., 1989, Guided search: an alternative to the feature integration model of visual search, *J. Exp. Psychol.: Human Percept. and Perf.* **15**: 419-433.
Wolfe, J. M., 1994, Guided search 2.0: a revised model of visual search, *Psychonom. Bull. Rev.* **1**: 202-238.
Wurtz, R. H., Richmond, B. J., and Newsome, W. T., 1984, Modulation of cortical visual processing by attention, perception, and movement, In: *Dynamic Aspects of Neocortical Functions.*, G. M. Edelman W. E. Gall and W. M. Cowan, eds., Wiley, New York.
Yantis, S., and Jonides, J., 1984, Abrupt visual onsets and selective attention: evidence from visual search, *J. Exp. Psychol.: Human Percept. and Perf.* **10**: 601-621.

Chapter 15

Cognitive Processing and Models of Reading

Erik D. Reichle[1, 2] and Keith Rayner[3]

[1]*Dept. of Psychology, Carnegie Mellon University, Pittsburgh, PA 15213*

[2] Now at: *Dept. of Psychology, Learning Research and Development Center, University of Pittsburgh, 3939 O'Hara Street, Pittsburgh, PA 15260, PH: (412) 624-7457, FX: (412) 624-9149, EM: reichle+@pitt.edu*

[3] *Dept. of Psychology, University of Massachusetts, Amherst, MA 01003, PH: (413) 545-2175, FX: (413) 545-0996, EM: rayner@psych.umass.edu*

> On the outside, the reader has rotated his eyes only a few millimeters....But on the inside, there has been a rapid succession of intricate events. Clearly, this succession could only be the product of a complex information processing system...It contains components that are asked to perform amazing feats with amazing rapidity, and precisely in concert. –Gough (1972, p. 341)

Reading involves the orchestration of many information-processing stages; as the eyes move across the printed page, the visual features of the text are converted into orthographic and phonological patterns, which are then used to guide further language processing so that the content of the text can be understood. One can gain an appreciation of the complexity of these processes by examining the table of contents in a psychology of reading textbook; minimally, one finds chapters devoted to the topics of eye movement control, word recognition, semantic and syntactic analyses, and discourse processing (Just and Carpenter, 1987; Perfetti, 1985; Rayner and Pollatsek, 1989; Taylor and Taylor, 1983).

Because this volume focuses on models of the visual system, this chapter will not review all of the models that have been proposed to explain the various components of reading. Instead, we will only discuss those models that have attempted to explain the interface between vision and low-level language processing; that is, models that specify some combination of the following components of reading: Eye movement control, visuospatial attention, and/or the visual processing of words. Likewise, although the title of this chapter may suggest that it will cover general aspects of visual cognition (e.g., scene perception), our discussion will be limited to reading. This decision was partially due to limitations in space, but it was also based on the fact that reading has been investigated more extensively than other tasks (for a recent review, see Rayner, 1998). Moreover, reading is arguably the most important and complex visual skill for which humans are not biologically programmed.

This chapter will be organized into four sections: The first provides a general overview of the reading process and reviews the ways in which various models have attempted to specify the relationship between the aforementioned components of reading. The second section provides a brief discussion of how models of reading have been used to address questions of practical importance. The third section evaluates the models in light of what has recently been learned from cognitive neuroscience. Finally, the chapter closes with some thoughts about how the models might be used to guide future research.

15.1 READING MODELS

15.1.1 Reading and Eye Movements

Any discussion of reading models must begin with a brief overview of what has been learned about visual processing during reading[1]. Contrary to one's subjective impressions, the eyes do not move smoothly across the printed page during reading (Ciuffreda and Bassil, 1990). Instead, the eyes make short and rapid movements, called *saccades* (Bahill and Stark, 1979; Ciuffreda, 1990), that move the eyes forward about 6-8 character spaces on average (McConkie, Kerr, Reddix, and Zola, 1988; Rayner, 1978, 1998). Since saccade size depends upon character spaces and not visual angle, character spaces is the appropriate metric (Morrison and Rayner, 1981). Saccades take 20-50 ms to complete depending upon the length of the movement. Between saccades, the eyes remain stationary for brief periods of time (typically 200-250 ms) called *fixations* (Rayner, 1978, 1998). Because visual information is only extracted from the printed page during fixations, reading is similar to a slide show in which short segments of text are displayed for approximately a quarter of a second. However, it is important to note that within a given reader there is considerable variability in that some fixations will be shorter than 100 ms and some will be longer than 400 ms.

Other constraints also severely curtail how much visual information can be processed during each fixation. First, because visual acuity is maximal in the center of the retina and rapidly decreases towards the periphery, fine visual discriminations can only be made within the *fovea*, or central 2° of vision. As a result, the visual features that make up individual letters can only be encoded from a very narrow window of vision. The practical significance of this fact is that it is necessary to fixate most words so that they can be identified. Indeed, there is considerable evidence that as the angular disparity between the fovea and the retinal image of a word increases, the word becomes increasingly difficult to identify (Rayner and Bertera, 1979; Rayner and Morrison, 1981).

A second set of constraints stems from the time required to both plan and execute saccades, and to identify individual words. Experiments in which subjects move their eyes to visual targets indicate that the *saccadic latency*, or the time needed to plan and execute a saccade, is approximately 180-200 ms (Becker and Jürgens, 1979). This result is seemingly at odds with the intuitively appealing idea that word recognition might drive eye movements during reading because most estimates indicate that lexical access requires 100-300 ms to complete (Sereno, Rayner, and Posner, 1998; Rayner and Pollatsek, 1989; Schilling, Rayner, and Chumbley, 1998). On first appearances, therefore, it is not immediately obvious how the identification of one word can be the signal to begin planning a saccade to the next.

The solution to this quandary is that words can sometimes be partially processed in the *parafovea*, or region of the retina that extends 5° on either side of the fovea. McConkie and Rayner (1975, 1976) demonstrated the importance of parafoveal processing using the *moving-window paradigm*, which is illustrated in Fig. 15.1. In this paradigm, the letters outside of a "window" spanning a given number of character spaces is distorted in some way (e.g., replaced with Xs). By varying the size of the window and making its location contingent upon where the reader is looking, it is possible to ascertain the *perceptual span*, or region from which useful visual information can be encoded. With English text, readers can progress at a more-or-less normal rate when the window extends 14-15 character spaces to the right and 3-4 character spaces to the left (i.e., the left word boundary) of the fixation point. However, word encoding probably does not extend more than seven characters to the right of fixation; beyond this distance, only low-spatial frequency information about letter shape (e.g., descenders vs. ascenders; letters vs. blank spaces) is extracted from the page. The left-right asymmetry reflects covert attention and is language specific; with

Hebrew text (which is read from right to left), the perceptual span extends asymmetrically to the left of fixation (Pollatsek, Bolozky, Well, and Rayner, 1981).

Normal Text

```
the link between language and eye movements
      *
the link between language and eye movements
            *
the link between language and eye movements
                   *
```

Moving Window: 2 Words

```
XXX link between XXXXXXXX XXX XXX XXXXXXXXX
      *
XXX XXXX between language XXX XXX XXXXXXXXX
            *
XXX XXXX XXXXXXX language and XXX XXXXXXXXX
                   *
```

Moving Window: 4 Letters Left, 14 Letters Right

```
XXe link between langXXXX XXX XXX XXXXXXXXX
      *
XXX XXXX between language aXX XXX XXXXXXXXX
            *
XXX XXXX XXXXXXn language and eye XXXXXXXXX
                   *
```

Figure 15.1. The top panel shows three successive fixations within a segment of normal text. Asterisks below each line indicate the locations of fixations. The middle and bottom panels depict the paradigm in which a moving "window" of normal text is displayed at a location that is contingent upon where the eyes are looking. The middle panel shows a 2-word moving window; that is, both the fixated word and the word to its right are displayed normally, and the letters in the remaining words are replaced by Xs. In the bottom panel, the window extends four character spaces to the left of fixation and 14 character spaces to the right of fixation.

Other experiments have shown that orthographic (Rayner, 1975; Balota, Pollatsek, and Rayner, 1985) and phonological (Pollatsek, Lesch, Morris, and Rayner, 1992) processing of a word can begin prior to the word being fixated. These results suggest that, during normal reading, the *parafoveal preview* of a word can reduce the duration of the subsequent fixation on the word, which is one measure of the time needed for identification (Schilling et al., 1998). Parafoveal preview can also produce skipping because words that can be identified in the parafovea do not have to be fixated and can therefore be skipped.

One indicator of the inherent difficulty of reading (even for skilled readers) is reflected in the fact that 10-15% of saccades move the eyes back to previous parts of the text. These backward movements, called *regressions*, are thought to result from problems with both linguistic processing and oculomotor error. The former hypothesis is supported by findings that regressions can be induced with structurally difficult "garden path" sentences; because such sentences often lead to incorrect syntactic analyses, the reader must make a regression and then re-interpret the sentence (Frazier and Rayner, 1982). The hypothesis that regressions are sometimes due to simple motor error is supported by the finding that when the eyes fixate near the end of a word, they often move back a few character spaces (O'Regan, 1990). This presumably happens because the eyes overshot their intended target (near the middle of the word) and a second fixation location affords a better place from which to see the word. This interpretation is consistent with the finding that identification is most rapid if a word is fixated just to the left of its center, on the *optimal viewing position* (O'Regan, 1990, 1992).

To investigate the components of reading, researchers typically have subjects read sentences or passages of text while an eye tracker records the locations and durations of individual fixations (Rayner, 1979a). Because an average college-level reader can read approximately 300 words per minute (Rayner and Pollatsek, 1989), this technique produces a staggering amount of data. To make sense of such data, they are usually reduced to *word-based measures*, which are across-subject averages that reflect how often and for how long individual words were fixated. A number of word-based measures are standard (Inhoff and Radach, 1998). The first is the *gaze duration*, which is defined as the sum of all fixations on a word, excluding any fixations after the eyes have left the word. Two related measures are the *first-fixation duration* and the *single-fixation duration*; the former is the duration of the first fixation on a word, while the latter is the time spent on a word conditional upon the word being fixated exactly once. These indices are typically reported along with indices of how often a word was fixated: The probability of a word being skipped, fixated once, and fixated more than once before moving to another word (i.e., a *refixation*).

The word-based measures provide a complete record of where and when fixations occurred. These two parameters (where vs. when) also provide a useful framework for organizing a discussion of reading models because much of the controversy surrounding reading concerns the determinants of where and how long the eyes remain fixated. The models that have been developed to explain eye movement control can be separated into two classes: Those in which eye movements are determined primarily by oculomotor factors (*oculomotor models*), and those in which eye movements are guided by some form of cognitive control (*processing models*).

15.1.2 Oculomotor Models

The oculomotor models share two basic principles. The first is that the visual properties of the text that is being read (e.g., word length) and operating characteristics of the visual and oculomotor systems (e.g., visual acuity, saccade accuracy) largely determine where the eyes will fixate. The second common principle is that fixation durations are largely determined by where the eyes have fixated.

Much of the motivation for the oculomotor models was due to the seminal work of McConkie and his colleagues (McConkie et al., 1988; McConkie, Kerr, Reddix, Zola, and Jacobs, 1989; McConkie, Zola, Grimes, Kerr, Bryant, and Wolff, 1991; Radach and McConkie, 1998). They expanded on the observation that readers typically fixate between the beginning and the middle of a word, the *preferred viewing location* (Rayner, 1979b)[2]. In several analyses of the patterns of readers' eye movements when they read a large corpus of text, they showed that the *landing site distributions* (i.e., distributions of the locations of individual fixations, collapsed across subjects and words of a given length) are very systematic. As Fig. 15.2 indicates, the distributions are approximately Gaussian in shape, but with truncated tails. The means of the distributions also tend to be near the centers of the words. However, the means shift from the word endings towards the word beginnings as the distance between the *launch site* (i.e., the location of the fixation prior to the saccade) and the landing site increases. In addition, the landing sites become less variable as the distance between launch site and landing site decreases, and as the fixation duration on the launch site increases.

McConkie and his colleagues (1988, 1989, 1991) believed that the pattern of results could be accounted for by the following: First, the shape of the distributions reflect random noise in the oculomotor system, while the missing tails presumably reflect instances in which the eyes under/overshot their intended targets. In addition to the random noise in the motor system, the system is also biased to make saccades approximately seven character spaces in length; because of this bias, longer saccades tend to undershoot their targets, and shorter saccades tend to overshoot their targets. This *systematic range error* thus causes the distributions to shift towards the beginnings of words as the launch site becomes more distant from the intended saccade target. However, when the oculomotor system is given more time to plan its saccades, its accuracy increases, and the systematic range error is reduced.

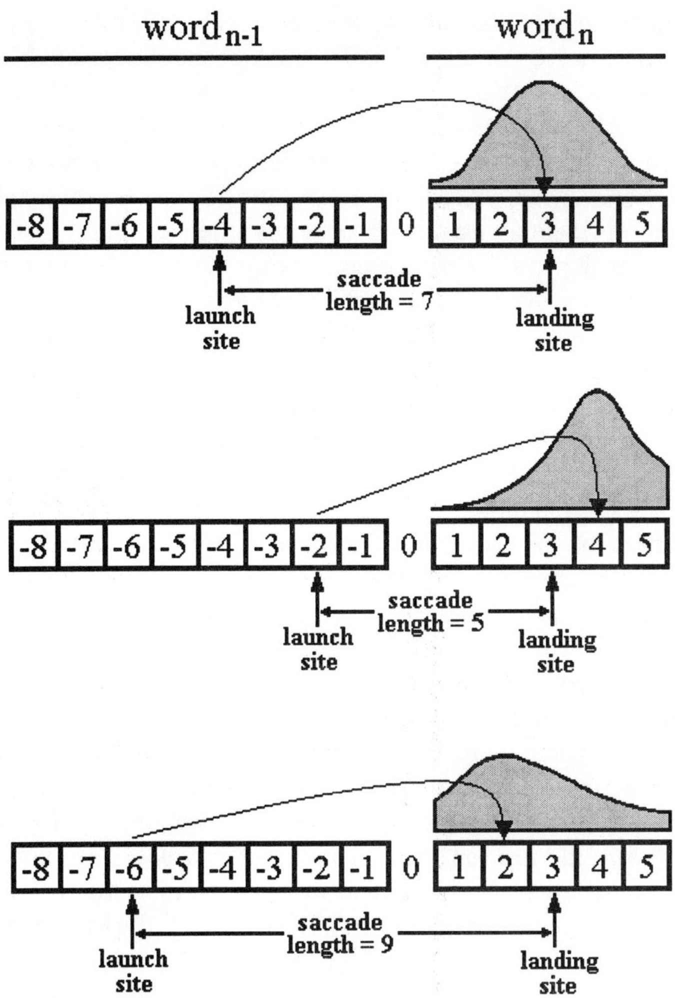

Figure 15.2. Landing site distributions as a function of saccade length. In all three panels, $word_{n-1}$ (the launch site) and $word_n$ (the landing site) are depicted as rectangles, with the character spaces represented by numbers (as per convention, the space to the left of $word_n$ is denoted as zero). The distributions are approximately Gaussian in shape and centered on $word_n$, which suggests that the eyes are directed towards the center of $word_n$. However, because the oculomotor system is biased towards making saccades that are seven character spaces in length (as in the top panel), there is a systematic range error: The eyes tend to overshoot closer targets (middle panel) and undershoot more distant targets (bottom panel).

These relationships among saccade length, launch site duration, and saccadic accuracy led to the development of some very precise mathematical descriptions of how these variables affect the landing site distributions during reading (McConkie, Kerr, and Dyre, 1994). Although there have also been attempts to provide similar mathematical descriptions of fixation durations (McConkie et al., 1994; McConkie and Dyre, 2000; see also Brysbaert and Vitu, 1998), these accounts are generally little more than precise descriptions of the data, and do not explain why fixation durations are modulated during reading. Moreover, because these descriptions address the "where" and "when" questions independently, they fail to explain why the durations of fixations are related to their spatial locations.

Another, more process-oriented oculomotor theory is the *strategy-tactics model* (O'Regan, 1990, 1992; O'Regan and Lévy-Schoen, 1987). The core of this model is based on the finding that words are identified most rapidly if they are fixated slightly to the left of center, on the *optimal viewing position*. This finding, combined with the fact that words are also less likely to be refixated if they are initially viewed from this position, suggests that readers adopt a "strategy" of directing their eyes from word to word, in an attempt to fixate each word's optimal viewing position. This process describes the "risky" variant of the basic reading strategy; with the "careful" strategy, people also adopt the following within-word "tactic:" If the eyes do not land near the optimal viewing position, then immediately move the eyes to the other end of the word. Doing this ensures that each word either will be viewed from its optimal position (in the case of single fixations) or will be viewed from two different locations (in the case of refixations).

Because the within-word tactics are guided by visual factors (e.g., word length), the model predicts that linguistic variables (e.g., word frequency) should: (a) only modulate fixation durations when there is a single long fixation or when the fixation is the second of the two, and (b) not modulate refixation probabilities. Unfortunately for the strategy-tactics model, neither prediction was confirmed (Rayner, Sereno, and Raney, 1996); word frequency effects were evident in the first of two fixations, and refixations were more likely on low-frequency words than on high-frequency words. In addition, neither fixation durations nor frequency effects on single-fixations varied as a function of landing position, which suggests that the optimal viewing position may be much less important in normal reading than in the identification of single words when they are presented in isolation.

Although the strategy-tactics model was not originally implemented as a computer program, variants of the model were recently implemented (Reilly and O'Regan, 1998), so that different word-targeting strategies could be evaluated with respect to how well they handle the findings related to landing-site distributions (McConkie et al., 1988). The word-targeting strategies included several oculomotor (e.g., *word-by-word, target long words*, and *skip short words*) and language-based strategies (e.g., *skip high-frequency words* and an *attention-shift strategy*; see the next section, below). The target-long-words strategy fit the landing-site distributions better than the other strategies, and the two language-based strategies fared rather poorly overall. On this basis, Reilly and O'Regan proposed that the language-processing models cannot provide an adequate account of eye-movement control during reading. We shall return to this point later in this chapter.

The final oculomotor model is Suppes' (1990, 1994) *minimal-control model.* This model assumes that reading is largely a low-level, automatic process, and that neither fixation durations nor saccade lengths are affected by linguistic or cognitive factors, but are instead only affected by the physical layout of the text. The model consists of a small number of axioms that describe fixation duration distributions and a random-walk process that determines where the eyes will move.

In the minimal-control model, fixation durations are a function of the number of operations (which are never specified in the model) that must be completed during each fixation. The durations are exponentially distributed in cases where a single operation must be completed; in cases requiring two operations, the durations are described by the convolution of two independent exponential distributions. Finally, fixation times are independent of both earlier processing and the current text content. Thus, the model stipulates that variability in fixation durations is not due to variability in the duration of the underlying cognitive processing, but instead reflects the probabilistic nature of the processing.

The minimal-control model's saccade axioms are as follows: First, if the processing within a region of regard has been completed, then move the eyes to the next word; otherwise, remain in the same location. Second, if processing has not finished and memory for the prior region of regard has decayed, then regress back to the prior region. Third, if perceptual processing of the upcoming word has finished from the current location, then skip the upcoming word. Finally, each saccade is independent of earlier processing and previous saccades.

Unfortunately, the minimal-control model has only been used to simulate eye movements during an arithmetic task, and not during reading (Suppes, 1990; Suppes, Cohen, Laddaga, Anliker, and Floyd, 1982, 1983). It is therefore difficult to evaluate the model. However, it is clear that the model only makes predictions on the level of individual words, and hence cannot account for landing site distributions (McConkie et al., 1988) or the optimal viewing position effects (O'Regan, 1990). The model also fails to account for many other factors that are acknowledged by Suppes (1994) to affect eye movements during reading.

In the final analysis, the oculomotor models have provided important insights into the nature of visual processing and eye movement control during reading. The phenomena that these models were designed to explain are important benchmarks for evaluating any model of reading. The models are limited because, in concentrating on oculomotor factors, they did not incorporate language-processing factors, which have been shown in the past three decades of research to have important effects on eye movement control during reading. Rather than being incorrect, however, the models should probably be viewed as being incomplete in that they do provide partial solutions to the problem of understanding reading.

15.1.3 Processing Models

In contrast to the oculomotor models, the processing models share an assumption that cognitive (i.e., linguistic) processing at least partially guides eye movements during reading. Generally speaking, the decision about how long to fixate is determined by the on-going linguistic processing of the text, while the decision about where to fixate is determined by a combination of linguistic, visual, and oculomotor factors.

One of the first processing models was *Reader* (Carpenter and Just, 1983; Just and Carpenter, 1980, 1987; Thibadeau, Just, and Carpenter, 1982). The goal in developing this model was to build a system that could simulate the entire reading process. In other words, the model was to account for low-level encoding of visual features, lexical processing, semantic and syntactic analysis, and the schema-guided comprehension and abstraction of key ideas that normally occurs during reading. As might be expected given the goals of the modeling effort, Reader posits a tight link between eye movements and cognitive processing, based on the following assumptions: First, the *immediacy hypothesis* stipulates that each word is processed (to the farthest extent possible) as it is fixated. The second, *eye-mind hypothesis* stipulates that the eyes remain fixated until processing has been completed. Both the durations and locations of individual fixations are thus determined by the immediate processing of the information being fixated.

The Reader model was implemented as a production-system program. *Productions* are procedural condition-action pairs (i.e., if-then statements) that perform operations on units of declarative knowledge. For example, the production:

IF <letter$_1$ = "c" & letter$_2$ = "a" & letter$_3$ = "t"> THEN <word = "cat">

is used to encode the word percept "cat," so that the meaning of the word can then be retrieved from semantic memory. In the Reader model, the productions are activation-based; that is, productions direct activation towards units of declarative knowledge which have thresholds that must be exceeded if the information is to be "active" in working memory (and thereby satisfy the conditions of other productions). The values of these thresholds are adjusted to modulate the cost associated with using each production. For example, the threshold of those productions that mediate lexical access are adjusted to reflect each word's normative frequency of occurrence, so that low-frequency words take longer to identify (and are consequently fixated longer) than high-frequency words. Also, in the most recent version of the model (*CC-Reader*; Just and Carpenter, 1992), the amount of activation that is available to support processing is limited, but is an adjustable parameter, so that individual differences in working memory capacity (i.e., activation) can be used to simulate individual differences in reading ability.

The major strength of the Reader model is its comprehensiveness. As mentioned above, the model was developed to simulate the entire reading process. The model therefore does reasonably well in simulating a number of language-related reading phenomena, such as word-frequency effects, increased reading times on lexically ambiguous words, and the processing difficulties which are found with syntactically ambiguous sentences. However, because of the model's complexity (Reader consists of 225 productions; Just and Carpenter, 1987), it lacks the conciseness and controllability of other computational models (e.g., the inner workings of the model can only be described verbally). It is also difficult to evaluate the model's performance because it depends upon the complex interplay of productions, several free parameters, and the regression weights (on several independent variables; e.g., whether or not a word is the first in a sentence) that are necessary to convert production cycles (arbitrary units of time) into processing time. Furthermore, the model only makes predictions about the locations of fixations at the level of individual words, using a composite measure (*gazes*) that averages fixation durations and skips. This means that, apart from word-length effects, the model fails to account for the phenomena that are explained by the oculomotor models (e.g., landing site distributions).

In addition to the above shortcomings, the Reader model has been criticized because of the immediacy and eye-mind assumptions. With respect to the former, there is considerable evidence that the lexical processing of a word is often initiated before the word has been directly fixated (i.e., parafoveal preview; Balota et al., 1985; McConkie and Rayner, 1975; Pollatsek et al., 1992; Rayner, 1975). With respect to the eye-mind hypothesis, it is now known that the normative frequency of $word_n$ can modulate how long the eyes remain on $word_{n+1}$ (Rayner and Duffy, 1986; Rayner, Sereno, Morris, Schmauder, and Clifton, 1989). These *spillover effects* indicate that the eyes often leave a word before the processing of that word is complete, contrary to the eye-mind assumption.

Thus, even if eye movements during reading are (at least partially) guided by language processing, the assumptions of the Reader model contain some gross over-simplifications of how this process probably occurs. Consequently, a second class of processing models, the *attention-shift models* (Reilly, 1993; Reilly and O'Regan, 1998), was proposed to explain the eye-mind connection. Although these models maintain a link between linguistic processing and eye movement control, the two are not as tightly coupled.

The central axioms of the attention-shift models were originally proposed by Morrison (1984). These principles are as follows: First, the identification of $word_n$ (which is currently being fixated) causes a "spotlight" of attention (Julesz, 1991) to move to $word_{n+1}$. This shift of attention signals the oculomotor system to start planning a saccade to $word_{n+1}$. If the word-recognition system is still trying to identify $word_{n+1}$ when the saccadic program has been completed, then the saccade will be executed and the eyes will move to $word_{n+1}$. If $word_{n+1}$ is recognized before the program completes, however, then the saccade can be canceled. If this happens, attention shifts to $word_{n+2}$, and the oculomotor system begins programming a saccade to $word_{n+2}$ so that $word_{n+1}$ is skipped. Thus, the two key assumptions of Morrison's model are that: (a) attention moves serially, from word to word; and (b) saccades can be programmed in parallel.

Morrison's (1984) assumption about the parallel programming of saccades followed Becker and Jürgen's (1979) demonstration that saccadic programming is completed in two stages: An initial, labile stage that is subject to cancellation, and an ensuing, non-labile stage in which the program cannot be canceled. Their work showed that, if the oculomotor system begins programming a saccade while another saccadic program is in its labile stage of development, then the first program is aborted. However, if the second program is initiated while the first saccadic program is in its non-labile stage, then both saccades will be executed, which typically results in a very short fixation between the two saccades.

By incorporating the known functional characteristics of the oculomotor system into his model, Morrison (1984) was able to provide an elegant account for a number of findings in the reading literature. For example, the model proposes a parafoveal preview process to explain skipping. Furthermore, because short, high-frequency words tend to be identified more easily than long, low-frequency words, the model explains why the former are skipped more often (and are fixated for shorter periods of time) than the latter.

However, despite the model's many successes, it is in many ways still incomplete. For example, because attention moves from word to word, Morrison's (1984) model cannot explain refixations; that is, each word is either fixated exactly once or skipped. Another limitation is that the amount of processing that occurs on a word when it is in the parafovea is determined by the saccadic latency. This is the case because, in the model, eye movements are tightly linked to shifts of attention, so that the time during which parafoveal processing can occur is equal to the amount of time that is necessary to program a saccade to the parafoveal word. Consequently, the model fails to predict that preview benefit will be modulated by foveal processing difficulty. And, for the same reason, the model also fails to predict spillover effects. Both of these shortcomings are problematic because, as already mentioned, spillover effects do occur, and because the size of the preview benefit does decrease as the foveal word becomes more difficult to process (Henderson and Ferreira, 1990; Kennison and Clifton, 1995).

One way to circumvent the limitations of Morrison's (1984) model is to augment it with a deadline for the completion of lexical processing (Henderson and Ferreira, 1990; Sereno, 1992). That is, if word identification is not completed within a certain period of time, attention does not shift to the next word, but instead remains on the current word, resulting in a refixation. This leads to the prediction (which has not been supported empirically; Schilling et al., 1998) that the first of two fixations should be longer than single fixations (because the processing deadline must be reached). An alternative to the deadline assumption is that problems with higher-order linguistic processing somehow signal the eyes to remain on the current word (Pollatsek and Rayner, 1990; Rayner and Poltatsek, 1989); unfortunately, the way in which this occurs has not been well specified.

The most recent attention-shift model is *E-Z Reader* (Pollatsek, Rayner, Fischer, and Reichle, 1999; Rayner, Reichle, and Pollatsek, 1998; Reichle, Pollatsek, Fisher, and Rayner, 1998). Like it's predecessor (Morrison, 1984), E-Z Reader was not designed to be a detailed model of oculomotor control or language processing. The model instead provides a framework for understanding how the two types of processing guide eye movements during reading. However, unlike Morrison's (1984) model, E-Z Reader has been implemented as a stochastic computational model so that it can be used to run simulations and make precise quantitative predictions. In the model, five processes determine the "when" and "where" of eye movements: A familiarity check, the completion of lexical access, the labile and non-labile stages of saccadic programming, and the actual saccades. These processes are shown schematically in Fig. 15.3.

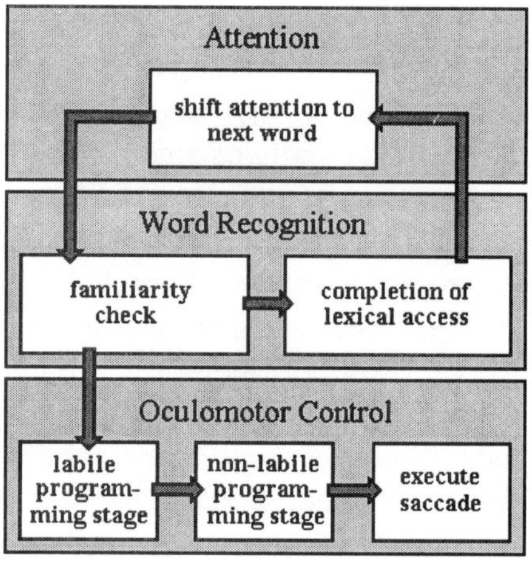

Figure 15.3. A schematic diagram of the E-Z Reader model (Reichle et al., 1998, 1999). The initial stage of word recognition (i.e., the familiarity check) signals the oculomotor system to begin planning a saccade to the next word, while the completion of lexical access causes attention to shift to the next word. Attention shifts are thus de-coupled from saccadic programming. The saccadic programs are completed in two stages: An initial, labile stage that is subject to cancellation, and a subsequent, non-labile stage that cannot be canceled. The completion of the non-labile stage is immediately followed by a saccade.

The first two processes (the familiarity check and completion of lexical access) correspond to the early and late stages of lexical access, respectively, and are conceptualized as being the tasks performed by the word-recognition system. The *familiarity check* is a quick assessment of whether or not lexical access is imminent. The completion of the familiarity check also signals the oculomotor system to cancel any pending labile saccades and to begin programming a saccade to the next word. The *completion of lexical access* refers to the final stage of word identification, in which a word's identity is determined. The completion of lexical access also causes attention to shift to the next word so that it can be processed. The mean times required to complete both stages of lexical processing are determined for each word as a function of its: (a) normative frequency of occurrence, and (b) predictability within the local sentence context. This is done using Eqs. 15.1 and 15.2:

$$t(f_n) = \{f_b - [f_m * \ln(frequency_n)]\} * (1 - \theta * predictability_n) \quad (15.1)$$

$$t(lc_n) = \Delta * \{f_b - [f_m * \ln(frequency_n)]\} * (1 - predictability_n) \quad (15.2)$$

In the equations, $t(f_n)$ and $t(lc_n)$ are the mean durations (in ms) of the familiarity check and completion of lexical access, respectively. f_b and f_m are free parameters that specify the linear relationship between the process durations and the natural log of the normative frequency of occurrence of $word_n$ (as tabulated by Francis and Kuĉera, 1982). θ is a free parameter that attenuates how much $word_n$'s predictability decrements its familiarity check duration, and the parameter Δ ensures that the time required for the completion of lexical access is some multiple of the familiarity check duration. Finally, the actual process durations, $T(f_n)$ and $T(lc_n)$, are random deviates which are sampled from gamma distributions having means equal to $t(f_n)$ and $t(lc_n)$, respectively, and having standard deviations equal to one third of the means.

Because words require more time to identify when they are viewed in the parafovea (Rayner and Morrison, 1981), an additional assumption of E-Z Reader is that the rate of lexical processing is modulated by *foveal eccentricity*, or distance between the center of the word being processed and the fovea (i.e., the character position that is being fixated), as follows:

$$duration(x) = duration_0 * \varepsilon^x \qquad (15.3)$$

where $duration(x)$ is the time required to complete lexical processing at a foveal eccentricity of x character spaces, $duration_0$ is time required to complete lexical processing when the eyes are fixated on the center of $word_n$, and ε is a free parameter.

The *labile* and *non-labile* stages of saccadic programming refer to the motor commands that are necessary to move the eyes, contingent upon whether or not the commands can be canceled. Again, this distinction follows the work of Becker and Jürgens (1979) and allows the model to explain skipping. In the model, the durations of the labile and non-labile programming stages, $T(m)$ and $T(M)$, respectively, are random deviates which are sampled from gamma distributions with fixed means, $t(m)$ and $t(M)$, respectively, and standard deviations equal to one third of the means.

In addition to word skipping, E-Z Reader accounts for many findings in the reading literature. For example, the model was fitted to Schilling et al.'s (1998) sentence corpus and was able to predict the mean values of six word-based measures (gaze, first-, and single-fixation durations, and the probability of skipping, making a single fixation, and refixating) for five frequency classes of words. E-Z Reader also generated gaze duration and first-fixation duration distributions that closely corresponded to those that were observed in the data. The model also replicated the frequency effects reported by Schilling et al. (1998) on target words, and produced both spillover effects and preview effects that were comparable to those reported in the literature. Finally, the model predicted a processing cost (i.e., gaze duration increase) on $word_n$ for skipping $word_{n+1}$, as well as a similar cost on $word_n$ for having skipped $word_{n-1}$. Although both predictions were confirmed in the data, the latter is particular noteworthy because it had not been previously reported in the literature.

The most recent version of E-Z Reader (Rayner, Reichle, and Pollatsek, 2000; Reichle, Rayner, and Pollatsek, 1999) has incorporated the assumptions of McConkie and his colleagues (McConkie et al., 1988) to explain how visual factors and characteristics of the oculomotor system affect eye movement control in reading (see Fig. 15.2). In the augmented model, saccades are always directed towards the optimal viewing position of the intended target word, but motor error (both random and systematic) cause the eyes to deviate from their targets. The fixation duration on the launch site also modulates the size of the systematic range error so that it decreases as the oculomotor system is given more time to plan its movements. The length of the saccade that is actually executed is given by:

$$Saccade = PSL + SRE + random\ error \tag{15.4}$$

where *PSL* is the programmed saccade length (in character spaces) and *SRE* is the systematic range error:

$$SRE = (\psi - PSL)\ [\Omega_b - \ln(fixation\ duration) / \Omega_m] \tag{15.5}$$

In Eq. 15.5, Ω_b and Ω_m are free parameters that specify how much the natural log of the launch site fixation duration modulates the systematic range error, and Ψ is a parameter that determines the saccade length for which the oculomotor system neither undershoots nor overshoots its intended target. In Eq. 15.4, the *random error* component is sampled from a Gaussian distribution with $\mu_{random\ error} = 0$ and $\sigma_{random\ error}$ given by:

$$\sigma_{random\ error} = \beta_b + (\beta_m * PSL) \tag{15.6}$$

The random error component of the saccade thus increases linearly with the programmed saccade length, as determined by the parameters β_b and β_m.

Because foveal eccentricity is defined in terms of character spaces, E-Z Reader predicts a parabolic refixation function; that is, refixations are most likely following initial fixations near word boundaries, and least likely following fixations near the optimal viewing position (McConkie et al., 1988; Rayner et al., 1996). This happens because fixations near the optimal viewing position generally allow the familiarity check to be completed very rapidly, so that the eye movement system begins programming an interword sacccade (to the next word) rather than the default intraword saccade (which would result in a refixation). E-Z Reader thus counters Reilly and O'Regan's (1998) claims that attention-shift models cannot explain the patterns of landing site distributions that are normally observed (McConkie et al., 1988).

The scope of the E-Z Reader model makes it a viable framework for thinking about the relationship between language processing and oculomotor control. For example, recent efforts[3] have been successful in modeling the effects of higher-order language processing. With minimal additional assumptions, the model can simulate the interaction between the presence/absence of disambiguating context and the meaning dominance of lexically ambiguous words (Binder and Rayner, 1998; Dopkins, Morris, and Rayner, 1992; Duffy, Morris, and Rayner, 1988; Rayner and Duffy, 1986; Rayner and Frazier, 1989).

Nonetheless, like the other models reviewed thus far, E-Z Reader approximates certain aspects of reading behavior at the expense of others. For example, Kennedy (1998, 2000; see also Inhoff, Starr, and Shindler, 2000) reports data suggesting that two words can sometimes be encoded in parallel, contrary to the assumption in E-Z Reader that attention shifts serially from word to word. Also, in contrast to the CC-Reader model (Just and Carpenter, 1987, 1992), E-Z Reader does not account for many of the effects of higher-level language processing (e.g., syntactic analysis). However, because many of these effects typically occur when the reader is having difficulty understanding the text that is being read, the model should be viewed as the "default" reading process; that is, the model simulates what happens during reading when higher-level linguistic processing is running smoothly.

One final point of contention concerns the distinction between the familiarity check and the completion of lexical access. This distinction was originally conceptualized as being broadly consistent with the *activation-verification model* of lexical access (Paap, Newsome, McDonald, and Schvaneveldt, 1982), with the familiarity check corresponding to an initial, activation stage of lexical access, and the completion of lexical access corresponding to a later, verification stage. However, the model's account of lexical-ambiguity effects may make it necessary to re-conceptualize the stages as lexical access followed by the integration of the lexical representation into the sentence representation.

Another processing model (Reilly, 1993) is remarkably similar to E-Z Reader in that it, too, was designed to be a framework for thinking about how language processing (i.e., lexical access) guides eye movements during reading. However, unlike E-Z Reader, Reilly's model (dubbed *ASM*, for attentional-shift model) is implemented as a pair of interacting connectionist (i.e., PDP; McClelland and Rumelhart, 1986; Rumelhart and McClelland, 1986) networks. The model's architecture (which is shown in Fig. 15.4) consists of two single-layer networks that were trained using a back-propagation learning algorithm (Rumelhart, Hinton, and Williams, 1986): One network is responsible for word recognition, and the other is responsible for programming saccades. As each word is identified, the lexical-encoding network signals attention to shift to the next word, so that it can be processed. Attention shifts cue the saccadic-programming network to begin programming a saccade to the next word. The time needed to complete lexical access and saccadic programming are determined by the number of cycles that the two networks need to settle into stable activation patterns. One other important feature of the model is that the visual input to the word recognition and oculomotor modules conforms to what is known about acuity limitations of the retina. That is, the patterns of activation representing the visual features of individual letters are more "degraded" (i.e., smaller values and less precise localization of these values across the array of input units) for letters that are encoded further away from the fovea, and for letters that share visual features. The model thus provides a natural way to explain why word identification becomes more difficult the further the word is from the fovea (Morrison and Rayner, 1981).

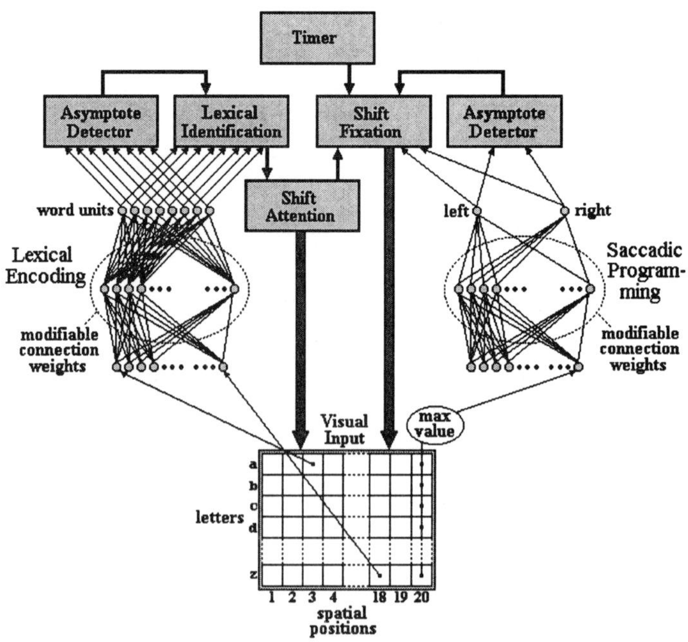

Figure 15.4. See figure legend on next page.

Figure 15.4. (See figure on previous page). A schematic diagram of the ASM model (Reilly, 1993). The model's input consists of an array representing the 20 possible spatial locations (position 8 is the center of the fovea) of 26 letters. The "Saccadic Programming" network uses the maximal value from each spatial position to compute the direction (left vs. right) and amplitude of the next saccade. The "Lexical Encoding" network uses the activation values of each letter from the central 16 spatial positions to determine the identity of each word (which are represented as unique 8-bit patterns by the word units). The "Asymptote Detectors" determine when the networks have settled into stable patterns, and provide an index of processing time. The lexical identification of one word causes attention to shift to the next, which modifies the visual input (represented by the thick shaded arrows) by reducing the activation values of unattended spatial positions. Attention shifts also enable saccades, which are executed after the "Saccadic Programming" network settles or after a fixed time interval (determined by the "Timer"). Saccades also modify the visual input by boosting the activation of letters in the next word.

Although Reilly (1993) does not provide a detailed account of the ASM model's performance, it is clear that the model simulates many of the basic eye-movement phenomena that are observed during reading. For instance, the model generates mean fixation durations and saccade lengths that are in close agreement to values typically observed. Moreover, because saccadic programming is prone to error, words are sometimes refixated. Low-frequency words are also fixated longer than high-frequency words, and because the model sometimes encodes two words in parallel, it will sometimes skip short, high-frequency words. The model also predicts preview effects; when $word_n$ is fixated, the absence of a parafoveal preview of $word_{n+1}$ will slow the processing of that word when it is later fixated, but facilitates the processing of $word_n$. Finally, one could make the argument that the model's primary strength lies in the fact that it has been implemented as a connectionist system. To the extent that connectionism offers a valid means to understand how cognitive processes might emerge from their underlying neural substrates (for a balanced discussion of this issue, see Clark, 1991), Reilly's (1993) model may represent real progress towards understanding the neural basis of eye movement control during reading.

The final processing model, *Mr. Chips* (Legge, Klitz, and Tjan, 1997; Klitz, Legge, and Tjan, 2000), represents a very different approach to understanding the inter-relationships among visual processing, word recognition, and eye movement control during reading. The model was proposed as a means to evaluate how an *ideal-observer* (a reader with perfect lexical knowledge and a well specified goal) would move his/her eyes so as to maximize reading speed. The model does this using three pieces of information: First, the input from a "retina" that encodes a small number of letters in the fovea and limited information about the presence/absence of letters in the periphery. The second source of information is knowledge about the relative frequencies with which words occur in text. Finally, the model knows the likelihood of making a saccadic error of each particular size. These three sources of knowledge are depicted in Fig. 15.5.

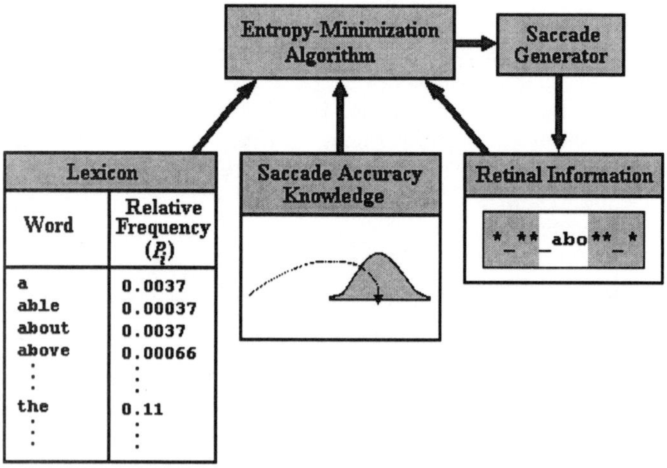

Figure 15.5. A schematic diagram of the Mr. Chips model (Legge et al., 1997; Klitz et al., 2000). The model uses the following information to compute the saccade length that will minimize the uncertainty about the identity of the current word: (1) the relative frequencies of words; (2) the accuracy of saccades; and (3) "Retinal Information" from the current fixation. The "retina" consists of two regions: a fovea, in which letters can be identified, and a parafovea, in which letters can be discriminated from blank spaces. (In the figure, the retina is represented by a rectangle; the fovea corresponds to the white area, and the parafovea corresponds to the gray area). The entropy-minimization algorithm computes the optimal saccade length, and then an error-prone "Saccade Generator" executes the saccade so that new letter information can be encoded.

The Mr. Chips model attempts to use this information to identify each word in a text using the fewest saccades possible. To do this, the model calculates the expected uncertainty that is associated with being able to identify a word for saccades of each possible length. It then executes a saccade that minimizes this uncertainty. For example, imagine that the model has the following partial information about a word: It is five letters long and begins with "abo" (see Fig. 15.5). The model uses this information in conjunction with its lexical knowledge to calculate conditional probabilities of the letter string being each of the words that meet these constraints. This is done using:

$$p_i = P_i \sum_j P_j \tag{15.7}$$

where p_i is the conditional probability of the letter string being $word_i$, P_i is the absolute probability of the letter string being $word_i$, and the P_js are the absolute probabilities of the letter string being each of the other candidate words. In the Fig. 15.5 example, the conditional probability that "abo--" is "about" is $p = 0.849$.

The conditional probabilities are then used to compute the conditional entropy, or degree of uncertainty, H, that would result from a saccade of a given length, L, under the assumption that the letter string is $word_i$:

$$H(L, word_i) = -\sum_i p_i \log_2(p_i) \tag{15.8}$$

For example, from the current fixation, the entropy associated with the letter string is *H(0, abo--)* = 0.613. (Smaller entropy values represent less uncertainty about the identity of a word, so that identification occurs with certainty when the entropy value associated with a letter string equals zero.) A saccade of $L = 1$ would reveal one letter, which, given the model's lexical knowledge, must be either "u" or "v." If the letter is "u," then the conditional probability of the word being "about" is $p = 1$, and the conditional entropy would be reduced to *H(1, about)* = 0. Likewise, if the letter is "v," then the conditional entropy would also be reduced to *H(1, above)* = 0.

After calculating the conditional entropies for each possible saccade length, Mr. Chips computes a probability-weighted average to determine the expected entropy associated with a saccade of each given length. This is done using:

$$H(L) = -\sum_i p_i H(L, word_i) \tag{15.9}$$

In the example, $H(L) = 0$ for $L = 1\text{-}5$. Because of saccadic error, however, each saccade of intended length L has an associated landing-site distribution, $P_L(x)$, which specifies the probability of making a saccade of actual length x. The model uses this knowledge to calculate the entropy associated with each saccade length, L, averaged across possible landing sites. This gives the expected uncertainty, H_L, of making a saccade of intended length L:

$$H_L = \sum_i P_L(x) H(x) \tag{15.10}$$

Finally, the model makes the saccade that minimizes H_L, and thereby maximizes the probability of identifying the word. In cases where two or more possible saccades yield the same expected entropy, Mr. Chips executes the longest saccade possible, so as to maximize reading speed.

Because Mr. Chips was designed to investigate how lexical knowledge and restrictions on visual encoding affect the lengths of saccades and the locations of fixations, the model does not answer the "when" question of eye movement control. The model does show several emergent properties that are consistent with the literature, however. For example, the model predicts saccades that have a mean length of seven character spaces (McConkie et al., 1988) tend to be directed towards the optimal viewing position (O'Regan, 1990). The model also predicts parafoveal preview because the left-most letters of upcoming words are often identified before the words are directly fixated. Although it does not seem plausible that readers compute the expected amount of information to be gained from each possible saccade length so as to make the saccade that maximizes this gain, Klitz et al. (2000) acknowledge that their model "is not intended as an exact model of how humans perform a task, but rather establishes an upper bound (i.e., a level of competence) for human performance." The model has been used in this manner to account for the erratic eye movements that are often observed when persons with *scotomata* (i.e., blind spots in the visual field) read text (Klitz et al., 2000). Rather than being maladaptive, these erratic patterns of eye movements seem to provide a way of maximizing the rate of information extraction. Finally, the Mr. Chips algorithm is well approximated by the simple heuristic of left-justifying the target word in the high-resolution part of vision, so that, on some level, the model is psychologically plausible.

Collectively, it is unequivocal that the processing models extend the theoretical coverage of the oculomotor models by attempting to specify how *the* key component of reading—word recognition—influences and is influenced by vision and the oculomotor system[4]. This is important because many linguistic variables have well documented effects on eye movements during reading (for reviews, see Rayner and Duffy, 1988; and Rayner and Sereno, 1994). Indeed, much of the interest surrounding the use of the eye-tracking methodology is that it affords a relatively non-intrusive way to study language processing on-line (see Endnote 1). Of course, not all of the processing models are equally successful in handling the phenomena addressed by the oculomotor models. Table 1 summarizes how well the different reading models handle the phenomena that were discussed in this chapter. In the next section, we will briefly discuss how the various reading models are currently being used to find answers to applied (e.g., engineering, clinical, etc.) questions.

Table 1. A comparison of the reading models[a] with respect to reading-related phenomena.[b]

Reading Phenomena	Oculomotor Models			Processing Models				
	Target Long Words	Strategy-Tactics	Minimal Control	CC-Reader	Attention-Shift	E-Z Reader	ASM	Mr. Chips
Landing-Site Distributions	Yes	Yes	No	No	No	Yes	Yes	No
Systematic Range Errors	Yes	Yes	No	No	No	Yes	No	No
Viewing-Position Effects	Ltd	Yes	No	No	No	Yes	Yes	Ltd
Word-based Measures	No	No	Ltd	Ltd	Ltd	Yes	Ltd	No
Frequency Effects	No	No	No	Ltd	Ltd	Yes	Ltd	No
Parafoveal Preview	No	No	Ltd	No	Ltd	Yes	Ltd	Ltd
Spillover Effects	No	No	No	No	No	Yes	No	No
Fixation Costs for Skipping	No	No	No	No	No	Yes	Ltd	No

[a]The primary references for the reading models are: (a) *Target Long Words* (Reilly and O'Regan, 1998); (b) *Strategy-Tactics* (O'Regan, 1990, 1992); (c) *Minimal-Control* (Suppes, 1990, 1994); (d) *CC-Reader* (Just and Carpenter, 1980, 1987, 1992; Thibadeau et al., 1982); (e) *Attention-Shift* (Morrison, 1984); (f) *E-Z Reader* (Reichle et al., 1998, 1999); (g) *ASM* (Reilly, 1993); (h) *Mr. Chips* (Legge et al., 1997; Klitz et al., 2000).

[b]*Yes* indicates that a model can handle a result; *No* indicates that it cannot; *Ltd* indicates that the prediction is limited in some way (e.g., the model predicts a parafoveal preview benefit, but the benefit is not modulated by foveal processing difficulty).

15.2 APPLICATIONS

The models that were reviewed in the first section of this chapter have obviously furthered our understanding of the reading process and—more specifically—how eye movements are related to vision and language. Although a complete understanding of the reading process would in itself be invaluable, the models have also proved useful in more immediate and tangible ways.

For example, among reading researchers, there has been a debate about how to analyze eye movements (see Inhoff and Radach, 1998). Because *microsaccades* (i.e., eye movements within approximately 25 ms and of 1-

20' of arc, and therefore smaller than a single character space) and *micronystagmus* (i.e., physiological tremors having an amplitude of less than 0.3' of arc) introduce uncertainties into eye-movement data, the algorithms that are used to calculate fixation durations sometimes lead to discrepancies. One possible solution to this problem is to use the reading models to make inferences about the intended locations of saccades, so that these predictions can be used to guide the analysis of the eye-movement data. Salvucci and Anderson (1998) have shown that this type of procedure can be used to markedly increase the reliability and speed with which eye-movement data can be analyzed.

This logic behind using a model of reading to make inferences about saccade targets has also been used to engineer more intelligent gaze-control interfaces, or systems that allow users to execute human-computer interactions with their eyes (Salvucci and Anderson, 2000). Needless to say, this has the potential for significantly improving the quality of life for many people with severe physical disabilities.

The reading models are also important analytical tools for understanding *dyslexia*, or reading-specific impairments that cannot be accounted for by factors such as low intelligence, inadequate instruction, lack of motivation, etc. (Rayner and Pollatsek, 1989; also see Chapter 19 of this volume). Although visual, attentional, and oculomotor deficits may cause reading problems (Ciuffreda, 1994; Ciuffreda and Tannen, 1995; Fischer and Biscaldi, 1998), and despite the fact that dyslexics often make irregular eye movements while reading (e.g., longer fixations, shorter saccades, and more regressions; Rayner, 1986), the claim that dyslexia is generally caused by oculomotor problems (Pavlidis, 1981, 1983) has received little support (Hyönä and Olson, 1994; for a review, see Rayner, 1986). There is instead a growing consensus that dyslexia largely stems from language-processing deficits, and that the faulty eye movements of most dyslexics reflect this more fundamental problem (Rayner, 1985, 1986; however, for recent evidence supporting the contrary view, see Eden, Stein, Wood, and Wood, 1993; and Eden et al., 1996). Indeed, the eye movements of dyslexics appear normal when they are given materials to read that are appropriate for their level of reading skill (Pirozzolo and Rayner, 1978), which suggests that it is the language-processing components of reading, and not the eye movements per se, which make reading difficult for dyslexics. Nevertheless, it may still be a useful exercise (both in terms of understanding reading impairments and providing additional constraints on the models) to consider how the different reading models might account for the eye movements of people with various reading deficits.

For example, the Mr. Chips model (Legge et al., 1997) was designed to investigate how different levels of visual impairment might affect eye movements during reading. In this capacity, the model has been successful (Klitz et al., 2000). A comparison of the model's performance to that of a human in a reading task[5] with a simulated scotoma indicated that, in contrast to the model, the human had difficulty integrating information across central scotomata more than a single character-space in size. The human was instead more proficient using visual information from only a single side of the scotoma. In contrast to the model, the human was also more proficient using the purely visual strategy of placing the region of normal vision over all of the character spaces, rather than using lexical knowledge to winnow down the possible identities of letter strings. Although this purely visual strategy sacrificed saccade length, it markedly increased reading speed. These analyses, therefore, suggest that the seemingly erratic eye movements of readers with scotomata may be adaptive; although their eye movements do not allow a maximal amount of information to be extracted from the page with each new fixation, the time needed to extract the information is minimized.

Of course, this raises the question of whether reading difficulties that are caused by visual and/or oculomotor deficits might be alleviated by rapidly presenting the text one word at a time (in the same spatial location on a computer screen) so that eye movements are not necessary. When this is done using a fixed rate of presentation (e.g., 250 ms per word), readers can comprehend sentences and short passages normally (Juola, Ward, and McNamara, 1982), but have difficulty understanding longer texts (Masson, 1983). This difficulty largely stems from the fact that, under such conditions, the readers cannot make regressions back to those earlier places in the text, and hence have difficulty whenever their comprehension falters. Unfortunately, making the text presentation rate self-paced (i.e., text is advanced or reversed via button presses) only provides a partial solution to this problem; although the reader can go back to earlier parts of the text, the reading speed is slowed by the need to make manual responses, which require considerably more time to execute than do saccades.

The more general question of whether or not people can learn to read more effectively by being trained to move their eyes in a specific manner has also been hotly debated in the context of speed-reading. Misconceptions about the validity of speed-reading continue to exist (especially among lay people) despite the fact that the best available evidence indicates that, as reading speed increases, comprehension rapidly falters (Just, Carpenter, and Masson, 1982; also reported in Just and Carpenter, 1987). Depending upon the difficulty of the material being read, college students normally read 250-350 words per minute (Rayner and Pollatsek, 1989). As with dyslexia, however, this issue could be examined by using the reading models to both determine the maximal attainable reading speed and to gain a better understanding of the constraints on reading speed. For example, the E-Z Reader model (which was fit to the mean data from thirty college students reading 48 sentences, excluding those trials in which regressions were made) predicts an average reading speed of approximately 315 words per minute. One could perhaps adjust the model's parameter values, however, to simulate the word identification performance of highly skilled individuals (see Eqs. 15.1 and 15.2). Likewise, one could include different assumptions about saccade-targeting strategies (i.e., the default assumption that saccades are directed towards the optimal viewing position) and/or saccadic accuracy (Eq. 15.5) in the model. Of course, these new assumptions and parameter values must not violate known constraints; for example, the lower bound for lexical access must be at least 100 ms because it takes approximately this long for neural transmission to propagate information from the retina to the occipital cortex (Rayner and Pollatsek, 1989). Likewise, other factors (e.g., minimal saccadic latencies, visual acuity limitations, etc.) are less subject to variation, and hence place hard constraints on the maximal attainable reading speed. Still, such an exercise might be informative with respect to understanding how these variables (in the limit) jointly determine reading speed.

Finally, to illustrate how a model of reading is being used to understand impairments in language processing, many of the core principles of CC-Reader (Just and Carpenter, 1992) were incorporated into a model of language comprehension to test the hypothesis that aphasia may be due to limitations in the verbal working memory resources that are necessary to support linguistic processing (Haarmann, Just, and Carpenter, 1997). The model simulated normal and aphasic levels of comprehension accuracy for sentences of varying syntactic difficulty. It also made a specific prediction about aphasic performance that was confirmed, thereby supporting the model's main assumption and furthering our understanding of aphasia.

Of course, the preceding examples were only mentioned to show the tremendous potential for using models to understand the reading process. If the reading models provide a theoretical framework for understanding the link between eye movements and cognition, then they may offer important new insights about normal reading and its development. In the final section of this chapter, we will attempt to frame these issues within the context of what has been learned about reading from cognitive neuroscience.

15.3 NEURAL SYSTEMS

The current consensus is that most cognitive operations are supported by groups of cortical and subcortical areas that are organized into large-scale networks (Mesulam, 1990, 1998; Posner and Raichle, 1997). Therefore, it should not be surprising that complex tasks like reading (which consists of many cognition operations) are themselves supported by several of these large networks. In the specific case of reading, these include (at a minimum) the systems that mediate vision, visuospatial attention, eye-movement control, and language. In this section, we will provide a brief overview of these systems, and also speculate about how the language-processing system might direct the system that is responsible for programming and executing saccades.

The most natural place to begin our analysis of the neural systems underlying reading is the printed page. Visual processing of the text begins in the retina, and then progresses by way of the optic nerve to the optic chiasm to the optic tract. From here, the visual-processing stream splits into two pathways. The first projects to the lateral geniculate nucleus, and then the occipital cortex; the second innervates several subcortical structures, including one that is known to play a key role in eye movements—the superior colliculus (Leigh and Zee, 1999; Sparks and Mays, 1990). On the basis of numerous electrophysiology (i.e., single cell recording) experiments with non-human primates, it has been estimated that 30 or more distinct cortical areas are involved in vision (Felleman and Van Essen, 1991; Maunsell and Newsome, 1987; Van Essen and DeYoe, 1995), although many of these areas perform functions that are less central to reading (e.g., color perception). Feature extraction (which is necessary for the perception of letters) is thought to be completed within the primary visual and extrastriate cortices (Grill-Spector et al., 1998). This has led to the proposal that the left medial extrastriate cortex is intrinsically involved in the recognition of word forms (Peterson, Fox, Posner, Mintun, and Raichle, 1989; Peterson, Fox, Snyder, and Raichle, 1990; Pugh et al., 2000), although this issue is still unresolved (see Posner, Abdullaev, McCandliss, and Sereno, 1999a, 1999b; and Price, 1997).

The visual-processing stream leaves the occipital cortex via two anatomically and functionally distinct pathways (Maunsell and Newsome, 1987; Sagi and Julesz, 1985; Ungerleider and Mishkin, 1985; Van Essen and DeYoe, 1995). The ventral, or "what," pathway extends along the inferior temporal cortices, and is thought to play an important role in the feature integration and object recognition (Ishai, Ungerleider, Martin, Schouten, and Haxby, 1999; Tanaka, 1996). Because words are in some sense visual objects, it is not surprising that the ventral system has also been implicated in the recognition of word forms (Cohen et al., 2000; Poldrack, Desmond, Glover, and Gabrieli, 1998).

The dorsal, or "where," pathway is thought to represent spatial information, such as the relative positions of two or more objects (Ungerleider, Courtney, and Haxby, 1998; Ungerleider and Haxby, 1994). (However, the dorsal system may also provide an interface between perception and action; Goodale and Milner, 1992.) The dorsal pathway has also been implicated in visuospatial attention; the region around the intraparietal sulcus (i.e., the parietal eye fields) is thought to be a central component of the visuospatial attention network. The other components include the superior colliculus (part of the mid-brain), the pulvinar nucleus of the thalamus, and a region that includes the precentral sulcus/gyrus and posterior tip of the superior frontal sulcus (i.e., the frontal eye fields) (Corbetta, Miezin, Schulman, and Peterson, 1993; Goldberg, 1994; Leigh and Zee, 1999; Luna et al., 1998; Kim et al., 1999; Posner and Raichle, 1997; Rafal and Robertson, 1995; Sweeney et al., 1996). Recent neuroimaging and electrophysiology work suggests that this network is involved in both covert and overt shifts of visuospatial attention, and that covert attention is probably represented in motor (eye movement) coordinates (Corbetta, 1998; Kim et al., 1999). This attention network also modulates object analysis in the ventral visual-processing pathway (Corbetta, 1998), as well as perceptual processing in the striate and extrastriate cortices (Somers, Dale, Seiffert, and Tootell, 1999).

Turning now to the higher-level, linguistic components of reading, a number of neuroimaging experiments indicate that the left inferior frontal gyrus (*Broca's area*) and the posterior part of the left superior and middle temporal gyri (*Wernicke's area*) are the two major language-processing areas. Both areas are engaged by a variety of receptive and expressive language tasks, including: (a) reading (Bavelier et al., 1997; Binder et al., 1997; Just, Carpenter, Keller, Eddy, and Thulborn, 1996); (b) speech comprehension (Binder et al., 1997; Caplan, Alpert, and Waters, 1999; Schlosser, Aoyagi, Fulbright, Gore, and McCarthy, 1998; Stromswold, Caplan, Alpert, and Rauch, 1996); and (c) speech production (Bookheimer et al., 1997; Müller et al., 1997). Although their exact functional roles are not known, it has been suggested that Broca's area is involved in articulatory-syntactic processing, and that Wernicke's area supports lexical-semantic processing (Mesulam, 1990). Because Wernicke's area receives input from the primary auditory cortex, its role in lexical processing may be to bind the phonological representations of words to their semantic representations (which are distributed elsewhere in the associative cortex; Mesulam, 1998).

Because the same language system is used to understand written and spoken language, one of the central questions in reading research has been:

How are the characters on a printed page converted into linguistic-based codes? The results of several recent neuroimaging experiments suggest that the left angular gyrus (which is located in the posterior part of the inferior parietal lobule) plays a critical role in computing grapheme-to-phoneme correspondences (Horwitz, Rumsey, and Donohue, 1998; Pugh et al., 2000). Because the left angular gyrus lies at the juncture of the extrastriate cortex and Wernicke's area, it is ideally situated to convert the orthographic forms of words into their phonological counterparts.

A recent meta-analysis (Posner et al., 1999a, 1999b) provides compelling evidence that word-form processing can be completed within the time window that is necessary for it to function as a signal to initiate saccadic programming. The results of a recent ERP experiment, for example, indicate that certain aspects of lexical processing (e.g., word frequency) can be discerned within 100-160 ms (Sereno, Rayner, and Posner, 1998). This would leave plenty of time (up to 80 ms) to signal the oculomotor system to begin programming a saccade (if one assumes a 180-ms saccadic latency and a 300-ms fixation). Unfortunately, the results of this study do not show that word identification drives eye movements, nor does the study suggest how this happens (if this interpretation of the data is correct).

One possibility (which is admittedly very speculative, but none-the-less consistent with the E-Z Reader model of reading; Reichle et al., 1998) is depicted by the numbered event sequences in Figs. 15.6 and 15.7. Figure 15.6 depicts the eye movements that might occur as $word_n$ and $word_{n+1}$ are identified during reading, the cognitive processes (as specified by E-Z Reader) which give rise to this pattern of eye movements, and the cortical and subcortical systems in which may underlie these cognitive processes. Fig. 15.7 shows both where in the brain these neural systems are localized, and how processing is coordinated among these systems.

In the sequence of events depicted in Figs. 15.6 and 15.7, the visual form of $word_n$ hits the retina and, after approximately 70 ms, has been processed by the primary visual cortex (1). The individual letter features are then integrated at successively higher levels as visual processing cascades from the striate to the extrastriate cortex (2). After approximately 100-160 ms, $word_n$'s orthographic form has been identified in the left extrastriate cortex (2) and/or left inferior temporal gyrus (3), and its phonological representation has been assembled in the left angular gyrus (4). Up to this point in time, both the eyes and attention have been on $word_n$. With the identification of $word_n$, however, the parietal eye fields (5) disengage attention. This allows the pulvinar nucleus of the thalamus (6) to move the attention "spotlight" forward, so that the frontal eye fields (7) and superior colliculus (8) can use low-spatial frequency information (e.g., word length) from the primary visual cortex to begin programming a saccade to $word_{n+1}$.

This saccadic program takes approximately 180 ms to complete. During this time, the processing of $word_n$ continues; its orthographic (2 and 3) and/or phonological form(s) (4) are used to activate the word's meaning via connections from Wernicke's area (9) to various parts of the associative cortex. If the meaning is accessed before the saccadic program has been completed, then the pulvinar (6) enhances the processing of $word_{n+1}$, and a preview benefit ensues. Otherwise, a saccade is executed by the neural circuitry of the brainstem (labeled "BS" in Fig. 15.7; see Leigh and Zee, 1999) and the eyes move to $word_{n+1}$.

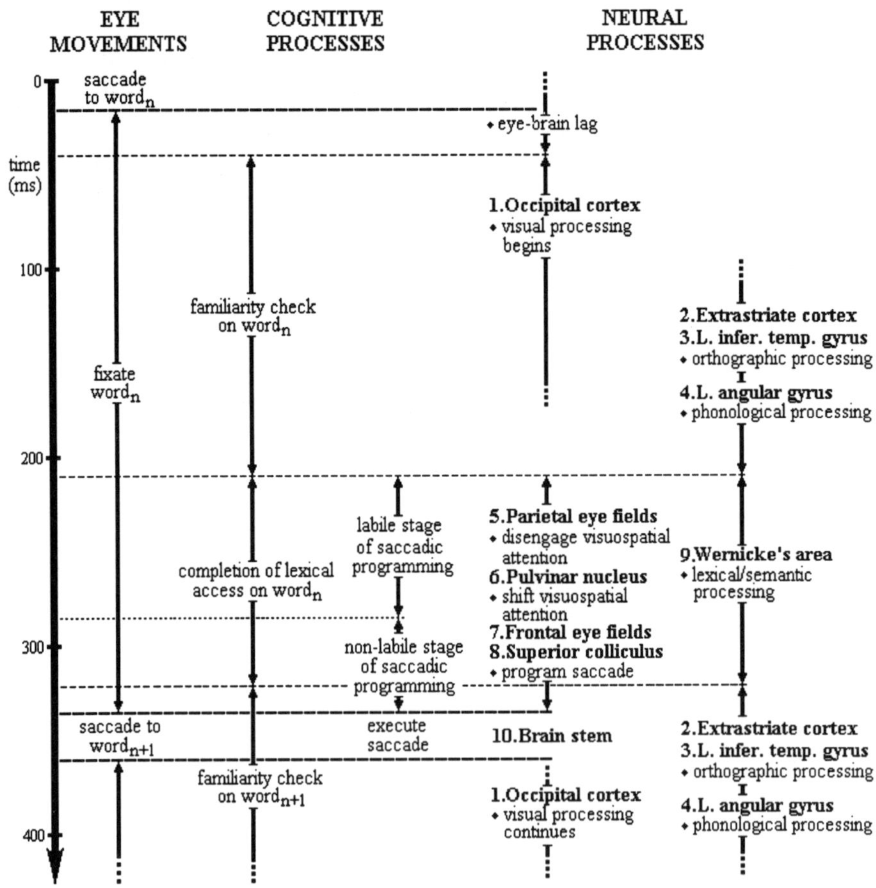

Figure 15.6. The time course of eye movements and cognitive processing (as specified by E-Z Reader; Reichle et al., 1998) and the cortical and subcortical regions which may mediate these processes.

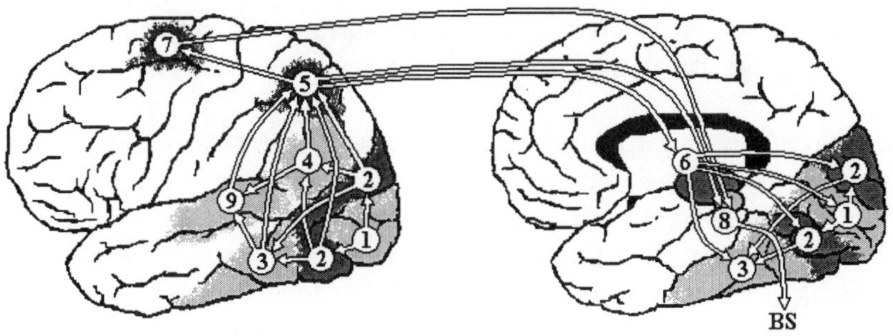

Figure 15.7. Sagittal views of the left lateral (left panel) and medial (right panel) cortical, thalamic (pulvinar nucleus), and mid-brain (superior colliculus) structures that support eye movements during reading. The numbers correspond to the following structures: (1) primary visual cortex (Brodmann's Area [BA] 17); (2) extrastriate cortex (BAs 18 and 19); (3) inferior temporal gyrus (BAs 20 and 37); (4) posterior inferior parietal lobule (i.e., angular gyrus; BA 39); (5) intraparietal sulci (i.e., parietal eye fields, BAs 7 and 40); (6) pulvinar nucleus of the thalamus; (7) superior prefrontal and posterior superior frontal gyri (i.e., frontal eye fields; BAs 6 and 8); (8) superior colliculus; and (9) posterior middle and superior temporal gyri (i.e., Wernicke's area; BAs 21 and 22). The pathway exiting the superior colliculus leads to the brainstem saccade generator, labeled "BS." (Although Fig. 15.7 only shows the left hemisphere, the right hemisphere homologues of structures 1, 2, 5, 6, 7, 8, and 40 are also components of the visuospatial, attention, and oculomotor networks.)

15.4 FUTURE RESEARCH

As Brewer (1972) noted nearly three decades ago: "an adequate model of reading will have to include some mechanism to allow an interaction between the higher linguistic processing and the control of eye fixations" (p. 360). The work that was reviewed in this chapter suggests that Brewer's assertion was well founded. To explain the relationship between language processing and eye movements during reading, any viable model of reading will (almost by definition) have to include an account of language processing. Although considerable progress has been made towards understanding the individual components of reading (e.g., word recognition, visuospatial attention, oculomotor control), the real challenge will be to develop a theoretical framework to understand their interactions. Indeed, the central assumption of processing models (i.e., cognition drives eye movements during reading) has yet to be directly investigated using the techniques of cognitive neuroscience. To date, there is no conclusive evidence of a direct functional link between the word recognition area(s) and the attention and/or eye-movement systems. However, if the processing models are correct, one should be able to demonstrate direct communication between language processing and eye-movement systems during normal reading.

The current reading models were developed primarily around the eye-tracking methodology; by studying the patterns of eye movements during reading, theorists were able to make inferences about the cognitive operations that occur during reading. Yet, the last decade has witnessed unprecedented theoretical and methodological advances in the field of cognitive neuroscience. Consequently, it is almost certain that advances in cognitive neuroscience will guide the development of the next generation of reading models. Like eye movement data in the past, the discoveries of today will provide important guideposts for models of the future.

15.5 ACKNOWLEDGMENTS

Preparation of this chapter was supported by Grant HD26765. The second author was also supported by a Research Scientist Award from the National Institute of Mental Health (MH01255). We would like to thank Ken Ciuffreda, Vaibhav Diwadkar, George Hung, Timothy Keller, Robert Mason, Rebecca Sandak, and Matt Starr for helpful comments on earlier versions of this chapter. Address correspondence to: Erik Reichle, Department of Psychology, Learning Research and Development Center, University of Pittsburgh, 3939 O'Hara St., Pittsburgh, PA 15260; or via e-mail at: reichle@pitt.edu.

15.6 ENDNOTES

1. Although a variety of behavioral tasks (e.g., pronunciation, lexical decision, semantic judgements) have been used to study reading, the endeavor has benefited tremendously from the advent of eye trackers and *eye-contingent paradigms*. This technology and methodology allows the experimenter to control what the reader sees by making the material that is displayed on the computer screen contingent upon where the eyes are looking. The advantages of this approach are that reading can be studied on-line, in an ecologically valid fashion, and that the experimenter can maintain precise moment-by-moment control over the materials being read. Most of the findings reported in this chapter come from experiments that have used the eye-tracking methodology. (For an overview of the eye-tracking methodology, see Rayner, 1979a.)
2. The preferred viewing location differs from the optimal viewing position in that it is the location where the eyes actually land in a word whereas the optimal viewing position is the position in a word that results in the easiest identification of the word. The preferred

viewing location is a bit to the left of the optimal viewing position. Studies dealing with the optimal viewing position have typically involved isolated word identification, whereas those dealing with the preferred viewing location have typically involved reading.

3. These unpublished simulation results were completed by the authors in collaboration with Alexander Pollatsek.
4. A number of word-recognition models have been proposed (Brown, 1991; Bullinaria, 1997; McClelland and Rumelhart, 1981; Paap et al., 1982; Plaut, McClelland, Seidenberg, and Patterson, 1996; Seidenberg, 1989; Seidenberg and McClelland, 1989). These models generally explain how orthography maps onto phonology and/or meaning, and how this process is affected by lexical variables (e.g., normative frequency, grapheme-phoneme regularity, etc.). Unfortunately, these models are limited in two ways: First, the entry point into these models is usually some highly abstract representation of the orthography that bears little resemblance to the features that one might expect to be encoded by the visual system. Second, the models are generally fit to data from paradigms other than natural reading (e.g., lexical decision latencies). The models therefore say very little about the relationships among vision, eye movements, and word identification. One interesting exception to this is McClelland's (1986) *programmable blackboard* model of reading. This model was designed to examine how fixation locations and visual acuity restrictions affect the model's word recognition performance.
5. These results are open to alternative interpretations because the task was not natural reading (and thus did not actually require eye movements), but instead required the subject to read text on a computer monitor that was displayed through a stationary nine-character "window." The text was manually advanced via pressing keys that moved the text forward 1-9 character spaces (or backwards 1-3 character spaces), and a mask covering 1, 3, or 5 character spaces was placed over the center of the viewing window to occlude letters in scotomata conditions.

15.7 REFERENCES

Bahill, T. and Stark, L., 1979, The trajectories of saccadic eye movements, *Sci. Am.* **240**: 108-117.

Balota, D. A., Pollatsek, A., and Rayner, K., 1985, The interaction of contextual constraints and parafoveal visual information in reading, *Cognit. Psychol.* **17**: 364-390.

Bavelier, D., Corina, D., Jezzard, P., Padmanabhan, S., Clark, V. P., Karni, A., Prinster, A., Braun, A., Lalwani, A., Rauschecker, J. P., Tuner, R., and Neville, H., 1997, Sentence reading: A functional MRI study at 4 Tesla, *J. Cogn. Neurosci.* **9**: 664-686.

Becker, W. and Jürgens, R., 1979, An analysis of the saccadic system by means of double step stimuli, *Vis Res* **19**: 967-983.

Binder, J. R., Frost, J. A., Hammeke, T. A., Cox, R. W., Rao, S. M., and Prieto, T., 1997, Human brain language areas indentified by functional magnetic resonance imaging, *J. Neurosci.* **17**: 353-362.

Binder, K. S. and Rayner, K., 1998, Contextual strength does not modulate the subordinate bias effect: Evidence from eye fixations and self-paced reading, *Psychonom. Bull. Rev.* **5**: 271-276.

Bookheimer, S. Y., Zeffiro, T. A., Blaxton, T., Malow, B. A., Gaillard, W. D., Sato, S., Kufta, C., Fedio, P., and Theodore, W. H., 1997, A direct comparison of PET activation and electocortical stimulation mapping for language localization, *Neurology.* **48**: 1056-1065.

Brewer, W. F., 1972, Is reading a letter-by-letter process?: A discussion of Gough's paper, in: *Language by Ear and Eye*, J. F. Kavanagh and I. G. Mattingly, eds., MIT Press, Cambridge, pp. 359-365.

Brown, P., 1991, DEREK: The direct encoding routine for evoking knowledge, in: *Basic Processes in Reading: Visual Word Recognition*, D. Besner and G. W. Humphreys, eds., Erlbaum, Hillsdale, pp. 104-147.

Brysbaert, M. and Vitu, F., 1998, Word skipping: Implications for theories of eye movement control in reading, in: *Eye Guidance in Reading and Scene Perception*, G. Underwood, ed., Elsevier, Oxford, pp. 125-147.

Bullinaria, J. A., 1997, Modeling reading, spelling, and past tense learning with artificial neural networks, *Brain Lang.* **59**: 236-266.

Caplan, D., Alpert, N., and Waters, G., 1999, PET studies of syntactic processing with auditory sentence presentation, *Neuroimage* **9**: 343-351.

Carpenter, P. A. and Just, M. A., 1983, What your eyes do while your mind is reading, in: *Eye Movements in Reading: Perceptual and Language Processes*, K. Rayner, ed., Academic Press, New York, pp. 275-307.

Ciuffreda, K. J., 1990, Etymology of the word 'saccade', *Ophthal. Physiol. Opt.* **10**: 307.

Ciuffreda, K. J., 1994, Reading eye movements in patients with oculomotor disturbances, in: *Eye Movements in Reading*, J. Ygge and G. Lennerstrand, eds., Pergamon Press, Oxford.

Ciuffreda, K.J. and Bassil, N., 1990, Translation of "Essay on the Physiology of Reading" (by E. Javal), *Ophthal. Physiol. Opt.* **10**: 381-384.

Ciuffreda, K. J. and Tannen, B., 1995, *Eye Movement Basics for the Clinician*, C. V. Mosby, St. Louis.

Clark, A., 1991, *Microcognition: Philosophy, Cognitive Science, and Parallel Distributed Processing*, MIT Press, Cambridge.

Cohen, L., Dehaene, S., Naccache, L., Lehéricy, S., Dehaene-Lambertz, G., Hénaff, M.-A., and Michel, F., 2000, The visual word form area: Spatial and temporal characterization of an initial stage of reading in normal subjects and posterior split-brain patients, *Brain* **123**: 291-307.

Corbetta, M., 1998, Frontoparietal cortical networks for directing attention and the eye to visual locations: Identical, independent, or overlapping neural systems?, *Proc. Nat. Acad. Sci. USA.* **95**: 831-838.

Corbetta, M., Miezin, F. M., Schulman, G. L., and Petersen, S. E., 1993, A PET study of visuospatial attention, *J. Neurosci.* **13**: 1202-1226.

Dopkins, S., Morris, R. K., and Rayner, K., 1992, Lexical ambiguity and eye fixations during reading: A test of competing models of lexical ambiguity resolution, *J. Mem. Lang.* **31**: 461-477.

Duffy, S. A., Morris, R. K., and Rayner, K., 1988, Lexical ambiguity and fixation times in reading, *J. Mem. Lang.* **27**: 429-446.

Eden, G. F., Stein, J. F., Wood, H. M., and Wood, F. B., 1993, Differences in eye movements and reading problems in dyslexic and normal children, *Vis. Res.* **34**: 1345-1358.

Eden, G. F., VanMeter, J. W., Rumsey, J. M., Maisog, J. M., Woods, R. P., and Zeffiro, T. A., 1996, Abnormal processing of visual motion in dyslexia revealed by functional brain imaging, *Nature* **382**: 66-69.

Felleman, D. J. and Van Essen, D. C., 1991, Distributed hierarchical processing in the primate cerebral cortex, *Cereb. Cortex.* **1**: 1-47.

Fischer, B. and Biscaldi, M., 1999, Saccadic eye movements in dyslexia, in: *Reading and Dyslexia: Visual and Attentional Processes*, J. Everatt, ed., Routledge, New York, pp 91-121.

Francis, W. N. and Ku era, H., 1982, *Frequency analysis of English usage: Lexicon and grammar*, Houghton-Mifflin, Boston.

Frazier, L. and Rayner, K., 1982, Making and correcting errors during sentence comprehension: Eye movements in the analysis of structurally ambiguous sentences. *Cognit. Psychol.* **14**: 178-210.

Goldberg, M. E., 1994, The cortical control of saccadic eye movements, in: *Eye Movements in Reading*, J. Ygge and G. Lennerstrand, eds., Pergamon Press, Oxford, pp. 51-64.

Goodale, M. A. and Milner, A. D., 1992, Separate visual pathways for perception and action, *Trends in Neurosci.* **15**: 20-25.

Gough, P. B., 1972, One second of reading, in: *Reading by Ear and Eye*, J. F. Kavanagh and I. G. Mattingly, eds., MIT Press, Cambridge, pp. 331-358.

Grill-Spector, K., Kushnir, T., Hendler, T., Edelman, S., Itzchak, Y., and Malach, R., 1998, A sequence of object-processing stages revealed by fMRI in the human occipital lobe, *Human Brain Map.* **6**: 316-328.

Haarmann, H., Just, M. A., and Carpenter, P. A., 1997, Aphasic sentence comprehension as a resource deficit: A computational approach, *Brain Lang.* **59**: 76-120.

Henderson, J. M. and Ferreira, F., 1990, Effects of foveal processing difficulty on the perceptual span in reading: Implications for attention and eye movement control, *J. Exp. Psychol. Learn. Mem. Cogn.* **16**: 417-429.

Horwitz, B., Rumsey, J. M., and Donohue, B. C., 1998, Functional connectivity of the angular gyrus in normal reading and dyslexia, *Proc. Nat. Acad. Sci. USA.* **95**: 8939-8944.

Hyönä, J. and Olson, R. K., 1994, Dyslexic and normal readers' eye movement patterns in reading, visual search and tracking, in: *Eye Movements in Reading*, J. Ygge and G. Lennerstrand, eds., Pergamon Press, Oxford, pp. 233-244.

Inhoff, A.W. and Radach, R., 1998, Definition and computation of oculomotor measures in the study of cognitive processes, in: *Eye Guidance in Reading and Scene Perception*, G. Underwood, ed., Elsevier, Oxford, pp. 29-53.

Inhoff, A.W., Starr, M., and Shindler, K.L., 2000, Is the processing of words during eye fixations in reading strictly serial?, *Percept. Psychophys.*, in press.

Ishai, A., Ungerleider, L. G., Martin, A., Schouten, J. L., and Haxby, J. V., 1999, Distributed representation of objects in the human ventral visual pathway, *Proc. Nat. Acad. Sci. USA* **96**: 9379-9384.

Julesz, B., 1991, Early vision and focal attention, *Rev. Mod. Physics* **63**: 735-772.

Juola, J. F., Ward, N. J., and McNamara, T., 1982, Visual search and reading of rapid serial presentations of letter strings, words, and text, *J. Exp. Psychol. Gen.* **111**: 208-227.

Just, M. A. and Carpenter, P. A., 1980, A theory of reading: From eye fixations to comprehension, *Psychol. Rev.* **87**: 329-354.

Just, M. A. and Carpenter, P. A., 1987, *The Psychology of Reading and Language Comprehension*, Allyn and Bacon, Boston.

Just, M. A. and Carpenter, P. A., 1992, A capacity theory of comprehension: Individual differences in working memory, *Psychol. Rev.* **99:** 122-149.

Just, M. A., Carpenter, P. A., Keller, T. A., Eddy, W. F., and Thulborn, K. R., 1996, Brain activation modulated by sentence comprehension, *Science.* 274: 114-116.

Just, M. A., Carpenter, P. A., and Masson, M. E. J., 1982, What eye movements tell us about speed reading and skimming, *Eye-Lab Technical Report*, Carnegie Mellon University, Pittsburgh.

Kennedy, A., 1998, The influence of parafoveal words on foveal inspection time: Evidence for a processing trade-off, in: *Eye Guidance in Reading and Scene Perception*, G. Underwood, ed., Elsevier, Oxford, pp. 149-179.

Kennedy, A., 2000, Parafoveal processing in word recognition, *Q. J. Exp. Psychol.* **53A**: 429-455.

Kennison, S. M. and Clifton, C., 1995, Determinants of parafoveal perview benefit in high and low working memory capacity readers: Implications for eye movement control, *J. Exp. Psychol. Learn. Mem. Cogn.* **21**: 68-81.

Kim, Y.-H., Gitelman, D. R., Nobre, A. C., Parrish, T. B., LaBar, K. S., and Mesulam, M. M., 1999, The large-scale neural network for spatial attention displays multifunctional overlap but differential asymmetry, *Neuroimage.* **9**: 269-277.

Klitz, T. S., Legge, G. E., and Tjan, B. S., 2000, Saccade planning in reading with central scotomas: Comparison of human and ideal performance, in: Reading as a Perceptual Process, A. Kennedy, R. Radach, D. Heller, and J. Pynte, eds., Elsevier, Oxford, manuscript in press.

Legge, G. E., Klitz, T. S., and Tjan, B. S., 1997, Mr. Chips: An ideal-observer model of reading, *Psychol. Rev.* **104**: 524-553.

Leigh, R. J. and Zee, D. S., 1994, *The Neurology of Eye Movements*, Oxford University Press, New York.

Luna, B., Thulborn, K. R., Strojwas, M. H., McCurtain, B. J., Berman, R. A., Genovese, C. R., and Sweeney, J. A., 1998, Dorsal cortical regions subserving visually guided saccades in humans: An fMRI study, *Cereb. Cortex.* **8**: 40-47.

Masson, M. E. J., 1983, Conceptual processing of text during skimming and rapid sequential reading, *Mem. Cognit.* **11**:262-274.

Maunsell, J. H. R. and Newsome, W. T., 1987, Visual processing in monkey extrastriate cortex, *Annu. Rev. Neurosci.* **10**: 363-401.

McClelland, J. L., 1986, The programmable blackboard model of reading, in: *Parallel Distributed Processing: Explorations in the Microstructure of Cognition, Vol. 2: Psychological and Biological Models*, McClelland, J. L., Rumelhart, D. E., and the PDP Research Group, eds., MIT Press, Cambridge, pp 122-169.

McClelland, J. L. and Rumelhart, D. E., 1981, An interactive activation model of context effects in letter perception: Part I. An account of basic findings, *Psychol. Rev.* **88**: 375-407.

McClelland, J. L., Rumelhart, D. E., and the PDP Research Group, 1986, *Parallel Distributed Processing: Explorations in the Microstructure of Cognition, Vol. 2: Psychological and Biological Models*, MIT Press, Cambridge.

McConkie, G. W. and Dyre, B. P., 2000, Eye fixation durations in reading: Models of frequency distributions, in: *Reading as a Perceptual Process*, A. Kennedy, R. Radach, D. Heller, and J. Pynte, eds., Elsevier, Oxford, manuscript in press.

McConkie, G. W., Kerr, P. W., and Dyre, B. P., 1994, What are 'normal' eye movements during reading: Toward a mathematical description, in: *Eye Movements in Reading*, J. Ygge and G. Lennerstrand, eds., Pergamon Press, Oxford, pp. 315-328.

McConkie, G. W., Kerr, P. W., Reddix, M. D., and Zola, D., 1988, Eye movement control during reading: I. The location of initial eye fixations on words, *Vis. Res.* **28**: 1107-1118.

McConkie, G. W., Kerr, P. W., Reddix, M. D., Zola, D., and Jacobs, A. M., 1989, Eye movement control during reading: II. Frequency of refixating a word, *Percept. Psychophys* **46**: 245-253.

McConkie, G. W. and Rayner, K., 1975, The span of the effective stimulus during a fixation in reading, *Percept. Psychophys.* **17**: 578-586.

McConkie, G. W. and Rayner, K., 1976, Asymmetry of the perceptual span in reading. *Bull. Psychonom. Soc.* **8**: 365-368.

McConkie, G. W., Zola, D., Grimes, J., Kerr, P. W., Bryant, N. R., and Wolff, P. M., 1991, Children's eye movements during reading, in: *Vision and Visual Dyslexia*, J. F. Stein, ed., Macmillan Press, London, pp. 251-262.

Mesulam, M. M., 1990, Large-scale neurocognitive networks and distributed processing for attention, language, and memory, *Ann. Neurol.* **28**: 597-613.

Mesulam, M. M., 1998, From sensation to cognition, *Brain* 121: 1013-1052.

Morrison, R. E., 1984, Manipulation of stimulus onset delay in reading: Evidence for parallel programming of saccades, *J. Exp. Psychol. Hum. Percept. Perf.* **10**: 667-682.

Morrison, R.E. and Rayner, K., 1981, Saccade size in reading depends upon character spaces and not visual angle, *Percept. Psychophys.* **30**: 395-396.

Müller, R.-A., Rothermel, R. D., Behen, M. E., Muzik, O., Mangner, T. J., and Chugani, H. T., 1997, Receptive and expressive language activations for sentences: a PET study, *Neuroreport.* **8**: 3767-3770.

O'Regan, J. K., 1990, Eye movements and reading, in: *Eye Movements and Their Role in Visual and Cognitive Processes*, E. Kowler, ed., Elsevier, Amsterdam, pp. 395-453.

O'Regan, J. K., 1992, Optimal viewing position in words and the strategy-tactics theory of eye movements in reading, in: *Eye Movements and Visual Cognition: Scene Perception and Reading*, K. Rayner, ed., Springer-Verlag, New York, pp. 333-354.

O'Regan, J. K. and Lévy-Schoen, A., 1987, Eye movement strategy and tactics in word recognition and reading, in: *Attention and Performance, XII: The Psychology of Reading*, M. Coltheart, ed., Erlbaum, Hillsdale, pp. 363-383.

Paap, K. R., Newsome, S. L., McDonald, J. E., and Schvaneveldt, R. W., 1982, An activation-verification model for letter and word recognition: The word-superiority effect, *Psychol. Rev.* **89**: 573-594.

Pavlidis, G. T., 1981, Do eye movements hold the key to dyslexia?, *Neuropsychologia*, **19**: 57-64.

Pavlidis, G. T., 1983, The 'dyslexic syndrome' and its objective diagnosis by erratic eye movements, in: *Eye Movements in Reading: Perceptual and Language Processing*, K. Rayner, ed., Academic Press, New York, pp. 441-466.

Perfetti, C. A. , 1985, *Reading Ability*, Oxford University Press, New York.

Peterson, S. E., Fox, P. T., Posner, M. I., Mintun, M., and Raichle, M. E., 1989, Positron emission tomographic studies of the cortical anatomy of single-word processing, *Nature* **331**: 585-589.

Peterson, S. E., Fox, P. T., Snyder, A. Z., and Raichle, M. E., 1990, Activation of extrastriate and frontal cortical areas by visual words and word-like stimuli, *Science.* **249**: 1041-1044.

Pirozzolo, F. J. and Rayner, K., 1978, Disorders of oculomotor scanning and graphic orientation in developmental Gerstmann syndrome, *Brain Lang.* **5**: 119-126.

Plaut, D. C., McClelland, J. L., Seidenberg, M. S., and Patterson, K., 1996, Understanding normal and impaired word reading: Computational principles in quasi-regular domains. *Psychol. Rev.* **103**: 56-115.

Poldrack, R. A., Desmond, J. E., Glover, G. H., and Gabrieli, J. D. E., 1998, The neural basis of visual skill learning: An fMRI study of mirror reading, *Cereb. Cortex.* **8**: 1-10.

Pollatsek, A., Bolozky, S., Well, A. D., and Rayner, K., 1981, Asymmetries in the perceptual span for Israeli readers, *Brain Lang.* **14:** 174-180.

Pollatsek, A., Lesch, M., Morris, R. K., and Rayner, K., 1992, Phonological codes are used in integrating information across saccades in word identification and reading, *J. Exp. Psychol. Hum. Percept. Perf.* **18**: 148-162.

Pollatsek, A. and Rayner, K., 1990, Eye movements and lexical access in reading, in: *Comprehension Processes in Reading*, D. A. Balota, G. B. Flores d' Arcais, and K. Rayner, eds., Erlbaum, Hillsdale, pp. 143-164.

Pollatsek, A., Rayner, K., Fischer, M. H., and Reichle, E. D., 1999, Attention and eye movements in reading, in: *Reading and Dyslexia: Visual and Attentional Processes*, J. Everatt, ed., Routledge, New York, pp. 179-209.

Posner, M. I., Aldullaev, Y. G., McCandliss, B. D., and Sereno, S. C., 1999a, Anatomy, circuitry and plasticity of word reading, in: *Reading and Dyslexia: Visual and Attentional Processes*, J. Everatt, ed., Routledge, New York, pp. 137-162.

Posner, M.I., Abdullaev, Y.G., McCandliss, B.D., and Sereno, S.C., 1999b, Neuroanatomy, circuitry and plasticity of word reading, *Neuroreport.* **10**: R12-R23.

Posner, M. I. and Raichle, M. I., 1997, *Images of Mind*, Scientific American Library, New York.

Price, C. J., 1997, Functional anatomy of reading, in: *Human Brain Function*, R. S. J. Frackowiak, K. J. Friston, C. D. Frith, R. J. Dolan, and J. C. Mazziotta, eds., Academic Press, San Diego, pp. 301-328.

Pugh, K. R., Mencl, W. E., Shaywitz, B. A., Shaywitz, S. E., Fulbright, R. K., Constable, R. T., Skudlarski, P., Marchione, K. E., Jenner, A. R., Fletcher, J. M., Liberman, A. M., Shankweiler, D. P., Katz, L., Lacadie, C., and Gore, J. C., 2000, The angular gyrus in developmental dyslexia: Task-specific differences in functional connectivity with posterior cortex, *Psychol. Sci.* **11**: 51-56.

Radach, R. and McConkie, G. W., 1998, Determinants of fixation positions in words during reading, in: *Eye Guidance in Reading and Scene Perception*, G. Underwood, ed., Elsevier, Oxford, pp. 77-100.

Rafal, R. and Robertson, L., 1995, The neurology of visual attention, in: *The Cognitive Neurosciences*, M. S. Gazzaniga, ed., MIT Press, Cambridge, pp. 625-648.

Rayner, K., 1975, The perceptual span and peripheral cues in reading, *Cognit. Psychol.* **7**:65-81.

Rayner, K., 1978, Eye movements in reading and information processing, *Psychol. Bull.* **85**: 618-660.

Rayner, K., 1979a, Eye movement and cognitive psychology: On-line computer approaches to studying visual information processing, *Beh. Res. Meth. Instrument.* **11**: 164-171.

Rayner, K., 1979b, Eye guidance in reading: Fixation locations within words, *Perception* **8**:21-30.

Rayner, K., 1985, Do faulty eye movements cause dyslexia?, *Dev. Neuropsychol.* **1**: 3-15.

Rayner, K., 1986, Eye movements and the perceptual span: Evidence for dyslexic typology, in: *Dyslexia: Its Neuropsychology and Treatment*, G. T. Pavlidis and D. F. Fisher, eds., Wiley, New York, pp. 111-130.

Rayner, K., 1998, Eye movements in reading and information processing: 20 years of research, *Psychol. Bull.* **124**: 372-422.

Rayner, K. and Bertera, J.H., 1979, Reading without a fovea, *Science* **206**: 468-469.

Rayner, J. and Duffy, S. A., 1986, Lexical ambiguity and fixation times in reading: Effects of word frequency, verb complexity, and lexical ambiguity, *Mem. Cognit.* **14**: 191-201.

Rayner, K. and Duffy, S. A., 1988, On-line comprehension processes and eye movements in reading, in: *Reading Research: Advances in Theory and Practice*, M. Daneman, G. E. MacKinnon, and T. G. Waller, eds., Academic Press, New York, pp. 13-66.

Rayner, K. and Frazier, L., 1989, Selection mechanisms in reading lexically ambiguous word, *J. Exp. Psychol. Learn. Mem. Cogn.* **15**: 779-790.

Rayner, K. and Morrison, R. E., 1981, Eye movements and identifying words in parafoveal vision, *Bull. Psychonom. Soc.* **17**: 135-138.

Rayner, K. and Pollatsek, A., 1989, *The Psychology of Reading*, Prentice Hall, Englewood Cliffs.

Rayner, K., Reichle, E. D., and Pollatsek, A., 1998, Eye movement control in reading: An overview and model, in: *Eye Guidance in Reading and Scene Perception*, G. Underwood, ed., Elsevier, Oxford, pp. 243-268.

Rayner, K., Reichle, E. D., and Pollatsek, A., 2000, Eye movement control in reading: Updating the E-Z Reader model to account for initial fixation locations and refixations, in: *Reading as a Perceptual Process*, A. Kennedy, R. Radach, D. Heller, and J. Pynte, eds., Elsevier, Oxford, in press.

Rayner, K. and Sereno, S. C., 1994, Eye movements in reading: Psycholinguistic studies, in: *Handbook of Psycholinguistics*, M. Gernsbacher, ed., Academic Press, New York, pp. 57-82.

Rayner, K., Sereno, S. C., Morris, R. K., Schmauder, A. R., and Clifton, C., 1989, Eye movements and on-line language comprehension processes [Special issue], *Lang. Cognit. Process.* **4**: 21-49.

Rayner, K., Sereno, S. C., and Raney, G. E., 1996, Eye movement control in reading: A comparison of two types of models, *J. Exp. Psychol. Hum. Percept. Perf.* **22**: 1188-1200.

Reichle, E. D., Pollatsek, A., Fisher, D. L., and Rayner, K., 1998, Toward a model of eye movement control in reading, *Psychol. Rev.* **105**: 125-157.

Reichle, E. D., Rayner, K., and Pollatsek, A., 1999, Eye movement control in reading: Accounting for initial fixation locations and refixations within the E-Z Reader model, *Vis. Res.* **39**: 4403-4411.

Reilly, R., 1993, A connectionist framework for modeling eye-movement control in reading, in: *Perception and Cognition: Advances in Eye Movement Research*, G. d'Ydewalle and Van Rensbergen, eds., North-Holland, Amsterdam, pp. 191-212.

Reilly, R. and O'Regan, J. K., 1998, Eye movement control in reading: A simulation of some word-targeting strategies, *Vis. Res.* **38**: 303-317.

Rumelhart, D. E., Hinton, G. E., and Williams, R. J., 1986, Learning internal representations by error propagation, in: *Parallel Distributed Processing: Explorations in the Microstructure of Cognition, Vol. 1: Foundations*, D. E. Rumelhart, J. L. McClelland, and the PDP Research Group, eds., MIT Press, Cambridge, pp. 318-362.

Rumelhart, D. E., McClelland, J. L., and the PDP Research Group, 1986, *Parallel Distributed Pocessing: Explorations in the Mcrostructure of Cgnition, Vl. 1: Foundations*, MIT Press, Cambridge.

Sagi, D. and Julesz, B., 1985, "Where" and "what" in vision, *Science.* **228**: 1217-1219.

Salvucci, D. D. and Anderson, J. R., 1998, Tracing eye movement protocols with cognitive process models, in: *Proceedings of the Twentieth Annual Conference of the Cognitive Science Society*, Erlbaum, Hillsdale, pp. 923-928.

Salvucci, D. D. and Anderson, J. R., 2000, Intelligent gaze-added interfaces, in: *Human Factors in Computing Systems: CHI 2000 Conference Proceedings*, ACM Press, New York, pp. 273-280, in press.

Schilling, H. E. H., Rayner, K., and Chumbley, J. I., 1998, Comparing naming, lexical decision, and eye fixation times: Word frequency effects and individual differences, *Mem. Cognit.* **26**: 1270-1281.

Schlosser, M. J., Aoyagi, N., Fulbright, R. K., Gore, J. C., and McCarthy, G., 1998, Functional MRI studies of auditory comprehension, *Human Brain Map.* **6**: 1-13.

Seidenberg, M. S., 1989, Visual word recognition and pronunciation: A computational model and its implications, in: *Lexical Representation and Process*, W. Marslen-Wilson, ed., MIT Press, Cambridge, pp. 25-74.
Seidenberg, M. S. and McClelland, J. L., 1989, A distributed, developmental model of word recognition and naming, *Psychol. Rev.* **96**: 523-568.
Sereno, S. C., 1992, Early lexical effects when fixating a word in reading, in: *Eye Movements and Visual Cognition: Scene Perception and Reading*, K. Rayner, ed., Springer-Verlag, New York, pp. 304-316.
Sereno, S. C., Rayner, K., and Posner, M. I., 1998, Establishing a time-line of processing during reading: Evidence from eye movements and event-related potentials, *Neuroreport* **9**: 2195-2200.
Somers, D. C., Dale, A. M., Seiffert, A. E., and Tootell, R. B. H., 1999, Functional MRI reveals spatially specific attentional modulation in human primary visual cortex, *Proc. Nat. Acad. Sci. USA.* **96**: 1663-1668.
Sparks, D. L. and Mays, L., 1990, Signal transformations required for the generation of saccadic eye movements, *Annu. Rev. Neurosci.* **13**: 309-336.
Stromswold, K., Caplin, D., Alpert, N., and Rauch, S., 1996, Localization of syntactic comprehension by Positron Emission Tomography, *Brain Lang.* **52**: 452-473.
Suppes, P., 1990, Eye-movement models for arithmetic and reading performance, in: *Eye Movements and Their Role in Visual and Cognitive Processes*, E. Kowler, ed., Elsevier, Amsterdam, pp. 455-477.
Suppes. P., 1994, Stochastic models of reading, in: *Eye Movements in Reading*, J. Ygge and G. Lennerstrand, eds., Pergamon Press, pp. 349-364.
Suppes, P., Cohen, M., Laddaga, R., Anliker, J., and Floyd, H., 1982, Research on eye movements in arithmetic performance, in: *Cognition and Eye Movements*, R. Groner and P. Fraisse, eds., North-Holland, Amsterdam, pp. 57-73.
Suppes, P., Cohen, M., Laddaga, R., Anliker, J., and Floyd, H., 1983, A procedural theory of eye movements in doing arithmetic, *J. Math. Psychol.* **27**: 341-369.
Sweeney, J. A., Mintun, M. A., Kwee, S., Wiseman, M. B., Brown, D. L., Rosenberg, D. R., and Carl, J. R., 1996, Positron emission tomography study of voluntary saccadic eye movements and spatial working memory, *J. Neurophysiol.* **75**: 454-468.
Tanaka, K., 1996, Inferotemporal cortex and object vision, *Annu. Rev. Neurosci.* **19**: 109-139.
Taylor, I. and Taylor, M. M., 1983, *The Psychology of Reading*, Academic Press, New York.
Thibadeau, R., Just, M. A., and Carpenter, P. A., 1982, A model of the time course and content of reading, *Cognit. Sci.* **6**: 157-203.
Ungerleider, L. G., Courtney, S. M., and Haxby, J. V., 1998, A neural system for human visual working memory, *Proc. Nat. Acad. Sci. USA.* **95**: 883-890.
Ungerleider, L. G. and Haxby, J. V., 1994, 'What' and "where' in the human brain, *Curr. Opin. Neurobiol.* **4**: 157-165.
Ungerleider, L. G. and Mishkin, M., 1982, Two cortical visual systems, in: *Analysis of Visual Behavior*, D. J. Ingle, M. A. Goodale, and R. J. W. Mansfield, eds., MIT Press, Cambridge, pp. 548-586.
Van Essen, D. C. and De Yoe, E. A., 1995, Concurrent processing in the primate visual cortex, in: *The Cognitive Neurosciences*, M. S. Gazzaniga, ed., MIT Press, Cambridge, pp. 383-400.

Chapter 16

Model of Visual Perceptual Space

Vasudevan Lakshminarayanan

School of Optometry, University of Missouri – St. Louis, 8001 Natural Bridge Road, St. Louis, MO 63121, PH: (314) 516-6533, FX: (314) 516-6712, EM: vengu@umsl.edu

16.1 INTRODUCTION

Space is a fundamental concept in physics and psychology. We consider space around us to be 3-dimensional and Euclidean. Can the same be said about the representation of this world in our perceptual system? This 3-dimensional manifold[1] of objects we are surrounded by have form and localization, in addition to characteristic qualities of color and brightness. (A brief definition of superscripted terms are given in the Notes (Section 16.4) at the end of the chapter). In a visual sensation, we are aware not only of a distribution of colors and brightness but also that certain of these qualities are combined into objects which have a definite geometric form and localization in a 3-dimensional space. This (the sensed space) is the visual space. The visual space is the final product of a long series of transformations/processes from the retina to the brain and it is coherent, self-organized, dynamic and complex (Indow 1991). Phenomenologically, it is articulated into individual objects, backgrounds and the "self". The self is a percept that is due to visual and proprioceptive experiences. Other visual percepts arise from stimuli in physical space. Luneberg (1947, 1950), in his classic work on binocular vision, defined visual space as a space obtained by transforming, using a specific mapping, a physical space into a cyclopean egocentered space with orthogonal coordinates.

The projection of the external object space onto the retina is well understood in terms of the optics of the eye. The transformations between the retinal mapping and subsequent remapping onto the first visual cortex area V1 can be approximately described as a log conformal transform[2] (Schwartz 1980). However, the transformations that occur between V1 and the ultimate level at which perception occurs is poorly understood. Since relevant anatomical/physiological substrates are unknown, we are left with psychophysical and modeling approaches which could help us identify some constraints that operate on the transformation process and may give us an insight into the characteristics of the ultimate perceptual representation. Hence, it is worthwhile to study the geometry of visual space. Historically, attempts to determine the geometry of the visual world date back to Euclid, when it was known that simple Euclidean geometric principles did not seem to hold in visual perception (See Rosar, 1985 for interesting discussions). Therefore, a recurrent theme has been the existence of a unified geometry of the visual field. It has been suggested that a major cause for the development of "objective" geometries is the discrepancies between our perception and "common sense" (Caelli 1981; see Pedoe (1976) for a detailed discussion of geometries and the arts).

16.1.1 Properties of Visual Space

As described by Indow and Watanabe (1988), visual space (VS) is "the most comprehensive percept that includes all objects appearing in front of the perceived self " and is the substrate of our visual experience. This is a well-defined theoretical formalism in sensory psychology. The global structure of visual space is discussed in detail by Indow (1991). They can be summarized as follows:

1. VS has three major directions, which correspond to the three dimensions in X, the physical world.
2. Percepts in VS are hierarchically related to each other. Each percept is localized with respect to other percepts which act as its framework.
3. We can perceive geometrical properties in VS.
4. How VS is structured depends on stimulus conditions in X.
5. VS is frameless in the sense that stimuli are presented in a homogenized ganzfeld.

Luneberg postulated that VS is a Riemannian space of constant curvature, based on his analysis of parallel and distance alley experiments under binocular viewing. In his mathematical model of binocular visual space perception, he applied the model to data from classical experiments on the arrangement of light stimuli in a dark room. In particular, he used data from Blumenfeld's (1913) finding that two rows of light points extending away from an observer and arranged so that they seem parallel (parallel alley) differ from those which appear equidistant (distance alley). The apparent phenomena are primarily determined by a relationship between the position of light stimulus and that of an observer, which may be called frameless – a ganzfeld (Indow, 1974). With the knowledge that a binocular geometry was not adequately represented by a Euclidean coordinate system, Luneberg used bipolar and modified bipolar coordinates to express the position of a point in three-space in a way consistent with properties of binocular vision. This model, which describes VS in terms of hyperbolic geometry, has been modified by Blank (19598, 1978) and has been tested by a number of investigators. A review of the literature is given by Suppes et al. (1989). In orthogonal sensory coordinates, the line element ds in VS is given in terms of sensory coordinates α,β and γ by (Indow 1979):

$$ds^2 = \frac{d\alpha^2 + d\beta^2 + d\gamma^2}{\left[1 + \frac{1}{4}\kappa\left(\alpha^2 + \beta^2 + \gamma^2\right)\right]^2} \tag{16.1}$$

where κ refers to the curvature of the space. If κ=0, we have the visual Euclidean space; κ > 0, hyperbolic space and κ < 0, an elliptical space. Therefore, if it were frameless, VS can be described in terms of a Riemannian space of constant curvature. It should be noted that this approach is not measurement theoretic and is not a qualitative approach to the axioms of VS. It is strictly a conceptual-theoretical approach based on psychophysical data.

16.1.2 Mathematical Background

Here we explain the differential geometric terms and their meaning. A Euclidean space is one in which the distance between any two points $P_1(x_1, y_1, z_1)$ and points $P_2(x_2, y_2, z_2)$ is given by

$$\begin{aligned} ds^2 &= (x_1 - x_2)^2 + (y_1 - y_2)^2 + (z_1 - z_2)^2 \\ &= dx^2 + dy^2 + dz^2 \end{aligned} \tag{16.2}$$

The extremum distance is given by a geodesic[3], a straight line in this case.

Consider a space which is not Euclidean, e.g., a curved space. Here, the distance between two points requires the notion of a metric tensor[4]. Usually, the metric tensor $g_{i,j}(x)$ will be symmetric between the indices i and j; and, the distance between two points is given by $g_{i,j}dx^i dx^j$ and if $g_{i,j} = \delta_{i,j}$, the Kronecker delta[5], we get the Euclidean space. The metric $g_{i,j}$ will depend on the x^i-s. If one considers a vector v^i at a point in the curved space and transports the vector to a neighboring point, the change in the vector, δv^i, consists of two parts: an ordinary change and a change due to the curved nature of the space. The second part, the "parallel transport", is proportional to the vector itself at the starting point and the extent of the distance displaced. The proportionality constant is the "affine connection" Γ (also known as Christoffel symbol) and the total change is given by (Einstein's summation convention – summation over repeated indices – is used throughout this chapter):

$$\delta v^i = dv^i + \Gamma^i_{j,k} v^j dx^k \tag{16.3}$$

and the affine connection is related to the metric as

$$\Gamma^i_{j,k} = g^{in}\Gamma_{jkin}$$
$$= \frac{1}{2} g^{in}\left(\frac{\partial g_{kn}}{\partial x^j} + \frac{\partial g_{jn}}{\partial x^k} - \frac{\partial g_{jk}}{\partial x^n}\right) \tag{16.4}$$

Once the metric $g^{i,j}$ is known, the entire geometric properties of the space can be determined. If we consider a space where Γ and the metric $g^{i,j}$ are not related and the connection Γ^i_{jk} is not symmetric under exchange of j and k, we call the space non-Riemannian. Since the connection need not be symmetric in terms of j and k, we can decompose Γ^i_{jk} into symmetric and anti-symmetric parts in j and k. The second part introduces "torsion"[6] into the space. The reader is referred to Spivak (1979) and Adler et al. (1965) for a good description of differential geometry.

16.2 A VISUAL SPACE FORMALISM

As mentioned previously, we approach the problem of perceptual space based on detailed psychophysical experiments (DeValois et al., 1990, 1989, 1988). We examined the constraints in visual spatial discriminations that reflect the mapping of VS. A detailed formalism of this non-Euclidean, Riemannian VS is given by Lakshminarayanan et al. (2000).

16.2.1 Experimental Foundations

DeValois et al. (1988, 1989, 1990) investigated the extent to which observers can successfully make discriminations based upon the relative positions of features under a variety of object transformations. Subjects were required to discriminate small effects in the relative positions of a single component within a 2-dimensional simple pattern presented in the fronto-parallel plane. The experiments could be divided into four general classes:

Class I: Discrimination of simple patterns with simple size scaling. Here comparisons involved discriminations of 1- and 2-dimensional patterns with relative sizes scaled to unity or in a 2:1 ratio in either one or two dimensions.

Class II: Discriminations among similar patterns with rotations. Here, similar patterns were either reflected about a vertical axis, inverted about a horizontal axis or rotated about the line of sight.

Class III: Comparisons of position defined by dissimilar features. In this class, the component features themselves differed between reference and comparison patterns. Only their relative positions remained comparable.

Class IV: Discriminations among similar patterns with transformations. Here, the test patterns were transformed under a set of transformation conditions.

Weber fractions for judgments of feature relative positions and orientations for simple 2-d stimuli were measured. Assuming the rigidity hypothesis and that wholeness of objects are preserved in the mapping from physical space to visual space, we can surmise the following:

1. Magnification, translation, inversion, reflection and rotation of objects in the fronto-parallel plane do not adversely affect performance.
2. Discriminations along one axis do not depend upon position in the perpendicular axis.
3. Positions to be compared can be equivalently defined by any of several features.
4. Performance does not depend on length ratios.
5. Certain classes of axis transformations have little effect on performance (ex., Cartesian to polar) but other classes of axis transformations (ex., linear to log) have significant adverse effect on performance.

16.2.2 The Model

Based on the conclusion from experimental results, a model of the perceptual space can be built wherein:

1. There is an orthogonal, labeled coordinate system;
2. Simple linear transformation of scale and orientation are allowable;
3. Some complex transformations are allowed, while others are not.

16.2.2.1 Affine Connection

Consider a rigid 2-dimensional object in physical space undergoing a general displacement. We assign a vector v_α to each such point in the physical space X. If this is mapped onto V_α in the visual space VS, the two vectors are related by

$$V_\alpha = f_{\alpha\beta} v^\beta \tag{16.5}$$

where $f_{\alpha\beta}(x,y)$ is an arbitrary covariant tensor[7]. If it were an orthogonal, non-singular transformation, then we have

$$f_{\alpha\beta} f^{\beta\gamma} = \delta^\gamma_\alpha \tag{16.6}$$

The explicit form of the tensor $f_{\alpha\beta}$ can be determined depending upon the characteristics of the retinal stimulus. (For example, $f_{\alpha\beta}$ was modeled based on the output of Reichardt motion detectors by Zhang and Wu (1990)). However, in the following analysis, the form of $f_{\alpha\beta}$ is not necessary.

As the object in the physical space moves, V_α traces out a curve *S* in VS. We can write

$$\begin{aligned} \frac{dV_\alpha}{dS} &= \frac{\partial V_\alpha}{\partial x^\beta}\frac{\partial x^\beta}{\partial S} \\ &= \frac{\partial f_{\alpha\gamma}}{\partial x^\beta} f^{\gamma\lambda} V_\lambda \frac{\partial x^\beta}{\partial S} \end{aligned} \tag{16.7}$$

from which it follows that the perceptual space is a Riemannian manifold of affine connection

$$\Gamma^\lambda_{\alpha\beta} = \frac{\partial f_{\alpha\gamma}}{\partial x^\beta} f^{\gamma\lambda} \tag{16.8}$$

Since any affine space is characterized by its connection Γ, the metrical structure of VS can be derived.

16.2.2.2 Length of a Vector

Does the space allow an orthogonal coordinate system? To answer this question, we need to consider the length of vectors V^α and v^α, and see if they are equal.

$$V^\alpha = f^{\alpha\beta} v_\beta \tag{16.9}$$

$$\begin{aligned} V^\alpha V_\alpha &= f^{\alpha\beta} v_\beta f_{\alpha\gamma} v^\gamma \\ &= v_\beta v^\beta \end{aligned} \tag{16.10}$$

Thus, all length orthogonal transformations preserve the vector lengths and vice versa.

16.2.2.3 The Metric of VS

The visual space is a metric space because we can make estimates of length. The metric tensor is related to the affine connection by Eq. 16.4. The affine connection is also given by Eq. 16.8. If we define F to be the determinant of the tensor f, we can get

$$f^{\rho\lambda} = \frac{1}{F}\frac{\partial F}{\partial f_{\lambda\rho}} \tag{16.11}$$

$$\begin{aligned} \frac{\partial f_{\lambda\rho}}{\partial x^\beta} f^{\rho\lambda} &= \frac{\partial f_{\lambda\rho}}{\partial x^\beta}\left(\frac{1}{F}\frac{\partial F}{\partial f_{\lambda\rho}}\right) \\ &= \frac{1}{F}\frac{\partial F}{\partial x^\beta} \\ &= \frac{\partial \log |F|}{\partial x^\beta} \end{aligned} \tag{16.12}$$

The affine connection coefficient is given by a well-known result (Adler et al., 1965)

$$\Gamma^{\lambda}_{\lambda\beta} = \frac{1}{2}\frac{\partial \log |G|}{\partial x^{\beta}} \tag{16.13}$$

where G is the determinant of the metric tensor g_{ij}. This gives us

$$\frac{\partial \log |F|}{\partial x^{\beta}} = \frac{1}{2}\frac{\partial \log |G|}{\partial x^{\beta}} \tag{16.14}$$

which results in

$$G = (kF)^2 \tag{16.15}$$

where k is a constant that relates the scaling from physical space to visual space.

16.2.2.4 Riemannian Curvature

The Riemannian curvature tensor $R^{\alpha}{}_{\eta\beta\gamma}$ is equal to zero for the space whose connection is given above. By definition, the Riemannian curvature tensor is a tensor that is constructed from the metric tensor of the space and its first and second derivatives:

$$R^{\alpha}_{\eta\beta\gamma} = \left(\Gamma^{\alpha}_{\beta\eta}\right)_{,\gamma} - \left(\Gamma^{\alpha}_{\eta\gamma}\right)_{,\beta} + \Gamma^{\alpha}_{\tau\gamma}\Gamma^{\tau}_{\beta\eta} - \Gamma^{\alpha}_{\tau\beta}\Gamma^{\tau}_{\gamma\eta} \tag{16.16}$$

where the notation $X_{,x}$ denotes differentiation of X with respect to x. Substituting the expression for Γ into the expression for R we can get the required result. Since

$$f_{\alpha\beta} = \frac{\partial f}{\partial x^{\alpha}\partial x^{\beta}}$$

we can consider the subscripts of f as both tensor indices and as a differentiation. This implies that we can alter the order:

$$\begin{aligned} R^{\alpha}_{\eta\beta\gamma} &= \left(f^{\alpha\beta}{}_{,\gamma} + f^{\alpha\mu} f^{\tau\rho} f_{\tau\mu\gamma}\right) f_{\beta\rho\eta} \\ &\quad - \left(f^{\alpha\rho}{}_{,\beta} + f^{\alpha\mu} f^{\tau\rho} f_{\tau\mu\beta}\right) f_{\gamma\rho\eta} \end{aligned} \tag{16.17}$$

From Eq. 16.6, the inverse of $f_{\alpha\rho}$ is $f^{\alpha\rho}$. Therefore, we have

$$f^{\alpha\rho}{}_{,\gamma} f_{\rho\beta} = -f^{\alpha\rho} f_{\rho\beta\gamma} \tag{16.18}$$

and hence the assertion that the Riemannian tensor is zero can be proved (since the quantities inside each of the parentheses in Eq. 16.17 are equal to zero). The Gaussian curvature[8], the so-called Ricci tensor $R_{\eta\gamma}$ which can be obtained by contracting[9] α,β in $R^{\alpha}{}_{\eta\beta\gamma}$, is also identically zero.

This result implies that the space is flat with no intrinsic curvature. This suggests that the difference of changes in the vector components is also equal to zero. Parallel transport of vectors in this space is path independent and uniquely defined. Zhang and Wu (1990) interpret this result to mean that in motion perception, we can unambiguously compare local velocities among all points in the space even at large separations. This implies that the output of the motion detectors in response to a rigid moving object as constant vectors and are therefore "computed" to yield percept of a single rigid object in motion.

16.2.2.5 Orientation of Vectors in VS

If the vector v_α in the physical space is taken around a closed loop, it will come back to the original starting point. Here we will show that when each point in the physical space is mapped on to VS, V_α will also span a closed loop such that initial and final vectors V_α and V_α' will coincide, preserving orientation.

This can be shown by realizing that when a vector is parallel transported, it undergoes a rotation. The change in the orientations of V_α after one complete loop is given by

$$\partial W = \oint dx^{\alpha}(S) \int_0^S \Gamma^{\beta}_{\alpha\gamma}(S') V_\beta(S') \frac{dx^{\gamma}}{dS'} dS' \tag{16.19}$$

where the integral on the right gives the infinitesimal change along the curve for a small segment and $dx^\alpha(S)$ is the corresponding measure. Substituting for the affine connection from Eq. 16.8, we obtain

$$\partial W = \oint dx^{\alpha}(S) \int_0^S \frac{\partial f_{\alpha\rho}}{\partial x^{\gamma}} f^{\rho\beta} V_\beta(S') \frac{dx^{\gamma}}{dS'} dS' \tag{16.20}$$

However, $f^{\rho\beta}V_\beta(S') = v^\rho$ which is a constant since it is a vector in physical space.

$$\partial W = v^\rho \oint dx^\alpha(S) \int_0^S \frac{\partial f_{\alpha\rho}}{\partial x^\gamma}\frac{dx^\gamma}{dS'}dS'$$

$$= v^\rho \oint dx^\alpha(S)\left[\delta f_{\alpha\beta}(S)\right] \tag{16.21}$$

$$= 0$$

The closed loop integral vanishes since the integrand includes differences between successive points in the closed loop. This implies that V_α does indeed come back to the original configuration. This result indicates that the visual space admits orientation in much the same way as physical space in spite of the non-Euclidean nature of the affine connection.

16.2.2.6 Geodesics in VS

The geodesic of an affine space is given by

$$\frac{d^2X^\lambda}{dS^2} + \Gamma^\lambda_{\alpha\beta}\frac{dX^\alpha}{dS}\frac{dX^\beta}{dS} = 0 \tag{16.22}$$

where S is some parameter (which can be arc length). Substituting for $\Gamma^\lambda{}_{\alpha\beta}$, we get

$$\frac{d^2X^\lambda}{dS^2} + \frac{df_{\alpha\rho}}{dx^\beta}f^{\rho\lambda}\frac{dX^\alpha}{dS}\frac{dX^\beta}{dS} = 0 \tag{16.23}$$

Multiplying by $f_{\sigma\lambda}$ we can show that the above is an exact differential given by

$$\frac{d}{dS}\left(f_{\sigma\alpha}\frac{dX^\alpha}{dS}\right) = \frac{d^2f_\sigma}{dS^2} = 0 \tag{16.24}$$

which has the solution

$$f_\sigma = a_\sigma S + b_\sigma \tag{16.25}$$

where a_σ and b_σ are constants determined from the initial conditions (position and direction of a particular geodesic). This defines a "straight line" in VS, which may not be a straight line in the physical space. This property may conceivably be used to study certain classes of visual illusions such as the Herring illusion (Zhang and Wu 1990).

Degenerating points are those points at which F=0 where $f_{\alpha\beta}$ is not defined. It is a known fact from differential geometry that all geodesics converge toward the degenerating direction. The degenerating direction is the eigenvector corresponding to the degenerate eigenvalue zero. With the Hering (and similar) illusions, the physically straight line appears perceptually bent in the presence of background (which determines the perceptual setting). In the Hering illusion, consider the background radiating lines to be degenerating curves with degenerating directions perpendicular to them. Since, at the intersection, the test line is not along the degenerating directions, perceptual distortion appears and hence the illusion.

16.2.2.7 Allowed and Non-Allowed Transformations

Since we are dealing with real physical spaces, we can write $f_{\alpha\beta}$ as functions of x and y. If $f_{\alpha\beta}$ is realizable as $\frac{\partial f}{\partial x^\alpha \partial x^\beta}$, then it is possible to investigate the form of the scalar f. Consider the solutions to the equation F=0 or $\det(f_{\alpha\beta}) = 0$. These are nothing but degenerating points. They occur when

$$f_{xy}{}^2 = f_{xx} f_{yy} \tag{16.26}$$

Here the subscript denotes differentiation by the subscripted variable. If we could write the scalar as log q, for some q, then

$$\left(qq_{xy} - q_x q_x\right)^2 = \left(qq_{xx} - q_x{}^2\right)\left(qq_{yy} - q_y{}^2\right) \tag{16.27}$$

where Eq. 16.26 can be satisfied formally, Eq. 16.27 cannot be, in general. This is because the left hand side of Eq. 16.27 is always positive as long as the q-s are real functions, whereas the right hand side can become negative if for instance, $q_x^2 > qq_{xx}$.

Using this, it is possible to consider some allowable forms of f(x,y). If f is separable in x and y, then, both additive and multiplicative forms can be considered. If $f(x,y) = f_1(x) + f_2(y)$, then forms such as $f(x,y) = ax + b + f_2(y)$, $f(x,y) = ay + b + f_1(x)$, $f(x,y) = ax + by + c$ are allowed. Here a, b, c are constants and f_1 and f_2 are arbitrary functions. In the multiplicative form, we have $f(x,y) = f_1(x)f_2(y)$ and we can have $f(x,y) = C \exp(ax+by)$, $f_1(x) = a_0 + a_1x + a_2x^2 + \dots$, $f_2(y) = = b_0 + b_1y + b_2y^2 + \dots$, $f(x,y) = C \sin(ax) \sin(by)$ with $ax = n\pi/4$ and $by = m\pi/4$. Similarly, simple rotations like $x \to x' = x\cos\theta + y\sin\theta$ and/or $y \to y' = -x\sin\theta + y\cos\theta$ leave Eq. 16.26 invariant. Similar results hold for scaling $x \to x' = ax, y \to y' = by$ while power law $x \to x' = ax^n, y \to y' = by^m$ or exponential transformation $x \to x' = \exp(ax)$, $y \to y' = \exp(by)$ do not leave Eq. 16.26 invariant.

16.3 DISCUSSION AND CONCLUSIONS

A Riemannian 2-dimensional manifold described by a special metric which incorporates certain features and defines a class of perceptual spaces has been developed. We suggest that there may be an infinite set of realizable VS, all of which obey certain transformation rules. The specific character of the perceptual representation is solely determined by the characteristics of the retinal image and the transformation rules. Thus, the geometry of the perceptual representation is not independent of the characteristics of the stimulus pattern. However, all such geometries must conform to a specifiable set of basic criteria. We can also postulate that:

1. No single simple manifold can be defined which will adequately describe every spatial visual pattern.
2. It should be possible, at least in principle, to specify the rule by which a given pattern gives rise to a particular perceptual representation.
3. If the rules are known, the unique spatial percepts that result from any given geometric pattern should be predictable. Thus, geometric visual illusions should follow from the application of the spatial transformation rules to the particular stimulus patterns which give rise to them.
4. The prediction of these unique percepts should not depend upon any assumptions about the ecological significance of or meaning attached to a particular feature or stimulus.

We have described a Riemannian affine space with zero curvature. This seems to contradict other studies, including Luneberg's. As Suppes (1995) points out, there are many things fundamentally wrong with Luneberg's approach to VS. For example, there is a complete lack of stability once we assume a space of constant curvature. The results for curvature vary considerably from experiment to experiment, and even from day to day for an individual subject doing the same experiment. Foley (1964) in an experimental study pointed out the Desarguesian property of VS, namely, the curvature of the space cannot vary from point to point. Wagner (1985) dealt with how observers judged distance. He found that two different models gave description of VS. In the first model, VS is an affine transformed version of a Euclidean physical space, similar to the one presented here. In the second model, distances viewed as vectors were broken down into an in-depth component and a frontal component relative to the observer. The in-depth component of this vector was found to be contracted by a constant amount in VS. Yamazaki (1987) modeled parallel and distance alley experiments as geodesics in a non-Riemannian VS. He could explain distance alley experiments as being Riemannian, while the parallel alley results indicate that the affine connection is non-Riemannian. In order to explain this, he used a theorem due to Eisenhart (1927), which states that a space can be divided such that some vector fields behave differently from the rest. Because of the presence of "torsion", closed curves in physical space will not be mapped onto closed curves in VS and orientations will not be preserved. It is possible that the distortions can be explained within the framework of the Riemannian model, however, this is for future work.

According to Suppes (1995), "Perhaps the most important feature of VS that must be taken into account in a thorough analysis is that VS is more like a physical system than a geometrical one". By this he means that the most important feature of VS is that it is context dependent. It is well known that perceptual judgments of symmetry or of congruence are affected by context. This implies that because of such data, we should abandon the idea of a single unified VS. Moreover, multiple geometrical models of VS should be required because different contexts lead to different models. MacLeod and Willen (1995) have noted that the concept of a perceptual space as a global and unified structure is not supported, because the relevant neurobiology does not support a unified conception. However, attempts have been made to incorporate neurophysiology into such modeling. As a example, Carlton (1988) has used Gabor function representation of the visual cortical receptive fields to link perceived mental rotations of a 3-dimensional object with the sequence of internal activities in the visual cortex (brain state space). See Lakshminarayanan and Santhanam (1995) for a discussion.

In physics, the concept of space-time as a continuum may need revision, since quantum theory implies that a continuum model may be untenable particularly at scales smaller than the Planck length (Wheeler 1990). In vision, the concept of perceptual/visual space is also currently in flux and seems to need radical revision. To quote Suppes (1995), "The goal of having a unified structure of visual space adequate to account for all important experimental results seems mistaken. A pluralistic and fragmentary approach seems required".

How can we get a unified picture of VS? An elegant way seems to be through the use of "Geometric psychology", a term coined by Hoffman (1980, 1984). Hoffman (1966, 1968, 1970, 1977) has formulated what is termed as the Lie Transformation Groups/Neuropsychology (LTG/NP) approach to study certain perceptual processing such as visual constancies. Hoffman's approach is to use the Lie Transformation Groups[10] to provide a metalanguage for perceptual processes (see also Dodwell 1970). There is some experimental evidence for the application of this formalism to pattern or contour extraction (Caelli 1977). Mathematically, if there is a situation involving LTG and differentiable manifolds, as in the case with VS, then one has to involve the so-called fiber bundles. A fiber bundle (a vector field or a vector bundle) is a set of vectors associated with all points on a surface or manifold which can potentially generate all possible paths or trajectories on that manifold. According to Hoffman, the structure of the manifold will determine the basic paths, trajectories, or orbits that can occur. This determines the basic gestalt properties of VS. The vector field operations generate or admit certain continuous groups, and these constrain visual processing. The structure of perception can then be thought of as a fiber bundle with the visual space as a base manifold and the affine connection as the base connection. The cross section of the fiber bundle will simply be the visual scene. Discussion of this idea is beyond the scope of this paper. To quote Dirac (1954), "The moral of the story is that one should have faith in a theory that is beautiful. If a theory fails to agree with experiment, its basic principles may still be correct, and the discrepancy may be due to some detail that will get cleared up in the future". There is plenty of scope for further research in both theory and experimentation to deduce the rich structure of visual space.

16.4 NOTES

1. A manifold is any set that could be continuously parameterized. Any surface, any hyperplane in an n-dimensional space, is a manifold. A manifold which allows differentiation is called a differentiable manifold. A Riemannian manifold, also known as a Riemannian space, is a differentiable manifold on which a "distance function" - metric is defined.

2. A conformal transformation is a transformation which preserves angles. For a transformation to be conformal, it must be an analytic function and have a non-zero derivative.

3. Given any two points on a surface, a geodesic is the shortest path on the surface connecting the two points.

4. A metric tensor, also known as a Riemannian metric, is a symmetric, positive-definite tensor, g_{ij}, that tells us how to compute the distance between two points in any given space. Its components can be viewed as multiplication factors which must be placed in front of the differential displacements dx_I in a generalized Pythagorean theorem $ds^2 = g_{11}dx^1d^1 + g_{12}dx^1d^2 + g_{21}dx^2d^1 + g_{22}dx^2d^2 + \cdots$

5. The Kronecker delta, δ_{ij}, is defined to be 1 when $i = j$ and equal to zero when $i \neq j$.

6. Torsion refers to the rate of change of osculating plane of a space curve. It can be thought of as the rate of change of curvature or the screw action of a curve, and is a local geometric feature of the curve. The torsion tensor can be defined as $T^l_{jk} = -\left(\Gamma^l_{jk} - \Gamma^l_{kj}\right)$.

7. A covariant tensor is a tensor having a specific transformation property. If A_i is any set of quantities which transform as $A_i = \frac{\partial x_j}{\partial x'_i} A'_j$, then A_i is a covariant tensor.

8. The Gaussian curvature is an intrinsic property of a space independent of the coordinate system used to describe it. For a sphere, Gaussian curvature is $\kappa = \frac{1}{a^2}$ where a is the radius. For Euclidean space, $\kappa = 0$.

9. The contraction of a tensor is obtained by setting unlike indices equal and summing according to the Einstein summation convention.

10. A transformation group is a set of transformations having the mathematical property of a group, namely: that the combination of any two transformations successively applied is a transformation in the set, the transformations obey associativity, there is an inverse (canceling) transformation and a zero (null) transformation. Sophus Lie, a Norwegian mathematician extensively studied continuous transformation groups and the resultant theory is known as Lie groups or Lie algebra.

16.5 REFERENCES

Adler, R., Bazan, M., Schiffer, M., 1965, *Introduction to general relativity*, McGraw Hill, New York.

Blank, A. A., 1958, Analysis of experiments in binocular space perception, *J. Opt. Soc. Am.* **48**: 911-925.

Blank, A. A., 1978, Metric geometry in human binocular perception – theory and fact, In: Leewenberg, E. and Buffart, H., (Eds) *Formal theories of visual perception*, Wiley, New York.

Blumenfeld, W., 1913, Unterschugen über die schienbare grosse in sehraume, *Zeitschrift for psyhcologie und physiologie der sinnesorgane*, **65**: 241-404.

Caelli, T., 1977, Psychophysical interpretations and experimental evidence for the Hoffman LTG/NP theory of perception. *Cahiers de Psychologie*, **20**: 107-134.

Caelli, T. 1981, *Visual perception theory and practice*, Pergamon Press, Oxford, U.K.

Carlton, E. H., 1988, Connection between internal representation of rigid transformation and cortical activity patterns, *Biol. Cybernetics*, **59**: 419-429.

DeValois, K. K., Lakshminarayanan, V., Nygaard, R., Schlussel, S., Sladkey, J., 1988, Relative position comparisons between dissimilar patterns, paper TUR5, *Technical Digest Series, Vol II*, Optical Society of America, Washington, D.C.

DeValois, K. K., Lakshminarayanan, V., Nygaard, R., Schlussel, S., Sladkey, J., 1989, Perceptual space as an elastic manifold, *Inves. Opthal. Vis. Sci.* ARVO supplement **30**: 486.

DeValois, K. K., Lakshminarayanan, V., Nygaard, R., Schlussel, S., Sladkey, J., 1990, Discrimination of relative spatial position, *Vision Res.* **30**: 1649-1660.

Dirac, P. A. M., 1954, Letter, *Sci. Monthly*, **79**(#4).

Dodwell, P., 1970, *Visual pattern recognition*, Holt, Reinhart and Winston, New York.

Eisenhart, L. P., 1927, *Non-Riemannian geometry*, American Mathematical Society Colloquia Publications, Vol 8, American Mathematical Society, Providence, R.I.

Foley, J. M., 1964, Desarguesian property in visual space, *J. Opt. Soc. Am* **54**: 684-692.

Hoffman, W. C., 1966, The Lie algebra of visual perception, *J. Math. Psych.*, **3**: 65-98, errata: *ibid* **4**: 348-349.

Hoffman, W. C., 1968, The neuron as a Lie group germ and a Lie group product, *Quat. J. Applied. Math.*, **25**: 423-441.

Hoffman, W. C., 1970, Higher visual perception as a prolongation of the basic Lie transformation group, *Math. Biosci.*, **6**: 437-471.

Hoffman, W. C., 1977, An informal, historical description (with biography) of the LTG/NP, *Cahiers de Psychologie*, **20**: 139-150.

Hoffman, W. C., 1980, Subjective geometry and geometric psychology, *Mathematical Modeling*, **1**: 349-367.

Hoffman, W. C., 1984, Figural synthesis by vector fields: Geometric neuropsychology, In: Dodwell, P., Caelli, T., (Eds) *Figural Synthesis*, Erlbaum, Hillsdale, NJ.

Indow, T., 1974, On geometry of frameless binocular perceptual space, *Psychologia*, **17**: 50-63.

Indow, T., 1979, Alleys in visual space, *J. Math. Psychol.* **19**: 221-258.

Indow, T., Watanabe, T., 1988, Alleys on an extensive apparent fronto-parallel plane: A second experiment, *Perception* **17**: 647-666.

Indow, T., 1991, A critical review of Luneberg's model with regard to global structure of visual space, *Psychol. Rev.* **98**: 430-453.

Lakshminarayanan, V., Santhanam, T. S., 1995, Representation of Rigid Stimulus transformations by cortical activity patterns, In: Luce, D. R., D'Zmura, M., Hoffman, D., Ivesron, G. J., Romney, A. K., (Eds), *Geometric representations of perceptual phenomena*, Lawrence Erlbaum, Mahwah, N.J., pp. 61-68.

Lakshminarayanan, V., Parthasarathy, R., DeValois, K. K., 2000, A generalized perceptual space, *Neurol. Res.* **22**: 699-702.

Luneberg, R. K., 1947, *Mathematical analysis of binocular vision*, Princeton University Press, Princeton, N.J.

Luneberg, R. K., 1950, The metric of binocular visual space, *J. Opt. Soc. Am.* **40**: 627-642.

MacLeod, D. A., Willer, J. D., 1995, Is there a visual space?, In: Luce, D. R., D'Zmura, M., Hoffman, D., Ivesron, G. J., Romney, A. K., (Eds), *Geometric representations of perceptual phenomena*, Lawrence Erlbaum, Mahwah, N.J., pp. 47-60.

Pedoe, D., 1976, *Geometry and the liberal arts*, Peregrine, New York.

Schwartz, E. L., 1980, Computational anatomy and functional architecture of striate cortex: A spatial mapping approach to perceptual coding, *Vision Res.* **20**: 645-669.

Spivak, M., 1979, *A comprehensive introduction to differential geometry*, Publish or Perish, Inc., Berkeley, CA.

Suppes, P., Krantz, D. H., Luce, D. R., Tversky, A., 1989, *Foundations of measurements, Vol II: Geometrical, threshold and probabilistic representations,* Academic Press, New York.

Suppes, P., 1995, Some fundamental problems in the theory of visual space, In: Luce, D. R., D'Zmura, M., Hoffman, D., Ivesron, G. J., Romney, A. K., (Eds), *Geometric representations of perceptual phenomena*, Lawrence Erlbaum, Mahwah, N.J., pp. 37-45.

Wagner, M., 1985, The metric of visual space, *Perception and Psychophysics* **38**: 483-495.

Wheeler, J., 1990, Information, physics, quantum: The search for the links: In: Zurek, W. (Ed), *Complexity, entropy and physics of information*, Addison- Wesley, Reading, MA, pp. 3-28.

Yamazaki, T., 1987, Non-Riemannian approach to geometry of visual space: An approach to affinely connected geometry to visual alleys and horopter, *J. Math. Psychol.* **31**: 270-298.

Zhang, J., Wu, S., 1990, Structures of visual perception, *Proc. Natl. Acad. Sci. USA* **87**: 7819-7823.

V

CLINICAL SYSTEM MODELS

Chapter 17

Model-Based Understanding of Clinical Vergence Testing

Kenneth J. Ciuffreda[1] and George K. Hung[2]

[1] *Dept. of Vision Sciences, State University of New York, State College of Optometry, 33 West 42nd St. New York, NY 10036; PH: (212) 780-5132, FX: (212) 780-5124; EM: kciuffreda@sunyopt.edu*

[2] *Dept. of Biomedical Engineering, Rutgers University, 617 Bowser Rd., Piscataway, NJ 08854-8014, PH: (732) 445-4137, FX: (732) 445-3753, EM: shoane@rci.rutgers.edu*

17.1 INTRODUCTION

Both symmetric and asymmetric disparity vergence have been investigated extensively over the past 50 years to gain a more complete understanding of the sensory, motor, and perceptual processes subserving and controlling binocular *fusion*, which is the primary goal of the vergence system (Schor and Ciuffreda, 1983; Ciuffreda and Tannen, 1995). Pure symmetric vergence can be elicited when a target is displaced in depth precisely along one's egocentric midline. In this case, the *retinal disparity* is equally distributed between the two eyes (Fig. 17.1A). The response consists of a saccade-free, relatively slow and smooth, symmetric disparity vergence tracking movement in depth. On the other hand, asymmetric vergence can be elicited when a target is displaced in depth anywhere except precisely along the egocentric midline, as is true under most naturalistic conditions. In this case, the initial retinal disparity is unequally distributed between the two eyes (Fig. 17.1B). However, now the response consists of a rapid saccade and a relatively slower disparity vergence movement, with the conjugate saccade functioning to shift the eyes laterally such that the retinal disparity is once again symmetrically distributed between the two eyes for symmetric disparity vergence to correct the residual bifixation error within foveal Panum's fusional areas (PFA). Furthermore, there is the special case of line-of-sight asymmetric vergence, in which the target is moved along the line-of-sight of one eye (Fig. 17.1C). Here too both a saccade and a disparity vergence movement are executed, as described earlier for the more general case of asymmetric vergence (Fig. 17.1B).

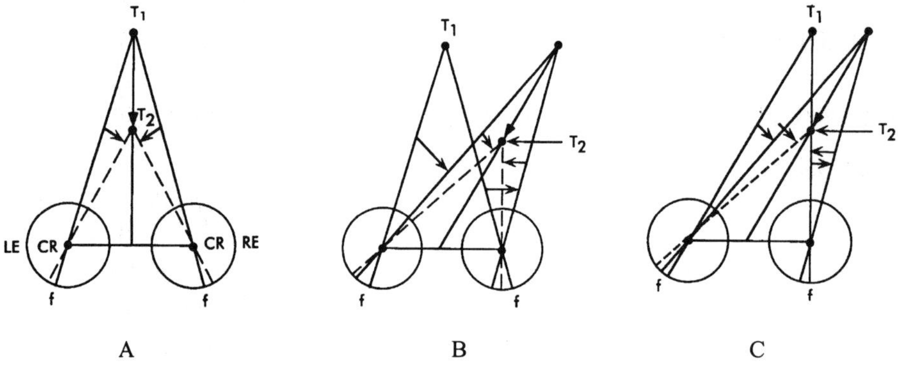

Figure 17.1. Symmetrical versus asymmetrical disparity stimuli and overall response patterns. A. Symmetric vergence. B. Asymmetric vergence. C. Line-of-sight asymmetric vergence. T_1, initially fixated target; T_2, subsequently fixated target; f, fovea; *CR*, center of rotation of the eye; *LE*, left eye; *RE*, right eye. Reprinted from Ciuffreda and Tannen (1995) with permission of Harcourt Health Sciences.

In this chapter, the static model-based aspects of clinical disparity vergence testing and the underlying saccade-vergence eye movements will be considered. Although much of the clinical testing to be discussed incorporates symmetric disparity stimuli to drive the vergence system directly (Benjamin, 1998), and the accommodative system indirectly (Ciuffreda, 1991, 1998), once fusion is disrupted and the disparity vergence system is rendered open-loop, asymmetric vergence becomes dominant. (Also see Chapter 11 in this volume). An understanding of the mechanisms involved requires conceptualization of the components that comprise the vergence response. This will be conducted in a systematic manner by first discussing and detailing a quantitative model of the system, then describing the clinical diagnostic testing, and concurrently relating these vergence tests and the clinical vergence deficits they may uncover to the underlying oculomotor model control structure. In essence, oculomotor-related clinical signs and symptoms of binocular vergence disorders may now be understood in terms of abnormality of specific disparity vergence and accommodative components in the static model.

To assist the reader in understanding the clinical terms denoted by italics, a glossary is provided at the end of this chapter.

17.2 BACKGROUND

17.2.1 Static Model of the Vergence and Accommodative Systems

A comprehensive and homeomorphic static, or steady-state, quantitative model of the vergence and accommodative systems and its motor interactions has been developed by Hung, Ciuffreda, and Rosenfield (1996). It incorporates disparity, blur, proximal, and tonic inputs to each system. This model and its earlier versions have been useful for furthering our understanding of a wide range of basic normal mechanisms and abnormal clinical conditions (Hung, 2001).

The latest version of the model is presented in Fig. 17.2, with model parameter values presented in Table 17.1. Progressing from left to right in Fig 17.2, it may be seen that the accommodative (upper) and vergence (lower) negative feedback control loops have similar component control structures.

Table 17.1

Static Model Values for Accommodation and Vergence

Accommodation	Vergence
DOF = ± 0.15 D	PFA = ± 5.0 min of arc
ACG = 10	VCG = 150
AC = 0.80 D/MA	CA = 0.37 MA/D
ABIAS = 0.61 D	VBIAS = -0.29 MA
ADAPTA = 4	ADAPTV = 9
APG = 2.10	VPG = 0.067
PDG = 0.212	PDG = 0.2.12

Input. The input or stimulus change for accommodation (AS; target distance in *diopters*, D) and disparity vergence (VS; target distance in meter angles, MA, or *prism diopters*, PD) sum with the negative feedback response of the respective system at that moment. The resultant difference represents the initial system error. The input for the proximal branch is perceived target distance, with such perceptually-derived information not having a separate feedback loop but rather inputting directly and simultaneously into both the accommodative and vergence negative feedback loops.

Under normal binocular, closed-loop viewing conditions (i.e., with blur and retinal disparity feedback present), blur and disparity provide the primary motor drives to their respective systems, with their crosslinks providing a secondary drive to the fellow system. The proximal drive adds to the final steady-state vergence (VR) and accommodative (AR) responses by only 0.4% and 4%, respectively. However, this concordant tertiary proximal information is still quite important, as it influences overall responsivity by providing perceptual cue reinforcement derived from the perceived depth information. Lastly, the tonic inputs to each system have little motor-response effect in

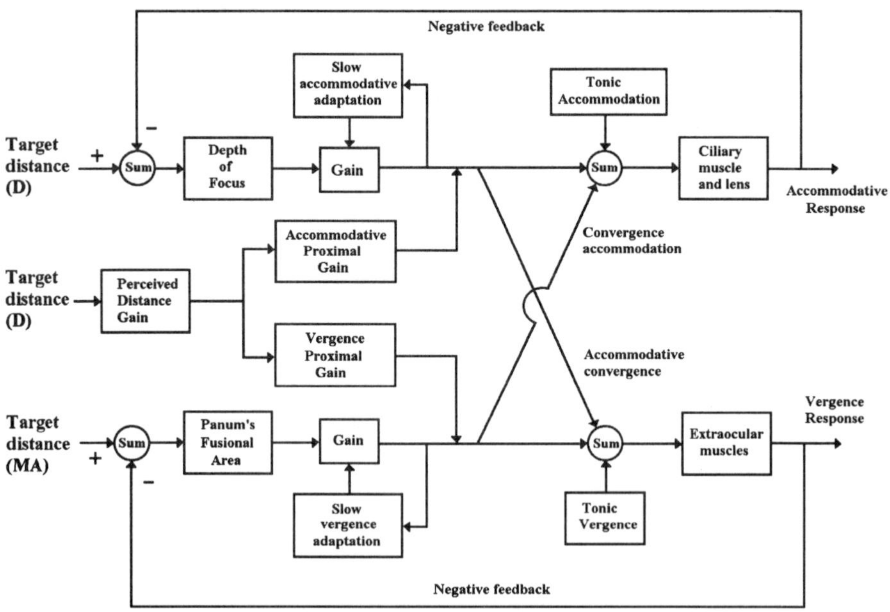

Figure 17.2. Conceptual version of Hung, Ciuffreda, and Rosenfield (1996) interactive static model of accommodation and vergence. Reprinted from Ong and Ciuffreda (1997) with permission of Optom. Extension Program Foundation Press, Santa Ana, CA.

visually-normal individuals, especially at near, under naturalistic binocular viewing conditions.

Threshold "Deadspace" Operator (DSP). This represents the depth-of-focus for accommodation (in diopters) and Panum's fusional areas for disparity vergence (in minutes of arc). DSP allows small neurosensory-based accommodative (AE) and vergence (VE) system errors (i.e., retinal defocus and retinal disparity, respectively) to be tolerated without adverse perceptual consequences (i.e., blur and diplopia, respectively). If the input error exceeds its threshold level, this error information proceeds to drive the respective system.

Gain. The accommodative (ACG=AR-ABIAS/AE-DSP) and vergence (VCG=VR-VBIAS/VE-DSP) controller gains represent the experimentally-derived, open-loop, internal neurological controller gains of the respective systems. The final system error signal, which equals the initial system error minus the deadspace threshold value, is multiplied by this gain element. Its output provides the primary neurological control signal and drive to formulate the final steady-state motor response. The output of the controller gain is then input to three other components (see next three components below).

Adaptive Gain. Although typically regarded as a dynamic model element, following sustained nearwork it may bias either the final static open-loop or closed-loop system response. However, under non-sustained viewing conditions, its value is zero. Hence, it will not be further considered in this chapter.

Cross-Link Gain. The cross-link gain (AC for accommodation and CA for vergence) multiplies the output of the direct ACG or VCG pathway, respectively. It provides a secondary drive to the fellow system, as mentioned earlier. For accommodation, this represents the effective accommodative-convergence to accommodation (AC/A) ratio, whereas for *convergence* it represents the effective convergence-accommodation to convergence (CA/C) ratio.

Tonic Input. Tonic input for accommodation (ABIAS; in diopters) and vergence (VBIAS; in meter angles) reflects midbrain baseline neural innervation. Although tonic terms may have substantial effects on the response amplitude when both systems are rendered open-loop (i.e., with visual feedback rendered ineffective), they have negligible influence on the overall closed-loop near response and only modest influence on the closed-loop far response (Hung & Semmlow, 1980). This is shown in Eq. 17.1 with respect to monocular blur-driven accommodation, where

$$AR = (AS - DSP) * \frac{ACG}{1 + ACG} + ABIAS * \frac{1}{1 + ACG} \qquad (17.1)$$

For a typical value of ABIAS = 1 diopter and ACG = 9, the effect of ABIAS on AR would only be 0.1 diopter. This relative lack of effect is even more dramatic for disparity vergence, with its much higher controller gain value, as shown in Eq. 17.2 with accommodation open-loop, where

$$VR = (VS - DSP) * \frac{VCG}{1 + VCG} + VBIAS * \frac{1}{1 + VCG} \tag{17.2}$$

For a typical value of VBIAS = 1 MA and VCG = 149, the effect of VBIAS on VR would only be 0.007 MA.

Summing Junction. The direct controller gain output is also sent to the summing junction, where it adds with the cross-link and tonic inputs, both of which have only modest influence on the fellow system, to formulate the final combined signal to drive the respective system.

Peripheral Apparatus. The output of the summing junction proceeds to cortical and subcortical centers related to accommodation and to vergence to formulate the respective aggregate neural signals (Ciuffreda, 1991, 1998; Ciuffreda and Tannen, 1995). It then advances to innervate the appropriate peripheral apparatus, that is the ciliary muscle and crystalline lens complex for accommodation and the extraocular muscles for vergence.

Output. These motor changes are then fed back to the initial summing junction via their respective negative feedback pathways. If a relatively large residual error remains, the cycle is repeated, until an acceptably small and stable steady-state error for both systems is attained.

17.2.2 Clinical Vergence Testing

In the early part of the 20th century, the development of clinical test procedures to assess sensory and motor aspects of the vergence system began in optometric clinical and academic facilities (Sheard, 1917). Concurrently, test values were ascertained in large groups of asymptomatic and symptomatic clinical patients, especially as related to nearwork activities (Sheard, 1917; Skeffington, 1928). Normative values evolved for each test, which segregated these two clinic populations: patients having symptoms typically fell outside the range of values established for the asymptomatic individuals. Thus, such normative test values could henceforth be used as a valuable diagnostic tool (Benjamin, 1998). Furthermore, with the development of oculomotor-based *vision therapy* (Skeffington, 1928; Peckham, 1928), these norms also possessed therapeutic value because vision therapy could now be justified on a sound clinical and motor learning basis (Ciuffreda, in press). Therapy would be continued, until the established norms were attained, and then maintained by

Table 17.2

Relationship of Clinical Vergence Test, Vergence Model Component, and Clinical Vergence Deficit

Test	Component	Deficit
Four base-out test	Disparity vergence (central field open-loop)	Suppression scotoma in strabismus
Cover test	Disparity vergence (central and peripheral field open-loop) Accommodative vergence crosslink gain (AC)	Large phoria /strabismus/paresis
Distance phoria	Tonic vergence (VBIAS) Accommodative vergence crosslink gain (AC) Disparity vergence (central and peripheral field open-loop)	Symptomatic large phoria / intermittent strabismus/paresis
Near phoria	Accommodative vergence crosslink gain (AC) Disparity vergence (central and peripheral field open-loop) Perceived / proximal vergence gain	Symptomatic large phoria / intermittent strabismus/paresis
Prism vergence ranges	Vergence controller gain (VCG) Extraocular muscles (PLANT)	Convergence insufficiency Convergence excess Large phoria Extraocular muscle paresis
Near point of convergence	Vergence controller gain (VCG) Extraocular muscles (PLANT)	Convergence insufficiency Large exophoria at near Extraocular muscle paresis

the patient for a sufficient period of time. Such clinical findings first demonstrated the remarkable degree of neural plasticity in the vergence oculomotor system, and, in turn, justified the use of specific interactive vergence and accommodative therapeutic regimens designed to attain normal static and dynamic symmetric and asymmetric vergence function (Ciuffreda, in press). Furthermore, with the development of quantitative models, the individual abnormal vergence and accommodative model components could now be ascertained, and vision therapy directed specifically towards each component's normalization (Hung et al, 1986; Ciuffreda, in press).

Gross abnormalities of vergence are typically found in patients with *strabismus* (Figs. 17.2 and 17.3; Table 17.2). This abnormal sensorimotor ocular condition is associated with either an anomalous central field, open-loop, disparity vergence response or an anomalous closed-loop, full-field disparity vergence response (Ciuffreda and Tannen, 1995). Binocular *suppression* is commonly found in conjunction with strabismus (Fig. 17.3). It is a key factor in understanding the anomalous asymmetric fusional vergence responses observed in some strabismic patients due to presence of a binocular *suppression scotoma* (Ciuffreda and Tannen, 1995).

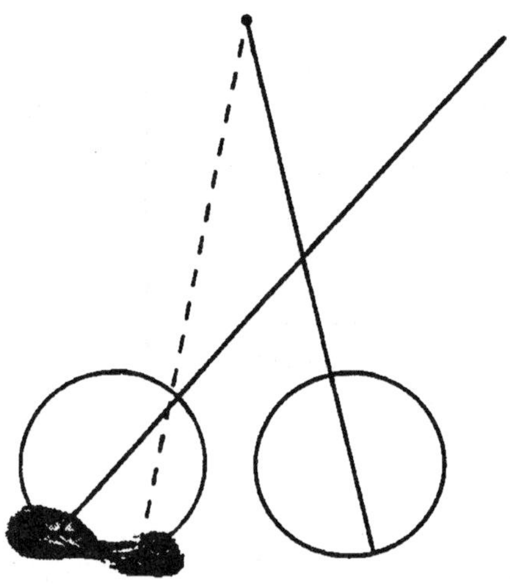

Figure 17.3. Binocular suppression scotoma (stippled region) in the deviated eye. Left esotropia. Solid lines project from the fovea of each eye. Small solid circle represents the test target. Reprinted from Ciuffreda and Tannen (1995) with permission of Harcourt Health Sciences.

17.3 CLINICAL DIAGNOSTIC VERGENCE TESTING

17.3.1 Four Base-Out Test

This test directly involves careful detection for the presence of an asymmetric vergence response to assess the neurosensory binocular status of the strabismic patient. It is used diagnostically to detect for the presence of a central binocular suppression scotoma in the deviated eye of *esotropic* patients (Fig. 17.3) (von Noorden and Maumenee, 1967). This sensory-based, binocular suppression scotoma is adaptive in nature at the cortical level (Blake and Lehmkule, 1976) to prevent the occurrence of visually-debilitating *diplopia* and *visual confusion* (Ciuffreda and Tannen, 1995). It effectively acts to open-loop the disparity vergence system over the central field by cortically suppressing the retinal disparity information over that region in the deviated eye, with this occurring in the model representation between the summing junction and Panum's fusional areas (Figs. 17.2 and 17.3, and Table 17.2).

In essence, the test involves careful scrutiny of the two eyes immediately upon interposition of a 4 prism diopter base-out ophthalmic prism before one eye, which deviates the incoming light (approximately 2 degrees) towards the prism base and projects the deviated image in the direction of the prism apex. This procedure is repeated in the fellow eye. Since the prism does not distribute the retinal disparity equally interocularly (i.e., a 4 prism diopter change in the eye with the prism, and zero in the fellow eye), the vergence stimulus and hence vergence response are asymmetric in nature in a normal individual with the prism placed before either eye (Figs. 17.4A and B). In contrast, in a strabismic patient with abnormal binocular vision and presence of a central binocular suppression scotoma, either only a conjugate versional movement (i.e., a saccade) (Fig. 17.4C) or no response (Fig. 17.4D) will be found, depending upon which eye the prism is interposed.

Figure 17.4. (See next page). The Four Base-Out Test. (A) On placing the prism over the right eye, a rapid leftward movement occurs during refixation with the right eye. This indicates absence of foveal suppression in the right eye. (B) A subsequent very slow fusional movement of the left eye is observed to correct for the image displacement. This indicates absence of foveal suppression in the left eye. The combined movements of (A) and (B) represent asymmetric vergence. (C) In another patient, the left eye remains turned out after a prism is placed over the right eye. Absence of the secondary asymmetric vergence fusional movement of the left eye indicates a foveal suppression scotoma of the left eye; the image has been shifted within a binocularly nonfunctioning retinal area. (D) To confirm this diagnosis, the prism is placed over the left eye. Neither eye will move under these circumstances, since the prism has merely displaced the image within the binocular suppression scotoma, and hence no stimulus for asymmetric vergence exists for refixation. Reprinted from von Noorden and Maumenee (1967) with permission of Harcourt Health Sciences.

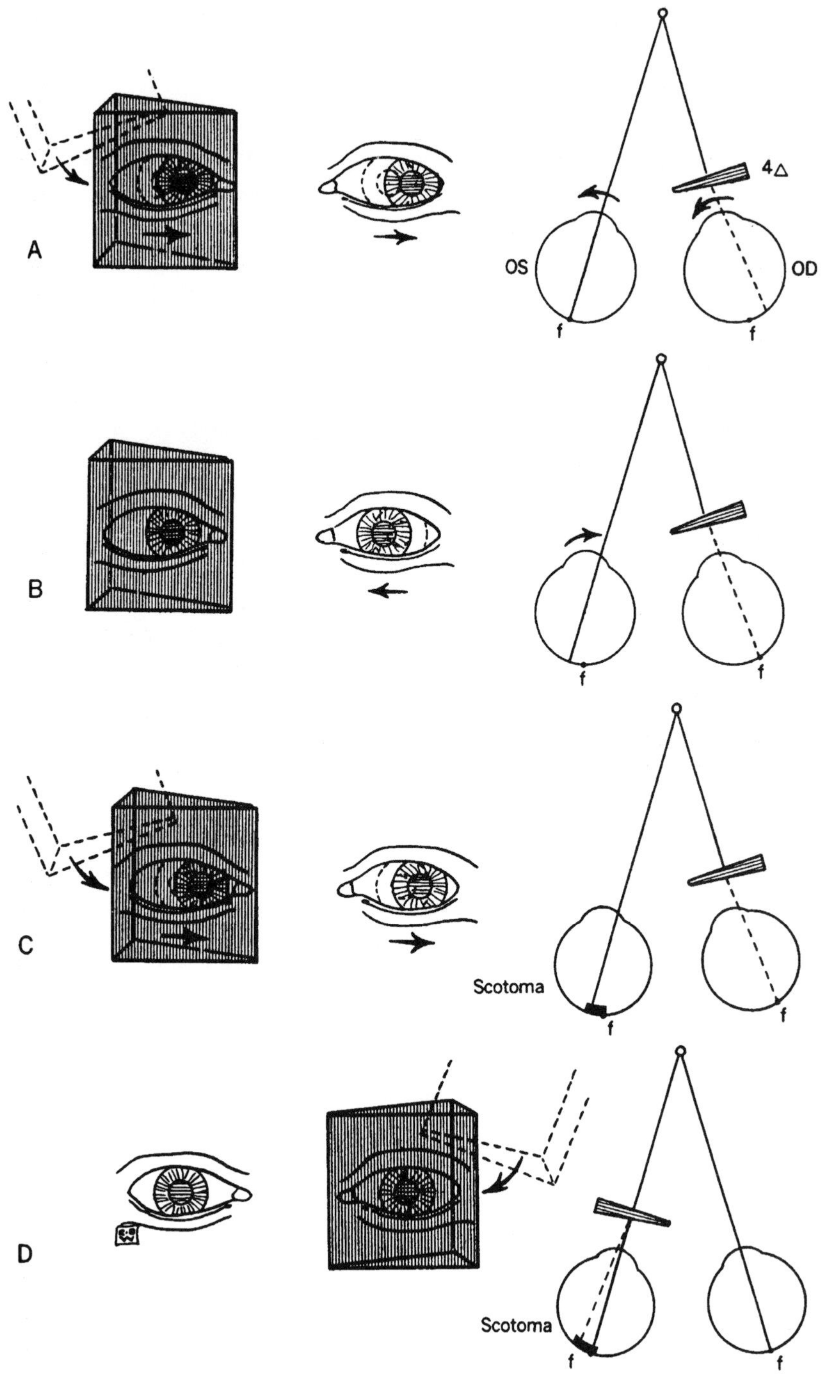

4Δ
A
OS
OD
f
f
B
f
f
C
Scotoma
f
f
D
Scotoma
f
f

17.3.2 Cover Test Measurement and Fusional Recovery

One of the most common tests performed in ophthalmic clinical practice is the "unilateral cover test", which is used to detect the presence of an oculomotor deviation when binocular sensory fusion is disrupted (Table 17.2) (Benjamin, 1998; von Noorden and Maumenee, 1967). This test is critical in the diagnosis and treatment of binocular vision disorders, such as a large *phoria, extraocular muscle paresis,* or intermittent strabismus.

The cover test assesses the position of the eyes when disparity vergence is open-loop both centrally and peripherally, and hence the main drive to the vergence system now comes from the accommodative system via its crosslink, AC; perceived distance and proximal vergence gain may influence the near response in a secondary manner by introducing a constant bias effect; and, the influence of tonic vergence (VBIAS) is tertiary at near, but may be of somewhat more importance at distance in assessment of the phoria (Table 17.2). In essence, while having the patient binocularly view an object placed along the midline, an occluder is first placed fully over one eye for up to 15 seconds to allow for dissipation and decay of the fusional vergence response adaptive component. Furthermore, this effectively renders the disparity vergence system open-loop, with the final eye position dictated by the accommodative vergence drive (i.e., AC/A ratio) via the AC model crosslink component. Then the occluder is quickly removed to restore binocular viewing and provide closed-loop retinal disparity feedback. The initial direction and magnitude of the movement in the uncovered eye is denoted (Fig. 17.5). This procedure is then repeated with the fellow eye.

Three possible motor responses may be found. (1) If the patient manifests *orthophoria*, and hence the occluded eye remains stationary, then no movement will be observed when the occluder is removed. (2) However, if the patient exhibits a phoria such that the eye under cover shifts in an exponentially decelerating manner to the fusion-free, phoria position (Fig. 17.2), then eye movements will be observed when the occluder is removed. Since an asymmetric vergence stimulus condition is present immediately upon removal of the occluder, as all of the retinal disparity is present in the deviated eye rather than being symmetrically distributed between the two eyes, an asymmetric vergence, "fusional recovery" motor response will be executed. At distance, the phoria position will approximate the tonic vergence level, or VBIAS (O'Shea et al, 1988) (Fig. 17.2). (3) And, if the patient has an intermittent strabismus, so that when the occluder is introduced, the eye under cover again exponentially shifts in approximately 10 to 15 seconds to the fusion-free strabismic position, then two responses are possible when the occluder is initially removed: (a) if the patient does not either reflexively or volitionally regain binocular motor fusion, the eye will remain in the deviated

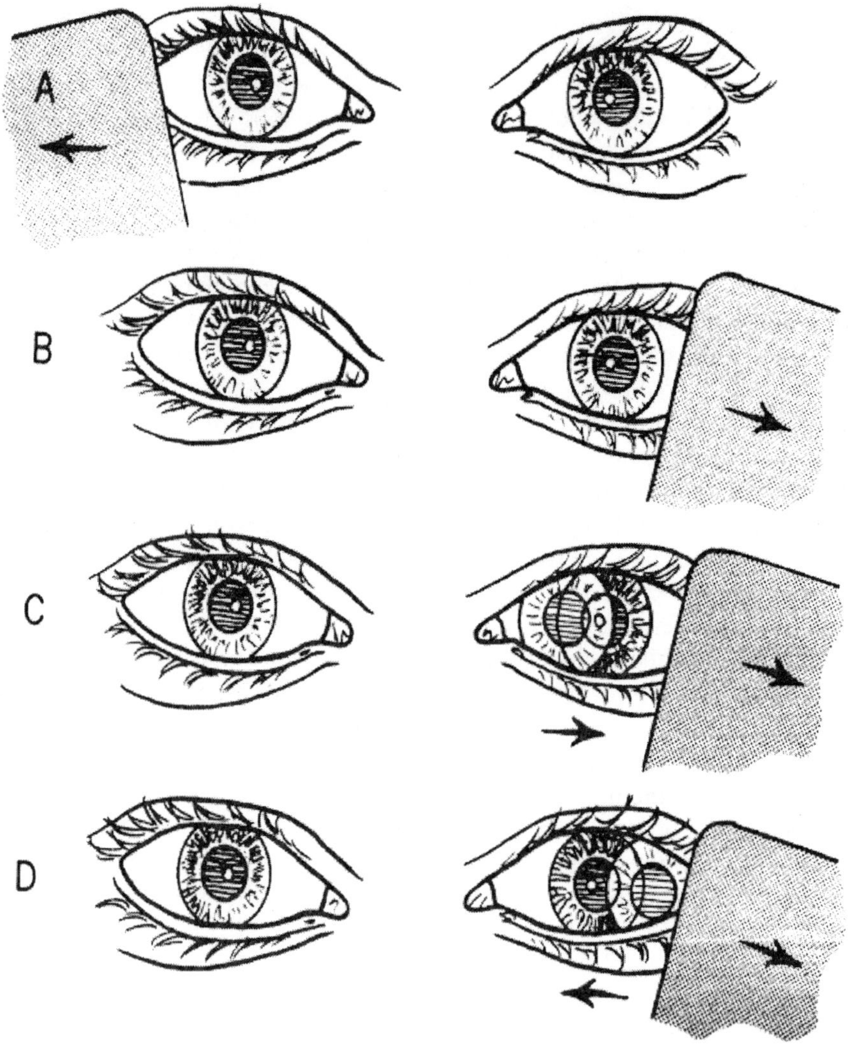

Figure 17.5. The Unilateral Cover Test. (A) The cover has been removed from the right eye, and no movement of the right eye can be detected. (B) The cover has been removed from the left eye, and no movement of the left eye can be detected. (C) When uncovered, the left eye moves outward to fixate. *Esophoria*. (D) When uncovered, the left eye moves inward to fixate. *Exophoria*. Reprinted from von Noorden and Maumenee (1967) with permission of Harcourt Health Sciences.

strabismic position, or (b) if the patient does regain binocular motor fusion, an asymmetric vergence "fusional recovery" motor response will take place, as all of the retinal disparity is present in the deviated eye rather than being symmetrically distributed between the two eyes (Ciuffreda et al, 2001).

17.3.3 Prism Disparity Vergence Ranges

Another important clinical test used in the diagnosis and treatment of binocular vision disorders is the measurement of prism vergence ranges at both distance and near. This assesses the motor limits of crossed and uncrossed disparity vergence, with the accommodative stimulus level remaining fixed (Ciuffreda, 1992). These tests examine the integrity of the vergence controller gain (VCG), and secondarily the vergence plant (i.e., extraocular muscles) (Table 17.2). A higher controller gain results in greater accuracy in the vergence response; however, abnormally-high gain results in *convergence excess,* and abnormally-low gain results in *convergence insufficiency*. A functional plant is needed to provide a full range of vergence movements, and hence a mild extraocular muscle paresis would produce restriction.

Basically, a pair of variable power ophthalmic prisms with bases in opposite directions (e.g., bases out, or templeward) is placed before the eyes and increasing amounts of fusional vergence stimulus demand via the prisms (i.e., crossed convergent or uncrossed divergent retinal disparity; horizontal, vertical, or cyclorotary in nature) (Ciuffreda and Tannen, 1995) is introduced with respect to a fixed target positioned along the midline. Hence, the initial stimulus and response are symmetric in nature. And, the magnitude of the vergence response is dictated by the vergence controller gain, or VCG, as well as the integrity of the plant (see Fig. 17.2 and Table 17.2). As the disparity vergence demand increases, the patient will eventually note some target blur due to the concurrently increased vergence accommodation from the crosslink gain (CA) as it exceeds the depth-of-focus of the eye (Benjamin, 1998) (Fig. 17.2); and with yet further increases in prism vergence demand will subsequently note the onset of diplopia as fusion can no longer be maintained. Once diplopia is consistently present, and hence the disparity vergence system is now rendered open-loop, the non-fixating eye will begin to shift slowly and exponentially to the fusion-free phoria position. After a few seconds, the vergence demand is then gradually reduced optically via the prisms, until binocular fusion is regained. The "fusional recovery" movement is accomplished with asymmetric vergence, as once again the retinal disparity is markedly asymmetric.

17.3.4 Near Point of Convergence

The symmetric and asymmetric responses as described above are also found in this important diagnostic test (Benjamin, 1998; Ciuffreda, 1992). And, as was true for the prism vergence ranges described above, this test reflects vergence controller gain (VCG) and extraocular muscle (i.e., plant) integrity (Table 17.2). However, now a physical (versus optical) target is moved in space slowly along the midline towards the patient, thus producing a symmetric vergence stimulus and response, to determine the maximum convergence ability. Once this point in physical space is exceeded, fusion will no longer be possible and hence become disrupted, and the non-dominant eye will shift exponentially to the disparity vergence, open-loop, phoria position (generally outward). Diplopia may or may not be reported. As the target is then moved away from the patient, at a certain point an asymmetric vergence movement will be executed to regain fusion, as the retinal disparity is markedly asymmetric.

17.4 CONCLUSIONS

Clearly, our interactive vergence and accommodation model, with its specific subcomponents, can be used successfully to describe and understand the diagnostic tests used by optometrists and others in their overall clinical armamentarium for a variety of binocular vision dysfunctions (Table 17.2). While such model-based conceptualization of these tests represents a major step forward as compared with their early origins over 75 years ago, such a notion must have been in the pioneering minds of Sheard (1917), Skeffington (1928), Peckham (1928) and others, as only if one could understand the basic normal and abnormal control structure could progress and advances take place in the equally important areas of prevention and treatment. Future directions are two-fold: to develop a comprehensive and quantitative interactive oculomotor model of binocular vision system dysfunctions and, (2) to perform computer simulations on the above to develop more effective vision therapy, and perhaps even surgical, paradigms that would "target" specific abnormal model component structures.

17.5 REFERENCES

Benjamin, W.J. (ed.), 1998, *Borish's Clinical Refraction*, W.B. Saunders Company, Philadel., PA.

Blake, R., and Lehmkule, S.W., 1976, On the site of strabismic suppression, *Invest. Ophthal. Vis. Sci.* **15:** 660-663.

Ciuffreda, K.J., 1991, Accommodation and its anomalies, in *Vision and Visual Dysfunction: Visual Optics and Instrumentation*, Vol. 1, W.N. Charman ed., Macmillan, London, pp. 231-279.

Ciuffreda, K.J., 1992, Components of clinical near vergence testing, *J. Behav. Optom.* **3:** 3-13.

Ciuffreda, K.J., 1998, Accommodation, the pupil, and presbyopia, in *Borish's Clinical Refraction*, W.J. Benjamin ed., W.B. Saunders Company, Philadel., PA, pp. 77-120.

Ciuffreda, K.J, The scientific basis for and efficacy of optometric vision therapy in non-strabismic accommodative and vergence disorders, *Optometry*, in press.

Ciuffreda, K.J., Suchoff, I.B., Kapoor, N., Jackowski, M.M., and Wainapel, S.F., 2001, Normal vision function, in *Downey and Darling's Physiological Basis of Rehabilitation Medicine*, Gonzalez, E.G. et al, eds., Butterworth-Heinemann, Boston, pp. 241-261.

Ciuffreda, K.J., and Tannen, B., 1995, *Eye Movement Basics for the Clinician*, Mosby Yearbook, St. Louis.

Hung, G.K., 2001, *Models of Oculomotor Control*, Singapore, World Scientific Publishing Co. Inc.

Hung, G.K., Ciuffreda, K.J., and Rosenfield, M., 1996, Proximal contribution to a linear static model of accommodation and vergence, *Ophthal. Physiol. Opt.* **16:** 31-41.

Hung, G.K., Ciuffreda, K.J., and Semmlow, J.L., 1986, Static vergence and accommodation: population norms and orthoptic effects, *Doc. Ophthal.* **62:** 165-179.

Hung, G.K., and Semmlow, J.L., 1980, Static behavior of accommodation and vergence: computer simulation of an interactive dual-feedback system, IEEE Trans. Biomed. Eng. **27:** 439-447.

Ong, E., and Ciuffreda, K.J., 1997, *Accommodation, Nearwork, and Myopia*, OEP Foundation Press, Santa Ana, CA.

O'Shea, W.F., Ciuffreda, K.J., Fisher, S.K., Tannen, B., and Super, P., 1988, The relationship between the distance heterophoria and tonic vergence, *Amer. J. Optom. Physiol. Opt.* **65:** 787-793.

Peckham, R.H., 1928, *The Modern Treatment of Binocular Imbalance*, Shur-On Standard Optical, New York.

Schor, C.M., and Ciuffreda, K.J., (eds.), 1983, *Vergence Eye Movements: Basic and Clinical Aspects*, Butterworth, Boston.

Sheard, C., 1917, *Dynamic Ocular Tests*, Lawrence Press, Columbus, Ohio.

Skeffington, A.M., 1928, *Procedure in Ocular Examination*, A.J. Fox Co., Chicago.

von Noorden, G.K., and Maumenee, A.E., 1967, *Atlas of Strabismus*, Mosby, St. Louis.

Glossary of Clinical Terms	
Convergence	The inward-directed turning of the lines-of-sight toward each other; this is in contrast to divergence, the outward turning of the lines-of-sight away from each other.
Convergence excess	Abnormal increase in convergence at near relative to that at distance under fusion-free conditions.
Convergence insufficiency	Abnormal decrease in convergence at near relative to that at distance under fusion-free conditions.
Diopter	A unit of ophthamic lens power; one diopter focuses light from infinity at a distance of one meter.
Diplopia	Similar images falling on non-corresponding retinal points, and hence projecting to different visual directions; "double" vision; non-fused images.
Esophoria	An inward lateral deviation of the eye in the fusion-free state accomplished either by prismatic dissociation or occlusion of one eye.
Esotropia	A manifest inward lateral deviation of one eye.
Exophoria	An outward lateral deviation of the eye in the fusion-free state accomplished either by prismatic dissociation or occlusion of one eye.
Fusion	Higher-order cortical integration of the left and right eyes' images; haplopia.
Orthophoria	Lack of deviation of the eye in the fusion-free state.
Paresis	Partially-paralyzed extraocular muscle.
Phoria	A deviaiton of the eye (inward, outward, upward, downward, or cyclorotatory in nature) in the fusion-free state (i.e., typically either with one eye occluded or with prismatic dissociation).
Prism diopter	A unit of ophthalmic prism power; one prism diopter deviates light from infinity 1 cm at 1m; 1.745 prism diopters equal 1 degree.
Retinal disparity	The geometric angular difference at the eyes between the bifixation target and any other object in the visual field.
Scotoma	A relative or absolute blind area of the visual field.
Strabismus	An anomaly of binocular vision in which the visual axis of one eye fails to intersect the object of interest.

Glossary of Clinical Terms (con't)	
Suppression	An anomaly of binocular vision in which part of the ocular image of the strabismic, deviated eye is prevented from contributing to the fused binocular percept.
Vision therapy	Highly specific, sequential, sensory-motor-perceptual stimulation paradigms and regimens aimed at normalizing binocular vision. It incorporates the use of blur (via lenses and target distance), disparity (via prisms and target distance), and proximity (via perceived target distance) stimuli; also referred to as v.t., vision training, visual training, visual therapy, "eye exercises", and orthoptics.
Visual confusion	Dissimilar images falling on foveal corresponding retinal points, and hence projecting to identical visual directions.

Chapter 18

Models of Refractive Error Development

George K. Hung[1] and Kenneth J. Ciuffreda[2]

[1]*Dept. of Biomedical Engineering, Rutgers University, 617 Bowser Rd., Piscataway, NJ 08854-8014, PH: (732) 445-4137, FX: (732) 445-3753, EM: shoane@rci.rutgers.edu;*

[2]*Dept. of Vision Sciences, State University of New York, State College of Optometry, 33 West 42nd St., New York, NY 10036; PH: (212) 780-5132, FX: (212) 780-5124; EM: kciuffreda@sunyopt.edu.*

18.1 INTRODUCTION

Clarity of the visual image is a vital component of ocular health. A common method for assessing image clarity is to measure distance visual acuity. The development of an uncorrected refractive error, however, reduces visual acuity, and in turn adversely impacts upon the quality of ocular health. This chapter discusses various analytical approaches taken in the understanding of refractive error development.

There are two main types of refractive error: hyperopia and myopia. Hyperopia, or farsightedness, occurs when the combined optical power of the cornea and the unaccommodated lens are less than that needed for the axial length of the eye, so that the retinal image is focused beyond the retina. On the other hand, myopia, or nearsightedness, occurs when the total ocular power of the eye exceeds that needed for its ocular length, so that the image is focused in front of the retina. Image clarity for the distant object in hyperopes can be attained by means of accommodation, or an increase in lens power, but at the expense of increased effort along with a reduced effective accommodative range. For myopes, however, image clarity cannot be attained with increased accommodation, and in fact, this would further degrade retinal-image clarity. Thus, myopia is associated with the more immediate concerns of everyday visual function.

Myopia is a worldwide public health concern (Goldschmidt, 1968), because it affects 25% of the adult population in the United States (Sperduto et al, 1983) and 75% or more of the adult population in Asian countries such as Taiwan (Lin et al, 1996). It can be corrected by optical means, but the estimated annualized cost to consumers in the United States for eye examinations and corrective lenses is $4.6 billion (Javitt and Chiang, 1994). Furthermore, the wearing of spectacles for myopia may restrict one's vocational and avocational options (Mahlman, 1982). Surgical techniques to reduce myopia are available, but they are expensive (Grosvenor and Goss, 1999). Furthermore, despite the continual developments and technological improvements over the past 20 years, there are still surgical and post-surgical risks, along with possible side effects such as long-term hazy vision (Javitt and Chiang, 1994). Moreover, surgery does not prevent the subsequent development of adult-onset myopia or other age-related refractive changes, for example, due to increased lens index and presbyopia (Javitt and Chiang, 1994). For these reasons, the slowing of myopic progression, as well as the prevention of its initial occurrence, has been of considerable interest to clinicians and scientists alike for decades.

To understand the fundamental mechanisms underlying refractive error development, both *genetic* and *environmental* factors must be examined (Ong & Ciuffreda, 1997; McBrien & Millodot, 1986; Gwiazda et al, 1993; Mutti et al, 1996; Jiang & Woessner, 1996; Grosvenor and Goss, 1999). Evidence for genetic influence is evident in the high correlation between refractive errors in twins (Kimura, 1965; Sorsby et al, 1962; Goss et al, 1988), and the higher prevalence of myopia in children whose parents were also myopic (Gwiazda et al, 1993). On the other hand, evidence for environmental influence comes from the very rapid increase in the prevalence of myopia in the Intuit, Japanese, Chinese, and Native Americans over the past 50 years (Young et al, 1969; Alward et al, 1985; Hosaka, 1988; Goh and Lam, 1994; and Lam et al, 1994; Woodruff and Samek, 1977), suggesting an association between their progressively greater amount of time spent on nearwork during formal schooling and higher rates of childhood myopia prevalence and progression (Pãssinen et al, 1989; Wu et al, 1999; Zhang et al, 2000). Clearly, both genetic and environmental factors are involved.

During infancy under normal genetic development, there is an inherent mismatch between the optical power of the cornea/lens and the axial length of the eyeball (Scammon & Armstrong, 1925). Yet, as the normal eye matures, the cornea/lens and ocular tunics begin to develop in concert to provide a relatively precisely focused image on the retina (Bennett & Rabbetts, 1989; Grosvenor and Goss, 1999). This process is called emmetropization (Yackle & Fitzgerald, 1999). Clearly, certain critical information is used to coordinate cornea/lens and axial growth. One of the most important cues for regulating axial growth appears to be retinal-image defocus (McBrien & Millodot, 1986; Ong & Ciuffreda, 1997; Wallman, 1997; Norton, 1999), which is dependent on the interaction of both cornea/lens and axial length. Cornea/lens growth and its consequent change in optical power will alter retinal-image defocus, but an appropriate change in the axial length growth rate will act to reduce this defocus, and in turn restore the balance between these two components. Since the basic growth of the cornea/lens is genetically-predetermined (Sorsby et al, 1962; Goss and Erickson, 1987; Goss and Jackson, 1993; Fledelius and Stubgaard, 1986), emmetropization involves only the regulation and modulation of axial length growth (McBrien and Millodot, 1986; Ong and Ciuffreda, 1997; Wallman, 1997; Norton, 1999).

Emmetropization also occurs under environmentally-induced conditions. This is seen in numerous studies which have attempted to determine the effect of various optically-based manipulations of retinal-image quality on induced ocular growth and refractive development. The findings have been mixed with respect to the resultant direction of refractive shift. Some manipulations produced a myopic shift. These included: prolonged nearwork (Goss and Wickham, 1993; Grosvenor & Goss, 1999), purposeful undercorrection for myopia (O'Leary, 2000), graded diffusers (Smith and Hung 2000), and black occluder contact lenses (Tigges et al, 1990; Iuvone et al, 1991). On the other hand, other manipulations resulted in a hyperopic shift. These included: very strong diffusers (O'Leary et al, 1992; Bradley et al, 1996), crystalline lens removal (Wilson et al, 1987), and initial imposition of graded diffusers (Smith & Hung, 2000). Finally, manipulations using plus or minus lenses in the chick (Schaeffel et al, 1990), tree shrew (Norton, 1999; Siegwart & Norton, 1999), and monkey (Smith & Hung, 1999) resulted in either hyperopic or myopic growth, respectively.

Thus, emmetropization occurs during both normal genetically-determined ocular growth and under environmentally-induced conditions, and it involves the regulation of axial length growth rate via some property of retinal-image defocus. It appears to be effective following temporary optical manipulations, such as that seen for imposed plus and minus lenses, in which the eye responds by changing its axial length growth rate appropriately to compensate for the optical defocus. However, it has been thought that with prolonged nearwork, emmetropization breaks down, so that the axial length continues to increase, thus resulting in the development of myopia.

The mechanism for the short-term emmetropization process appears to be relatively simple, since visual feedback related to retinal-image defocus could provide the requisite control signal to regulate the direction and magnitude of axial growth. However, such appropriate changes in growth rate occur even when the optic nerve is severed (Troilo et al, 1987; Wildsoet & Pettigrew, 1988) or the midbrain nuclei for controlling accommodation are lesioned (Troilo, 1989), thus precluding any central visual feedback mechanism. Moreover, since defocus blur per se is an even-error signal (Stark, 1968), it lacks the requisite directional sensitivity for controlling axial growth. For these reasons, the controlling mechanism for the short-term emmetropization process, and in turn the long-term development of myopia, has remained a puzzle for decades.

18.2 EARLIER MODELS

Various descriptive as well as quantitative models have been proposed to account for the emmetropization process and the development of refractive error (Medina, 1987; Medina and Fariza, 1993; Schaeffel & Howland, 1988; Bartmann & Schaeffel, 1994; Flitcroft, 1998; Blackie & Howland, 1999; Hung & Ciuffreda, 1999; Norton, 1999; Wick, 2000). However, these models have not been able to explain the fundamental underlying mechanism of refractive error development.

For example, Medina and Fariza model (Fig. 18.1a) proposed a model whose simulated output curve matched the experimental refractive error development timecourse. Thus, the time constant, k, was selected to be very large (about 10 years) to fit the refractive error vs. age curve (see Fig. 18.1b). However, the model lacked homeomorphic correspondence with physiological processes associated with refractive error development and emmetropization. And, as such, the model simply served as a curve fitting mechanism, but would not be able to respond to stimulus changes such as those employed in the experimental optical manipulations discussed above.

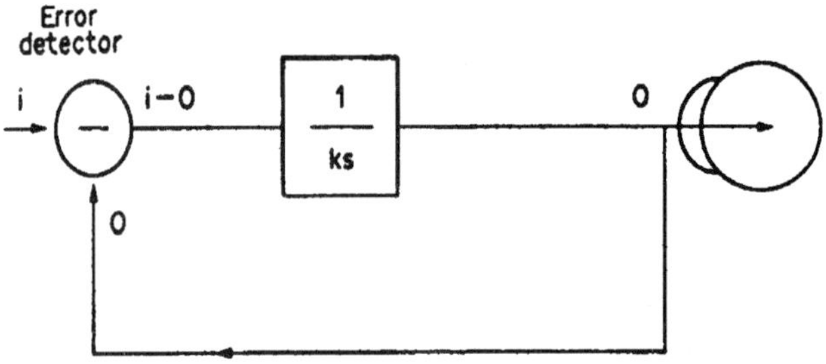

Figure 18.1(a). Servo mechanism of emmetropization proposed by Medina and Fariza (1993). The refractive error is controlled by a feedback system: i, input or commanded refraction; o, output or measured refraction; i - o, error. Transfer function F(s) = O(s) / I(s) = 1 / (1+ks) is of first-order in s, where s is a complex variable, k is the time constant, and O(s) and I(s) are the Laplace transforms of the temporal input and output function. Reprinted from Medina and Fariza (1993), pg.23, Fig. 1, with permission of Elsevier Science.

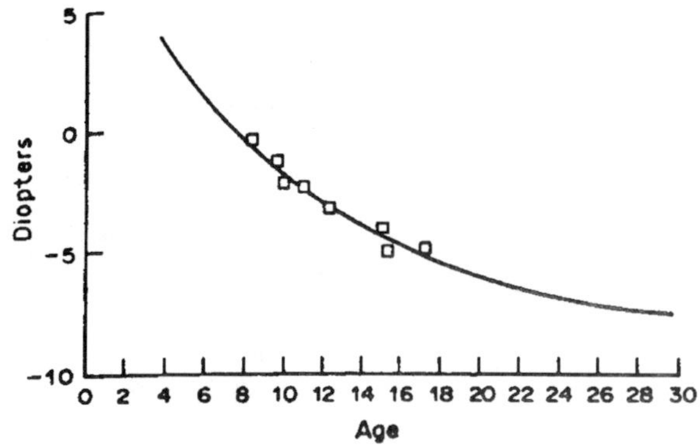

Figure 18.1(b). Refractive error vs age plot for experimental data (square symbols) and model (solid curve). Note that the time constant (estimated as the time to reach a change in exponential response that is 63% of the final difference in amplitude) for the curve is about 10 years. Reprinted from Medina and Fariza (1993), pg. 25, Fig. 3, with permission of Elsevier Science.

Schaeffel and Howland (1988) proposed a model in which two independent feedback loops were assumed to regulate ocular growth (Fig. 18.2). The accommodative loop operated under crystalline lens optical power feedback, whereas the retinal loop operated under local retinal feedback without accommodation. They presented simulation results showing that the accommodative feedback loop (Fig. 18.2, left loop) was responsible for the imposed-lens experimental results and also for the transient hyperopia observed after sectioning the optic nerve and lesioning the Edinger-Westphal nucleus. On the other hand, the retinal feedback loop (Fig. 18.2, right loop) was responsible for form-deprivation myopia, myopia in restricted retinal areas, myopia after optic nerve section, and recovery from form-deprivation myopia even with lesioning of the Edinger-Westphal nucleus. However, the model did not provide for interaction between accommodation and retinal feedback, which would have been expected to occur normally (Ciuffreda, 1991; 1998; also see Chapter 8, Models of Accommodation, in this volume). In addition, they had to vary the gains of both loops to simulate the experimental results. Yet, such gain variations should not have been permitted except in unusual cases, such as the simulation of an adaptive control system (Hung, 1992). Moreover, the increased degrees of freedom provided by gain adjustments in the two feedback loops may have resulted in an artificial, and not physiologically justified, matching of model and experimental data.

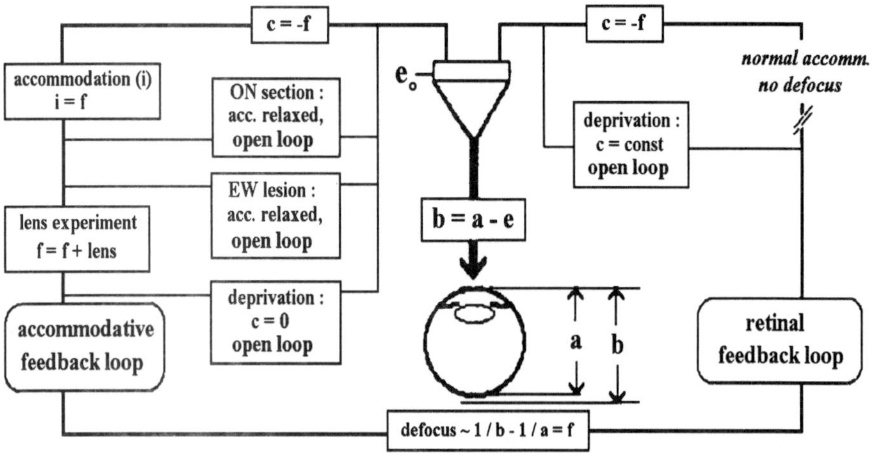

Figure 18.2. Model of refractive error development proposed by Schaeffel and Howland (1988) consists of independent accommodative and retinal feedback loops that regulate ocular growth. Symbols: a = control or reference axial length of the eye, b = current axial length, c = correction signal derived from either accommodative or retinal feedback, e_0 = initial difference between control and current axial length, EW = Edinger-Westphal nucleus, f = refractive state of the eye, lens = imposed lens optical power, ON section = optic nerve section, i = average amount of accommodation due to hyperopic defocus. Reprinted from Schaeffel and Howland (1988), pg. 2083, Fig. 2, with permission of Optical Society of America.

Flitcroft (1998) proposed a model of emmetropization and myopia (Fig. 18.3) using the dual-interactive accommodation and vergence feedback model by Hung and colleagues (1990, 1992, 1996, 1997). Refraction was defined as a function of the integral of the accommodative error, or blur, which could be obtained via solution of the accommodative and vergence interactive feedback equations (Hung, 1990). Simulations of the model using various initial refractions and a selected set of interactive accommodation-vergence model parameter values based on Hung (1996) was believed to demonstrate emmetropization of the response towards a constant refraction level (Fig. 18.3b). Analysis of the model, however, revealed that it simply generated decay curves that asymptote to a specified level rather than being an actual dynamic emmetropization mechanism that compensated for the imposed retinal defocus.

Flitcroft's (1998) model is based on the equation

$$r_{t+1} = r_t - k \int a_{error}(t)dt \qquad (18.1)$$

with

$$a_{error} = ae + r \qquad (18.2)$$

where r is refractive correction, t is time, k is gain of the emmetropization process, ae is accommodative error (or accommodative stimulus minus accommodative response), and a_{error} is net accommodative error. The accommodative error was obtained by the solution of the static interactive accommodative and vergence model equations provided by Hung (1996) and organized in a slightly different form by Flitcroft (1998). For simplicity in analyzing Eq. 18.1 without a loss of generality, the depth of focus was not included. For this analysis, time domain variables will be designated by lower case, whereas the Laplace transform variables will be designated by upper case. One can rewrite Eq. 18.1 as

$$dr = -k \int (ae + r)\, dt \qquad (18.3)$$

Taking the derivative of both sides of Eq. 18. 3 gives

$$\frac{dr}{dt} = -k \bullet (ae + r) \qquad (18.4)$$

Taking the Laplace transform of both sides of Eq. 18.4 gives

$$s \bullet R - r(0^+) = -k\,(AE + R) \qquad (18.5)$$

where s is the Laplace operator, and $r(0^+)$ equals the initial value of refraction, with the + symbol designating the value immediately after time 0. Let AE = ae/s, which represents a unit step of accommodative error. Re-arranging and solving for R gives

$$R = -k \bullet ae \bullet \frac{1}{s+k} \bullet \frac{1}{s} + \frac{r(0^+)}{s+k} \qquad (18.6)$$

Taking the partial fraction expansion of the product term in Eq. 18.6 gives

$$R = ae \bullet \left(\frac{1}{s+k} - \frac{1}{s} \right) + \frac{r(0^+)}{s+k} \tag{18.7}$$

Taking the inverse transform and re-arranging gives

$$r = \left[r(0^+) + ae \right] \bullet e^{-kt} - ae \tag{18.8}$$

Therefore, the refraction in Flitcroft's model is comprised of the sum of two terms: (1) an exponential decay term whose initial value depends on the initial refraction and the accommodative error, which then decays with a time constant of (1/k) towards a zero value; and (2) a constant bias term given by (-ae) which determines the final steady-state value of the simulation response (see Fig. 18.3b). Hence, instead of being a feedback process that acts on some fundamental property of the retinal-image defocus, the model simply sets the response to eventually arrive at the pre-selected (-ae) level. Moreover, the various initial refractions simply change the initial condition for the prescribed exponential decay of the response curve. Thus, although the model has some of the decay characteristics expected of an emmetropization model, it is clearly not an emmetropization model. Nevertheless, using this model, Flitcroft (1998) developed a relationship between refraction and oculomotor parameters. One of his conclusions based on the simulations was that "an appropriate spectacle correction … imposed after the model has reached a stable myopic refraction due to a high near work demand … has the effect of destabilising the refraction and causing a further myopic shift". On the other hand, recent human experimental results showed greater myopic progression in undercorrected (by approximately 0.75 D) than fully corrected myopic subjects (O'Leary, 2000). Also, animal experimental results (Schaeffel et al, 1990; Siegwart and Norton, 1999; Smith and Hung, 1999) demonstrated myopic growth of the eye with large negative lenses. Thus, both experimental results show myopic progression with large rather than small retinal defocus, which contrasts with Flitcroft's simulation results.

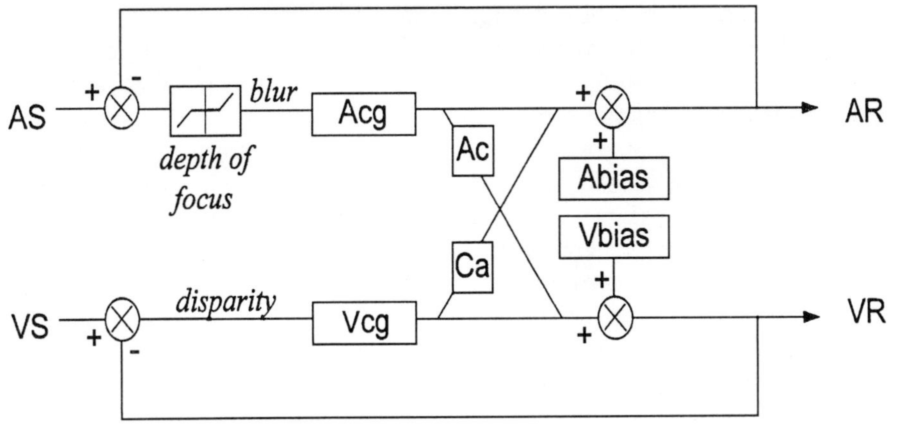

Figure 18.3a. Flitcroft (1998) model of emmetropization and myopia using the dual-interacting feedback model of Hung and colleagues (1980, 1990, 1992, 1996, 1997). AS and VS = accommodative and vergence demand. AR and VR = accommodative and vergence response. ACG and VCG = gains of the accommodative and vergence controllers. ABIAS and VBIAS = tonic level of accommodation and vergence. AC and CA = gain of accommodative convergence and convergence-accommodation crosslink. Reprinted from Flitcroft (1998), pg. 2870, Fig. 1, with permission of Elsevier Science.

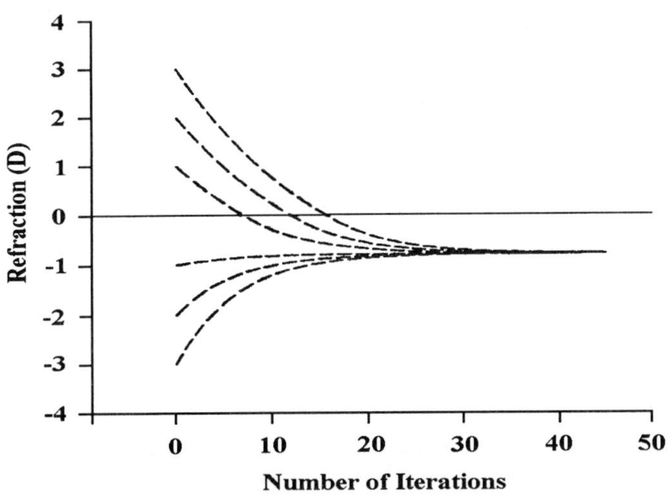

Figure 18.3b. Simulation of Flitcroft (1998) model for refraction as a function of number of interations (in arbitrary time units) for various initial refractions and a selected set of interactive accommodative-vergence model parameter values (based on Hung, 1996) believed to show emmetropization towards a fixed refraction value of about 1 D myopia. Reprinted from Flitcroft (1998), pg. 2874, Fig. 5a, with permission of Elsevier Science.

Hung and Ciuffreda (1999) proposed a descriptive block diagram model of refractive error development (Fig. 18.4a). The main concept of the model is that there are two pathways that interact, which can lead to the development of refractive error. First, a genetically-controlled pathway governs the growth of the cornea, lens, and axial length. A match between the optical and axial length components will result in emmetropia, whereas a mismatch will result in a refractive error. Both normal and abnormal preprogrammed growth are possible. Second, an environmentally-controlled pathway is driven by retinal defocus from the accommodative feedback loop. Subthreshold amounts of retinal defocus do not increase the axial length, whereas prolonged exposure to suprathreshold amounts of retinal defocus will increase the axial length. The genetically-preprogrammed axial length is summed with the environmentally-induced axial length to result in the overall axial length. The difference between cornea/lens optical power and the overall axial length optical power is the refractive error. Any refractive error will in turn serve as a portion of the accommodative stimulus in the accommodative feedback loop.

In addition, a more detailed quantitative version of this model was developed (Hung and Ciuffreda, 1999; Fig. 18.4b). In the basic accommodative (lower) loop, the difference between accommodative stimulus AS and response AR gives the accommodative error AE, which represents retinal defocus. This signal is input to deadspace element (±DSP) representing the depth-of-focus. The output of the deadspace element is input to the accommodative controller, which is represented by a dynamic element with gain ACG and time constant τ_C (= 4 sec). Other components added to the basic loop include a proximal element, an adaptive element, convergence accommodation crosslink gain, tonic accommodation (ABIAS), and lens plant with time constant τ_P (= 0.3 sec). Added to this basic loop is a long-term growth (upper) loop that is driven in part by the root-mean square (rms; equal to the average of the absolute value of the instantaneous response over a given time interval) of the accommodative error (AE), which can contribute to axial growth. Another part of the long-term growth loop is driven by genetic control of the cornea/lens and axial length (τ_E = 9 yrs; τ_F = 2 yrs; and τ_G = 9 years, for late-onset myopes; Hung and Ciuffreda, 1999).. The environmental (i.e., retinal-defocus induced) and genetically-controlled axial lengths are added together to provide the total axial length. Any mismatch between the power for the cornea/lens and the axial length results in a refractive error. Thus, this model provides a basis for both genetically-preprogrammed and environmentally-induced interactive development of refractive error.

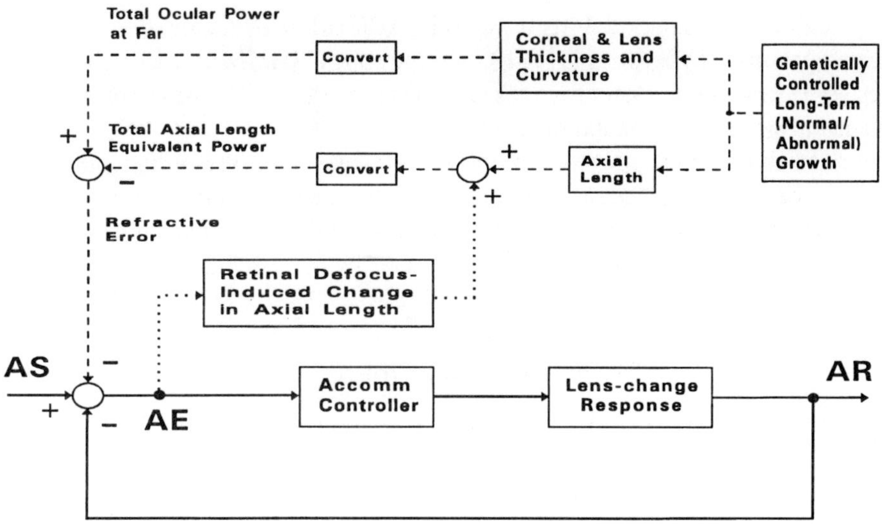

Fig. 18.4a.

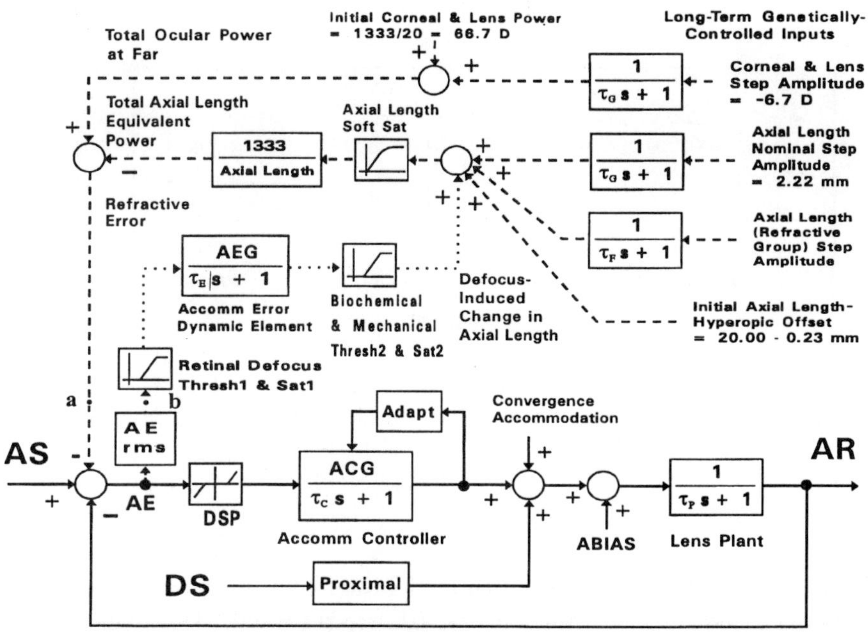

Fig. 18.4b.

Figure 18.4. *(a)* Descriptive block diagram of refractive error development (Hung and Ciuffreda, 1999). Solid-line pathway represents the normal accommodative system, whereas the dashed-line pathway represents refractive error development derived from both genetically-preprogrammed and retinal-defocus drives to refractive error. (b) Detailed block diagram of refractive error development. Reprinted from Hung and Ciuffreda (1999), pg. 43, Fig. 2A, and pg. 44, Fig. 2B, with permission of Swets and Zeitlinger.

The model was simulated using MATLAB/SIMULINK for the four refractive groups: hyperopes (HYP), emmetropes (EMM), early-onset myopes (EOM), and late-onset myopes (LOM) (Fig. 18.5). The time courses are consistent with experimental data.

Although this model provided important insight into the interactions between genetically-preprogrammed and environmental components, it nevertheless still did not uncover the fundamental underlying mechanism in the development of refractive error.

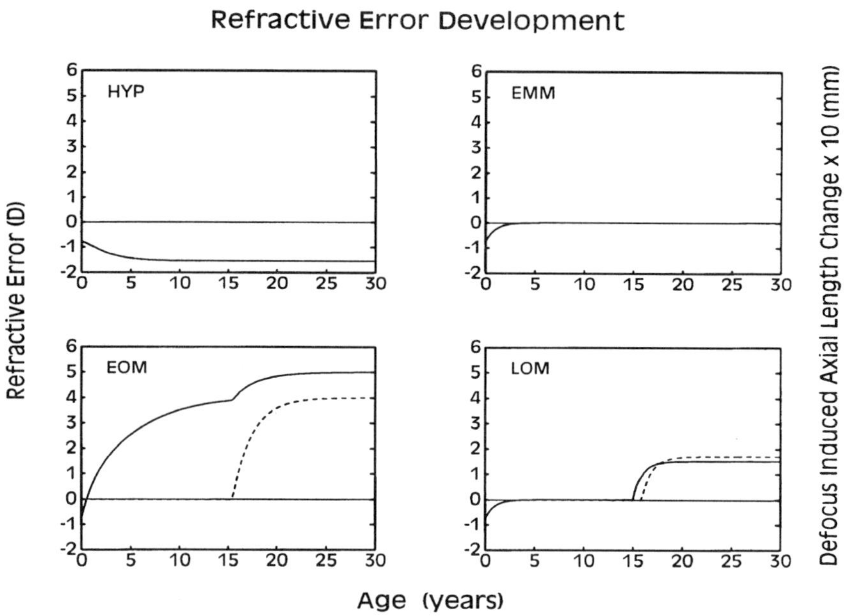

Figure 18.5. Simulation of model using SIMULINK showing long-term (30 yrs) timecourses for refractive error (solid) development and axial length change (dashed) in different refractive groups that are consistent with experimental data. Reprinted from Hung and Ciuffreda (1999), pg. 48, Fig. 5, with permission of Swets and Zeitlinger.

Wick (2000) proposed a conceptual model of emmetropization (Fig. 18.6) and myopia similar to that by Hung and Ciuffreda (1999). However, there are some important differences between the two models. Wick's model provides for separate paths for refractive error development in the two eyes; in addition, drug effects and clinical conditions such as anisometropic and astigmatism are also discussed. However, unlike Hung and Ciuffreda's model (1999), which was quantitatively simulated, Wick's model was only descriptive.

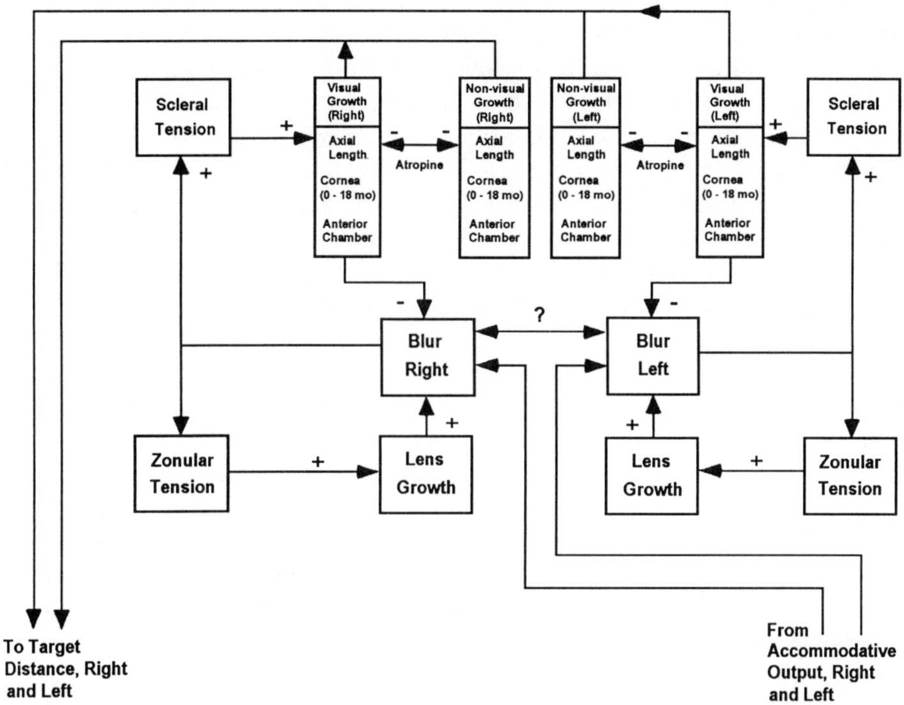

Figure18.6. Wick's (2000) conceptual model of emmetropization takes the form of dual intersecting feedback loops where genetically programmed ocular growth of each eye is altered by blur derived from interactions between accommodation and vergence. The potential influence of suppression on the response of the refractive state to blur is indicated by the crosslink between the blur mechanisms. In the visual growth mechanism, continued relative hyperopic blur (e.g., lag of accommodation at near) increases scleral tension and promotes axial elongation; lens growth is retarded by concurrent reduction in zonular tension. The resulting reduction in lens thickness (with increasing power) and increased axial length decrease the accommodative demand associated with near visual tasks. Visual growth feeds into and combines with genetically programmed ocular growth to result in the final refractive state." Reprinted from Wick (2000), pg. 49, Fig. 1, with permission of J. Optom. Vis. Devel.

Moreover, there are two problems with Wick's conceptual model. First, blur is derived from the combination of accommodative response, axial length, and lens growth, but not from the accommodative stimulus (see Fig. 18.6). Since retinal-image defocus is the difference between the accommodative stimulus and response, it is unclear how the blur signal could be appropriately obtained in the model. Second, his concept that increased scleral tension promotes axial length growth contradicts the findings of reduced scleral rigidity (Castrén and Pohjola, 1961a,b) and decreased tensile strength (Avetisov et al, 1984) in myopic individuals. In addition, instead of being the driving force for axial elongation, scleral tension is the consequence of active remodeling of the sclera during emmetropization (Reeder and McBrien, 1993; Phillips and McBrien, 1995; Siegwart and Norton, 1999). Further, Ong and Ciuffreda (1997) have argued, based on anatomical and biomechanical studies, that pressure on the sclera itself actually decreases during accommodation, and moreover, its effect is relatively small, thus precluding any direct mechanical role of accommodation in the process of axial elongation. These difficulties reduce the usefulness and general applicability of this model.

Each of these models has provided their own important insights into the overall interactive mechanisms in refractive error development. However, none of these models accounted for two important experimental findings. First, the imposition of plus or minus lenses results in a decrease and increase, respectively in the rate of ocular growth in young animals (Schaeffel et al, 1990; Norton, 1999; Siegwart and Norton, 1999; Smith and Hung, 1999). Second, the rate of ocular growth can be regulated even when the optic nerve is severed, thus precluding any accommodative feedback mechanism in the emmetropization process (Troilo et al, 1987; Wildwoet and Pettigrew, 1988). The development of a new theory, called the Incremental Retinal Defocus Theory, have been an attempt to resolve these long-statnding problems (see below).

18.3 INCREMENTAL RETINAL-DEFOCUS THEORY (IRDT)

To explore the underlying mechanism and to account for the apparently mixed experimental findings, a recent theory of refractive error development was formulated by us (Hung & Ciuffreda, 1999, 2000a-c). Two fundamental insights underlie our theory, which is called the Incremental Retinal-Defocus Theory (IRDT). First, local retinal-defocus <u>magnitude</u> is critical in the development of environmentlly-induced refractive error. The perceived blur is an even-error signal, which provides magnitude but not direction information regarding retinal defocus. Second, manipulations of the visual environment are effective in producing and/or modulating refractive error development mainly during the ocular growth and maturational period. This demonstrates the importance of a time-dependent element in providing the appropriate directional sense. It can be seen, however, that each insight alone is <u>insufficient</u> to provide a workable bidirectionally-sensitive theory. But, when these two are combined, they provide a coherent framework for a unifying theory of refractive error development. Our theory is based on the concept that the <u>change</u> in magnitude of retinal defocus during an <u>increment</u> of genetically-programmed axial length growth provides the critical information for directional modulation of growth rate. The term genetically-programmed is used to describe the normally-occurring ocular growth that has been pre-programmed genetically. This should be distinguished from environmentally-induced growth that is due to a change in retinal defocus. Both, however, involve neuromodulator release, with the environmentally-induced component acting to modulate the normal genetically-programmed release rate.

17.3.1 Basic Principles of the Theory

[1] *Neuromodulators Control Sensitivity to Changes in Retinal-Image Contrast*

In contrast to neurotransmitters, such as glutamate, acetylcholine, and GABA, which respond rapidly to retinal stimulation (Dowling, 1996), neuromodulators, such as dopamine, seratonin, and neuropeptides (Stone et al, 1989; Iuvone et al, 1991; Dowling, 1996), act over a longer period, and in addition, may cause changes in the neuronal synapses (Windhorst, 1996). An example of synaptic plasticity in the retina can be seen in the interplexiform cells in the retina (Dowling, 1996). These neurons, which contain dopamine, receive their inputs from the amacrine cells in the inner plexiform layer, and then send their outputs back to the horizontal cells in the outer plexiform layer (Werblin, 1973; Kolb, 1994; 1981; Dowling, 1996). Dopamine serves as a neuromodulator by altering the properties of the horizontal cell membrane and decreasing the flow of current across the membrane (Dowling, 1996; Windhorst, 1996). Moreover, because of the center-surround structure of the retina, the interplexiform neurons respond in a graded manner to local retinal-image contrast (Werblin, 1973; Kolb, 1994; Dowling, 1996).

We have proposed that feedback regulation provided by the interplexiform neurons from the inner to outer plexiform layers acts to maintain a relatively constant sensitivity to retinal-image contrast, and furthermore that interplexiform neuronal activity leads to a corresponding change in the neuromodulators (Hung and Ciuffreda, 2000a-c). Such feedback regulation is useful. It precludes the need for a memory mechanism to register and store previous levels of retinal defocus for the purposes of update and comparison. The release of neuromodulators results in synaptic changes in the horizontal cells (Dowling, 1996; Windhorst, 1996). This in turn alters retinal sensitivity to center-surround input, which helps to shift the steady-state operating level to permit responsivity to transient changes in local retinal-image contrast. Thus, the net rate of release of neuromodulators is <u>not</u> dependent on the absolute level of retinal defocus, but rather on the <u>change</u> in retinal-defocus magnitude. The release of neuromodulators also causes structural changes in the sclera via modulation of proteoglycan synthesis (Rada et al, 1992; Norton & Rada, 1995), wherein an increase in proteoglycan synthesis rate results in greater structural integrity of the sclera, and in turn, a decrease in axial growth rate relative to normal. Conversely, a decrease in proteoglyccan synthesis rate results in less structural integrity of the sclera, and in turn, an increase in axial growth rate relative to normal (Gottlieb et al, 1990; McBrien et al, 1999; Wilsoet, 1998; Christiansen & Wallman, 1991; Marzani & Wallman, 1997; Siegwart & Norton, 1999).

[2] *The Overall Mechanism for Regulating the Rate of Axial Length Growth*

Genetically-programmed mechanisms determine a baseline rate of neuromodulator release that is associated with normal axial growth rate. Retinal defocus-induced changes in the rate of neuromodulator release are superimposed onto this baseline level to result in changes relative to the normal axial growth rate. The net effect of the local-retinal mechanism, as discussed above, is that the change in retinal-defocus magnitude, and in turn the change in the rate of neuromodulator release, are in opposite directions with respect to the change in the rate of defocus-induced axial growth relative to normal. Thus, during an increment of genetically-programmed ocular growth, a change in retinal-defocus magnitude due to the incremental change in ocular geometry provides the directional information needed to modulate the rates of release of neuromodulators and proteoglycan synthesis, which in turn produce structural changes in the sclera for regulation of ocular growth (Siegwart & Norton, 1999; Wildsoet, 1998). For example, during an increment of genetically-programmed ocular growth (over days), if the retinal-defocus magnitude decreases, the axial growth rate increases. This results in relative myopic growth. On the other hand, if the retinal-defocus magnitude increases, the axial growth rate decreases. Hence, this results in relative hyperopic growth. These resultant axial growth rate changes are consistent with the emmetropization process. See Fig. 18.7 and the next section for details.

18.3.2 Applications of the Theory

This theory was tested under five critical experimental conditions to assess the generality of the proposed underlying mechanism. In addition, a MATLAB/SIMULINK model was constructed to demonstrate quantitatively the direct effect of change in retinal defocus via signal cascade through the retinal layers on scleral growth rate.

Lenses

During ocular development, the eye exhibits continuous genetically-programmed growth (Hung and Ciuffreda, 1999, 2000a-c). The imposition of a lens causes changes in retinal defocus, which acts to modulate the genetically-predetermined normal growth rate, and thereby alter overall axial length growth rate. This modulation can be illustrated by the following example. Consider the effect of introducing spherical lenses in front of the eye. The change in size of the blur circle during a small increment of normal genetically-programmed ocular growth for large imposed zero, minus, and plus lenses is shown schematically in Figs. 18.7a, b, and c, respectively. A neuromodulator, such as dopamine, maintains a certain level of neuronal activity related to retinal-image contrast by means of the local retinal feedback mechanism described earlier. The net effect is that the rate of neuromodulator release is dependent not on the absolute level of retinal-defocus magnitude, but rather on the change in retinal-defocus magnitude during the increment of genetically-programmed ocular growth, as was also mentioned earlier. For example, for a zero power lens, there is no change in the size of the blur circle. Thus, no additional neuromodulator is released,

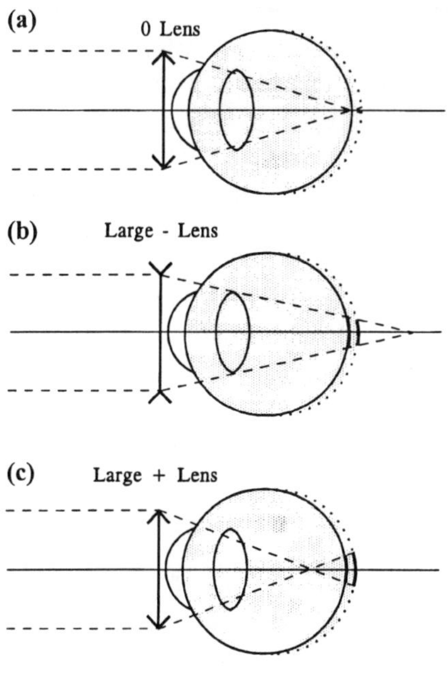

Figure 18.7. Below local blur threshold; Normal rate of neuromodulators; Normal growth rate. Dashed lines represent distant light rays. Dotted curves represent growth increments.

Decrease in local blur magnitude; Decrease in rate of neurotransmitters; Decrease in rate of proteoglycan synthesis Increase in axial growth rate relative to normal.

Increase in local blur magnitude; Increase in rate of neuromodulators; Increase in rate of proteoglycan synthesis; Decrease in axial growth rate relative to normal.

Reprinted from Hung and Ciuffreda (2000c), pg. 1094, Fig. 1, with permission of Bull. Math. Biol.

and the normal genetically-based incremental axial growth pattern of the young eye is maintained. With the introduction of a minus lens, however, the size of the blur circle is decreased during the growth increment; thus, the rates of neuromodulator release and in turn proteoglycan synthesis are decreased, thereby resulting in an increase in axial growth rate (Norton, 1999). On the other hand, with the introduction of a plus lens, the size of the blur circle is increased during the growth increment; thus, the rates of neuromodulator release and in turn proteoglycan synthesis are increased, thereby resulting in a decrease in axial growth rate (Norton, 1999). Hence, either a decrease or increase in mean retinal-defocus magnitude during an increment of genetically-programmed axial growth is proposed to cause a change in the rate of neuromodulator release, which in turn leads to biochemically-mediated structural changes in the sclera (Siegwart & Norton, 1999; Wildsoet, 1998), that are manifest as appropriate changes in the rate of axial growth and reflect the active emmetropization process.

Graded Diffusers

The IRDT theory can also be applied to recent experimental results on the effect of graded diffusers in monkeys (Smith & Hung, 2000). Although a diffuser can have complex optical effects (Smith & Atchison, 1997), its primary effect is to disperse or scatter the rays of light that are transmitted through the diffuser. This is schematically represented by a cone of light and seen in the figure as two lines representing the boundaries of the cone (Fig. 18.8a-c). The angle is increased for a stronger diffuser to represent its greater dispersional effect (Fig. 18.8c). This results in a very diffuse blur circle on the retina. Since accommodation would be quite imprecise under this condition, for simplicity, the focal point for the central ray of the cone of diffused light is set midway between the two incremental foveal positions along the visual axis. It can be shown that based on the geometrical configuration using ray tracing, for the weak diffuser with small dispersion (Fig. 18.8a), the decrease in retinal-defocus magnitude during the genetically-programmed incremental growth is relatively minor. Thus, the decrease in neuromodulator and proteoglycan synthesis is small, thereby resulting in a relatively small increase in axial growth rate. On the other hand, for the stronger diffusers with stronger dispersions (Figs. 18.8b, c), there is a progressively greater decrease in retinal-defocus magnitude. Therefore, the intermediate diffuser will result in an intermediate increase in axial growth rate, whereas the strong diffuser will result in the largest increase in axial growth rate. These findings are consistent with experimental results in animals (Smith & Hung, 2000).

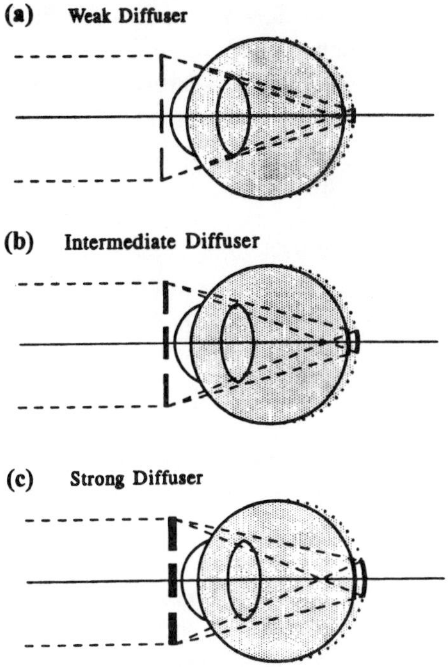

Figure 18.8. Note that the diffuser increases the effective dispersion of any ray of light, resulting in a blur circle. See text for details. Dashed lines represent distant light rays. Dotted curves represent growth increments. See text for details. Reprinted from Hung and Ciuffreda (2000c), pg. 1095, Fig. 2, with permission of Bull. Math. Biol.

Black Occluder

Under the more extreme condition of a full black occluder (Smith & Hung, 2000), there is complete absence of form vision along with a drastically reduced retinal luminance level. Our theory predicts that as a result of the absence of retinal signal-induced feedback regulation of horizontal cells in the dark (Dowling, 1996), there would likewise be a drastic reduction in the rate of neuromodulators, and in turn a substantial decrease in the rate of proteoglycan synthesis (Fig. 18.9a). This will result in a marked increase in the axial growth rate and the development of myopia, which is also consistent with experimental results in animals (Tigges et al, 1990; Iuvone et al, 1991).

Very Strong Diffuser or Removal of Crystalline Lens

For the condition of either a very strong diffuser (Fig. 18.9b) or the removal of the crystalline lens (Fig. 18.9c), there is a large initial increase in retinal-image blur magnitude. Our theory predicts that this will cause an initial increase in the rates of neuromodulators and proteoglycan synthesis, which will result in a decrease in the axial growth rate, and therefore relative hyperopia. However, the large blur magnitude, with its boundary rays being nearly parallel, does not change substantially during subsequent time increments. This lack of change in retinal-defocus magnitude will result in a normal rate of neuromodulator release, and therefore the subsequent axial growth rate will be nearly normal. Thus, the initial hyperopia is retained (Smith & Hung, 2000).

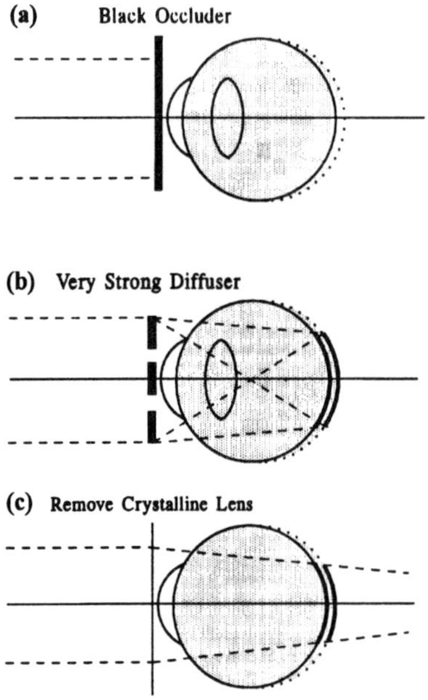

Figure 18.9. (a) Absence of retinal signal-induced feedback regulation in the dark; Low rate of neuromodulators; Low rate of proteoglycan synthesis; High rate of ocular growth. Dashed lines represent distant light rays. Dotted curves represent growth increments.

(b & c) Large initial increase in blur magnitude; Increase in rate of neuromodulators; Increase in rate of proteoglycan synthesis; Decrease in axial growth rate relative to normal; Subsequent growth rate near normal since change in blur magnitude is relatively small; Hence, remain hyperopic.

Reprinted from Hung and Ciuffreda (2000c), pg. 1096, Fig. 3, with permission of Bull. Math. Biol.

Transient Hyperopia

Transient hyperopia following the imposition of a diffuser can be explained as follows (Fig. 18.10). Prior to the imposition of the diffuser, the retinal-defocus magnitude is near or below the threshold level. Thus, there is a relatively small amount of retinal defocus-induced neuromodulator release and proteoglycan synthesis, and the growth rate is determined primarily by genetic factors. However, the sudden imposition of the diffuser results in an immediate increase in retinal-defocus magnitude. Thus, there is a transient increase in the rates of neuromodulators and proteoglycan synthesis. This results in a decrease in axial growth rate, or a transient relative hyperopia. However, subsequently, as the genetically-programmed growth continues, the condition becomes similar to that shown in Figs. 18.8a-c. Since the boundaries of the light bundles from the diffuser converge beyond the retina, there will now be an incremental decrease in retinal-defocus magnitude. The resultant decrease in the rates of neuromodulators and proteoglycan synthesis, and in turn an increase in axial growth rate, will first effectively null out the initial hyperopia and then eventually develop into myopia (see Figs. 18.8a-c). This can account for the transient nature of the initial hyperopia seen in some of the recent monkey experimental results (Smith & Hung, 2000).

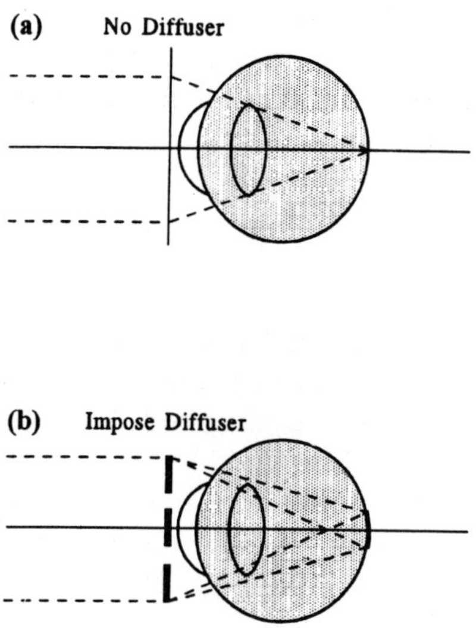

Figure 18.10. Prior to imposition of diffuser, below local blur threshold;

Normal rate of neuromodulators;

Normal rate of proteoglycan synthesis;

Normal growth rate.

Dashed lines represent distant light rays.

Imposition of diffuser results in sudden increase in local blur magnitude;

Thus, there is a transient increase in rate of neuromodulators;

Increase in rate of proteoglycan synthesis;

Decrease in axial growth rate relative to normal;

Resulting in an initial hyperopia;

Subsequent myopic effect during incremental growth first overcomes the initial hyperopia, and then develops into relative myopia.

Reprinted from Hung and Ciuffreda (2000c), pg. 1097, Fig. 4, with permission of Bull. Math. Biol.

The initial transient increase in retinal-defocus magnitude may also explain some recent findings of a greater effect of plus over minus lenses in controlling ocular growth. For example, Wildsoet and Collins (2000) found that imposition of plus/minus multifocal lenses in chicks resulted in a preference towards hyperopic ocular development, which indicates a preference of the plus over the minus lens. Also, Winauer et al (2000) found that chicks compensated to imposed plus lenses even when blurred with diffusers, thus suggesting a particular potency of myopic blur in changing ocular growth rate. These results can be explained by the fact that the imposition of either a plus or a minus lens will cause an initial transient increase in retinal-defocus magnitude. During a subsequent increment of genetically-programmed growth, the plus lens will cause an increase in retinal-defocus magnitude (see Fig. 18.7c), which augments the initial transient increase. On the other hand, the minus lens will cause a decrease in retinal-defocus magnitude (see Fig. 18.7b), which opposes the initial transient increase. The net result is a greater effect of plus over minus lenses during the initial phases of ocular development following such optical manipulations.

Prolonged Nearwork

Finally, the theory can be analyzed for the condition of prolonged nearwork, as in the case of the development of school myopia, wherein relatively small amounts of retinal defocus are present over extended periods of time (i.e., weeks or months) (Ong & Ciuffreda, 1995, 1997). This can be understood in terms of the interactions between two clinical measures: the dynamic nearwork-induced transient myopia (NITM) and the static normal accommodative stimulus/response (AS/R) function (Ciuffreda, 1991, 1998; Ciuffreda & Kenyon, 1983; Ong et al, 1993; Hung, 1998) (Fig. 18.11). NITM refers to the transitory myopic refractive shift found in distance viewing immediately following sustained nearwork (Ong and Ciuffreda, 1995, 1997). It is measured as the difference between post- (□ in Fig. 18.11) and pre-task (■ in Fig. 18.11) accommodative levels at distance. The AS/R function is a static s-shaped curve which shows a slight over-accommodation at distance and progressive under-accommodation at near with increased dioptric demand (Ciuffreda, 1991, 1998).

During nearwork, the accommodative response (AR) lags the accommodative stimulus (AS) (Fig. 18.11, point A; Fig. 18.12a). However, immediately following the nearwork and returning to the far-target viewing, AR exceeds AS more than usual due to the presence of NITM and its relatively slow decay back to the initial pre-task distance refractive state (Figs. 18.11 and 18.12b). This transient myopia can be conceptually regarded as an equivalent low-powered plus lens that is placed in front of the eye (Fig. 18.12c). The transient myopia remains for a period of time [30 sec or longer (Ong and Ciuffreda, 1995, 1997)]. In returning to nearwork within this slow decay period, the accommodative stimulus is now the net result of the negative optical power due to the distance of the target minus the small equivalent plus lens associated with the residual transient myopia, i.e., the non-decayed portion of the NITM. Thus, the net accommodative stimulus is slightly less (~ 0.25 D) than 4D (point B in Fig. 18.11) (Ong and Ciuffreda, 1997). The response to this reduced effective AS is a slightly reduced AR, and thereby a smaller AE is present. Therefore, the cumulative effect of such repeated, non-fully decayed NITM, which results in repeated transient decreases in retinal defocus at near, is conceptually similar to that of an imposed large static minus lens during an increment of genetically-determined ocular growth (Fig. 18.7b). By the earlier arguments, this results in a decrease in the net rate of release of neuromodulators, a decrease in proteoglycan synthesis, and in turn an increase in the rate of axial growth relative to normal, or relative myopic growth. It is somewhat ironic that rather than representing a failure of the emmetropization process, myopia development is actually a result of the emmetropization process that operates under the constraints of the AS/R function during increments of genetically-programmed ocular growth.

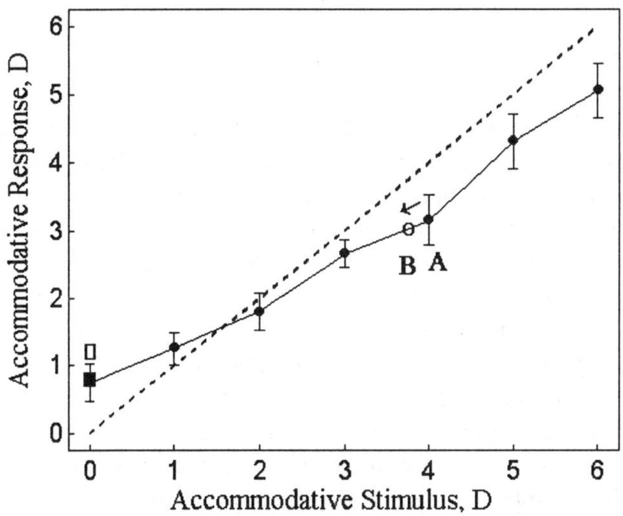

Figure 18.11. Plot of mean accommodative stimulus-response data for 10 visually-normal subjects. Symbols: ● = group mean accommodative response, error bars = ±SEM, ■ = initial pre-task refractive state at distance, □ = initial post-task NITM, data point at A = initial near response, and o = subsequent near response with superimposed non-decayed NITM. Adapted from Ong et al (1993), pg. 199, Fig. 6, with permission from Invest. Ophthal. Vis. Sci.

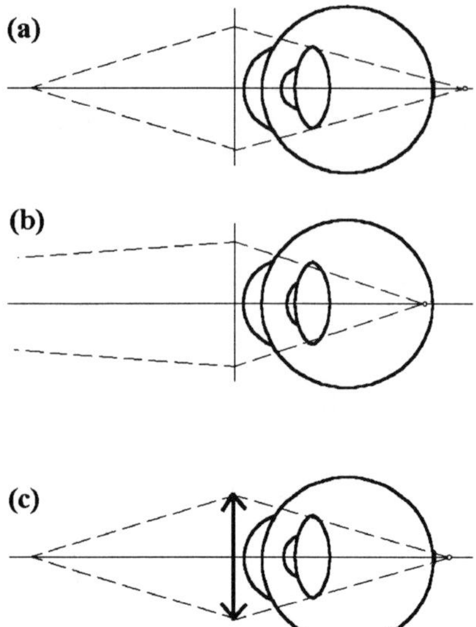

Figure 18.12. Nearwork results in lag of AR and associated hyperopic defocus (pt. A in AS/R curve in Fig. 18.11). Dashed lines represent light rays from object.

Subsequent viewing at far results in a small nearwork-induced transient myopia (NITM).

Plus lens represents equivalent amount of residual transient myopia. Far-to-near carry-over of NITM due to slow decay of crystalline lens power can be represented by an equivalent plus lens. Thus, when returning to nearwork, the effective AS is reduced slightly, and AR is also reduced (i.e., moved down from A to B in the AS/R curve in Fig. 18.11), resulting in a slight decrease in AE, or retinal defocus.

Adapted from Hung and Ciuffreda (2000c), pg. 1099, Fig. 5, with permission of Bull. Math. Biol.

The analysis above also explains why the increase in axial growth rate is primarily associated with prolonged nearwork rather than farwork. This results from a difference in their operating regions on the AS/R curve. It was noted that the net result of repeated NITM is a decrease in the accommodative stimulus, or a shift of AS to the left on the AS/R curve. For nearwork (AS = 4 D), this results in a decrease in retinal defocus <u>magnitude</u>. On the other hand, for farwork (AS $\leq$ 1.5 D), this results in an increase in retinal defocus <u>magnitude</u>. Thus, in contrast to the effect of nearwork discussed above, farwork is analogous to the imposition of a plus lens (see Fig. 18.7c), which according to IRDT, results in a relative decrease in axial growth rate. Furthermore, some individuals (hyperopes and emmetropes) may not develop myopia subsequent to nearwork, because they may have higher sensory blur thresholds for inducing axial length growth than the myopes (Hung & Ciuffreda, 1999; Flitcroft, 2000).

In addition, according to this theory, the rate of ocular growth is dependent on the <u>change</u> in retinal-defocus magnitude regardless of how it is generated. Therefore, in the absence of an increment of genetically-programmed ocular growth, retinal defocus-induced axial elongation due to prolonged nearwork can still occur as long as the individual exhibits susceptibility to the neurochemical influences on scleral growth. This may explain the finding of form deprived myopia in adolescent animals even after they are past the rapid juvenile growth phase (Troilo et al, 2000b), as well as the finding of axial-based permanent myopia in newly-trained adult microscopists (Adams and McBrien, 1992).

18.3.3 Basic Retinal Anatomy and Physiology of IRDT

Since the IRDT involves detailed aspects of retinal signal processing, a brief review of retinal neural signal transmission is provided below. Signals are transmitted in the retina through three types of neurons: photoreceptors, bipolar cells, and ganglion cells. The photoreceptors (rods and cones) are stimulated by light and relay their signals through bipolar cells, which in turn relay the information through the ganglion cells. The axons of the ganglion cells in the retina form the optic nerve, which transmits retinal-image information to the higher cortical centers. Bipolar cells also receive light stimulus information from neighboring, or surround, photoreceptors via lateral connections from horizontal cells in the outer plexiform layer. This center-surround organizational structure provides local retinal-image contrast information to the "sustained" ganglion cells, which respond to sustained contrast information. On the other hand, "transient" ganglion cells respond to <u>change</u> in the surround via amacrine cells in the inner plexiform layer. Thus, these neurons relay information regarding any change in retinal-defocus magnitude. In addition, interplexiform neurons from the inner-to-outer plexiform layer modulate the long-term sensitivity of horizontal cells to surround input. Thus, this feedback mechanism serves to adjust the steady-state sensitivity level to provide relatively constant sensitivity to changes in local contrast (Dowling, 1996).

18.4 QUANTITATIVE MATLAB/SIMULINK MODEL

A conceptual block diagram of the model is shown in Fig. 18.13a. It is based on the principle that the magnitude of retinal defocus can be represented by the difference in center and surround excitation. A <u>change</u> in this signal, and thus a change in retinal-defocus magnitude, provides the requisite sign for modulating ocular growth. The sensitivity to local retinal-image contrast is maintained at a relatively constant level by means of feedback regulation of horizontal cell gain provided by the interplexiform neurons. This precludes the need for a "memory mechanism" (Norton, 1999) for storing information regarding the immediately previous level of retinal-defocus magnitude, so that its change can be discerned. The release of neuromodulator in turn results in changes in the rate of scleral proteoglycan synthesis, which causes a change in scleral growth rate. This relative growth rate is added to the ongoing and normal genetically-programmed ocular growth rate to provide the overall axial length growth.

The detailed model is shown in Fig. 18.13b. The sustained pathway consists of the photoreceptor, bipolar, and sustained ganglion cells. It is modulated by surround signals via horizontal cells in the outer plexiform layer to provide local steady-state or sustained contrast information. The transient pathway also consists of photoreceptor, bipolar, and transient ganglion cells. However, it is modulated by surround signals via amacrine cells in the inner plexiform layer to provide information regarding local change or transients in contrast information. Feedback regulation is provided locally by the interplexiform neurons that receive signals for neuromodulator release in the inner plexiform layer and modulate the gain of horizontal cells in the outer plexiform layer to maintain a relatively constant sensitivity to change in local contrast. The center bipolar cell receives a signal derived from the difference between center and summed surround inputs, which represents the summated amount of retinal-image defocus across the overlapping, spatially-contiguous center and surround receptive field area. This signal is differentiated by neural circuitry in the inner plexiform layer, which most likely contains amacrine cells. This change is rectified, so that the "envelope" of the signal, which represents the overall change in retinal-defocus magnitude, drives the rate of neuromodulator release. The neuromodulator, or a cascade of neurochemicals related to the release of the neuromodulator (Wallman, 1997), passes through the choroid to reach the sclera. The transit of the neuromodulator through the choroid may result, at least in the monkey, in a volume change that is observed as a change in choroidal thickness (Curtin, 1985; Cheng et al, 1992; Wildsoet and Wallman, 1995; Marzani and Wallman, 1997; Hung et al, 2000b,c; Troilo et al, 2000a). This may explain why, as expected, choroidal thickness changes in the monkey are correlated with changes in retinal-defocus magnitude, but the optical change associated the thickness change is too small to account for any significant contribution towards full emmetropization (Hung et al, 2000b.c; Troilo et al, 2000a). On the other hand, the neuromodulator that reaches the sclera modifies proteoglycan synthesis to result in changes in ocular growth that does provide nearly full emmetropization, as described in the schematic model (Fig. 18.13a).

Model simulation responses to center and surround stimuli are shown in Figs. 18.14 a-d. The center stimulus representing sharp focus consists of a ± 1 amplitude (in arbitrary units representing change in luminance relative to the background level) peak-to-peak, 0.1 Hz, square-wave signal. The surround stimuli, representing varying degrees of retinal-image defocus, consists of the same square wave but modulated by different step levels over the time span of the simulation. Fig. 18.14a shows the various steps of modulation of the surround amplitude (solid) and the feedback-regulated change in gain of the horizontal cells. As noted above, this provides relatively constant sensitivity to changes in retinal-defocus magnitude. The pulse-like responses for the rates of neuromodulator release (solid) and proteoglycan synthesis (dashed) are shown in Fig. 18.14b. The change in proteoglycan synthesis rate in turn causes changes in the scleral growth rate relative to normal (Fig. 18.14c). Finally, the cumulative change in axial length relative to normal is shown in Fig. 18.14d. These results clearly demonstrate that the model is able to simulate the bi-directional aspects of choroidal and scleral axial length changes found experimentally.

Figure 18.13. (See next page). (a) Conceptual block diagram model of the retinal-defocus pathway for regulating sclera growth. The difference between center and surround excitation provides the retinal-defocus signal. The derivative of the signal drives the release of neuromodulators, which provides the feedback via interplexiform neurons to regulate horizontal cell gain. In addition, release of neuromodulators causes changes in the rate of proteoglycan synthesis, and in turn relative scleral growth rate. (b) Detail block diagram model depicting the regulation of scleral growth rate. The retinal layers (outer to inner) are arranged from left to right: photoreceptor, outer plexiform, bipolar, inner plexiform, and ganglion. The sustained pathway consists of center photoreceptor, center bipolar B, and sustained ganglion cell. Horizontal cell, whose gain is regulated by feedback via interplexiform cells, relays surround information to modify sustained ganglion output. The transient pathway consists of center photoreceptor, center bipolar A, and transient ganglion cell. Amacrine cell relay change in surround information to modify transient ganglion output. Center bipolar B signal consists of retinal-defocus information and passes through a rectifier, lowpass filters, and elements representing neuromodulator release, the choroid, and proteoglycan synthesis. This is inverted to provide relative scleral growth rate relative to normal. Adapted from Hung and Ciuffreda (2000c), pg. 1101, Figs. 6a,b, with permission of Bull. Math. Biol.

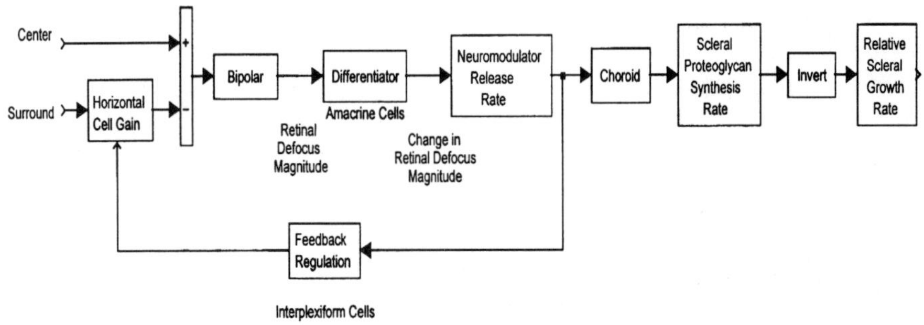

Fig. 18.13a.

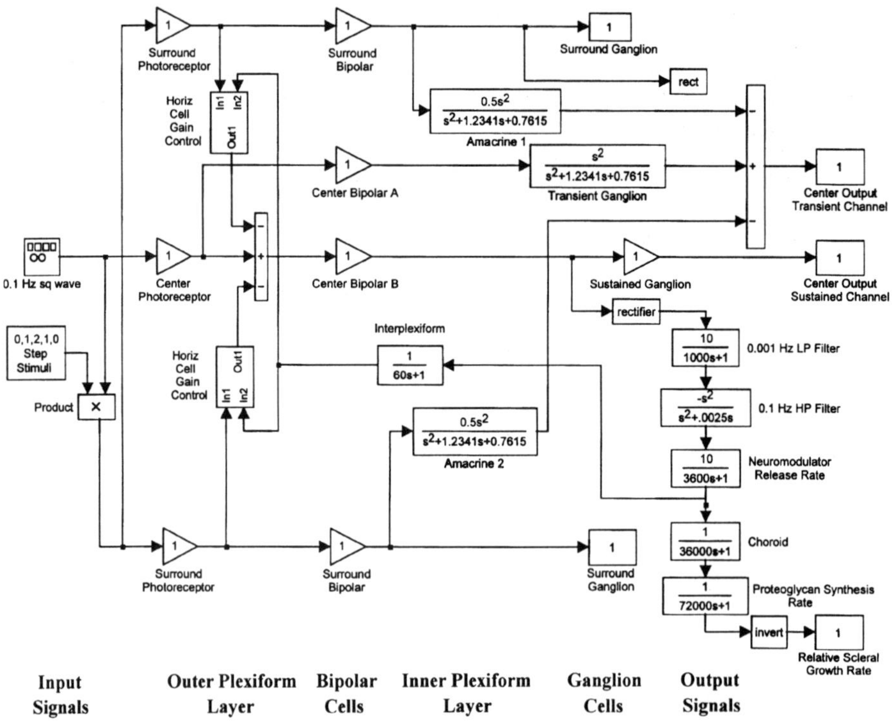

Fig. 18.13b.

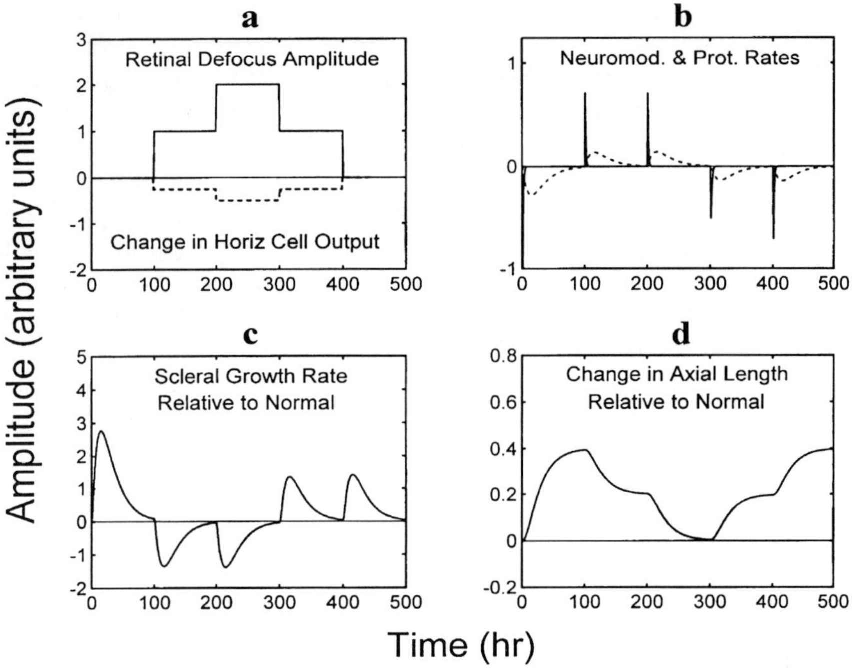

Figure 18.14. (a) Envelope of surround stimulus representing various levels of defocus (solid). Changes in horizontal cell output, which are regulated by interplexiform neuronal feedback, shows a complementary response to surround defocus (dashed). (b) Pulses of rates of neuromodulator release (solid) and proteoglycan synthesis (dashed) occur at the transitions of surround stimulus (see Fig. 18.14a). (c) Rate of scleral growth relative to normal follows the pulses in the rate of proteoglycan synthesis (see Fig. 18.14b). (d) Integration of scleral growth rate provides the change in axial length relative to normal. The direction of change is consistent with experimental findings and with the analysis provided by the schematic model. Adapted from Hung and Ciuffreda (2000c), pg. 1103, Fig. 7, with permission of Bull. Math. Biol.

18.5 SUMMARY AND CONCLUSIONS

Emmetropization appears to be governed by a relatively simple mechanism, which has been described by our newly proposed Incremental Retinal-Defocus Theory (Hung and Ciuffreda, 1999, 2000). The theory states that the rate of change of retinal defocus determines the rate of release of neuromodulators, which modulates rate of proteoglycan synthesis, and in turn regulates the rate of scleral growth. Schematic analysis of the theory has provided a clear explanation for the eye's ability to grow in the appropriate direction under a wide range of experimental conditions including lens, diffusers, black occluder, and removal of the crystalline lens. In addition, the theory has been able to explain how prolonged nearwork could lead to increased cumulative decrease in proteoglycan synthesis, and thereby increased axial growth and permanent myopia. The critical point is that the detection mechanism does not depend on the sign of the blur, but rather on the change in blur magnitude that is either environmentally-induced or results from an increment of genetically-programmed ocular growth. And, it is not necessary to invoke more complicated processes, such as sensing and analyzing of chromatic aberration, spherical aberration, spatial gradient of blur, or spatial frequency content (Ciuffreda, 1991, 1998). Thus, this unifying theory provides an understanding of the basic underlying retinal mechanism for detecting blur magnitude, and furthermore explains how the neurochemical signal is processed to modulate the rate of eye growth, and in turn the resultant development of axial myopia.

18.6 REFERENCES

Adams, D. W., and McBrien, N. A. 1992, Prevalence of myopia and myopic progression in a population of clinical microscopists, *Optom. Vis. Sci.* **69**: 467-473.

Alward, W.L., Bender, T.R., Demske, J.A., and Hall, D.B., 1985, High prevalence of myopia among young adult Yupik Eskimo, *Can. J. Ophthalmol.* **20**: 241-245.

Avetisov, E. S., Savitskaya, N. F., Vinetskaya, M. I., and Iomdina, E. N., 1984, A study of biochemical and biomechanical qualities of normal and myopic eye sclera in humans of different age groups, *Metab. Pediatr. Syst. Ophthalmol.* **7**: 183-188.

Bartmann, M., and Schaeffel, F., 1994, A simple mechanism for emmetropization without cues from accommodation or colour, *Vis. Res.* **34**: 873-876.

Bennett, A. G., and Rabbetts, R.B., 1989, *Clinical Visual Optics*, Butterworth-Heinemann, Woburn, MA, pg. 75.

Blackie, C. A. and, Howland, H. C., 1999, Extension of the Flitcroft model of emmetropization: inclusion of pupil size, *Ophthal. Physiol. Opt.* **19**: 112-125.

Bradley, D. V., Fernandes, A., Tigges, M., and Boothe, R.G., 1996, Diffuser contact lenses retard axial elongation in infant rhesus monkeys, *Vis. Res.* **36**: 509-514.

Castrén, J. A., and Pohjola, S., 1961a, Refraction and scleral rigidity, *Acta Ophthalmol.* **39**: 1011-1014.

Castrén, J. A., and Pohjola, S., 1961b, Scleral rigidity in puberty, *Acta Ophthalmol.* **39**: 1015-1019.

Cheng, H.-M., Omah, S. S., and Kwong, K. K., 1992, Shape of the myopic eye as seen with high-resolution magnetic resonance imaging, *Optom. Vis. Sci.* **69**: 698-701.

Christiansen, A. M. and Wallman, J., 1991, Evidence that increased scleral growth underlies visual deprivation myopia in chicks, *Invest. Ophthal. Vis. Sci.*, **32**: 2134-2150.

Ciuffreda, K. J., 1991, Accommodation and its anomalies, in: *Vision and Visual Dysfunction: Visual Optics and Instrumentation*, Vol. 1, W. N. Charman, ed., Macmillan, London, pp. 231-279.

Ciuffreda, K. J., 1998, Accommodation, pupil, and presbyopia, in: *Borish's Clinical Refraction,* W. J. Benjamin, ed,, W. B. Saunders Co, Philadelphia, PA, pp. 77-120.

Ciuffreda, K. J., Kenyon R. V., 1983, Accommodative vergence and accommodation in normals, amblyopes, and strabismics, in: *Vergence Eye Movements: Basic and Clinical Aspects,* C. M. Schor and K. J. Ciuffreda, eds., Butterworths, Boston, MA, pp. 101-173.

Ciuffreda, K. J., and Wallis, D. M., 1998, Myopes show increased susceptibility to nearwork aftereffects, *Invest. Ophthal. Vis. Sci.* **39**: 1797-1803.

Curtin, B. J., 1985, The etiology of myopia, in: *The Myopias: Basic Science and Clinical Management*, Harper & Row, Philadel. PA, pp. 61-151.

Dowling, J. E., 1996, Retinal processing of vision, in *Comprehensive Human Physiology: From Cellular Mechanisms to Integration, Vol. 1,* Greger R and U. Windhorst, eds., Springer-Verlag, Berlin, pp. 773-778.

Fledelius, H. C., and Stubgaard, M., 1986, Changes in refraction and corneal curvature during growth and adult life. A cross-sectional study, *Acta Ophthalmol.* **64**: 487-491.

Flitcroft, D. I., 1998, A model of the contribution of oculomotor and optical factors to emmetropization and myopia, *Vis. Res.*, **38**: 2869-2879.

Goh, W.S., and Lam, C.S., 1994, Changes in refractive trends and optical components of Hong Kong Chinese aged 19-39, *Ophthal. Physiol. Opt.* **14**: 378-382.

Goldschmidt, E., 1968, On the etiology of myopia - an epidemiological study. *Acta Ophthalmol.* **98 (suppl)**: 1-72.

Goss, D. A., and Erickson, P., 1987, Meridional corneal components of myopia progression in young adults and children, *Am. J. Optom. Physiol. Opt.* **64**: 475-481.

Goss, D. A., Hampton, M. J., and Wickham, M. G., 1988, Selected review on genetic factors in myopia, *J. Am. Optom. Assoc.* **59**: 875-884.

Goss, D. A., and Jackson, T. W., 1993, Cross-sectional study of changes in the ocular components in school children, *Appl. Opt.* **32**: 4169-4173.

Goss, D. A., and Wickham, M. G., 1995, Retinal-image mediated ocular growth as a mechanism for juvenile onset myopia and for emmetropization, *Doc. Ophthalmol.* **90**: 341-375.

Goss, D. A., and Winkler, R. L., 1983, Progression of myopia in youth: age of cessation, *Am. J. Optom. Physiol. Opt.* **60**: 651-658.

Gottlieb, M. D., Joshi, H. B., and Nickla, D. L., 1990, Scleral changes in chicks with form-deprived myopia, *Curr. Eye Res.* **9**: 1157-1165.

Grosvenor, T., and Goss, D. A., 1998, Role of the cornea in emmetropia and myopia, *Optom. Vis. Sci.* **75**: 132-145.

Grosvenor, T., and Goss, D. A., 1999, *Clinical Management of Myopia.* Butterworth-Heinemann, Boston, MA, pp. 49-62.

Gwiazda, J., Thorn F., Bauer J., and Held, R., 1993, Emmetropization and the progression of manifest refraction in children followed from infancy to puberty, *Clin. Vis. Sci.* **8**: 337-344.

Hosaka, A., 1988, Population studies - myopia experience in Japan, *Acta Ophthalmol (Supp) (Kbh).* **185**: 37-40.

Hung, G. K., 1990, Fixation disparity under open- and closed-loop accommodation, *Ophthal. Physiol. Opt.* **10**: 211-214.

Hung, G. K., 1992. Adaptation model of accommodation and vergence, *Ophthal. Physiol. Opt.* **12**: 319-326.

Hung, G. K., 1998, Sensitivity analysis of the stimulus-response function of a static nonlinear accommodation model, *IEEE Trans Biomed Engin.* **45**: 335-341.

Hung, G. K., and Ciuffreda, K. J., 1999, Model of refractive error development, *Cur. Eye. Res.*, **19**: 41-52.

Hung, G. K. and Ciuffreda, K. J., 2000a, Differential retinal-defocus magnitude during eye growth provides the appropriate direction signal, *Med. Sci. Monitor.* **6**: 791-795.

Hung, G. K., and Ciuffreda, K. J., 2000b, Quantitative analysis of the effect of near lens addition on accommodation and myopigenesis, *Cur. Eye. Res.* **20**: 293-312.

Hung, G. K., and Ciuffreda, K. J., 2000c, A unifying theory of refractive error development, *Bull. Math. Biol.* **62**: 1087-1108.

Iuvone, P. M., Tigges, M., Stone, R. A., Lambert, S., and Laties, A. M., 1991, Effect of apomorphine, a dopamine receptor agonist, on ocular refraction and axial elongation in primate model of myopia, *Invest. Ophthal. Vis. Sci.* **32**: 1674-1677.

Javitt, J. C., and Chiang, Y. P., 1994, The socioeconomic aspects of laser refractive surgery, *Arch. Ophthalmol.* **112**: 1526-1530.

Jiang, B. C. and Woessner, W. M., 1996, Increase in axial length is responsible for late-onset myopia, *Optom. Vis. Sci.* **73**: 231-234.

Kolb, H., 1994, The architecture of functional neural circuits in the vertebrate retina. The Proctor Lecture, *Invest. Ophthalmol. Vis. Sci.* **35**: 2385-2404.

Lam, C. S., Goh, W. S., Tang, Y. K., Tsui, K. K., Wong W. C., and Man, T. C., 1994, Changes in refractive trends and optical components of Hong Kong Chinese aged over 40 years, *Ophthal. Physiol. Opt.* **14**: 383-388.

Lin, L. L. K., Shih, Y. F., Lee, Y. C., Hung, P. T., and Hou, P. K., 1996, Changes in ocular refraction and its components among medical students - a 5-year longitudinal study, *Optom. Vis. Sci.* **73**: 495-498.

Kimura, T. 1965, Developmental change of the optical components in twins, *Acta Soc. Ophthalmol. Jpn.* **69**: 963-969.

Mahlman, H. E. 1982, *Handbook of Federal Vision Requirements and Information.* Professional Press, Chicago, IL, USA, pp. 8-18.

Marzani, D., and Wallman, J., 1997, Growth of the two layers of the chick sclera is modulated reciprocally by visual conditions, *Invest. Ophthal. Vis. Sci.* **38**: 1726-1739.

McBrien, N. A., Gentle, A., and Cottriall, C., 1999, Optical correction of induced axial myopia in the tree shrew: implications for emmetropization, *Optom. Vis. Sci.* **76**: 419-427.

McBrien, N. A., and Millodot, M., 1986, The effect of refractive error on the accommodative response gradient, *Ophthal. Physiol. Opt.* **6**: 145-149.

Medina, A., 1987, A model of emmetropization, the effect of corrective lenses, *Acta. Ophthalmol.* **65**: 585-571.

Medina, A., and Fariza, E., 1993, Emmetropization as a first-order feedback system, *Vis. Res.* **33**: 21-26.

Mutti, D.D., Zadnik K., and Adams, A.J., 1996, Myopia. The nature vs nurture debate goes on, *Invest. Ophthal. Vis. Sci.* **37**: 952-957.

Norton, T. T., 1999, Animal models of myopia: learning how vision controls the size of the eye, *Instit. Lab. Animal Res. Journal.* **40**: 59-77.

Norton, T. T., and Rada, J. A., 1995, Reduced extracellular matrix in mammalian sclera with induced myopia, *Vis. Res.* **35**: 1271-1281.

O'Leary, D. J., Chung, K. M., and Mohikin, N., 2000, Undercorrection causes more rapid progression of myopia in children, *Am. Acad. Optom. 2000 (Abstract)*, pg. 24.

O'Leary, D. J., Chung, K. M., and Othman, S., 1992, Contrast reduction without myopia induction in monkey, *Invest. Ophthal. Vis. Sci.* **33**: S712.

Ong, E., and Ciuffreda, K. J., 1995, Nearwork-induced transient myopia - a critical review, *Doc. Ophthalmol.* **91**: 57-85.

Ong, E., and Ciuffreda, K. J., 1997, *Accommodation, Nearwork, and Myopia,* Optometric Extension Program Foundation, Inc, Santa Ana, CA, pp. 76-96, 177-201.

Ong, E. , Ciuffreda, K. J., and Tannen, B., 1993, Static accommodation in congenital nystagmus, *Invest. Ophthal. Vis. Sci.* **34**: 194-204.

Pässinen, O., Hemminki, E., and Klemetti, A., 1989, Effect of spectacle use and accommodation on myopia progression: final results of a three-year randomised clinical trial among schoolchildren, *Br. J. Ophthalmol.* **73**: 547-551.

Phillips, J. R., and McBrien, N. A., 1995, Form deprivation myopia: elastic properties of the sclera. *Ophthal. Physiol. Opt.* **15**: 357-362.

Rada, J. A., McFarland, A. L., Cornuet, P. K., and Hassell, J. R., 1992, Proteoglycan synthesis by scleral chondrocytes is modulated by a vision dependent mechanism, *Cur. Eye Res.* **11**: 767-782.

Reeder, A. P., and McBrien, N. A., 1993, Biochemical changes in the sclera of tree shrew with high degrees of experimental myopia, *Ophthal. Physiol. Opt.* **13**: 105.

Rosenfield, M., and Gilmartin, B., 1998, Myopia and nearwork: causation or merely association?, in: *Myopia and Nearwork*, M. Rosenfield and B. Gilmartin, eds., Butterworth-Heinemann, Oxford, pp. 193-206.

Scammon, R. E., and Armstrong, E. L., 1925, On the growth of the human eyeball and optic nerve, *J. Comp. Neurol.* **38**: 165-219.

Schaeffel, F., and Howland, H. C., 1988, Mathematical model of emmetropization in the chicken, *J. Opt. Soc. Am. A* **5**: 2080-2086.

Siegwart, J. T. Jr., and Norton, T. T., 1999, Regulation of the mechanical properties of tree shrew sclera by the visual environment, *Vis. Res.* **39**: 387-407.

Smith, G., and Atchison, D. A., 1997, *The Eye and Visual Optical Instruments*, Cambridge Univ. Press, Cambridge, United Kingdom, pp. 274, 796.

Smith, E. L., and Hung, L. F., 1999, The role of optical defocus in regulating refractive development in infant monkeys, *Vis. Res.* **39**: 1415-1435.

Smith, E. L., and Hung, L. F., 2000, Form deprivation myopia in monkeys is a graded phenomenon, *Vis. Res.* **40**: 371-381.

Sorsby, A., Sheridan, M., and Leary, G. A., 1962, *Refraction and Its Components in Twins,* London: Her Majesty's Stationary Service.

Sperduto, R. D., Seigel, D., Roberts, J., and Rowland, M., 1983, Prevalence of myopia in the United States, *Arch. Ophthalmol.* **101**: 405-407.

Stark, L., 1968, *Neurological Control Systems, Studies in Bioengineering*, Plenum Press, New York, pp. 205-219.

Stone, R. A., Lin, T., and Laties, A. M., 1989, Retinal dopamine and form-deprivation myopia, *Proc. Natl. Acad. Sci.* **86**: 704-706.

Tigges, M., Tigges, J., Fernendes, A., Effers, H. M., and Gammon, J. A., 1990, Postnatal axial eye elongation in normal and visually deprived rhesus monkeys, *Invest. Ophthal. Vis. Sci.* **31**: 1035-1046.

Troilo, D., 1989, *The Visual Control of Eye Growth in Chicks,* Ph. D. Dissertation, Faculty of Biology, City University of New York, New York, NY.

Troilo, D., Gottlieb, M. D., and Wallman, J., 1987, Visual deprivation causes myopia in chicks with optic nerve section, *Cur. Eye Res.* **6**: 993-999.

Troilo, D., Nickla, D. L., and Wallman, J., 2000a, Choroidal thickness changes during altered eye growth and refractive state in a primate, *Invest. Ophthal. Vis. Sci.* **41**: 1249-1258.

Troilo, D., Nickla, D. L., and Wildsoet, C. F., 2000b, Form deprivation myopia in mature common Marmoset (Callitbrix jaccbus), *Invest. Ophthal. Vis. Sci.* **41**: 2043-2049.

Wallman, J., 1997,. Can myopia be prevented? in: *14th Biennial Research to Prevent Blindness Science Writers Seminar in Ophthalmology*, Research to Prevent Blindness, New York, pp. 50-52.

Werblin, F., 1973, Control of sensitivity of the retina, *Sci. Am.* **228 (1)**: 71-79.

Wick, B., 2000, On the etiology of refractive error - Parts I-III, *J. Optom. Vis. Devel.* **31**: 5-21, 48-63, 93-99.

Wildsoet, C. F., 1998, Structural correlates of myopia, in: *Myopia and Nearwork*, M. Rosenfield and B. Gilmartin, eds., Butterworth-Heinemann, Oxford, pp. 32-51.

Wildsoet, C. F., and Collins, M. J., 2000, Competing defocus stimuli of opposite sign produce opposite effects in eyes with intact and sectioned optic nerves in the chick, *Invest. Ophthal. Vis. Sci.* **41**: S738.

Wildsoet, C. F., and Pettigrew, J. D., 1988, Experimental myopia and anomalous eye growth patterns unaffected by optic nerve section in chickens: Evidence for local control of eye growth, *Clin. Vis. Sci.* **3**: 99-107.

Wildwoet, C. F., and Wallman, J., 1995, Choroidal and scleral mechanisms of compensation for spectacle lenses in chicks, *Vis. Sci.* **35**: 1175-1194.

Wilson, J. R., Fernandes, A., Chankler, C. V., Tigges, M., Boothe, R. G., and Gammon, J. A., 1987, Abnormal development of the axial length of aphakic monkey eyes, *Invest. Ophthal. Vis. Sci.* **28**: 2096-2099.

Winauer, J. A., Zhu, X., Park, T., and Wallman, J., 2000, Is myopic blur more important than sharp vision for positive-lens compensation? *Invest. Ophthal. Vis. Sci.* **41**: S136.

Windhorst, U., 1996, Specific networks of the cerebral cortex: functional organization and plasticity, in: *Comprehensive Human Physiology: From Cellular Mechanisms to Integration. Vol. 1*, R. Greger and U. Windhorst, eds., Springer-Verlag, Berlin, pp. 1105-1136.

Woodruff, M.E., and Samek M.J., 1977, A study of the prevalence of spherical equivalent refractive states and anisometropia in Amerind population in Ontario, *Can. J. Public Health.* **68**: 414-424.

Wu, M. M.-M., and Edwards, M. H., 1999, The effect of having myopic parents: An analysis of myopia in three generations, *Optom. Vis. Sci.* **76**: 387-392.

Yackle, K., and Fitzgerald, D. E., 1999, Emmetropization: an overview, *J. Behav. Optom.* **10**: 38-43.

Young, F.A., Leary, G.A., Baldwin, W.R., West, D.C., Box, R.A., Harris, E., and Johnson, C., 1969, The transmission of refractive errors within eskimo families, *Am. J. Optom. Arch. Am. Acad. Optom.* **46**: 676-685.

Zhang, M.-Z., Saw, S.-M., Hong, R.-Z., Fu, Z.-F., Yang, H., Shui, Y.-B., Yap, M. K. H., and Chew, S.-J., 2000, Refractive errors in Singapore and Xiamen, China - A comparative study in school children aged 6 to 7 years, *Optom. Vis. Sci.,* **77**: 302-308.

Chapter 19

Models of Reading Disability and Their Implications

Harold A. Solan

Dept. of Clinical Sciences, State University of New York, State College of Optometry, 33 West 42nd St. New York, NY 10036; PH: (212) 780-4993; FX: (212) 780-5124; EM: hsolan@sunyopt.edu

19.1 INTRODUCTION

Although reading disability has been discussed for over a century (Berlin, 1887; Hinshelwood, 1896, 1917), the nature of the disorder is still controversial. Children identified as being reading disabled (RD)[1] have average or better than average intelligence, abundant educational opportunities, normal sensory acuities (seeing and hearing), no frank brain damage (e.g., cerebral palsy), and no primary emotional disturbances. This exclusionary definition supports the impression that we are dealing with an unexpected inability to learn to read. There have been numerous and varied estimates of the incidence of specific reading disability, some as high as 15% to 25% (Badian, 1984; Johnson, 1988). Most primary grade teachers would agree that, in an average class of twenty-five, four or five children are

[1] Because the term *dyslexia* is abstract and the definition varies with the author, the term reading disability is preferred. This exclusionary definition represents a population whose standardized reading test scores are between 0.5 s.d. and 1.0 s.d. below the mean for their grade levels.

one year or more behind in reading. There appear to be quantitative differences between research and teacher identified RDs, the latter being significantly greater (Shaywitz, et al., 1990). By grade 4, a third of the class may be expected to read below the basic level. It is not unusual for these RD children to show some signs of a mild maturational lag such as ambilaterality, left-right directionality problems, clumsiness, short attention span, and a history of delay in language development.

Several models of reading disability are presented. It will be apparent that they are not mutually exclusive. Although each diagnostic approach is analyzed specifically, we are dealing with a single commonplace disorder, *reading disability*, viewed from different perspectives. Each of the topics in this chapter describes different mechanisms, nevertheless, all are relevant to understanding the nature and treatment of reading disability.

19.2 BACKGROUND

19.2.1 Fundamental Aspects of Reading Disability

19.2.1.1 The Nature of Reading Disability

To understand and appreciate the complexity of reading disability requires a review of some of the educational components of learning to read. Usually the average child's receptive memory and oral subsystems for processing and responding to spoken language are at age expecteds prior to receiving formal reading instruction. There are numerous models for primary reading instruction that are systematically used to develop a child's subsystems for decoding, blending sounds, recalling whole words and families of words, and comprehending meaning. Most bright children, whose perceptual readiness, visual imagery, intersensory integration, and sensory-motor skills have matured spontaneously, are receptive to the phonetics and orthographics (i.e., sounding and spelling), and learn to read equally well regardless of the methodology. They learn intuitively to integrate the subskills, generalize, and transfer previous learning to new learning. For example, after learning the words *bat* and *fig*, recognizing big, fat, bag, and fit occurs naturally. If spoken language has not been delayed and grammatical development has progressed normally for their age, basic comprehension is not usually a problem in primary grade children. Since linguistic skills and learning to read are developmental, reading materials and instructional strategies are programmed to lead toward the mastery of skills not yet acquired. On the other hand, students whose pattern of standardized reading scores fall between the $16^{th}$and 31st percentile (1.0 to 0.5 SD < Mean.) may be considered as RD and should be receiving varying amounts of supplementary instruction. A child in the lowest 15th percentile usually requires a psychoeducational evaluation and subsequent individualized instruction in reading and language development.

Reading disability can be very complex, and only an experienced reading diagnostician is capable of formulating a comprehensive educational program. The remedial instruction must be success-oriented to counteract the succession of previous failures. At the outset, the child must be taught to recognize his increasing mastery of the skills that are being acquired, a sensitivity that is frequently lacking in RD children. Supplementary instruction is most effective when the remedial teacher focuses upon the specific needs of the child, but maintains sufficient flexibility to modify the program as the child progresses. Not only is it important for the teacher to maintain orderly sequences of skill development, but there is also a need for the program to be meaningful to the child. Although a positive approach should be used in pointing out errors, at no time should the child be patronized.

The following example illustrates the complexity of evaluating the progress of remediation. Many of the children who are seen for remedial instruction are two years or more below grade level. If individual supplementary instruction were administered twice a week, it is probable that a minimum of two school years would be required for an entering grade 4 student to approach grade level. Initially, this 4th grade student has attended school for three years and has improved just one year in reading skills, from grade 1.1 to 2.1. Learning rate (LR) is 1:3 or 33%. If, during first year of supplementary individual instruction, the child gains a full year in reading, LR for the year will improve to 100%. When entering grade 5, the reading grade will be 3.1, still two years below grade level. The overall LR however, has improved to 50%, two years growth in four years. Usually, two years of intensive remediation, combined with better intrinsic motivation, improved attention, and concurrent diversified classroom educational experiences such as reading in the content areas, enable most students to progress more independently.

19.2.1.2 Reading Acquisition in the Primary Grades

The development of reading skills requires competency at two principal levels: first, learning to read and then, reading to learn (Satz, et al., 1970; Satz, et al., 1971). Instructional techniques are quite different for each; nevertheless subsequently the two approaches become complementary. In the primary grades, a core deficit in phonological processes can impede the acquisition of fluent reading. Although auditory and verbal processing skills are primary, certain visual components are necessary if the orthographic representations are to be processed automatically. Included are the transition in visual processing from low to high spatial frequencies which is essential for initial feature detection, visual discrimination, and integration of visual feature and pattern information with stored orthographic and phonological representations (Wolf & Bowers, 1999; Chase, 1996). Other perceptual and perceptual-motor skills that correlate with early reading include auditory-visual integration (Solan, Usprich, Mozlin, et al., 1983), rapid visual information processing, and visual-motor integration (Solan & Mozlin, 1986).

At this juncture, it seems appropriate to summarize briefly the basic processes and instructional practices in word and letter identification and early reading (Adams, 1995). It is important to appreciate that, although applying the rules of decoding appears to be a concrete experience, the act of learning to read is not a unitary task. The process involves complex theoretical concepts. A history of good verbal and language skills is considered an asset, but it is not a guarantee that a child will learn to read and comprehend easily. Similarly, efficient visual perception and visual information skills usually are necessary to process words accurately and automatically.

Because the *pros and cons* of teaching *phonics* have been so controversial, it is a logical topic to open the discussion of conventional teaching methodology. Acquiring a correspondence between visualizing and sounding letters is a basic decoding skill. As is the case in many cognitive alternatives, phonetic skills are necessary but not sufficient to become a good reader, and hardly a reason to result in *The Great Debate* (Chall, 1967). With hindsight, it is reasonable to propose that a linguistic program, which combines the whole word approach and phonics, would prove to be most beneficial. Unfortunately, this controversy took place for many years among educators (and book publishers!). Since there are significant differences between school populations and teachers, it is difficult to generalize whether phonic or whole word instruction is always better. Neither the students nor the programs are monolithic. A review of the literature suggests that although teaching phonics appears to have the advantage in the primary grades, ultimately the inclusion of meaning and language instruction results in superior reading. In the final analysis, each child must learn to read whole words effortlessly, automatically, *and visually* (Birch, 1962; Adams, 1995). To accomplish this goal, reading instruction must be adjusted to accommodate the diversity of individual reading abilities and disabilities. As children proceed from the primary to elementary, and then to middle grades, where the emphasis on learning to read is replaced by reading to learn, academic demands increase, and the need for proficient reading and study skills becomes more critical.

19.2.2 Remediation of Reading Disability

19.2.2.1 Effect of Increased Visual Attention

Sergeant (1996) proposed that visual attention is multidimensional with traits that involve both cognitive and psychophysiological variables. (See Chapter 14 in this volume.) The phylogenesis of the nervous system from which attention is derived may be traced back to the brain stem and basic neuron structures in reptiles out which the limbic system emerged prior to the complex development of the neocortex (Halperin, 1996). Nevertheless, the brain stem remains a key component of attention within the framework of the evolution of neocortical structures. However, there is no universal agreement as to how to define *attention.* Although attention requires arousal, activation, and vigilance, perceptual and cognitive processing skills also are necessary, albeit subsumed, components. Sometimes it is difficult to identify the precise constructs being measured in attention tests since they frequently include significant perceptual components such as visual memory and visually-guided motor skills (e.g., pencil and paper tests). For example, the Cancellation Test and WISC-R Coding Test, which are sometimes used to assess attention, are *timed*-integration tests with visual-motor components. Accuracy *and* automaticity are also necessary conditions for these tests, which begs the question: Should a speed/accuracy trade-off be acceptable in a visual attention test?

Furthermore, outcomes of attention tests often depend markedly upon the individual's ability to develop a rapid response strategy in a novel testing situation. Therefore, memory and executive function (strategies) provide additional complementary superordinate constructs, notwithstanding they too lack precise definitions and measurement validation. A universal consensus as to what is being measured appears to be found wanting. This is not a surprising observation since attention is primarily involved with early stages of information processing. Yet, our attention tests frequently measure the endpoint of a number of perceptual and cognitive processes. Fortunately, visual attention is malleable and can be clinically modified, provided that appropriate effort is put forth to solve the task. There remains, however, some concern about the differences in response between children and adults.

Two categories of attention are to be considered: selective and sustained attention. Although they deal with brief and protracted durations, respectively, they are not necessarily mutually exclusive (Halperin, 1996). In testing attention and reading, we must also be concerned about the necessity to inhibit irrelevant stimuli. In selective attention, this is especially the case in impulsive ADHD (attention deficit hyperactive disorder) individuals who have not learned to delay responses. In reading, visual attention shifts *within a fixation*, from the fixation point to the right parafoveal region (previewing), so that the next fixation can be programmed. Oculomotor readiness, sometimes called the *premotor theory of attention* (Hoffman and Subramaniam, 1995), depends upon this shift in visual attention that precedes executing refixation, since selective attention drives the saccades. Sustained attention is more closely related to the maintenance of vigilance which contributes to the high degree of automaticity required for efficient processing of information. Arousal is a necessary, but in the absence of memory and executive function, not a sufficient condition for information processing. However, when provided with the appropriate therapeutic techniques, significant improvements in attending are attainable if intact neural functioning is available, and the individual is experiencing correctable deficits in reading (Solan, Larson, et al., 2001). Clinical testing and experience should continue to be supported with controlled research using fMRI studies to explore the pathophysiology of attention and its relation to reading disability.

19.2.2.2 Effect of Improvement of Eye Movements

The study of eye movements in reading raises numerous questions. For example, which components of eye movements are of special interest? Do these components have normal expecteds? What factors could interfere with normal maturation of eye movements in an individual? Are poorly developed eye movements solely the result of poor reading skills? Are inefficient eye movements correctable, and if so which procedures are most effective? Over a lifetime, what are the effects of age on eye movement development and therapy?.

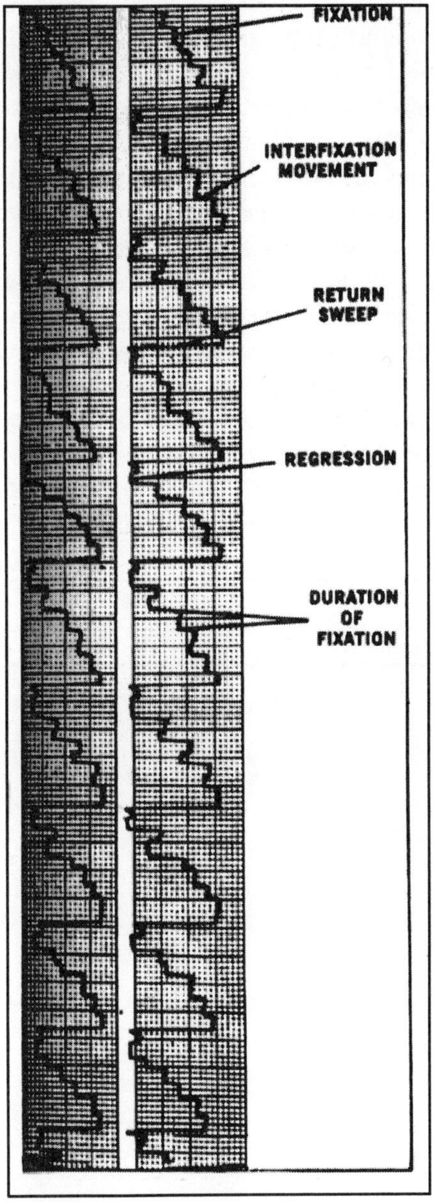

Figure 19.1. Schematic diagram of an eye movement graph showing how eye movement characteristics are represented. From Taylor, E.A.. *The Fundamental Reading Skill*, 1966. Courtesy of Charles C Thomas, Publish, Ltd., Springfield, IL.

Pavlidis (1983) objectively recorded eye movements of 12 normal readers and 12 dyslexics, age matched (Mean age, 12yrs.,2mos.), using 5 sequentially flashing red LEDs as non-verbal stimuli. Stimuli were flashed consecutively left-to-right, then right-to-left. Dyslexics showed a significantly larger number of regressions (reverse eye movements). In a subsequent eye movement study (Pavlidis, 1985) involving 13 dyslexics, 10 normal, and 16 retarded readers (e.g., low socioeconomic populations), subjects sequentially fixated 7 equidistantly spaced digits from left-to-right then right-to-left. Dyslexics made significantly more regressions than normal or retarded readers, although dyslexics and retarded readers had similar reading ages. The results were attributed to sequential processing and attentional deficits, as no true text processing occurred. Almost a decade elapsed before Eden et al. (1994), recording eye movements with *non-verbal* stimuli, also dispelled the notion that verbal and language disorders were the sole underlying defects in the control of eye movements frequently observed in RDs.

Just 56.5% of their disabled readers had poor phonologic awareness, and the percent of saccadic inaccuracies was significantly greater for RDs than NRs. Eye movement stability of RD children was poorer at all levels of binocular convergence magnitudes. Furthermore, oculomotor vergence amplitudes were lower for RDs than NRs. The outcome of this research lends strong support to the notion that deficits other than phonological play a significant role in RD, particularly oculomotor factors. In reading, the eye movement time course ideally forms a series of saccadic staircases shown in Fig. 19.1. As an individual views a word, there is no eye movement, and a *fixation* pause (sometimes called a *progression)* is recorded. Between fixations, a brief left to right interfixation movement known as a *saccade* is recorded. When a right to left eye movement is made, the fixation is considered to be a *regression.* The average number of words per fixation seen by a subject is called *span of recognition*. At the end of a line of print, a rapid *return sweep* to the beginning of the next line takes place. Table 19.1 provides acceptable

TABLE 19. 1

Eye – Movement Norms

	1st	2nd	3rd	4th	5th	6th	JrH	HS	Col
Fixations, Per 100 words	240	200	170	136	118	105	95	83	75
Regressions, Per 100 words	55	45	37	30	26	23	18	15	11
Average Span of Recognition, Words per fixation	.42	.50	.59	.73	.85	.95	1.05	1.21	1.33
Rate of Comprehension, Words per min.	75	100	138	180	216	235	255	296	340

This data was compiled by E.A. Taylor from over 5,000 eye-movement records (Taylor, 1966).

Table 19.1. Eye Movement Norms

From "Deficient eye-movement patterns in achieving high school students: Three case histories" by H.A. Solan, 1985, *Journal of Learning Disabilities,18*, pp. 66-70. Copyright 1985 by PRO-ED, Inc. Reprinted with permission.

norms for all age-grade levels, albeit without standard deviations[2]. Span of recognition is of special interest, since it contradicts the myth that good readers read in phrases. Individuals with college level reading efficiency may make an average of 75 fixations per 100 words, equivalent to a span of recognition of 1.33 words per fixation. Since the fovea, the locus of highest resolution in the retina, has an angular size of about 2 degrees (about 6 characters), 1.33 words per fixation is an acceptable estimate. Interpreting the nuances of an eye movement graph requires more experience than simply counting the lines. For example, return sweeps warrant careful observation. Some readers make two or more fixations within the return sweep as if they are reading the previous line from right to left. This highly inefficient performance can be corrected readily with oculomotor therapy. A slight overshoot or undershoot with subsequent re-fixation at the beginning of the next line is not unusual and generally does not contribute to oculomotor inefficiency.

Implied is the assumption that the subject has read the 100 word test selection with adequate comprehension, at least 70% (7 of 10 objective questions answered correctly). A decrease in the number of fixations and regressions is desirable. On the other hand, duration of fixation may vary in either direction. A reduction in duration could be the result of an individual making many very short fixations or a decrease in the number of fixations could result in longer duration of fixation (see Fig. 19.2). Of the two, the latter is more efficient and less fatiguing for the reader. Although these fundamental reading skills stress reading efficiency, improvement in rate of reading is also a prime concern and should not be overlooked in evaluating and treating individuals identified as inefficient readers. However, rate of reading may be influenced by the purpose in reading the article or book. For example, Eisenhower's *Crusade in Europe* would probably be read at a slower rate by an army field officer taking courses at the War College than an engineer or lawyer who is interested in obtaining an overview of the European phase of World War II.

Based on the research of Lovegrove, et al. (1986), Williams, et al. (1992), and Solan, et al. (1997), who had demonstrated that the magnocellular pathways were sensitive to short wavelengths of light, Solan, Ficarra, et al. (1998) postulated that blue filters would increase the intensity of the stimulus and reduce the transient response latency, which, in turn would restore the timing of the transient-sustained synchronization (see Table 19.2).

[2] For further information concerning eye movement norms, see Teachers Guide for Visagraph II, Taylor Associates, 200 – 2 East 2nd Street, Huntington Station, NY 11746.

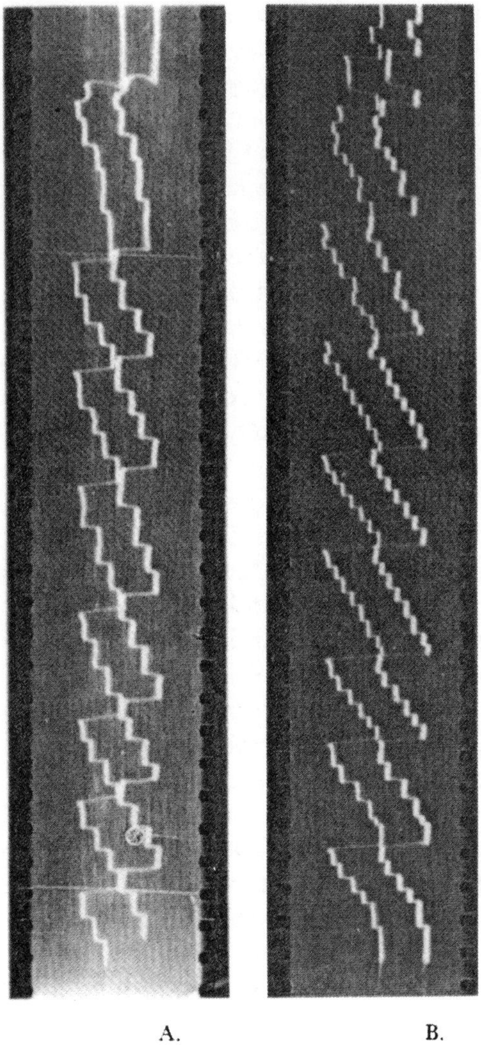

Figure 19.2. Objective evidence that readers who read at the same rate (421 words per minute) do not necessarily read with the same degree of efficiency. From E. A. Taylor, *The Fundamental Reading Skill*, 1966, page 48. Courtesy of Charles C Thomas, Publisher, Ltd., Springfield, IL.

They identified 54 RD and NR students, grades 4 – 6, as measured with the Gates-MacGinitie Reading Tests. Eye movement graphs of fixations, regressions, and rate of reading were recorded objectively with the Visagraph II, a commercially available infra-red recording device (see Fig. 19.3). Since text difficulties influence eye movement measures, it was necessary to equalize comprehension levels between the two groups. Therefore, all subjects were tested with selections at their respective independent reading levels (80% comprehension). Although the level of language difficulty for RDs had been decreased, the mean eye movement findings of RDs remained significantly poorer for fixations, regressions, and rate of reading as compared to NRs. However, when RDs, but not NRs, read selections through blue filters (short wavelength), their eye

TABLE 19.2
CHARACTERISTICS OF TRANSIENT AND SUSTAINED SUBSYSTEMS

Transient (M-cell)	Sustained (P-cell)
1. Most sensitive to low and middle spatial frequencies: large overall shapes	1. Most sensitive to high spatial frequencies: fine detail
2. High sensitivity to contrast	2. Low sensitivity to contrast.
3. Peripheral vision dominant.	3. Central (foveal) vision dominant.
4. Responds to onset and offset of stimulus. Short response persistence.	4. Response persists after stimulus. Longer response persistence.
5. Most sensitive to high temporal frequencies	5. Most sensitive to low temporal frequencies.
6. Responds to quickly moving target (early warning).	6. Sensitive to stationary or slowly moving targets.
7. Sensitive to short wavelengths (e.g., blue).	7. Sensitive to longer wavelengths (e.g., red).
8. Global analysis of incoming visual information	8. Identification of shapes and patterns.
9. Involved in perception of depth, flicker, motion, brightness discrimination.	9. Involved in processing color information
10. Prepares visual system for the input of slower detailed information which follows.	10. Responds subsequent to transient and is dependent upon transient output.

Table 19.2. Characteristics of Transient (M-cell) and Sustained (P-cell) Processing.

movement components improved significantly ($p< 0.01$) and were comparable to the normal readers. This study and prior research (Eden et al., 1994, 1995; Pavlidis, 1983, 1985) support the notion that poor oculomotor skills in RDs are not solely the consequence of language and phonologic disabilities.

Eye movements deficits in reading can be explained often as a magnocellular and/or an attentional deficit that has resulted in a mismatch between retinal signals and cortical signals in the motion sensitive areas of the extrastriate cortex. In theory, shifting attention to the right parafoveal region within a fixation in order to program the next fixation synchronizes the activation of brain structures that are intimately involved in eye movements.

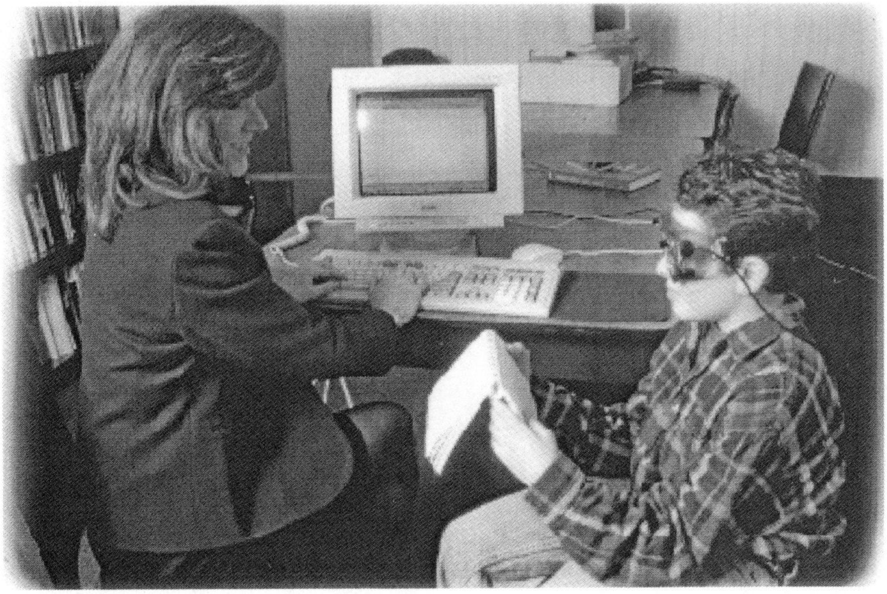

Figure 19.3. A child wearing infrared sensitive goggles being tested with the Visagraph II. Reproduced with permission. Taylor Associates, Huntington Station, New York 11746.

Solan, Larson, et al. (2001) recently compared the effect of eye movement and comprehension therapy of 31 grade 6, urban, reading disabled children using a controlled factorial crossover design. Initial reading levels, which averaged Grade Equivalent 4.1, were established from standardized reading tests, and eye movement tests were administered objectively with the Visagraph II. Half of the subjects received eye movement therapy first, and the other half received comprehension therapy first. The crossover took place after 12 hours of therapy that also stressed enhancement of visual attention, as suggested by Lennerstrand, Ygge, and Jacobson (1993). Although there was no difference between the two therapeutic groups (eye movement first and comprehension first) after either 12 or 24 hours of therapy, the within group improvements were statistically significant and educationally meaningful ($p<0.01$). The combined therapy rendered a 2.5 years improvement in reading comprehension and a 2.7 years average growth in eye movement skills in 7 months.

Improvement in oculomotor efficiency and reading comprehension is predicated on the ability of the subject to sustain effort visually and concentrate in reading and study type activities for extended periods of time. Fortunately, most learning-related vision disorders such as convergence insufficiency, convergence excess, and accommodative (focusing) disturbances are amenable to vision therapy and symptoms are promptly relieved. Vision therapy also provides the reader with sufficient binocular fusional reserves to compensate for transient vergence variations and accommodative changes necessary to maintain clear vision as the eyes sweep across the page when reading (Ciuffreda & Tannen, 1995). Hung (1989) reported that 2 dyslexic subjects showed reduced vergence velocities and fewer step-ramp and multiple-step responses than 2 normal controls. Buzzelli (1991) compared binocular vergence facility of 13 dyslexics with 13 normal readers attending middle school. Subjects were required to align a vertical picture (box-X-O) anaglyph slide at near as they alternately viewed the target through 16 p.d. base-out and 4 p.d. base in for 20 cycles. Dyslexics completed the task significantly more slowly than normal readers ($p=0.043$). A similar test was performed for binocular accommodation with $\pm$ 2.00 D. lenses. Although dyslexics performed more slowly than normal readers, differences were not statistically significant. Solan and Mozlin made some interesting observations in an unpublished study of several adults identified as having binocular dysfunctions and a slow reading gait, but good reading comprehension (1984). Eye movement graphs were completed before and after administering vision therapy. The therapy included improving quality and ranges of fusion, both convergence and divergence, and facility of accommodation (focusing). Although no specific eye movement therapy was rendered, the components of the graphs showed improvements. When Guided Reading, an oculomotor procedure which exposes the reading selection in a left to right moving horizontal aperture, was subsequently added to the therapeutic regimen to train reading efficiency while maintaining acceptable comprehension, eye movement components were normalized. That is, the number of fixations, regressions, and rate of reading with comprehension reached expected levels of performance. These results support the notion that traditional vision therapy not only is effective in correcting binocular anomalies, but also has a salutary effect on eye movement efficiency when reading.

The efficacy of concurrent vision therapy and eye movement therapy was reported by Solan (1985). The subjects included two achieving high school senior students whose standardized reading comprehension scores were at about the 50th percentile. The first subject presented with a convergence insufficiency (tendency of the eyes to diverge at the reading distance with reduced amplitude of binocular convergence) and sluggish accommodation (focusing) at the reading distance. The student reported persistent problems in sustaining effort in reading after 45 minutes. In addition, he was concerned that his slow reading rate, compared to normal expecteds, would be a handicap in college. The vision analysis and an objective recording of his eye movements with high school level selections confirmed his presenting complaints (see Tables 19.1 & 19.3). Vision therapy that stressed the improvement of binocular fusion, convergence, and accommodative facility was rendered. In addition, a reading efficiency program was

TABLE 19.3

Eye-Movement Photography

Case 1: R.T.	Before Training		After Training	
	Raw Score	Grade Score	Raw Score	Grade Score
Fixations (per 100 wds.)	153	4+	82	H.S.
Regressions (per 100 wds.)	255	5	1	Col.
Span of Recognition (Wds. per fix.)	.65	4+	1.22	H.S.
Rate of Comprehension (Wds. per min.)	143	3+	256	8
Comprehension	70%		90%	
Selection Level		H.S.		H.S.

Table 19.3. Eye Movement Skills of Case 1 before and after therapy.
From "Deficient eye-movement patterns in achieving high school students: Three case histories" by H.A. Solan, 1985, *Journal of Learning Disabilities,18*, pp. 66-70. Copyright 1985 by PRO-ED, Inc. Reprinted with permission.

administered that included: 1.Tachistoscopic training to improve rate of visual processing; 2. Guided Reading to improve oculomotor functioning, reduce number of fixations and regressions, and enlarge span of recognition while improving reading rate with comprehension; and 3. High interest timed reading selections designed to improve reading rate and comprehension. After completing this 30 hour supplementary program, the student's visual functioning was symptom free which enabled him to sustain effort comfortably in reading and study type activities for extended periods of time. Furthermore, he reported noticeable improvements in his reading rate and comprehension. Post therapy eye movement recordings revealed a significant reduction in fixations and regressions accompanied by increased rate of over 100 words per minute with improved comprehension.

A second high school senior showed a similar developmental reading efficiency disorder. She presented with headaches associated with reading that often caused her to withdraw from studying. Although very bright, she had serious problems in courses that required considerable reading such as English literature. The vision analysis revealed a binocular insufficiency and sluggish accommodation at the reading distance. Reading comprehension was 80%, but objective eye movement recordings resembled that of a grade 2 child. Span of recognition was about half a word per fixation. On the other hand, on a standardized reading test, she scored 92nd percentile in vocabulary and 47th percentile in comprehension. After ten hours of vision therapy to improve binocular functioning, accommodative facility, and reading efficiency training, she reported significant improvement in visual comfort. Eye movement recordings showed dramatic changes (see Fig. 19.4): Fixations improved from 198 to 47 per 100 words, regressions from 40 to 8 per 100 words, and rate from 113 to 533 words per minute, all with 90% comprehension (see Table 19.4). Although the need to maintain sustained effort comfortably in reading and study-type activities is indisputable, it does not predict that a binocular dysfunction will be found in all or even most individuals identified as reading disabled. On the other hand, it is not unusual that reading disability is accompanied with insufficient control over saccadic eye movements in reading (Lennerstrand, et al., 1993).

Of special interest is a reading efficiency program that was conducted for twenty older adults, ages 62 to 75 years, who were in good health and passed a thorough eye examination (Solan, Feldman, Tujak, 1995). Although they were not required to take a standardized reading comprehension test, eye movement recordings with comprehension were completed. Each participant was assigned to either the control group or reading efficiency training group. In the control group, eye movements were retested after the control period; they were then provided with reading

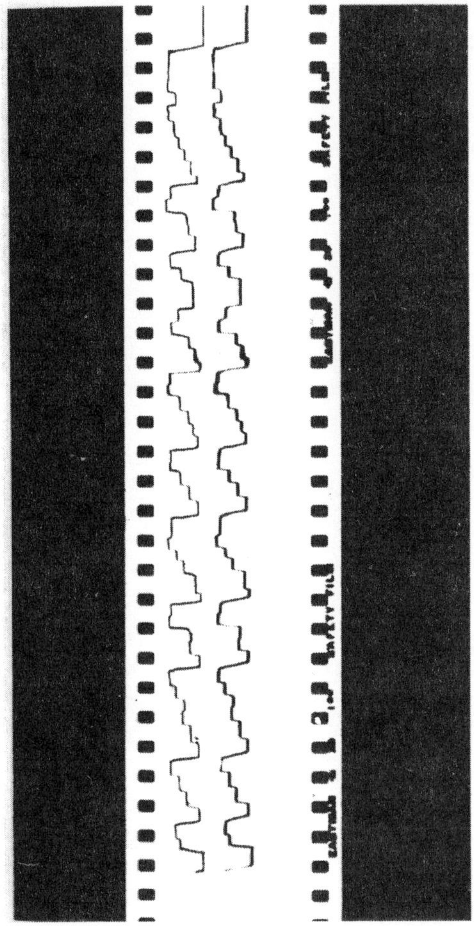

Figure 19.4. Graph of high school student after receiving Guided Reading training. (Before training graph is not available.) Table 4 lists eye movement components before and after training. From "Deficient eye-movement patterns in achieving high school students: Three case histories" by H.A. Solan, 1985, *Journal of Learning Disabilities,18*, pp. 66-70. Copyright 1985 by PRO-ED, Inc. Reprinted with permission..

efficiency training. Each group was retested at the conclusion of experimental intervention which consisted of rapid visual processing, oculomotor, and Guided Reading training. Results of the training revealed statistically significant ($p<0.05$) and clinically meaningful improvements in all aspects of reading efficiency including reduced number of fixations and regressions per 100 words, increased average span of recognition, and improved rate of reading without loss of comprehension after training. All of the participants provided the experimenters with positive feedback. They could read more efficiently with improved concentration for longer periods of time. Hence, training of reading eye movements with concurrent gains in reading ability appears to be independent of age.

TABLE 19.4

Eye-Movement Photography

Case 2: D.H.	Before Training		After Training	
	Raw Score	Grade Score	Raw Score	Grade Score
Fixations (per 100 words)	198	2	47	Col.+
Regressions (per 100 words)	40	2	8	Col.+
Span of Recognition, (Wds. per fix.)	.51	2	2.13	Col.+
Rate of Comprehension, (Wds. per min.)	113	2	533	Col.+
Comprehension	80%		90%	
Selection	H.S.		H.S.	

Table 19.4. Eye movement skills of Case 2 before and after therapy.
From "Deficient eye-movement patterns in achieving high school students: Three case histories" by H.A. Solan, 1985, *Journal of Learning Disabilities, 18*, pp. 66-70. Copyright 1985 by PRO-ED, Inc. Reprinted with permission.

19.3 MODELS

In the Introduction, the controversial nature of *reading disability* was emphasized. It was pointed out that the descriptor *dyslexia* was avoided because it was, in fact, too abstract. The development of models, therefore, becomes the vehicle of choice if we intend to conceptualize the essential attributes of reading disability. Interpreting the models as mutually exclusive, however, would be a mistake. Rather than *either/or*, it is much more productive to look for commonality of function.

19.3.1 Verbal / Visual Deficiency Model

The extent to which disordered verbal and phonological development (Vellutino, 1979; Galaburda, 1988), as compared to visual dysfunctions, contributes to problems in the acquisition of reading skills has not been resolved completely. Visual skills include refractive errors (Grisham & Simons, 1986), binocular anomalies (Simons & Grisham, 1987), perceptual and developmental lags (Solan & Ficarra, 1990), and inefficient eye movements (Eden, Stein, Wood, & Wood, 1994). Therefore, visual factors may interfere with an individual's reading skills in at least four different ways: seeing clearly, visual comfort, information processing, and reading efficiency. Experimental evidence (Eden, Stein, Wood, and Wood, 1995) supports the position that visual tasks are almost as useful as phonological tests to discriminate between normal (NR) and disabled readers (RD): 68% of the variations in reading of both NR and RD subjects could be predicted by combining the visual and phonological scores in a multiple regression analysis. After considering the complementary nature of the two sensory systems, the reasoning behind this artificial dichotomy appears to be moot. Visual sensory input is a function of sequential ocular fixation pauses. Reading is not a matter of instantaneous recognition of the contents of a visual span (Geyer, 1970). It involves the interaction of complex sensory and neural systems over time. That is, the verbal-language system that follows the visual input necessitates storage and cognitive feedback loops to provide the development of meaning. As we shall see, the value of visual as compared to verbal skills is not the only difference of opinion that has been generated among the attempts to maximize the quality of reading instruction.

19.3.2 Neuro-maturational Lag Model

The concept of *maturational lag* requires some clarification since there are numerous ramifications. In children and adolescents, certain physical and cognitive developmental characteristics mature according to expected longitudinal patterns. According to Bender (1958), *maturational lag* signifies slow and uneven neural differentiation in the absence of pathology or structural defect. To appreciate this term, the concept of *plasticity* needs to be understood. Plasticity intimates: "Being as yet unformed, but capable of being formed", a definition Bender borrowed from embryology. Maturational lags in children are associated with poor language skills, deficient visual information processing skills, delayed gross and fine motor skills, reading disability, uneven patterns of intellectual development, soft neurological signs[3], and delayed establishment of cortical dominance. These children usually show more diffuse functional disorders that are associated with developmental delay than is generally observed in adults who present with acquired brain damage. Therefore, therapeutic procedures administered to children stress habilitation, since the disorder is primarily developmental

[3] Soft neurological signs: Slow development of certain integrative skills that are not attributed to a specific cerebral area or neural system such as: coordination deficits, motor impersistence, dyspraxia, and motor overflow.

as compared to adults for whom *re*habilitation is required to restore previously acquired skills.

Satz et al. (1970, 1971) found that the nature of reading disorders varies as a function of age. They identified two sub-groups: Younger RDs, ages 7-8 years, who presented with visual-motor and auditory-visual integration problems; and older children with maturational delays who presented with disorders in language and visual conceptualization. They also found that, after children with early problems had "caught up", they subsequently were prone to lag in those skills which have a later ontogenetic development. Mattis et al. (1975) reported that 92% of his cohort of 113 subjects (ages 8 - 18 years) could be divided into three principal sub-groups: language disorder syndrome (39%), articulatory and graphomotor dyscoordination syndrome (37%), and visual perceptual disorder (16%).

It is important to recognize that, regardless of etiology, no single cognitive, sensory, or motor deficit or cluster of deficiencies which is sufficient to limit the development of reading skills in all children has been identified. Visual spatial deficiencies are frequently associated with RD in younger rather than older children. In the Mattis et al. (1975) study, it appeared that adequate development of language, motor speech blending fluency, and visual spatial perception are critical to learning to read. Delayed development of any one of these skills at the pre-school level may result in atypical reading skills. All three factors are subject to subsequent maturational lags that usually are apparent during K to grade two. These antecedents comprise non-reading skills that suggest reading disability is more than a reading disorder, *per se*. They are predictive of future reading disability if intervention is not initiated promptly. The findings lend support to the value of early neuropsychological and optometric evaluations to identify developmental delays in visual information, sensory-motor, and intermodal processing skills. It is critical that children receive therapeutic intervention when the acquisition of developmental skills crucial to the early stages of reading are in primary ascendency and prior to experiencing the consequences of academic failure.

19.3.3 Neurological Deficit Model

Some RDs reveal a history of biological precursors which suggests that a specific pathogenesis accounts for the individual's difficulty in learning to read. One should not conclude, however, that all RDs have abnormal brains, or that a single neural anomaly accounts for all reading disabilities. Galaburda (1988) postulated that the condition, for which he has preferred the term *dyslexia,* was primarily a linguistic deficiency that was not solely the product of a developmental delay. Since dyslexic individuals seemed to display incomplete cerebral dominance more frequently than normal readers (NRs), the presence of anatomical asymmetry warranted consideration.

Normally, the acquisition and control of language are lateralized in the left hemisphere of the brain. Therefore, a logical first step was to look for anatomical evidence of cerebral asymmetry, since it could have a significant impact on reading (Geschwind, et al., 1987). In the normal human brain, the *planum temporale* exhibits an easily detectable asymmetry. It is located in the sylvian fissure, and constitutes a substantial portion of the temporal-parietal speech and language region (see Fig. 19.5). The planum temporale is contiguous with Heschl's gyrus, the primary auditory area and Wernicke's area that mediates the understanding of language. Although the left hemisphere develops more slowly than the right, the left planum

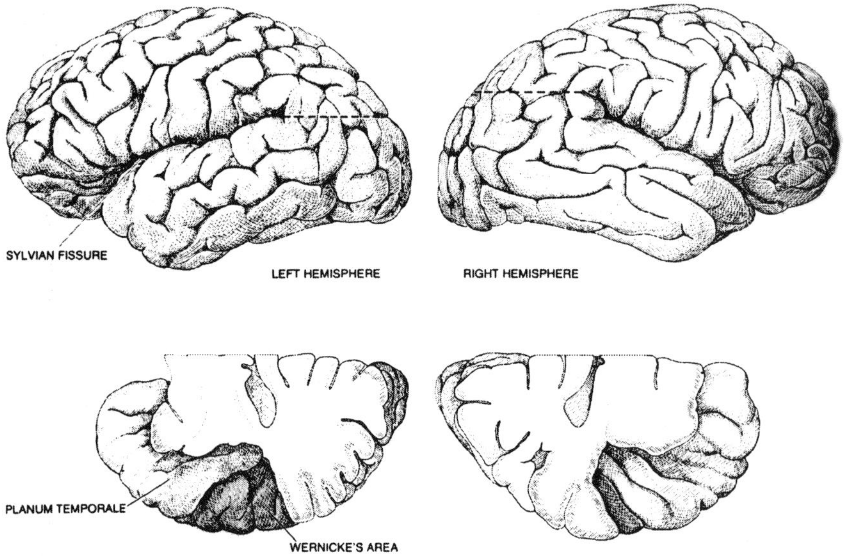

Figure 19.5. Anatomical Asymmetries. One asymmetry is readily observed in the intact brain: the Sylvian Fissure, which defines the upper margin of the temporal lobe, rises more sharply on the right side of the brain. Another asymmetry is found on the planum temporale, which forms the upper surface of the temporal lobe, and which can be seen only when the Sylvian fissure is opened. The posterior part of the planum temporale is usually much larger on the left side. The enlarged region is part of Wernicke's area which suggests that the asymmetry may be related to the linguistic dominance of the left hemisphere. From Specializations of the human brain by Norman Geschwind, 1979, *Scientific American*, 241(3), pp 180-199.

temporale is normally larger and has a greater number of neurons than the homologous area in the right hemisphere. Ordinarily, neural development is orderly and sequential. The earlier developing right hemisphere, on the other hand, usually produces neurons during the prenatal period in excess of the number of synaptic sites for neurons to target. In the absence of completing intrahemispheric or contralateral neuronal connections, a process of cell-death takes place. This asymmetrical reduction of excess neurons in the right hemisphere during corticogenesis contributes to normal development of left hemisphere dominance.

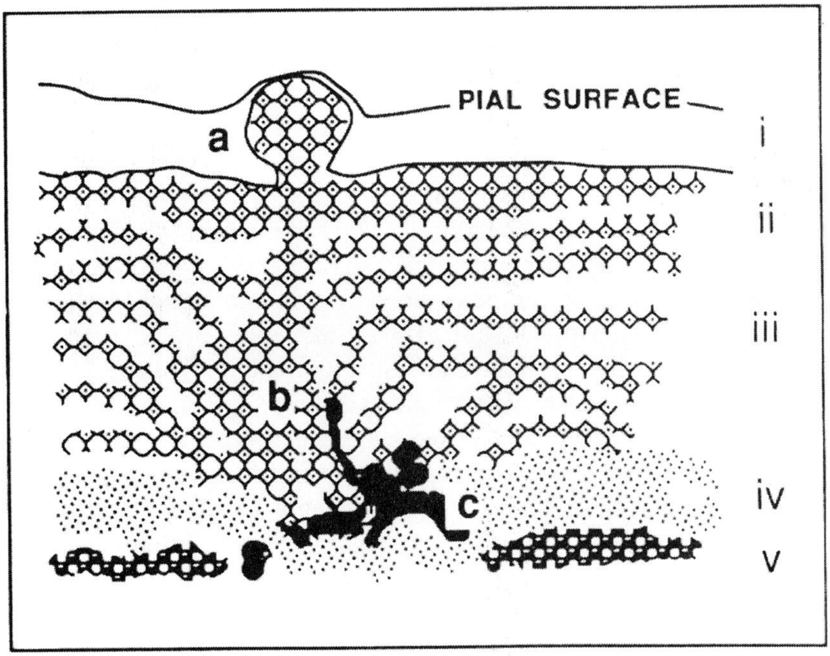

Figure 19.6. Semi-stylized, composite diagram of the types of cerebrocortical dysgenesis seen in brains of dyslexics. An ectopic collection of neurons is seen in layer i (a); disorganization of cortical lamination (dysplasia) is seen predominantly in the upper layers (b); abnormal architecture of cortical vessels often accompanies the dysplasia (c). From Galaburda, A.M., 1988, "The pathogenesis of childhood dyslexia", in: *Language, Communication, and the Brain*, F. Plum ed., Lippincott Williams and Wilkins, Philadelphia, PA. 19106, pp.127-137 Reprinted with permission.

To understand the dyslexic brain, it is first necessary to appreciate which aberrant developmental events are capable of decreasing the cell death in the right hemisphere. The developmental anomalies are the consequences of disordered neural migration, called *microdysgenesis,* primarily in the region of the left perisylvian cortex during the second half of gestation (Galaburda, et al., 1989). The aberrations result in abnormally placed neurons and distorted cortical architecture rarely seen in normal brains, but frequently in dyslexics (see Fig. 19.6). These relatively minor migration defects formerly were considered insignificant and were not thought to be associated with neurological impairment. However, due to the microdysgenesis in individuals destined to become dyslexic and the delay in left cortical development, many more neurons of the earlier developing right hemisphere cross the corpus callosum and randomly connect with left hemisphere targets. Therefore, the number of targets available for subsequent left intrahemispheric connections in the area associated with language development is reduced. Further, the pattern of these random aberrant connections is not discreet and produces more "noise" in the system.

Consequently, in the intact right hemisphere, competition to establish intrahemispheric connections is reduced. There is diminished cell death, and the size of the right planum temporale is increased. Symmetry replaces asymmetry. Yet, this may not be entirely a negative development. Indeed, potential for such an individual (e.g., da Vinci, Edison, and Einstein) to become gifted in certain visualization skills, normally attributed to the right hemisphere such as mathematics and painting, by this *Pathology of Superiority* should not be overlooked. Since speech and language are predominantly centered in the left temporal lobe of the brain, it would be misleading to ignore the impact of asymmetry on auditory and visual processing in reading disorders. Impairments in both systems have been implicated. Phonological skills are essential concomitants of reading, and the inability to process visual information rapidly can severely handicap perceptual analysis and may even interfere with phonemic discrimination (Chase, 1996).

19.3.4 Magnocellular Pathway Deficit Model

Recent resurgence of the role of visual processing in reading has, in part, been the result of temporal processing studies, especially those involving two interactive parallel pathways, magnocellular (M-cell) and parvocellular (P-cell). They are sometimes identified as transient and sustained, respectively. Dual (parallel) processing has its roots in the anatomy and physiology of the visual pathways (Shapley, 1992; Gross, 1992; Fellerman, 1991; Motter, 1991; Mishkin, 1972; Livingston, Rosen, Drislane, & Galaburda, 1991). Psychophysical studies reported by Breitmeyer (1980, 1993), Lovegrove, Martin and Slaghuis (1986), and Lehmkuhle, Garzia, Turner, et al. (1993) support the hypothesis for an M-cell abnormality in reading disability. If the two subsystems are not synchronized as a result of increased transient response latency, the neural image from the previous fixation is not masked (erased). This timing defect could produce a superimposition of successive inputs that results in a retinal overlay or smear. It is reasonable to postulate that this "noise" in the system could interfere with information processing, thereby causing the reader to refixate frequently and decrease reading efficiency. Specifically, Williams, LeCluyse, & Rock-Faucheux (1992), Solan, Brannan, Ficarra, & Byne ((1997) and Brannan, Solan, Ficarra, & Ong (1998) reported significant correlations between reading performance and short wavelength and reduced luminance in the early stages of visual temporal processing among disabled readers. These results are not surprising since a blue overlay is effective in increasing the relative contribution of the M-pathway (see Table 19.2) by reducing the transient response latency. In part, the normal paucity of short wavelength cones in the fovea (Lehmkuhle, 1993) also may be contributing to the outcome.

The introduction of functional magnetic resonance imaging (fMRI) to measure brain activity in locations and conditions selected to stimulate preferentially M pathways has added another dimension to objective testing of affected RD subjects. The schematic of the striate and extra-striate systems illustrates the roles of the M- and P- cell pathways in maintaining oculomotor efficiency (see Fig19.7). After entering the extrastriate cortex, the M-cell stream projects to the posterior parietal cortex (PPC) via the middle temporal (MT) and superior temporal (MST) areas. Whereas the MT areas specialize in the analysis of motion, a small area within the posterior parietal cortex, the lateral intraparietal area (LIP), serves as a sensory-motor integrating bridge to the frontal eye fields in the premotor area of the brain, the source of voluntary saccadic movements. The LIP carries not only sensory information related to the targets for eye movements but also signals related to the planning of eye movements (Andersen 1995). Eden, Van Meter, et al. (1996) observed increasing evidence that RDs exhibit visual processing abnormalities in addition to deficits in phonological awareness. Using fMRI measurements, their data revealed subtle differences in the regional functional organization of the cortical visual system in area V5/MT. Demb, Boynton, & Heeger (1998) also reported reduced brain activity in RDs compared to controls in the primary visual cortex (V1) and extrastriate areas MT and motion sensitive MT+.

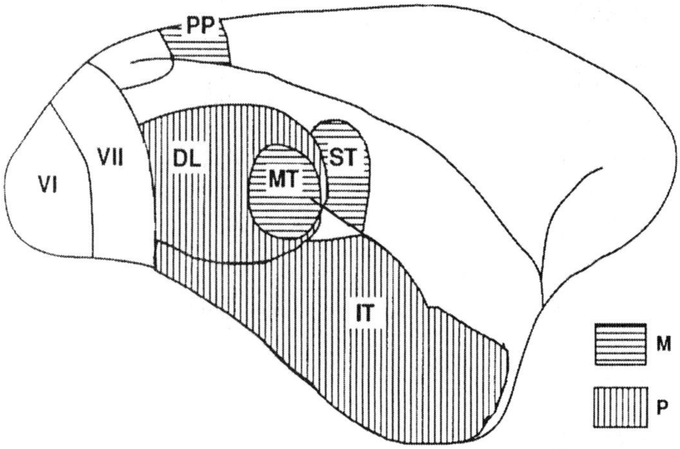

Figure 19.7. Schematic of striate and extrastriate cortex. Horizontal hatching denotes areas related to M-cell pathway, vertical hatching related to P-cell pathway. PP = posterior parietal, DL = dorsal lateral, MT = middle temporal, ST = superior temporal, IT = inferotemporal. Modified from Bassi C. and Lehmkuhle S. Clinical implications of parallel visual pathways. *J Am Optom Assoc* 1990;61:98-110. Copyright© 1990 by the American Optometric Association. Reprinted with permission.

These results support the hypothesis for an M pathway abnormality in RDs and imply strong relationships between the integrity of the M pathway, visual motion perception, and reading disability.

The pathophysiology of reading disability apparently is more complex than originally thought, extending well beyond the classically defined language areas of the brain to the visual processing areas and centers controlling eye movements (Eden & Zeffiro, 1996). The linkage of phonological, visual, and eye movement deficits in RDs suggests the potential for a common antecedent, a *temporal processing deficit* (Eden, Van Meeter, 1996; Stein & Walsh, 1997). Furthermore, since the M-cell pathways are more sensitive than P-cell pathways to global analysis of incoming visual information (i.e. low spatial frequencies), perception of movement, high temporal frequencies, and low contrast, the consequences of poor visual processing efficiency is primarily associated with magno system deficits (see Table 19.2). Therefore, M pathway disorders in disabled readers may interfere with initial perceptual identification and mental representation skills in the early stages of visual processing in reading. In normally achieving readers, the magnocellular pathways "set the table" by providing global pattern information prior to the arrival of the final visual details via the parvo channel.

The vulnerability of the M pathway as compared to the P pathway has led to considerable speculation about why the M pathway is selectively compromised. Since the number of M-cells is significantly less than P-cells in the foveal region, visual deficits with the former may be manifested earlier. Although we may continue to expect considerable variations between individuals, any alteration of the order and timing (i.e., synchronization) of the relative contributions or processing speed of the M- and P- pathways can result in a visual deficit. Since auditory and visual processing are dependent upon temporal processing, synchronization of M- and P-cell processing subsystems plays a significant role in learning to read (Chase, 1996; Eden, et al., 1995; Chase and Tallel, 1990).

19.3 5 Role of Models in the Understanding and Remediation of Reading Disability

At this juncture, it should be apparent that we are dealing with a central nervous system disorder with variable expressivity. The most obvious expression is in the verbal/visual deficiency model, since reading disability represents an overt clinical outcome. In this model, the contributions of two complementary neurosensory pathways, auditory and visual, are addressed. Audition and vision are both sensory responses to their respective stimuli; however, visual processing is also synchronized with a motor component that involves eye movements. In either case, the stimuli must be processed ultimately as language. It is reasonable to expect that the antecedents of reading disability would include not only the potential for an auditory, but also a visual processing and/or oculomotor deficit. Comprehensive reading therapy must stress development of phonological, word and sentence meaning, and other required cognitive skills. When deficits exist, rapid sensory processing, efficient eye movements, and the capacity to sustain effort and attention visually in reading and study type activities for extended periods of time should be enhanced with optometric vision therapy (Solan, 1985; Solan, Larson, et al., 2001).

Audition and vision embrace parallel hierarchies of sensory processing skills, such as attention, figure ground, discrimination, memory, analysis, and synthesis, that have been associated with reading disabilities. Normally, there is an orderly ontogenetic development of these perceptual skills with auditory preceding visual. Almost 4 decades ago, Birch (1962) postulated:

Reading disability may stem from the inadequate development of appropriate hierarchical organization of sensory systems, and so, at least in part, be the product of the failure of the visual system's hierarchical dominance. Failure for such dominance to occur will result in a pattern of functioning inappropriate for the development of reading skills. Additionally, Solan, et al. (1983) investigated intersensory liaison. They reported that the ability to equate a temporally distributed auditory stimulus to a spatially distributed visual response in the primary grades distinguished good from poor readers. The study also established that temporal - spatial integration within the visual modality is related to the development of early reading skills. Lest we adopt the position that "one model fits all", it has been necessary to consider several alternative, albeit interactive, paradigms.

In the absence of frank brain damage, clinical evidence and experience supports the notion that the maturation of auditory and visual perceptual and attentional skills required for reading takes place more slowly in some children than others. Selective and sustained attention involves executive function that originates in the prefrontal cortex, a brain system that has evolved relatively recently and is more likely to be vulnerable to genetic and environmental variations. The left hemisphere phonological and language processing systems discussed earlier are similarly vulnerable to developmental insult (Pennington, 1991). Sometimes even minor developmental maturational lags can have a profound effect on information processing, reading, and academic achievement, especially in the primary grades at which time children are learning to respond accurately with improved automaticity. Accuracy is a necessary, but not a sufficient condition for efficient learning. Fortunately, these developmental anomalies are amenable to specific processing therapy that routinely improves readiness for learning (Solan, Larson, et al., 2001). Neuromaturational lags distinguish themselves differently at different ages. Middle grade students who have overcome their primary grade decoding difficulties may have problems with understanding main ideas, figurative language, and identifying inferences. Academic performance may suffer selectively or generally. In either case, the students usually are characterized as having the potential to perform at a higher level. Secondary effects such as lowered motivation may result from failure and poor self - image. The condition sometimes produces compensatory behavior such as clowning in class that compounds the presenting problem.

In some individuals, the true *dyslexics*, a specific neuro-pathogenesis as described earlier may account for an acute failure in learning to read initially. Progress, especially development of the verbal-phonological-language sequence, occurs more slowly. In practice, this diagnosis often is made in hindsight, because it is not usually possible to identify anatomical anomalies *a priori*. That is, the diagnosis is based upon the individual's performance. Educational interventions that use multisensory techniques have been helpful. One such approach is the Orton-Gillingham program which emphasizes visual, auditory, kinesthetic, and tactile processing (VAKT), however it must be rendered by an experienced reading therapist (Birsh, 1999). Because the RD child requires more time to decode, the need to provide vision therapy to ameliorate visual processing and visual functional deficits continues to be important.

Finally, there appears to be abundant evidence that defects in the magnocellular system alter its characteristics and contribute to reading disabilities (see Table 19.1). Furthermore, prior investigations support the notion that at least some magnocellular deficits and the effects of intervention can be measured objectively (Solan, et al. 1998; Demb, et al. 1998; Eden, et al. 1996). Current studies are concerned with measuring the effects of training oculomotor efficiency and visual attention on reading disability. Subsequent investigations will use *f*MRI and/or coherent motion analysis to measure M-pathway defects before and after therapy.

19.4 CONCLUSION/SUMMARY

As we enter the 21st century, we are still searching for *the* cause of reading disability in children. Since a universal consensus of the definition of reading disability is lacking, finding specific antecedents approaches an "impossible dream". For example, Shaywitz et al. (1990) reported that there was no significant difference in the prevalence of reading disability between boys and girls in grades 2 or 3. These results contradicted the current notion promulgated by educators that a significant gender discrepancy, about 3 boys for each girl, exists. Careful analysis of the research, however, revealed that Shaywitz and her colleagues had defined reading disability as 1.5 SD below the IQ predicted score. To be identified as RD, a child with an IQ of 100 (50th percentile) would have to be reading at about the 10th percentile. A classroom teacher probably would use 25th percentile, about 2 years below grade level, as a criterion for RD, which would represent an entirely different population. Furthermore, some highly regarded educational psychologists consider IQ scores irrelevant to the definition of reading disability (Siegel, 1988). Although progress has been made, the resolution of basic as well as therapeutic controversies are necessary prerequisites to improve the quality of education for this special group of educationally handicapped individuals.

At this juncture, it appears that efficient reading skills entail the ability of the brain to interpret, associate, and unify inputs from different sensory modalities, both temporal and spatial. That is, an individual's information processing skills are largely dependent upon sensory integration, although exactly how remains elusive (Ayres, 1989).

The reductionist approach may have the value of gaining a more precise understanding of the molecular causes of reading disability, and this research should be encouraged. However, our clinical and educational procedures should also be guided by the significant contributions of an impressive array of professionals. They include experimental and clinical psychology and psychiatry, neurology, optometry, endocrinology, epidemiology, and educational specialists.

19.5 FUTURE DIRECTIONS

We must not stray too far from our primary goal, the analysis, treatment, and management of reading disability that has been abundantly documented. During the past decade, clinicians on the "front line" have realized, albeit gradually, that the future of sound clinical diagnosis and therapy will be highly dependent upon tools provided by advanced technology. Further research in the neurobiological antecedents and the neural behavior of reading and reading disability should be encouraged. It would be particularly valuable to develop a better understanding of the role of biological plasticity. Finally, we must learn more about the acquisition of the skills required for reading and the systems that constitute efficient visual information processing. The fulfilment of this idealized cooperative partnership between bio-engineers and professionals who render care to individuals identified as reading disabled will redound to the benefit of society.

19.6 REFERENCES

Adams, M. J., 1995, *Beginning to Read: Thinking and Learning about Print,* The MIT Press, Cambridge, MA, pp. 31-54.

Andersen, R. A., 1995, Coordinate transformations and motor planning in posterior parietal cortex, in: *The Cognitive Neurosciences,* M. S. Gazzaniga, ed., The MIT Press, Cambridge, MA, Ch. 33, pp. 519-532.

Ayres, A. J., 1989, *Sensory Integration and Praxis Tests,* Western Psychological Services, Los Angeles.

Badian, N.A., 1984, Reading disability in an epidemiological context: Incidence and environmental correlates, *J Learn Disabil.* **17**: 129-136.

Bender L., 1958, Specific reading disability as a maturational lag, in: *Psychopathology of Communications,* P. H. Hoch & J. Zubin. eds., Grune & Stratton, New York, Ch. 11, pp.155-176.

Berlin, V. P., 1887, *Eine Besondere Art der Wortblindheit.* Stuttgart, Weisbaden.

Birch, H.G., 1962, Dyslexia and the maturation of visual function, in: *Reading Disability*, J. Money, ed., Johns Hopkins University, Baltimore, pp. 161-169.

Birsh, J.R., 1999, *Multisensory Teaching of Basic Language Skills*, Paul H. Brookes, Baltimore.

Brannan, J. R., Solan, H. A., Ficarra, A. P., & Ong, E., 1998, Effect of luminance on visual evoked potential amplitude in normal and disabled readers, *Optometry Vision Sci.* **75**: 279-283.

Breitmeyer, B. G., 1980, Unmasking visual masking: A look at the "why" behind the veil of the "how", *Psychol Rev.* **87**: 52-69.

Breitmeyer, B. G., 1993, The roles of sustained (P) and transient (M) channels in reading and reading disability, in: *Facets of Dyslexia and its Remediation,* S. F. Wright & R. Groner, eds., Elsevier Science Publishers, Amsterdam, pp.13-31.

Buzzelli, A. R., 1991, Stereopsis, accommodative, and vergence facility: Do they relate to dyslexia? *Optom Vis Sci.* **68**: 842-846.

Chall, J., 1967, *Learning to Read: The Great Debate*, McGraw-Hill, New York.

Chase, C. H., 1996, A visual deficit model of developmental dyslexia, in: *Developmental Dyslexia: Neural, Cognitive, and Genetic Mechanisms,* C. H. Chase, G. D. Rosen, & G. F. Sherman, eds., York Press, Baltimore, Ch. 7, pp.127-156.

Chase, C. H. & Tallal, P., 1990, A developmental, interactive activation model of the word superiority effect, *J Exper Child Psychol.* **49**: 448-487.

Ciuffreda, K. J. & Tannen, B., 1995, *Eye Movement Basics for the Clinician,* Mosby, St. Louis.

Demb, J. B., Boynton, G. M., & Heeger, D. J., 1998, Functional magnetic resonance imaging of early visual pathways in dyslexia, *J Neurosci.* **18**: 6939-6951.

Eden, G. F., Stein, J. F., Wood, M. H., Wood, F. B., 1994, Differences in eye movements and reading problems in dyslexia and normal children, *Vision Res.* **4**: 1345-1358.

Eden, G. F., Stein, J. F., Wood, M. H., Wood, F. B., 1995, Verbal and visual problems in reading disability, *J Learn Disabil.* **28**: 272-290.

Eden, G. F., Van Meter, J. W., Rumsey, J. M., Maisog, J., et al., 1996,. Abnormal processing of visual motion in dyslexia revealed by functional brain imaging, *Nature.* **382**: 66-69.

Eden, G.F. & Zeffiro, T.A..,1996, Looking beyond the reading deficits in dyslexia, a vision deficit, *J NIH Res.* **8**: 31-35.

Fellerman, D. J. & Van Essen, D. C., 1991, Distributed hierarchical processing in primate visual cortex, *Cereb Cortex.* **1**: 1-47.

Galaburda, A. M., 1988, Pathogenesis of childhood dyslexia, in: *Language, Communications and the Brain,* F. Plum, ed., Raven Press, New York, pp. 127-137.

Galaburda, A. M., Rosen, F. D., Sherman, G. F., 1989, The neural origin of developmental dyslexia: implications for medicine, neurology, and cognition, in: *From Reading to Neurons*, A. M. Galaburda,, ed., The MIT Press, Cambridge, MA, Ch. 14, pp. 377-385.

Geshwind, N. & Galaburda, A. M., 1987, *Cerebral Lateralization: Biological Mechanisms, Association, and Pathology*, The MIT Press, Boston, pp. 20-47.

Geyer, J. J., 1970, Models of perceptual processes in reading, in: *Theoretical Models and Processes of Reading,* H. Singer & R. B. Rudel, eds., International Reading Association., Newark, Del, pp. 47-94.

Grisham, J. D. & Simons, H. D., 1986, Refractive error and the reading process: a literature analysis, *J Am Optom Assoc.* **57**: 44-55.

Gross, C. G., 1992, Visual function of inferotemporal cortex, in: *Handbook of Sensory Physiology,* R. Jung, ed., Springer-Verlag, Berlin, Vol 7 (3B), Ch.23, pp. 451-482.

Halperin, J. M., 1996, Conceptualizing, describing, and measuring components of attention, in: *Attention, Memory, and Executive Function*, G. R. Lyon & N. A. Krasnegor, eds., Brookes Publishing, Baltimore, pp. 119-136.

Hinshelwood, J.,1896. A case of dyslexia: a peculiar form of word blindness, *Lancet.* **2**: 1451.

Hinshelwood, J., 1917, *Congenital Word Blindness*, H. K. Lewis, London:

Hoffman, J.E. & Subramaniam, B., 1995, The role of visual attention in saccadic eye movements, *Perception and Psychophysics.* **57**: 787-795.

Hung, G. K., 1989, Reduced vergence response velocities in dyslexics: a preliminary report, *Ophthal Physiol Opt.* **9**: 420-423.

Johnson, D., 1988, Specific developmental disabilities of reading writing, and mathematics, in: *Learning Disabilities: Proceedings of the National Conference*, J. F. Kavanagh & T. J. Truss, Jr.,.eds., York Press, Parkton, MD, pp. 79-163.

Lehmkuhle, S, Garzia, R. P., Turner, L., Hash, T., Baro, J.A., et al., 1993, A defective visual pathway in children with reading disability, *New Engl J Med.* **328**: 989-996.

Lehmkuhle, S., 1993, Nurological basis of visual processes in reading, in: *Visual Processes in Reading and Reading Disabilities,* D. M. Willows, R. Kruk, & E.Corcos, eds., Lawrence Erlbaum, Hillsdale, NJ, pp.77-94

Lennerstrand, G., Ygge, J., Jacobson, C., 1993, Control of binocular eye movements in normals and dyslexics, in: *Temporal Information Processing in the Nervous System,* P. Tallal, A. M. Galaburda, R. R.. Llinas, & C. von Euller, eds., New York Academy of Sciences, New York, pp. 231-239.

Livingstone, M. S., Rosen, G., Drislane, F., & Galaburda, A., 1991, Physiological and anatomical evidence for magnocellular defect in developmental dyslexia, *Proc National Acad Sci, USA.* **88**: 7943-7947.

Lovegrove, W. J., Martin, F., & Slaghuis, W., 1986, A theoretical and experimental case for a visual deficit in specific reading disability, *Cognitive Neuropsych.* **3**: 225-267.

Mattis, S., French J. H., & Rapin, I., 1975, Dyslexia in children and young adults: Three independent neuro psychological syndromes, *Dev Med Child Neurol.* **17**: 150-163.

Mishkin, M., 1972, Cortical visual areas and their interconnections, in: A. G. Karczmar & J. C. Eccles, *Brains and Human Behavior,* A. G. Karczmar & J. C. Eccles, eds, Springer-Verlag, Berlin, pp.185-208.

Motter, B. C., 1991, Beyond extrastriate cortex: The parietal visual system, in: *The Neural Basis of Visual Function. Vision and Visual function,* Vol.4, A. G. Leventhal, ed., CRC Press, Boca Raton, FL, pp. 371-381

Pavlidis, G. Th., 1983, The "dyslexia syndrome" and its objective diagnosis by erratic eye movements, in: *Eye Movements in Reading: Perceptual and Language Processes, K. Rayner, ed.,* Academic Press, New York, pp. 441-456.

Pavlidis, G. Th., 1985, Eye movement differences between dyslexics, normal, and retarded readers while sequentially fixating digits. *Amer J Optom & Physiol Optics.* **62**: 820-832.

Pennington, B. F., 1990, *Diagnosing Learning Disorders: A Neurological Framework*, The Guilford Press, New York, pp. 3-22.

Satz P., Rardin D., & Ross, J., 1971, An evaluation of a theory of specific developmental dyslexia. *Child Dev.* **27**: 2009-2021.

Satz, P. & Sparrow, S., 1970, Specific developmental dyslexia: A theoretical reformulation, in: *Specific Reading Disability: Advances in Theory and Method*, D. J. Bakker & P. Satz, eds., University of Rotterdam Press, Rotterdam, pp.17-40.

Sergeant, J., 1996,.A theory of attention: An information processing perspective, in: *Attention, Memory, and Executive Function,* R..Lyon & N. A. Krasnegor, eds., Brookes Publishing, Baltimore, pp.57-69.

Shapley, R., 1992, Parallel retinocortical channels: X and Y and M. Applications of parallel processing in vision. in: *Applications of Parallel Processing in Vision*, J. R. Brannan, ed., Amsterdam: Elsevier Science Publishers, Amsterdam, pp. 3-36.

Shaywitz, S.E., Shaywitz, B.A., Fletcher, J.M., Escobar, M.D. 1990, Prevalence of reading disability in boys and girls, *JAMA.* **269**: 998-1002.

Siegel, L. S., 1988,.Evidence that IQ scores are irrelevant to the definition and analysis of reading disability, *Can J Psychol.* **42**: 201-215.

Simons, H. D. & Grisham, J. D., 1987, Binocular anomalies and reading problems., *J Am Optom Assoc.* **58**: 578-587.

Solan, H. A., Larson, S., Shelley-Tremblay, J., Ficarra, A., Silverman, M., 2001, Role of visual attention in cognitive control of oculomotor readiness in reading disabled students, *J Learning Disabil,* **34**: 107-118 .

Solan, H. A., Ficarra, A. P., Brannan, J. R., Rucker, F., 1998, Eye movement efficiency in normal and reading disabled elementary school children: effects of varying luminance and wavelength, *J Am Optom Assoc.* **69**: 455-464.

Solan, H. A., Brannan, J. R., Ficarra, A. P., & Byne, R., 1997, Transient and sustained processing: effects of varying luminance and wavelength on reading comprehension, *J Am Optom Assoc.* **68**: 503-510.

Solan, H. A., Feldman, J., & Tujak, L., 1995, Developing visual and reading efficiency in older adults, *Optometry Vision Sci.* **72**: 139-145.

Solan H. A. & Ficarra, A. P., 1990, A study of perceptual and verbal skills of disabled readers in grades 4, 5, 6, *J Am Optom Assoc.* **61**: 628-634.

Solan, H. A., 1985, Deficient eye-movement patterns in achieving high school students: Three case histories, *J Learn Disabil.* **18**: 66-70.

Solan, H. A., Usprich, C., Mozlin, R., Ali, S., Fitzpatrick, C., 1983, The auditory-visual integration test: Intersensory or temporal spatial? *J Am Optom Assoc.* **54**: 607-616.

Solan, H. A. & Mozlin, R. 1986, The correlation of perceptual motor maturation to readiness and reading in kindergarten and the primary grades, *J Am Optom Assoc.* **57**: 28-34.

Stein, J. F. & Walsh, J., 1997, To see but not to read; the magnocellular theory of dyslexia, *Trends Neurosci.* **20**: 147-152.

Taylor, E. A., 1966, *The Fundamental Reading Skill,* Charles C Thomas, Springfield, IL.

Vellutino, F. R., 1979, *Dyslexia: Theory and Research.*, The MIT Press, Cambridge, MA., pp. 232-311.

Williams, M. C., LeCluyse, K., Rock-Faucheux, A., 1992, Effective intervention for reading disability, *J Am Optom Assoc.* **63**: 411-417.

Wolf, M. & Bowers, P. G., 1999, The double deficit hypothesis for the double deficit dyslexias, *J Educ Psychol.* **91**: 415-438.

Chapter 20

Nystagmus Basics:
Normal Models that Simulate Dysfunction

Louis F. Dell'Osso

Dept. of Neurology, University Hospitals, Cleveland, OH 44106, PH: (216) 421-3224, FX: (216) 231-3461, EM: lfd@po.cwru.edu

20.1 INTRODUCTION

Instabilities of the ocular motor system may take either of two forms, nystagmus or saccadic oscillations. Nystagmus (an involuntary oscillation of the eyes) is caused by instabilities in subsystems responsible for slow eye movements, whereas saccadic oscillations stem from subsystems responsible for generating saccadic eye movements. Nystagmus may exhibit either a pendular (sinusoidal) or a jerk waveform. In jerk waveforms, the slow phases displace the eye and fovea away from the target and the fast (saccadic) phases attempt to refoveate it. An important indication of the underlying mechanism for a particular type of nystagmus is the shape of the slow phase. Linear (constant velocity) slow phases of jerk nystagmus stem from tonic imbalances in any of the ocular motor subsystems. Decelerating (decreasing velocity) slow phases of jerk nystagmus are due to failure of gaze holding mechanisms, either central (common neural integrator) or peripheral (extraocular plant). Finally, accelerating (increasing velocity) slow phases of jerk nystagmus indicate a basic instability in a subsystem that causes it to "run away." Pendular nystagmus indicates a resonant-frequency oscillation of a subsystem.

Identifying the waveform is the first step towards understanding the mechanisms involved in generating each type of nystagmus; one must also identify where in the ocular motor system (i.e., which subsystem, not the anatomical site) the instability arises and how the oscillation affects other ocular motor responses. Because of the complexity of the system, identifying the cause of the oscillation requires the development of a detailed, quantitative model. In contrast, attempts to do so by studying specific neurophysiological sites has not yielded the insights necessary for a thorough understanding of how the ocular motor system functions and how it deals with specific dysfunction.

In this chapter, I will discuss an approach to modelling the ocular motor system that is primarily based on function, dysfunction, and system-level responses rather than on specific neuroanatomy or neurophysiology. Although the latter two are incorporated into system models as much as possible (based on current knowledge), the absence of specific neuroanatomical or neurophysiological data is not sufficient to preclude a necessary hypothetical function from inclusion in the model. For example, both the common neural integrator and the local, resettable neural integrator were hypothesized and included in models long before they were either located or neural networks were hypothesized to simulate their operation. As we learn more about neurophysiology, it is becoming clear that the original expectations of modelers, namely that neurophysiological signals exist that parallel the functional signals of their models, were probably too optimistic (Robinson, 1994). Rather, it is more likely that most signals carried by neural interconnections are composites of several such functional signals that may defy decomposition into recognizable parts. Indeed, the signals in the hidden layers of even simple neural networks are usually functionally unrecognizable. That being the case, modelling on both the system and the neuronal levels should be pursued, with neither waiting for confirmation from the other. Modelling of specific subpopulations of neurons will continue to elucidate their particular behavior and possibly suggest where they fit into the overall system. However, it is doubtful that prediction of system function will come out of such investigations. The performance of any complex feedback system, including the ocular motor system, is determined by the myriad interconnections of its functional building blocks and not by the performance of any individual block – that is the lesson of negative feedback and its raison d'être.

It will come as no surprise to anyone familiar with modelling that a multitude of solutions exists for each particular input/output relationship to be simulated; simulations are not unique. Thus, within a complex system comprised of many interconnected subsystems, such as the ocular motor system, it is not *which* simulation of each block that is important to overall behavior but rather, *how* the blocks interact (i.e., how they are interconnected). It is this fact that has guided our modelling of ocular motor function and dysfunction. The models consist of distributed functional blocks with associated delays. One need only choose the particular model for each of the subsystems that performs the required tasks of that subsystem and insert it into the total system model. If the system model is designed properly, it will function correctly with any of a number of subsystem realizations. If a particular subsystem choice proves unsatisfactory, another can be substituted without compromising system behavior.

The goals of this type of ocular motor system modelling are: to arrive at a *robust* model capable of simulating most (and perhaps eventually all) normal behavior; to add to that model the ability to simulate a wide range of specific dysfunctions (e.g., different types of nystagmus or saccadic oscillations); and to use such a model to help understand both normal and abnormal behavior, in a pedagogical setting and in a medical setting. To model is to be forced to quantify the "hand waving" and to learn what works and what does not. The model embodies our thoughts about how a system functions in a mathematical framework and, as it becomes more complex, its emergent properties, which we have neither built in nor anticipated, are what teach us more about system function. Toward that end, we have arrived at a point in this ambitious journey where we are developing a single ocular motor model that exhibits many normal and abnormal behaviors, such as nystagmus. At present (three years from its inception), they include: saccadic responses to step, pulse, and pulse-step target changes; short-latency corrective saccades, pursuit responses to ramp and step-ramp target changes; saccadic hypo- and hypermetria; macro saccadic oscillations; gaze-evoked nystagmus; muscle/gaze-paretic nystagmus in myasthenia gravis or other types of palsies; fixation and saccadic responses during latent/manifest latent nystagmus (LMLN); and fixation, saccadic and pursuit responses during several of the forms of congenital nystagmus (CN).

20.2 HISTORICAL OVERVIEW

There have been many models proposed to simulate specific dysfunction in a limited ocular motor context, using parts of normal models as a foundation (Collins, 1975; Robinson, 1973; Stark, 1968). In some ways, each contributed to our understanding of ocular motor function and dysfunction. Different types of ocular motor abnormalities have one thing in common; they cause unwanted motion of the *moving oculocentric coordinate system* (i.e., the retina and its fovea-centric base) upon which world images serve as inputs to the ocular motor system. In normal models, retinal image motion is usually presumed to be due to target motion that is unmatched by the eyes. That built-in presumption cannot be made in the presence of nystagmus or saccadic oscillations. Similarly, in normal models, it is presumed that all pulses from the pulse generator are fully integrated by the common neural integrator; again, as several types of dysfunction demonstrate, that cannot be the case. We make the more reasonable presumption that individuals with either congenital or acquired ocular motor dysfunction have substantially the same ocular motor systems found in normal individuals. Therefore, structures and functions required to simulate the ocular motor system's responses in the presence of experiments of nature (i.e., the abnormalities causing dysfunction) must also be present in the normal system. Current evidence supports these assumptions, and the models arrived at by simulating dysfunction more closely represent the actual physiological system than do the simplistic and, many times, erroneous models designed to simulate only a limited range of normal behavior (e.g., see the discussion of the need for a resettable neural integrator and for control of the common neural integrator in section 20.2.3 of this chapter).

Most ocular motor system models of version eye movements are of unilateral (containing bidirectional (left and right) signals), yoked control architecture that move the two eyes together in a coordinated manner in both directions. However, the actual architecture of the physiological system is bilateral (each side containing only unidirectional signals), yoked, independent control that allows both conjugate and disjunctive eye movements (Dell'Osso, 1994). There are many reasons for this, including conservation of computing power, simplicity, engineering "elegance", and the assumption of conjugacy (i.e., a single signal drives both eyes to move similarly). Unfortunately, the latter may not always be the case in normal individuals and certainly is not in those with specific abnormalities (e.g., achiasma or LMLN). Despite the limitations of such models, they are still valuable tools that aid in our understanding of ocular motor control. The models of version discussed in this chapter, with the exception of the Alexander's law model, have bidirectional, unilateral, and binocular architecture. Their outputs represent either both eyes (presuming conjugacy) or, in some types of dysfunction, the fixating eye. Finally, bidirectional, unilateral, and binocular models can also be made to drive two separate eyes, producing a combination of the two architectures (see section 20.2.4 of this chapter). Obviously, models of vergence do not presume conjugacy (see also Chapter 11, Models of Saccade-Vergence Interactions, by G. Hung and K.J. Ciuffreda in this volume).

20.2.1 Congenital Nystagmus

Congenital nystagmus is the most common type of infantile nystagmus; it may appear at birth or in early infancy (also, see Chapter 21 on the clinical aspects of CN in this volume). CN may exhibit a number of idiosyncratic waveforms that arise from either pendular or jerk nystagmus instabilities (or both) (Dell'Osso and Daroff, 1975). The first ocular motor system model of dysfunction was that of the saccadic and pursuit responses of individuals with CN (Dell'Osso, 1967; 1968; 1970). That dual-mode model (see Fig. 20.1) was designed and simulated on a Burroughs B5500 computer using Analog Algol software. It incorporated modified, contemporary models of the saccadic and smooth pursuit subsystems (Young and Stark, 1963a; b) and accurately simulated physiological responses both of normals and a CN subject to pulses, steps, pulse-steps, ramps, step-ramps, and sinusoidal target inputs. Because the purpose of the model was to demonstrate how a *normal* ocular motor system could properly respond to specific target inputs despite an internal nystagmus oscillation, the latter was simulated by an asymmetric triangular waveform with no attempt to suggest either its anatomical site or physiological mechanism. Data from an individual with CN revealed "normal" responses to all of the above stimuli; that is, with the exception of the CN superimposed on each ocular motor response, the underlying responses of this individual were equivalent to those of normal subjects in gains, latencies, and components of the responses.

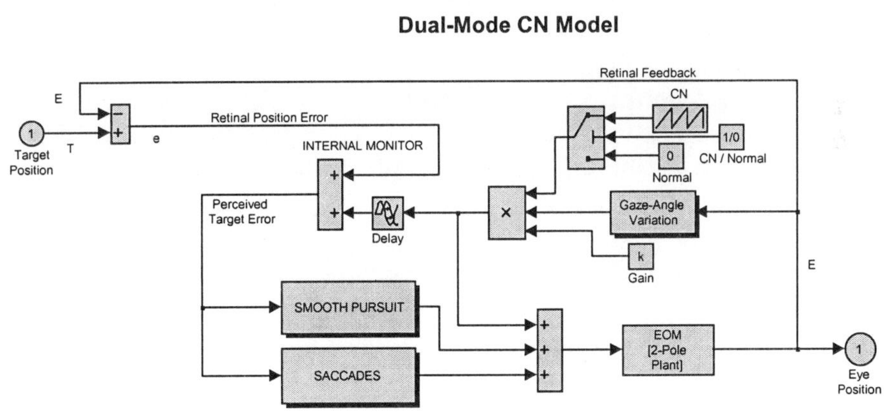

Figure 20.1. *A* dual-mode model used to demonstrate how a rudimentary internal monitor of efferent commands allows a normal ocular motor system to respond properly to target inputs in the presence of an internal oscillation (here, congenital nystagmus). In this and the following Figures, T – target, E – eye, e – retinal error, CN – congenital nystagmus, and drop shadows indicate subsystems (their absence indicates a basic Simulink building block).

Once the CN oscillation was introduced into the model, it became clear that none of the observed physiological behavior could be simulated unless the model was able to extract true target position and velocity from the oscillating retinal error signal (i.e., the target signal minus the eye signal). Because individuals with CN perceive a clear and stable world despite the continuously oscillating images presented to them by their retinas (i.e., they have no oscillopsia — the illusory motion of the world), the CN must have been canceled out. Thus, in the model some *internal copy* of the CN motor signal was needed to cancel out the oscillations of the eye to simulate accurately the physiological responses. To relieve the model of its 'oscillopsia,' it was necessary to add a delayed copy of the CN motor command to the retinal error signal. That allowed 'perception' of true target inputs and simulation of accurate responses, despite the ongoing CN. In effect, the model was using an *efference copy* signal of motor output to reconstruct true target information. This simple use of a motor command was a radical departure from the existing models of normal ocular motor behavior that functioned without the benefit of efference copy. A later study demonstrated the adequacy of perceived motion in stimulating smooth pursuit. (Yasui and Young, 1975) It was the simple basis for what was to become the *internal monitor*, an essential component of subsequent system models of ocular motor function and dysfunction. Despite the fact that neither the two subsystems in this model nor its ocular motor plant were physiologically accurate, the broad range of simulations it produced in the presence of an internal oscillation were precise enough that such model robustness would not be exceeded in over thirty years of research.

Later studies into the genesis of CN identified specific waveforms that fell into two categories (Dell'Osso and Daroff, 1975). This suggested that there were at least two sources for the observed instabilities. One was a pair of jω-axis poles for sinusoidal CN (possibly in the smooth pursuit system) (Dell'Osso et al., 1972), and the other, a right-half plane pole for the accelerating slow phases of jerk CN (possibly in the common neural integrator circuitry) (Dell'Osso and Daroff, 1981). The singularities represented by poles on the complex s-plane correspond to roots of the characteristic equation of the differential equation that defines the system under study. Pairs of jw-axis poles define a sinusoidal oscillation whose frequency is given by the value on that axis. A singularity in the right-half plane corresponds to a positive exponential output, causing the system to run away. Beginning about two decades after the above system model, several attempts were made to model CN waveforms (Harris, 1995; Optican and Zee, 1984; Tusa et al., 1992). These models used the suggested right-half plane pole in the neural integrator circuitry to generate CN-like jerk waveforms with accelerating slow phases.

One of the above models (Optican and Zee, 1984) postulated a "reversed" (i.e., opposite sign) velocity pathway associated with the neural integrator. That was based on: the misperception that, despite abundant evidence to the contrary, individuals with CN pursued moving targets in the wrong direction and, similar to albinos (who usually have CN), they also had some misrouting of their temporal retinal fibers. There is no evidence for such maldevelopment in the great majority of individuals with CN and, given the excellent ability to pursue and use their vestibulo-ocular reflex (Dell'Osso, 1967; 1986; Dell'Osso et al., 1992a; b; c; Kurzan and Büttner, 1989), they appear to have normal ocular motor system structures and interconnections. Although the model simulated some CN waveforms, it also produced behaviors not exhibited by individuals with CN. Integral to the model was the hypothesis that the slow phases were generated from pulse-step mismatches in the "preceding" fast phases. That is, the CN resulting from a large fast phase or voluntary saccade would be predicted to be greater than that from a smaller one. However, this hypothesis is incorrect because first, it is the CN *slow phase* that is responsible for the genesis of the oscillation; the fast phase *follows* it and is in fact a response to its generated error. Some CN fast phases simply *brake* the slow phases, while others also attempt to refoveate the target. Second, the magnitude of CN during fixation of static targets is a function of gaze angle itself, *not* of the amplitude of the saccade used to acquire a given gaze angle. In fact, for a given individual, the CN at 15° right gaze is the same *regardless* of how that gaze angle was achieved, including either a slow pursuit or head movement to that gaze angle. The latter would contain no pulse-step mismatch. Thus, the very basis for their model was inconsistent with recorded physiological data taken from individuals with CN.

Despite evidence that the pendular and jerk waveforms arise from different sites and types of instability, Optican and Zee's 1984 model hypothesized a single source. However, attempts to simulate pendular waveforms using their model failed to match physiological data, and the use of specific non-linearities tailored to produce each waveform greatly weakened that approach. Finally, their model suggested that there could be two distinct version null angles in CN, a condition that had been reported in: individuals with LMLN and adducting-eye fixation that was mistaken for CN, or in CN with a latent component, where alternation in the fixating eye results in a null shift. In many hundreds of recordings made of individuals with CN, I have never observed more than one *static* version null angle (vergence nulls and asymmetric (a)periodic alternating nulls also exist in some individuals). The absence of a pursuit system in the model precluded testing of the pursuit responses in the presence of CN, and no attempt was made to provide a mechanism for target *foveation*, the hallmark of CN waveforms. (Dell'Osso and Daroff, 1975) Besides ruling out the putative mechanisms suggested, one of the positive contributions of their model was its demonstration of how the circuitry of the neural integrator might be involved in generating accelerating nystagmus slow phases. It remains to be demonstrated how that could occur in individuals with gaze-holding failure, due to leaky neural integrators, in addition to CN (Dell'Osso et al., 1993).

A subsequent attempt to model CN (Tusa et al., 1992) was unfortunately based on three patients whose nystagmus was atypical for CN, but instead closely resembled vestibular nystagmus (Raphan et al., 1979). *Fixation attempt*, which is known to elicit CN, *suppressed* the nystagmus of these patients, who also had oscillopsia; both of these signs conform to vestibular nystagmus, not CN. Most of the other clinical characteristics of these patients were antithetical to CN. Based on the eye-movement and clinical data, the consensus was that they did not have CN but rather a 'congenital,' primary vestibular abnormality with, perhaps, a variable CN-like instability. The model of the primarily linear nystagmus of these unusual patients was an extension of the above CN model but had optokinetic, pursuit, and fixation subsystems. However, the latter included the abnormal, reversed velocity loop around the neural integrator from the previous model with a "switch" to turn it *off* rather than *on* with fixation attempt. This is opposite to what is exhibited in CN. Attempts to assess the patients' pursuit and vestibulo-ocular 'gains' failed to take foveation into account and were therefore inaccurate. This led to simulations in the respective subsystems that may have looked like those of the patients but were conceptually flawed. To the student of ocular motor system modelling, parts of the model's performance were both interesting and instructive, and may have application in a CN model.

A third model of CN (Harris, 1995) used many of the concepts of the preceding two models but tried to avoid the problems described above. One contention was that the neural-integrator time constant in CN is short. That was based on EOG measurements of vestibular nystagmus in the dark and the assumption that velocities just before and after fast phases could provide an estimate of the time constant. Because of the unreliability of EOG and the failure to take into account the slow-phase velocities of the ever-present CN waveforms, this is a questionable method that led to an improbable conclusion. Individuals with CN can maintain foveation of eccentric targets for periods up to 400 msec per CN cycle, which is hardly indicative of a leaky neural integrator. The bottom line is that the waveforms produced by this model also depend on the size of the preceding saccades, a characteristic *not* representative of CN. Also, the method used to generate pendular waveforms produced only weak, damped pendular oscillations, and not the continuous and varied pendular oscillations of CN. Furthermore, it was suggested that they were more likely to be generated in the dark than in the light. This would come as a surprise to those whose CN is both strong and pendular under both conditions. However, the model does exhibit interesting characteristics regarding nulls and their shifts with eye velocity, as well as the conditions for 'oscillopsia.'

20.2.2 Saccadic Pulses

The internal monitor postulated in the dual-mode CN model was to reappear in attempts to explain other types of dysfunction, such as saccadic pulses (previously called macro square-wave jerks) (Dell'Osso et al., 1975). Although the model proposed to account for the generation of saccadic pulses required efference copy for the short-latency return saccade, responses from such a model were not presented. It is presumed that the same logic that produces the short-latency corrective saccades following dysmetric voluntary saccades could be used for the second saccade of a saccadic pulse.

20.2.3 Gaze-Evoked Nystagmus

Gaze-evoked nystagmus (GEN) is an *acquired* form of jerk nystagmus not present in the primary position but appearing when gaze is directed laterally. It is the most common form of acquired nystagmus, appearing often as a form of drug-induced nystagmus. The first attempt to model an acquired form of nystagmus was a model of GEN; it also required an internal monitor (Abel et al., 1978a). It was simulated on a Systron-Donner SD/80 analog computer containing 95 operational amplifiers, 85 digitally settable potentiometers, and 10 diode function generators, as well as relays and logic circuitry. This saccadic model was also the first to use the *local, resettable* neural integrator to generate the pulse of activity necessary for saccades (see Fig. 20.2). For several years prior to this use, I had postulated its presence in models developed in our lab. This was based on observations that nature appears to choose redundancy and separation of function over engineering 'elegance' and that dysfunction affecting gaze-holding did *not* affect saccades. The GEN model showed that it simply was not possible to use the previous commonly accepted scheme whereby the output signal from the *common* neural integrator was used for that purpose. In the presence of partial or total absence of the gaze-holding ability of the common neural integrator, its use to generate saccades was inconsistent with the observed normal saccades generated by individuals with GEN. A separate, local, resettable integrator was needed to turn off the individual burst required for each normal saccade or fast phase of nystagmus. Two non-linearities in the pulse generator determined the pulse height and width respectively for saccades of differing amplitudes. When the local integrator's output matched the appropriate value of the width non-linearity, it turned off the pulse and was reset to zero.

Another departure from commonly accepted dogma sprang from modelling GEN. The model predicted that individual populations of neural integrator cells were under neural control, so that they only integrated pulses to that point where their output would match the desired gaze angle. The populations are simulated in the model as two neural integrators whose proportional contributions were adjusted using gains p1 and p2. Integrator control circuitry and provisions for either neural-integrator leakiness or saturation were incorporated into the model (see Fig. 20.2).

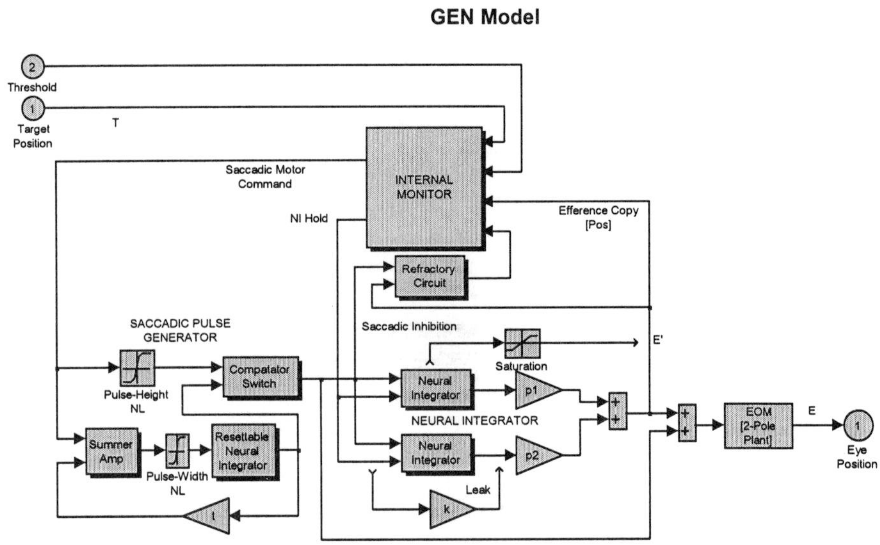

Figure 20.2. A saccadic-system model used to simulate gaze-evoked nystagmus (GEN) employing a more complex internal monitor and a population of neural-integrator cells. In this and the following Figures, Pos – position, NI – neural integrator, NL – non-linearity, and E' – efference copy of eye position.

In addition to a wide range of normal saccades, the model simulated several varieties of GEN, depending on the degree of neural-integrator leakiness, percentage of neuronal pool that was leaky, or the amount of saturation exhibited by the neural integrator (Abel et al., 1978a). Figure 20.3 shows a simulation of GEN produced using our more recent *ocular motor system* model (see below). Note that despite the GEN, a normal corrective saccade is made during the 20° refixation, the centripetal slow phases increase in velocity as gaze becomes more lateral, and there is no GEN in the primary position (consistent with clinical observation).

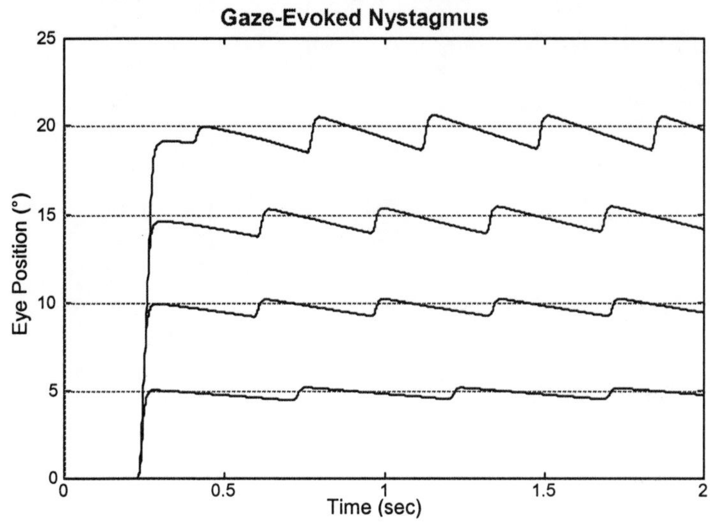

Figure 20.3. A simulation of GEN produced by our ocular motor system model. Shown are fixation at 0° and saccades to 5, 10, 15, and 20°.

20.2.4 Muscle-Paretic Nystagmus in Myasthenia Gravis

Muscle-paretic (gaze-paretic) nystagmus is caused by a peripheral failure to maintain the tonic signal for eye position. A model of myasthenia gravis, also simulated on the SD/80 analog computer (Abel et al., 1980), included both the stunted saccades of the myasthenic eye plant and the overdriven saccades of the normal eye (behind cover). Either a tonic deficit (leak) or paresis (saturation) in the plant, as shown in Fig. 20.4, simulated different myasthenic responses, including muscle-paretic nystagmus (Schmidt et al., 1980). Muscle-paretic nystagmus is similar to GEN, but the leak is peripheral (in the ocular motor plant) rather than central (in the common neural integrator). Also part of the model was provision for increasing central saccadic gain as an adaptation to the plant deficit and, furthermore, simulating the saccadic hypermetria and macro saccadic oscillations that would result from a Tensilon test that transiently eliminated the plant deficit. Both the internal monitor functions, the local, resettable integrator in the pulse generator, and neural control of the common neural integrators were integral parts of that model. This was a system model that simulated normal and abnormal saccades and muscle-paretic nystagmus; it did not have a pursuit system.

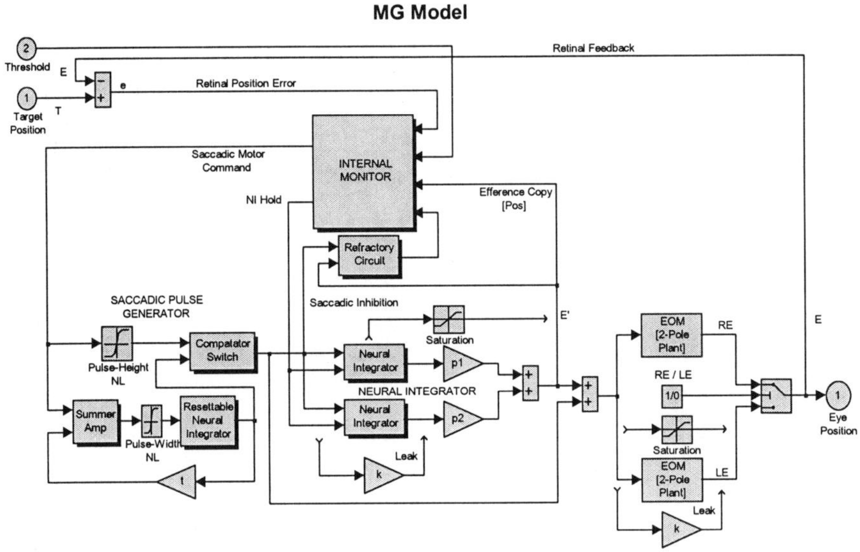

Figure 20.4. A saccadic-system model used to simulate myasthenia gravis (MG) built upon the GEN model of Fig. 20.2 and including provisions to lesion the ocular motor plant. In this and the following Figures, RE – right eye and LE – left eye.

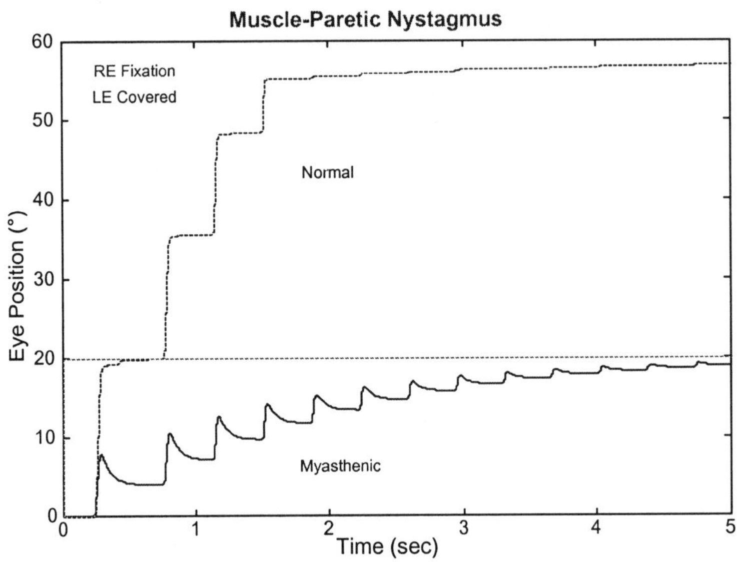

Figure 20.5. A simulation of muscle-paretic nystagmus in MG produced by our ocular motor system model. Shown are the saccadic responses of both the covered normal LE and fixating myasthenic RE to a target shift from 0 to 20°.

As Fig. 20.4 shows, one eye (here, the left) could be made weak either by a leak or saturation (similar to those shown for the common neural integrators). Also, either eye could be made the fixating eye by means of a switch. The model simulated the various types of myasthenic responses (including muscle-paretic nystagmus) (Abel et al., 1980). A myasthenic response simulated by our recent ocular motor system model is shown in Fig. 20.5. Here, the reduced gain, hypometric saccades of the fixating myasthenic eye are reflected in the covered, normal eye as high-gain saccades that bring that eye far beyond the target.

20.2.5 Alexander's Law and Vestibular Nystagmus

Alexander's law (Alexander, 1912) states that the amplitude of (vestibular) jerk nystagmus increases as gaze is directed towards the fast phase. To simulate this variation in a nystagmus caused by a tonic imbalance, a bilateral, push-pull model was constructed on the SD/80 analog computer (Doslak et al., 1979; 1982). The importance of Alexander's law stems from the observation that it applies not only to vestibular nystagmus but also to LMLN, and furthermore, that it may play a role in the variation about the null angle of CN. The mechanism proposed in the model was based on steady-state data derived from experiments on normals using caloric stimulation. It predicted linear slow phases whose velocities increased linearly with gaze angle in the direction of the fast phases. This was in agreement with the linear slow phases measured in vestibular nystagmus. However, in another study of normals and patients with vestibular lesions, it suggested that neural-integrator leakiness was responsible for Alexander's law (Robinson et al., 1984). Although this mechanism was not modeled, it predicts decelerating slow phases due to the leaky common neural integrator. However, as the authors pointed out, the amount of curvature to be expected in the slow-phase waveform, based on the known values of time constants, is so low as to be undetectable in most cases (i.e., they look linear). The second mechanism(Robinson et al., 1984) is the more parsimonious of the two, because it requires only portions of the normal ocular motor system, and without the need for the brain stem circuitry proposed in the Doslak et al. model. (Doslak et al., 1979; 1982) Further study of the linearity of vestibular slow phases is needed. Both mechanisms need to be tested within a robust ocular motor system model.

20.2.6 Periodic Alternating Nystagmus

Periodic alternating nystagmus (PAN) is a jerk nystagmus (usually acquired vestibular nystagmus but may be congenital also). The main characteristic of acquired PAN is the periodic reversal of the nystagmus direction with short intervals (usually <15 sec) of either no nystagmus or pendular nystagmus interspaced between the longer (usually ~90 sec) intervals of jerk nystagmus. Congenital APAN (Asymmetric (a)Periodic Alternating Nystagmus) differs by exhibiting longer and usually unequal periods of jerk nystagmus (several to >10 min) with short intervals (0 to ~20 sec) of either pendular nystagmus or no nystagmus. Also, the waveforms are those of CN. It has been suggested that the congenital form is underdiagnosed and may exceed 35% of CN patients (Shallo-Hoffmann et al., 1999). Acquired PAN has been modeled as a limit-cycle involving the brain stem neural networks that generate the slow phases of vestibular and optokinetic nystagmus, an adaptive network that normally suppresses prolonged, inappropriate nystagmus, and the inability to use retinal-error velocity information (Leigh et al., 1981). Although this was not an ocular motor system model in the sense described above, it was a model of system properties of the combined optokinetic and vestibulo-ocular subsystems with associated adaptation circuitry that normally produced post-rotary and optokinetic after-nystagmus. By depriving the model of retinal-error velocity information and raising the gain of the eye-velocity efference copy signal, the model was made to oscillate in a manner that simulated the PAN cycle. Justification for the higher gain was the observation that loss of retinal-error velocity information in patients results in a higher vestibulo-ocular gain. The resulting post-rotary nystagmus reversal grows until the system oscillates. By strategically stimulating the model at an appropriate point in the limit-cycle with an impulse of head velocity, the model's oscillation was made to stop and then increase slowly until the limit-cycle was reached. Application of such a stimulus to one of the patients stopped the PAN for 20 minutes. This model yielded valuable insights into both the interactions of these two subsystems and their adaptive circuitry; it also reaffirmed the need for efference copy in producing normal ocular motor behavior.

It has been postulated that the asymmetric (a)periodic alternating nystagmus (APAN) seen in many cases of CN is the result of a shifting neutral zone (Daroff and Dell'Osso, 1974). It is quite possible that these individuals also failed to calibrate their vestibular-optokinetic adaptation circuitry properly, and their CN is modulated by the same type of limit-cycle that is found in acquired PAN, albeit in a more asymmetrical manner.

20.2.7 Endpoint and Rebound Nystagmus

Endpoint nystagmus is a nystagmus present in normals that develops in lateral gaze, usually after prolonged fixation (Abel et al., 1978b). It is a jerk nystagmus with linear, centripetal slow phases and centrifugal fast phases. A system model of endpoint nystagmus (including saccadic, pursuit, and optokinetic subsystems) attributed its development and form to slow, drift velocities in lateral gaze (Eizenman et al., 1990). Reductions in drift velocity during fixation, as opposed to in the dark, were attributed to the pursuit system. However, in the absence of a moving target, it is more probable that the fixation subsystem was responsible.

Rebound nystagmus develops after prolonged lateral gaze followed by a return to primary position; it beats in the opposite direction of the preceding endpoint nystagmus and can be elicited in the light (Shallo-Hoffmann et al., 1990). A combination of a velocity bias and a null shift has been hypothesized as the cause of rebound nystagmus. The velocity bias causes the null to move, and a decrease in the time constant of the common neural integrator determines its exact position (Gordon et al., 1986). This putative mechanism needs to be tested within an ocular motor system model.

Ocular Motor System Block Diagram

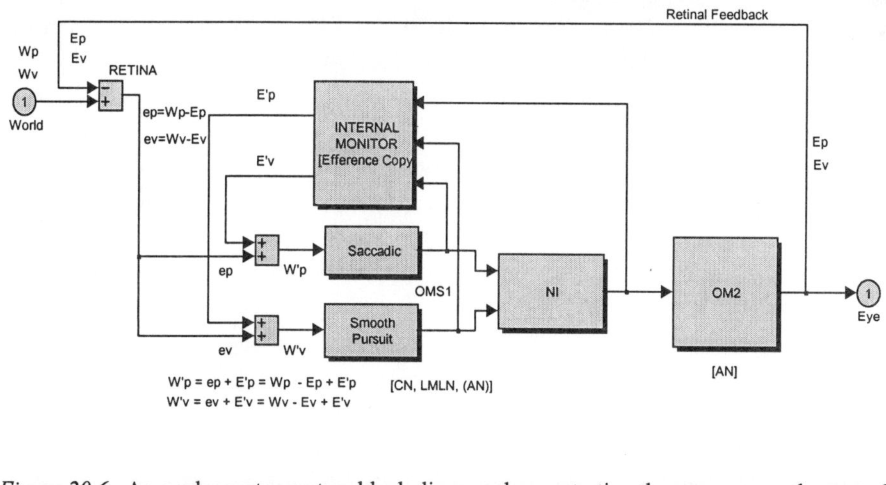

Figure 20.6. An ocular motor system block diagram demonstrating the presence or absence of oscillopsia based on the site of an internal oscillation. In this and the following Figures, W – world (target or background), OMS1 – distributed portions of the ocular motor system within efference copy loops, OMS2 – distributed portions of the ocular motor system outside of efference copy loops, LMLN – latent/manifest latent nystagmus, AN – acquired nystagmus, and p or v following a variable's symbol indicates position or velocity, respectively.

20.2.8 Oscillopsia

The topic of oscillopsia is intimately connected to ocular motor dysfunction, and therefore must be considered in our analysis of models of dysfunction (Dell'Osso et al., 1997). Oscillopsia is the perception of motion of the world when there is no actual physical motion. It may be the perception of a unidirectional 'spinning' of the world or a bidirectional oscillation. The first type can be experienced by normal individuals who are spun around in one direction for several minutes and then stopped, and the second type by vibrating one's finger as it is pressed on the side of one eye. It is an extremely debilitating symptom that interferes with standing, walking, reading. and many other basic functions. Some types of oscillations (usually congenital) are not associated with oscillopsia, while most acquired types are. Ocular motor models of these conditions should demonstrate the same properties. That is, the signals that represent reconstructed target position and velocity (i.e., *perceived* position and velocity) should either accurately reflect true target parameters (no oscillopsia) or be contaminated with the eye oscillation signal (oscillopsia). An ocular motor system block diagram (Fig. 20.6) demonstrates how lesions in different parts of the ocular motor system (arbitrarily divided into two parts, OM1 and OM2) may or may not produce oscillopsia. OM1 and OM2 are not meant to represent specific anatomical sites; rather, they are distributed *functional* subsystems. Thus, if a nystagmus position or velocity signal, N, is introduced within OM1 such that both the actual E and efference copy E' become the desired E + N, the perceived world signal, W', is given by e + E' = W – (E + N) + (E + N) = W. That is, there is no oscillopsia. If, on the other hand, N is introduced in OM2, the efference copy signal, E', would not contain N and W' = W – N and oscillopsia results. It is hypothesized that oscillations that do not cause oscillopsia lie within the efference copy loops used by the ocular motor system to calculate target parameters, while those that do cause oscillopsia lie outside of those loops.

20.3 CURRENT MODELS

In this chapter, I have concentrated on discussions of system models that fulfill the definition of being robust. These are models that: simulate a wide range of normal responses; simulate the responses to the same stimuli that are exhibited by individuals with specific abnormalities; and, are insensitive to internal changes or errors. Such models may be limited to specific ocular motor subsystems (e.g., saccadic, pursuit, etc.), but do not simply generate waveforms. The latter type of 'model' is more an intellectual exercise for students of control systems than a true simulation of the behavior of an ocular motor system that contains a dysfunction. Furthermore, the specific alterations to the normal system necessary to simulate the desired abnormal behavior should be both physiologically possible and consistent with the known pathophysiology of the abnormality being simulated (e.g., a neural-integrator lesion, leak, or saturation to produce gaze-evoked nystagmus). Experience with system models of dysfunction has shown that, although many proposed subsystem models, functional blocks, or small neural networks might suffice for specific, limited input/output simulations, one cannot easily predict which will not also adversely affect other, even seemingly unrelated, system functions. It is also difficult to prove that one such model is more accurate (i.e., physiological) than the others. The most stringent, and ultimately the best, test of any subsystem model is to place it within a more robust system model and demonstrate that the resulting ocular motor system model can simulate both normal and abnormal behavior when presented with a wide range of stimuli.

We employ this method of "evolving" a model after each subsystem change or addition and, although tedious, it ensures that a particular solution to a desired simulation of dysfunction is compatible with the rest of the system; if it is not, the proposed solution is discarded (i.e., made "extinct"). In addition to being a *fail-safe* methodology, this "hands-on" evolution provides satisfying support for any incipient feelings of omnipotence the modeler might harbor. However, the transient highs resulting from a successful modification to simulate a desired specific behavior are short lived, only to be dashed by the feelings of despair and incredulity when the model fails to successfully accomplish the next (and there *always* is a next) supposedly simple simulation. The claim that much more than this (i.e., the ocular motor system) was completed in seven days, is the humbling thought that keeps the modeler in his/her place. The model must always be considered a work in progress, a working hypothesis, and *never* a finished product. The goal is to change it for the better, not defend any specific part of it.

The remainder of this chapter will concentrate on the development of a robust ocular motor system model that has been used to simulate physiological responses of individuals with either of two disorders, CN or LMLN.

20.3.1 Congenital Nystagmus

Recently, a preliminary simulation of one of the pendular waveforms of CN was incorporated into a *normal* model of saccades and smooth pursuit (Dell'Osso and Jacobs, 1998). This model was constructed using Simulink (a component of the MATLAB environment). The preliminary CN model is shown in Fig. 20.7 combined with a dual-mode LMLN model into a multi-modal model. It was capable of simulating normal physiological saccades, short-latency corrective saccades, saccadic dysmetria, macro saccadic oscillations, and smooth pursuit. As Fig. 20.8 shows, a modified Robinson pursuit model (Robinson et al., 1986) was used, and our pulse generator was an updated version of those shown in Figs. 20.2 and 20.4, including a local, resettable neural integrator. The common neural integrator consists of a leaky neural integrator with a 25-second time constant (from normal data in the dark) within a positive feedback loop that allows sustained gaze holding in the light.

Incorporated into the internal monitor (shown in Fig. 20.9) were means to detect target changes, reconstruct target position and velocity, generate braking saccades, (Dell'Osso and Daroff, 1976) make saccadic logic and size decisions, enable and time saccades, provide saccadic blanking, and generate neural integrator control signals. As the drop shadows indicate, each of these functional blocks is itself a complex subsystem consisting of internal blocks and basic Simulink functions. Their specific design is less important than how well they perform the functions required of them and how they interact with the other blocks.

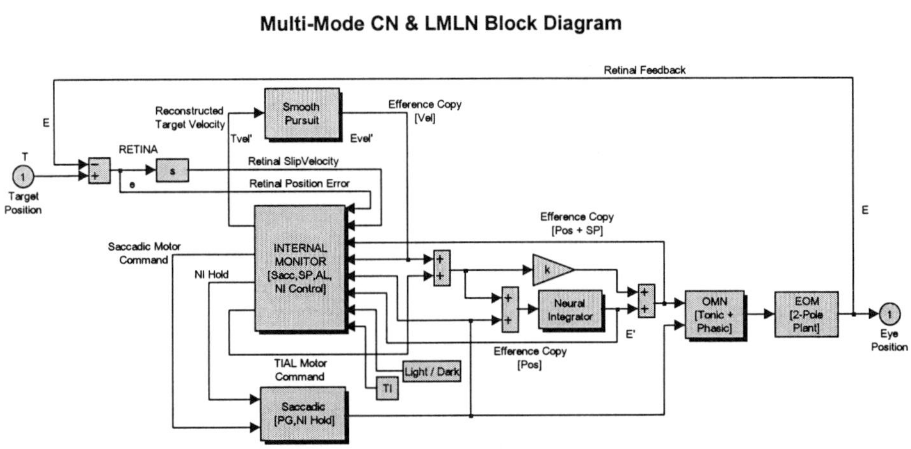

Figure 20.7. A multi-mode block diagram including simulations for CN and LMLN. In this and the following Figures, Tvel' – reconstructed target velocity, Evel' – reconstructed eye velocity equal to efference copy of eye velocity (Vel) signal, SP – smooth pursuit, TI – tonic imbalance, and TIAL – tonic imbalance modified by Alexander's law.

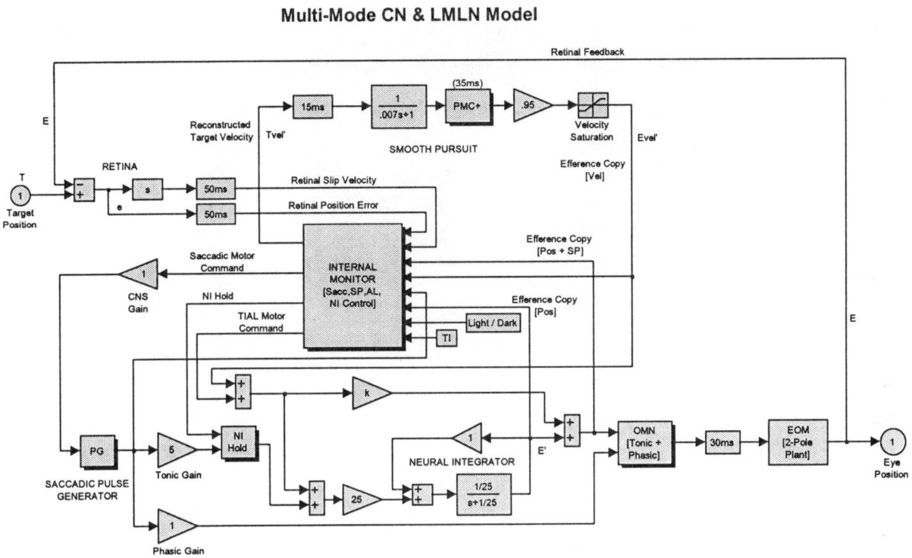

Figure 20.8. A multi-mode model of CN and LMLN containing a modified Robinson pursuit model (capable of oscillating), a saccadic pulse generator containing pulse-height and pulse-width non-linearities and a resettable neural integrator, a leaky common neural integrator with provision to overcome the leak, and distributed delays within and between major subsystems.

Shown in Fig. 20.10 are the model's responses to step changes in target position during simulation of orthometria, hypometria, and hypermetria as well as fixation with macro saccadic oscillations. In addition to generating normal pursuit with a damped oscillation (see Fig. 20.11), the smooth pursuit system was capable of oscillating, producing a sinusoidal *velocity* oscillation that was hypothesized to be the source of pendular CN. When the model was allowed to generate braking saccades, the pendular oscillation of the smooth pursuit system became a pseudopendular oscillation of the eyes, one of the waveforms that are pathognomonic for CN (see Fig. 20.11). Note that although the model programmed 1° braking saccades, the actual braking saccades were truncated by the oppositely directed CN slow phases. The model was able to make normal saccadic and pursuit responses, while oscillating in a pendular manner with two braking saccades per cycle; the latter is also shown in Fig. 20.11. This model was the first step in an attempt to model all of the CN waveforms and their hypothetical sources using a robust normal ocular motor system model as a foundation.

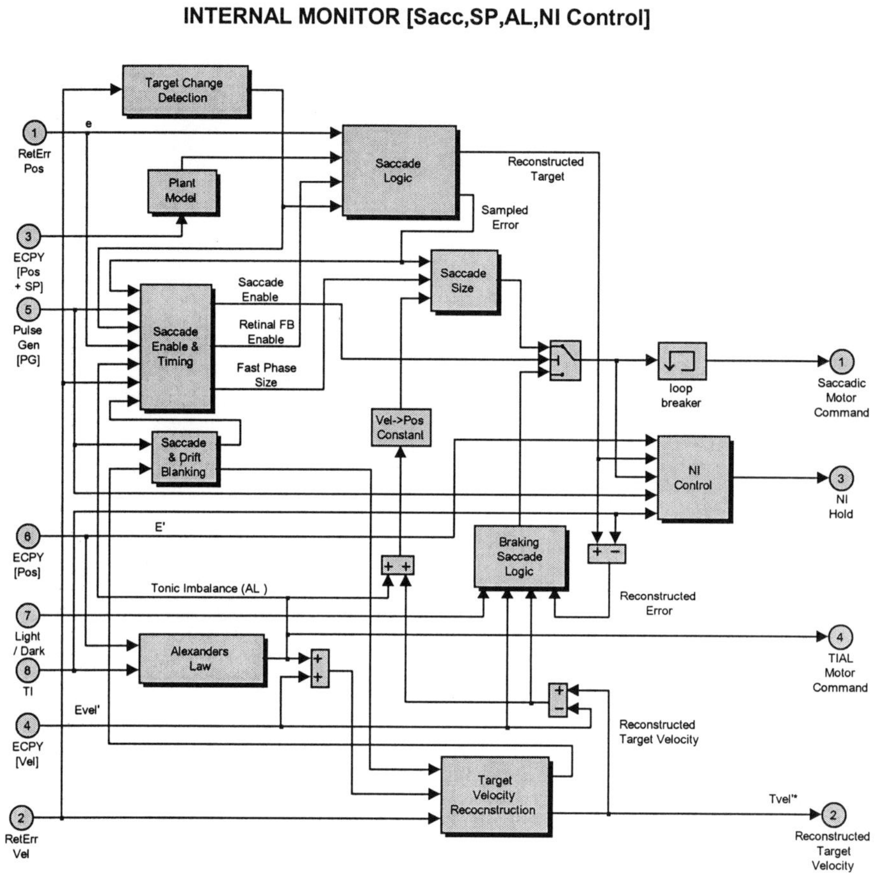

Figure 20.9. Functional organization of the Internal Monitor of the multi-mode CN and LMLN model. Target changes are detected and used by the saccadic logic and timing blocks to aid in the generation of saccadic responses. Both target position and velocity are reconstructed for use by the saccadic and pursuit subsystems, respectively; both are also used by the braking saccade logic block that determines the timing and amplitudes of braking saccades. Any existing tonic imbalance produces an Alexander's law variation that is used in calculating target velocity and sent as an output to the neural integrator. Finally, the reconstructed and efference copy signals are used by the neural integrator control block to determine what portions of saccadic pulses are to be integrated.

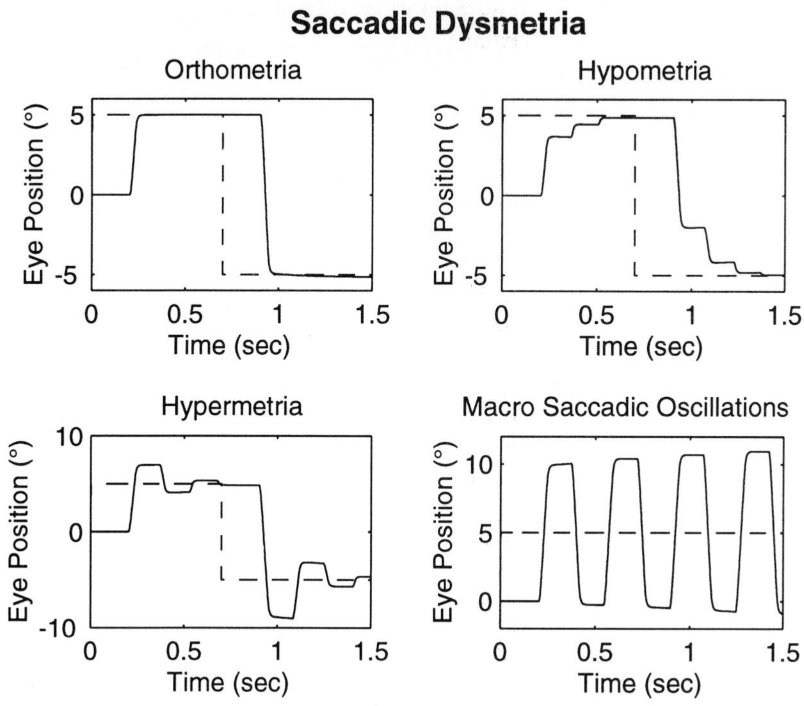

Figure 20.10. Simulations of saccadic dysmetria including orthometric, hypometric, and hypermetric saccades and macro saccadic oscillations.

Beginning with our preliminary model of CN and including the functions required to simulate LMLN (see below), we proceeded to add the ability to produce foveating fast phases to the braking saccades (Jacobs and Dell'Osso, 2000). At present, in addition to the simulations listed above for both the preliminary CN model and the dual-mode LMLN model, this *ocular motor system* model (still under development) simulates fixation, voluntary saccades, short-latency corrective saccades, pulse-step responses, and smooth pursuit of ramps and step-ramps (Rashbass stimuli) (Rashbass, 1961) of both normals and individuals with either pendular with foveating saccades or pseudopendular with foveating saccades CN waveforms. It also simulates GEN and MG (see Figs. 20.3 and 20.5), since it contains the neural-integrator and plant elements (leaks and saturations) of these earlier models.

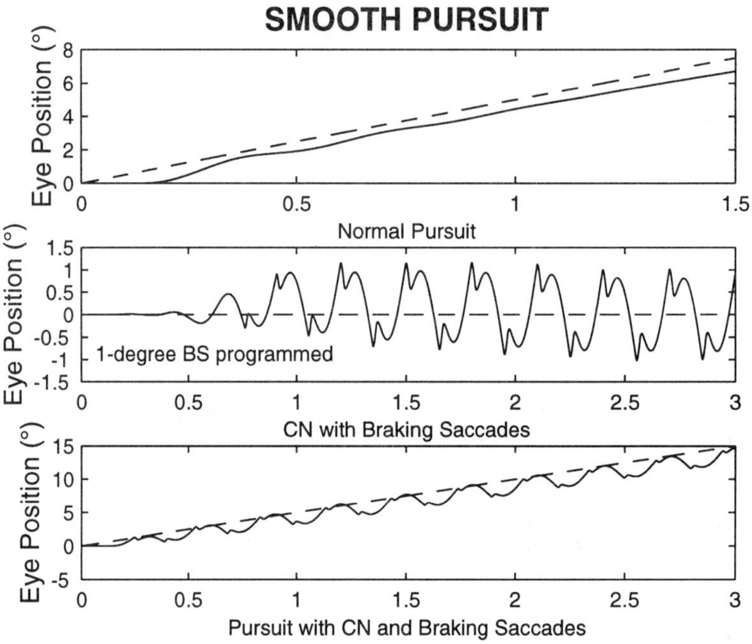

Figure 20.11. Simulations of normal smooth pursuit, fixation with pseudopendular CN, and smooth pursuit with pseudopendular CN.

20.3.2 Latent/Manifest Latent Nystagmus

Latent/manifest latent nystagmus is the second most common type of infantile nystagmus. It may be present in individuals with strabismus (a misalignment of the eyes) and is a jerk nystagmus whose direction is that of the fixating eye. That is, jerk right when the right eye is fixating and jerk left when the left eye is fixating. Pure latent nystagmus (LN) is very rare(Ciuffreda, 1977) and is only present when one eye is occluded. More commonly, manifest latent nystagmus (MLN) is present with both eyes open but only one fixating. (Dell'Osso et al., 1979) The linear slow phases of MLN usually become decelerating when LN is induced by occlusion of one eye. We expanded the capabilities of our preliminary CN model into the multi-modal model of Figs 20.7—9 to model LMLN. We hypothesized that a tonic imbalance was the driving force that produced the linear slow phases of LMLN. They are corrected by foveating fast phases in the direction of the fixating eye and defoveating fast phases when the slow phase velocity exceeds that necessary for good visual acuity (Dell'Osso et al., 1995). The multi-modal model was able to simulate: normal voluntary and corrective saccades; fixation; voluntary and corrective saccades during LMLN (of both types); the automatic transition from one type of LMLN to the other as gaze angle changed (based on the change in slow-phase velocity due to Alexander's law); the eye movements seen during the alternate cover test; and the LMLN changes seen during fixation with the adducting eye (Jacobs and Dell'Osso, 1999). Figure 20.12 shows a simulation of LMLN including saccades to different gaze angles, as well as the automatic transition of the LN from defoveating to foveating fast phases and of the LMLN from foveating to defoveating fast phases. Note that despite the LMLN, larger saccadic refixations include corrective saccades which may be suppressed, allowing the LMLN slow phases to bring the eye onto the target. As the increased size and number of inputs and outputs suggests, the Internal Monitor of Fig. 20.9 has grown considerably from its beginnings as a delay and summing junction. In this model, in addition to the functions listed above for the preliminary CN model, the circuitry provides Alexander's law variation to nystagmus slow phases. There are extensive interconnections between the different functional networks to ensure that they continue to work synergistically as we add function. We have endeavored to maintain backwards compatibility with our previous models to ensure no loss of function.

20.4 SYSTEM MODELS AND NEUROPHYSIOLOGY

As stated in the beginning of this chapter, these models of dysfunction and normal function are based on properties of the ocular motor system as a whole, as determined from responses to various controlled inputs. They neither depend on specific neurophysiological data nor do they ignore that which exists. However, when a function is deemed necessary for the model's performance, it suggests that somewhere within the physiological system that function, or a similar one, is being performed and that related signals might be found if one is looking for them. The latter statement comes with the caveat, as discussed in the Introduction, that such signals may be buried within composite signals and be undetectable. At present, these system models consist of functional blocks that are functions of time alone, whereas we know that the physiological system also makes use of spatial maps in its calculations of eye or target parameters (e.g., the retina, superior colliculus, visual cortex, etc.). Thus, some of the model's functional blocks can be replaced with more physiological spatio-temporal blocks as they are developed. I expect that, as in the past, neurophysiological researchers will continue to search for and elucidate the functions of specific brain stem areas known to be involved with ocular motor control and we, the modelers, will incorporate those findings into specific functional blocks.

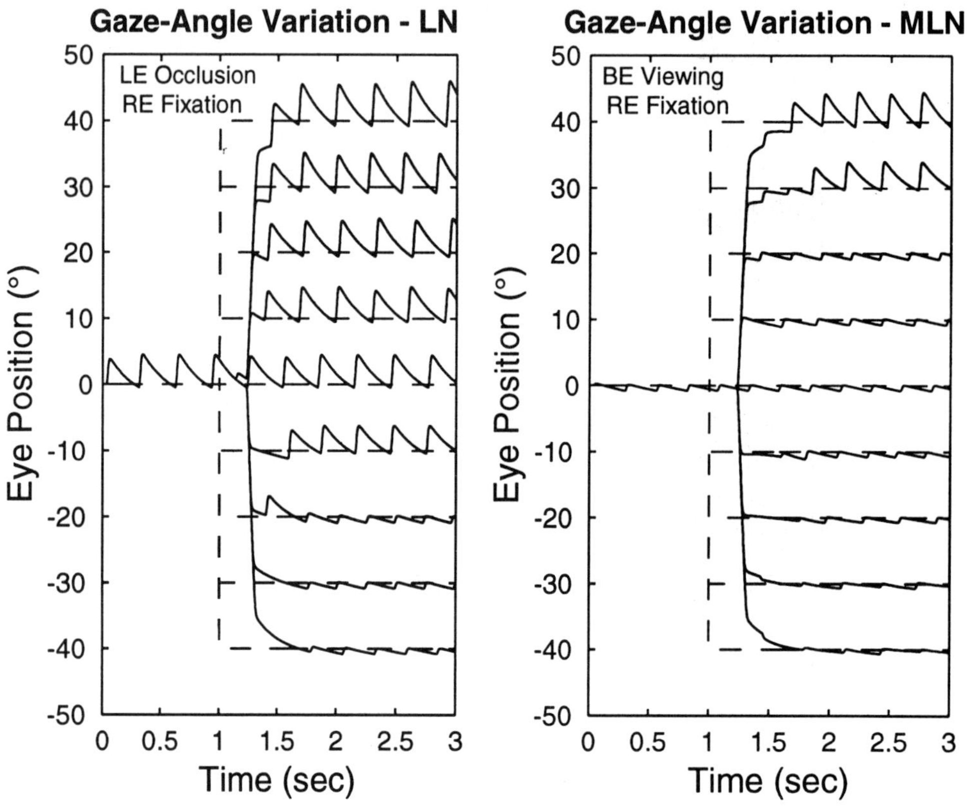

Figure 20.12. Simulations of LMLN showing the variation of the slow-phase velocity with gaze angle and the automatic transition from defoveating to foveating (during LN) and foveating to defoveating (during MLN) fast phases. For LN, the LE was occluded and the RE fixated the target; for MLN, both eyes (BE) were viewing (i.e., neither occluded) and the RE fixated the target. The target either remained at 0° or changed to one of the lateral positions indicated at 1 sec.

20.5 IMPLICATIONS OF THE MODELS

20.5.1 Basic

The most important findings that have come from the modelling of ocular motor *dysfunction* have been the elucidation of *normal* ocular motor function and architecture. Models of CN, square-wave pulses in multiple sclerosis, GEN, muscle-paretic nystagmus in myasthenia gravis, PAN, and LMLN revealed the necessity for an internal monitor to reconstruct target position and velocity from efference copy of motor commands and the retinal error signals. Modelling GEN demonstrated the need for a local, resettable neural integrator in the pulse generator and also the need for neural control of the percentage of the pulse that is integrated by the common neural integrator. The latter was also required by models of LMLN when fast phases became defoveating, and by saccadic pulses. Our current model of CN demonstrates how normal fast-phase generating mechanisms may be used to generate both the braking and foveating saccades of many CN waveforms. It also demonstrates how efference copy allows normal function of both the saccadic and smooth pursuit systems despite the presence of complex oscillations. Finally, modelling the uniocular saccades seen in achiasmatic Belgian sheepdogs required bilateral, yoked, independent control architecture of the ocular motor system, suggesting that each eye and muscle can be independently activated under the proper (perhaps, abnormal) circumstances. Although this is unusual in normal humans, its demonstration in abnormal human ocular motor behavior supports the hypothesis that the underlying architecture is uniocular and unimuscular with a strong yoking overlay. Models that do not contain these elements are severely limited in their ability to simulate the variety of ocular motor responses that the physiological system is capable of making.

20.5.2 Clinical

The PAN model described above demonstrated how the normal velocity storage mechanism enters into a limit cycle if not properly calibrated, and the model predicted a method to temporarily stop the PAN. Consideration of efference copy in ocular motor models led to the hypothesis that oscillopsia in acquired nystagmus occurred when the site of the dysfunction was outside efference-copy loops. The LMLN model demonstrated how individuals with LMLN could use their saccadic system to defoveate the target in an effort to utilize the slower tail ends of the slow phases to achieve better visual acuity. The CN model demonstrated how, by means of braking and foveating saccades, individuals with CN improve their foveation and therefore, their acuity.

20.6 FUTURE DIRECTIONS

20.6.1 Short Term

Because our ocular motor system model is a work in progress, some of the short-term plans listed below may have already been accomplished at the time of publication of this book. With regard to the simulation of LMLN, we are assessing the effects of background on the waveform and will include this in the model. Similarly, the effects of total darkness without a target will be simulated. For the CN simulation, a fixation subsystem will be added to achieve extended foveation of targets for the different waveforms. We will simulate the jerk waveforms and their variation with gaze angle and eye-velocity, and based on our data, the transitions of some individuals from pendular waveforms to jerk waveforms. Also, the model's emergent property of spontaneous bias shifts (which mimic physiological CN) will be studied to provide for control of that phenomenon. Finally, we plan to expand the types of dysfunction that the model can simulate by adding saccadic oscillations (e.g., flutter, flutter dysmetria, psychogenic nystagmus, and square-wave jerks/oscillations) and other types of nystagmus (e.g., rebound, Brun's, and acquired pendular nystagmus).

20.6.2 Long Term

As computing power continues to grow, we will be able to increase dramatically the complexity of our models and concurrently demand more from them. Additional specific behaviors will be simulated in more accurate ways. Duplication of unilateral, yoked control architecture into the bilateral, yoked, independent control architecture that defines the physiological system will become both feasible and the accepted norm in well-tested ocular motor models. When specific subsystems become totally specified by neurophysiological studies, their counterparts in the system model can be replaced by neural networks that accomplish and extend the simulation of those subsystems and of the model as a whole. Also, by appropriate time scaling, models will incorporate the plasticity exhibited by the physiological system and be better able to simulate the results of specific dysfunction. This includes both its slow development (in some cases) and the slow adaptations that take place within the ocular motor system to overcome the dysfunction. It will keep us busy, and it will take us much longer than seven days.

20.7 ACKNOWLEDGMENTS

This work emanated from the Ocular Motility Laboratory, Veterans Affairs Medical Center and the Departments of Neurology and Biomedical Engineering, Case Western Reserve University, Cleveland, OH. As the reference list indicates, many of the models that resulted from studies in our laboratory were the result of interdisciplinary collaborative efforts. I would like to specifically acknowledge the modelling contributions of L.A. Abel, M.J. Doslak, and J.B. Jacobs and the medical and physiological contributions of R.B. Daroff, B.T. Troost, D. Schmidt, and R.J. Leigh.

20.8 REFERENCES

Abel L. A., Dell'Osso L. F., Daroff R. B., 1978a, Analog model for gaze-evoked nystagmus, *IEEE Trans Biomed Engng.* **BME-25:** 71-75.

Abel L. A., Parker L., Daroff R. B., Dell'Osso L. F., 1978b, Endpoint nystagmus, *Invest Ophthalmol Vis Sci.* **17:** 539-544.

Abel L. A., Dell'Osso L. F., Schmidt D., Daroff R. B., 1980, Myasthenia gravis: Analogue computer model, *Exp Neurol.* **68:** 378-389.

Alexander G., 1912, Die Ohrenkrankhieten im Kindesalter, in: *Handbuch der Kinderheilkunde*, M. Pfaundler and A. Schlossman, eds., Vlg FCW Vogel, Leipzig, pp. 84-96.

Ciuffreda K. J., 1977, *Movements in amblyopia and strabismus. (Ph.D Dissertation)*, School of Optometry, Univ. of California,, Berkeley.

Collins C. C., 1975, The human oculomotor control system, in: *Basic Mechanisms of Ocular Motility and their Clinical Implications*, G. Lennerstrand and P. Bach-y-Rita, eds., Pergamon Press, Oxford, pp. 145-180.

Daroff R. B., Dell'Osso L. F., 1974, Periodic alternating nystagmus and the shifting null, *Can J Otolaryngol.* **3:** 367-371.

Dell'Osso L. F., 1967, A model for the horizontal tracking system of a subject with nystagmus, *Proc 20th Ann Conf EMB.* **9:** 24.2.

Dell'Osso L. F., 1968, *A Dual-Mode Model for the Normal Eye Tracking System and the System with Nystagmus. (Ph.D. Dissertation)*, Electrical Engineering (Biomedical), University of Wyoming, Laramie, pp. 1-131.

Dell'Osso L. F., 1970, A dual-mode model for the normal eye tracking system and the system with nystagmus, *IEEE Trans Biomed Engng.* **BME:** 87.

Dell'Osso L. F., Gauthier G., Liberman G., Stark L., 1972, Eye movement recordings as a diagnostic tool in a case of congenital nystagmus, *Am J Optom Arch Am Acad Optom.* **49:** 3-13.

Dell'Osso L. F., Daroff R. B., 1975, Congenital nystagmus waveforms and foveation strategy, *Doc Ophthalmol.* **39:** 155-182.

Dell'Osso L. F., Troost B. T., Daroff R. B., 1975, Macro square wave jerks, *Neurology.* **25:** 975-979.

Dell'Osso L. F., Daroff R. B., 1976, Braking saccade--A new fast eye movement, *Aviat Space Environ Med.* **47:** 435-437.

Dell'Osso L. F., Schmidt D., Daroff R. B., 1979, Latent, manifest latent and congenital nystagmus, *Arch Ophthalmol.* **97:** 1877-1885.

Dell'Osso L. F., Daroff R. B., 1981, Clinical disorders of ocular movement, in: *Models of Oculomotor Behavior and Control*, B. L. Zuber, ed., CRC Press Inc, West Palm Beach, pp. 233-256.

Dell'Osso L. F., 1986, Evaluation of smooth pursuit in the presence of congenital nystagmus, *Neuro ophthalmol.* **6:** 383-406.

Dell'Osso L. F., Van der Steen J., Steinman R. M., Collewijn H., 1992a, Foveation dynamics in congenital nystagmus I: Fixation, *Doc Ophthalmol.* **79:** 1-23.

Dell'Osso L. F., Van der Steen J., Steinman R. M., Collewijn H., 1992b, Foveation dynamics in congenital nystagmus II: Smooth pursuit, *Doc Ophthalmol.* **79:** 25-49.

Dell'Osso L. F., Van der Steen J., Steinman R. M., Collewijn H., 1992c, Foveation dynamics in congenital nystagmus III: Vestibulo-ocular reflex, *Doc Ophthalmol.* **79:** 51-70.

Dell'Osso L. F., Weissman B. M., Leigh R. J., Abel L. A., Sheth N. V., 1993, Hereditary congenital nystagmus and gaze-holding failure: The role of the neural integrator, *Neurology.* **43:** 1741-1749.

Dell'Osso L. F., 1994, Evidence suggesting individual ocular motor control of each eye (muscle), *J Vestib Res.* **4:** 335-345.

Dell'Osso L. F., Leigh R. J., Sheth N. V., Daroff R. B., 1995, Two types of foveation strategy in 'latent' nystagmus. Fixation, visual acuity and stability, *Neuro Ophthalmol.* **15:** 167-186.

Dell'Osso L. F., Averbuch-Heller L., Leigh R. J., 1997, Oscillopsia suppression and foveation-period variation in congenital, latent, and acquired nystagmus, *Neuro Ophthalmol.* **18:** 163-183.

Dell'Osso L. F., Jacobs J. B., 1998, A preliminary model of congenital nystagmus (CN) incorporating braking saccades, *Invest Ophthalmol Vis Sci.* **39:** S149.

Doslak M. J., Dell'Osso L. F., Daroff R. B., 1979, A model of Alexander's law of vestibular nystagmus, *Biol Cyber.* **34:** 181-186.

Doslak M. J., Dell'Osso L. F., Daroff R. B., 1982, Alexander's law: A model and resulting study, *Ann Otol Rhinol Laryngol.* **91:** 316-322.

Eizenman M., Cheng P., Sharpe J. A., Frecker R. C., 1990, End-point nystagmus and ocular drift: An experimental and theoretical study, *Vision Res.* **30:** 863-877.

Gordon S. E., Hain T. C., Zee D. S., Fetter M., 1986, Rebound nystagmus, *Soc Neurosci Abstr.* **12:** 1091.

Harris C. M., 1995, Problems in modeling congenital nystagmus: Towards a new model, in: *Eye Movement Research: Mechanisms, Processes and Applications*, J. M. Findlay, R. Walker and R. W. Kentridge, eds., Elsevier, Amsterdam, pp. 239-253.

Jacobs J. B., Dell'Osso L. F., 1999, A dual-mode model of latent nystagmus. ARVO abstracts, *Invest Ophthalmol Vis Sci.* **40:** S962.

Jacobs J. B., Dell'Osso L. F., 2000, A model of congenital nystagmus (CN) incorporating braking and foveating saccades. ARVO abstracts, *Invest Ophthalmol Vis Sci.* **41:** S701.

Kurzan R., Büttner U., 1989, Smooth pursuit mechanisms in congenital nystagmus, *Neuro ophthalmol.* **9:** 313-325.

Leigh R. J., Robinson D. A., Zee D. S., 1981, A hypothetical explanation for periodic alternating nystagmus: Instability in the optokinetic-vestibular system, *Ann NY Acad Sci.* **374:** 619-635.

Optican L. M., Zee D. S., 1984, A hypothetical explanation of congenital nystagmus, *Biol Cyber.* **50:** 119-134.

Raphan T., Matsuo V., Cohen B., 1979, Velocity storage in the vestibuloocular reflex arc (VOR), *Exp Brain Res.* **35:** 229-248.

Rashbass C., 1961, The relationship between saccadic and smooth tracking eye movements, *J Physiol (Lond).* **159:** 326-338.

Robinson D. A., 1973, Models of saccadic eye movement control systems, *Kybernetik.* **14:** 71-83.

Robinson D. A., Zee D. S., Hain T. C., Holmes A., Rosenberg L. F., 1984, Alexander's law: Its behavior and origin in the human vestibulo-ocular reflex, *Ann Neurol.* **16:** 714-722.

Robinson D. A., Gordon J. L., Gordon S. E., 1986, A model of smooth pursuit eye movements, *Biol Cyber.* **55:** 43-57.

Robinson D. A., 1994, Implications of neural networks for how we think about brain function, in: *Movement Control*, P. Cordo and S. Harnad, eds., Cambridge University Press, Cambridge, pp. 42-53.

Schmidt D., Dell'Osso L. F., Abel L. A., Daroff R. B., 1980, Myasthenia gravis: Saccadic eye movement waveforms, *Exp Neurol.* **68:** 346-364.

Shallo-Hoffmann J., Schwarze H., Simonsz H. J., Mühlendyck H., 1990, A reexamination of end-point and rebound nystagmus in normals, *Invest Ophthalmol Vis Sci.* **31:** 388-392.

Shallo-Hoffmann J., Faldon M., Tusa R. J., 1999, The incidence and waveform characteristics of periodic alternating nystagmus in congenital nystagmus, *Invest Ophthalmol Vis Sci.* **40:** 2546-2553.

Stark L., 1968, *Neurological Control Systems (Studies in Bioengineering)*, Plenum Press, New York, pp. 1-428.

Tusa R. J., Zee D. S., Hain T. C., Simonsz H. J., 1992, Voluntary control of congenital nystagmus, *Clin Vis Sci.* **7:** 195-210.

Yasui S., Young L. R., 1975, Perceived visual motion as effective stimulus to pursuit eye movement system, *Science.* **190:** 906-908.

Young L. R., Stark L., 1963a, A discrete model for eye tracking movements, *IEEE Trans Military Elect MIL 7.* 13-115.

Young L. R., Stark L., 1963b, Variable feedback experiments testing a sampled data model for eye tracking movements, *IEEE Trans Prof Tech Group Human Factors Electron.* **HFE:** 38-51.

Chapter 21

Multi-Sensory Feedback Therapy for Oculomotor Dysfunction

Kenneth J. Ciuffreda[1], Barry Tannen[2], and Daniella Rutner[3]

[1] *Dept. of Vision Sciences, State Univ. of New York, State College of Optometry, 33 West 42nd St. New York, NY 10036; PH: (212) 780-5132, FX: (212) 780-5124; EM: kciuffreda@sunyopt.edu*

[2] *Dept. of Clinical Sciences, State University of New York, State College of Optometry, 33 West 42nd St. New York, NY 10036; PH: (212) 780-5132, FX: (212) 780-5124; EM: btannenod@aol.com*

[3] *Dept. of Vision Sciences, State Univ. of New York, State College of Optometry, 33 West 42nd St. New York, NY 10036; PH: (212) 780-5132, FX: (212) 780-5124; EM: dsharvit@sunyopt.edu*

21.1 BACKGROUND

Biofeedback refers to the process of gaining voluntary control over some bodily function by immediate use of information regarding its physiological state. Thus, the individual is provided immediate information from biological processes normally beyond their awareness, thereby facilitating the regulation of these same functions (Brown, 1977; Wolf, 2001). The experimental applications of this technique have continued to grow, and the method has been used to control such disparate physiological processes as blood pressure, heart rate, muscle tension, brain wave activity, galvanic skin response, intestinal motility, and skin temperature (Fuller, 1978; Lee et al, 1977; Spilker, 1991). Biofeedback training has been used in the medical treatment of hypertension, cardiac arrhythmias, neuromuscular dysfunction, circulatory insufficiency, migraine, autonomic dysfunction, and chronic anxiety. The general biofeedback therapy paradigm is three-fold: (1) to provide information related to system variation/state that the individual learns to control, (2) to withdraw this information gradually, once control is well-learned, and (3) to provide occasional "reinforcement" using the prior auxillary feedback information to maintain the newly-learned ability.

The clinical application of biofeedback therapy for vision disorders is one of the relatively more recent areas of investigation. It has been used to achieve better control over such diverse ocular functions as visual scanning and accommodation, to reduce the effects of such pathological conditions as blepharospasm and glaucoma, and to improve fixational ability in strabismus, amblyopia, and nystagmus (Ciuffreda and Goldrich, 1983; Rothberg and Surwit, 1981; Halperin and Yolton, 1986). More specifically, oculomotor auditory biofeedback has been used in the treatment of a variety of specific oculomotor conditions (Table 21.1). In addition, oculomotor auditory feedback has been used experimentally with visually-normal individuals. The results provide insight into the normal auditory/visual interactive feedback process, as will be described later.

Table 21.1 **Therapeutic Uses of Oculomotor Auditory Feedback**

- To improve fixation in nystagmus (congenital and acquired)
- To improve fixation, reading eye movements, and visual scanning strategies in acquired brain injury (head trauma and stroke)
- To change from eccentric to centric (i.e., foveal) fixation in amblyopic eyes
- To embed and stabilize fixational eccentric viewing in central scotomas
- To improve reading and visual scanning ability in central scotomas and hemianopias
- To attain bifixation in strabismus
- To improve inefficient oculomotor skills of a functional origin in children
- To increase reading eye movement efficiency in normal and slow readers
- To improve reading eye movement ability in gross pathological oculomotor dysfunctions (e.g., nystagmus, saccadic intrusions, muscle paresis)

In this chapter, we will overview three types of oculomotor feedback --- auditory, visual, and tactile --- and discuss both their clinical and laboratory uses, as we and others have done for over 20 years. We will first consider each form of feedback in isolation, and then in combination in a multi-sensory manner. And, henceforth, we will generally use the word "feedback" rather than "biofeedback", as the former is more general, and the latter has historically had a negative connotation.

21.2 AUDITORY FEEDBACK

Auditory feedback refers to feedback in which the information about the bodily physiological function of interest is transmitted to and received by the auditory system of the individual. More specifically, oculomotor auditory feedback refers to the use of a tonal signal related to and correlated with changes in eye position that can be used to monitor one's oculomotor status (Fig. 21.1). By a trial-and-error process, one has to learn and develop the necessary auditory/oculomotor transform. The potential benefits of this therapy are three-fold: (1) improved visual acuity, (2) enhanced cosmetic aspects, and (3) reduced psychological effects subsequent to improvements in 1 and 2 above (Ciuffreda and Goldrich, 1983).

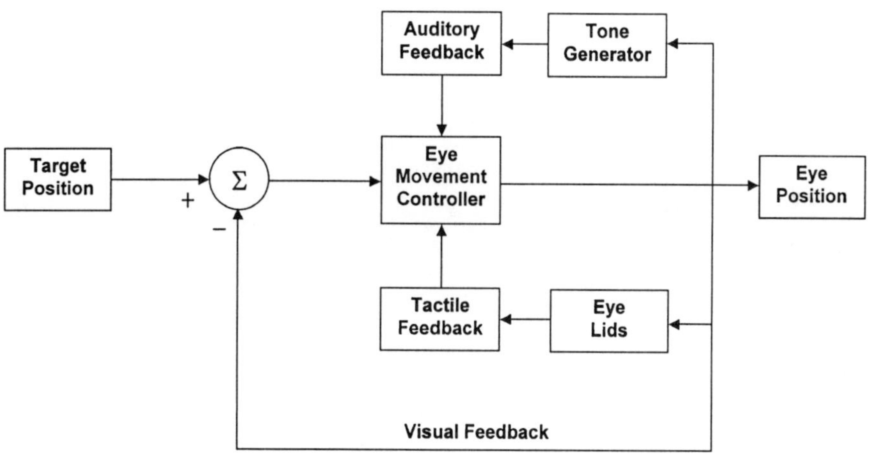

Figure 21.1. Block diagram of multi-sensory oculomotor feedback model showing visual, auditory, and tactile loops.

21.2.1 Instrumentation

The infrared limbal reflection technique (Ciuffreda and Tannen, 1995) has been used to record horizontal (sometimes also vertical) eye position of one eye during viewing of a test target with auditory feedback available. Such eye movement systems typically have a resolution of 0.25 degrees, a linear range of $\pm$10 degrees, and a bandwidth of at least dc to 100 Hz.

The specially designed and fabricated auditory feedback component receives electrical signals from the eye movement monitoring system by means of an adjustable level detector integrated circuit whose output signal controls a voltage-regulated audio oscillator having a frequency range of 500-5000 Hz. The output signal is amplified and played back to the individual through a loudspeaker, as schematically represented in Fig. 21.2. Quality of the tone reflects the individual's eye movements. For example, in the case of attempted fixation on a stationary target, the steadily-fixating eye of a normal individual would produce a relatively unwavering tone of constant pitch; in contrast, in a patient with continuously unsteady fixation (e.g., nystagmus), a distinctive wavering tone would be produced with pitch variation reflecting the position, amplitude, frequency, and velocity of the abnormal eye movements. The resolution of the audio component of the system is approximately 0.5 degrees.

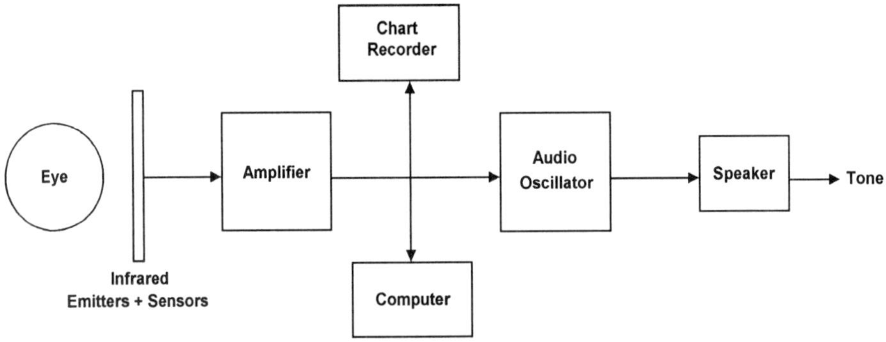

Figure 21.2. A schematic representation of an oculomotor auditory feedback system.

An example of such a system is presented in Fig. 21.3. This device dubbed "EyeTone" was recently commercially developed by the senior author for use within the clinic environment, as well as in the home environment for patients with nystagmus and other types of oculomotor dysfunctions to reinforce and extend the oculomotor therapy performed weekly in the clinic (Hall and Ciuffreda, in press; IOTA Eye Trace Systems, Sundsvall, Sweden; none of the authors has a financial interest in the device or company).

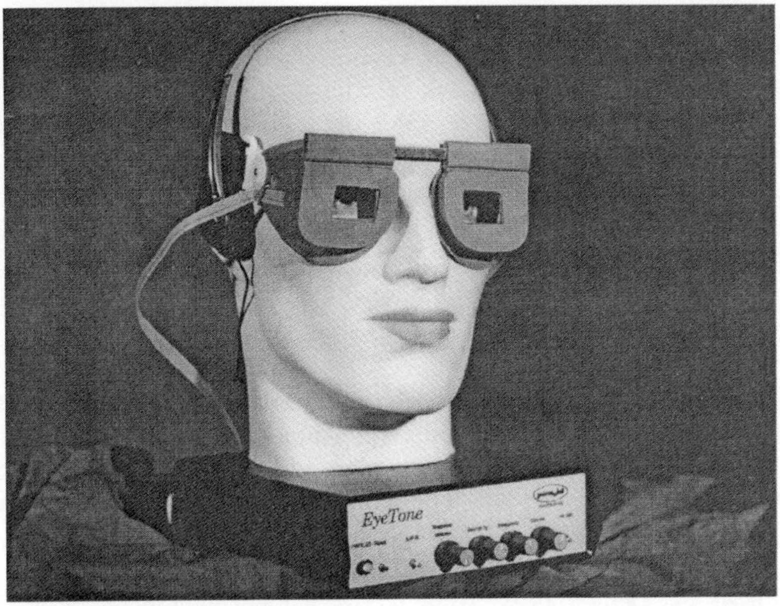

Figure 21.3. Commercially-available clinic and home training auditory feedback system developed by the senior author. Components include infrared emitter/sensor eye monitoring goggles, control unit, and privacy headphones.

21.2.2 Protocol

The instruction set is dependent upon the oculomotor abnormality and goal of therapy. In the case of nystagmus, cerebellar disease, and macular degeneration, in which fixation exhibits varying degrees of saccade-related unsteadiness, the instruction is to "learn to do 'something' to reduce the jerkiness of the tone". Hence, by trial-and-error, the individual learns to develop an internal, higher-level, cognitively-based control strategy and related trigger mechanism to override and suppress the unsteady eye movements, at least transiently (i.e., up to a few minutes), to view some object of interest with improved contrast and visual resolution. In patients manifesting reading dysfunctions, however, the goal is to shift the eyes across the line of print rhythmically and discretely in a step-like manner from word-to-word. The successful individual would hear a rhythmical sequence of tones varying progressively in pitch as their eyes moved sequentially across the line of print. And, if one were training smooth pursuit, the desired tonal variation should be extremely smooth and staccato-free, just like the sound of a slide trombone shifting in a glissando manner up and down a musical scale. Thus, the goal of auditory feedback training is to have the tone reflect time-optimal, accurate, and efficient eye movement patterns for whatever oculomotor system and task is being trained.

Training is typically conducted in 45 minute sessions, at least once per week, using the general feedback paradigm described earlier. If a home training device is available, then the clinic training is supplemented by the home training for 15-30 minutes each day. Home-based training can prove invaluable, as it provides daily rather than only weekly reinforcement.

21.2.3 Laboratory and Clinical Findings

21.2.3.1 Normal Vision

Surprisingly few studies have been conducted exploring the use of oculomotor auditory feedback in eyes with unimpaired vision function to assess the normal visual/auditory oculomotor interaction. Smith (1964) was the first to determine the extent to which auditory feedback could supplement and/or replace visual feedback used in the normal control of eye position. Monocular horizontal eye position was monitored with an infrared eye movement system (Ciuffreda and Tannen, 1995), either with or without auditory feedback, during four different conditions of fixation in which the quality and level of visual feedback were gradually reduced: (1) fixation on a small spot of light, which provided fine foveal visual feedback; (2) fixation between two small spots of light separated by 2.4 degrees, which provided parafoveal visual feedback; (3) fixation to the right (1.2 degrees) of a small spot of light, which provided somewhat less parafoveal visual feedback than in condition 2; and, (4) fixation in total darkness, which provided no visual feedback related to accuracy of eye position, and presumably eye position was maintained by proprioceptive feedback from the eye muscles and by visual memory (Ciuffreda and Tannen, 1995).

Smith found that the total time of accurate fixation increased, and mean eye position error decreased, under conditions 2, 3, and 4 when auditory feedback was made available. Under condition 1, however, in which normal visual feedback was present, fixation was actually slightly worse when the auditory feedback was added. Under conditions 2, 3, and 4, eye position control progressively decreased as visual feedback decreased, both with and without the presence of auditory feedback. Smith concluded that auditory feedback alone was sufficient for relatively good maintenance of eye position, especially under conditions in which visual feedback was not optimal.

In a later study, Kirschen (1977) found that tracking of a smoothly-moving target at high frequencies with the normal eye of amblyopic patients could be adversely affected when auditory feedback was provided. Thus, both Smith and Kirschen provided evidence that eye position control in normally-functioning eyes may actually become slightly worse when auditory feedback was added in the presence of high quality, normal, foveal visual feedback. Hence, under optimal conditions of visual feedback, the addition of correlated auditory information appeared to distract and possibly compete with vision. Only when visual feedback was degraded and non-optimal did the supplementary auditory information provide an enhancement effect.

There have been two more recent investigations of the normal visual/auditory oculomotor interaction. Hung et al (1988) monitored both horizontal and vertical, monocular, midline eye position using the infrared reflection technique (Ciuffreda and Tannen, 1995) in normal subjects under 3 conditions: (1) in the light, (2) in the dark, and (3) in the dark with two-dimensional auditory feedback, each for 100 second test periods. In the latter two conditions, if the subject's eye movements exceeded either ± 1 degree horizontally or ± 2 degrees vertically, the error-related auditory information would be activated. This consisted of a continuous, variable pitch tone horizontally and a 5 Hz discrete "clicking" tone vertically. The results are presented in Fig. 21.4 for one subject. Fixational ability was best in the light, worst in the dark, and at an intermediate level in the dark with the correlated auditory information present; percent time on target was 100, 50, and 80%, respectively. Thus, addition of auditory information related to eye stability enhanced considerably that found in the dark alone in the absence of a visual reference, with only extraocular muscle proprioceptive and visual memory

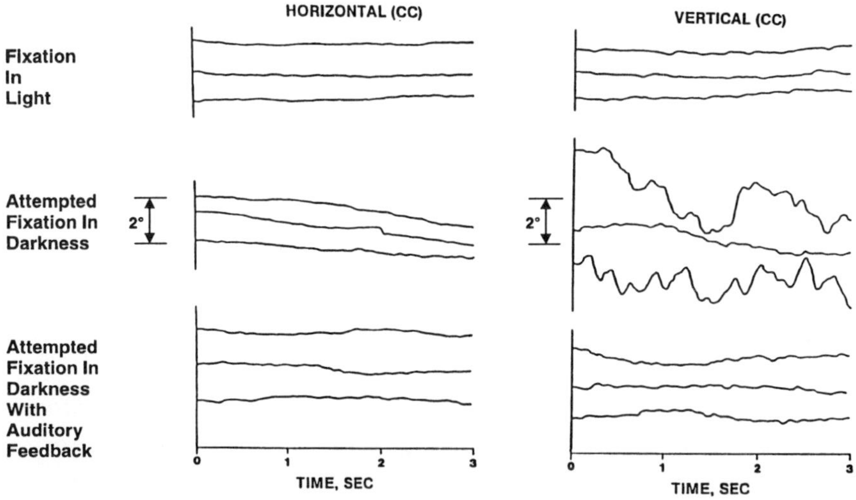

Figure 21.4. Horizontal (left) and vertical (right) eye movement traces for subject CC under the three test conditions. Three representative records are displayed for each test condition. Upward deflections represent leftward and upward eye movements. From Hung et al, 1988 with permission of Assoc. for Research in Vision and Ophthalmology.

information presumably available. Similar results were found in the dark at different bifixation distances for the vergence eye movement system (Shelhamer et al, 1994).

21.2.3.2 Nystagmus

Nystagmus refers to abnormally-large, involuntary, rhythmic movement of the eyes. It can be of either a congenital or acquired nature, in the latter case typically in conjunction with a variety of neurological diseases and oscillopsia (Ciuffreda and Tannen, 1995).

The first to use oculomotor auditory biofeedback in individuals with nystagmus was Abadi et al (1979, 1980, 1981). Using a simple infrared-based, eye monitoring home training device, they found improvement in both eye movements and related sensory functions, such as contrast sensitivity due to reduced smearing of the retinal image, after a few months of training. Others

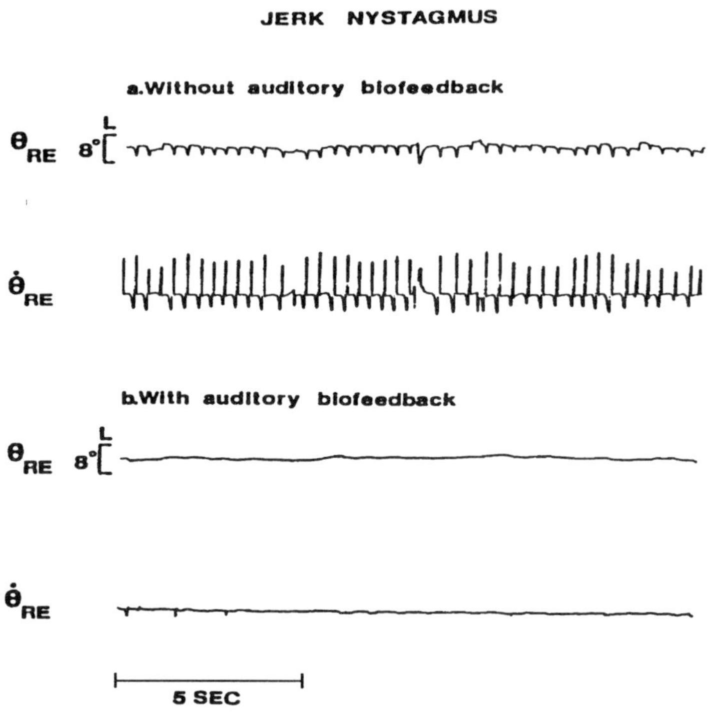

Figure 21.5. Right eye horizontal position and velocity as a function of time. Binocular viewing. Midline fixation. (a) Without auditory feedback, constant jerk nystagmus is present. (b) With auditory feedback during successful inhibition of nystagmus in absence of significant ocular drift. Symbols: θ = eye position, $\dot{\theta}$ = eye velocity, RE-right eye, L-leftward direction. Reprinted from Ciuffreda et al. (1980) with permission of Amer. Optom. Assoc.

followed over the next two decades with similar results (Abplanalp and Bedell, 1983, 1987; Ishikawa et al, 1985; Kirschen, 1983; Mezawa et al, 1990).

At the same time, our own laboratory independently began investigating this new and exciting therapeutic paradigm in adults, with even better results than the Abadi group. Perhaps this was the result of performing therapy in the more demanding and controlled clinic environment, with the addition of constant verbal feedback and encouragement of the researchers.

Our earliest recording in a nystagmat is shown in Fig. 21.5 (Ciuffreda et al, 1980). This was an optometry student in whom we conducted preliminary testing of the eye movement recording and feedback system, as well as test paradigm. He was not a patient, and, in fact, was not particularly motivated to participate in this early testing phase. Much to our surprise, after only 15 minutes of training, he was indeed able to reduce his nystagmus remarkably well for several seconds. We have since been able to demonstrate a pronounced training effect in nystagmus patients with as little as two minutes of auditory feedback, although more typically it takes 1 or 2 thirty minute sessions for a consistent and positive response to begin to evolve.

The results of our initial study are presented in Table 21.2 (Ciuffreda, 1982 a, b; Ciuffreda and Hung, 1988). The oculomotor findings in one of the best subjects are shown in Fig. 21.6. In Fig. 21.6A, pre-/post-treatment fixational

Table 21.2 **Results of Eye Movement Auditory Feedback Training**

Patient No.	Visual Acuity Initial/Final (20/)	Mean Amplitude (deg) Nystagmus Without/With Biofeedback	Mean Frequency (Hz) Nystagmus Without/With Biofeedback	Mean Peak Slow-Phase Velocity (deg/sec) Nystagmus Without/With Biofeedback
1	50^{-2} / 40^{-1}	2.2 / 0.4	2.6 / 1.8	20.0 / 2.8
2	30^{+3} / 15^{-1}	1.2 / 0.2	2.8 / 1.4	27.7 / 1.3
3	50^{+1} / 40^{+1}	1.0 / 0.2	3.4 / 1.5	7.0 / 1.2
4	100 / 50	3.8 / 0.4	2.6 / 2.4	22.0 / 2.5
5	25^{-3} / 15^{-1}	0.8 / 0.2	3.1 / 2.4	12.0 / 2.8
$\overline{X}$	50/30	1.8 / 0.3	2.9 / 1.9	17.7 / 2.1

results are presented. Clearly, with the additional audio information related to changes in horizontal eye position, the subject was able to reduce his nystagmus considerably for relatively long periods of time (up to 2 minutes). Of course, a critical question relates to how well the reduced nystagmus can be maintained, once the audio information is withdrawn. This is demonstrated in Fig. 21.6B. Again, the reduced nystagmus could be maintained for up to 2 minutes. Table 21.2 presents the visual acuity and eye movement results for each of the 5 subjects tested. Snellen visual acuity improved by 1 or 2 lines concurrent with the reduced nystagmus. All subjects could now volitionally reduce their nystagmus either with or without the auditory feedback present.

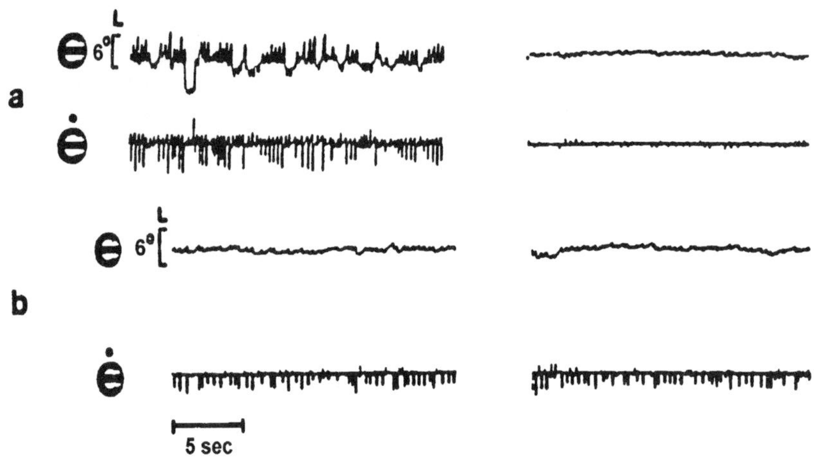

Figure 21.6. (a) Eye position (θ) and velocity ($\dot{\theta}$) as a function of time. Fixation without (left) and with (right) auditory feedback. Nystagmus is reduced with audio cues. (b) Eye position (θ) and velocity ($\dot{\theta}$) as a function of time; fixation with auditory feedback provided (left) and then withdrawn (right). Reduced nystagmus is maintained after auditory feedback withdrawal. Reprinted from Ciuffreda and Hung (1988) with permission of Measurements & Data Corporation.

We have continued these studies both in the clinic and research laboratory over the past 20 years, in which we have successfully tested and trained over 100 patients with nystagmus. The following results were recently obtained in one of our adult patients with congenital nystagmus, who performed home therapy with the device shown in Fig. 21.3 for 2 months (Fig. 21.7 and Table 21.3) (Rydberg et al, 1999). There was a marked reduction of the nystagmus concurrent with improvement in both suprathreshold visual acuity and threshold contrast sensitivity, again due to the reduced retinal-image motion and correlated reduction of retinal-image smearing across the fovea. We have noted another interesting phenomenon in a few patients who have maintained periodic auditory feedback therapy sessions over a several year period. Now, without attempting to employ any conscious strategy to reduce the nystagmus, and in the absence of any auditory feedback, their habitual nystagmus has become considerably reduced and barely perceptible to most. We speculate that with repeated conscious effort both in the clinic and at home, their cognitively-based strategy has now become automatic and reflexive. Thus, while surgical results for nystagmus show gradual reduction of the initial positive effect, non-invasive auditory feedback produces a gradual enhancement of the initial positive effect.

Table 21.3 - **Vision Function and Fixational Eye Movements Before and After Oculomotor Feedback Therapy In an Adult Nystagmat**

Parameters			Before Training	After Training
Vision Function	Visual Acuity	RE	0.8 - 1.0	1.0
		LE	0.5 / 0.65	0.5 / 0.65
		OU	0.8 - 1.0	1.0
	Contrast Sensitivity	OU	Reduced at all spatial frequencies	Improved at all spatial frequencies
Eye Movements	Peak Velocity	RE	158.3 deg/sec	35.1 deg/sec
		LE	142.6 deg/sec	38.8 deg/sec
	Amplitude	RE	4.0 deg	0.9 deg
		LE	3.8 deg	1.0 deg
	Frequency	OU	2.2 Hz	0.9 Hz

Key: RE - right eye; LE - left eye; OU - binocular; visual acuity decimal notation

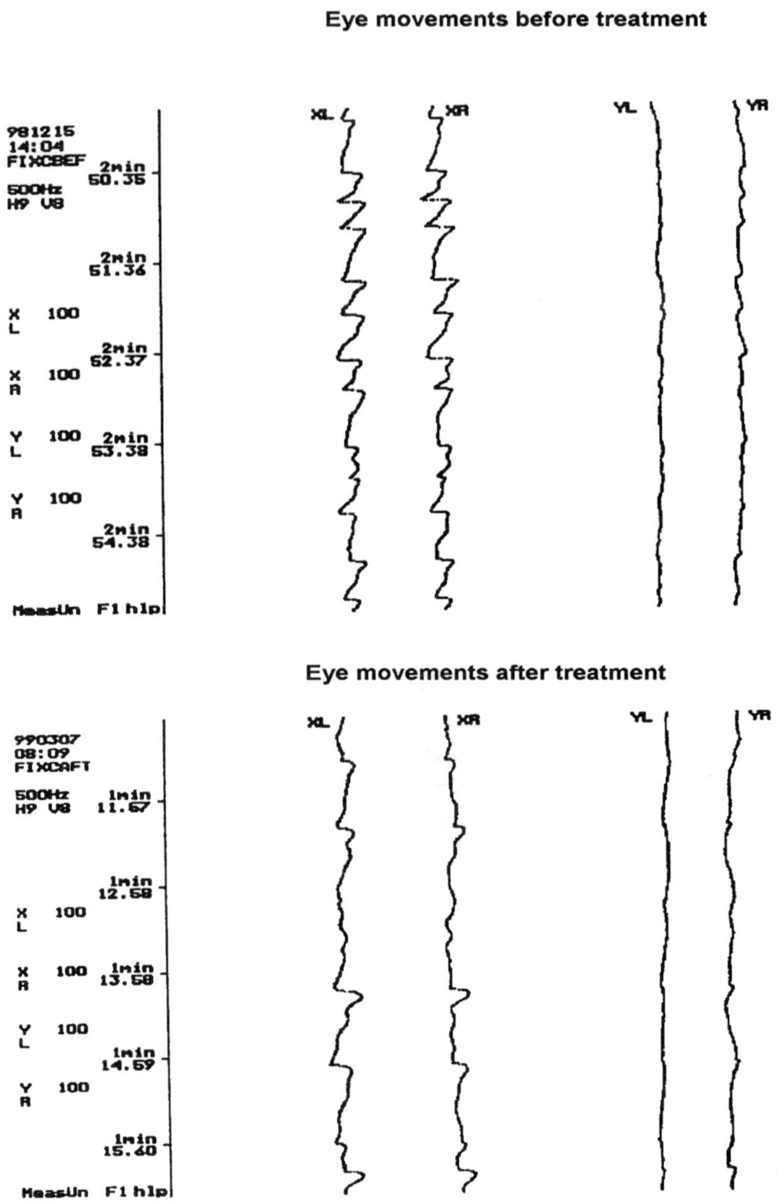

Figure 21.7. Binocular horizontal (XL, XR) and vertical (YL, YR) eye position as a function of time during midline fixation before and after 2 months of home-based, oculomotor auditory feedback using the EyeTone system (see Fig. 21. 3) in an adult patient with congenital nystagmus.

A new area of our investigation is the use of auditory feedback in young children, typically 4-10 years of age. We have found that some can understand and respond favorably by 8-10 years of age (Ciuffreda and Tannen, 1995). This is consistent with a recent study by Alvero et al (1998) in children with nystagmus aged 8-16 years. However, it is the youngest children in which both motor (e.g., eye movements) and sensory (e.g., visual acuity) improvements would be expected to be maximal and thus therapy more beneficial due to their greater degree of visual system plasticity. Also see Section 21.4.

21.2.3.3 Macular Degeneration

Macular degeneration is a common cause of legal blindness and reading disability, especially in the elderly (Hall and Ciuffreda, in press). It results from vascular insufficiency at the fovea and macular region. Typically, a central scotoma, or blind area, results. Such individuals must eccentrically view objects or words to be fixated with a non-foveal region of the retina having reduced visual resolution and non-primary visual direction, thus resulting in grossly inefficient fixational and saccadic scanning eye movement patterns. For example, reading rates in these patients rarely exceed 100 words per minute, which is far below the norm of 200-300 words (Taylor, 1966).

Our recent findings in this area clearly demonstrate that these patients, just like the nystagmats, can respond favorably and rapidly to oculomotor auditory feedback therapy primarily consisting of basic fixation and saccadic tracking, as well as reading of adult level text (Hall and Ciuffreda, in press). With only a total of 1.5 hours of training distributed over a period of two weeks, group mean reading rate significantly improved by over 20%. Nine out of the 10 elderly subjects tested demonstrated a significant gain in their reading ability and related oculomotor pattern. This is clearly seen in Fig. 21.8. Reading eye movements over the 5 second example presented normalized considerably and became more "staircase-like". With a longer and more intense treatment plan, perhaps in conjunction with conventional vision therapy to improve visual scanning ability and perceptual span (Taylor, 1966), even better results would be anticipated.

21.2.3.4 Slow and Normal Readers

Can one alter chronically inefficient reading eye movement patterns? Furthermore, can one alter and improve reading eye movement patterns in normal readers? Although early thinking on this notion was negative (Carmichael and Dearborn, 1947), results from our laboratory clearly refuted this old idea (Fayos and Ciuffreda, 1998). Twelve visually-normal adult

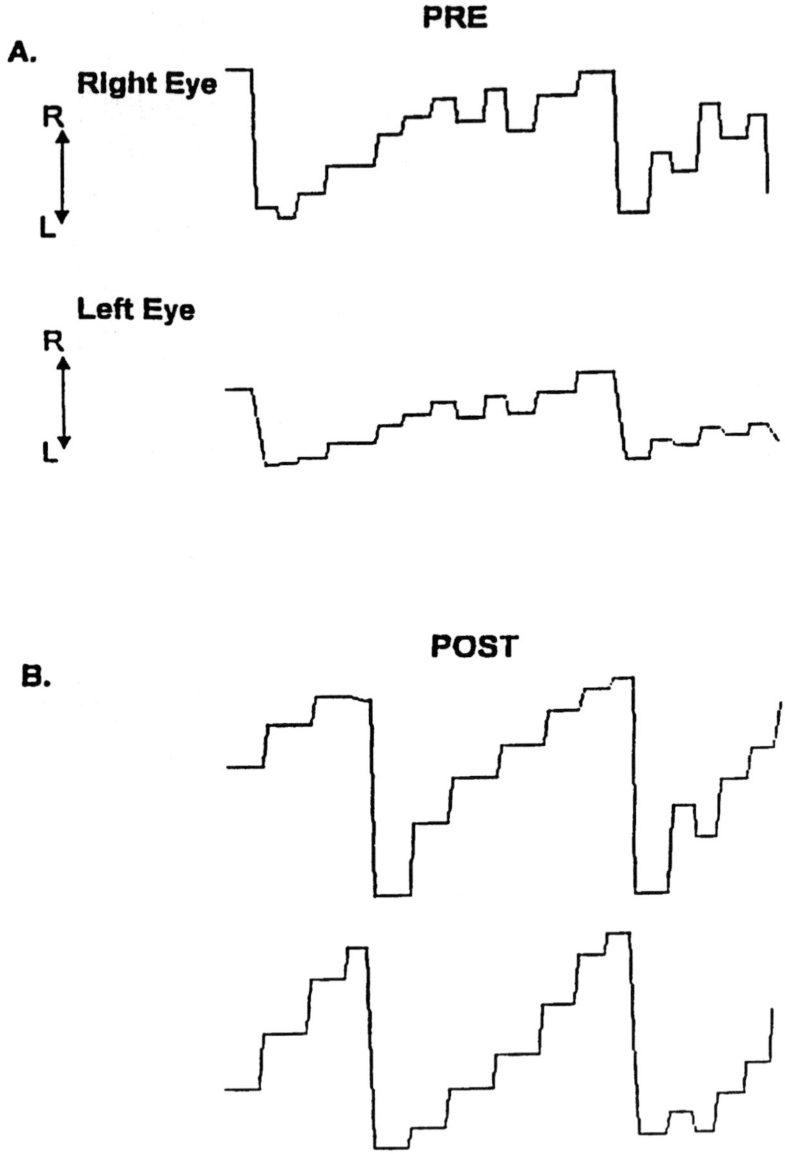

Figure 21.8. (A) Pre-training and (B) post-training Visagraph system reading eye movement recordings, from one observer (MNWI) with macular degeneration (MD), who exhibited a substantial increase in reading rate after training. Five second record length. R = rightward and L = leftward. Reprinted from Hall and Ciuffreda, with permission of Optometric Extension Program Foundation, Inc.

subjects with a range of low (100-200 words per minute) and normal (>200 words per minute) reading rates received four 30 minute auditory feedback training sessions distributed over a two-week period. Reading material consisted of adult classical novels. They were instructed to move their eyes rhythmically and rapidly, only making one saccade per fixational change, including the large right-to-left return-sweep saccade which shifts the eyes back to the next line of print, as per the desired tonal variations. Reading eye movements were objectively recorded immediately before and after training without the auditory feedback present. Results are presented in Fig. 21.9. All but one of the 12 subjects exhibited moderate to large improvement in reading rate following the training.

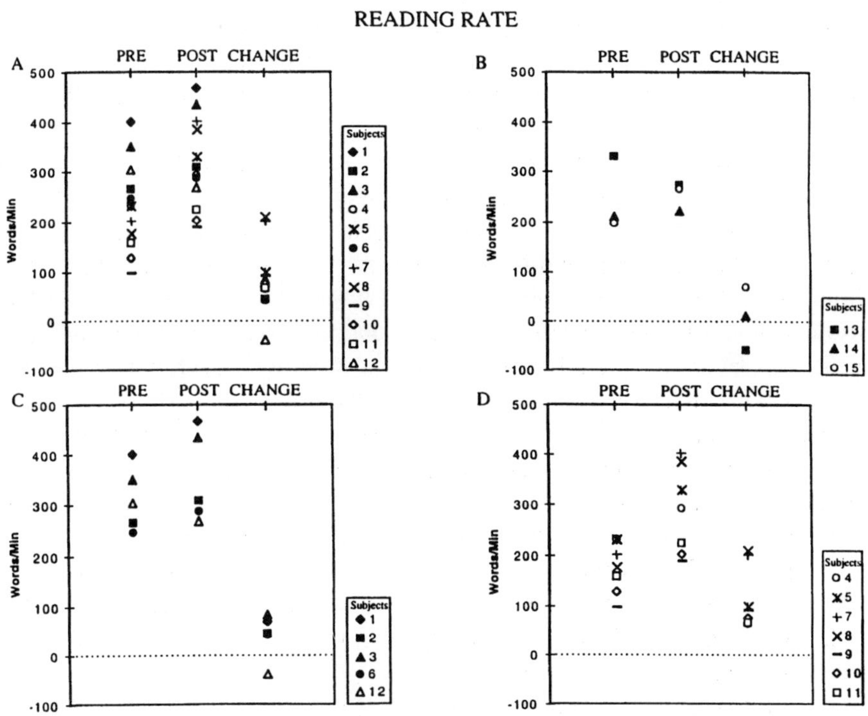

Figure 21.9. Reading rate from Ober2:Visagraph eye movement recording system prior to training (PRE) and after training (POST). The change in pre- and post-training raw scores are also shown (CHANGE). Experimental subjects (A), control subjects (B), experimental subjects subgrouped by Ober2:Visagraph grade level 12 or higher (C), or below grade level 12 (D). Reprinted from Fayos and Ciuffreda (1998) with permission of Optometric Extension Program Foundation, Inc.

The findings clearly demonstrated a rapid and easily learned ability of the subjects to alter and improve their reading patterns, thus resulting in increased reading speed, in contrast to long-standing conventional wisdom. It appears that by repeatedly hearing their reading eye movement patterns, subjects were able to minimize regressions, as well as properly sequence progressive movements to develop a more natural rhythm demanded for good reading (Fletcher, 1993; Robinson, 1977).

21.2.3.5 Strabismus and Amblyopia

Strabismus is an anomaly of binocular vision, in which the visual axis of one eye fails to intersect the object of interest due to presence of either a horizontal and/or vertical oculomotor deviation (Ciuffreda and Tannen, 1995). Presence of strabismus causes functional disruption of normal binocular vision due to occurrence of binocular suppression (See Chapter 17 in this volume), which if present most or all of the time will lead to amblyopia ("lazy eye") in the deviated eye (Ciuffreda et al, 1991).

Typically, this frank oculomotor deviation is corrected with either optometric vision therapy (Ciuffreda, in press) or eye muscle surgery (Caloroso and Rouse, 1993). However, auditory feedback has also been successful in many cases of strabismus and related vergence dysfunctions (Afanador, 1982; Hirons and Yolton, 1978; Inverso and Larsen, 1981; Letourneau, 1976; Letourneau and Ludlam, 1976; Letourneau and Giroux, 1984; Scheiman et al, 1983; Van Brocklin et al, 1981), even in those with disease-related macular scotoma (Goldrich, 1982). However, with the strabismic patient, the tonal feedback is not continuous. Use of discrete "deadband" electronics are configured such that when the eyes are perfectly aligned and bifixating the test object, the tone is extinguished; and, when misaligned, the tone is present. The patient is instructed to "learn to do something" to keep the tone off.

Figure 21.10 presents recent recordings in one teenage patient with constant, alternating strabismus (exotropia, or outward lateral deviation of the eye). After just four 30 minute weekly sessions, he was able to control his binocular eye position rapidly, repeatedly, and at will to maintain bifixation on the near test target. And, following this short period of therapy, the strabismus is now rarely manifest in his everyday activities, and in fact could not be elicited in our most recent clinical evaluation.

Auditory feedback has also been used successfully in training monocular central fixation in amblyopic eyes (i.e., an anomaly of monocular vision in which visual acuity is reduced due to early abnormal visual experience including suppression and contrast deprivation) (Ciuffreda et al, 1991), especially with respect to steadiness (Kirschen, 1977, 1991; Flom et al, 1978; Schor and Hallmark, 1978). This technique has been used in conjunction with both afterimage and entoptic image training (Ciuffreda et al, 1991). With this combined feedback protocol, the patient can visualize where in space the eye is fixating relative to the intended target via cortical image projection, as well as hear the degree of oculomotor unsteadiness via the auditory feedback, with the goal being steady, foveal, monocular fixation.

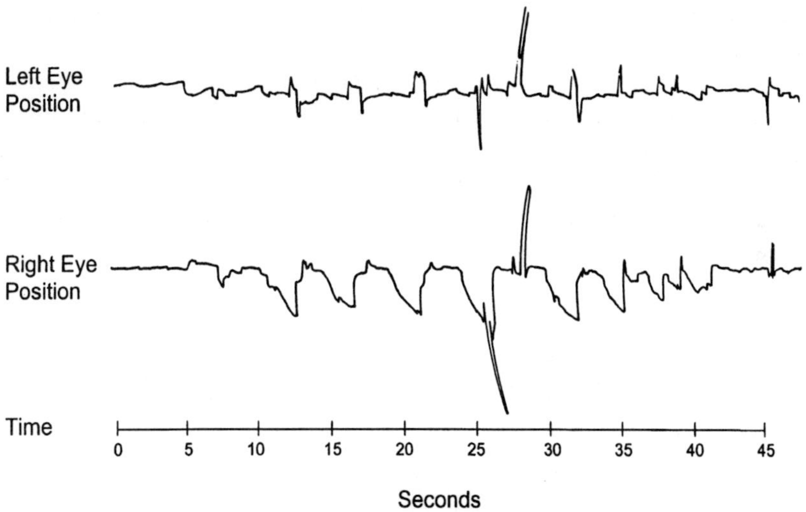

Figure 21.10. Binocular horizontal eye position as a function of time in a constant, alternating exotropic patient. Demonstrates repeated volitional asymmetric fusional vergence responses to auditory feedback information. Up is left, and down is right for the right eye, and the reverse holds for the left eye. Large deflections are artifacts due to blinks.

21.2.3.6 Blindness

In Section 21.2.3.1, it was demonstrated that an excessive amount of feedback related to eye position may be distracting rather than enhancing to motor performance. Thus, especially in normal observers, but also in well-trained nystagmats (personal observation of K.C. and B.T.), inclusion of auditory feedback in conjunction with visual feedback can result in non-optimal performance. In light of the above with regard to auditory/visual oculomotor interactions, it was decided to test the extreme --- blind observers who had no residual vision and related visual feedback for years, or perhaps even never had vision in their life, using oculomotor auditory feedback alone.

Recently, we completed such a study (Hall and Ciuffreda, submitted). Twelve blind adults with visual experience ranging from 0-28 years and a blindness duration of 6-55 years were tested. An infrared eye movement system with auditory feedback related to horizontal eye position was employed. An ABA experimental design was used: (A) subjects were asked to "look" straight ahead and try to keep their eyes as steady as possible; this was done for three 30-second periods; (B) auditory feedback was then added with the same instruction set and test periods; and, (C) feedback was terminated, and the instruction and test periods were as before without auditory feedback.

Ten of the twelve subjects were able to reduce their baseline eye movements by at least 25% both during and immediately following the feedback (Fig. 21.11). Two important correlations were found: eye stability improvement both during, and immediately following, the auditory feedback, with each being positively correlated with duration of visual experience. Clearly, blind individuals demonstrated a remarkable ability to use the auditory information to learn to increase eye stability very rapidly, thus revealing age-independent, higher-level, motor plasticity. This is consistent with that proposed earlier with regard to auditory/vision interactions: the poorer the visual feedback, the more effective the auditory feedback.

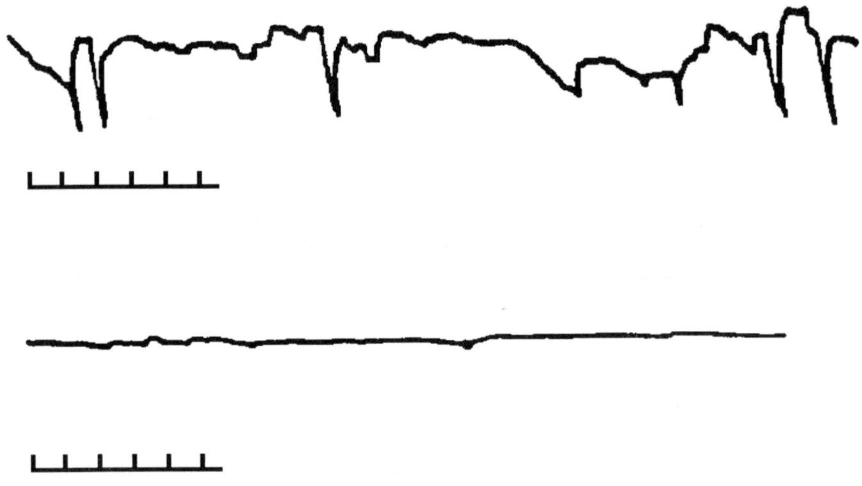

Figure 21.11. --- Horizontal eye movements in the right eye of a blind subject before (top) and immediately after (bottom) three, 30 second auditory feedback training sessions. Up = leftward, and down = rightward. Timing marks = 1 second. Large deflections are artifacts due to blinks.

21.3 TACTILE FEEDBACK

For the present discussion, tactile feedback refers to the sensory, or afferent, information derived from the somatosensory system, which is either directly related to or correlated with one's oculomotor status (Fig. 21.1). It is derived from the eyeball and surrounding tissues, especially the palpebral conjunctiva which lines the inner portions of the eyelids, via diffuse trigeminal nerve innervation (CNV). This tactile feedback results from lid sensations created by the presence of either a hard/rigid contact lens (preferred) or a soft contact lens moving on the eye during one's normal complement of eye movements (Ciuffreda and Tannen, 1999).

The use of contact lenses in nystagmus patients has been described periodically in the clinical literature for nearly 50 years (Abrams, 1955; Allen and Davies, 1983; Stevenson, 1975), in which hard and soft corneal lenses, as well as large molded scleral lenses, have been used. However, the reasons attributed for the improvements over that found with spectacles were typically of an optical nature: (1) less optical aberrations, (2) enlarged retinal image (in refractive myopes), and increased peripheral visual field. Any or all of the above would improve the quality of the retinal image, and hence provide a visual input of higher fidelity for subsequent visual processing.

However, in one early clinic report (Allen and Davies, 1983), the notion of lid-based tactile feedback was advanced. More recent evidence based upon objectively-recorded eye movements confirmed that lid-derived, tactile feedback indeed plays a significant role in this phenomenon (Abadi, 1979; Dell `Osso et al, 1988; Matsubayashi et al, 1991). When hard contact lenses were used in place of spectacles, both oculomotor and related sensory improvements were evident: nystagmus amplitude decreased (Fig. 21.12), and both visual acuity and contrast sensitivity increased. This appears to represent a more general trigeminal nerve phenomenon; for example, gentle touching of the

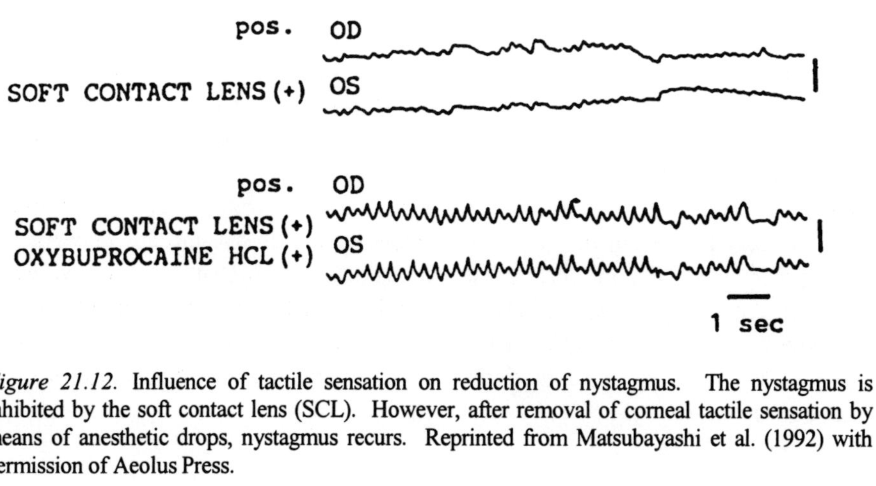

Figure 21.12. Influence of tactile sensation on reduction of nystagmus. The nystagmus is inhibited by the soft contact lens (SCL). However, after removal of corneal tactile sensation by means of anesthetic drops, nystagmus recurs. Reprinted from Matsubayashi et al. (1992) with permission of Aeolus Press.

upper eyelid and the forehead also reduced the nystagmus (Dell'Osso et al, 1990; Sheth et al, 1995); further, use of a topical anesthetic to block these lid sensations restored the full nystagmus complement. This in turn seems to represent a yet more general phenomenon of afferent/somatosensory influence on the oculomotor control pathways, as effects were also found as a result of either neck vibration (Ozawa et al, 1981) or acupuncture involving the neck musculature (Blekher et al, 1998; Ishikawa et al, 1987).

21.4 VISUAL FEEDBACK

For the present discussion, visual feedback refers to sensory information derived from the visual system, which is either directly related to or correlated with one's oculomotor status. Clearly, in all but totally-blind individuals, some degree of visual feedback related to eye position in space, typically with respect to an object of interest, is present (Fig. 21.1). However, in individuals with foveally-based eye diseases (e.g., macular degeneration; see Section 21.2.3.3), the quality of the visual feedback may be compromised. And, in patients with nystagmus (see Section 21.2.3.2), the chronically-increased retinal-image motion and consequent retinal-image degradation leads to compromised visual input for later stages of visual processing.

Clinically, visual feedback-based therapy has evolved into two basic forms: afterimages and conventional vision therapy. Flash-induced, foveal afterimages (Stohler, 1973), in which changes in position of the perceived cortically-based afterimage reflect changes in eye position, have been used by some clinicians for years with limited success. In patients with nystagmus, a constantly oscillating image would be perceived. The instruction would be to attempt to foveate a target and try to "do something" to reduce the oscillation of the afterimage. However, the procedure is cumbersome, and the afterimage typically short-lived (<1 minute). Vision therapy has also been used in these cases (Healy, 1952, 1962; Leung et al, 1996; Ordonez and Ciuffreda, 1997; Pearlman, 1976). Basically, it involves a series of highly specific, sequential, sensory-motor-perceptual stimulation paradigms and regimens aimed at normalizing sensory and motor aspects of binocular vision, incorporating the use of various lenses, prisms, and test distances to manipulate the blur, disparity, and proximal information, respectively (Ciuffreda, in press). Visual feedback takes the form of blur, disparity/diplopia, and/or perceived distance/direction. Vision therapy has met with reasonable success. However, both forms of visual feedback-based training are difficult to implement with very young children (i.e., less than 10 or so years of age) due to technical problems, low level of maintained interest, and difficulty conceptualizing some of the task demands.

Recently, we have developed a novel form of external visual feedback, with a goal towards treating very young children with nystagmus (5-10 years of age) (Rutner and Ciuffreda, submitted) (patent pending) in whom oculomotor auditory feedback is neither conceptually nor practically understood and appreciated. This system, which consists of an eye movement device and a modified combination television/video monitor, is shown schematically in Fig. 21.13. In essence, the child's horizontal eye movements control the video aspect of the combined video/audio presentation consisting of a female storyteller conveying a well-known children's story. The spartan video image consists only of the storyteller's head and shoulders, which is restricted to a small region located at the bottom of the television monitor. The remainder of the screen is blank to reduce distractions. The horizontal eye position of one eye is continuously monitored using an infrared eye movement system. The child is told that if his/her eyes are "steady", the storyteller will be both seen and heard; however, if the eyes are not "steady", the video portion will be extinguished, until the eyes are once again steady and gaze is maintained at the monitor location where the storyteller's face was initially present. Essentially, voltage signals from the eye movement system are transmitted to an adjustable deadband controller, which allows the experimenter to control the angular extent of allowable horizontal eye movement. If the designated range is exceeded, it will extinguish the video component only. Once the appropriate eye movement control is regained, the video aspect is restored. During the course of training, the deadband width is gradually reduced to increase fixational difficulty level.

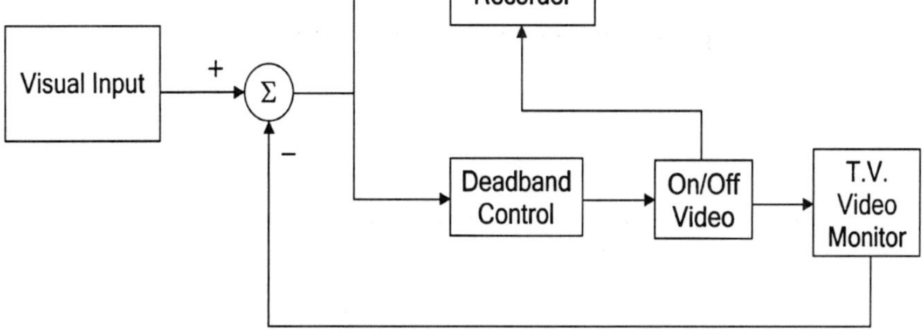

Figure 21.13. Block diagram of the visual feedback eye movement training system developed by two of the authors (D.R. and K.C.).

Thus far, we have found this form of oculomotor visual feedback to be readily understood by both visually-normal young children in our initial control tests (Fig. 21.14), as well as in a few young children with congenital nystagmus. We plan on continuing this training in additional young nystagmats, with the hope that attainment of such oculomotor control will transfer to their normal everyday activities, thereby resulting in improved vision and cosmetic aspects.

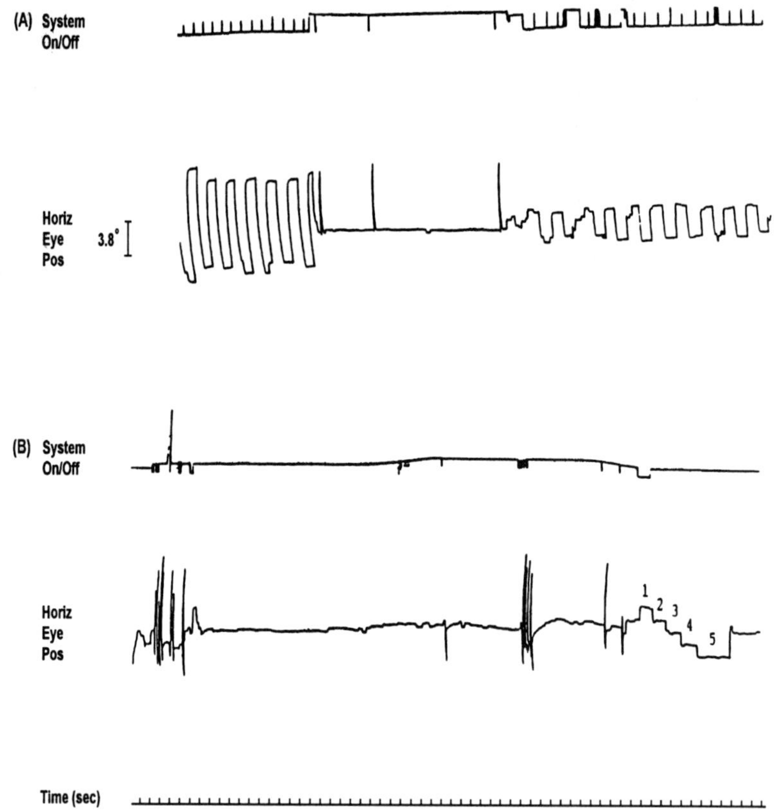

Figure 21.14. A) Visually-normal eight-year-old demonstrating volitional ability to turn on and off video aspect of video-tape with his lateral eye movements when deadband set for either 7.0 (left) or 3.5 (right) degrees; center portion shows steady midline fixation. (B) Same as above showing steady midline fixation and video aspect present, followed by calibration (#'s 1-5). For video system, up is on, and down is off; for eye movement system, up is left, and down is right. Large deflections are artifacts due to blinks.

21.5 MULTI-SENSORY FEEDBACK THERAPY FOR NYSTAGMUS AND OTHER OCULOMOTOR DYSFUNCTIONS

Based on the above information, we have developed a comprehensive, multi-sensory, three component, two-phase therapeutic paradigm for nystagmus (Ciuffreda and Tannen, 1999) (Table 21.4) (Fig. 21.1). This paradigm can also be applied to other forms of oculomotor dysfunctions. Phase One consists of oculomotor auditory feedback alone, with initial emphasis on fixational ability. This is then extended to basic version (saccades and pursuit), simulated reading, and vergence. In Phase Two, various types of visual and tactile feedback are added, both in isolation and in conjunction with the other forms of feedback. For example, the patient might be trained to use combined visual (e.g., very low contrast target discrimination) and tactile (e.g., rigid corneal contact lens) feedback, especially for home therapy.

Table 21.4 **Nystagmus Multi-Sensory Feedback Clinical Therapy Protocol**

- **Auditory feedback (phase 1 and 2)**
  - fixation
  - version (saccades and pursuit)
  - reading (stimulated)
  - vergence
- **Visual feedback (phase 2)**
  - version and vergence eye movement training
  - sustained afterimages
  - very low contrast target discrimination
  - fine line separation discrimination
- **Tactile feedback (phase 2)**
  - digital assessment
  - rigid contact lenses

We have successfully performed our multi-sensory therapy in the clinic and at home on cooperative patients ranging in age from 10 to 70 years. While we have also attempted Phase One oculomotor auditory feedback in a few children as young as 4 years of age, they did not seem to grasp the relatively difficult concept that the tone was related to their eye movements, perhaps in part because they did not fully appreciate and understand their own oculomotor dysfunction. In essence, they could not conceptualize and formulate the requisite auditory/oculomotor transform. In the future, new test paradigms will be developed to facilitate the understanding and appreciation this critical conceptual link in young children. For example, use of video-based cartoon test stimuli in conjunction with remote video-based monitoring of eye movements is a possible scenario, with visibility or even clarity of the cartoon video aspect contingent upon an appropriate level of oculomotor control. See Section 21.4, where our initial attempts in this area are discussed.

One area that requires careful attention is the long-term efficacy of oculomotor auditory feedback therapy, both in isolation as well as in conjunction with the visual and tactile feedback paradigms. We have informally tested patients who had successfully completed primarily the auditory feedback treatment up to 2 years earlier. They reported using the learned strategy during specific situations in their activities of daily living. Furthermore, they could invoke it readily during both subjective and objective testing of their eye movements in the clinic and laboratory, respectively. However, formal testing over relatively long post-treatment time intervals (i.e., 3-5 years) is required to assess the retention ability more reliably and quantitatively, especially with respect to its multi-sensory component contribution to the overall enhanced oculomotor performance. Clearly, a large-scale clinical trial in both adults and children is essential.

21.6 POSSIBLE NEURAL MECHANISMS

Evidence abounds that both single- and multi-sensory feedback training and therapy can be effective in conditioning, and hence modifying, oculomotor behavior in both normal and visually-abnormal individuals, at times in a remarkably rapid manner. Although details of the neural mechanisms have yet to unfold, psychophysical and electrophysiological investigations over the past decade have provided some insights. Based on this valuable new information, as well as our own experiences, we propose the following possible, albeit incomplete, scenarios.

A common aspect to feedback therapy for oculomotor dysfunction is relaxation, which might include components such as visual imagery, thinking about some pleasant experience, etc., with all of one's attention directed to the specific task at hand. In a recent paper by Critchley et al (2001), PET brain imaging was performed as normal volunteers either relaxed in isolation or with visual feedback present. In the former condition, the left anterior cingulate, globus pallidus, and parietal cortex were activated; in contrast, in the later condition, the entire anterior cingulate cortex and cerebellar vermis were, in addition, activated. This provided evidence for distinct neural mechanisms being involved in the central control of autonomic bodily responses and their modulation by cognitive intent. Such an interpretation is consistent with our own findings and ideas about oculomotor feedback therapy, as well as the broader field of medical "biofeedback" therapy.

Another aspect of interest is the rapidity with which an intended oculomotor behavior can be conditioned and ultimately changed. Although we have objectively documented eye movements in a few nystagmats in which auditory feedback was effective within only a few minutes, such was the norm in our blind population. However, this may not be too surprising in the blind group. First, it is consistent with the early results of Smith (1964), as well as those of Hung et al (1988): as visual feedback became less effective, auditory feedback become more effective. Certainly, the blind subjects represent the visual extreme. Second, there is evidence for cortical plasticity to become manifest after only one second of stimulation (Kapadia et al, 1994); furthermore, it was argued that such "plasticity" really represented normal dynamic "learning" in response to specific changes in sensory stimulation. And, third, there is evidence that other brain areas may effectively "recruit" from the visual cortex, hence resulting in crossmodal reorganization (Kujala et al, 1992; Alho et al, 1993; Cohen et al, 1997; Buchel et al, 1998). Thus, there may be additional brain resources for processing and responding to auditory information in blind individuals.

And, other more general processes may be involved in the non-blind patients tested. First, it has been found that synchronous and relatively widely expansive brain oscillations occur when performing specific detailed activities, such as directing attention to sensorimotor integrated activities, thus enhancing global response saliency (Murthy and Fetz, 1992; Munk et al, 1996). And, second, under certain conditions, correlated visual and auditory input produces neural facilitation resulting in reduced oculomotor saccadic response latency (Hughes et al, 1998). This effect has been attributed to the multi-sensory afferent stimuli converging upon the multi-modal superior colliculus, which in turn influences subsequent oculomotor control centers.

21.7 CONCLUSIONS

Feedback control represents a basic biological concept (McFarland, 1971; Milsum, 1966; Stark, 1968; Toates, 1975). However, in the present context, it does not refer to the use of one's naturally-occurring feedback information and mechanism alone, but rather to exploitation of the process by making readily available new and novel forms of information from the various senses that are either correlated with or related to the oculomotor system's state and/or motor variation. This novel information can come in the form of auditory, tactile, and visual feedback. Clearly, such information was capable of being used to exert considerable influence upon and control over an individual's normal and abnormal oculomotor function. However, despite widespread evidence for its rapid and effective employment as a possible mode of therapy in the clinician's armamentarium, these novel forms of oculomotor feedback have not been used to their full potential, due in part to the tradition of clinical conservatism, as well as both the theoretical and technical demands required in many cases of the clinician/technician. However, as we proceed in this new millenium emphasizing medical economics and non-traditional approaches to health care (e.g., acupuncture and yoga), multi-sensory feedback therapy for oculomotor dysfunctions will likely be embraced on a much broader scale.

21.8 ACKNOWLEDGMENTS

We thank the following individuals for their collaboration, support, and free exchange of ideas in our research activities in multi-sensory feedback therapy over the years: K. Engber, M. Ferri, B. Fayos, O. Franzen, S. Goldrich, E. Hall, Y. Han, G. Hung, G. Lennerstrand, T. Lofman, A. Martins and IOTA EyeTrace, J. Ober, J. Orzuchowski and A. Rydberg.

21.9 REFERENCES

Abadi, R.V., 1979, Visual performance with contact lenses and congenital idiopathic nystagmus, *Brit. J. Physiol. Opt.* **33:** 32-37.

Abadi, R.V., Carden, C., and Simpson, J., 1979, Controlling abnormal eye movements, *Vis. Res.* **19:** 961-963.

Abadi, R.V., Carden, C., and Simpson, J., 1980, A new treatment for congenital nystagmus, *Brit. J. Ophthal.* **64:** 2-6.

Abadi, R.V., Carden, C., and Simpson, J., 1981, Listening for eye movements, *Ophthal. Physiol. Opt.* **1:** 19-27.

Abrams, B.S., Correcting nystagmus with corneal lenses, 1955, *Optom. Wkly.* **46:** 809-812.

Abplanalp, P., and Bedell, H., 1983, Biofeedback therapy in rehabilitative optometry, *The Rehab. Optom. J.* **1:** 11-14.

Abplanalp, P., and Bedell, H., 1987, Visual improvement in an albinotic patient with alternation of congenital nystagmus, *Amer. J. Optom. Physiol. Opt.* **64:** 944-951.

Afanador, A.J., 1982, Auditory biofeedback and intermittent exotropia, *J. Amer. Optom. Assoc.* **53:** 481-483.

Alho, K., Kujala, T., Paavilainen, P., Summala, H., Naatanen, R., 1993, Auditory processing in visual brain areas of the early blind: evidence from event-related potentials, Electroenceph. Clin. Neurophysiol. **86:** 418-427.

Allen, E.D., and Davies, P.D., 1983, Role of contact lenses in the management of congenital nystagmus. *Brit. J. Ophthal.* **67:** 834-836.

Blekher, T., Yamada, T., Yee, R.D., and Abel, L.A., 1998, Effects of acupuncture on foveation characteristics in congenital nystagmus, *Brit. J. Ophthal.* **82:** 115-120.

Brown, B., 1977, *Stress and the Art of Biofeedback*, Harper and Row, New York.
Buchel, C., Price, C., Frackowiak, R.S.J., and Friston, K., 1998, Different activation patterns in the visual cortex of late and congenitally blind subjects, *Brain*, **121:** 409-419.
Caloroso, E.E., and Rouse, M.W., 1993, *Clinical Management of Strabismus,* Butterworth-Heinemann, Boston.
Carmichael L., and Dearborn, W.F., 1947, *Reading and Visual Fatigue,* Houghton Mifflin, Boston.
Ciuffreda, K.J., The scientific basis for and efficacy of optometric vision therapy in non-strabismic accommodative and vergence disorders, *Optometry*, in press.
Ciuffreda, K.J., and Hung, G.K., 1988, Eye movement auditory biofeedback: "hearing" abnormal eye movements, *Med. Electron.,* December, pp. 66-70.
Ciuffreda, K.J., and Goldrich, S.G., 1983, Oculomotor biofeedback therapy, *Internat. Rehab. Med.* **5:** 111-117.
Ciuffreda, K.J., Goldrich, S., and Neary, C., 1980, Auditory biofeedback as a potentially important new tool in the treatment of nystagmus, *J. Amer. Optom. Assoc.* **51:** 615-617.
Ciuffreda, K.J., Goldrich, S.G., and Neary, C., 1982a, Control of nystagmus using eye movement auditory biofeedback, in *Functional Basis of Ocular Motility Disorders,* Lennerstrand, G. et al eds, Pergamon Press, New York, pp. 147-150.
Ciuffreda, K.J., Goldrich, S.G., and Neary, C., 1982b, Use of eye movement auditory biofeedback in the control of nystagmus, *Amer. J. Optom. Physiol. Opt.* **59:** 396-409.
Ciuffreda, K.J., Levi, D.M., and Selenow, A., 1991, *Amblyopia: Basic and Clinical Aspects,* Butterworth-Heinemann, Boston.
Ciuffreda, K.J., and Tannen, B., 1995, *Eye Movement Basics for the Clinician,* Mosby Yearbook, St. Louis.
Ciuffreda, K.J., and Tannen, B., 1999, Training of nystagmus: a multi-sensory approach, *J. Behav. Optom.* **10:** 63-66.
Cohen, L.G., Celnik, P., Pascual-Leone, A. et al, 1997, Functional relevance of cross-modal plasticity in blind humans, Nature, **389:** 180-183.
Critchley, H.D., Melmed, R.N., Featherstone, E., Mathias, C.J., and Dolan, R.J., 2001, Brain activity during biofeedback relaxation: a functional neuroimaging investigation, *Brain,* **124:** 1003-1012.
Dell 'Osso, L.F., Leigh, R.J., and Daroff, R.B., 1990, Suppression of congenital nystagmus by cutaneous stimulation, *Neuro-ophthal.* **11:** 173-175.
Dell 'Osso, L.F., Traccis, S., Abel, L.A., and Erzurum, S.I., 1988, Contact lenses and congenital nystagmus, Clin. Vision Sci. **3:** 229-232.
Fayos, B., and Ciuffreda, K.J., 1998, Oculomotor auditory biofeedback training to improve reading efficiency, *J. Behav. Optom.* **9:** 143-152.
Fletcher, J., 1993, Eye movement rhythmicity and reading comprehension, *J. Learn. Disab.* **26:** 683-687.
Flom, M.C., Kirschen, D.G., and Bedell, H.E., 1978, Control of unsteady, eccentric fixation in amblyopic eyes by auditory biofeedback of eye position, *Invest. Ophthal. Vis. Sci.,* **19:** 1371-1381.
Fuller, G., 1978, Current status of biofeedback in clinical practice, *Amer. Psychol.* **33:** 39-48.
Goldrich, S.G., 1982, Oculomotor biofeedback therapy for exotropia, *Amer. J. Optom. Physiol. Opt.* **59:** 306-317.
Hall, E.C., and Ciuffreda, K.J., Eccentric viewing training in macular disease using ocular motor biofeedback, *J. Behav. Optom.,* in press.
Hall, E.C., and Ciuffreda, K.J., submitted, Fixational ocular motor control is plastic despite visual deprivation.
Halperin, E., and Yolton, R.L., 1986, Ophthalmic applications of biofeedback, *Amer. J. Optom. Physiol. Opt.* **63:** 985-998.
Healy, E., 1952, Nystagmus treated by orthoptics, Amer. J. Orthopt. **2:** 53-55.
Healy, E., 1962, Nystagmus treated by orthoptics: a second report, *Amer. J. Orthopt.* **12:** 89-91.

Hirons, R., and Yolton, R.L., 1978, Biofeedback treatment of strabismus: case studies, *J. Amer. Optom. Assoc.* **49:** 875-882.
Hughes, H.C., Nelson, M.D., and Aronchick, D.M., 1998, Spatial characteristics of visual-auditory summation in human saccades, *Vision Res.* **38:** 3955-3963.
Hung, G.K., Ciuffreda, K.J., Carley, C.A., Fang, P., and Menditto, S., 1988, Auditory biofeedback to control vertical and horizontal eye movements in the dark, *Invest. Ophthal. Vis. Sci.,* **29:** 1860-1865.
Ishikawa, S., Ozawa, H., and Fujiyama, Y., 1987, Treatment of nystagmus by acupuncture, *Proc. 6th Annual Meeting of the International Neuro-Ophthalmology Society,* Hakone, Japan, pp. 227-232.
Ishikawa, S., Tanakadate, A., Nabatame, K., and Ishu, M., 1985, Biofeedback treatment of congenital nystagmus, *Neuro-ophthalmol. Jpn.* **2:** 58-64.
Inverso, M., and Larsen, T., 1981, Biofeedback-enhanced vision training for strabismus, O.D. thesis, Pacific University, College of Optometry, Forest Grove, OR.
Kapadia, M.K., Gilbert, C.D., and Westheimer, G., 1994, A quantitative measure for short-term cortical plasticity in human vision, *J. Neurosci.,* **14:** 451-457.
Kirschen, D.G., 1977, Visual acuity and oculomotor control in amblyopia, Ph.D. dissertation, Univ. Calif. (Berkeley), School of Optometry.
Kirschen, D.G., 1983, Auditory biofeedback in the control of congenital nystagmus, *Amer. J. Optom. Physiol. Opt.* **60:** 364-368.
Kirschen, D.G., 1991, Biofeedback therapy for amblyopia, in *Problems in Optometry: Amblyopia,* Vol. 3, No. 2, R. London ed., Lippincott, J.B., Philadelphia, pp. 369-375.
Kujala, T., Alho, K., Paavilainen, P., Summala, H., and Naatanen, R., 1992, Neural plasticity in processing of sound localization by the early blind: an event-related potential study, Electroenceph. Clin. Neurophysiol. **84:** 469-472.
Lee, R.M., Baldwin, S.E., and Lee, J.A., 1977, Clinical uses of biofeedback: a review of recent research, *Henry Ford Hosp. Med. J.* **25:** 99-118.
Letourneau, J.E., 1976, Application of biofeedback and behavior modification techniques in visual training, *Amer. J. Optom. Physiol. Opt.* **53:** 187-190.
Letourneau, J.E., and Giroux, R., 1984, Using biofeedback in an exotropic aphake, *J. Amer. Optom. Assoc.* **55:** 909-910.
Letourneau, J.E., and Ludlam, W.M., 1976, Biofeedback reinforcement in the training of limitation of gaze, *Amer. J. Optom. Physiol. Opt.* **53:** 672-676.
Leung, V., Wick, B., and Bedell, H.E., 1996, Multifaceted treatment of congenital nystagmus: a report of 6 cases, *Optom. Vis. Sci.* **73:** 114-124.
Matsubayashi, K., Fukushima, M., and Tabuchi, A., 1992, Application of soft contact lenses for children with nystagmus, *Neuro-ophthalmol.* **12:** 47-52.
McFarland, D.J., 1971, *Feedback Mechanisms in Animal Behavior,* Academic Press, New York.
Mezawa, M., Ishikawa, S., and Ukai, K., 1990, Changes in waveform of congenital nystagmus associated with biofeedback treatment, *Brit. J. Ophthal.* **74:** 472-476.
Milsum, J.H., 1966, *Biological Control Systems Analysis,* McGraw-Hill, New York.
Munk, M.H., Roelfsema, P.R., Konig, P., Engel, A.K., and Singer, W., 1996, Role of reticular activation in the modulation of intracortical synchronization, *Science,* **272:** 271-274.
Murthy, V.N., and Fetz, E.E., 1992, Coherent 25- to 35-Hz oscillations in the sensorimotor cortex of awake behaving monkeys, *Proc. Natl. Acad. Sci. USA,* **89:** 5670-5674.
Ordonez, X.P., and Ciuffreda, K.J., 1997, Reading eye movements in patients with nystagmus and proposed therapeutic paradigms, *J. Behav. Optom.* **8:** 99-103.
Ozawa, H., Fujiyama, Y., and Ishikawa, S., 1983, The effect of vibrations on intensity of nystagmus and body movement, *Agressologie,* **24:** 229-230.
Pearlman, C.A., 1976, A case study, *J. Amer. Optom. Assoc.* **47:** 396-398.
Robinson, G.M., 1977, Rhythmic organization in speech processing, *J. Exp. Psych: Human Percept. Perform.* **3:** 83-87.

Rotberg, M.H., and Surwit, R.S., 1981, Biofeedback techniques in the treatment of visual and ophthalmologic disorders, *Biofeed. Self-Regul.* **6:** 375-388.

Rydberg, A., Han, Y., and Ciuffreda, K.J., 1999, Eye movement auditory biofeedback therapy for patients with nystagmus, *Trans. 25th Meeting of the European Strabismological Assoc.*, Jerusalem, Israel, pp. 330-334.

Scheiman, M.M., Peli, E., and Libassi, D., 1983, Auditory biofeedback used to enhance convergence insufficiency, *J. Amer. Optom. Assoc.* **54:** 1001-1003.

Schor, C.M. and Hallmark, W., 1978, Slow control of eye position in strabismic amblyopia, *Invest. Ophthal. Vis. Sci.*, **17:** 577-581.

Shelhamer, M., Merfield, D.M., and Mendoza, J.C., 1994, Vergence can be controlled by audio feedback and influence downward ocular deviation, *Exp. Br. Res.* **101:** 169-172.

Sheth, N.V., Dell 'Osso, L.F., Leigh, R.J., Van Doren, C.L., and Peckham, H.P., 1995, The effects of afferent stimulation on congenital nystagmus foveation periods, *Vision Res.* **35:** 2371-2382.

Smith, W.M., 1964, Control of eye fixation by auditory feedback, *Psychon. Sci.* **1:** 233-234.

Spilker, B., 1991, Special modalities: biofeedback, radiation therapy, hyperthermia, hypnosis, electroconvulsive therapy, apheresis, and hyperbaric oxygen, in *Guide to Clinical Trials*, B. Spilker ed., Raven Press, New York, pp. 653-656.

Stark, L., 1968, *Neurological Control Systems,* Plenum Press, New York.

Stevenson, R.W.W., 1975, The application of contact lenses to nystagmus, *Optics (Scotland),* June: 4-8.

Stohler, T., 1973, Afterimage treatment in nystagmus, *Amer. Orthopt. J.* **23:** 65-67.

Taylor, E.A., 1966, *The Fundamental Reading Skill,* Charles C. Thomas, Springfield, Ill.

Toates, F.M., 1975, *Control Theory in Biology and Experimental Psychology,* Hutchinson Educ. Ltd., London.

Van Brocklin, M.D., Vasche, T.R., Hirons, R., and Yolton, R.L., 1981, Biofeedback enhanced strabismus therapy, *J. Amer. Optom. Assoc.* **52:** 731-736.

Wolf, S.L., 2001, Biofeedback, in *Downey and Darling's Physiological Basis of Rehabilitation Medicine,* E.G. Gonzalez et al eds., Butterworth-Heinemann, Boston, pp. 747-759.

SUBJECT INDEX